T0296256

CAMBRIDGE LIBRARY COLLECTION

Books of enduring scholarly value

Life Sciences

Until the nineteenth century, the various subjects now known as the life sciences were regarded either as arcane studies which had little impact on ordinary daily life, or as a genteel hobby for the leisured classes. The increasing academic rigour and systematisation brought to the study of botany, zoology and other disciplines, and their adoption in university curricula, are reflected in the books reissued in this series.

Handbook of the New Zealand Flora

Sir Joseph Dalton Hooker (1817–1911), botanist, explorer, and director of the Royal Botanical Gardens at Kew, is chiefly remembered as a close friend and colleague of Darwin, his publications on geographical distribution of plants supporting Darwin's theory of evolution by natural selection. In 1839 Hooker became an assistant surgeon on HMS *Erebus* during Ross' Antarctic expedition. The boat wintered along the New Zealand coast, Tasmania and the Falkland Islands, enabling Hooker to collect over 700 plant species. Drawing heavily on Hooker's illustrated *Flora Novae Zelandiae* (1854–1855), this two-volume work (1864–1867) contains a comprehensive list of New Zealand plant species as well as those of the Chatham, Kermadec, Auckland, Campbell and Macquarrie Islands. As the first major study of New Zealand flora, Hooker's handbook remained the authority on the subject for half a century. Volume 1 begins Hooker's exhaustive list of species encountered during his three-year voyage.

Cambridge University Press has long been a pioneer in the reissuing of out-of-print titles from its own backlist, producing digital reprints of books that are still sought after by scholars and students but could not be reprinted economically using traditional technology. The Cambridge Library Collection extends this activity to a wider range of books which are still of importance to researchers and professionals, either for the source material they contain, or as landmarks in the history of their academic discipline.

Drawing from the world-renowned collections in the Cambridge University Library, and guided by the advice of experts in each subject area, Cambridge University Press is using state-of-the-art scanning machines in its own Printing House to capture the content of each book selected for inclusion. The files are processed to give a consistently clear, crisp image, and the books finished to the high quality standard for which the Press is recognised around the world. The latest print-on-demand technology ensures that the books will remain available indefinitely, and that orders for single or multiple copies can quickly be supplied.

The Cambridge Library Collection will bring back to life books of enduring scholarly value (including out-of-copyright works originally issued by other publishers) across a wide range of disciplines in the humanities and social sciences and in science and technology.

Handbook of the New Zealand Flora

A Systematic Description of the Native Plants of New Zealand

VOLUME 1

JOSEPH DALTON HOOKER

CAMBRIDGE UNIVERSITY PRESS

Cambridge, New York, Melbourne, Madrid, Cape Town,
Singapore, São Paolo, Delhi, Tokyo, Mexico City

Published in the United States of America by Cambridge University Press, New York

www.cambridge.org
Information on this title: www.cambridge.org/9781108030397

© in this compilation Cambridge University Press 2011

This edition first published 1864
This digitally printed version 2011

ISBN 978-1-108-03039-7 Paperback

This book reproduces the text of the original edition. The content and language reflect the beliefs, practices and terminology of their time, and have not been updated.

Cambridge University Press wishes to make clear that the book, unless originally published by Cambridge, is not being republished by, in association or collaboration with, or with the endorsement or approval of, the original publisher or its successors in title.

NOTICE.

The Second Part, containing the remaining Orders of Cryptogamic Plants, with Index, and Catalogues of Native Names and of Naturalized Plants, will appear shortly.

HANDBOOK

OF THE

NEW ZEALAND FLORA.

HANDBOOK

OF THE

NEW ZEALAND FLORA:

A SYSTEMATIC DESCRIPTION

OF THE

Native Plants

OF

NEW ZEALAND

AND THE

CHATHAM, KERMADEC'S, LORD AUCKLAND'S, CAMPBELL'S, AND MACQUARRIE'S ISLANDS.

BY

J. D. HOOKER, M.D., F.R.S. L.S. & G.S.,

AND HONORARY MEMBER OF THE PHILOSOPHICAL INSTITUTE OF CANTERBURY, NEW ZEALAND.

PUBLISHED UNDER THE AUTHORITY OF THE GOVERNMENT OF NEW ZEALAND.

LONDON:
LOVELL REEVE & CO., 5, HENRIETTA STREET, COVENT GARDEN.
1864.

PRINTED BY JOHN EDWARD TAYLOR,
LITTLE QUEEN STREET, LINCOLN'S INN FIELDS.

TO HIS EXCELLENCY

SIR GEORGE GREY, K.C.B., D.C.L. OXON.,

ETC. ETC. ETC.,

GOVERNOR AND COMMANDER-IN-CHIEF OF THE

COLONY OF NEW ZEALAND,

WHO, THROUGHOUT A LONG AND DISTINGUISHED CAREER,

IN THE COLONIES OF

SOUTH AUSTRALIA, NEW ZEALAND, AND THE CAPE OF GOOD HOPE,

HAS BEEN THE LIBERAL ENCOURAGER OF EVERY SCIENTIFIC UNDERTAKING,

THIS WORK

IS GRATEFULLY DEDICATED,

BY HIS EXCELLENCY'S VERY FAITHFUL SERVANT,

J. D. HOOKER.

ROYAL GARDENS, KEW,

July, 1864.

PREFACE.

The desirability of publishing a compendious account of the plants of New Zealand having been represented to the Colonial Government by Dr. Knight, F.L.S., Auditor-General, and other gentlemen interested in the Natural History of the Islands, and in the development of their resources, that Government was pleased to entrust me with the preparation of such a work, and to place at my disposal the necessary funds for its publication, including a liberal remuneration for my services. I was at the same time instructed to make Mr. Bentham's 'Hongkong Flora' my guide as to the form of the work and method of describing the plants, and to adhere in these and in all other matters to the plan* recommended by Sir W. J. Hooker for publishing in a uniform series Floras of all the British Colonies. The title 'Handbook of the New Zealand Flora' is adopted in accordance with the wishes of its promoters.

Though as complete as the materials at my disposal enable me to make this book, it is still imperfect as to the descriptions of several Orders of Flowering plants; whilst with regard to the Flowerless, it is tolerably complete in the Orders Ferns and Lycopods only; of the others, the islands no doubt possess twice as many Mosses and Jungermannias as have hitherto been discovered; and I have been able to offer but a meagre sketch of the Fungi, of the lower tribes of Algæ, and of the more minute and especially crustaceous Lichens that grow on rocks and on the bark of trees. It must, however, be many years before the multitudinous New Zealand genera and species of these very obscure and imperfectly-known tribes of plants are fully known, and I have thought it best to give good descriptions of all the commoner and most conspicuous only, which can be easily found and studied, and to treat more generally the rarer and more obscure, of which I have but imper-

* See 'Natural History Review,' 1861, p. 255.

fect specimens, and which, were they perfect, could only be satisfactorily examined in a living state. I have, however, omitted no species known to me as a native of New Zealand, whether from books or collections. By adopting this plan I have, I hope, made this portion of the 'Handbook' a fair introduction to the study of the Orders of New Zealand Flowerless plants, and a safe guide to the principal species; and I also hope that this will prove to be a more useful way of treating so very abstruse a subject, than would systematic descriptions, of equal length and pretensions to accuracy, of all the obscure and supposed species, whether common or scarce, perfect or imperfect.

In the course of preparing this work, I have re-examined most of the materials described in my 'Flora Novæ-Zelandiæ;'* these consisted of the collections of Banks and Solander, and of Forster, contained in the British Museum, and of those of the Cunninghams, Colenso, Sinclair, Bidwill, Dieffenbach, Raoul, Lyall, and my own, all preserved in the 'Hookerian Herbarium.' Since the publication of that work, little of novelty has been added to the Flora of the Northern Island, but very many interesting discoveries have been made in the Middle Island, adding fully one-third to the previously known number of New Zealand Flowering Plants.

Much remains to be done towards the Botany of the Northern Island especially; of the whole province Taranaki, nothing is known; and except the Ruahine range, by Colenso, no mountain region has been approximately well explored. Then too of the outlying islands, as the Kermadec and Chatham Islands, very little is known, and of Bounty or Antipodes Island nothing, whilst much remains to be collected on Lord Auckland's group, Campbell's Island and Macquarrie Island. The materials are still wanting for a comparison of the volcanic mountains of the Northern Island with the primitive or other mountains of the Middle Island, a comparison essential to make before the geological or climatic relations of the flora of either island can be ascertained. These subjects and those of the geographical distribution of New Zealand plants, and of the apparently recent development of many of its species by variation from others still existing in the islands, are, however, foreign to a purely systematic Handbook, and I shall hope to take them up when this is finished.

In the 'Flora Novæ-Zelandiæ,' I have detailed at length the labours

* This, which forms the second part of the 'Botany of the Antarctic Expedition of Sir J. Ross,' was published in 1854–5, in two volumes, 4to, with 130 plates, coloured (including 1060 species), of New Zealand plants.

of all my predecessors, whether as collectors or authors, up to the year 1850, and shall therefore only briefly recapitulate them here.

In August 1769, SIR JOSEPH BANKS and his companion DR. SOLANDER visited the islands in Captain Cook's first voyage, and collected in Poverty Bay, Tegadoo, Tolaga, Opuragi, the Thames river, Bay of Islands, Queen Charlotte's Sound and Admiralty Bay. They obtained about 360 Flowering plants and ferns, had folio drawings made of most of them, and excellent manuscript descriptions. These MSS., together with about 200 engraved plates, were, I believe, all prepared for the press, and are preserved in the British Museum, but have never been published.

In 1772 Captain Cook again visited New Zealand, accompanied by the two FORSTERS, REINWOLD and GEORGE (father and son), and by DR. SPARRMAN; they collected at Dusky Bay and Queen Charlotte's Sound. Their herbarium amounted to only about 160 species of Flowering plants and ferns. Of these, 150 are published in Forster's 'Florulæ Insularum Australium Prodromus,' and a few others in his 'Characteres Generum,' and 'De Plantis Esculentis Insularum Oceani Australis Commentatio Botanica.' The specimens were distributed to various museums, and being often carelessly named, much confusion has crept into descriptive works.

In 1777, Captain Cook, during his third voyage, visited New Zealand. On this occasion MR. ANDERSON, his surgeon, was the botanical collector, who obtained very little indeed, and nothing of any importance.

In 1791, Captain Vancouver arrived in Dusky Bay, on his way to survey the coasts of North-West America, having with him as surgeon MR. ARCHIBALD MENZIES, a very assiduous collector of Flowerless plants, who procured many species of *Filices*, *Musci*, and *Hepaticæ*, most of which are described at length, and beautifully illustrated in Hooker's 'Musci Exotici,' and in Hooker and Greville's 'Icones Filicum.'

In 1822, Captain Duperrey visited the islands in the French discovery corvette 'Coquille,' when one of his officers, the late ADMIRAL D'URVILLE, made excellent collections.

In 1827, CAPTAIN (afterwards ADMIRAL) DUMONT D'URVILLE again visited New Zealand in the same ship, renamed the 'Astrolabe,' accompanied by an able naturalist, M. LESSON, when additional botanical collections were made in Cook's Straits, the Thames river, and the Bay of Islands. The materials of this voyage (containing upwards of 200 Flowering plants and ferns) were published by M. A. Richard, in his 'Essai d'une Flore de la Nouvelle-Zélande,' with folio plates (Paris,

1832). Some of Forster's plants, together with extracts from his MSS., preserved in the Paris Museum, were also published in this work.

In 1825, Mr. Charles Fraser, then superintendent of the Sydney Botanical Gardens, landed for one day in the Bay of Islands, made a small collection of dried plants. He, however, procured more living ones, some of which were amongst the first plants of the islands which were introduced into European gardens.

In 1826, and again in 1838, Allan Cunningham, the eminent Australian botanist and explorer, made extensive botanical explorations in the northern parts of the Northern Island, chiefly at the Bay of Islands; and in 1833, his brother, Richard Cunningham (Fraser's immediate successor in the Sydney gardens), was sent in H.M.S. Buffalo, to procure timber for the Government of Australia. The results of the labours of the brothers, and especially of Allan, whose arduous exertions in the islands led to his untimely death at Sydney in 1839, added considerably to the known Flora, and were collected by Allan into his 'Floræ Novæ-Zelandiæ Præcursor,' which was published by Sir W. J. Hooker, partly in his 'Companion to the Botanical Magazine,' vol. ii.; and partly in the 'Annals and Magazine of Natural History,' vols. i. ii. iii.

The herbarium of the Cunninghams, which has lately been presented by its possessor, R. Heward, Esq., F.L.S., to the Royal Gardens, Kew, had been lent to me by its liberal possessor during the preparation of the 'Flora Novæ-Zelandiæ,' and I have again consulted it during the preparation of this work.

In 1840 and 1841, the French frigate 'L'Aube,' and in 1842–3 another, the 'Allier,' made a lengthened sojourn at the islands; during those occasions M. Raoul, a very intelligent medical officer, diligently explored Banks's Peninsula and the Bay of Islands, making excellent collections at the former locality especially; most of the new species discovered were published first in the 'Annales des Sciences Naturelles' (ser. iii. vol. ii. p. 113) by MM. Raoul and Decaisne, and more recently were described and figured in a beautiful work, entitled, 'Choix de Plantes de la Nouvelle-Zélande,' which further contains thirty plates, and an enumeration of all then known New Zealand plants. The collections are preserved in the Paris Museum, and a set has been communicated to Sir W. Hooker's herbarium.

In 1841, the Antarctic Expedition visited the Bay of Islands, when, accompanied by my friend the Rev. W. Colenso, and by Dr. A. Sinclair during a part of the time, I was enabled to explore the neighbourhood very fully, and to add largely to the Cryptogamic Flora.

In 1847-9, CAPTAIN STOKES, R.N., in H.M.S. Acheron, surveyed the coast of New Zealand; he was accompanied by Dr. Lyall, who made very large and excellent collections, especially of Flowerless plants, on various parts of the coast, but chiefly of the Middle Island.

The other collectors to whom I am principally indebted for the materials published in the 'Flora Novæ-Zelandiæ,' are, firstly, the REV. W. COLENSO, who, during many successive years, has collected throughout the whole length of the Northern Island, with great care and skill, discovering more new and interesting plants (especially on the Ruahine Range, Tongariro, Hikurangi, etc.) than any botanist since Banks and Solander. In every respect Mr. Colenso is the foremost New Zealand botanical explorer, and the one to whom I am most indebted for specimens and information. The late DR. ANDREW SINCLAIR, R.N., F.L.S., formerly Colonial Secretary, a man of great attainments in many ways, certainly ranks second to Mr. Colenso. He collected very copiously in the Bay of Islands, the Auckland districts, and in the Nelson mountains, and was engaged in a botanical exploration of the Southern Alps in company with Mr. Haast, when he was drowned in the Rangitata river. His loss has been a very great one, whether as a botanist or as an enthusiastic and liberal patron of science.

The first alpine collections were made by my late friend J. T. BIDWILL, Esq., of Sydney, who was the earliest explorer of the interior of the Northern Island, and in 1839 ascended the lofty active volcano Tongariro, incurring considerable danger; at a later period he was the first explorer of the Southern Alps, making extensive and very important collections on the Nelson mountains, which were transmitted with copious notes to Sir W. Hooker.

The same mountains have been still better explored by DR. MUNRO, who has added many beautiful alpine species to the 'New Zealand Flora,' and sent an excellent herbarium of Nelson plants to Sir W Hooker.

In 1840, DR. DIEFFENBACH visited many parts of the Northern Island and northern part of the Middle Island, and is the first person who ascended Mount Egmont. His collections, which are however most scanty, compared with the great extent of interesting ground he passed over, were also communicated to Sir W. Hooker's herbarium. MR. LOGAN, MR. EDGERLEY, the REV. W. TAYLOR, the late GENERAL BOLTON, CAPT. HAULTAINE, CAPT. DRURY, R.N., MR. JOLLIFFE, CAPT. D. ROUGH, and MR. STEPHENSON, have all contributed interesting and important collections which are embodied in the 'Flora Novæ-Zelandiæ.'

Since 1854, as I have before observed, no addition of importance has been made to our knowledge of the Flowering plants of the Northern Island; which I greatly regret, as much remains to be done in all the mountain districts in collecting the *Gramineæ*, *Cyperaceæ*, and *Cryptogamia* everywhere, and in observing the habits and characters of the species of *Veronica*, *Coprosma*, *Astelia*, and many other genera. As regards Flowerless plants, two valuable papers by MR. KNIGHT and MR. MITTEN, on some of the Lichens of Auckland,* and by MR. RALPHS† on the Tree-ferns, are almost the only published contributions made since that date to our knowledge of its Flora.

The Middle Island, on the other hand, has furnished several diligent explorers and many capital discoveries. DR. MUNRO has made further excellent collections on the Nelson mountains, as have DR. SINCLAIR and CAPT. ROUGH. To W. T. LUKE TRAVERS, Esq., F.L.S., of Canterbury, I am indebted for many fine plants discovered in the alpine ranges of Canterbury, Nelson, and Marlborough, during various excursions. Many of these being ticketed as to elevation, are the more valuable. His observations on the spread of introduced plants are extremely interesting, and will, I hope, be fully followed up.‡

The great opportunities enjoyed by the distinguished geologist and explorer, JULIUS HAAST, ESQ., F.L.S., F.G.S., Government Geologist of Canterbury, have been used to the best advantage in the furtherance of botanical science, he having contributed more new species to the Flora of the islands than any collector since Mr. Colenso. I am indebted to him also for a series of maps, notes, and observations, especially respecting the ranges of the mountain plants, including the most alpine species hitherto discovered, which have been of great service. It is difficult to imagine how, with so many and such arduous duties as surveyor and geologist, Mr. Haast can have personally effected so much for botany as he has done, and I anticipate that his method of making complete collections on each mountain and on each line of march, will eventually do much to develop the extremely curious subject of the variations of New Zealand plants. Mr. Haast has further called my attention to the labours of his assistant MR. WILLIAM YOUNG, who has made several interesting discoveries, more particularly amongst the Grasses and Sedges of the alpine regions.

From the Otago province I have an excellent herbarium of Dunedin

* Transactions of the Linnean Society of London, xxiii. 99 and 101.

† Journal of Linnean Society of London, Bot. iii. 163.

‡ See 'Natural History Review,' January, 1864.

plants, made by Dr. Lauder Lindsay, F.L.S.; and more recently very extensive and valuable collections, containing much novelty, from the Alps of the interior and west coasts, by another eminent geologist, Dr. Hector, F.G.S., Government Geologist, and Mr. Buchanan, his assistant. The most important of these last collections arrived whilst the sheets of this work were passing through the press, and have materially delayed its publication; for the discoveries which they contained seemed to me to be of sufficient importance to render it desirable that they should be embodied in the portions that had already been printed, which had to be recalled for the purpose. Again, since the completion of the Flowering plants, I have received two more contributions from these surveyors, including various new discoveries and new habitats, which must be reserved for the Supplement.

To render this Handbook more complete, I have included in it the plants of the outlying islands properly belonging to the New Zealand group. They are the following:—

Chatham Islands, whence I have a very few plants collected by Dr. Dieffenbach. The splendid *Myosotidium nobile* inhabits this group, which is well worthy of a careful exploration. This, through the liberality of Mr. Travers, has been done by his son, who, he tells me, has returned from the group with considerable collections.

Kermadec Islands. One of these, Sunday or Raoul Island, was visited by Captain Denham, in H.M.S. Herald, and botanized on by Mr. MacGillivray, naturalist, and his assistant, Mr. Milne; its Flora, though characteristic of New Zealand, is more tropical than the latter, containing the widely-diffused *Metrosideros polymorpha,* and several tropical Ferns of the Pacific islands. I published a list of its plants in the Linnean Society's Journal, Botany, vol. i. p. 125.

Lord Auckland's Group and **Campbell's Island** were explored by Dr. Lyall and myself, during the stay there of the Antarctic Expedition, in the year 1840. Our collections amounted to 370 species, and are published in the first volume of the 'Antarctic Flora,' with 80 plates of 150 species. Lord Auckland's group had been visited in the previous year by Admiral D'Urville's Antarctic Expedition, but the collections made by his naturalists, MM. Hombron and Jacquinot, were extremely small. The Cryptogamia alone are described, by M. Montagne, in a work entitled 'Voyage au Pôle Sud,' Bot. Crypt., 8vo, Paris, 1845, with a folio atlas of 20 plates: figures of some of the Flowering plants and Ferns have likewise been published in the same form, but without descriptions. Lord Auckland's group was also visited, in the same year, by Commodore Wilkes's Expedition, when very few, if any, plants appear to have been collected; and more recently by GEN. BOLTON, who added several species to the Flora of the group.

Considering how many beautiful plants different from those of New Zealand these islands contain, it is obvious that they deserve a very close and careful botanical scrutiny.*

Macquarrie's Island. A few plants from this desolate spot were sent to Sir W. Hooker by MR. FRASER, when Superintendent of the Sydney Botanic Garden.

Of Bounty or Antipodes Islands nothing whatever seems to be botanically or geographically known.

The Flora of Lord Howe's Islands is intermediate in character (as the islands are in position) between that of New Zealand and Australia, but much more nearly allied to the latter; whilst Norfolk Island, which should perhaps have been included in the New Zealand group, is much more tropical and may have equal claims to rank botanically with the New Caledonian or Fiji Islands.

Of the 303 New Zealand genera of Flowering plants described in this part, about 252 (containing 222 species and 51 representatives) are common to Australia; 174 (containing 11 species and 32 representatives) to South America; 31 are peculiar to the group (comprising 59 species), and 6 (with 20 species) are found in the Pacific islands and elsewhere, but not in Australia or South America.

Again, of the 935 species of Flowering plants, 677 are peculiar to the islands; 222 are Australian; and 111 American. There are further 51 Australian representative and 32 American representative species.

Comparing New Zealand with Europe, these countries have 115 genera and 58 species in common, the latter including many water-plants, and several land-plants which are doubtful natives. Of these European genera, the shrubby Veronicas and *Ligusticum* are the only ones that appear to be vastly more numerous in New Zealand than in Europe.

It remains for me to apologize for many imperfections that will be

* Of about one hundred Flowering plants, natives of these small groups, no less than twenty-seven are hitherto unknown in New Zealand proper, including three genera and twelve most conspicuous and singular species, viz., *Ligusticum latifolium* and *antipodum*, *Pleurophyllum* two species, *Celmisia vernicosa*, *Gentiana* two species, *Plantago Antarctica*, *Chiloglottis cornuta*, *Anthericum Rossii*, *Rostkovia* two species.

found in this work, and to express a hope that these will be pointed out to me when discovered, so that they may be corrected in a future edition. Of the two principal sources of error in any work of the nature of a Systematic Flora, one is unavoidable, and that is the impossibility of deciding, in many cases, as to what should be regarded a species and what a variety; in my case this difficulty is greatly enhanced by my having only dried specimens to examine. But this is not all; for it is now admitted that one and the same species may be represented by two or more permanently distinct forms in one district, in other districts by but one of these forms, and in still other districts by forms which unite the characters of the most distinct forms of the first district; and moreover, that these forms are usually permanent under cultivation. It hence follows that the several characters will have different values in the estimation of the observers in each district, and that there must always be differences of opinion regarding the claims of such forms to take specific rank. The other great source of error is of more real importance, as it relates to facts and not at all to opinions; it is, that in examining dried specimens, important and constant characters are often overlooked, unimportant and transient ones exaggerated, and that errors accumulate in the successive process of examining so many organs, in applying technical terms to them, and in describing, transcribing, printing, and even in correcting the press. The number of these errors is always great in works which, like the Phænogamic Part of this Handbook, consist of descriptions of plants, two-thirds of which have been examined and described by one author alone, and it is to succeeding observers that I must look for their detection and correction.

Royal Gardens, Kew:
June 30, 1864.

OUTLINES OF BOTANY,

TO ACCOMPANY THE COLONIAL FLORAS.

FROM BENTHAM'S 'FLORA AUSTRALIENSIS.'

CHAP. I. DEFINITIONS AND DESCRIPTIVE BOTANY.

1. The principal object of a **Flora** of a country, is to afford the means of *determining* (*i. e.* ascertaining the name of) any plant growing in it, whether for the purpose of ulterior study or of intellectual exercise.

2. With this view, a Flora consists of descriptions of all the wild or native plants contained in the country in question, so drawn up and arranged that the student may identify with the corresponding description any individual specimen which he may gather.

3. These descriptions should be *clear*, *concise*, *accurate*, and *characteristic*, so as that each one should be readily adapted to the plant it relates to, and to no other one; they should be as nearly as possible arranged under *natural* (184) divisions, so as to facilitate the comparison of each plant with those nearest allied to it; and they should be accompanied by an *artificial key* or index, by means of which the student may be guided step by step in the observation of such peculiarities or *characters* in his plant, as may lead him, with the least delay, to the individual description belonging to it.

4. For descriptions to be clear and readily intelligible, they should be expressed as much as possible in ordinary well-established language. But, for the purpose of accuracy, it is necessary not only to give a more precise technical meaning to many terms used more or less vaguely in common conversation, but also to introduce purely technical names for such parts of plants or forms as are of little importance except to the botanist. In the present chapter it is proposed to define such *technical* or *technically limited* terms as are made use of in these Floras.

5. At the same time mathematical accuracy must not be expected. The forms and appearances assumed by plants and their parts are infinite. Names cannot be invented for all; those even that have been proposed are too numerous for ordinary memories. Many are derived from supposed resemblances to well-known forms or objects. These resemblances are differently appreciated by different persons, and the same term is not only differently applied by two different botanists, but it frequently happens that the same writer is led on different occasions to give somewhat different meanings to the same word. The botanist's endeavours should always be, on the one hand, to make as near an approach to precision as circumstances will allow, and on the other hand to avoid that prolixity of detail and overloading with technical terms which tends rather to confusion than clearness. In this he will be more or less successful. The aptness of a botanical description, like the beauty of a work of imagination, will always vary with the style and genius of the author.

§ 1. *The Plant in General.*

6. The **Plant,** in its botanical sense, includes every being which has *vegetable life,* from the loftiest tree which adorns our landscapes, to the humblest moss which grows on its stem, to the mould or fungus which attacks our provisions, or the green scum that floats on our ponds.

7. Every portion of a plant which has a distinct part or *function* to perform in the operations or phenomena of vegetable life is called an **Organ.**

8. What constitutes *vegetable life*, and what are the functions of each organ, belong to *Vegetable Physiology;* the microscopical structure of the tissues composing the organs, to *Vegetable Anatomy;* the composition of the substances of which they are formed, to *Vegetable Chemistry;* under *Descriptive and Systematic Botany* we have chiefly to consider the forms of organs, that is, their *Morphology*, in the proper sense of the term, and their general structure so far as it affects classification and specific resemblances and differences. The terms we shall now define belong chiefly to the latter branch of Botany, as being that which is essential for the investigation of the Flora of a country. We shall add, however, a short chapter on Vegetable Anatomy and Physiology, as a general knowledge of both imparts an additional interest to and facilitates the comparison of the characters and affinities of the plants examined.

9. In the more perfect plants, their organs are comprised in the general terms **Root, Stem, Leaves, Flowers,** and **Fruit.** Of these the three first, whose function is to assist in the growth of the plant, are *Organs of Vegetation;* the flower and fruit, whose office is the formation of the seed, are the *Organs of Reproduction.*

10. All these organs exist, in one shape or another, at some period of the life of most, if not all, *flowering plants*, technically called *phænogamous* or *phanerogamous plants;* which all bear some kind of flower and fruit in the botanical sense of the term. In the lower classes, the ferns, mosses, fungi, moulds or mildews, seaweeds, etc., called by botanists *cryptogamous plants*, the flowers, the fruit, and not unfrequently one or more of the organs of vegetation, are either wanting, or replaced by organs so different as to be hardly capable of bearing the same name.

11. The observations comprised in the following pages refer exclusively to the flowering or phænogamous plants. The study of the cryptogamous classes has now become so complicated as to form almost a separate science. They are therefore not included in these introductory observations, nor, with the exception of ferns, in the present Flora.

12. **Plants** are

Monocarpic, if they die after one flowering-season. These include *Annuals*, which flower in the same year in which they are raised from seed; and *Biennials*, which only flower in the year following that in which they are sown.

Caulocarpic, if, after flowering, the whole or part of the plant lives through the winter and produces fresh flowers another season. These include *Herbaceous perennials*, in which the greater part of the plant dies after flowering, leaving only a small perennial portion called the Stock or Caudex, close to or within the earth; *Undershrubs, suffruticose* or *suffrutescent* plants, in which the flowering branches, forming a considerable portion of the plant, die down after flowering, but leave a more or less prominent perennial and woody base; *Shrubs* (*frutescent* or *fruticose plants*), in which the perennial woody part forms the greater part of the plant, but branches near the base, and does not much exceed a man's height; and *Trees* (*arboreous* or *arborescent plants*) when the height is greater and forms a woody *trunk*, scarcely branching from the base. *Bushes* are low, much branched shrubs.

13. The terms *Monocarpic* and *Caulocarpic* are but little used, but the other distinctions enumerated above are universally attended to, although more useful to the gardener than to the botanist, who cannot always assign to them any precise character. Monocarpic plants, which require more than two or three years to produce their flowers, will often, under certain circumstances, become herbaceous perennials, and are generally confounded with them. Truly perennial herbs will often commence flowering the first year, and have then all the appearance of annuals. Many tall shrubs

and trees lose annually their flowering branches like undershrubs. And the same botanical species may be an annual or a perennial, a herbaceous perennial or an undershrub, an undershrub or a shrub, a shrub or a tree, according to climate, treatment, or variety.

14. Plants are usually *terrestrial*, that is, growing on earth, or *aquatic*, *i. e.* growing in water; but sometimes they may be found attached by their roots to other plants, in which case they are *epiphytes* when simply growing upon other plants without penetrating into their tissue, *parasites* when their roots penetrate into and derive more or less nutriment from the plant to which they are attached.

15. The simplest form of the perfect plant, the annual, consists of—

(1) The **Root,** or descending axis, which grows downwards from the stem, divides and spreads in the earth or water, and absorbs food for the plant through the extremities of its branches.

(2) The **Stem,** or ascending axis, which grows upwards from the root, branches and bears first one or more leaves in succession, then one or more flowers, and finally one or more fruits. It contains the tissues or other channels (217) by which the nutriment absorbed by the roots is conveyed in the form of *sap* (192) to the leaves or other points of the surface of the plant, to be *elaborated* or *digested* (218), and afterwards redistributed over different parts of the plant for its support and growth.

(3) The **Leaves,** usually flat, green, and horizontal, are variously arranged on the stem and its branches. They *elaborate* or *digest* (218) the nutriment brought to them through the stem, absorb carbonic acid gas from the air, exhaling the superfluous oxygen, and returning the assimilated sap to the stem.

(4) The **Flowers,** usually placed at or towards the extremities of the branches. They are destined to form the future seed. When perfect and complete, they consist: 1st, of a *pistil* in the centre, consisting of one or more *carpels*, each containing the germ of one or more seeds; 2nd, of one or more *stamens* outside the pistil, whose action is necessary to *fertilize* the pistil or enable it to ripen its seed; 3rd, of a *perianth* or *floral envelope*, which usually encloses the stamens and pistil when young, and expands and exposes them to view when fully formed. This complete perianth is double; the outer one, called *Calyx*, is usually more green and leaf-like; the inner one, called the *Corolla*, more conspicuous, and variously coloured. It is the perianth, and especially the corolla, as the most showy part, that is generally called the flower in popular language.

(5) The **Fruit,** consisting of the pistil or its lower portion, which persists or remains attached to the plant after the remainder of the flower has withered and fallen off. It enlarges and alters more or less in shape or consistence, becomes a *seed-vessel*, enclosing the seed until it is ripe, when it either opens to discharge the seed or falls to the ground with the seed. In popular language the term *fruit* is often limited to such seed-vessels as are or look juicy and eatable. Botanists give that name to all seed-vessels.

16. The herbaceous perennial resembles the annual during the first year of its growth; but it also forms (usually towards the close of the season), on its *stock* (the portion of the stem and root which does not die), one or more *buds*, either exposed, and then popularly called *eyes*, or concealed among leaves. These buds, called *leaf-buds*, to distinguish them from *flower-buds* or unopened flowers, are future branches as yet undeveloped; they remain dormant through the winter, and the following spring grow out into new stems bearing leaves and flowers like those of the preceding year, whilst the lower part of the stock emits fresh roots to replace those which had perished at the same time as the stems.

17. Shrubs and trees form similar leaf-buds either at the extremity of their branches, or along the branches of the year. In the latter case these buds are usually *axillary*, that is, they appear in the *axil* of each leaf, *i. e.* in the angle formed by the leaf and the branch. When they appear at any other part of the plant they are called *adventitious*. If these buds by producing roots (19) become distinct plants before separating from the parent, or if adventitious leaf-buds are produced in the place of flowers or seeds, the plant is said to be *viviparous* or *proliferous*.

§ 2. *The Root.*

18. **Roots** ordinarily produce neither buds, leaves, nor flowers. Their branches, called *fibres* when slender and long, proceed irregularly from any part of their surface.

19. Although roots proceed usually from the base of the stem or stock, they may also be produced from the base of any bud, especially if the bud lie along the ground, or is otherwise placed by nature or art in circumstances favourable for their development, or indeed occasionally from almost any part of the plant. They are then often distinguished as *adventitious*, but this term is by some applied to all roots which are not in prolongation of the original radicle.

20. **Roots** are

fibrous, when they consist chiefly of slender fibres.

tuberous, when either the main root or its branches are thickened into one or more short fleshy or woody masses called *tubers* (25).

taproots, when the main root descends perpendicularly into the earth, emitting only very small fibrous branches.

21. The stock of a herbaceous perennial, or the lower part of the stem of an annual or perennial, or the lowest branches of a plant, are sometimes underground and assume the appearance of a root. They then take the name of *rhizome*. The rhizome may always be distinguished from the true root by the presence or production of one or more buds, or leaves, or scales.

§ 3. *The Stock.*

22. The **Stock** of a herbaceous perennial, in its most complete state, includes a small portion of the summits of the previous year's roots, as well as of the base of the previous year's stems. Such stocks will increase yearly, so as at length to form dense tufts. They will often preserve through the winter a few leaves, amongst which are placed the buds which grow out into stems the following year, whilst the under side of the stock emits new roots from or amongst the remains of the old ones. These perennial stocks only differ from the permanent base of an undershrub in the shortness of the perennial part of the stems and in their texture usually less woody.

23. In some perennials, however, the stock consists merely of a branch, which proceeds in autumn from the base of the stem either aboveground or underground, and produces one or more buds. This branch, or a portion of it, alone survives the winter. In the following year its buds produce the new stem and roots, whilst the rest of the plant, even the branch on which these buds were formed, has died away. These *annual stocks*, called sometimes *hybernacula*, *offsets*, or *stolons*, keep up the communication between the annual stem and root of one year and those of the following year, thus forming altogether a perennial plant.

24. The stock, whether annual or perennial, is often entirely underground or root-like. This is the *rootstock*, to which some botanists limit the meaning of the term *rhizome*. When the stock is entirely root-like, it is popularly called the *crown* of the root.

25. The term *tuber* is applied to a short thick, more or less succulent rootstock or rhizome, as well as to a root of that shape (20), although some botanists propose to restrict its meaning to the one or to the other. An Orchis tuber, called by some a *knob*, is an annual tuberous rootstock with one bud at the top. A potato is an annual tuberous rootstock with several buds.

26. A *bulb* is a stock of a shape approaching to globular, usually rather conical above and flattened underneath, in which the bud or buds are concealed, or nearly so, under *scales*. These scales are the more or less thickened bases of the decayed leaves of the preceding year, or of the undeveloped leaves of the future year or of both. Bulbs are annual or perennial, usually underground or close to the ground, but occasionally buds in the axils of the upper leaves become transformed into bulbs. Bulbs are said to be *scaly* when their scales are thick and loosely imbricated, *tunicated* when the scales are thinner, broader, and closely rolled round each other in concentric layers.

27. A *corm* is a tuberous rootstock, usually annual, shaped like a bulb, but in which the bud or buds are not covered by scales, or of which the scales are very thin and membranous.

§ 4. *The Stem.*

28. **Stems** are

erect, when they ascend perpendicularly from the root or stock; *twiggy* or *virgate*, when at the same time they are slender, stiff, and scarcely branched.

sarmentose, when the branches of a woody stem are long and weak, although scarcely climbing.

decumbent or *ascending*, when they spread horizontally, or nearly so, at the base, and then turn upwards and become erect.

procumbent, when they spread along the ground the whole or the greater portion of their length; *diffuse*, when at the same time very much and rather loosely branched.

prostrate, when they lie still closer to the ground.

creeping, when they emit roots at their nodes. This term is also frequently applied to any rhizomes or roots which spread horizontally.

tufted or *cæspitose*, when very short, close, and many together from the same stock.

29. Weak climbing stems are said to *twine*, when they support themselves by winding spirally round any object; such stems are also called *voluble*. When they simply climb without twining, they support themselves by their leaves, or by special clasping organs called *tendrils* (169), or sometimes, like the Ivy, by small root-like excrescences.

30. *Suckers* are young plants formed at the end of creeping, underground rootstocks. *Scions*, *runners*, and *stolons*, or *stoles*, are names given to young plants formed at the end or at the nodes (31) of branches or stocks creeping wholly or partially above-ground, or sometimes to the creeping stocks themselves.

31. A *node* is a point of the stem or its branches at which one or more leaves, branches, or leaf-buds (16) are given off. An *internode* is the portion of the stem comprised between two nodes.

32. **Branches** or **leaves** are

opposite, when two proceed from the same node on opposite sides of the stem.

whorled or *verticillate* (in a *whorl* or *verticil*), when several proceed from the same node, arranged regularly round the stem; *geminate*, *ternate*, *fascicled*, or *fasciculate*, when two, three, or more proceed from the same node on the same side of the stem. A tuft of fasciculate leaves is usually in fact an axillary leafy branch, so short that the leaves appear to proceed all from the same point.

alternate, when one only proceeds from each node, one on one side and the next above or below on the opposite side of the stem.

decussate, when opposite, but each pair placed at right angles to the next pair above or below it; *distichous*, when regularly arranged one above another in two opposite rows, one on each side of the stem; *tristichous*, when in three rows, etc. (92).

scattered, when irregularly arranged round the stem; frequently, however, botanists apply the term *alternate* to all branches or leaves that are neither opposite nor whorled.

secund when all start from or are turned to one side of the stem.

33. **Branches** are *dichotomous*, when several times forked, the two branches of each fork being nearly equal; *trichotomous*, when there are three nearly equal branches at each division instead of two; but when the middle branch is evidently the principal one, the stem is usually said to have two opposite branches; *umbellate*, when divided in the same manner into several nearly equal branches proceeding from the same point. If, however, the central branch is larger than the two or more lateral ones, the stem is said to have opposite or whorled branches, as the case may be.

34. A *culm* is a name sometimes given to the stem of Grasses, Sedges, and some other Monocotyledonous plants.

§ 5. *The Leaves.*

35. The ordinary or perfect **Leaf** consists of a flat *blade* or *lamina*, usually green, and more or less horizontal, attached to the stem by a stalk called a *footstalk* or *petiole*. When the form or dimensions of a leaf are spoken of, it is generally the blade that is meant, without the petiole or stalk.

36. The end by which a leaf, a part of the flower, a seed, or any other organ, is

attached to the stem or other organ, is called its *base*, the opposite end is its *apex* or summit, excepting sometimes in the case of anther-cells (115).

37. **Leaves** are

sessile, when the blade rests on the stem without the intervention of a petiole.

amplexicaul or *stem-clasping*, when the sessile base of the blade clasps the stem horizontally.

perfoliate, when the base of the blade not only clasps the stem, but closes round it on the opposite side, so that the stem appears to pierce through the blade.

decurrent, when the edges of the leaf are continued down the stem so as to form raised lines or narrow appendages, called *wings*.

sheathing, when the base of the blade, or of the more or less expanded petiole, forms a vertical sheath round the stem for some distance above the node.

38. Leaves and flowers are called *radical*, when inserted on a rhizome or stock, or so close to the base of the stem as to appear to proceed from the root, rhizome, or stock; *cauline*, when inserted on a distinct stem. Radical leaves are *rosulate* when they spread in a circle on the ground.

39. **Leaves** are

simple and *entire*, when the blade consists of a single piece, with the margin nowhere indented, *simple* being used in opposition to *compound*, *entire* in opposition to *dentate*, *lobed*, or *divided*.

ciliate, when bordered with thick hairs or fine hair-like teeth.

dentate or *toothed*, when the margin is only cut a little way in, into what have been compared to teeth. Such leaves are *serrate*, when the teeth are regular and pointed like the teeth of a saw; *crenate*, when regular and blunt or rounded (compared to the battlements of a tower); *serrulate* and *crenulate*, when the serratures or crenatures are small; *sinuate*, when the teeth are broad, not deep, and irregular (compared to bays of the coast); *wavy* or *undulate*, when the edges are not flat, but bent up and down (compared to the waves of the sea).

lobed or *cleft*, when more deeply indented or divided, but so that the incisions do not reach the midrib or petiole. The portions thus divided take the name of *lobes*. When the lobes are narrow and very irregular, the leaves are said to be *laciniate*. The spaces between the teeth or lobes are called *sinuses*.

divided or *dissected*, when the incisions reach the midrib or petiole, but the parts so divided off, called *segments*, do not separate from the petiole, even when the leaf falls, without tearing.

compound, when divided to the midrib or petiole, and the parts so divided off, called *leaflets*, separate, at least at the fall of the leaf, from the petiole, as the whole leaf does from the stem, without tearing. The common stalk upon which the leaflets are inserted is called the *common petiole* or the *rhachis*; the separate stalk of each leaflet is a *petiolule*.

40. Leaves are more or less marked by *veins*, which, starting from the stalk, diverge or branch as the blade widens, and spread all over it more or less visibly. The principal ones, when prominent, are often called *ribs* or *nerves*, the smaller branches only then retaining the name of *veins*, or the latter are termed *veinlets*. The smaller veins are often connected together like the meshes of a net, they are then said to *anastomose*, and the leaf is said to be *reticulate* or *net-veined*. When one principal vein runs direct from the stalk towards the summit of the leaf, it is called the *midrib*. When several start from the stalk, diverge slightly without branching, and converge again towards the summit, they are said to be *parallel*, although not mathematically so. When 3 or 5 or more ribs or nerves diverge *from the base*, the leaf is said to be 3-*nerved*, 5-*nerved*, etc., but if the lateral ones diverge from the midrib a little above the base, the leaf is *triplinerved*, *quintuplinerved*, etc. The arrangement of the veins of a leaf is called their *venation*.

41. The **Leaflets, Segments, Lobes,** or **Veins** of leaves are

pinnate (feathered), when there are several succeeding each other on each side of the midrib or petiole, compared to the branches of a feather. A pinnately lobed or divided leaf is called *lyrate* when the terminal lobe or segment is much larger and broader than the lateral ones, compared, by a stretch of imagination, to a lyre; *run-*

cinate, when the lateral lobes are curved backwards towards the base of the leaf; *pectinate*, when the lateral lobes are numerous, narrow, and regular, like the teeth of a comb.

palmate or *digitate*, when several diverge from the same point, compared to the fingers of the hand.

ternate, when three only start from the same point, in which case the distinction between the palmate and pinnate arrangement often ceases, or can only be determined by analogy with allied plants. A leaf with ternate lobes is called *trifid*. A leaf with three leaflets is sometimes improperly called a ternate leaf: it is the leaflets that are ternate; the whole leaf is *trifoliolate*. Ternate leaves are leaves growing three together.

pedate, when the division is at first ternate, but the two outer branches are forked, the outer ones of each fork again forked, and so on, and all the branches are near together at the base, compared vaguely to the foot of a bird.

42. Leaves with pinnate, palmate, pedate, etc., leaflets, are usually for shortness called *pinnate, palmate, pedate*, etc., *leaves*. If they are so cut into segments only, they are usually said to be *pinnatisect, palmatisect, pedatisect*, etc., although the distinction between segments and leaflets is often unheeded in descriptions, and cannot indeed always be ascertained. If the leaves are so cut only into lobes, they are said to be *pinnatifid, palmatifid, pedatifid*, etc.

43. The teeth, lobes, segments, or leaflets, may be again toothed, lobed, divided, or compounded. Some leaves are even three or more times divided or compounded. In the latter case they are termed *decompound*. When twice or thrice pinnate (*bipinnate* or *tripinnate*), each primary or secondary division, with the leaflets it comprises, is called a *pinna*. When the pinna of a leaf or the leaflets of a pinna are in pairs, without an odd terminal pinna or leaflet, the leaf or pinna so divided is said to be *abruptly pinnate;* if there is an odd terminal pinna or leaflet, the leaf or pinna is *unequally pinnate* (*imparipinnatum*).

44. The number of leaves or their parts is expressed adjectively by the following numerals, derived from the Latin :—

uni-,	bi-,	tri-,	quadri-,	quinque-,	sex-,	septem-,	octo-,	novem-,	decem-,	multi-
1-,	2-,	3-,	4-,	5-,	6-,	7-,	8-,	9-,	10-,	*many-*

prefixed to a termination, indicating the particular kind of part referred to. Thus—

unidentate, bidentate, multidentate, mean one-toothed, two-toothed, many-toothed, etc.

bifid, trifid, multifid, mean two lobed, three-lobed, many-lobed, etc.

unifoliolate, bifoliolate, multifoliolate, mean having one leaflet, two leaflets, many leaflets, etc.

unifoliate, bifoliate, multifoliate, mean having one leaf, two leaves, many leaves, etc.

biternate and *triternate*, mean twice or thrice ternately divided.

unijugate, bijugate, multijugate, etc., pinnæ or leaflets, mean that they are in one, two, many, etc., pairs (*juga*).

45. **Leaves** or their parts, when **flat,** or any other flat organs in plants, are

linear, when long and narrow, at least four or five times as long as broad, falsely compared to a mathematical line, for a linear leaf has always a perceptible breadth.

lanceolate, when about three or more times as long as broad, broadest below the middle, and tapering towards the summit, compared to the head of a lance.

cuneate, when broadest above the middle, and tapering towards the base, compared to a wedge with the point downwards; when very broadly cuneate and rounded at the top, it is often called *flabelliform* or *fan-shaped*.

spathulate, when the broad part near the top is short, and the narrow tapering part long, compared to a spatula or flat ladle.

ovate, when scarcely twice as long as broad, and rather broader below the middle, compared to the longitudinal section of an egg; *obovate* is the same form, with the broadest part above the middle.

orbicular, oval, oblong, elliptical, rhomboidal, etc., when compared to the corresponding mathematical figures.

transversely oblong, or *oblate*, when conspicuously broader than long.

falcate, when curved like the blade of a scythe.

46. Intermediate forms between any two of the above are expressed by combining two terms. Thus, a *linear-lanceolate* leaf is long and narrow, yet broader below the middle, and tapering to a point; a *linear-oblong* one is scarcely narrow enough to be called linear, yet too narrow to be strictly oblong, and does not conspicuously taper either towards the summit or towards the base.

47. The *apex* or *summit* of a leaf is

acute or *pointed*, when it forms an acute angle or tapers to a point.

obtuse or *blunt*, when it forms a very obtuse angle, or more generally when it is more or less rounded at the top.

acuminate or *cuspidate*, when suddenly narrowed at the top, and then more or less prolonged into an *acumen* or *point*, which may be acute or obtuse, linear or tapering. Some botanists make a slight difference between the *acuminate* and *cuspidate* apex, the acumen being more distinct from the rest of the leaf in the latter case than in the former; but in general the two terms are used in the same sense, some preferring the one and some the other.

truncate, when the end is cut off square.

retuse, when very obtuse or truncate, and slightly indented.

emarginate or *notched*, when more decidedly indented at the end of the midrib; *obcordate*, if at the same time approaching the shape of a heart with its point downwards.

mucronate, when the midrib is produced beyond the apex in the form of a small point.

aristate, when the point is fine like a hair.

48. The base of the leaf is liable to the same variations of form as the apex, but the terms more commonly used are *tapering* or *narrowed* for acute and acuminate, *rounded* for obtuse, and *cordate* for emarginate. In all cases the petiole or point of attachment prevent any such absolute termination at the base as at the apex.

49. A leaf may be *cordate* at the base whatever be its length or breadth, or whatever the shape of the two lateral lobes, called *auricles* (or *little ears*), formed by the indenture or notch, but the term *cordiform* or *heart-shaped* leaf is restricted to an ovate and acute leaf, cordate at the base, with rounded auricles. The word auricles is more particularly used as applied to sessile and stem-clasping leaves.

50. If the auricles are pointed, the leaf is more particularly called *auriculate;* it is moreover said to be *sagittate*, when the points are directed downwards, compared to an arrow-head; *hastate*, when the points diverge horizontally, compared to a halbert.

51. A *reniform* leaf is broader than long, slightly but broadly cordate at the base, with rounded auricles, compared to a kidney.

52. In a *peltate* leaf, the stalk, instead of proceeding from the lower edge of the blade, is attached to the under surface, usually near the lower edge, but sometimes in the very centre of the blade. The peltate leaf has usually several principal nerves radiating from the point of attachment, being, in fact, a cordate leaf, with the auricles united.

53. All these modifications of division and form in the leaf pass so gradually one into the other that it is often difficult to say which term is the most applicable—whether the leaf be toothed or lobed, divided or compound, oblong or lanceolate, obtuse or acute, etc. The choice of the most apt expression will depend on the skill of the describer.

54. **Leaves,** when **solid, Stems, Fruits, Tubers,** and other parts of plants, when not flattened like ordinary leaves, are

setaceous or *capillary*, when very slender like bristles or hairs.

acicular, when very slender, but stiff and pointed like needles.

subulate, when rather thicker and firmer like awls.

linear, when at least four times as long as thick; *oblong*, when from about two to about four times as long as thick, the terms having the same sense as when applied to flat surfaces.

ovoid, when egg-shaped, with the broad end downwards, *obovoid* if the broad end is upwards; these terms corresponding to *ovate* and *obovate* shapes in flat surfaces.

globular or *spherical*, when corresponding to *orbicular* in a flat surface. *Round* applies to both.

turbinate, when shaped like a top.

conical, when tapering upwards; *obconical*, when tapering downwards, if in both cases a transverse section shows a circle.

pyramidal, when tapering upwards; *obpyramidal*, when tapering downwards, if in both cases a transverse section shows a triangle or polygon.

fusiform, or spindle-shaped, when tapering at both ends; *cylindrical*, when not tapering at either end, if in both cases the transverse section shows a circle, or sometimes irrespective of the transverse shape.

terete, when the transverse section is not angular; *trigonous*, *triquetrous*, if the transverse section shows a triangle, irrespective in both cases of longitudinal form.

compressed, when more or less flattened laterally; *depressed*, when more or less flattened vertically, or at any rate at the top; *obcompressed* (in the achenes of *Compositæ*), when flattened from front to back.

articulate or *jointed*, if at any period of their growth (usually when fully formed and approaching their decay, or in the case of fruits when quite ripe) they separate, without tearing, into two or more pieces placed end to end. The joints where they separate are called *articulations*, each separate piece an *article*. The name of *joint* is, in common language, given both to the articulation and the article, but more especially to the former. Some modern botanists, however, propose to restrict it to the article, giving the name of *joining* to the articulation.

didymous, when slightly two-lobed, with rounded obtuse lobes.

moniliform, or beaded, when much contracted at regular intervals, but not separating spontaneously into articles.

55. In their consistence **Leaves** or other organs are

fleshy, when thick and soft; *succulent* is generally used in the same sense, but implies the presence of more juice.

coriaceous, when firm and stiff, or very tough, of the consistence of leather.

crustaceous, when firm and brittle.

membranous, when thin and not stiff.

scarious or *scariose*, when very thin, more or less transparent and not green, yet rather stiff.

56. The terms applied botanically to the consistence of solids are those in general use in common language.

57. The mode in which unexpanded leaves are disposed in the leaf-bud is called their *vernation* or *præfoliation*; it varies considerably, and technical terms have been proposed to express some of its varieties, but it has been hitherto rarely noticed in descriptive botany.

§ 6. *Scales, Bracts, and Stipules.*

58. **Scales** (*Squamæ*) are leaves very much reduced in size, usually sessile, seldom green or capable of performing the respiratory functions of leaves. In other words, they are organs resembling leaves in their position on the plant, but differing in size, colour, texture, and functions. They are most frequent on the stock of perennial plants, or at the base of annual branches, especially on the buds of future shoots, when they serve apparently to protect the dormant living germ from the rigour of winter. In the latter case they are usually short, broad, close together, and more or less *imbricated*, that is, overlapping each other like the tiles of a roof. It is this arrangement as well as their usual shape that has suggested the name of *scales*, borrowed from the scales of a fish. Imbricated scales, bracts, or leaves, are said to be *squarrose*, when their tips are pointed and very spreading or recurved.

59. Sometimes, however, most or all the leaves of the plant are reduced to small scales, in which case they do not appear to perform any particular function. The name of *scales* is also given to any small broad scale-like appendages or reduced organs, whether in the flower or any other part of the plant.

60. **Bracts** (*Bracteæ*) are the upper leaves of a plant in flower (either all those of the flowering branches, or only one or two immediately under the flower), when differ-

ent from the stem-leaves in size, shape, colour, or arrangement. They are generally much smaller and more sessile. They often partake of the colour of the flower, although they very frequently also retain the green colour of the leaves. When small they are often called *scales*.

61. *Floral leaves* or *leafy bracts* are generally the lower bracts on the upper leaves at the base of the flowering branches, intermediate in size, shape, or arrangement, between the stem-leaves and the upper bracts.

62. *Bracteoles* are the one or two last bracts under each flower, when they differ materially in size, shape, or arrangement from the other bracts.

63. **Stipules** are leaf-like or scale-like appendages at the base of the leaf-stalk, or on the node of the stem. When present there are generally two, one on each side of the leaf, and they sometimes appear to protect the young leaf before it is developed. They are however exceedingly variable in size and appearance, sometimes exactly like the true leaves except that they have no buds in their axils, or looking like the leaflets of a compound leaf, sometimes apparently the only leaves of the plant; generally small and narrow, sometimes reduced to minute scales, spots or scars, sometimes united into one opposite the leaf, or more or less united with, or *adnate* to the petiole, or quite detached from the leaf, and forming a ring or sheath round the stem in the axil of the leaf. In a great number of plants they are entirely wanting.

64. *Stipellæ*, or secondary stipules, are similar organs, sometimes found on compound leaves at the points where the leaflets are inserted.

65. When scales, bracts, or stipules, or almost any part of the plant besides leaves and flowers are stalked, they are said to be *stipitate*, from *stipes*, a *stalk*.

§ 7. *Inflorescence and its Bracts.*

66. The **Inflorescence** of a plant is the arrangement of the flowering branches, and of the flowers upon them. *An Inflorescence* is a flowering branch, or the flowering summit of a plant above the last stem-leaves, with its branches, bracts, and flowers.

67. A single flower, or an inflorescence, is *terminal* when at the summit of a stem or leafy branch, *axillary* when in the axil of a stem-leaf, *leaf-opposed* when opposite to a stem-leaf. The inflorescence of a plant is said to be *terminal* or *determinate* when the main stem and principal branches end in a flower or inflorescence (not in a leaf-bud), *axillary* or *indeterminate* when all the flowers or inflorescences are axillary, the stem or branches ending in leaf-buds.

68. A *Peduncle* is the stalk of a solitary flower, or of an inflorescence; that is to say, the portion of the flowering branch from the last stem-leaf to the flower, or to the first ramification of the inflorescence, or even up to its last ramifications; but the portion extending from the first to the last ramifications or the axis of inflorescence is often distinguished under the name of *rhachis*.

69. A *Scape* or *radical Peduncle* is a leafless peduncle proceeding from the stock, or from near the base of the stem, or apparently from the root itself.

70. A *Pedicel* is the last branch of an inflorescence, supporting a single flower.

71. The branches of inflorescences may be, like those of stems, opposite, alternate, etc. (32, 33), but very often their arrangement is different from that of the leafy branches of the same plant.

72. **Inflorescence** is

centrifugal, when the terminal flower opens first, and those on the lateral branches are successively developed.

centripetal, when the lowest flowers open first, and the main stem continues to elongate, developing fresh flowers.

73. Determinate inflorescence is usually centrifugal. Indeterminate inflorescence is always centripetal. Both inflorescences may be combined on one plant, for it often happens that the main branches of an inflorescence are centripetal, whilst the flowers on the lateral branches are centrifugal; or *vice versâ*.

74. An **Inflorescence** is

a *Spike*, or *spicate*, when the flowers are sessile along a simple undivided axis or rhachis.

a *Raceme*, or *racemose*, when the flowers are borne on pedicels along a single undivided axis or rhachis.

a *Panicle*, or *paniculate*, when the axis is divided into branches bearing two or more flowers.

a *Head*, or *capitate*, when several sessile or nearly sessile flowers are collected into a compact head-like cluster. The short, flat, convex or conical axis on which the flowers are seated, is called the *receptacle*, a term also used for the torus of a single flower (135). The very compact flower-heads of *Compositæ* are often termed *compound flowers*.

an *Umbel*, or *umbellate*, when several branches or pedicels appear to start from the same point and are nearly of the same length. It differs from the head, like the raceme from the spike, in that the flowers are not sessile. An umbel is said to be *simple*, when each of its branches or *rays* bears a single flower; *compound*, when each ray bears a *partial umbel* or *umbellule*.

a *Corymb*, or *corymbose*, when the branches and pedicels, although starting from different points, all attain the same level, the lower ones being much longer than the upper. It is a flat-topped or *fastigiate* panicle.

a *Cyme*, or *cymose*, when branched and centrifugal. It is a centrifugal panicle, and is often corymbose. The central flower opens first. The lateral branches successively developed are usually forked or opposite (dichotomous or trichotomous), but sometimes after the first forking the branches are no longer divided, but produce a succession of pedicels on their upper side forming apparently unilateral centripetal racemes; whereas if attentively examined, it will be found that each pedicel is at first terminal, but becomes lateral by the development of one outer branch only, immediately under the pedicel. Such branches, when in bud, are generally rolled back at the top, like the tail of a scorpion, and are thence called *scorpioid*.

a *Thyrsus*, or *thyrsoid*, when cymes, usually opposite, are arranged in a narrow pyramidal panicle.

75. There are numerous cases where inflorescences are intermediate between some two of the above, and are called by different botanists by one or the other name, according as they are guided by apparent or by theoretical similarity. A spike-like panicle, where the axis is divided into very short branches forming a cylindrical compact inflorescence, is called sometimes a spike, sometimes a panicle. If the flowers are in distinct clusters along a simple axis, the inflorescence is described as an *interrupted* spike or raceme, according as the flowers are nearly sessile or distinctly pedicellate; although when closely examined the flowers will be found to be inserted not on the main axis, but on a very short branch, thus, strictly speaking, constituting a panicle.

76. The *catkins* (*amenta*) of *Amentaceæ*, the *spadices* of several Monocotyledons, the *ears* and *spikelets* of Grasses are forms of the spike.

77. **Bracts** are generally placed singly under each branch of the inflorescence, and under each pedicel; bracteoles are usually two, one on each side, on the pedicel or close under the flower, or even upon the calyx itself; but bracts are also frequently scattered along the branches without axillary pedicels; and when the differences between the bracts and bracteoles are trifling or immaterial, they are usually all called bracts.

78. When three bracts appear to proceed from the same point, they will, on examination, be found to be really either one bract and two stipules, or one bract with two bracteoles in its axil. When two bracts appear to proceed from the same point, they will usually be found to be the stipules of an undeveloped bract, unless the branches of the inflorescence are opposite, when the bracts will of course be opposite also.

79. When several bracts are collected in a whorl, or are so close together as to appear whorled, or are closely imbricated round the base of a head or umbel, they are collectively called an *Involucre*. The bracts composing an involucre are described under the names of *leaves*, *leaflets*, *bracts*, or *scales*, according to their appearance. *Phyllaries* is a useless term, lately introduced for the bracts or scales of the involucre of *Compositæ*. An *Involucel* is the involucre of a partial umbel.

80. When several very small bracts are placed round the base of a calyx or of an

involucre, they have been termed a *calycule*, and the calyx or involucre said to be *calyculate*, but these terms are now falling into disuse, as conveying a false impression.

81. A *Spatha* is a bract or floral leaf enclosing the inflorescence of some Monocotyledons.

82. *Paleæ*, *Pales*, or *Chaff*, are the inner bracts or scales in *Compositæ*, *Gramineæ*, and some other plants, when of a thin yet stiff consistence, usually narrow and of a pale colour.

83. *Glumes* are the bracts enclosing the flowers of *Cyperaceæ* and *Gramineæ*.

§ 8. *The Flower in General.*

84. A *complete* **Flower** (15) is one in which the calyx, corolla, stamens, and pistils are all present; a *perfect* flower, one in which all these organs, or such of them as are present, are capable of performing their several functions. Therefore, properly speaking, an *incomplete* flower is one in which any one or more of these organs is wanting; and an *imperfect* flower, one in which any one or more of these organs is so altered as to be incapable of properly performing its functions. These imperfect organs are said to be *abortive* if much reduced in size or efficiency, *rudimentary* if so much so as to be scarcely perceptible. But, in many works, the term *incomplete* is specially applied to those flowers in which the perianth is simple or wanting, and *imperfect* to those in which either the stamens or pistil are imperfect or wanting.

85. A **Flower** is

dichlamydeous, when the perianth is double, both calyx and corolla being present and distinct.

monochlamydeous, when the perianth is single, whether by the union of the calyx and corolla, or the deficiency of either.

asepalous, when there is no calyx.

apetalous, when there is no corolla.

naked, when there is no perianth at all.

hermaphrodite or *bisexual*, when both stamens and pistil are present and perfect.

male or *staminate*, when there are one or more stamens, but either no pistil at all or an imperfect one.

female or *pistillate*, when there is a pistil, but either no stamens at all, or only imperfect ones.

neuter, when both stamens and pistil are imperfect or wanting.

barren or *sterile*, when from any cause it produces no seed.

fertile, when it does produce seed. In some works the terms *barren*, *fertile*, and *perfect* are also used respectively as synonyms of *male*, *female*, and *hermaphrodite*.

86. The flowers of a plant or species are said collectively to be *unisexual* or *diclinous* when the flowers are all either male or female.

monœcious, when the male and female flowers are distinct, but on the same plant.

diœcious, when the male and female flowers are on distinct plants.

polygamous, when there are male, female, and hermaphrodite flowers on the same or on distinct plants.

87. A head of flowers is *heterogamous* when male, female, hermaphrodite, and neuter flowers or any two or three of them, are included in one head; *homogamous*, when all the flowers included in one head are alike in this respect. A spike or head of flowers is *androgynous* when male and female flowers are mixed in it. These terms are only used in the case of very few Natural Orders.

88. As the scales of buds are leaves undeveloped or reduced in size and altered in shape and consistence, and bracts are leaves likewise reduced in size, and occasionally altered in colour; so the parts of the flower are considered as leaves still further altered in shape, colour, and arrangement round the axis, and often more or less combined with each other. The details of this theory constitute the comparatively modern branch of botany called *Vegetable Metamorphosis*, or *Homology*, sometimes improperly termed *Morphology* (8).

89. To understand the arrangement of the floral parts, let us take a *complete* flower, in which moreover all the parts are *free* from each other, *definite* in number, *i. e.* always the same in the same species, and *symmetrical* or *isomerous*, *i. e.* when each whorl consists of the same number of parts.

90. Such a complete symmetrical flower consists usually of either four or five whorls of altered leaves (88), placed immediately one within the other.

The **Calyx** forms the outer whorl. Its parts are called *sepals*.

The **Corolla** forms the next whorl. Its parts, called *petals*, usually *alternate* with the sepals; that is to say, the centre of each petal is immediately over or within the interval between two sepals.

The **Stamens** form one or two whorls within the petals. If two, those of the outer whorl (the *outer stamens*) alternate with the petals, and are consequently opposite to, or over the centre of the sepals; those of the inner whorl (the *inner stamens*) alternate with the outer ones, and are therefore opposite to the petals. If there is only one whorl of stamens, they most frequently alternate with the petals; but sometimes they are opposite the petals and alternate with the sepals.

The **Pistil** forms the inner whorl; its carpels usually alternate with the inner row of stamens.

91. In an axillary or lateral flower the *upper* parts of each whorl (sepals, petals, stamens, or carpels) are those which are next to the main axis of the stems or branch, the *lower* parts those which are furthest from it; the intermediate ones are said to be *lateral*. The words *anterior* (front) and *posterior* (back) are often used for lower and upper respectively, but their meaning is sometimes reversed if the writer supposes himself in the centre of the flower instead of outside of it.

92. The number of parts in each whorl of a flower is expressed adjectively by the following numerals derived from the Greek:—

mono-,	di-,	tri-,	tetra-,	penta-,	hexa-,	hepta,	octo-,	ennea-,	deca-, etc.,	poly-
1-,	2-,	3-,	4-,	5-,	6-,	7-,	8-,	9-,	10-,	*many-*

prefixed to a termination indicating the whorl referred to.

93. Thus, a **Flower** is

disepalous, *trisepalous*, *tetrasepalous*, *polysepalous*, etc., according as there are 2, 3, 4, or many (or an indefinite number of) sepals.

dipetalous, *tripetalous*, *polypetalous*, etc., according as there are 2, 3, or many petals.

diandrous, *triandrous*, *polyandrous*, etc., according as there are 2, 3, or many stamens.

digynous, *trigynous*, *polygynous*, etc., according as there are 2, 3, or many carpels.

And generally (if symmetrical), *dimerous*, *trimerous*, *polymerous*, etc., according as there are 2, 3, or many (or an indefinite number of) parts to each whorl.

94. Flowers are *unsymmetrical* or *anisomerous*, strictly speaking, when any one of the whorls has a different number of parts from any other; but when the pistils alone are reduced in number, the flower is still frequently called symmetrical or isomerous, if the calyx, corolla, and staminal whorls have all the same number of parts.

95. Flowers are *irregular* when the parts of any one of the whorls are unequal in size, dissimilar in shape, or do not spread regularly round the axis at equal distances. It is however more especially irregularity of the corolla that is referred to in descriptions. A slight inequality in size or direction in the other whorls does not prevent the flower being classed as *regular*, if the corolla or perianth is conspicuous and regular.

§ 9. *The Calyx and Corolla, or Perianth.*

96. The **Calyx** (90) is usually green, and smaller than the corolla; sometimes very minute, rudimentary, or wanting, sometimes very indistinctly whorled, or not whorled at all, or in two whorls, or composed of a large number of sepals, of which the outer ones pass gradually into bracts, and the inner ones into petals.

97. The **Corolla** (90) is usually coloured, and of a more delicate texture than the calyx, and, in popular language, is often more specially meant by the *flower*. Its petals are more rarely in two whorls, or indefinite in number, and the whorl more rarely broken than in the case of the calyx, at least when the plant is in a natural state. *Double flowers* are in most cases an accidental deformity or monster in which the ordinary number of petals is multiplied by the conversion of stamens, sepals, or even carpels into petals, by the division of ordinary petals, or simply by the addition of supernumerary ones. Petals are also sometimes very small, rudimentary, or entirely deficient.

98. In very many cases, a so-called *simple perianth* (15) (of which the parts are usually called *leaves* or *segments*) is one in which the sepals and petals are similar in form and texture, and present apparently a single whorl. But if examined in the young bud, one half of the parts will generally be found to be placed outside the other half, and there will frequently be some slight difference in texture, size, and colour, indicating to the close observer the presence of both calyx and corolla. Hence much discrepancy in descriptive works. Where one botanist describes a simple perianth of six segments, another will speak of a double perianth of three sepals and three petals.

99. The following terms and prefixes, expressive of the modifications of form and arrangement of the corolla and its petals, are equally applicable to the calyx and its sepals, and to the simple perianth and its segments.

100. The Corolla is said to be *monopetalous* when the petals are united, either entirely or at the base only, into a cup, tube, or ring; *polypetalous* when they are all free from the base. These expressions, established by a long usage, are not strictly correct, for *monopetalous* (consisting of a single petal) should apply rather to a corolla really reduced to a single petal, which would then be on one side of the axis; and *polypetalous* is sometimes used more appropriately for a corolla with an indefinite number of petals. Some modern botanists have therefore proposed the term *gamopetalous* for the corolla with united petals, and *dialypetalous* for that with free petals; but the old-established expressions are still the most generally used.

101. When the petals are partially united, the lower entire portion of the corolla is called the *tube*, whatever be its shape, and the free portions of the petals are called the *teeth*, *lobes*, or *segments* (39), according as they are short or long in proportion to the whole length of the corolla. When the tube is excessively short, the petals appear at first sight free, but their slight union at the base must be carefully attended to, being of importance in classification.

102. The **Æstivation** of a corolla, is the arrangement of the petals, or of such portion of them as is free, in the unexpanded bud. It is

valvate, when they are strictly whorled in their whole length, their edges being placed against each other without overlapping. If the edges are much inflexed, the æstivation is at the same time *induplicate; involute*, if the margins are rolled inward; *reduplicate*, if the margins project outwards into salient angles; *revolute*, if the margins are rolled outwards; *plicate*, if the petals are folded in longitudinal plaits.

imbricate, when the whorl is more or less broken by some of the petals being outside the others, or by their overlapping each other at least at the top. Five-petaled imbricate corollas are *quincuncially* imbricate when one petal is outside, and an adjoining one wholly inside, the three others intermediate and overlapping on one side; *bilabiate*, when two adjoining ones are inside or outside the three others. Imbricate petals are described as *crumpled* (*corrugate*) when puckered irregularly in the bud.

twisted, *contorted*, or *convolute*, when each petal overlaps an adjoining one on one side, and is overlapped by the other adjoining one on the other side. Some botanists include the twisted æstivation in the general term *imbricate;* others carefully distinguish the one from the other.

103. In a few cases the overlapping is so slight that the three æstivations cannot easily be distinguished one from the other; in a few others the æstivation is variable, even in the same species, but, in general, it supplies a constant character in species, in genera, or even in Natural Orders.

104. In general shape the **Corolla** is

tubular, when the whole or the greater part of it is in the form of a tube or cylinder.

campanulate, when approaching in some measure the shape of a cup or bell.

urceolate, when the tube is swollen or nearly globular, contracted at the top, and slightly expanded again in a narrow rim.

rotate or *stellate*, when the petals or lobes are spread out horizontally from the base, or nearly so, like a wheel or star.

hypocrateriform or *salver-shaped*, when the lower part is cylindrical and the upper portion expanded horizontally. In this case the name of *tube* is restricted to the cylindrical part, and the horizontal portion is called the *limb*, whether it be divided to the base or not. The orifice of the tube is called its *mouth* or *throat*.

infundibuliform or *funnel-shaped*, when the tube is cylindrical at the base, but enlarged at the top into a more or less campanulate limb, of which the lobes often spread horizontally. In this case the campanulate part, up to the commencement of the lobes, is sometimes considered as a portion of the tube, sometimes as a portion of the limb, and by some botanists again described as independent of either, under the name of *throat* (*fauces*). Generally speaking, however, in campanulate, infundibuliform, or other corollas, where the lower entire part passes gradually into the upper divided and more spreading part, the distinction between the *tube* and the *limb* is drawn either at the point where the lobes separate, or at the part where the corolla first expands, according to which is the most marked.

105. Irregular corollas have received various names according to the more familiar forms they have been compared to. Some of the most important are the

bilabiate or *two-lipped* corolla, when, in a four- or five-lobed corolla, the two or three upper lobes stand obviously apart, like an *upper lip*, from the two or three lower ones or *underlip*. In *Orchideæ* and some other families the name of lip, or *labellum*, is given to one of the divisions or lobes of the perianth.

personate, when two-lipped, and the orifice of the tube closed by a projection from the base of the upper or lower lip, called a *palate*.

ringent, when very strongly two-lipped, and the orifice of the tube very open.

spurred, when the tube or the lower part of the petal has a conical hollow projection, compared to the spur of a cock; *saccate*, when the spur is short and round like a little bag; *gibbous*, when projecting at any part into a slight swelling; *foveolate*, when marked in any part with a slight glandular or thickened cavity.

resupinate or *reversed*, when a lip, spur, etc which in allied species is usually lowest, lies uppermost, and *vice versâ*.

106. The above terms are mostly applied to the forms of monopetalous corollas, but several are also applicable to those of polypetalous ones. Terms descriptive of the special forms of corolla in certain Natural Orders, will be explained under those Orders respectively.

107. Most of the terms used for describing the forms of leaves (39, 45) are also applicable to those of individual petals; but the flat expanded portion of a petal, corresponding to the blade of the leaf, is called its *lamina*, and the stalk, corresponding to the petiole, its *claw* (*unguis*). The stalked petal is said to be *unguiculate*.

§ 10. *The Stamens.*

108. Although in a few cases the outer stamens may gradually pass into petals, yet, in general, **Stamens** are very different in shape and aspect from leaves, sepals, or petals. It is only in a theoretical point of view (not the less important in the study of the physiological economy of the plant) that they can be called altered leaves.

109. This usual form is a stalk, called the *filament*, bearing at the top an *anther* divided into two pouches or *cells*. These anther-cells are filled with *pollen*, consisting of minute grains, usually forming a yellow dust, which, when the flower expands, is scattered from an opening in each cell. When the two cells are not closely contiguous, the portion of the anther that unites them is called the *connectivum*.

110. The filament is often wanting, and the anther sessile, yet still the stamen is perfect; but if the anther, which is the essential part of the stamen, is wanting, or does not contain pollen, the stamen is imperfect, and is then said to be *barren* or *sterile* (without pollen), *abortive*, or *rudimentary* (84), according to the degree to which the imperfection is carried. Imperfect stamens are often called *staminodia*.

111. In unsymmetrical flowers, the stamens of each whorl are sometimes reduced in number below that of the petals, even to a single one, and in several Natural Orders they are multiplied indefinitely.

112. The terms *monandrous* and *polyandrous* are restricted to flowers which have really but one stamen, or an indefinite number respectively. Where several stamens are united into one, the flower is said to be *synandrous*.

113. **Stamens** are

monadelphous, when united by their filaments into one cluster. This cluster either forms a tube round the pistil, or, if the pistil is wanting, occupies the centre of the flower.

diadelphous, when so united into two clusters. The term is more especially applied to certain *Leguminosæ*, in which nine stamens are united in a tube slit open on the upper side, and a tenth, placed in the slit, is free. In some other plants the stamens are equally distributed in the two clusters.

triadelphous, pentadelphous, polyadelphous, when so united into three, five, or many clusters.

syngenesious, when united by their anthers in a ring round the pistil, the filaments usually remaining free.

didynamous, when (usually in a bilabiate flower) there are four stamens in two pairs, those of one pair longer than those of the other.

tetradynamous, when (in *Cruciferæ*) there are six, four of them longer than the two others.

exserted, when longer than the corolla, or even when longer than its tube, if the limb be very spreading.

114. An **Anther** (109) is

adnate, when continuous with the filament, the anther-cells appearing to lie their whole length along the upper part of the filament.

innate, when firmly attached by their base to the filament. This is like an adnate anther, but rather more distinct from the filament.

versatile, when attached by their back to the very point of the filament, so as to swing loosely.

115. Anther-cells may be *parallel* or *diverging* at a less or greater angle; or *divaricate*, when placed end to end so as to form one straight line. The end of each anther-cell placed nearest to the other cell is generally called its *apex* or *summit*, and the other end its *base* (36); but some botanists reverse the sense of these terms.

116. Anthers have often, on their connectivum or cells, appendages termed *bristles* (setæ), *spurs, crests, points, glands*, etc., according to their appearance.

117. Anthers have occasionally only one cell: this may take place either by the disappearance of the partition between two closely contiguous cells, when these cells are said to be *confluent*; or by the abortion or total deficiency of one of the cells, when the anther is said to be *dimidiate*.

118. Anthers will open or *dehisce* to let out the pollen, like capsules, in *valves, pores*, or *slits*. Their dehiscence is *introrse*, when the opening faces the pistil; *extrorse*, when towards the circumference of the flower.

119. Pollen (109) is not always in the form of dust. It is sometimes collected in each cell into one or two little wax-like masses. Special terms used in describing these masses or other modifications of the pollen will be explained under the Orders where they occur.

§ 11. *The Pistil.*

120. The carpels (91) of the **Pistil**, although they may occasionally assume, rather more than stamens, the appearance and colour of leaves, are still more different in shape and structure. They are usually sessile; if stalked, their stalk is called a *podocarp*. This stalk, upon which each separate carpel is supported above the receptacle, must not be confounded with the *gynobasis* (143), upon which the whole pistil is sometimes raised.

121. Each carpel consists of three parts:

1. The **Ovary**, or enlarged base, which includes one or more cavities or *cells*, containing one or more small bodies called *ovules*. These are the earliest condition of the future seeds.

2. the **Style**, proceeding from the summit of the ovary, and supporting—

3. the **Stigma**, which is sometimes a point (or *punctiform* stigma) or small head (a *capitate* stigma) at the top of the style or ovary, sometimes a portion of its surface more or less lateral and variously shaped, distinguished by a looser texture, and covered with minute protuberances called *papillæ*.

122. The style is often wanting, and the stigma is then sessile on the ovary, but in

the perfect pistil there is always at least one ovule in the ovary, and some portion of stigmatic surface. Without these the pistil is imperfect, and said to be *barren* (not setting seed), *abortive*, or *rudimentary* (84), according to the degree of imperfection.

123. The ovary being the essential part of the pistil, most of the terms relating to the number, arrangement, etc., of the carpels, apply specially to their ovaries. In some works each separate carpel is called a pistil, all those of a flower constituting together the *gynœcium;* but this term is in little use, and the word *pistil* is more generally applied in a collective sense. When the ovaries are at all united, they are commonly termed collectively a compound ovary.

124. The number of carpels or ovaries in a flower is frequently reduced below that of the parts of the other floral whorls, even in flowers otherwise symmetrical. In a very few genera, however, the ovaries are more numerous than the petals, or indefinite. They are in that case either arranged in a single whorl, or form a head or spike in the centre of the flower.

125. The terms *monogynous*, *digynous*, *polygynous*, etc. (with a pistil of one, two, or more parts), are vaguely used, applying sometimes to the whole pistil, sometimes to the ovaries alone, or to the styles or stigmas only. Where a more precise nomenclature is adopted, the flower is

monocarpellary, when the pistil consists of a single simple carpel.

bi-, *tri-*, etc., to *poly-carpellary*, when the pistil consists of two, three, or an indefinite number of carpels, whether separate or united.

syncarpous, when the carpels or their ovaries are more or less united into one compound ovary.

apocarpous, when the carpels or ovaries are all free and distinct.

126. A *compound* ovary is

unilocular or *one-celled*, when there are no partitions between the ovules, or when these partitions do not meet in the centre so as to divide the cavity into several cells.

plurilocular or *several-celled*, when completely divided into two or more cells by partitions called *dissepiments* (*septa*), usually vertical and radiating from the centre or axis of the ovary to its circumference.

bi-, *tri-*, etc., to *multi-locular*, according to the number of these cells, two, three, etc., or many.

127. In general the number of cells or of dissepiments, complete or partial, or of rows of ovules, corresponds with that of the carpels, of which the pistil is composed. But sometimes each carpel is divided completely or partially into two cells, or has two rows of ovules, so that the number of carpels appears double what it really is. Sometimes again the carpels are so completely combined and reduced as to form a single cell, with a single ovule, although it really consist of several carpels. But in these cases the ovary is usually described as it appears, as well as such as it is theoretically supposed to be.

128. In apocarpous pistils the styles are usually free, each bearing its own stigma. Very rarely the greater part of the styles, or the stigmas alone, are united, whilst the ovaries remain distinct.

129. Syncarpous flowers are said to have

several styles, when the styles are free from the base.

one style, with several branches, when the styles are connected at the base, but separate below the point where the stigmas or stigmatic surfaces commence.

one simple style, with several stigmas, when united up to the point where the stigmas or stigmatic surfaces commence, and then separating.

one simple style, with a branched, lobed, toothed, notched, or entire stigma (as the case may be), when the stigmas also are more or less united. In many works, however, this precise nomenclature is not strictly adhered to, and considerable confusion is often the result.

130. In general the number of styles, or branches of the style or stigma, is the same as that of the carpels, but sometimes that number is doubled, especially in the stigmas, and sometimes the stigmas are dichotomously or pinnately branched, or *penicillate*, that is, divided into a tuft of hair-like branches. All these variations sometimes make it a difficult task to determine the number of carpels forming a compound ovary, but the point is of considerable importance in fixing the affinities of plants, and, by careful

consideration, the real as well as the apparent number has now in most cases been agreed upon.

131. The *Placenta* is the part of the inside of the ovary to which the ovules are attached, sometimes a mere point or line on the inner surface, often more or less thickened or raised. *Placentation* is therefore the indication of the part of the ovary to which the ovules are attached.

132. Placentas are

axile, when the ovules are attached to the axis or centre, that is, in plurilocular ovaries, when they are attached to the inner angle of each cell; in unilocular simple ovaries, which have almost always an excentrical style or stigma, when the ovules are attached to the side of the ovary nearest to the style; in unilocular compound ovaries, when the ovules are attached to a central protuberance, column, or axis rising up from the base of the cavity. If this column does not reach the top of the cavity, the placenta is said to be *free* and *central*.

parietal, when the ovules are attached to the inner surface of the cavity of a one-celled compound ovary. Parietal placentas are usually slightly thickened or raised lines, sometimes broad surfaces nearly covering the inner surface of the cavity, sometimes projecting far into the cavity, and constituting partial dissepiments, or even meeting in the centre, but without cohering there. In the latter case the distinction between the one-celled and the several-celled ovary sometimes almost disappears.

133. Each **Ovule** (121), when fully formed, usually consists of a central mass or *nucleus* enclosed in two bag-like *coats*, the outer one called *primine*, the inner one *secundine*. The *chalaza* is the point of the ovule at which the base of the nucleus is confluent with the coats. The *foramen* is a minute aperture in the coats over the *apex* of the nucleus.

134. **Ovules** are

orthotropous or *straight*, when the chalaza coincides with the base (36) of the ovule, and the foramen is at the opposite extremity, the axis of the ovule being straight.

campylotropous or *incurved*, when the chalaza still coinciding with the base of the ovule, the axis of the ovule is curved, bringing the foramen down more or less towards that base.

anatropous or *inverted*, when the chalaza is at the apex of the ovule, and the foramen next to its base, the axis remaining straight. In this, one of the most frequent forms of the ovule, the chalaza is connected with the base by a cord, called the *raphe*, adhering to one side of the ovule, and becoming more or less incorporated with its coats, as the ovule enlarges into a seed.

amphitropous or *half-inverted*, when the ovule being as it were attached laterally, the chalaza and foramen at opposite ends of its straight or curved axis are about equally distant from the base or point of attachment.

§ 12. *The Receptacle and Relative Attachment of the Floral Whorls.*

135. The **Receptacle** or *torus* is the extremity of the peduncle (above the calyx), upon which the corolla, stamens, and ovary are inserted. It is sometimes little more than a mere point or minute hemisphere, but it is often also more or less elongated, thickened, or otherwise enlarged. It must not be confounded with the receptacle of inflorescence (74).

136. *A Disk*, or *disc*, is a circular enlargement of the receptacle, usually in the form of a cup (*cupular*), of a flat disk or quoit, or of a cushion (*pulvinate*). It is either immediately at the base of the ovary within the stamens, or between the petals and stamens, or bears the petals or stamens or both on its margin, or is quite at the extremity of the receptacle, with the ovaries arranged in a ring round it or under it.

137. The disk may be *entire*, or *toothed*, or *lobed*, or *divided* into a number of parts, usually equal to or twice that of the stamens or carpels. When the parts of the disk are quite separate and short, they are often called glands.

138. *Nectaries*, are either the disk, or small deformed petals, or abortive stamens, or appendages at the base of petals or stamens, or any small bodies within the flower which do not look like petals, stamens, or ovaries. They were formerly supposed to

supply bees with their honey, and the term is frequently to be met with in the older Floras, but is now deservedly going out of use.

139. When the disk bears the petals and stamens, it is frequently adherent to, and apparently forms part of, the tube of the calyx, or it is adherent to, and apparently forms part of, the ovary, or of both calyx-tube and ovary. Hence the three following important distinctions in the relative insertion of the floral whorls.

140. Petals, or as it is frequently expressed, flowers, are

hypogynous (*i.e.* under the ovary), when they or the disk that bears them are entirely free both from the calyx and ovary. The ovary is then described as *free* or *superior*, the calyx as *free* or *inferior*, the petals as being *inserted on the receptacle.*

perigynous (*i. e.* round the ovary), when the disk bearing the petals is quite free from the ovary, but is more or less combined with the base of the calyx-tube. The ovary is then still described as *free* or *superior*, even though the combined disk and calyx-tube may form a deep cup with the ovary lying in the bottom ; the calyx is said to be *free* or *inferior*, and the petals are described as *inserted on the calyx.*

epigynous (*i. e.* upon the ovary), when the disk bearing the petals is combined both with the base of the calyx-tube and the base outside of the ovary ; either closing over the ovary so as only to leave a passage for the style, or leaving more or less of the top of the ovary free, but always adhering to it above the level of the insertion of the lowest ovule (except in a very few cases where the ovules are absolutely suspended from the top of the cell). In epigynous flowers the ovary is described as *adherent* or *inferior*, the calyx as *adherent* or *superior*, the petals as *inserted on* or *above the ovary.* In some works, however, most epigynous flowers are included in the perigynous ones, and a very different meaning is given to the term *epigynous* (144), and there are a few cases where no positive distinction can be drawn between the epigynous and perigynous flowers, or again between the perigynous and hypogynous flowers.

141. When there are no petals, it is the insertion of the stamens that determines the difference between the hypogynous, perigynous, and epigynous flowers.

142. When there are both petals and stamens,

in hypogynous flowers, the petals and stamens are usually free from each other, but sometimes they are combined at the base. In that case, if the petals are distinct from each other, and the stamens are monadelphous, the petals are often said to be *inserted on* or *combined with the staminal tube ;* if the corolla is gamopetalous and the stamens distinct from each other, the latter are said to be *inserted in the tube of the corolla.*

in perigynous flowers, the stamens are usually inserted immediately within the petals, or alternating with them on the edge of the disk, but occasionally much lower down within the disk, or even on the unenlarged part of the receptacle.

in epigynous flowers, when the petals are distinct, the stamens are usually inserted as in perigynous flowers ; when the corolla is gamopetalous, the stamens are either free and hypogynous, or combined at the base with (inserted in) the tube of the corolla.

143. When the receptacle is distinctly elongated below the ovary, it is often called a *gynobasis*, *gynophore*, or *stalk of the ovary.* If the elongation takes place below the stamens or below the petals, these stamens or petals are then said to be *inserted on the stalk of the ovary*, and are occasionally, but falsely, described as *epigynous.* Really epigynous stâmens (*i. e.* when the filaments are combined with the ovary) are very rare, unless the rest of the flower is epigynous.

144. An *epigynous disk* is a name given either to the thickened summit of the ovary in epigynous flowers, or very rarely to a real disk or enlargement of the receptacle closing over the ovary.

145. In the relative position of any two or more parts of the flower, whether in the same or in different whorls, they are

connivent, when nearer together at the summit than at the base.

divergent, when further apart at the summit than at the base.

coherent, when united together, but so slightly that they can be separated with little or no laceration ; and one of the two cohering parts (usually the smallest or least important) is said to be *adherent* to the other. Grammatically speaking, these two terms convey nearly the same meaning, but require a different form of phrase ; practically however it has been found more convenient to restrict *cohesion* to the union of parts of the same whorl, and *adhesion* to the union of parts of different whorls.

connate, when so closely united that they cannot be separated without laceration. Each of the two connate parts, and especially that one which is considered the smaller or of the least importance, is said to be *adnate* to the other.

free, when neither coherent nor connate.

distinct is also used in the same sense, but is also applied to parts distinctly visible or distinctly limited.

§ 13. *The Fruit.*

146. The **Fruit** (15) consists of the ovary and whatever other parts of the flower are *persistent* (*i. e.* persist at the time the seed is ripe), usually enlarged, and more or less altered in shape and consistence. It encloses or covers the seed or seeds till the period of maturity, when it either opens for the seed to escape, or falls to the ground with the seed. When stalked, its stalk has been termed a *carpophore*.

147. Fruits are, in elementary works, said to be *simple* when the result of a single flower, *compound* when they proceed from several flowers closely packed or combined in a head. But as a fruit resulting from a single flower, with several distinct carpels, is compound in the sense in which that term is applied to the ovary, the terms *single* and *aggregate*, proposed for the fruit resulting from one or several flowers, may be more appropriately adopted. In descriptive botany a fruit is always supposed to result from a single flower unless the contrary be stated. It may, like the pistil, be syncarpous or apocarpous (125); and as in many cases carpels united in the flower may become separate as they ripen, an apocarpous fruit may result from a syncarpous pistil.

148. The involucre or bracts often persist and form part of aggregate fruits, but very seldom so in single ones.

149. The receptacle becomes occasionally enlarged and succulent; if when ripe it falls off with the fruit, it is considered as forming part of it.

150. The adherent part of the calyx of epigynous flowers always persists and forms part of the fruit; the free part of the calyx of epigynous flowers or the calyx of perigynous flowers, either persists entirely at the top of or round the fruit, or the lobes alone fall off, or the lobes fall off with whatever part of the calyx is above the insertion of the petals, or the whole of what is free from the ovary falls off, including the disk bearing the petals. The calyx of hypogynous flowers usually falls off entirely or persists entirely. In general a calyx is called *deciduous* if any part falls off. When it persists it is either enlarged round or under the fruit, or it withers and dries up.

151. The corolla usually falls off entirely; when it persists it is usually withered and dry (*marcescent*), or very seldom enlarges round the fruit.

152. The stamens either fall off, or more or less of their filaments persists, usually withered and dry.

153. The style sometimes falls off or dries up and disappears; sometimes persists, forming a point to the fruit, or becomes enlarged into a wing or other appendage to the fruit.

154. The *Pericarp* is the portion of the fruit formed of the ovary, and whatever adheres to it exclusive of and outside of the seed or seeds, exclusive also of the persistent receptacle, or of whatever portion of the calyx persists round the ovary without adhering to it.

155. Fruits have often external appendages called *wings* (alæ), *beaks*, *crests*, *awns*, etc., according to their appearance. They are either formed by persistent parts of the flower more or less altered, or grow out of the ovary or the persistent part of the calyx. If the appendage be a ring of hairs or scales round the top of the fruit, it is called a *pappus*

156. Fruits are generally divided into *succulent* (including *fleshy*, *pulpy*, and *juicy* fruits) and *dry*. They are *dehiscent* when they open at maturity to let out the seeds, *indehiscent* when they do not open spontaneously but fall off with the seeds. Succulent fruits are usually indehiscent.

157. The principal kinds of succulent fruits are

the *Berry*, in which the whole substance of the pericarp is fleshy or pulpy, with the exception of the outer skin or rind, called the *Epicarp*. The seeds themselves are usually immersed in the pulp; but in some berries, the seeds are separated from the pulp by the walls of the cavity or cells of the ovary, which forms as it were a thin inner skin or rind, called the *Endocarp*.

the *Drupe*, in which the pericarp, when ripe, consists of two distinct portions, an outer succulent one called the *Sarcocarp* (covered like the berry by a skin or epicarp), and an inner dry endocarp called the *Putamen*, which is either *cartilaginous* (of the consistence of parchment) or hard and woody. In the latter case it is commonly called a *stone*, and the drupe a *stone-fruit*. When the putamen consists of several distinct stones or nuts, each enclosing a seed, they are called *pyrenes*, or sometimes *kernels*.

158. The principal kinds of dry fruits are

the *Capsule* or *Pod*,* which is dehiscent. When ripe the pericarp usually splits longitudinally into as many or twice as many pieces, called *valves*, as it contains cells or placentas. If these valves separate at the line of junction of the carpels, that is, along the line of the placentas or dissepiments, either splitting them or leaving them attached to the axis, the dehiscence is termed *septicidal*; if the valves separate between the placentas or dissepiment, the dehiscence is *loculicidal*, and the valves either bear the placentas or dissepiments along their middle line, or leave them attached to the axis. Sometimes also the capsule discharges its seeds by *slits*, *chinks*, or *pores*, more or less regularly arranged, or bursts irregularly, or separates into two parts by a horizontal line; in the latter case it is said to be *circumsciss*.

the *Nut* or *Achene*, which is indehiscent and contains but a single seed. When the pericarp is thin in proportion to the seed it encloses, the whole fruit (or each of its lobes) has the appearance of a single seed, and is so called in popular language. If the pericarp is thin and rather loose, it is often called an *Utricle*. A *Samara* is a nut with a wing at its upper end.

159. Where the carpels of the pistil are distinct (125) they may severally become as many distinct berries, drupes, capsules, or achenes. Separate carpels are usually more or less compressed laterally, with more or less prominent inner and outer edges, called *sutures*, and, if dehiscent, the carpel usually opens at these sutures. A *Follicle* is a carpel opening at the inner suture only. In some cases where the carpels are united in the pistil they will separate when ripe; they are then called *Cocci* if one-seeded.

160. The peculiar fruits of some of the large Orders have received special names, which will be explained under each Order. Such are the *siliqua* and *silicule* of Cruciferæ, the *legume* of Leguminosæ, the *pome* of *Pyrus* and its allies, the *pepo* of Cucurbitaceæ, the *cone* of Coniferæ, the *grain* or *caryopsis* of Gramineæ, etc.

§ 14. *The Seed.*

161. The **Seed** is enclosed in the pericarp in the great majority of flowering plants, called therefore *Angiosperms*, or *angiospermous plants*. In *Coniferæ* and a very few allied genera, called *Gymnosperms*, or *gymnospermous plants*, the seed is naked, without any real pericarp. These truly gymnospermous plants must not be confounded with *Labiatæ*, *Boragineæ*, etc., which have also been falsely called gymnospermous, their small nuts having the appearance of seeds (158).

162. The seed when ripe contains an *embryo* or young plant, either filling or nearly filling the cavity, but not attached to the outer skin or the seed, or more or less immersed in a mealy, oily, fleshy, or horn-like substance, called the *albumen*, or *perisperm*. The presence or absence of this albumen, that is, the distinction between *albuminous* and *exalbuminous* seeds, is one of great importance. The embryo or albumen can often only be found or distinguished when the seed is quite ripe, or sometimes only when it begins to germinate.

163. The shell of the seed consists usually of two separable *coats*. The outer coat, called the *testa*, is usually the principal one, and in most cases the only one attended to in descriptions. It may be hard and *crustaceous*, woody or bony, or thin and *mem-*

* In English descriptions, *pod* is more frequently used when it is long and narrow; *capsule*, or sometimes *pouch*, when it is short and thick or broad.

branous (skin-like), dry, or rarely succulent. It is sometimes expanded into *wings*, or bears a tuft of hair, cotton, or wool, called a *coma*. The inner coat is called the *tegmen*.

164. The *funicle* is the stalk by which the seed is attached to the placenta. It is occasionally enlarged into a membranous, pulpy, or fleshy appendage, sometimes spreading over a considerable part of the seed, or nearly enclosing it, called an *aril*. A *strophiole* or *caruncle* is a similar appendage proceeding from the testa by the side of or near the funicle.

165. The *hilum* is the scar left on the seed where it separates from the funicle. The *micropyle* is a mark indicating the position of the foramen of the ovule (133).

166. The **Embryo** (162) consists of the *Radicle* or base of the future root, one or two *Cotyledons* or future seed-leaves, and the *Plumule* or future bud within the base of the cotyledons. In some seeds, especially where there is no albumen, these several parts are very conspicuous, in others they are very difficult to distinguish until the seed begins to germinate. Their observation, however, is of the greatest importance, for it is chiefly upon the distinction between the embryo with one or with two cotyledons that are founded the two great classes of phænogamous plants, *Monocotyledons* and *Dicotyledons*.

167. Although the embryo lies loose (unattached) within the seed, it is generally n some determinate position with respect to the seed or to the whole fruit. This position is described by stating the direction of the radicle next to or more or less remote from the *hilum*, or it is said to be *superior* if pointing towards the summit of the *fruit*, *inferior* if pointing towards the base of the *fruit*.

§ 15. *Accessory Organs.*

168. Under this name are included, in many elementary works, various external parts of plants which do not appear to act any essential part either in the vegetation or reproduction of the plant. They may be classed under four heads: *Tendrils* and *Hooks*, *Thorns* and *Prickles*, *Hairs*, and *Glands*.

169. **Tendrils** (*cirrhi*) are usually abortive petioles, or abortive peduncles, or sometimes abortive ends of branches. They are simple or more or less branched, flexible, and coil more or less firmly round any objects within their reach, in order to support the plant to which they belong. *Hooks* are similar holdfasts, but of a firmer consistence, not branched, and less coiled.

170. **Thorns** and **Prickles** have been fancifully called the weapons of plants. A *Thorn* or *Spine* is the strongly pointed extremity of a branch, or abortive petiole, or abortive peduncle. A *Prickle* is a sharply pointed excrescence from the epidermis, and is usually produced on a branch, on the petiole or veins of a leaf, or on a peduncle, or even on the calyx or corolla. When the teeth of a leaf or the stipules are pungent, they are also called *prickles*, not *thorns*. A plant is *spinous* if it has thorns, *aculeate* if it has prickles.

171. **Hairs**, in the general sense, or the *indumentum* (or clothing) of a plant, include all those productions of the epidermis which have, by a more or less appropriate comparison, been termed *bristles*, *hairs*, *down*, *cotton*, or *wool*.

172. Hairs are often branched. They are said to be *attached by the centre*, if parted from the base, and the forks spread along the surface in opposite directions; *plumose*, if the branches are arranged along a common axis, as in a feather; *stellate*, if several branches radiate horizontally. These stellate hairs have sometimes their rays connected together at the base, forming little flat circular disks attached by the centre, and are then called *scales*, and the surface is said to be *scaly* or *lepidote*.

173. The *Epidermis*, or outer skin, of an organ, as to its surface and indumentum, is

smooth, when without any protuberance whatever.

glabrous, when without hairs of any kind.

striate, when marked with parallel longitudinal lines, either slightly raised or merely discoloured.

furrowed (*sulcate*) or *ribbed* (*costate*) when the parallel lines are more distinctly raised.

rugose, when wrinkled or marked with irregular raised or depressed lines.

umbilicate, when marked with a small round depression.

umbonate, when bearing a small boss like that of a shield.

viscous, *viscid*, or *glutinous*, when covered with a sticky or clammy exudation.

scabrous, when rough to the touch.

tuberculate or *warted*, when covered with small, obtuse, wart-like protuberances.

muricate, when the protuberances are more raised and pointed but yet short and hard.

echinate, when the protuberances are longer and sharper, almost prickly.

setose or *bristly*, when bearing very stiff erect straight hairs.

glandular-setose, when the setæ or bristles terminate in a minute resinous head or drop. In some works, especially in the case of *Roses* and *Rubus*, the meaning of *setæ* has been restricted to such as are glandular.

glochidiate, when the setæ are hooked at the top.

pilose, when the surface is thinly sprinkled with rather long simple hairs.

hispid, when more thickly covered with rather stiff hairs.

hirsute, when the hairs are dense and not so stiff.

downy or *pubescent*, when the hairs are short and soft; *puberulent*, when slightly pubescent.

strigose, when the hairs are rather short and stiff, and lie close along the surface all in the same direction; *strigillose*, when slightly strigose.

tomentose or *cottony*, when the hairs are very short and soft, rather dense and more or less intricate, and usually white or whitish.

woolly (*lanate*), when the hairs are long and loosely intricate, like wool. The wool or tomentum is said to be *floccose* when closely intricate and readily detached, like fleece.

mealy (*farinose*), when the hairs are excessively short, intricate and white, and come off readily, having the appearance of meal or dust.

canescent or *hoary*, when the hairs are so short as not readily to be distinguished by the naked eye, and yet give a general whitish hue to the epidermis.

glaucous, when of a pale bluish-green, often covered with a fine bloom.

174. The meanings here attached to the above terms are such as appear to have been most generally adopted, but there is much vagueness in the use practically made of many of them by different botanists. This is especially the case with the terms *pilose*, *hispid*, *hirsute*, *pubescent*, and *tomentose*.

175. The name of **Glands** is given to several different productions, and principally to the four following:—

1. Small wart-like or shield-like bodies, either sessile or sometimes stalked, of a fungous or somewhat fleshy consistence, occasionally secreting a small quantity of oily or resinous matter, but more frequently dry. They are generally few in number, often definite in their position and form, and occur chiefly on the petiole or principal veins of leaves, on the branches of inflorescences, or on the stalks or principal veins of bracts, sepals, or petals.

2. Minute raised dots, usually black, red, or dark-coloured, of a resinous or oily nature, always superficial, and apparently exudations from the epidermis. They are often numerous on leaves, bracts, sepals, and green branches, aud occur even on petals and stamens, more rarely on pistils. When raised upon slender stalks they are called *pedicellate* (or *stipitate*) *glands*, or *glandular hairs*, according to the thickness of the stalk.

3. Small, globular, oblong or even linear vesicles, filled with oil, imbedded in the substance itself of leaves, bracts, floral organs, or fruits. They are often very numerous, like transparent dots, sometimes few and determinate in form and position. In the pericarp of *Umbelliferæ* they are remarkably regular and conspicuous, and take the name of *vittæ*.

4. Lobes of the disk (137), or other small fleshy excrescences within the flower, whether from the receptacle, calyx, corolla, stamens, or pistil.

Chap. II. Classification, or Systematic Botany.

176. It has already been observed (3) that descriptions of plants should, as nearly as possible, be arranged under natural divisions, so as to facilitate the comparison of each plant with those most nearly allied to it. The description of plants here alluded to are descriptions of *species;* the natural divisions of the Flora refer to natural *groups of species.*

177. A **Species** comprises all the individual plants which resemble each other sufficiently to make us conclude that they are all, or *may have been* all, descended from a common parent. These individuals may often differ from each other in many striking particulars, such as the colour of the flower, size of the leaf, etc., but these particulars are such as experience teaches us are liable to vary in the seedlings raised from one individual.

178. When a large number of the individuals of a species differ from the others in any striking particular they constitute a **Variety.** If the variety generally comes true from seed, it is often called a *Race.*

179. A *Variety* can only be propagated with certainty by grafts, cuttings, bulbs, tubers, or any other method which produces a new plant by the development of one or more buds taken from the old one. A *Race* may with care be propagated by seed, although seedlings will always be liable, under certain circumstances, to lose those particulars which distinguish it from the rest of the species. A real *Species* will always come true from seed.

180. The known species of plants (now near 100,000) are far too numerous for the human mind to study without classification, or even to give distinct single names to To facilitate these objects, an admirable system, invented by Linnæus, has been universally adopted, viz. one common substantive name is given to a number of species which resemble each other more than they do any other species; the species so collected under one name are collectively called a **Genus,** the common name being the *generic* name. Each species is then distinguished from the others of the same genus by the addition of an adjective epithet or *specific name.* Every species has thus a botanical name of two words. In Latin, the language usually used for the purpose, the first word is a substantive and designates the genus; the second, an adjective, indicates the species.

181. The genera thus formed being still too numerous (above 6,000) for study without further arrangement, they have been classed upon the same principles; viz. genera which resemble each other more than they do any other genera, have been collected together into groups of a higher degree called **Families** or **Natural Orders,** to each of which a common name has been given. This name is in Latin an adjective plural, usually taken from the name of some one *typical* genus, generally the best known, the first discovered, or the most marked (e.g. *Ranunculaceæ* from *Ranunculus*). This is however for the purpose of study and comparison. To speak of a species, to refer to it and identify it, all that is necessary is to give the generic and specific names.

182. Natural Orders themselves (of which we reckon near 200) are often in the same manner collected into **Classes**; and where Orders contain a large number of genera, or genera a large number of species, they require further classification. The genera of an Order are then collected into minor groups called *Tribes*, the species of a genus into *Sections*, and in a few cases this intermediate classification is carried still further. The names of these several groups the most generally adopted are as follows, beginning with the most comprehensive or highest:—

Classes.
Subclasses or *Alliances.*
Natural Orders or Families.
Suborders.
Tribes.
Subtribes.
Divisions.
Subdivisions.
Genera.
Subgenera.
Sections.
Subsections.
Species.
Varieties.

183. The characters (3) by which a species is distinguished from all other species of

the same genus are collectively called the *specific character* of the plant; those by which its genus is distinguished from other genera of the Order, or its Order from other Orders, are respectively called the *generic* or *ordinal* character, as the case may be. The *habit* of a plant, of a species, a genus, etc., consists of such general characters as strike the eye at first sight, such as size, colour, ramification, arrangement of the leaves, inflorescence, etc., and are chiefly derived from the organs of vegetation.

184. Classes, Orders, Genera, and their several subdivisions, are called *natural* when, in forming them, all resemblances and differences are taken into account, valuing them according to their evident or presumed importance; *artificial*, when resemblances and differences in some one or very few particulars only are taken into account independently of all others.

185. The number of species included in a genus, or the number of genera in an Order, is very variable. Sometimes two or three or even a single species may be so different from all others as to constitute the entire genus; in others, several hundred species may resemble each other so much as to be all included in one genus; and there is the same discrepancy in the number of genera to a Family. There is moreover, unfortunately, in a number of instances, great difference of opinion as to whether certain plants differing from each other in certain particulars are varieties of one species or belong to distinct species; and again, whether two or more groups of species should constitute as many sections of one genus, or distinct genera, or tribes of one Order, or even distinct Natural Orders. In the former case, as a species is supposed to have a real existence in nature, the question is susceptible of argument, and sometimes of absolute proof. But the place a group should occupy in the scale of degree is very arbitrary, being often a mere question of convenience. The more subdivisions upon correct principles are multiplied, the more they facilitate the study of plants, provided always the main resting-points for constant use, the Order and the Genus, are comprehensive and distinct. But if every group into which a genus can be divided be erected into a distinct genus, with a substantive name to be remembered whenever a species is spoken of, all the advantages derived from the beautiful simplicity of the Linnæan nomenclature are gone.

CHAP. III. VEGETABLE ANATOMY AND PHYSIOLOGY.

§ 1. *Structure and Growth of the Elementary Tissues.*

186. If a very thin slice of any part of a plant be placed under a microscope of high magnifying power, it will be found to be made up of variously shaped and arranged ultimate parts, forming a sort of honeycombed structure. These ultimate parts are called *cells*, and form by their combination the *elementary tissues* of which the entire plant is composed.

187. A cell in its simplest state is a closed membranous sac, formed of a substance permeable by fluids, though usually destitute of visible pores. Each cell is a distinct individual, separately formed and separately acting, though cohering with the cells with which it is in contact, and partaking of the common life and action of the tissue of which it forms a part. The membranes separating or enclosing the cells are also called their *walls*.

188. Botanists usually distinguish the following tissues:—

(1) *Cellular tissue*, or *parenchyma*, consists usually of thin-walled cells, more or less round in form, or with their length not much exceeding their breadth, and not tapering at the ends. All the soft parts of the leaves, the pith of stems, the pulp of fruits, and all young growing parts, are formed of it. It is the first tissue produced, and continues to be formed while growth continues, and when it ceases to be active the plant dies.

(2) *Woody tissue*, or *prosenchyma*, differs in having its cells considerably longer than broad, usually tapering at each end into points and overlapping each other. The cells are commonly thick-walled; the tissue is firm, tenacious, and elastic, and constitutes

the principal part of wood, of the inner bark, and of the nerves and veins of leaves, forming, in short, the framework of the plant.

(3) *Vascular tissue*, or the *vessels* or *ducts* of plants, so called from the mistaken notion that their functions are analogous to those of the vessels (veins and arteries) of animals. A *vessel* in plants consists of a vertical row of cells, which have their transverse partition-walls obliterated, so as to form a continuous tube. All phænogamous plants, as well as ferns and a few other cryptogamous plants, have vessels, and are therefore called *vascular plants*; so the majority of cryptogams having only cellular tissue are termed *cellular plants*. Vessels have their sides very variously marked; some, called *spiral vessels*, have a spiral fibre coiled up their inside, which unrolls when the vessel is broken; others are marked with longitudinal slits, cross bars, minute dots or pits, or with transverse rings. The size of vessels is also very variable in different plants; in some they are of considerable size and visible to the naked eye in cross sections of the stem, in others they are almost absent or can only be traced under a strong magnifier.

189. Various modifications of the above tissues are distinguished by vegetable anatomists under names which need not be enumerated here as not being in general practical use. *Air-vessels*, *cysts*, *turpentine-vessels*, *oil-reservoirs*, etc., are either cavities left between the cells, or large cells filled with peculiar secretions.

190. When tissues are once formed, they increase, not by the general enlargement of the whole of the cells already formed, but by *cell-division*, that is, by the division of young and vitally active cells, and the enlargement of their portions. In the formation of the embryo, the first cell of the new plant is formed, not by division, but around a segregate portion of the contents of a previously existing cell, the embryo-sac. This is termed *free cell-formation*, in contradistinction to cell-division.

191. A young and vitally active cell consists of the *outer wall*, formed of a more or less transparent substance called *cellulose*, permeable by fluids, and of ternary chemical composition (carbon, hydrogen, and oxygen); and of the *cell-contents*, usually viscid or mucilaginous, consisting of *protoplasm*, a substance of quaternary chemical composition (carbon, hydrogen, oxygen, and nitrogen), which fills an important part in cell-division and growth. Within the cell (either in the centre or excentrical) is usually a minute, soft, subgelatinous body called the *nucleus*, whose functions appear to be intimately connected with the first formation of the new cell. As this cell increases in size, and its walls in thickness, the protoplasm and watery cell-sap become absorbed or dried up, the firm cellulose wall alone remaining as a permanent fabric, either empty or filled with various organized substances produced or secreted within it.

192. The principal organized contents of cells are

sap, the first product of the digestion of the food of plants; it contains the elements of vegetable growth in a dissolved condition.

sugar, of which there are two kinds, called *cane-sugar* and *grape-sugar*. It usually exists dissolved in the sap. It is found abundantly in growing parts, in fruits, and in germinating seeds.

dextrine, or vegetable mucilage, a gummy substance, between mucilage and starch.

starch or *fecula*, one of the most universal and conspicuous of cell-contents, and often so abundant in farinaceous roots and seeds as to fill the cell-cavity. It consists of minute grains called *starch-granules*, which vary in size and are marked with more or less conspicuous concentric lines of growth. The chemical constitution of starch is the same as that of cellulose; it is unaffected by cold water, but forms a jelly with boiling water, and turns blue when tested by iodine. When fully dissolved it is no longer starch, but dextrine.

chlorophyll, very minute granules, containing nitrogen, and coloured green under the action of sunlight. These granules are most abundant in the layers of cells immediately below the surface or epidermis of leaves and young bark. The green colouring matter is soluble in alcohol, and may thus be removed from the granules.

chromule, a name given to a similar colouring matter when not green.

wax, *oils*, *camphor*, and *resinous* matter, are common in cells or in cavities in the tissues between the cells, also various mineral substances, either in an amorphous state or as microscopic crystals, when they are called *Raphides*.

§ 2. *Arrangement of the Elementary Tissues, or Structure of the Organs of Plants.*

193. Leaves, young stems, and branches, and most parts of phænogamous plants, during the first year of their existence consist anatomically of

1, a *cellular system*, or continuous mass of cellular tissue, which is developed both vertically as the stem or other parts increase in length, and horizontally or laterally as they increase in thickness or breadth. It surrounds or is intermixed with the fibro-vascular system, or it may exist alone in some parts of phænogamous plants, as well as in cryptogamous ones.

2, a *fibro-vascular system*, or continuous mass of woody and vascular tissue, which is gradually introduced vertically into, and serves to bind together, the cellular system. It is continued from the stem into the petioles and veins of the leaves, and into the pedicels and parts of the flowers, and is never wholly wanting in any phænogamous plant.

3, an *epidermis*, or outer skin, formed of one or more layers of flattened (horizontal), firmly coherent, and usually empty cells, with either thin and transparent or thick and opaque walls. It covers almost all parts of plants exposed to the outward air, protecting their tissues from its immediate action, but is wanting in those parts of aquatic plants which are constantly submerged.

194. The epidermis is frequently pierced by minute spaces between the cells, called *Stomates*. They are oval or mouth-shaped, bordered by *lips*, formed of two or more elastic cells so disposed as to cause the stomate to open in a moist, and to close up in a dry state of the atmosphere. They communicate with intercellular cavities, and are obviously designed to regulate evaporation and respiration. They are chiefly found upon leaves, especially on the under surface.

195. When a phænogamous plant has outlived the first season of its growth, the anatomical structure of its stem or other perennial parts becomes more complicated and very different in the two great classes of phænogamous plants called *Exogens* and *Endogens*, which correspond with very few exceptions to the two classes Dicotyledons and Monocotyledons (167), founded on the structure of the embryo. In Exogens (Dicotyledons) the woody system is placed in concentric layers between a central *pith* (198, 1), and an external separable *bark* (1985). In Endogens (Monocotyledons) the woody system is in separate small bundles or fibres running through the cellular system without apparent order, and there is usually no distinct central pith, nor outer separable bark.

196. The anatomical structure is also somewhat different in the different organs of plants. In the **Root,** although it is constructed generally on the same plan as the stem, yet the regular organization, and the difference between Exogens and Endogens, is often disguised or obliterated by irregularities of growth, or by the production of large quantities of cellular tissue filled with starch or other substances (192). There is seldom, if ever, any distinct pith, the concentric circles of fibro-vascular tissue in Exogens are often very indistinct or have no relation to seasons of growth, and the epidermis has no stomates.

197. In the **Stem** or branches, during the first year or season of their growth, the difference between Exogens and Endogens is not always very conspicuous. In both there is a tendency to a circular arrangement of the fibro-vascular system, leaving the centre either vacant or filled with cellular tissue (pith) only, and a more or less distinct outer rind is observable even in several Endogens. More frequently, however, the distinction is already very apparent the first season, especially towards its close. The fibro-vascular bundles in Endogens usually anastomose but little, passing continuously into the branches and leaves. In Exogens the circle of fibro-vascular bundles forms a more continuous cylinder of network emitting lateral offsets into the branches and leaves.

198. The Exogenous stem, after the first year of its growth, consists of

1, the *pith*, a cylinder of cellular tissue, occupying the centre or longitudinal axis of the stem. It is active only in young stems or branches, becomes dried up and compressed as the wood hardens, and often finally disappears, or is scarcely distinguishable in old trees.

2, the *medullary sheath*, which surrounds and encases the pith. It abounds in spiral vessels (188, 3), and is in direct connection, when young, with the leaf-buds and branches, with the petioles and veins of leaves, and other ramifications of the system. Like the pith, it gradually disappears in old wood.

3, the *wood*, which lies immediately outside the medullary sheath. It is formed of woody tissue (188, 2), through which, in most cases, vessels (188, 3) variously disposed are interspersed. It is arranged in annual concentric circles (211), which usually remain active during several years, but in older stems the central and older layers become hard, dense, comparatively inactive, and usually deeper coloured, forming what is called *heart-wood* or *duramen*, the outer, younger, and usually paler-coloured living layers constituting the *sapwood* or *alburnum*.

4, the *medullary rays*, which form vertical plates, originating in the pith, and, radiating from thence, traverse the wood and terminate in the bark. They are formed of cellular tissue, keeping up a communication between the living portion of the centre of the stem and its outer surface. As the heart-wood is formed, the inner portion of the medullary rays ceases to be active, but they usually may still be seen in old wood, forming what carpenters call the *silver grain*.

5, the *bark*, which lies outside the wood, within the epidermis. It is, like the wood, arranged in annual concentric circles (211), of which the outer older ones become dry and hard, forming the *corky layer* or *outer bark*, which, as it is distended by the thickening of the stem, either cracks or is cast off with the epidermis, which is no longer distinguishable. Within the corky layer is the *cellular*, or *green*, or *middle bark*, formed of loose thin-walled pulpy cells containing chlorophyll (192); and which is usually the layer of the preceding season. The innermost and youngest circle, next the young wood, is the *liber* or *inner bark*, formed of long tough woody tissue called *bast-cells*.

199. The Endogenous stem, as it grows old, is not marked by the concentric circles of Exogens. The wood consists of a *matrix* of cellular tissue irregularly traversed by vertical cords or bundles of woody and vascular tissue, which are in connection with the leaves. These vascular bundles change in structure and direction as they pass down the stem, losing their vessels, they retain only their bast- or long wood-cells, usually curving outwards towards the rind. The old wood becomes more compact and harder towards the circumference than in the centre. The epidermis or rind either hardens so as to prevent any increase of diameter in the stem, or it distends, without increasing in thickness or splitting or casting off any outer layers.

200. In the **Leaf**, the structure of the petioles and principal ribs or veins is the same as that of the young branches of which they are ramifications. In the expanded portion of the leaf the fibro-vascular system becomes usually very much ramified, forming the smaller veins. These are surrounded and the interstices filled up by a copious and very active cellular tissue. The majority of leaves are horizontal, having a differently constructed upper and under surface. The cellular stratum forming the upper surface consists of closely set cells, placed vertically, with their smallest ends next the surface, and with few or no stomates in the epidermis. In the stratum forming the under surface, the cells are more or less horizontal, more loosely placed, and have generally empty spaces between them, with stomates in the epidermis communicating with these intercellular spaces. In vertical leaves (as in a large number of Australian plants) the two surfaces are nearly similar in structure.

201. When leaves are reduced to scales, acting only as protectors of young buds, or without taking any apparent part in the economy of vegetable life, their structure, though still on the same plan, is more simple; their fibro-vascular system is less ramified, their cellular system more uniform, and there are few or no stomates.

202. Bracts and floral envelopes, when green and much developed, resemble leaves in their anatomical structure, but in proportion as they are reduced to scales or transformed into petals, they lose their stomates, and their systems, both fibro-vascular and cellular, become more simple and uniform, or more slender and delicate.

203. In the stamens and pistils the structure is still nearly the same. The fibro-vascular system, surrounded by and intermixed with the cellular tissue, is usually simple in the filaments and style, more or less ramified in the flattened or expanded parts, such as the anther-cases, the walls of the ovary, or carpellary leaves, etc. The pollen consists of granular cells variously shaped, marked, or combined, peculiar forms being constant in the same species, or often in large genera, or even Orders. The stigmatic portion of the pistil is a mass of loosely cellular substance, destitute of epidermis, and

usually is in communication with the ovary by a channel running down the centre of the style.

204. Tubers, fleshy thickenings of the stem or other parts of the plant, succulent leaves or branches, the fleshy, woody, or bony parts of fruits, the albumen, and the thick fleshy parts of embryos, consist chiefly of largely developed cellular tissue, replete with starch or other substances (192), deposited apparently in most cases for the eventual future use of the plant or its parts when recalled into activity at the approach of a new season.

205. Hairs (171) are usually expansions or processes of the epidermis, and consist of one or more cells placed end to end. When thick or hardened into prickles, they still consist usually of cellular tissue only. Thorns (170) contain more or less of a fibro-vascular system, according to their degree of development.

206. Glands, in the primary sense of the word (175, 1), consist usually of a rather loose cellular tissue without epidermis, and often replete with resinous or other substances.

§ 3. *Growth of the Organs.*

207. Roots grow in length constantly and regularly at the extremities only of their fibres, in proportion as they find the requisite nutriment. They form no buds containing the germ of future branches, but their fibres proceed irregularly from any part of their surface without previous indication, and when their growth has been stopped for a time, either wholly by the close of the season, or partially by a deficiency of nutriment at any particular spot, it will, on the return of favourable circumstances, be resumed at the same point, if the growing extremities be uninjured. If during the dead season, or at any other time, the growing extremity is cut off, dried up, or otherwise injured, or stopped by a rock or other obstacle opposing its progress, lateral fibres will be formed on the still living portion ; thus enabling the root as a whole to diverge in any direction, and travel far and wide when lured on by appropriate nutriment.

208. This growth is not however by the successive formation of terminal cells attaining at once their full size. The cells first formed on a fibre commencing or renewing its growth, will often dry up and form a kind of terminal cap, which is pushed on as cells are formed immediately under it ; and the new cells, constituting a greater or lesser portion of the ends of the fibres, remain some time in a growing state before they have attained their full size.

209. The roots of Exogens, when perennial, increase in thickness like stems by the addition of concentric layers, but these are usually much less distinctly marked ; and in a large number of perennial Exogens and most Endogens the roots are annual, perishing at the close of the season, fresh adventitious roots springing from the stock when vegetation commences the following season.

210. The Stem, including its branches and appendages (leaves, floral organs, etc.), grows in length by additions to its extremity, but a much greater proportion of the extremity and branches remains in a growing and expanding state for a much longer time than in the case of the root. At the close of one season, leaf-buds or seeds are formed, each containing the germ of a branch or young plant to be produced the following season. At a very early stage of the development of these buds or seeds, a commencement may be found of many of the leaves it is to bear ; and before a leaf unfolds, every leaflet of which it is to consist, every lobe or tooth which is to mark its margin, may often be traced in miniature, and thenceforth till it attains its full size, the branch grows and expands in every part. In some cases however the lower part of a branch and more rarely (*e.g.* in some *Meliaceæ*) the lower part of a compound leaf attains its full size before the young leaves or leaflets of the extremity are yet formed.

211. The perennial stem, if exogenous (198), grows in thickness by the addition every season of a new layer or ring of wood between the outermost preceding layer and the inner surface of the bark, and by the formation of a new layer or ring of bark within the innermost preceding layer and outside the new ring of wood, thus forming a succession of concentric circles. The sap elaborated by the leaves finds its way, in a manner not as yet absolutely ascertained, into the *cambium-region*, a zone of tender thin-walled cells connecting the wood with the bark, by the division and enlargement of which new

cells (190) are formed. These cells separate in layers, the inner ones constituting the new ring of wood, and the outer ones the new bark or liber. In most exogenous trees, in temperate climates, the seasons of growth correspond with the years, and the rings of wood remain sufficiently distinct to indicate the age of the tree; but in many tropical and some evergreen trees, two or more rings of wood are formed in one year.

212. In endogenous perennial stems (199), the new wood or woody fibre is formed towards the centre of the stem, or irregularly mingled with the old. The stem consequently either only becomes more dense without increasing in thickness, or only increases by gradual distention, which is never very considerable. It affords therefore no certain criterion for judging of the age of the tree.

213. Flowers have generally all their parts formed, or indicated by protuberances or growing cells at a very early stage of the bud. These parts are then usually more regularly placed than in the fully developed flower. Parts which afterwards unite are then distinct, many are present in this rudimentary state which are never further developed, and parts which are afterwards very unequal or dissimilar are perfectly alike at this early period. On this account flowers in this very early stage are supposed by some modern botanists to be more *normal*, that is, more in conformity to a supposed type; and the study of the early formation and growth of the floral organs, called *Organogenesis*, has been considered essential for the correct appreciation of the affinities of plants. In some cases, however, it would appear that modifications of development, not to be detected in the very young bud, are yet of great importance in the distinction of large groups of plants, and that Organogenesis, although it may often assist in clearing up a doubtful point of affinity, cannot nevertheless be exclusively relied on in estimating the real value of peculiarities of structure.

214. The flower is considered as a *bud* (*flower-bud*, *alabastrum*) until the perianth expands, the *period of flowering* (*anthesis*) is that which elapses from the first expanding of the perianth, till the pistil is set or begins to enlarge, or, when it does not set, until the stamens and pistil wither or fall. After that, the enlarged ovary takes the name of *young fruit*.

215. At the close of the season of growth, at the same time as the leaf-buds or seeds are formed containing the germ of future branches or plants, many plants form also, at or near the bud or seed, large deposits, chiefly of starch. In many cases,—such as the tubers of a potato or other root-stock, the scales or thickened base of a bulb, the albumen or the thick cotyledons of a seed,—this deposit appears to be a store of nutriment, which is partially absorbed by the young branch or plant during its first stage of growth, before the roots are sufficiently developed to supply it from without. In some cases, however, such as the fleshy thickening of some stems or peduncles, the pericarps of fruits which perish long before *germination* (the first growth of the seed), neither the use nor the cause of these deposits has as yet been clearly explained.

§ 4. *Functions of the Organs.*

216. The functions of the Root are,—1. To fix the plant in or to the soil or other substance on which it grows. 2. To absorb nourishment from the soil, water, or air, into which the fibres have penetrated (or from other plants in the case of parasites), and to transmit it rapidly to the stem. The absorption takes place through the young growing extremities of the fibres, and through a peculiar kind of hairs or absorbing organs which are formed at or near those growing extremities. The transmission to the stem is through the tissues of the root itself. The nutriment absorbed consists chiefly of carbonic acid and nitrogen or nitrogenous compounds dissolved in water. 3. In some cases roots secrete or exude small quantities of matter in a manner and with a purpose not satisfactorily ascertained.

217. The stem and its branches support the leaves, flowers, and fruit, transmit the crude sap, or nutriment absorbed by the roots and mixed with previously organized matter, to the leaves, and re-transmit the assimilated or elaborated sap from the leaves to the growing parts of the plant, to be there used up, or to form deposits for future use (204). The transmission of the ascending crude sap appears to take place chiefly through the elongated cells associated with the vascular tissues, passing from one cell to another by a process but little understood, but known by the name of *endosmose*.

218. Leaves are functionally the most active of the organs of vegetation. In them is chiefly conducted digestion or *Assimilation*, a name given to the process which accomplishes the following results:—1. The chemical decomposition of the oxygenated matter of the sap, the absorption of carbonic acid, and the liberation of pure oxygen at the ordinary temperature of the air. 2. A counter-operation by which oxygen is absorbed from the atmosphere and carbonic acid is exhaled. 3. The transformation of the residue of the crude sap into the organized substances which enter into the composition of the plant. The exhalation of oxygen appears to take place under the influence of solar heat and light, chiefly from the under surface of the leaf, and to be in some measure regulated by the stomates; the absorption of oxygen goes on always in the dark, and in the daytime also in certain cases. The transformation of the sap is effected within the tissues of the leaf, and continues probably more or less throughout the active parts of the whole plant.

219. The Floral Organs seldom contribute to the growth of the plant on which they are produced; their functions are wholly concentrated on the formation of the seed with the germ of a future plant.

220. The Perianth (calyx and corolla) acts in the first instance in protecting the stamens and pistils during the early stages of their development. When expanded, the use of the brilliant colours which they often display, of the sweet or strong odours they emit, has not been adequately explained. Perhaps they may have great influence in attracting those insects whose concurrence has been shown in many cases to be necessary for the due transmission of the pollen from the anther to the stigma.

221. The pistil, when stimulated by the action of the pollen, forms and nourishes the young seed. The varied and complicated contrivances by which the pollen is conveyed to the stigma, whether by elastic action of the organs themselves, or with the assistance of wind, of insects, or other extraneous agents, have been the subject of numerous observations and experiments of the most distinguished naturalists, and are yet far from being fully investigated. Their details, however, as far as known, would be far too long for the present outline.

222. The fruit nourishes and protects the seed until its maturity, and then often promotes its dispersion by a great variety of contrivances or apparently collateral circumstances, *e.g.* by an elastic dehiscence which casts the seed off to a distance; by the development of a pappus, wings, hooked or other appendages, which allows them to be carried off by winds, or by animals, etc., to which they may adhere; by their small specific gravity, which enables them to float down streams; by their attractions to birds, etc., who taking them for food drop them often at great distances, etc. Appendages to the seeds themselves also often promote dispersion.

223. Hairs have various functions. The ordinary indumentum (171) of stems and leaves indeed seems to take little part in the economy of the plant besides perhaps some occasional protection against injurious atmospheric influences, but the root-hairs (216) are active absorbents, the hairs on styles and other parts of flowers appear often materially to assist the transmission of pollen, and the exudations of glandular hairs (175, 2) are often too copious not to exercise some influence on the phenomena of vegetation. The whole question, however, of vegetable exudations and their influence on the economy of vegetable life, is as yet but imperfectly understood.

CHAP. IV. COLLECTION, PRESERVATION, AND DETERMINATION OF PLANTS.

224. Plants can undoubtedly be most easily and satisfactorily examined when freshly gathered. But time will rarely admit of this being done, and it is moreover desirable to compare them with other plants previously observed or collected. *Specimens* must, therefore, be selected for leisurely observation at home, and preserved for future reference. A collection of such specimens constitutes a *Herbarium*.

225. A botanical **Specimen**, to be perfect, should have *root, stem, leaves, flowers* (both open and in the bud), and *fruit* (both young and mature). It is not, however, always possible to gather such complete specimens, but the collector should aim at

completeness. Fragments, such as leaves without flowers, or flowers without leaves, are of little or no use.

226. If the plant is small (not exceeding 15 in.) or can be reduced to that length by folding, the specimen should consist of the whole plant, including the principal part of the root. If it be too large to preserve the whole, a good flowering-branch should be selected, with the foliage as low down as can be gathered with it; and one or two of the lower stem-leaves or radical leaves, if any, should be added, so as to preserve as much as possible of the peculiar aspect of the plant.

227. The specimens should be taken from healthy uninjured plants of a medium size. Or if a specimen be gathered because it looks a little different from the majority of those around it, apparently belonging to the same species, a specimen of the more prevalent form should be taken from the same locality for comparison.

228. For bringing the specimens home, a light portfolio of pasteboard, covered with calico or leather, furnished with straps and buckles for closing, and another for slinging on the shoulder, and containing a few sheets of stout coarse paper, is better than the old-fashioned tin box (except, perhaps, for stiff prickly plants and a few others). The specimens as gathered are placed between the leaves of paper, and may be crowded together if not left long without sorting.

229. If the specimen brought home be not immediately determined when fresh, but dried for future examination, a note should be taken of the time, place, and situation in which it was gathered; of the stature, habit, and other particulars relating to any tree, shrub, or herb of which the specimen is only a portion; of the kind of root it has; of the colour of the flower; or of any other particulars which the specimen itself cannot supply, or which may be lost in the process of drying. These memoranda, whether taken down in the field, or from the living specimen when brought home, should be written on a label attached to the specimen or preserved with it.

230. To dry specimens, they are laid flat between several sheets of bibulous paper, and subjected to pressure. The paper is subsequently changed at intervals, until they are dry.

231. In laying out the specimen, care should be taken to preserve the natural position of the parts as far as consistent with the laying flat. In general, if the specimen is fresh and not very slender, it may be simply laid on the lower sheet, holding it by the stalk and drawing it slightly downwards; then, as the upper sheet is laid over, if it be slightly drawn downwards as it is pressed down, it will be found, after a few trials, that the specimen will have retained a natural form with very little trouble. If the specimen has been gathered long enough to have become flaccid, it will require more care in laying the leaves flat and giving the parts their proper direction. Specimens kept in tin boxes, will also often have taken unnatural bends which will require to be corrected.

232. If the specimen is very bushy, some branches must be thinned out, but always so as to show where they have been. If any part, such as the head of a thistle, the stem of an *Orobanche*, or the bulb of a Lily, be very thick, a portion of what is to be the under side of the specimen may be sliced off. Some thick specimens may be split from top to bottom before drying.

233. If the specimen be succulent or tenacious of life, such as a *Sedum* or an *Orchis*, it may be dipped in boiling water *all but the flowers*. This will kill the plant at once, and enable it to be dried rapidly, losing less of its colour or foliage than would otherwise be the case. Dipping in boiling water is also useful in the case of Heaths and other plants which are apt to shed their leaves during the process of drying.

234. Plants with very delicate corollas may be placed between single leaves of very thin unglazed tissue-paper. In shifting these plants into dry paper the tissue-paper is not to be removed, but lifted with its contents on to the dry paper.

235. The number of sheets of paper to be placed between each specimen or sheet of specimens, will depend, on the one hand, on the thickness and humidity of the specimens; on the other hand, on the quantity and quality of the paper one has at command. The more and the better the paper, the less frequently will it be necessary to change

it, and the sooner the plants will dry. The paper ought to be coarse, stout, and unsized. Common blotting-paper is much too tender.

236. Care must be taken that the paper used is well dried. If it be likewise hot, all the better; but it must then be very dry; and wet plants put into hot paper will require changing very soon, to prevent their turning black, for hot damp without ventilation produces fermentation, and spoils the specimens.

237. For pressing plants, various more or less complicated and costly presses are made. None is better than a pair of boards the size of the paper, and a stone or other heavy weight upon them if at home, or a pair of strong leather straps round them if travelling. Each of these boards should be double, that is, made of two layers of thin boards, the opposite way of the grain, and joined together by a row of clenched brads round the edge, without glue. Such boards, in deal, rather less than half an inch thick (each layer about 2½ lines) will be found light and durable.

238. It is useful also to have extra boards or pasteboards the size of the paper, to separate thick plants from thin ones, wet ones from those nearly dry, etc. Open wooden frames with cross-bars, or frames of strong wire-work lattice, are still better than boards for this purpose, as accelerating the drying by promoting ventilation.

239. The more frequently the plants are shifted into dry paper the better. Excepting for very stiff or woody plants, the first pressure should be light, and the first shifting, if possible, after a few hours. Then, or at the second shifting, when the specimens will have lost their elasticity, will be the time for putting right any part of a specimen which may have taken a wrong fold or a bad direction. After this the pressure may be gradually increased, and the plants left from one to several days without shifting. The exact amount of pressure to be given will depend on the consistence of the specimens and the amount of paper. It must only be borne in mind that too much pressure crushes the delicate parts, too little allows them to shrivel, in both cases interfering with their future examination.

240. The most convenient specimens will be made, if the drying-paper is the same size as that of the herbarium in which they are to be kept. That of writing-demy, rather more than 16 inches by 10½ inches, is a common and very convenient size. A small size reduces the specimens too much, a large size is both costly and inconvenient for use.

241. When the specimens are quite dry and stiff, they may be packed up in bundles with a single sheet of paper between each layer, and this paper need not be bibulous. The specimens may be placed very closely on the sheets, but not in more than one layer on each sheet, and care must be taken to protect the bundles by sufficient covering from the effects of external moisture or the attacks of insects.

242. In laying the specimens into the herbarium, no more than one species should ever be fastened on one sheet of paper, although several specimens of the same species may be laid side by side. And throughout the process of drying, packing, and laying in, great care must be taken that the labels be not separated from the specimens they belong to.

243. To examine or dissect flowers or fruits in dried specimens it is necessary to soften them. If the parts are very delicate, this is best done by gradually moistening them in cold water; in most cases, steeping them in boiling water or in steam is much quicker. Very hard fruits and seeds will require boiling to be able to dissect them easily.

244. For dissecting and examining flowers in the field, all that is necessary is a penknife and a pocket-lens of two or three glasses from 1 to 2 inches focus. At home it is more convenient to have a mounted lens or simple microscope, with a stage holding a glass plate, upon which the flowers may be laid; and a pair of dissectors, one of which should be narrow and pointed, or a mere point, like a thick needle, in a handle; the other should have a pointed blade, with a sharp edge, to make clean sections across the ovary. A compound microscope is rarely necessary, except in cryptogamic botany and vegetable anatomy. For the simple microscope, lenses of ¼, ½, 1, and 1½ inches focus are sufficient.

245. To assist the student in *determining* or ascertaining the name of a plant belonging to a Flora, analytical tables should be prefixed to the Orders, Genera, and

Species. These tables should be so constructed as to contain, under each bracket, or equally indented, two (rarely three or more) alternatives as nearly as possible contradictory or incompatible with each other, each alternative referring to another bracket, or having under it another pair of alternatives further indented. The student having a plant to determine, will first take the general table of Natural Orders, and examining his plant at each step to see which alternative agrees with it, will be led on to the Order to which it belongs; he will then compare it with the detailed character of the Order given in the text. If it agrees, he will follow the same course with the table of the genera of that Order, and again with the table of species of the genus. But in each case, if he finds that his plant does not agree with the detailed description of the genus or species to which he has thus been referred, he must revert to the beginning and carefully go through every step of the investigation before he can be satisfied. A fresh examination of his specimen, or of others of the same plant, a critical consideration of the meaning of every expression in the characters given, may lead him to detect some minute point overlooked or mistaken, and put him into the right way. Species vary within limits which it is often very difficult to express in words, and it proves often impossible, in framing these analytical tables, so to divide the genera and species, that those which come under one alternative should absolutely exclude the others. In such doubtful cases both alternatives must be tried before the student can come to the conclusion that his plant is not contained in the Flora, or that it is erroneously described.

246. In those Floras where analytical tables are not given, the student is usually guided to the most important or prominent characters of each genus or species, either by a general summary prefixed to the genera of an Order or to the species of the genus, for all such genera or species; or by a special summary immediately preceding the detailed description of each genus or species. In the latter case this summary is called a *diagnosis*. Or sometimes the important characters are only indicated by italicizing them in the detailed description.

247. It may also happen that the specimen gathered may present some occasional or accidental anomalies peculiar to that single one, or to a very few individuals, which may prevent the species from being at once recognized by its technical characters. It may be useful here to point out a few of these anomalies which the botanist may be most likely to meet with. For this purpose we may divide them into two classes, viz.:

1. *Aberrations from the ordinary type or appearance of a species for which some general cause may be assigned.*

A bright, light, and open situation, particularly at considerable elevations above the sea, or at high latitudes, without too much wet or drought, tends to increase the size and heighten the colour of flowers, in proportion to the stature and foliage of the plant.

Shade, on the contrary, especially if accompanied by richness of soil and sufficient moisture, tends to increase the foliage and draw up the stem, but to diminish the number, size, and colour of the flowers.

A hot climate and dry situation tend to increase the hairs, prickles, and other productions of the epidermis, to shorten and stiffen the branches, rendering thorny plants yet more spinous. Moisture in a rich soil has a contrary effect.

The neighbourhood of the sea, or a saline soil or atmosphere, imparts a thicker and more succulent consistence to the foliage and almost every part of the plant, and appears not unfrequently to enable plants usually annual to live through the winter. Flowers in a maritime variety are often much fewer, but not smaller.

The luxuriance of plants growing in a rich soil, and the dwarf stunted character of those crowded in poor soils, are too well known to need particularizing. It is also an everyday observation how gradually the specimens of a species become dwarf and stunted as we advance into the cold damp regions of the summits of high mountain-ranges, or into high northern latitudes; and yet it is frequently from the want of attention to these circumstances that numbers of false species have been added to our Enumerations and Floras. Luxuriance entails not only increase of size to the whole plant, or of particular parts, but increase of number in branches, in leaves, or leaflets of a compound leaf; or it may diminish the hairiness of the plant, induce thorns to grow out into branches, etc.

Capsules which, while growing, lie close upon the ground, will often become larger, more succulent, and less readily dehiscent, than those which are not so exposed to the moisture of the soil.

Herbs eaten down by sheep or cattle, or crushed underfoot, or otherwise checked in their growth, or trees or shrubs cut down to the ground, if then exposed to favourable circumstances of soil and climate, will send up luxuriant side-shoots, often so different in the form of their leaves, in their ramification and inflorescence, as to be scarcely recognizable for the same species.

Annuals which have germinated in spring, and flowered without check, will often be very different in aspect from individuals of the same species, which, having germinated later, are stopped by summer droughts or the approach of winter, and only flower the following season upon a second growth. The latter have often been mistaken for perennials.

Hybrids, or crosses between two distinct species, come under the same category of anomalous specimens from a known cause. Frequent as they are in gardens, where they are artificially produced, they are probably rare in nature, although on this subject there is much diversity of opinion, some believing them to be very frequent, others almost denying their existence. Absolute proof of the origin of a plant found wild, is of course impossible; but it is pretty generally agreed that the following particulars must always co-exist in a *wild hybrid*. It partakes of the characters of its two parents; it is to be found isolated, or almost isolated, in places where the two parents are abundant; if there are two or three, they will generally be dissimilar from each other, one partaking more of one parent, another of the other; it seldom ripens good seed; it will never be found where one of the parents grows alone.

Where two supposed species grow together, intermixed with numerous intermediates bearing good seed, and passing more or less gradually from the one to the other, it may generally be concluded that the whole are mere varieties of one species. The beginner, however, must be very cautious not to set down a specimen as intermediate between two species, because it appears to be so in some, even the most striking characters, such as stature and foliage. Extreme varieties of one species are connected together by transitions in all their characters, but these transitions are not all observable in the same specimens. The observation of a single intermediate is therefore of little value, unless it be one link in a long series of intermediate forms, and, when met with, should lead to the search for the other connecting links.

2. *Accidental aberrations from the ordinary type, that is, those of which the cause is unknown.*

These require the more attention, as they may sometimes lead the beginner far astray in his search for the genus, whilst the aberrations above-mentioned as reducible more or less to general laws, affect chiefly the distinction of species.

Almost all species with coloured flowers are liable to occur occasionally with them all white.

Many may be found even in a wild state with double flowers, that is, with a multiplication of petals.

Plants which have usually conspicuous petals will occasionally appear without any at all, either to the flowers produced at particular seasons, or to all the flowers of individual plants, or the petals may be reduced to narrow slips.

Flowers usually very irregular, may, on certain individuals, lose more or less of their irregularity, or appear in some very different shape. Spurs, for instance, may disappear, or be produced on all instead of one only of the petals.

One part may be occasionally added to, or subtracted from, the usual number of parts in each floral whorl, more especially in regular polypetalous flowers.

Plants usually monœcious or diœcious may become occasionally hermaphrodite, or hermaphrodite plants may produce occasionally unisexual flowers by the abortion of the stamens or of the pistils.

Leaves cut or divided where they are usually entire, variegated or spotted where they are usually of one colour, or the reverse, must also be classed amongst those accidental aberrations which the botanist must always be on his guard against mistaking for specific distinctions.

INDEX OF TERMS, OR GLOSSARY.

The Figures refer to the Paragraphs of the Outlines.

Par.
Palate 105
Palea, paleæ 82
Paleaceous = of a chaffy consistence.
Palmate 41, 42
Palmatifid, palmatisect . 42
Panicle, paniculate . . 74
Papillæ 122
Pappus 155
Parallel veins 40
Parasite 14
Parenchyma 188
Parietal 132
Pectinate 41
Pedate 41, 42
Pedatifid, pedatisect . 42
Pedicel 70
Pedicellate = on a pedicel.
Peduncle 68
Pedunculate = on a peduncle.
Peltate 52
Penicillate 130
Penta- (5 in composition) 92
Pepo 160
Perennials 12
Perfect flower . . . 84
Perfoliate 37
Perianth . 15, 98, 202, 220
Pericarp 154
Perigynous 140
Perisperm 162
Persistent 146
Personate 105
Petal 90
Petiole 35
Petiolule 39
Phænogamous, phanerogamous 10
Phyllaries 79
Phyllodium = a flat petiole with no blade.
Pilose 173
Pinna 43
Pinnate 41, 42
Pinnatifid, pinnatisect . 42
Pistil . 15, 90, 120, 203, 221
Pistillate 85
Pith 198
Placenta, placentation . 131
Plant 6
Plicate 102
Plumose 172
Pumule 166
Pluri- = *several*, in composition.
Plurilocular 126

Par.
Pod 158
Podocarp 120
Pollen 109, 119
Poly- (*many*, or an indefinite number, in composition) . . . 92
Polyadelphous . . . 113
Polyandrous 92, 112
Polygamous 86
Polygynous . . . 92, 125
Polypetalous 100
Pome 160
Posterior 91
Præfoliation 57
Preservation of specimens 224
Prickles 170
Primine 133
Procumbent 28
Proliferous 17
Prosenchyma 188
Prostrate 28
Protoplasm 191
Pubescent, puberulent . 173
Pulvinate (cushion-shaped) 136
Punctiform = like a point or dot.
Putamen 157
Pyramidal 54
Pyrenes 157

Quadri- (4 in composition) 44
Quincuncial 102
Quinque- (5 in composition) 44
Quintuplinerved . . . 40

Race 178
Raceme, racemose . . 74
Rachis 39, 68
Radical 38
Radicle 166
Raphe 134
Raphides 192
Receptacle 74, 135
Reduplicate 102
Regular 95
Reniform 51
Resupinate 105
Reticulate 40
Retuse 47
Revolute 102
Rhachis 39, 68
Rhaphe 134
Rhizome 21, 24

Par.
Rhomboidal 45
Ribs 40
Ribbed 173
Ringent 105
Root 15, 18, 196, 207, 216
Rootstock 24
Rostrate = beaked.
Rosulate 38
Rotate 104
Rudimentary 84
Rugose 173
Runcinate 41
Runner 30

Saccate 105
Sagittate 50
Salver-shaped . . . 104
Samara 158
Sap 192
Sapwood 198
Sarcocarp 157
Sarmentose 28
Scabrous 173
Scales . . 58, 59, 172, 201
Scaly bulb 26
Scaly surface 172
Scape 69
Scariose, scarious . . 55
Scattered 32
Scion 30
Scorpioid cyme . . . 74
Section 182
Secund 32
Secundine 133
Seed 161
Segment 39
Sepals 90
Septem- (7 in composition) 44
Septicidal 158
Septum = partition . . 126
Serrate, serrulate . . 39
Sessile 37
Seta, setæ (bristles) . . 173
Setaceous (bristle-like) . 54
Setose (bearing bristles) 173
Sex- (6 in composition) 44
Sheathing 37
Shrubs 12
Silicule, siliqua . . . 160
Silver grain 198
Simple 39
Sinuate 39
Sinus 39
Smooth 173
Spadix 76
Spatha 81

GLOSSARY OF TERMS.

CLASSIFICATIONS OF THE ORDERS AND GENERA.

I. KEY, CHIEFLY ADAPTED FROM DR. LINDLEY'S 'VEGETABLE KINGDOM.'
II. KEY, ACCORDING TO THE LINNÆAN CLASSES.
III. ARRANGEMENT AND CHARACTERS OF THE ORDERS ACCORDING TO THE NATURAL SYSTEM, AS ADOPTED IN THE PRESENT WORK.

THE following Keys are intended to facilitate the student's endeavours to determine the names of New Zealand plants. I have tried to make them as simple as possible by avoiding the use of more technical terms than necessary, and by employing in many cases characters taken from the general habit of the plants. None of these Keys can, however, be used, without some previous study of the elements of structural botany;* for the terms employed have each an exact meaning, which cannot safely be guessed at. The amount of study required depends much upon whether the student's powers of observation and of reasoning are good and accurate; but no amount of ability will obviate the absolute necessity of observing the characters of plants carefully and accurately, and clearly understanding the application of the terms used in defining these characters; and I would remind both teachers and students, that it is now a generally received opinion, that no subject is so well suited as systematic botany, to quicken the observing powers, and to improve the reasoning faculties of the young; and I believe that a little training in the use of these Keys alone, will sharpen the intellect of the quickest to a remarkable degree, and materially improve that of the dullest.

So many New Zealand plants are variable, have minute or unisexual flowers, or are otherwise difficult of determination, that by one key alone the student may fail to find out his plant; he must then try by means of the others; but there are a few New Zealand plants, which, as it appears to me, no system of keys will enable an uninstructed student to find out; just as there are idioms and expressions in languages that no grammar will teach.

All plants are naturally divisible into two great primary groups:—Flowering (Phænogamic) and Flowerless (Cryptogamic). To the first belong all

* These can be obtained from the excellent outlines of Botany by G. Bentham, Esq., P.L.S., prefixed to this work.

such as have more or less obvious flowers, and in which fertilization is effected by pollen (shed by the stamens) falling on the stigma of the pistil, which contains the ovules, or on the ovule itself. The effect of this fertilization is, that the ovule ripens into a seed, which consists of one or more integuments enclosing an embryo or rudimentary plant. This embryo, again, consists of distinct parts, from which in germination the stem and leaves are developed upwards, and the root downwards. Cryptogamic plants have no such apparatus, no obvious flowers, no stamens nor pistil nor ovule, nor have their seeds any distinguishable integuments or embryo. Their fertilization is effected in a very different manner, by most minute organs, extremely difficult to discover; and they are propagated chiefly by minute spores, or microscopic globular or angular bodies, usually without integument, and never containing an embryo. When the spore germinates, it is by growth from any point of its surface. As a rule, all commonly recognized trees, shrubs, and herbs belong to Phænogams, or flowering plants, whilst Cryptogams include Ferns, Lycopods, Mosses, Hepaticæ, Lichens, Fungi, and Algæ; of these the Ferns and Lycopods have leaves and branches more or less resembling those of Phænogams (but never flowers); the remainder are successively less and less like flowering plants, till we arrive at some of the lowest forms of Fungi and Algæ, which are with difficulty to be distinguished from the lowest forms of animal life.

The only New Zealand Phænogam that can be mistaken for a Cryptogam is *Lemna* (p. 277); and the Cryptogams which most nearly approach Phænogams are Lycopods, which have much of the habit of miniature Conifers, and are often furnished with cones also: these resemblances are, however, mainly in appearance.

The Cryptogams are a special study of great difficulty; and it will take time and research, and the collection of many specimens, before the student can make much progress in a knowledge of any of the Cryptogamic Orders except Ferns. I confess to having found very great difficulty in so describing the plants of any Cryptogamic Orders that a student shall readily name a specimen by this book; and with regard to Fungi, and certain divisions of the other Orders, as the crustaceous Lichens, freshwater, filamentous, unicellular and other Algæ, no descriptions alone will suffice either to give the uninstructed student a general acquaintance with the subject, or to enable him to name an isolated species. To understand these plants, good magnified drawings are essential; and indeed the same may be said for all departments of Cryptogamic botany after the Lycopods. I would recommend the student who is disposed to study Mosses, Hepaticæ, Lichens, Algæ, and Fungi, to take up one of these Orders at a time, to dissect, magnify,* and draw the organs of fructification (in all stages) of a good many species, before attempting to name any by this book. Should he have access to the 'Botany of the Antarctic Voyage,' he will find in all its parts (Antarctic Flora, Flora of New

* For this purpose, a good "simple microscope," with powers of $\frac{1}{2}$ in., $\frac{1}{4}$ in., $\frac{1}{10}$ in., and $\frac{1}{20}$ in. (the 2 latter Coddington lenses), is sufficient; such may be had of Ross, optician, Featherstone Buildings, Holborn, for about £4. 10*s*. Triangular-pointed (glovers') needles stuck in a handle, and a small keen-edged knife (such as are used in eye-operations) are the best instruments for ordinary purposes to dissect with; they are easily sharpened on a whetstone.

Zealand, and Flora of Tasmania) figures of a vast number of New Zealand species, for very many of these are widely diffused over the southern hemisphere.

Of New Zealand flowering plants, the following are more or less remarkable for some peculiarity of habit, habitat, locality, etc.

1. More or less aquatic species will be found in the following genera and Orders :—

Ranunculus, p. 3.
Montia, p. 27.
Elatine, p. 28.
Myriophyllum, p. 66.
Callitriche, p. 68.
Hydrocotyle, p. 85.
Limosella, p. 204.
Utricularia, p. 222.
Typhaceæ, p. 276.
Naiadeæ, p. 277.
Scirpus, p. 299.
Eleocharis, p. 300.
Isolepis, p. 301.
Cladium, p. 303.
(*Azolla*, which somewhat resembles a flowering plant, is a Cryptogam.)

2. Leafless plants, or plants provided with scale-like or very reduced leaves only, will be found in the following genera :—

Discaria, p. 43.
Carmichaelia, p. 48.
Rubus, p. 54.
Viscum, 108.
Cuscuta, p. 199.
Veronica (§ 4, *a*), p. 205.
Utricularia, p. 222.
Salicornia, p. 233.
Cassytha, p. 239.
Exocarpus, p. 246.
Dactylanthus, p. 255.
Gastrodia, p. 263.
Prasophyllum, p. 272.
Lemna, p. 277.
Juncus, p. 288.
Leptocarpus, p. 294.
Schœnus, p. 297.
Scirpus, p. 299.
Eleocharis, p. 300.
Cladium, p. 301.
Lepidosperma, p. 307.

3. Parasitic plants growing on branches (exclusive of Epiphytes) :—

Loranthaceæ, p. 106.
Cuscuta, p. 199.
Cassytha, p. 239.

Euphrasia, p. 219: the English species in germination attaches itself to the roots of grasses, from which it derives its nourishment. *Santalum*, p. 247, also should be observed in a young state, for it belongs to a family of which some species are parasites.

4. Plants with milky juice :—

Microseris, p. 164.
Crepis, p. 164.
Taraxacum, p. 165.
Sonchus, p. 165.
Picris, p. 165.
Wahlenbergia, p. 169.
Colensoa, p. 170.
Lobelia, p. 171.
Sapota, p. 185.
Parsonsia, p. 187 (slightly).
Convolvulus, p. 197. (do.)
Euphorbia, p. 247.
Epicarpurus, p. 250.

5. The leaves, etc., of species of the following genera and Orders are aromatic, fetid, or otherwise odorous when bruised; and probably of others with which I am not acquainted :—

Drimys, p. 10.
Rutaceæ, p. 38.
Myrtaceæ, p. 69.
Umbelliferæ, p. 84.
Araliaceæ, p. 99.
Coprosma, p. 110.
Olearia, p. 123.
Brachycome, p. 137.
Cotula, p. 140.
Mentha, p. 225.
Chenopodium, p. 229.
Laurineæ, p. 238.
Monimiaceæ, p. 239.
Santalum, p. 247.
Piper, p. 254.
Coniferæ, p. 255.

6. The following are *usually* sea-side genera or species (to which probably many others may be added) :—

Lepidium oleraceum, p. 14.
Hymenanthera, p. 18.
Spergularia rubra, p. 25.
Linum monogynum, p. 35.
Corynocarpus, p. 46.
Tillæa moschata, p. 61.
Metrosideros tomentosa, p. 72.
Sicyos, p. 82.
Mesembryanthemum, p. 83.
Tetragonia, p. 83.
Apium, p. 89.
Meryta, p. 104.
Coprosma Baueriana, p. 112.
C. petiolata, p. 113.
Cassinia retorta, p. 145.

Senecio lautus, p. 160.	Myoporum, p. 224.	Carumbium, p. 248.
S. odoratus, p. 160.	Plantago Brownii, p. 227.	Triglochin, p. 278.
Selliera radicans, p. 173.	Pisonia, p. 229.	Ruppia, p. 279.
Scævola gracilis, p. 173.	Chenopodium, p. 229.	Juncus maritimus, p. 289.
Samolus, p. 185.	Suæda, p. 231.	Scirpus maritimus, p. 300.
Sapota, p. 185.	Atriplex, p. 231.	Desmoschœnus, p. 303.
Convolvulus Soldanella, p. 198.	Salsola, p. 232.	Carex pumila, p. 315.
Vitex, p. 223.	Salicornia, p. 233.	Spinifex, p. 322.
Veronica elliptica, p. 209.	Pimelea arenaria, p. 244.	Zoysia, p. 324.
Avicennia, p. 224.	Euphorbia, p. 247.	Festuca littoralis, 341.

I. KEY TO THE NATURAL ORDERS, ETC., OF FLOWERING PLANTS.

(*Chiefly adapted from Lindley's 'Vegetable Kingdom.'*)

Class I. Dicotyledons. Stem, when perennial, with pith, rings of wood, and separable bark. Leaves with branching netted veins. Leaflets or lobes of the perianth 4 or 5, or multiples of 4 or 5 (in the third division the perianth is often absent or imperfect). Embryo with 2 opposite cotyledons; radicle elongating in germination.

The exceptions to each of the foregoing characters are very numerous, but a little practice, and the tact and knowledge that practice alone can give, will enable the student to decide at once whether almost any New Zealand flowering plant belongs to this or the following Class.

Class II. Monocotyledons. Stem, when perennial, without rings of wood or separable bark; the wood-bundles being isolated and scattered, apparently promiscuously through the cellular-tissue. Leaves (never opposite in New Zealand, often sheathing at the base) with parallel veins joined by straight cross-veinlets. Leaflets or lobes of the perianth 3 or 6, rarely 4; in grasses and sedges the perianth is imperfect or 0, and the flowers enclosed in dry imbricating scales. Embryo with one cotyledon or two alternate ones; radicle not elongating in germination, but giving off root-fibres.

Palms and Cordylines are the only New Zealand trees of this class. *Rhipogonum* and *Freycinetia* the only shrubs. *Rhipogonum* and *Callixene* have somewhat netted veins.

Class III. Cryptogamia. See p. lxvi.

CLASS I. DICOTYLEDONS.

I. Flowers having both calyx and corolla, the latter polypetalous.

A. Polyandrous. Stamens more than 20.

§ Ovary inferior. Leaves usually opposite.

Leaves with translucent dots. Shrubs or trees	Myrtaceæ, p. 69.
Leaves fleshy. Creeping or trailing herbs	Ficoideæ, p. 83.

§§ Ovary superior.

† Leaves stipulate.

Carpels free. Anthers 2-celled. Leaves compound .	Rosaceæ, p. 53.
Carpels free or combined. Anthers 1-celled. Leaves simple	Malvaceæ, p. 29.
Carpels combined. Anthers 2-celled. Leaves simple .	Tiliaceæ, p. 32.

†† Leaves exstipulate.

Stamens perigynous. Carpels free	Rosaceæ, p. 53.
Stamens hypogynous. Carpels free, many	Ranunculaceæ, p. 1.
Stamens hypogynous. Carpels free, few	Drimys, p. 10.
Stamens hypogynous. Leaves with transparent dots .	Hypericineæ, p. 28.
Stamens hypogynous. Leaves without dots	Tiliaceæ, p. 32.

B. OLIGANDROUS. Stamens fewer than 20.

§ Ovary inferior (or apparently so).

† Flowers umbelled or capitate.

Herbs. Stamens 5. Fruit of 2 separable carpels . . UMBELLIFERÆ, p. 84.
Shrubs or trees. Fruit of 2 or more combined carpels. ARALIACEÆ, p. 99.
Herbs. Stamens 1 or 2 . . . . ACÆNA, p. 55.

†† Flowers not umbelled nor capitate.

Leaves stipulate . . . . RHAMNEÆ, p. 42.
Leaves exstipulate.

α. Style 1; stigma simple.

Petals 4, imbricate. Stamens perigynous . . . ONAGRARIEÆ, p. 75.
Petals 4 or 5, valvate. Stamens epipetalous . . LORANTHACEÆ, p. 106.
Petals 4 or 5. Stamens alternate with petals.
Cells of fruit 1-seeded . . . . CORNEÆ, p. 104.
Cells of fruit many-seeded . . . . SAXIFRAGEÆ, p. 57.

β. Styles or stigmas 2 or more, or stigma divided.

Cells of fruit 1-seeded. Shrubs. Leaves alternate GRISELINIA, p. 104.
Cells of fruit 1-seeded. Herbs . . . . . . HALORAGEÆ, p. 64.
Cells of fruit many-seeded . . . . SAXIFRAGEÆ, p. 57.

§§ Ovary superior.

† Leaves stipulate.

1. Carpels solitary, 2- or more seeded. Flowers irregular LEGUMINOSÆ, p. 47.
2. Carpels several, free, 1-seeded. Flowers regular . ROSACEÆ, p. 53.
3. Carpels combined into a 1- or more celled ovary.

α. Ovary 1-celled. Ovules on its walls (parietal).

Climbing shrubs . . . . PASSIFLOREÆ, p. 81.
Viscid, glandular herbs. Flowers regular . . . DROSERACEÆ, p. 62.
Herbs or shrubs, not glandular nor climbing.
Flowers regular or irregular . . . . . . VIOLARIEÆ, p. 15.

β. Ovary 1-celled. Ovules fixed to its base.

Sepals 4 or 5. Herbs. Leaves opposite . . . CARYOPHYLLEÆ, p. 22.
Sepals 2. Herbs. Leaves opposite or alternate . PORTULACEÆ, p. 26.

γ. Ovary 2- or more celled.

Stamens hypogynous. Leaves opposite (water-herbs) ELATINE, p. 28.
Stamens hypogynous. Leaves alternate. Herbs GERANIACEÆ, p. 35.
Stamens 5, perigynous, opposite the petals. Ovules
1 in each cell. Shrubs . . . . RHAMNEÆ, p. 42.
Stamens perigynous, 5 alternate, or 10 opposite
and alternate with the petals. Ovules many . SAXIFRAGEÆ, p. 57.

†† Leaves exstipulate.

1. Carpels several, free.

Herbs. Leaves opposite. Carpels 4 or 5 . . . TILLÆA, p. 61.
Herbs. Leaves alternate. Carpels 5 or more . . RANUNCULACEÆ, p. 1.
Shrubs or herbs. Leaves opposite, simple. Carpels
5; styles very short . . . . CORIARIEÆ, p. 46.
Tree. Leaves alternate, simple, aromatic . . . . DRIMYS, p. 10.

2. Carpel solitary, 1-celled.

Tree. Stamens 5, hypogynous. Ovule 1, pendulous. PENNANTIA, p. 41.
Tree. Stamens 5, hypogynous. Ovules 2 or more . DRIMYS, p. 10.
Stamens 5, perigynous, alternating with scales. Ovule 1 CORYNOCARPUS, p. 46.
Stamens 10, perigynous. Ovules 2 or more . . . LEGUMINOSÆ, p. 47.
Stamens 4–5, epipetalous . . . . MYRSINE, p. 185.

3. Carpels combined into a 1- or more-celled ovary.

Glandular herbs. Ovary 1-celled, many-ovuled . . DROSERA, p. 63.
Herbs, not glandular. Ovary 1-celled, many-ovuled CARYOPHYLLEÆ, p. 22.
Ovary 2- or more celled. Stamens hypogynous.
Herbs. Sepals 4. Stamens 6. Ovary 2-celled . CRUCIFERÆ, p. 10.
Shrubs and trees. Sepals 5. Stamens 5, free.
Ovary 2–5-celled . . . . PITTOSPORUM, p. 18.

Herbs. Sepals 5. Stamens 5. Ovary 5-celled . LINUM, p. 34.
Herbs. Sepals 5. Stamens 10. Ovary 5-celled GERANIACEÆ, p. 35.
Ovary 3- or more celled. Stamens inserted at the base of a tumid disk, or perigynous.
Shrubs. Leaves with transparent dots. Sepals 4 or 5. Stamens 4, 5, 8 or 10 . . . . . RUTACEÆ, p. 38.
Shrubs and trees. Sepals 4 or 5. Stamens 4 or more. Petals lobed or cut . . . . . . TILIACEÆ, p. 32.
Trees, leaves pinnate. Sepals 4 or 5. Anthers 8–10, within a fleshy tube . . . . DYSOXYLUM, p. 40.
Tree ; leaves with glandular serratures. Calyx 5-lobed. Stamens 5, equal . . . . . . IXERBA, p. 59.
Small herb. Calyx 5-lobed. Petals linear. Stamens 5, unequal . . . . STACKHOUSIA, p. 42.

II. Flowers having both calyx and corolla, the latter monopetalous.
§ Ovary inferior.
Flowers minute, numerous, in involucrate heads . . . . COMPOSITÆ, p. 121.
Flowers numerous, in globose not involucrate heads . . . ACÆNA, p. 55.
Flowers not collected into many-flowered heads.
Leaves opposite and stipulate, or whorled . . . . . . RUBIACEÆ, p. 110.
Leaves alternate, exstipulate.
Stamens 2 ; filaments cohering with the style . . . STYLIDIEÆ, p. 166.
Stamens 5, inserted at the mouth of the corolla, alternate with its lobes . . . . ALSEUOSMIA, p. 109.
Stamens 5, opposite the lobes of the corolla . . . . SAMOLUS, p. 185.
Stamens 5, epigynous or inserted at the base of the corolla . . . . CAMPANULACEÆ, p. 169.
§§ Ovary superior. Corolla regular.
† Ovary and fruit very deeply 2–4-lobed; lobes 1-celled, 1-seeded.
Leaves alternate. Stamens perigynous. Ovary 3-lobed . STACKHOUSIA, p. 42.
Leaves opposite. Stamens nearly straight, epipetalous. Ovary 4-lobed . . . . LABIATÆ, p. 225.
Leaves alternate. Stamens epipetalous. Ovary 4-lobed . BORAGINEÆ, p. 191.
Leaves alternate. Stamens epipetalous. Ovary 2-lobed . DICHONDRA, p. 199.
†† Ovary not deeply lobed.
‡ Leaves alternate or radical (0 in *Cuscuta* of *Convolvulaceæ*).
Ovary 1-celled. Stamens epipetalous.
Herbs. Sepals 2 . . . . PORTULACEÆ, p. 26.
Shrubs ; leaves with glandular dots. Sepals 4 or 5 MYRSINE, p. 183.
Herbs ; leaves not dotted. Sepals 4 or 5 . . . SAMOLUS, p. 185.
Ovary 2- or more celled. Stamens epipetalous.
Minute tufted herbs. Stamens 2 . . . . . . PYGMÆA, p. 217.
Herbs. Stamens 4, filaments very long . . . PLANTAGO, p. 226.
Shrub or tree. Leaves alternate . . . . . SAPOTA, p. 185.
Leafless climbing herb . . . . CUSCUTA, p. 199.
Stamens 4 or 5, without alternating scales.
Shrubs or trees. Anthers 1-celled. Ovary 5–10-celled . . . . ERICEÆ, p. 173.
Climbing or trailing herbs. Anthers 2-celled. Ovary 2–4-celled . . . . CONVOLVULACEÆ, p. 197.
Erect shrubs. Corolla-lobes imbricate. Anthers 2-celled. Ovary 2–4-celled . . . . . . VERBENACEÆ, p. 223.
Erect herbs. Corolla-lobes plaited. Anthers 2-celled. Ovary 2-celled . . . . SOLANEÆ, p. 200.
Stamens 6 or more. Anthers 1-celled . . . . MALVACEÆ, p. 29.
Ovary 5-celled. Stamens 10, hypogynous . . . ERICEÆ, p. 173.

‡‡ Leaves opposite. Stamens epipetalous.
Stamens 5. Climbing shrubs . . . . PARSONSIA, p. 187.
Stamens 5. Herbs, very bitter . . . . GENTIANEÆ, p. 189.
Stamens 5. Erect or prostrate shrubs . . . . . . LOGANIACEÆ, p. 188.
Stamens 2 or 4. Herbs or shrubs . . . . SCROPHULARINEÆ, p. 200.
Stamens 5. Sepals 2. Herbs . . . . PORTULACEÆ, p. 26.

§§§ Ovary superior. Corolla irregular, 2-lipped.
Leaves opposite.
Ovary 4-lobed to the base. Filaments curved downwards. Herbs . . . . LABIATÆ, p. 225.
Ovary 4-lobed to the base. Filaments curved upwards. Shrubs or trees . . . . VERBENACEÆ, p. 223.
Ovary not deeply 4-lobed.
Ovary 1-celled, nearly 2-celled from the 2 projecting placentas . . . . RHABDOTHAMNUS, p. 221.
Ovary 2-celled . . . . SCROPHULARINEÆ, p. 200.
Leaves alternate, or 0 in some *Lentibularieæ*.
Shrub. Stamens 4. Leaves with pellucid dots . . . MYOPORUM, p. 224.
Herbs. Stamens 2. Leaves linear capillary or 0 . . UTRICULARIA, p. 222.

III. Flowers with a single perianth or 0, the calyx or corolla or both being absent.
§ Perianth single.
† Ovary inferior. These marked * are all provided with a double perianth, but the calyx has so obscure a limb, that they may naturally be sought for in this division.
a. Trees or shrubs.
Parasitic shrubs. Leaves opposite, exstipulate . . . *LORANTHACEÆ, p. 106.
Tree. Leaves opposite or alternate or 0, exstipulate. Flowers hermaphrodite . . . . SANTALUM, p. 247.
Tree. Leaves alternate, exstipulate, very large, long-petioled. Flowers capitate, unisexual . . . . . MERYTA, p. 104.
Trees. Leaves alternate, stipulate. Flowers unisexual FAGUS, p. 249.
Shrub. Leaves alternate, small, stipulate . . . . POMADERRIS, p. 43.
β. Herbs (flowers unisexual in all).
Prostrate or climbing, with tendrils . . . . . . *SICYOS, p. 82.
Aquatic. Leaves opposite or whorled . . . . . . HALORAGEÆ, p. 64.
Scapigerous. Leaves radical. Flowers unisexual . . *GUNNERA, p. 67.
Tuberous root parasite. Stems or scapes scaly . . . DACTYLANTHUS, p. 255.
†† Ovary superior.
a. Leaves stipulate.
Spiny shrub or tree, often leafless. Calyx valvate . . DISCARIA, p. 43.
Herbs and shrubs. Stipules membranous, sheathing the stem . . . . POLYGONEÆ, p. 235.
Herbs and shrubs. Stipules free. Flowers unisexual. URTICEÆ, p. 250.
Tree. Stipules free. Flowers unisexual. Stamens 6 or more . . . . CARUMBIUM, p. 248.
β. Leaves exstipulate.
1. Carpels many, free.
Stamens hypogynous . . . . RANUNCULACEÆ, p. 1.
Stamens perigynous . . . . MONIMIACEÆ, p. 239.
2. Carpels solitary or ovary 1-celled.
Leaves 0.
Fleshy, jointed, maritime herb . . . . SALICORNIA, p. 233.
Twining, slender, parasitic herb . . . . . . CASSYTHA, p. 239.
Small shrub. Branches grooved. Fruit red . . EXOCARPUS, p. 246.
Leaves opposite.
Trees or shrubs. Leaves opposite and alternate. Stamens 4 or 5 . . . . SANTALUM, p. 247.

Herbs. Flowers hermaphrodite, minute, white, in axillary fascicles. . . . . ALTERNANTHERA, p. 234.
Herbs. Flowers hermaphrodite, minute, green, in axillary spikelets . . . . CHENOPODIUM, p. 229.
Herbs. Flowers unisexual, minute, green. Stamens 4, opposite perianth-lobes . . . . . . PARIETARIA, p. 252.
Herbs. Flowers hermaphrodite. Stamens 4 or 5, alternate with perianth-lobes. Ovules many CARYOPHYLLEÆ, p. 22.
Herbs. Flowers hermaphrodite. Stamen 1. Seed 1 SCLERANTHUS, p. 234.

Leaves alternate.

Shrubs or trees.

Leaves large, long petioled. Stamens 6–10, hypogynous . . . . PISONIA, p. 229.
Leaves opposite and alternate. Stamens 4 or 5 at the base of deciduous perianth-lobes . SANTALUM, p. 247.
Leaves without transparent dots. Stamens 4, on middle or top of deciduous perianth-lobes . PROTEACEÆ, p. 240.
Leaves with transparent dots. Stamens 4 or 5, on base of perianth-lobes . . . . MYRSINE, p. 183.
Bark tough. Stamens 2–4, on top of perianth-tube . . . . THYMELEÆ, p. 241.
Stamens 6–15, perigynous. Anthers with upturned valves . . . . LAURINEÆ, p. 238.
Tree. Leaves pinnate. Flowers unisexual. Stamens 6–8. . . . . ALECTRYON, p. 45.
Flowers minute, green. Perianth 4- or 5-partite CHENOPODIACEÆ, p. 229.

3. Ovary 2- or 3-celled.

Herb, juice milky. Leaves alternate . . . . . . EUPHORBIA, p. 247.
Shrubs or trees. Leaves opposite. Stamens 2 . . OLEA, p. 186.
Shrubs or trees. Leaves alternate. Stamens 4, on perianth segments . . . . PROTEACEÆ, p. 240.
Spinous shrub or tree. Leaves opposite or 0. Stamens 4 or 5, perigynous . . . . DISCARIA, p. 43.
Rigid shrub. Leaves alternate. Stamens 5, hypogynous . . . . *Pittosporum rigidum*, p. 20.
Tree. Leaves alternate. Flowers unisexual. Stamens 6–8 . . . . DODONÆA, p. 44.
Herbs. Sepals 4. Stamens 6 . . . . CRUCIFERÆ, p. 10.

§§ Perianth wholly wanting.

† Leaves opposite, serrate, stipulate . . . . ASCARINA, p. 253.
Leaves opposite, exstipulate. Stamen 1. Water-herb . CALLITRICHE, p. 68.
Leaves alternate. Juice milky. Maritime herb . . . EUPHORBIA, p. 247.
Leaves alternate, very aromatic. Shrub . . . . . . PIPER, p. 254.
Leaves alternate, fleshy. Spikes slender. Herb . . . PEPEROMIA, p. 254.
Leaves reduced to scales. Root-parasite . . . . DACTYLANTHUS, p. 255.
Leaves various, very coriaceous, or minute imbricating scales. Trees and shrubs . . . . CONIFERÆ, p. 255.

CLASS II. MONOCOTYLEDONS.

§ Perianth superior, of 6 leaflets in 2 rows.

Flowers very irregular. Anther 1, adnate to the style . ORCHIDEÆ, p. 260.
Flowers regular. Stamens 3 . . . . LIBERTIA, p. 274.
Flowers regular. Stamens 6 . . . . HYPOXIS, p. 275.

§§ Perianth inferior; segments petaloid or herbaceous.

Perianth petaloid. Fruit a 1–3-celled, 3- or more seeded berry or capsule . . . . LILIACEÆ, p. 280.
Perianth herbaceous. Fruit a 1-seeded drupe . . . . . ARECA, p. 288.
Perianth herbaceous. Fruit of 3, 4, or 6 1-seeded, free or connate carpels . . . . NAIADEÆ, p. 277.

§§§ Perianth inferior, of 6 dry glumaceous leaflets.
Flowers panicled or capitate. Anthers 2-celled. Capsule 3-valved . . . . JUNCEÆ, p. 288.
Flowers in the axils of glumes, disposed in spikelets. Anthers 1-celled. Fruit a nut or 1–3-celled capsular utricle RESTIACEÆ, p. 293.

§§§§ Perianth 0 or incomplete.
α. Flowers in dense heads or cylindric catkins or spikes.
Climbing, shrubby. Leaves long, prickly . . . . . . FREYCINETIA, p. 275.
Erect, water or marsh plants. Heads spherical . . . SPARGANIUM, p. 276.
Erect, water or marsh plants. Catkins cylindric . . . TYPHA, p. 276.
Floating or submerged water-plants . . . . NAIADEÆ, p. 277.
β. Flowers spiked or axillary and solitary.
Water-herbs of various habit . . . . NAIADEÆ, p. 277.
Rush-like plants. Anthers 1-celled . . . . RESTIACEÆ, p. 293.
γ. Flowers in the axils of imbricating glumes, arranged in spikelets.
Perianth 0, or of 6 small leaflets. Anthers 1-celled . . RESTIACEÆ, p. 293.
Perianth 0, or of minute scales or bristles. Anther 2-celled. Stems solid. Leaf-sheaths closed . . . . CYPERACEÆ, p. 296.
Perianth 0, or of 1–3 most minute scales. Stamens usually 3, 2-celled. Stems hollow, jointed. Leaf-sheaths split to the base . . . . GRAMINEÆ, p. 317.

CLASS III. CRYPTOGAMS. *See p.* lxvi.

II. KEY TO THE GENERA OF NEW ZEALAND FLOWERING PLANTS, ARRANGED UNDER THE LINNÆAN CLASSES.

The Linnæan Classes are:—

1. *Flowers perfect, having stamens in the same flower with the pistils.*

Class		
Class I.	MONANDRIA . .	1 stamen in each flower.
II.	DIANDRIA . . .	2 stamens, free from one another.
III.	TRIANDRIA . . .	3 stamens, free from one another.
IV.	TETRANDRIA . .	4 stamens, free, equal in height.
V.	PENTANDRIA . .	5 stamens, free.
VI.	HEXANDRIA . .	6 stamens, free, equal in height.
VII.	HEPTANDRIA .	7 stamens, free.
VIII.	OCTANDRIA . .	8 stamens, free.
IX.	ENNEANDRIA .	9 stamens, free.
X.	DECANDRIA . .	10 stamens, free.
XI.	DODECANDRIA .	12 to 19 stamens, free.
XII.	ICOSANDRIA . .	20 or more stamens on the calyx (perigynous).
XIII.	POLYANDRIA . .	20 or more stamens on the receptacle (hypogynous).
XIV.	DIDYNAMIA . .	4 stamens, 2 long and 2 short.
XV.	TETRADYNAMIA	6 stamens, 4 long and 2 short.
XVI.	MONADELPHIA .	Filaments united into 1 set at the base or higher.
XVII.	DIADELPHIA . .	Filaments united into 2 sets, usually 9 united and 1 free (*Leguminosæ*).
XVIII.	POLYADELPHIA.	Filaments united into 3 or more sets.

XIX. SYNGENESIA . . Anthers united, but filaments free.
XX. GYNANDRIA . . Stamens and style consolidated.

2. *Stamens and pistils usually in different flowers.*

XXI. MONŒCIA . . . Male and female flowers on one plant.
XXII. DIŒCIA . . . . Male and female flowers on different plants.
XXIII. POLYGAMIA . . Male and female flowers separate or together, and flowers sometimes hermaphrodite.
XXIV. CRYPTOGAMIA . Fructification concealed.

CLASS I. MONANDRIA.

Tufted herb. Leaves opposite, subulate . . . . SCLERANTHUS, p. 234.
Water herb. Leaves opposite . . . . CALLITRICHE, p. 68.
Leafless, fleshy, jointed, seaside herb . . . . SALICORNIA, p. 233.
Small alpine herb. Leaves subulate, alternate . . . . . . ALEPYRUM, p. 295.
Minute, floating, green, scale-like fronds . . . . LEMNA, p. 277.
Prostrate herbs. Leaves pinnate. Flowers in a globose head . ACÆNA, p. 55.

Besides the above, *Orchideæ* (see XX.) are usually monandrous, but the anther is adnate to the style, and a few *Cyperaceæ* and species of *Chenopodium* have sometimes but one stamen.

CLASS II. DIANDRIA.

§ Perianth inferior, double.
Ovary 1-celled. Flowers irregular . . . . UTRICULARIA, p. 222.
Ovary 2-celled. Flowers regular or irregular . . . . SCROPHULARINEÆ, p. 200.
§§ Perianth inferior, single.
Trees. Leaves opposite . . . . OLEA, p. 186.
Shrubs; bark tough. Leaves alternate . . . . PIMELEA, p. 242.
Tufted herbs. Leaves subulate . . . . GAIMARDIA, p. 295.
Erect herbs. Leaves alternate or opposite . . . . . . CHENOPODIUM, p. 229.
Prostrate herbs. Leaves pinnate. Flowers capitate . . . ACÆNA, p. 55.
§§§ Perianth 0. Herb. Leaves alternate . . . . PEPEROMIA, p. 254.
Perianth 0. Shrub, aromatic. Leaves alternate . . . PIPER, p. 254.
§§§§ Perianth superior.
Tufted herb. Leaves imbricate. Flowers solitary . . . DONATIA, p. 58.
Prostrate herbs. Leaves pinnate. Flowers capitate . . . ACÆNA, p.55.
Herb. Leaves simple, radical. Flowers spiked or panicled GUNNERA, p. 67.

(See *Lemna* and *Salicornia* in I., *Forstera* and *Helophyllum* in XX.)

CLASS III. TRIANDRIA..

§ 1. Flowers in spikelets. Perianth incomplete or 0.—Grasses, sedges, etc. Leaves sheathing
Stem terete, hollow. Sheaths split to the base. Anthers 2-celled, versatile . . . . GRAMINEÆ, p. 317.
Stem solid, often 3-gonous. Sheaths not split. Anthers 2-celled, adnate . . . . CYPERACEÆ, p. 296.
Stem solid, terete. Sheaths entire. Anthers 1-celled . . RESTIACEÆ, p. 293.
§ 2. Flowers in globose, peduncled heads. Leaves pinnate . . ACÆNA, p. 55.
§ 3. Flowers not in spikelets nor heads. Perianth single.
Herbs. Perianth superior, of 6 spreading pieces . . . . LIBERTIA, p. 274.
Tree. Perianth superior. Calyx-limb 5–9-lobed . . . MERYTA, p. 104.
Herbs. Perianth inferior, of 5 green leaflets . . . . . . CHENOPODIUM, p. 229.
§ 4. Flowers various. Perianth double.
Herb. Leaves opposite. Sepals and petals 3 or 4, inferior ELATINE, p. 28.
Herb. Leaves opposite. Sepals 2. Petals 5, inferior . . MONTIA, p. 27.
Herb. Leaves linear, imbricate. Stamens 2 or 3, epigynous DONATIA, p. 58.
Herbs. Leaves radical. Stamens 2 or 3, epigynous . . GUNNERA, p. 67.

CLASS IV. TETRANDRIA.

§ 1. Perianth double, inferior. Corolla monopetalous.

Leaves opposite.

Maritime tree. Leaves entire, downy below . . . . AVICENNIA, p. 224.

Marsh or water, minute herbs . . . . { GLOSSOSTIGMA, p. 203. LIMOSELLA, p. 204.

Herb very aromatic, prostrate. Ovary 4-lobed . . . . MENTHA, p. 225.

Herbs. Leaves densely 4-fariously imbricate . . . . . LOGANIA, p. 188.

Herbs, bitter. Corolla campanulate or rotate . . . . GENTIANEÆ, p. 189.

Leaves alternate or radical.

Shrubs. Leaves entire, with transparent dots. Stamens opposite petals . . . . MYRSINE, p. 183.

Tree. Stamens epipetalous, alternating with scales . . SAPOTA, p. 185.

Shrub. Leaves serrate, with transparent dots . . . . MYOPORUM, p. 224.

Scapigerous herbs. Flowers spiked or capitate. Filaments very long, flexuose . . . . PLANTAGO, p. 226.

§ 2. Perianth double, inferior. Corolla polypetalous.

Leaves opposite.

Small herbs . . . . TILLÆA, p. 61.

Shrubs and trees. Petals lobed . . . . ARISTOTELIA, p. 33.

Leaves alternate or radical.

Shrubs or herbs. Flowers white, small, racemed . . . LEPIDIUM, p. 13.

Shrubs. Leaves with transparent dots . . . . MYRSINE, p. 183.

Climbing shrub. Flowers with a fringe inside the petals PASSIFLORA, 81.

Herbs. Leaves with glandular hairs . . . . DROSERA, p. 63.

§ 3. Perianth double, superior. Corolla monopetalous.

Leaves opposite; stipules 0. Style 1 . . . . LORANTHACEÆ, p. 107.

Leaves opposite, stipuled. Styles 2 . . . . RUBIACEÆ, p. 110.

Leaves alternate; stipules 0. Style 1 . . . . ALSEUOSMIA, p. 109.

Leaves whorled . . . . { ASPERULA, 121. GALIUM, 120.

Leaves 0. Stems jointed. Parasites . . . . VISCUM, p. 103.

§ 4. Perianth double, superior. Corolla polypetalous.

Herbs. Flowers minute . . . . HALORAGEÆ, p. 64.

Parasitic shrub . . . . LORANTHUS, p. 106.

§ 5. Perianth single or 0.

Leaves opposite (or 0 in Discaria *and* Viscum).

Herbs. Leaves subulate . . . . COLOBANTHUS, p. 24.

Tree or shrub, spiny. Branches opposite . . . . . DISCARIA, p. 43.

Parasitic shrubs. Stamens opposite perianth . . . { TUPEIA, p. 108. VISCUM, p. 108.

Erect tree. Stamens opposite perianth-lobes . . . . SANTALUM, p. 247.

Herbs. Leaves broad. Flowers small, green, clustered . CHENOPODIUM, p. 229.

Leaves alternate.

a. Perianth inferior.

Tufted, alpine. Leaves linear. Perianth tubular. Stamens on the tube . . . . DRAPETES, p. 245.

Trees and shrubs. Stamens on the 4 lobes of the perianth . . . . PROTEACEÆ, p. 240.

Herbs. Flowers minute, green . . . . CHENOPODIUM, p. 229.

Aquatic herbs. Leaves floating or submerged . . . NAIADEÆ, p. 277.

Tall, grass-like plants. Flowers in spikelets . . . . GAHNIA, p. 305.

β. Perianth superior.

Prostrate herb. Leaves pinnate. Flowers in a globose head . . . . ACÆNA, p. 55.

Fleshy maritime herbs. Leaves simple . . . . . . TETRAGONIA, p. 83.

Erect tree. Flowers cymose . . . . SANTALUM, p. 247.

CLASS V. PENTANDRIA.

§ 1. Perianth, double, inferior. Corolla monopetalous.

Leaves opposite.

Herbs. Sepals 2. Petals united at the base only . . . PORTULACEÆ, p. 26.
Herbs, bitter. Calyx 4- or 5-cleft. Corolla campanulate GENTIANEÆ, p. 189.
Prostrate small shrubs. Leaves connate or stipulate . . LOGANIACEÆ, p. 188.
Climbing shrubs. Leaves exstipulate . . . . PARSONSIA, p. 187.
Erect tree. Stamens alternating with scales . . . . SAPOTA, p. 185.

Leaves alternate.

Shrubs. Stamens epipetalous, anthers 1-celled. . . . . ERICEÆ, p. 173.
Shrubs. Leaves with transparent dots. Stamens opposite corolla-lobes . . . . MYRSINE, p. 183.
Herb. Leaves linear, not dotted. Stamens opposite corolla-lobes . . . . SAMOLUS, p. 185.
Climbing herbs. Corolla plicate. Ovary not lobed, 2-celled . . . . CONVOLVULACEÆ, p. 197.
Herbs. Peduncles supra-axillary. Berry 2-celled . . SOLANEÆ, p. 200.
Hispid herbs. Ovary 4-lobed. Fruit of 4 little nuts . BORAGINEÆ, p. 191.
Minute, glabrous herb. Ovary 2- or 3-lobed . . . . STACKHOUSIA, p. 42.

Leaves 0. Climbing slender herb . . . . CUSCUTA, p. 199.

§ 2. Perianth double, inferior. Corolla polypetalous.

Leaves opposite.

Herbs. Sepals 5. Ovary 1-celled . . . . CARYOPHYLLEÆ, p. 22.
Herbs. Sepals 2. Ovary 1-celled . . . . PORTULACEÆ, p. 26.
Shrub. Sepals 5. Ovary 3-celled . . . . ARISTOTELIA, p. 33.
Herb. Sepals 5. Ovary 5-celled . . . . PELARGONIUM, p. 37

Leaves alternate or radical.

Herbs. Leaves radical, glandular. Ovary 1-celled . . DROSERA, p. 63.
Herbs. Leaves subulate, glabrous. Ovaries free . . . MYOSURUS, p. 3.
Herb. Leaves linear, glabrous. Ovary 5-celled . . . LINUM, p. 34.
Minute herb. Leaves linear, glabrous. Ovary 3-celled . STACKHOUSIA, p. 42.
Shrubs. Leaves with transparent dots . . . . . . MYRSINE, p. 183.
Shrubs and trees. Ovary 2-celled . . . . PITTOSPORUM, p. 18.
Shrubs and trees. Stamens short. Ovary 1-celled; ovules several . . . . MELICYTUS, p. 17.
Tree. Stamens long, without alternate scales. Ovary 1-celled; ovule 1 . . . . PENNANTIA, p. 41.
Tree. Stamens short, with alternating scales. Ovary 1-celled; ovule 1 . . . . CORYNOCARPUS, p. 46.
Tree. Stamens long. Ovary 5-celled; ovules few . . IXERBA, p. 59.

§ 3. Perianth double, superior. Corolla monopetalous. (Leaves always alternate.)

Herb. Corolla regular. Stamens opposite corolla-lobes . SAMOLUS, p. 185.
Herbs. Corolla regular or irregular. Stamens alternate with corolla-lobes . . . . CAMPANULACEÆ, p. 169.
Shrubs. Corolla regular. Stamens alternate with corolla-lobes . . . . ALSEUOSMIA, p. 109.

§ 4. Perianth double, superior. Corolla polypetalous. (Leaves always alternate.)

Shrubs or trees. Flowers cymose. Ovary-cells many-ovuled SAXIFRAGEÆ, p. 57.
Shrubs or trees. Flowers umbelled. Ovary-cells 1-ovuled . ARALIACEÆ, p. 99.
Shrubs. Leaves exstipulate. Flowers solitary or few. Ovary-cells 1-ovuled . . . . CORNEÆ, p. 104.
Shrubs. Leaves stipulate. Stamens opposite petals. Ovary-cells 3, 1-ovuled . . . . POMADERRIS, p. 43.
Herbs. Flowers umbelled. Ovary-cells 2, 1-ovuled . . UMBELLIFERÆ, p. 84.

§ 5. Perianth single, inferior.
Herbs. Leaves opposite, exstipulate . . . . CARYOPHYLLEÆ, p. 22.
Herb. Leaves opposite, stipulate. Flowers small, white ALTERNANTHERA, p. 234.
Herb. Leaves subulate, radical . . . . MYOSURUS, p. 3.
Herbs. Leaves cauline, Flowers small, clustered, green . CHENOPODIUM, p. 229.
Shrub, spiny. Leaves or branches opposite . . . . . DISCARIA, p. 43.
Shrub. Leaves alternate . . . . , *Pittosporum rigidum*, p. 20.

§ 6. Perianth single, superior.
Shrub. Leaves stipulates . . . . POMADERRIS, p. 43.
Tree. Leaves very broad. Flowers capitate . . . . . . MERYTA, p. 104.
Tree. Leaves opposite and alternate. Flowers cymose . . SANTALUM, p. 247.
Herbs. Leaves alternate, fleshy . . . . TETRAGONIA, p. 83.

CLASS VI. HEXANDRIA.

§ 1. Perianth double, or, if single, of 6 pieces, inferior.
Herbs or shrubs. Perianth of 6 pieces. Style 1 . . . . LILIACEÆ, p. 280.
Marsh herb. Perianth small, green, of 6 pieces. Styles 3 or 6 . . . . TRIGLOCHIN, p. 278.
Rushes. Perianth small, of 6 dry leaflets. Stigmas 3 . . JUNCEÆ, p. 288.
Climbing shrubs, with opposite compound leaves . . . . CLEMATIS, p. 1.
Herbs and shrubs. Leaves alternate; stipules sheathing . POLYGONEÆ, p. 235.
Shrub. Leaves opposite. Petals 4, lobed . . . . ARISTOTELIA, p. 33.

§ 2. Perianth double, superior . . . . CARPODETUS, p. 59.

§ 3. Perianth single or 0.
Tree. Leaves large, broad. Flowers capitate . . . . MERYTA, p. 104.
Herb. Leaves radical, linear. Flower small . . . . . MYOSURUS, p. 3.
Climbing shrubs, with opposite leaves . . . . CLEMATIS, p. 1.
Succulent herbs. Leaves alternate . . . . TETRAGONIA, p. 83.
Tree. Leaves opposite and alternate. Flowers small, cymose SANTALUM, p. 247.
Herbs and shrubs. Leaves alternate. Stipules sheathing . POLYGONEÆ, p. 235.
Grass-like tall sedge, with harsh cutting foliage . . . . GAHNIA, p. 305.

CLASS VII. HEPTANDRIA.

Pisonia (see Class XVI.) is the only usually heptandrous New Zealand plant, but some few other genera, with indefinite stamens, are occasionally so.

CLASS VIII. OCTANDRIA.

§ 1. Perianth double, inferior. Corolla polypetalous.
Shrubs or trees. Leaves opposite, simple or pinnate . . . WEINMANNIA, p. 60.
Tree, aromatic. Leaves alternate, simple . . . . DRIMYS, p. 10.
Shrubs. Leaves with pellucid dots . . . . RUTACEÆ, p. 38.

§ 2. Perianth double, superior. Corolla polypetalous.
Ovary 4-celled, many-ovuled . . . . ONAGRARIEÆ, p. 75.
Ovary 2–4-celled; cells 1-ovuled . . . . HALORAGEÆ, p. 64.

§ 3. Perianth single, inferior.
Climbing shrubs. Leaves opposite, compound . . . . . CLEMATIS, p. 1.
Herbs or shrubs. Leaves alternate . . . . POLYGONEÆ, p. 235.
Tree. Leaves pinnate . . . . ALECTRYON, p. 45.

§ 4. Perianth single, superior.
Prostrate or climbing, fleshy herb . . . . TETRAGONIA, p. 83.
Shrub or tree. Leaves membranous . . . . FUCHSIA, p. 75.

CLASS IX. ENNEANDRIA.

A very few New Zealand plants, with indefinite stamens, are occasionally enneandrous.

CLASS X. DECANDRIA.

Stamens 10, free, in the same flower with the pistil. (Perianth inferior in all.)

§ 1. Perianth double. Corolla monopetalous.

Shrub. Ovary 5-celled . . . . { GAULTHERIA, p. 174. / PERNETTYA, p. 176.

§ 2. Perianth double. Corolla polypetalous.

Leaves opposite.

Herbs. Leaves entire, exstipulate. Ovary 1-celled . . CARYOPHYLLEÆ, p. 22.
Herbs and shrubs. Leaves entire, exstipulate. Carpels 5. CORIARIA, p. 46.
Herbs. Leaves lobed, stipulate . . . . GERANIACEÆ, p. 35.
Shrubs or tree. Leaves compound or simple . . . . SAXIFRAGEÆ, p. 57.

Leaves alternate or fascicled.

Shrubs or trees. Leaves pinnate. Flowers yellow . . SOPHORA, p. 52.
Herbs, acid. Leaves 3-foliolate . . . . OXALIS, p. 38.
Tree. Leaves oblong, not dotted; aromatic . . . . DRIMYS, p. 10.
Shrub. Leaves linear, with pellucid dots . . . . . PHEBALIUM, p. 39.
Herbs. Leaves lobed or cut . . . . GERANIUM, p. 35.

§ 3. Perianth single, inferior . . . . CLEMATIS, p. 1.

CLASS XI. DODECANDRIA. These will be found in the two following Classes.

CLASS XII. ICOSANDRIA. (Including dodecandrous perigynous plants.)

§ 1. Perianth superior.

Shrubs or trees. Leaves with pellucid dots . . . . . . MYRTACEÆ, p. 69.
Maritime herb. Leaves opposite, very fleshy . . . MESEMBRYANTHEMUM, p. 83.

§ 2. Perianth inferior.

Leaves compound. Carpels many, distinct . . . . ROSACEÆ, p. 53.
Leaves simple. Ovary 1-celled, 1-ovuled . . . . LAURINEÆ, p. 238.

CLASS XIII. POLYANDRIA. Perianth inferior in all. (Including dodecandrous hypogynous plants.)

Herbs or climbing shrubs. Carpels many, distinct . . . . RANUNCULACEÆ, p. 1.
Herbs. Carpels 5, cohering. Styles 5 . . . . . . HYPERICUM, p. 28.
Trees and shrubs. Ovary 3–6-celled . . . . TILIACEÆ, p. 32.
Maritime herb with milky juice . . . . EUPHORBIA, p. 247.

CLASS XIV. DIDYNAMIA. (Perianth inferior and irregular in all.)

Leaves 3–5-foliolate . . . . VITEX, p. 223.

Leaves simple.

Ovary not lobed, 2-celled, many-ovuled . . . . . . SCROPHULARINEÆ, p. 200.
Ovary not lobed, 1-celled, many-ovuled . . . . RHABDOTHAMNUS, p. 221.
Ovary 4-lobed. Stamens curved downwards . . . . SCUTELLARIA, p. 226.
Ovary 4-lobed. Stamens curved upwards . . . . . . TEUCRIDIUM, p. 224.
Ovary 4-lobed. Stamens nearly straight . . . . . MENTHA, p. 225.

CLASS XV. TETRADYNAMIA.

Herbs. Leaves alternate or radical . . . . CRUCIFERÆ, p. 10.

CLASSES XVI.—XVIII. MONADELPHIA, DIADELPHIA, AND POLYADELPHIA.

§ 1. Perianth double. Corolla polypetalous, or petals united at the base only. Ovary superior.

Leaves opposite.

Herbs. Leaves quite entire, exstipulate, with pellucid dots HYPERICUM, p. 28.

Herbs. Leaves lobed, stipulate, without dots . . . . GERANIUM, p. 35.
Leaves alternate or 0 *in* Carmichælia.
Herbs, shrubs, and trees; bark tough. Anthers 1-celled . MALVACEÆ, p. 29.
Herbs, acid. Leaves 3-foliolate . . . . OXALIS, p. 38.
Herb or undershrub. Leaves simple, entire . . . . . LINUM, p. 34.
Tree. Leaves pinnate. Flowers regular, racemed . . . DYSOXYLUM, p. 40.
Herbs and shrubs. Leaves 0, or pinnate or 3-foliolate. Flower irregular . . . . LEGUMINOSÆ, p. 47.
Herbs. Leaves lobed, stipuled . . . . GERANIACEÆ, p. 35.

§ 2. Perianth single.
Spiny maritime undershrub . . . . SALSOLA, p. 232.
Herb. Leaves opposite, stipulate. Flower small, white . ALTERNANTHERA, p. 234.
Tree. Leaves very large . . . . PISONIA, p. 229.

CLASS XIX. SYNGENESIA.

Flowers small, collected in involucrate heads . . . . COMPOSITÆ, p. 121.
Flowers not collected in involucrate heads.
Shrub. Leaves opposite. Corolla 2-labiate, inferior . . RHABDOTHAMNUS, p. 221.
Twining shrub. Leaves opposite. Corolla regular. Stamens epipetalous . . . . PARSONSIA, p. 187.
Shrubs and herbs. Leaves alternate, stipulate. Stamens hypogynous . . . . VIOLARIEÆ, p. 15.
Shrubs and herbs. Leaves alternate. Stamens epigynous or epipetalous . . . . CAMPANULACEÆ, p. 169.

CLASS XX. GYNANDRIA.

Herbs. Perianth irregular, superior. Anther 1 . . . . . . ORCHIDEÆ, p. 260.
Herbs. Perianth regular or irregular, superior. Anthers 2 . . STYLIDIEÆ, p. 166.
Shrubs, twining. Perianth regular, inferior. Anthers 5 . . . PARSONSIA, p. 187.

CLASSES XXI.—XXIII.

§ 1. Perianth double, inferior. Corolla monopetalous.
Tree. Leaves alternate. Stamens 4 or 5, alternating with scales . . . . SAPOTA, p. 185.

§ 2. Perianth double, inferior. Corolla polypetalous.
Tree. Leaves opposite, pinnate . . . . ACKAMA, p. 60.
Shrubs or trees. Leaves opposite, simple . . . . ARISTOTELIA, p. 33.
Climbing prickly shrub. Leaves opposite, 3–5-foliolate . . RUBUS, p. 54.
Shrubs and trees. Ovary 2- or 3-celled, many-ovuled . . PITTOSPORUM, p. 19.
Tree. Ovary 1-celled, 1-ovuled . . . . PENNANTIA, p. 41.

§ 3. Perianth double, superior. Corolla monopetalous.
Leaves opposite or whorled.
Leaves opposite, stipuled or whorled (not aquatic) . . . RUBIACEÆ, p. 110.
Leaves opposite or whorled. Aquatic herbs. . . . . . MYRIOPHYLLUM, p. 66.
Leaves opposite, exstipulate. Parasitic shrubs . . . . LORANTHACEÆ, p. 107.
Leaves alternate.
Shrubs or trees. Leaves alternate. Flowers umbelled . ARALIACEÆ, p. 99.
Herbs. Leaves alternate or radical. Flowers umbelled . UMBELLIFERÆ, p. 84.
Shrubs. Leaves alternate. Flowers not umbelled . . CORNEÆ, p. 104.
Trailing herb with tendrils. Leaves angled . . . . . . SICYOS, p. 82.
Trees and shrubs; bark tough. Filaments united in a tube PLAGIANTHUS, p. 29.
Herbs, scapigerous. Anthers 2 adnate to style . . . . STYLIDIEÆ, p. 166.

§ 4. Perianth single, inferior, regular. Flowers not in the scales of spikelets
Leaves opposite.
Trees. Calyx 4-lobed. Stamens 2 . . . . OLEA, p. 186.

Herbs. Stamens 4, opposite 4 perianth lobes . . . . URTICEÆ, p. 250.
Climbing shrubs. Leaves compound . . . . CLEMATIS, p. 1.
Erect shrubs and trees. Carpels separate, 1-ovuled . . MONIMIACEÆ, p. 239.
Herb. Stamen 1. Styles 2 . . . . CALLITRICHE, p. 68.

Leaves alternate.

Tree. Leaves simple, exstipulate. Calyx 5-lobed . . . DODONÆA, p. 44.
Tree. Leaves simple, stipuled. Male perianth campanulate FAGUS, p. 248.
Unbranched tree. Leaves very large, pinnate . . . . . ARECA, p. 288.
Trees. Leaves simple. Anthers with recurved valves . LAURINEÆ, p. 238.
Herbs, shrubs and trees. Stamens 4 or 5, opposite perianth lobes . . . . URTICEÆ, p. 250.
Herbs. Male flower 4- or 5-partite. Female 2-partite . ATRIPLEX, p. 231.
Tree. Perianth of 2 peltate scales. Stamens numerous . CARUMBIUM, p. 248.
Climbing shrub. Leaves long, sheathing, prickly . . . FREYCINETIA, p. 275.
Herbs often large, epiphytic, usually scaly or silky. Perianth 6-lobed. Stamens 6 . . . . ASTELIA, p. 283.

Leaves 0.

Small, alpine, rigid, twiggy shrub . . . . EXOCARPUS, p. 246.
Slender twining parasite . . . . CASSYTHA, p. 239.

§ 6. Perianth single, regular, superior. (The perianth is double in most, but the calyx-limb being minute or absent it appears to be single.)

Leaves opposite.

Parasitic shrub. Leaves exstipulate. Flowers green . TUPEIA, p. 109.
Shrubs and herbs. Leaves stipulate. Stamens usually 4 RUBIACEÆ, p. 110.
Tree. Leaves opposite and alternate, exstipulate . . . SANTALUM, p. 247.

Leaves whorled . . . . MYRIOPHYLLUM, p. 66.

Leaves alternate or radical.

Herb. Leaves radical. Flowers small, spiked or racemed GUNNERA, p. 67.
Shrubs. Leaves exstipulate. Flowers not umbelled . . CORNEÆ, p. 104.
Shrubs and trees. Flowers umbelled or capitate . . . ARALIACEÆ, p. 99.
Herb. Flowers umbelled or capitate . . . . UMBELLIFERÆ, p. 84.
Parasite shrub. Leaves exstipulate. Flowers green . . TUPEIA, p. 109.
Erect tree. Flowers cymose . . . . SANTALUM, p. 247.
Trailing herb, with tendrils. Leaves angled . . . . SICYOS, p. 82.

§ 5. Perianth very incomplete or 0.

Leaves opposite.

Shrub. Leaves serrate. Flowers spiked . . . . . ASCARINA, p. 253.
Herb. Leaves entire. Flowers solitary . . . . CALLITRICHE, p. 68.

Leaves alternate or radical.

α. Herbs. Flowers in the axils of scales, arranged in spikelets. (Grasses, sedges, etc. Leaves sheathing, long, slender, or 0.)

Stems hollow, terete, jointed. Leaf-sheaths split to the base . . . . GRAMINEÆ, p. 317.
Stems solid, flat, angled or 3-gonous. Leaf-sheath entire CYPERACEÆ, p. 296.
Stems solid. Leaf-sheaths usually entire. Anthers 1-celled . . . . RESTIACEÆ, p. 293.

β. Herbs not grassy in foliage, of various habit.

Marsh herbs. Flowers capitate or in large catkins . TYPHACEÆ, p. 276.
Water plants. Flowers obscure . . . . NAIADEÆ, p. 277.
Land plants. Flowers in very slender catkins . . . . PIPERACEÆ, p. 254.
Land plants. Flowers small, green . . . . URTICEÆ, p. 250.
Fleshy root-parasite, with tuberous rhizome and stout scaly peduncles . . . . DACTYLANTHUS, p. 255.

γ. Trees and shrubs.

Stem climbing. Leaves sheathing, long, prickly . . FREYCINETIA, p. 275.
Shrub. Leaves very aromatic . . . . PIPER, p. 254.

Erect trees. Leaves stipulate. Ovary 3-celled . . . FAGUS, p. 248.
Trees or shrubs. Anthers solitary in the scales of a cone. Ovules solitary or few in the axils of scales, without ovary or perianth . . . . CONIFERÆ, p. 255.

CLASS XXIV. CRYPTOGAMIA.

(See p. lxvi.)

III. CLASSIFICATION OF THE NEW ZEALAND NATURAL ORDERS ACCORDING TO THE NATURAL SYSTEM, AS ADOPTED IN THE PRESENT WORK.

The number of New Zealand genera is so small, and many of them are so exceptional in characters, that they afford but an imperfect idea of the relations that subsist between the Orders they belong to; or between these and the same and other Orders in the world at large. The natural sequence of the Orders of plants and their relationships being determined by the characters of the majority of all the known genera which they severally include; this sequence and these relationships cannot be demonstrated by a small flora, like that of New Zealand, in which many large Orders and more small ones are unrepresented, or represented by one or two uncharacteristic genera only.

The following system is not altogether natural, but it is so in the main, and as much so as any hitherto devised. The primary divisions Phænogams and Cryptogams are perfectly natural and well-defined, and the Subclasses and Orders of Cryptogams follow, on the whole, a very natural sequence. The two Classes of Phænogams, Dicotyledons and Monocotyledons, are also natural and well defined, as are, on the whole, the Subclasses and Orders of Monocotyledons. With the Dicotyledons the case is different: the subdivisions Angiosperms and Gymnosperms are distinct and natural; but no natural arrangement of the Orders of Angiosperms has yet been devised. The Five Subclasses (*Thalamifloræ*, etc.) are only in so far natural, that each consists of Orders more or less naturally related to one another; for the cross affinities between certain Orders of all the Subclasses are very numerous, and in some cases so strong that the single technical character of the Subclass alone keeps them where they are. In the case of Subclass V., *Incompletæ*, matters are still worse, the Orders of this Subclass having no common relationship, but consisting of Thalamifloral, or Calycifloral, plants in which the perianth happens to be incomplete or absent. Therefore it must be borne in mind, that the Subclasses of *Angiospermæ* are merely artificial devices, to enable us to find our way to the Order we seek; and that these Orders are connected by so many cross affinities that no one has ever yet been able to arrange them in a linear series.

For further information the reader is referred to Lindley's 'Vegetable Kingdom,'* which contains an excellent history of this subject, besides being the best Encyclopædia of all that relates to the Natural Orders which we possess.

A. *Phænogamic or Flowering Plants.*

CLASS I. DICOTYLEDONS.

Stem, when perennial, furnished with pith, concentric layers of wood and bark. Leaves usually with netted venation. Organs of the flower generally 4 or 5 each, or multiples of those numbers. Seeds having an embryo with 2 cotyledons. In germination the radicle lengthens, and forks or branches.

The exceptions to one or other of these characters are too numerous to be indicated.

---

* A work that should be in every naturalist's library (one thick volume, 8vo, with 500 woodcuts), price 36*s.*, 3rd edition.

SUBDIVISION I. *ANGIOSPERMÆ.*

Ovules enclosed in an ovary, and the seeds in a seed-vessel.

SUBCLASS I. **Thalamifloræ.** Flowers with both calyx and corolla. Petals free, and stamens usually inserted immediately beneath the pistil or ovaries. Ovary always superior.

Exceptions: Petals 0 in *Clematis, Myosurus, Caltha,* some species of *Cruciferæ, Colobanthus,* and *Stellaria,* united at the base in some *Portulaceæ* and *Malvaceæ.*
Sepals absent in one *Pittosporum;* petaloid in *Clematis* and *Caltha.*
Stamens perigynous in some *Stellariæ* and *Colobanthus.*

§ 1. *Anthers adnate, opening by lateral slits. Pistil apocarpous.*

1. **Ranunculaceæ.** Sepals 3–10, often petaloid. Petals 0, or 5–20. Stamens indefinite. Fruit of many or few achenes or follicles.—Herbs or opposite-leaved climbers (p. 1).

Of the four genera, three have petaloid sepals and no petals.

2. **Magnoliaceæ.** Sepals and petals forming together three or more series, imbricate in æstivation. Carpel 1 or more.—Tree with alternate exstipulate leaves (p. 9).

§ 2. *Anthers opening towards the stigma (inwards). Pistil syncarpous. Placentas parietal (rarely axile in* Pittosporeæ*).*

3. **Cruciferæ.** Sepals and petals 4. Stamens 6, 4 longer than the others.—Herbs with alternate or rosulate exstipulate leaves (p. 10).

One *Lepidium* is rather shrubby. *Nasturtium* sometimes wants petals, and two or more of the stamens.

4. **Violarieæ.** Sepals and petals 5. Anthers 5, their connectives enlarged or produced upwards, often connate. Placentas usually 3.—Herbs or shrubs, with alternate, stipulate leaves (p. 15).

5. **Pittosporeæ.** Sepals, petals, and stamens 4 or 5. Placentas usually 2. Capsule coriaceous or woody, 2-valved.—Shrubs or trees, with usually opposite, evergreen, exstipulate leaves (p. 18).

Ovary sometimes 2–5-celled. Calyx apparently absent in one *Pittosporum.*

§ 3. *Pistil syncarpous, 1-celled. Placenta basal.*

6. **Caryophylleæ.** Sepals 4 or 5. Petals 4 or 5 or 0, free. Stamens 4 or 5, 8 or 10, hypogynous or perigynous.—Herbs with opposite, entire leaves. Flowers white or green (p. 22).

Petals absent in some *Stellariæ* and *Colobanthus.* Stamens perigynous in *Colobanthus.*

7. **Portulaceæ.** Sepals 2. Petals 5, usually united at the base. Stamens 5, usually opposite and adherent to the bases of the petals.—Herbs with opposite alternate or imbricate leaves. Flowers white (p. 26).

§ 4. *Pistil more or less syncarpous, 2- or more-celled. Placentas axile. Disk 0, or a raised torus.*

8. **Elatineæ.** Sepals and petals 2–5, all free, imbricate. Stamens definite, hypogynous, free. Ovary 2–5-celled; ovules many.—Small, creeping water-herb. Leaves opposite, stipulate, pellucid-dotted. Flowers minute, solitary, axillary (p. 28).

9 **Hypericineæ.** Sepals and petals 5, hypogynous, free, imbricate. Stamens indefinite, hypogynous, free or polyadelphous. Ovary 3–5-celled; styles 3–5, free or connate; ovules numerous.—Herbs shrubs or trees, with opposite, exstipulate pellucid-dotted leaves. Flowers yellow, in 3-chotomous cymes (p. 28).

10. **Malvaceæ.** Calyx-lobes 5, valvate. Petals 5, usually connate at the base and adnate to the staminal tube, contorted. Stamens indefinite, filaments monadelphous; anthers 1-celled. Ovary of 1 or more, free or connate carpels.—Herbs shrubs and trees, with often stellate down. Leaves alternate, stipulate (p. 29).

11. **Tiliaceæ.** Sepals 4 or 5, valvate. Petals 4 or 5, often lobed or cut, imbricate. Stamens numerous, on a raised torus, filaments free; anthers 2-celled, often with terminal pores. Ovary 2–10-celled.—Trees or shrubs. Leaves alternate or opposite, stipulate (p. 32).

12. **Lineæ.** Sepals 5, free, imbricate. Petals 5, free, contorted. Stamens 5, hypogynous; filaments united at the base into a cup. Ovary 3–5-celled; styles 3–5, free or connate.—Herbs. Leaves alternate, small, exstipulate. Flowers large, usually corymbose (p. 34).

13. **Geraniaceæ.** Sepals 5, free, imbricate. Petals 5, equal or unequal, free, imbricate. Stamens 10, hypogynous, all fertile or some without anthers; filaments free or united at the base. Ovary 3–5-lobed; cells usually 1-seeded.—Herbs. Leaves alternate or opposite, stipulate or exstipulate. Flowers usually axillary, solitary geminate or umbelled (p. 35).

SUBCLASS II. **Disciflorae.** Flowers with both calyx and corolla. Petals free, and stamens usually inserted upon the surface or at the base of a thickened hypogynous disk. Ovary rarely inferior.

Exceptions: Petals absent in *Dodonæa*, in one *Pomaderris*, *Discaria*, and *Alectryon*. Stamens hypogynous in *Pennantia*, *Coriaria*, and *Dodonæa*.
Ovary inferior in *Pomaderris*.

14. **Rutaceæ.** Sepals and petals 4 or 5, free, imbricate or valvate. Stamens 8 or 10, rising from the outer base of an hypogynous disk. Ovary of 4 or 5, free or united, 2-ovuled carpels, separating when ripe into as many 1-seeded 2-valved cocci (p. 38).

15. **Meliaceæ.** Calyx small, 4- or 5-lobed, imbricate. Petals 4 or 5, linear, usually valvate and adnate at the base with the staminal tube. Stamens united into a thick tube, usually inserted below an annular or tubular disk. Ovary 3–5-celled.—Trees. Leaves exstipulate, compound (p. 40).

16. **Olacineæ.** Calyx small, 4- or 5-lobed. Petals 4 or 5, free or connate, valvate. Stamens 4 or 5, hypogynous or surrounding an annular disk. Ovary 1-celled or imperfectly 2- or 3-celled; style 1; ovules 1–3, pendulous. —Shrubs or trees. Leaves alternate, exstipulate (p. 41).

17. **Stackhousieæ.** Calyx 5-lobed. Petals 5, linear, erect, free or connate above the base. Stamens 5, 2 shorter. Ovary 2–5-celled. Fruit of 2–5 globose cocci.—Herbs. Leaves alternate, small, quite entire. Flowers small, greenish, racemose (p. 42).

18. **Rhamneæ.** Calyx superior or inferior, 4- or 5-lobed. Petals 4 or 5, minute or 0. Stamens inserted on the edges of a disk, as many as the petals and opposite them, small, incurved. Ovary 3-celled; style 1, ovule in each cell 1, erect. Fruit of 3 cocci.—Shrubs or trees, with often stellate down. Leaves alternate, rarely opposite, stipulate, or 0. Flowers small (p. 42).

19. **Sapindaceæ.** Calyx 2–5-sepalled. Petals 0 in the New Zealand species. Stamens 5–8, hypogynous or inserted within a disk. Ovary 2- or 3-celled; style 1; ovules 1 or 2 in each cell, pendulous.—Trees. Leaves exstipulate, simple or compound. Flowers racemose (p. 44).

20. **Anacardiaceæ.** Calyx 3–7-lobed. Petals 5, imbricate. Stamens 5, inserted at the base of a lobed disk. Ovary 1-celled; style 1; ovule 1, erect or pendulous.—Shrubs or trees. Leaves usually alternate, exstipulate. Flowers usually small and panicled (p. 45).

---

21. **Coriarieæ.** Sepals 5. Petals 5, free, becoming fleshy after flowering. Stamens 10, hypogynous, all free, or 5 with the filaments adnate to the petals. Carpels 5–10, 1-celled, 1-ovuled, whorled round a fleshy disk; styles 5–10. Fruit of dry achenes enclosed in the fleshy petals.—Shrubs or herbs. Leaves opposite, exstipulate. Flowers racemose (p. 46).

This Order is a very anomalous one, whose affinities have never yet been discovered.

SUBCLASS III. **Calycifloræ.** Flowers with both calyx and corolla. Petals usually free, and stamens inserted on the tube of the calyx or top of the ovary, which is often inferior.

Exceptions: perianth apparently absent in some *Halorageæ*.

Petals absent in *Meryta*, *Fuchsia*, *Tetragonia* and some *Halorageæ*. Petals united at the base in *Acæna*, obscurely in *Tillæa*, united into a tubular corolla in some *Loranthi*.

Stamens hypogynous in some *Droseræ* and *Tillææ*.

§ 1. *Corolla very irregular or regular. Pistil apocarpous. Albumen* 0.

22. **Leguminosæ.** Calyx tubular or campanulate. Petals papilionaceous in the New Zealand species. Stamens 10, sheathing the ovary. Ovary of one 1-celled carpel. Fruit a legume (p. 47).

Carmichælia has a very exceptional pod.

23. **Rosaceæ.** Calyx tubular or expanded. Petals regular. Stamens numerous (free in *Acæna*). Carpels 2 or more. Fruit various (p. 53).

§ 2. *Corolla regular. Stamens definite. Pistil syncarpous or apocarpous, inferior or superior. Albumen fleshy.*

24. **Saxifrageæ.** Calyx inferior or superior, 5-cleft. Stamens 5 or 10. Pistil syncarpous; placentas axile; ovules numerous.—Herbs shrubs or trees. Leaves opposite or alternate, simple or compound (p. 57).

Stamens 2 or 3 in *Donatia*.

25. **Crassulaceæ.** Calyx 3- or 5-sepalled, inferior, free. Petals and stamens subhypogynous. Pistil apocarpous. Follicles 1–∞-seeded.—Small herbs. Leaves opposite, quite entire (p. 61).

26. **Droseraceæ.** Calyx 5-cleft, inferior, free. Petals and stamens 4

or 5, usually hypogynous. Ovary 1-celled, with parietal placentas; ovules numerous.—Small herbs. Leaves radical or alternate, covered with long glandular hairs (p. 62).

27. **Halorageæ.** Calyx-tube adnate to the ovary; limb 4-toothed or 0. Petals 2, 4, or 0. Stamens 2 or 4, epigynous. Ovary 1- 2- or 4-celled, with 1 pendulous ovule in each cell. Fruit small, indehiscent.—Herbs. Leaves radical or opposite or whorled. Flowers minute, often unisexual (p. 64).

Callitriche has no perianth, and 1 stamen in the male flower. *Myriophyllum* has an incomplete perianth in the male flowers, as have some *Gunneræ*.

§ 3. *Corolla regular. Pistil syncarpous, wholly inferior; ovules numerous, on axile placentas. Albumen* 0. *Petals* 0 *in one* Fuchsia.

28. **Myrtaceæ.** Calyx-lobes 4 or 5, valvate or imbricate. Stamens indefinite. Ovules few or many.—Trees or shrubs. Leaves evergreen, opposite, with pellucid dots (p. 69).

29. **Onagrarieæ.** Calyx-lobes 4, valvate. Stamens 8; ovules indefinite.—Herbs shrubs or trees. Leaves opposite or alternate (p. 75).

§ 4. *Corolla regular. Pistil syncarpous, wholly inferior; ovules few or numerous, on parietal placentas. Embryo straight.*

Ovary 1-celled, 1-ovuled, in the only New Zealand genus of *Cucurbitaceæ*.

30. **Passifloreæ.** Petals persistent with the sepals. Stamens definite, adnate with the stalk of the ovary; albumen fleshy.—Climber, with tendrils. Leaves alternate (p. 81).

31. **Cucurbitaceæ.** Flowers unisexual. Stamens 3 or 5, usually variously combined by their anthers. Ovary 1-celled and 1-ovuled in the only New Zealand genus; albumen 0.—Climbers or trailers, with tendrils. Leaves alternate (p. 82).

§ 5. *Corolla regular or* 0. *Pistil syncarpous, inferior; ovules few or numerous, on axile placentas. Albumen fleshy. Embryo curved.*

32. **Ficoideæ.** Calyx 3–∞-lobed. Petals in the New Zealand genera numerous or 0. Stamens definite or indefinite. Fruit an indehiscent drupe, or fleshy below and bursting at the top by many valves within the calyx.—Fleshy herbs. Leaves opposite or alternate (p. 83).

§ 6. *Petals regular, often small, deciduous. Stamens* 4 *or* 5, *epigynous. Pistil syncarpous. Ovary wholly inferior, with an epigynous disk; ovules solitary in each cell.*
Petals 0 in *Meryta*.

33. **Umbelliferæ.** Petals 5, usually imbricate. Stamens 5. Styles always 2. Fruit separating into 2 1-seeded carpels.—Herbs, rarely undershrubs. Leaves alternate, simple or compound. Flowers small, umbelled or capitate (p. 84).

34. **Araliaceæ.** Petals 5, usually valvate. Stamens 5. Styles 2–5. Fruit not separating, drupaceous or dry, 2–many-celled.—Shrubs or trees, rarely herbs. Leaves simple, or 1–7-foliolate (p. 99).

Meryta has no petals and anomalous inflorescence.

35. **Corneæ.** Petals 5, valvate. Stamens 5. Style 1. Ovary 1–2-celled. Fruit ovoid, 1–2-celled.—Shrubs or trees. Leaves simple (p. 104).

36. **Loranthaceæ.** Petals 4 or 5, often united into a tube. Stamens 4 or 5, inserted on the petals or free. Ovary 1-celled. Style 1. Parasitical herbs or shrubs. Leaves opposite or alternate (p. 106).

Tupeia is diœcious, and *Viscum* leafless.

SUBCLASS IV. **Corolliforæ** or **Monopetalæ.** Flowers with both calyx and corolla. Petals combined into a lobed corolla. Stamens inserted on the tube of the corolla.

Exceptions: Corolla absent in *Jasmineæ*.
Petals free or almost free in some *Campanulaceæ* and *Myrsineæ*.
Stamens epigynous in *Stylidieæ*, some *Campanulaceæ* and *Ericeæ*; hypogynous in some *Jasmineæ*.

§ 1. *Corolla epigynous, bearing the stamens.* (*See* Ericeæ *in* § 2.)

37. **Caprifoliaceæ.** Flowers panicled or solitary. Anthers free. Ovary 2-celled. Leaves opposite or alternate, exstipulate (p. 109).

38. **Rubiaceæ.** Flowers panicled capitate or solitary. Anthers free. Ovary 2-celled. Leaves opposite and stipulate, or whorled and exstipulate (p. 110).

39. **Compositæ.** Flowers collected in involucrate heads. Anthers combined. Ovary 1-celled; ovule erect.—Herbs shrubs or trees. Leaves usually alternate or radical (p. 121).

§ 2. *Corolla epigynous. Stamens epigynous, or inserted at the very base of the corolla.*

Stamens on the tube of the corolla in some *Ericeæ*.

40. **Stylidieæ.** Stamens 2, united with the style into one column, bearing the anthers at its top.—Herbs, usually small (p. 166).

41. **Campanulaceæ.** Stamens 5, all free or more or less united in a tube sheathing the style. Anthers 2-celled, opening by 2 slits.—Herbs (p. 169).

42. **Ericeæ.** Stamens 5 or 10, all free, hypogynous or epipetalous. Anthers 1-celled, or, if 2-celled, opening by terminal pores.—Small shrubs, rarely trees (p. 173).

§ 3. *Corolla hypogynous.*

a. Stamens either opposite the corolla-lobes or more than their number. Corolla regular.

43. **Myrsineæ.** Stamens opposite the corolla lobes, which are almost free in the New Zealand genus. Ovary and ovules as in *Primulaceæ*. Fruit an indehiscent berry.—Shrubs. Leaves with pellucid dots (p. 183).

44. **Primulaceæ.** Stamens opposite the corolla-lobes. Ovary 1-celled; ovules numerous, on a free central placenta. Fruit capsular.—Herbs. Leaves without pellucid dots (p. 185).

45. **Sapoteæ.** Stamens opposite the corolla-lobes, or more numerous. Ovary 2- or more celled.—Shrubs or trees. Leaves without pellucid dots (p. 185).

β. Stamens as many as and alternating with the corolla-lobes (fewer in Jasmineæ). *Corolla regular.*

Corolla absent in *Jasmineæ.*

46. **Jasmineæ.** Corolla absent in the New Zealand genus. Stamens 2. Ovary 2-celled.—Shrubs and trees. Leaves opposite, exstipulate (p. 186).

47. **Apocyneæ.** Stamens 5, often adhering to the stigma. Carpels 2, usually distinct.—Climbers, often with milky juice. Leaves usually opposite, exstipulate (p. 187).

48. **Loganiaceæ.** Stamens 4–5, anthers free. Ovary 2- or more celled; placentas axile.—Herbs or shrubs. Leaves opposite, often stipulate (p. 188).

49. **Gentianeæ.** Stamens 4 or 5, anthers free. Ovary 1-celled; placentas parietal.—Herbs, with bitter juice. Leaves opposite, quite entire, exstipulate (p. 189).

50. **Boragineæ.** Stamens usually 5, anthers free or conniving. Ovary 4-lobed to the base, 4-celled; cells 1-ovuled. Fruit of 4 small nuts.—Herbs, often hispid. Leaves alternate, quite entire, exstipulate (p. 191).

51. **Convolvulaceæ.** Stamens usually 5, anthers free. Ovary entire or 2-lobed, 2–4-celled; cells 1- or 2-ovuled.—Herbs, prostrate or climbing. Leaves alternate, exstipulate, 0 in *Cuscuta* (p. 197).

52. **Solaneæ.** Stamens 5, anthers often conniving or cohering, opening by pores or slits. Ovary entire, 2-celled; ovules many, on axile placentas.—Herbs or shrubs. Leaves alternate, exstipulate (p. 200).

(See 58, *Plantagineæ*, at the end of δ.)

γ. Stamens fewer in number than the lobes of the corolla. Corolla usually very irregular, nearly regular in many Veronicas *and* in Pygmæa. *Ovules numerous.*

53. **Scrophularineæ.** Stamens 2–4. Ovary 2-celled; ovules many, on axile placentas.—Herbs shrubs or small trees. Leaves opposite (p. 200).

54. **Gesneriaceæ.** Stamens 4. Ovary 2-celled; ovules many, on projecting parietal placentas.—A twiggy shrub. Leaves opposite (p. 221).

55. **Lentibularieæ.** Stamens 2. Ovary 1-celled; ovules many, on a free central placenta.—Small herbs, often aquatic (p. 222).

δ. Stamens as many or fewer than the lobes of the corolla. Corolla regular and irregular. Ovules few.

56. **Verbenaceæ.** Corolla regular or irregular. Stamens 2, 4, or 5. Ovary rarely lobed, 2- or 4-celled; cells 1- or 2-ovuled.—Herbs shrubs or trees. Leaves opposite or alternate, simple or compound, exstipulate (p. 223).

57. **Labiatæ.** Corolla irregular or nearly regular. Stamens 2 or 4. Ovary 4-lobed to the base, 4-celled; cells 1-ovuled. Fruit of 4 small nuts.—Herbs and shrubs. Leaves opposite, simple (p. 225).

58. **Plantagineæ.** Corolla regular, scarious, 4-lobed. Stamens 4, filaments very long, flexuose. Ovary entire, 2-celled; ovules on the septum.

Capsule circumsciss.—Herbs. Leaves radical. Flowers capitate or spiked, green or brownish (p. 226).

SUBCLASS V. **Incompletæ.** Perianth single or 0.

§ 1. *Flower usually hermaphrodite. Perianth single, enclosing the fruit. Stamens hypogynous or perigynous. Ovary* 1-*celled*, 1-*ovuled* (3-*ovuled in some* Amaranthaceæ).

59. **Nyctagineæ.** Perianth elongate, tubular. Stamens hypogynous. Style 1. Ovule erect. Embryo folded; albumen scanty.—Shrubs trees and herbs (p. 228).

60. **Chenopodiaceæ.** Perianth 2–5-lobed or -partite. Stamens perigynous. Styles 2 or 3, or 2- or 3-fid; ovule 1, pendulous from a basilar cord. Embryo annular or spiral in mealy albumen.—Usually herbs (p. 229).

61. **Amaranthaceæ.** Perianth of 5 leaflets. Stamens perigynous, monadelphous. Style usually simple; ovules 1 or more, pendulous from basilar cords. Embryo annular in mealy albumen.—Usually herbs (p. 233).

62. **Paronychieæ.** Perianth 4- or 5-lobed or -partite. Stamens 1–10, perigynous. Style 2- or 3-fid; ovule 1, erect. Embryo annular in mealy albumen.—Herbs (p. 234).

63. **Polygoneæ.** Perianth 5- or 6-partite. Stamens 6–9, perigynous. Styles 2 or 3, very short; ovule erect. Embryo straight or curved in mealy albumen.—Herbs (p. 235).

64. **Laurineæ.** Perianth of 4–8-segments. Stamens 12–15, perigynous; anthers opening by recurved valves. Style 1; ovule 1, pendulous. Embryo with thick cotyledons; albumen 0.—Shrubs and trees, rarely herbs (p. 238).

§ 2. *Perianth single. Stamens perigynous. Carpels numerous*, 1-*celled*, 1-*ovuled.*

65. **Monimiaceæ.** Perianth 4–15-lobed; anthers opening by slits or recurved valves. Embryo small, in fleshy albumen.—Shrubs and trees. Leaves opposite (p. 239).

§ 3. *Flowers hermaphrodite. Perianth single, inferior (in flower at least), usually tubular. Stamens inserted on its lobes. Ovary* 1-*celled.*

66. **Proteaceæ.** Perianth of 4 narrow deciduous leaflets, often connate below. Stamens 4. Style 1; ovules 1 or more.—Shrubs or trees (p. 240).

67. **Thymeleæ.** Perianth tubular, 4-lobed. Stamens 1–4. Style 1; ovule 1, pendulous.—Shrubs with tough fibrous bark (p. 241).

68. **Santalaceæ.** Perianth 3–5-lobed, valvate. Stamens 3–6. Style simple or 3-fid; ovules 3–5, pendulous from a central placenta. Fruit often inferior, 1-seeded (p. 246).

§ 4. *Flowers unisexual. Perianth* 0, *or of scales* (*single in most* Urticeæ). *Stamens hypogynous or perigynous. Ovary* 3-*celled* (1-*celled in* Urticeæ).

69. **Euphorbiaceæ.** Fruit a 3-celled capsule.—Herbs shrubs or trees (p. 247).

70. **Cupuliferæ.** Fruit of 3-angled, 1-seeded nuts, contained in a hard 4-lobed involucre.—Trees. Leaves alternate, stipulate (248).

71. **Urticeæ.** Fruit a minute 1-seeded nut, enclosed by the perianth.—Herbs shrubs or trees (p. 250).

§ 5. *Flowers uni- bisexual. Perianth* 0. *Ovary* 1-*celled.*

72. **Chloranthaceæ.** Flowers uni- or bisexual. Stamens 1–3, epigynous. Ovary 1-celled; ovule 1, pendulous.—Herbs and shrubs. Leaves opposite (p. 253).

73. **Piperaceæ.** Flowers bisexual, minute, closely packed in slender spikes. Stamens usually 2, hypogynous. Ovary 1-celled; ovule 1, erect.—Herbs and shrubs (p. 254).

74. **Balanophoreæ.** Flowers unisexual. Stamens various. Ovary 1-celled; ovules solitary, pendulous.—Root parasites. Leafless or with scales on the stems (p. 255).

SUBDIVISION II. *GYMNOSPERMÆ.*

Ovules naked, not enclosed in an ovary.

75. **Coniferæ.** Anthers in the male, ovules in the female, inserted on scales, which often form catkins or cones.—Trees and shrubs (p. 255).

CLASS II. MONOCOTYLEDONS.

Stem, when perennial, without pith, bark, or rings of wood, but consisting of a cellular axis with scattered longitudinal vascular bundles. Veins of the leaves usually parallel, not netted, or if so, by parallel veinlets. Perianth when present usually 3- or 6-merous, the leaflets all petaloid; often absent, the flowers being contained in the axils of scales arranged in spikelets. Stamens usually 3 or 6. Embryo with 1 cotyledon, the plumule being developed in a cavity at its side, and the rootlets from its radicular end, which does not elongate.

SUBCLASS I. **Petaloideæ.** Perianth, when present, of 6 leaflets in 2 whorls; obscure or absent in *Pandaneæ*, *Typhaceæ*, and some *Naiadeæ*.

§ 1. *Perianth superior, of* 6 *coloured leaflets.*

1. **Orchideæ.** Perianth very irregular. Anthers 1 or 2, sessile in a stigmatiferous column. Ovary 1-celled (p. 260).

2. **Irideæ.** Perianth regular. Stamens 3. Ovary 3-celled (p. 273).

3. **Hypoxideæ.** Perianth regular. Stamens 6. Ovary 3-celled (p. 274).

§ 2. *Perianth incomplete or* 0 (*see* Lemna, Ruppia, *and* Zannichellia *in* § 3.)

4. **Pandaneæ.** Flowers unisexual, in dense spikes or catkins. Perianth 0, or imperfect. Stamens numerous. Ovaries 1-celled, usually numerous and connate, truncate; stigmas sessile.—Shrubby. Leaves with sheathing bases (p. 275).

5. **Typhaceæ.** Flowers unisexual, in dense spikes or catkins. Perianth 0, or of slender hairs. Stamens crowded. Ovaries crowded, tapering into a slender style, stigma lateral.—Marsh or water plants (p. 276).

§ 3. *Perianth of* 6 *inferior leaflets* (*absent in* Lemna, Ruppia, *and* Zannichellia).

6. **Naiadeæ.** Perianth of small green leaflets or 0. Stamens 1–6. Ovaries 1–6, free, each 1-ovuled.—Marsh or water plants (p. 277).

7. **Liliaceæ.** Flowers hermaphrodite. Perianth of 6 petaloid leaflets. Stamens 6. Ovary 3-celled, cells usually 2- or many-ovuled (p. 280).

8. **Palmeæ.** Flowers unisexual. Perianth of 6 coriaceous or fleshy leaflets. Stamens 6. Ovary 3-celled or ovaries 3, cells 1-ovuled (p. 287).

9. **Junceæ.** Flowers hermaphrodite. Perianth of 6 dry brown lanceolate leaflets. Stamens 3 or 6. Ovary 1- or 3-celled (p. 288).

SUBCLASS II. **Glumaceæ.** Perianth usually absent or of minute scales or hairs. Flowers in the axils of concave scales collected into spikelets.

The perianth is regular and evident in some *Restiaceæ*.

10. **Restiaceæ.** Perianth of 4–6 leaflets, or absent or reduced to scales. Anthers versatile, 1-celled. Ovary 1–3-celled, or of several free carpels.—Grass or rush-like plants. Leaves sheathing, sheath split to the base. Flowers unisexual (p. 293).

11. **Cyperaceæ.** Perianth 0, or of bristles or minute scales. Anthers terminal, 2-celled. Ovary 1-celled, 1-ovuled. Pericarp coriaceous. Embryo at the base of albumen.—Grass or rush-like herbs. Culms solid. Leaves sheathing, sheath entire (p. 296).

12. **Gramineæ.** Perianth 0, or of 2 minute scales. Stamens 3. Anthers versatile. Ovary 1-celled, 1-ovuled. Pericarp membranous, adhering firmly to the seed. Embryo at the side of the base of the albumen.—Herbs. Culms fistular, jointed. Leaves sheathing; sheaths split to the base (p. 317).

B. *Flowerless Plants*, answering to

CLASS III. CRYPTOGAMIA.

Plants cellular or vascular, without true stamen, pistil, or ovules. Organs of fructification often very minute, giving origin to microscopic spores by which the species are propagated. Spores germinating by microscopic threads, or by a prothallium. Fecundation (when known) effected by spermatozoids (not by pollen grains).

SUBCLASS I. **Acrogens.**—Plants usually furnished with distinct stem and leaves, the latter symmetrically arranged. Stems usually dichotomously branched; sometimes reduced to simple fronds or membranous green expansions, then furnished with a midrib. Fructification various.

1. **Filices.** Stems of cellular tissue traversed by spirally marked vessels, which are often collected into hard woody bundles. Fructification of very

minute capsules, full of microscopic spores, situated on the under surface of the frond, or on separate branches of the frond; rarely of larger capsules, which are confluent on the under surface of the frond (*Marattia*) or collected in simple or branched spikes (*Ophioglossum* and *Botrychium*).—Plants rarely or never aquatic. Vernation circinate, except in the tribe *Ophioglosseæ* (p. 344).

2. **Lycopodiaceæ.** Stem elongate, erect creeping or pendulous, its tissues similar to those of *Filices*. Leaves imbricate all round, or distichous or tetrastichous, usually small, flat or subulate. Fructification of capsules, which are axillary in the upper leaves or in the scales of a cone, sessile, 1–3-celled, bursting by 2 or 3 valves, full of microscopic spores, which are marked by 3 radiating lines.—Plants never aquatic. Vernation somewhat circinate (p. 387).

Phylloglossum differs from the rest of the Order in its fleshy roots, subulate radical leaves, and solitary scape, bearing one terminal cone.

3. **Marsileaceæ.** Aquatic (freshwater) plants, of various habit, furnished with spiral vessels, creeping or floating. Capsules of 2 kinds, very various in form and structure, situated on the roots or leaves or stems of the frond, 1- or many-locular (p. 392).

Azolla, the only New Zealand genus hitherto discovered, consists of floating bright red pinnate fronds, covered with minute imbricating leaves.

4. **Musci.** Erect or creeping, small, usually terrestrial plants, with distinct stem and leaves, without spirally marked vessels (spirally marked cells are found only in *Sphagnum*). Leaves always small, usually with a midrib, membranous or coriaceous. Fructification of two kinds; 1, more or less obovoid or ovoid, brown, sessile or stalked, erect or drooping capsules, which open by a lid, or rarely by 4 lateral slits, or not at all, and contain minute spores; 2, minute cylindric membranous sacs (antheridia), which are axillary or crowded at the tips of the branchlets, and contain spermatozoa.

5. **Jungermannieæ.** Plants all cellular, usually with the habit of *Musci*, but often forming flat continuous fronds with a stout midrib. Leaves without a midrib, usually distichous or secund, entire, 2- or more lobed. Fructification of two kinds, as in *Musci*, but the capsules are split from the top to the base into 4 diverging valves, and the spores are mixed with spiral filaments.

6. **Marchantieæ.** Leafless, wholly cellular plants, consisting of broad, green, rather thick, flat lobed fronds, with or without a midrib, closely appressed to the ground, and emitting rootlets from the under surface; cuticle porous. Fructification of two kinds: 1, capsules, usually symmetrically disposed on the under side of a peltate peduncled receptacle, which rises from the edge of the frond (rarely solitary and sessile), and contains spores mixed with spiral filaments; 2, antheridia, contained in sessile or peduncled peltate or discoid receptacles.

7. **Characeæ.** Freshwater plants. Stems leafless, branched dichotomously, and furnished with whorled branchlets, consisting of long parallel transparent tubes often coated with an opaque crust of carbonate of lime.

Fructification of two kinds; 1, lateral red "globules" composed of 8 3-angular scales which enclose a mass of jointed filaments; 2, axillary "nucules" surrounded by 5 spiral filaments, and filled with starch granules.

SUBCLASS II. **Thallogens.**—Plants usually without a distinct leafy stem, forming a flattened or cylindric, dichotomously branched or variously formed frond or thallus, or composed of articulated threads or simple cells variously disposed; vascular or spiral tissue 0 or extremely rare. Fructification imbedded in the substance of the thallus, very various.

8. **Lichenes.** Perennial, coriaceous, or rigid or crustaceous plants, all terrestrial, consisting of a thallus which is erect or appressed to the ground, or to rocks or trees, often reduced to mere scales or a powdery crust; substance always very dense, cellular externally; filamentous internally. Fructification of two kinds: 1, septate spores contained in tubes (asci) which are usually collected into hard peltate disks or shields formed of the upper surface of the thallus, but sometimes are imbedded in cracks of the thallus; 2, spermogones, or small sacs containing spermatia, which latter are supposed to be a form of spermatozoa; 3, pycnides, obscure organs, giving origin to spore-like bodies at their tips; 4, gonidia, or globose spore-like bodies imbedded in the filamentous substance of the thallus, and sometimes breaking through the cortical substance, and forming powdery masses called soredia and cyphella.

9. **Fungi.** Cellular, terrestrial or epiphytic or parasitic plants, presenting an infinite variety of forms, but never forming flat crusts or foliaceous expansions as the *Lichenes* and *Algæ*, frequently existing on animal matter and on living or dead foliage, often ephemeral, variously coloured, rarely green. Substance consisting of a congeries of cells or cellular filaments, usually soft or succulent, never containing gonidia. Fructification of microscopic spores attached to the outer cellular surface, or seated on the top of peculiar cells, or contained in asci as in *Lichenes*.

Most delicate spiral filaments have been found in a few genera.

10. **Algæ.** Cellular marine and freshwater plants, consisting of foliaceous variously often brightly coloured fronds, which are simple or branched or pinnately divided, with or without midrib, or reduced to cellular filaments, or to simple cells. Fructification of four kinds, free or imbedded in the tissue of the frond, either promiscuously or in separate sacs or vesicles; 1, zoospores, or minute bodies moving through the water by the motion of fine cilia, and requiring very high power of the microscope to render them visible; 2, spores of various forms, which are fertilized by antheridia; 3, antheridia, containing spermatozoa; 4, gonidia, or minute organs corresponding to the buds of higher plants.

HANDBOOK

OF THE

NEW ZEALAND FLORA.

CLASS I. DICOTYLEDONS.

ORDER I. RANUNCULACEÆ.

Herbs, with alternate or radical leaves (*Clematis* excepted). Flowers usually hermaphrodite.—Sepals 3–6, free, often petaloid, usually deciduous. Petals 5–10 or 0, sometimes spurred or deformed, often with a pit or scale towards the base, deciduous. Stamens hypogynous, usually very numerous; anthers adnate. Carpels numerous, free, on a torus which sometimes elongates. Fruit of few or many 1-seeded achenes, or many-seeded follicles. Seeds with copious fleshy albumen, and a minute embryo.

An Order abounding in all temperate climates, rarer in tropical. Many European and other genera have irregular flowers, and otherwise deviate from the New Zealand types; such are the cultivated Aconite, Larkspur, Columbine, etc. All the New Zealand genera are British.

Climbing shrubs with opposite compound leaves . . . . 1. CLEMATIS.
Herb. Leaves radical, linear. Sepals 5. Petals 0 . . . . 2. MYOSURUS.
Herbs. Sepals 5. Petals 5–20. Carpels 1-ovuled . . . . 3. RANUNCULUS.
Herb. Leaves radical, sagittate. Sepals petaloid. Petals 0. Carpels many-ovuled . . . . 4. CALTHA.

1. CLEMATIS, Linn.

Much branched, slender, climbing shrubs, with opposite compound leaves, and panicles of white or cream-coloured polygamous flowers.—Sepals 4–8, petaloid, valvate. Stamens 6 or more. Carpels many, 1-ovuled. Achenes indehiscent, the styles elongated into long feathered awns.

A large and widely diffused genus, of which some foreign species have blue or purple flowers, or herbaceous or erect stems, or entire leaves, or minute petals. The New Zealand species are very variable, passing one into the other; their flowers are almost unisexual, the males having no carpels, and the females few stamens. The anthers have no appendage at the tip as most Australian species have, *C. parviflora* alone having a very minute one.

Leaves simply 3-foliolate.

Flowers 1½–4 in. Sepals oblong. Leaflets 2–3 in. . . . . . . 1. *C. indivisa.*
Flowers 1–2 in. Sepals oblong. Leaflets 1–1½ in., lobulate . . 2. *C. hexasepala.*
Flowers ⅔–1¼ in. Sepals linear, with thick tomentum . . . . 3. *C. fœtida.*
Flowers ⅓–1 in. Sepals linear, with silky pubescence . . . . . 4. *C. parviflora.*

Leaves usually biternate, or 3-foliolate with 3-partite leaflets.

Flowers 1–1½ in. Sepals linear-oblong . . . . 5. *C. Colensoi.*

1. **C. indivisa,** *Willd.;—Fl. N. Z.* i. 6. A large, strong, woody climber, with trunk often ½ ft. diam. Leaves 3-foliolate, coriaceous, glabrous or downy; leaflets 1–4 in. long, varying from linear-oblong to ovate-cordate, all petioled, entire, rarely lobed. Flowers 1–4 in. diam., most abundantly produced, white, sweet-scented. Sepals 6–8, broad- or narrow-oblong. Anthers obtuse, oblong. Achenes very downy.—Hook. Bot. Mag. t. 4398 (a variety with leaflets lobed as in *C. hexasepala*); *C. integrifolia*, Forst.

Abundant throughout the islands, festooning trees, etc., especially on the skirts of forests, *Banks and Solander*, etc.

2. **C. hexasepala,** *DC.—C. Colensoi*, Hook. fil., Fl. N. Z. i. 6. t. 1. Smaller in all parts than *C. indivisa*, and best distinguished by its smaller narrow ovate-cordate, often lobulate leaflets, small flowers, and very narrow anthers. Leaves glabrous, coriaceous; leaflets 1–1½ in. long, coriaceous, ovate-lanceolate, serrate or lobulate, rarely entire. Peduncles pubescent. Flowers white, 1–2 in. diam. Sepals 6, broadly linear, obtuse, downy. Anthers long, linear, obtuse. Achenes pilose.—*C. hexapetala*, Forst.

Northern Island: sandy banks on the east coast, *Colenso*. **Middle** Island, *Forster*: Upper Awatere Valley, *Sinclair*. I have examined an authentic specimen of this, gathered and named by Forster, in Mus. Paris, and find that the plant which I formerly called *Colensoi* is the true *hexasepala*.

3. **C. fœtida,** *Raoul, Choix*, xxiv. *t.* 22;—*Fl. N. Z.* i. 7. Leaves as in *C. parviflora*, but coriaceous, larger, and glabrous below. Panicle densely tomentose, as are the sepals. Flowers small, ⅓–⅔ in. diam., fœtid (Raoul). Sepals 4–6, linear. Filaments slender; anthers shortly linear-oblong. Achenes silky.

Var. *β. depauperata*, Fl. N. Z. l. c. Leaflets narrow-linear, very small, ⅛–1¼ in. diam. Peduncles short, 1-flowered.

Common in the **Northern** and **Middle** Islands. Var. *β*. Lake Rotoatara, *Colenso*; Canterbury, *Travers*. The large leaflets and dense tomentum on the inflorescence distinguish this from *parviflora* and *Colensoi*, and the small flowers and narrow sepals from *hexasepala*.

4. **C. parviflora,** *A. Cunn.;—Fl. N. Z.* i. 7. A slender climber, more or less covered with fulvous pubescence. Leaves 3-foliolate; leaflets ⅓–1 in. long, ovate-cordate, usually broad, subacute, entire or lobed, pubescent chiefly beneath. Flowers ⅓–1 in. diam. Sepals covered with silky pubescence, narrow-linear. Filaments slender, anthers short, broad, with a minute swelling at the apex of the connective. Achenes silky.

Var. *β. depauperata*. Leaflets very small. Sepals narrowed into long slender points.—Perhaps a distinct species.

Northern Island: abundant on skirts of woods, *Banks and Solander*, etc. Var. *β*, Nelson, *Travers*.

5. **C. Colensoi,** *Hook. f.—C. hexasepala*, Fl. N. Z. i. 7 (not DC.). A

slender climber. Leaves generally biternate, or 3-foliolate with the leaflets 3-lobed or 3-partite. Leaflets small, $\frac{1}{4}$–$\frac{1}{2}$ in. long, more membranous than in *C. fœtida*, ovate-cordate, unequally toothed, lobed or almost 3-partite. Peduncles and pedicels slender. Flowers small, green, very sweet-scented, 1–$1\frac{1}{2}$ in. diam., male polyandrous; hermaphrodite 6–8-androus. Sepals silky. Anthers narrow-linear. Achenes glabrous or silky.—*C. hexasepala*, Lindl. Bot. Reg. xxxii. t. 44.

Var. *β. rutæfolia*, Fl. N. Z. i. 7. Leaves biternate or bipinnate, leaflets $\frac{1}{4}$ in. long.

In various parts of the **Northern** and **Middle** Islands. Var. *β*, common in the **Middle** Island. Allied to *C. hexasepala*, from which it is easily distinguished by its smaller size and small green flowers; and from *fœtida* and *parviflora*, by the broad sepals and narrow long anthers: it is certainly not the *C. hexapetala* of Forster, to which I gave the name of *C. Colensoi* in the New Zealand Flora.

2. MYOSURUS, Linn.

Small stemless annual herbs, with linear leaves and many 1-flowered scapes.—Sepals 5, gibbous or tubular or spurred at the base. Petals 0 in the N. Z. species. Stamens 5 or more. Carpels 1-ovuled. Achenes small, beaked, sessile and crowded on the torus, which elongates as they ripen.

A small genus, native of the temperate northern and southern hemispheres.

1. **M. aristatus,** *Benth.;—Fl. N. Z.* i. 8. About 1 in. high. Leaves $\frac{1}{20}$ in. broad. Flower minute, greenish, apetalous. Sepals with a spurred base. Stamens 5 or 6. Fruiting torus $\frac{1}{4}$–$\frac{1}{2}$ in. long, erect.

Northern Island: pebbly beach near Cape Palliser, *Colenso;* also a native of California, and of the Andes of Chili, at 11,500 ft. elevation, but not found in Australia, where the European (and only other known) species takes its place.

3. RANUNCULUS, Linn.

Herbs with petioled radical leaves, and yellow or white flowers.—Sepals 3–5, concave. Petals 5–20, with 1–3 glands or scales near the base. Achenes numerous, small, with short, straight or hooked styles, and one ascending ovule.

A very large genus in all temperate countries, rare in tropical; many are acrid and poisonous. Some of the N. Z. species are the finest known; all are very variable indeed.

1. *Stem erect. Leaves peltate. Flowers white or cream-coloured.*

Margin of leaf simply crenate	1. *R. Lyallii.*
Margin of leaf bicrenate and lobed at the base	2. *R. Traversii.*

2. *Stem erect, without creeping stolons. Leaves not peltate. Flowers yellow (white in* Buchanani). *Achenes not muricate.*

a. *Achenes tumid, often angled (not flattened, with thick margins).*

Leaves entire, 4–8 in. Petals 5–6. Achenes hirsute	3. *R. insignis.*
Leaves entire, 1–3 in. Petals 5–8. Achenes glabrous	4. *R. pinguis.*
Leaves deeply 3–7-lobed. Petals 10–15. Flowers many, $1\frac{1}{2}$ in. .	5. *R. nivicola.*
Leaves 3–5-lobed. Petals 10–12. Flowers few, $\frac{1}{2}$–$1\frac{1}{2}$ in. . . .	6. *R. geraniifolius.*
Glabrous. Leaves lobed or partite. Scape 1-flowered. *Flowers white*	7. *R. Buchanani.*
Glabrous, fleshy. Leaves multifid, cauline involucrate. Flowers many	8. *R. Haastii.*
Glabrous, fleshy. Leaves multifid. Scape naked, short, 1-flowered	9. *R. crithmifolius.*
Silky, short, stout. Leaves multifid. Scape stout, naked, 1-flowered	10. *R. sericophyllus.*
Pilose, slender. Leaves multifid. Scape slender, naked, 1-flowered	11. *R. Sinclairii.*

b. *Achenes flattened, border thickened; style short, hooked. Leaves hairy.*

Stem branched, leafy. Leaves lobed or pinnate. Peduncle slender, longer than the leaves. Sepals reflexed . . . . 12. *R. plebeius.*
Stem 0. Leaves all radical. Scapes slender, longer than the leaves. Sepals spreading . . . . 13. *R. lappaceus.*
Stem leafy. Scape shorter than the leaves, very stout, 1-flowered 14. *R. subscaposus.*

3. *Stem creeping, or with creeping stolons (unknown in* R. gracilipes). *Flowers yellow. Sepals reflexed? Achenes tumid, not muricate.—Marsh or water plants, all glabrous except* 19. *Scapes* 1-*flowered.*

Leaves 3–5-partite. Flowers ½–¾ in. Petals 5–10, narrow . . 15. *R. macropus.*
Leaves 3–5-partite. Flowers ¼–¾ in. Petals 5–8, linear . . . 16. *R. rivularis.*
Leaves 3-foliolate. Flowers ¼–⅓ in. Petals 5–8, spathulate . . 17. *R. acaulis.*
Leaves pinnate or 3–5-foliolate. Flowers ⅓–¾ in. Petals 8–10, obovate-cuneate . . . . 18. *R. gracilipes.*
Stem very stout, short. Leaves entire or lobed. Scape very short. Petals 10–15. . . . . 19. *R. pachyrrhizus.*

4. *Stem very slender, pilose. Achenes muricate.*

Flowers minute, almost sessile, opposite the leaves . . . . . . 20. *R. parviflorus.*

1. **R. Lyallii,** *Hook. f., n. sp.* An erect, very handsome, coriaceous plant, 2 to 4 ft. high, with paniculately branched many-flowered stem. Leaves peltate, on long stout petioles, glabrous; limb orbicular, very concave, thick and coriaceous, 15 in. diam., simply crenate; veins reticulated; cauline sessile, lobed and crenate; seedling-leaves not peltate, broadly rhomboid, with cuneate bases. Peduncles very numerous, villous, stout, erect, with linear-oblong bracts. Flowers waxy-white, 4 in. diam. Sepals 5, broad, pilose. Petals broadly cuneate, with an obscure oblong basilar gland. Stamens small, short; anthers oblong. Torus cylindric, hairy, lengthening after flowering. Achenes villous, oblique; style flexuose, subulate; edges compressed, not margined.

Middle Island: Milford Sound, *Lyall*; moist places in the Southern Alps, 2–3000 ft., *Travers, Sinclair, and Haast*; Otago, alt. 1–4000 ft., *Hector and Buchanan.* "Water Lily" of the shepherds. The most noble species of the genus. This and the following are the only known *Ranunculi* with peltate leaves.

2. **R. Traversii,** *Hook. f., n. sp.* Very similar to *R. Lyallii*, but smaller, the leaves 6 to 7 in. diam. and broadly twice or thrice crenate with deeper notches, and with two incisions near the base. Flowers cream-coloured.

Middle Island: moist gullies in Wurumui mountains, *Travers.* I follow Mr. Travers's opinion in distinguishing this species, which is certainly closely allied to the former. I have but very indifferent specimens.

3. **R. insignis,** *Hook. f. Fl. N. Z.* i. 8. *t.* 2. Erect, robust, paniculately branched, villous, often 4 ft. high, fulvous or rufous when dry. Leaves rounded-cordate, 4–8 in. broad, very coriaceous, crenate and lobed; petioles 6 in. long. Peduncles very numerous, stout, with (often opposite) linear-oblong bracts. Flowers golden, 1½ in. diam. Sepals 5–6, oblong, woolly at the back. Petals 5–6, obcordate, with 3 glands near the base. Achenes forming a small head upon an oblong pubescent torus, villous, tumid, with a slender, nearly straight style.

Northern Island: Ruahine range, Tongariro, and Hikurangi, *Colenso.*

4. **R. pinguis,** *Hook. f. Fl. Antarct.* i. 3. *t.* 1. Stout, rather fleshy,

erect, 4–10 in. high; stem simple, branched, more or less villous with scattered soft white hairs, or quite glabrous. Radical leaves on short (2–3 in.) stout petioles, fleshy, rounded reniform, deeply crenate-lobulate, 1–3 in. diam., veins reticulated, in young plants oblong or cuneate; cauline more or less cut. Scapes 1- or many-flowered, very stout or slender. Flowers ¾–1 in. diam., golden. Sepals linear-oblong, glabrous or hairy. Petals 5–8, obovate-cuneate, more or less retuse; glandular depressions 1–3 at the base, between the veins. Achenes numerous, crowded, forming a globose head, glabrous, sharply angled; style subulate, nearly straight.

Var. *α*. Scapes branched, many-flowered. Petals longer than sepals. Head of fruit small. —*R. Munroi*, *Fl. N. Z.* ii. 323.

Var. *β*. Much shorter, stouter, and more fleshy. Scapes 1- or few-flowered. Sepals as long as petals.—*R. pinguis*, H. f. l. c.

Middle Island: mountains of Nelson, summit of Macrae's Run, Tarndale, Discovery Peaks, and Wairau Gorge, alt. 4–5500 ft., *Munro*, *Sinclair*; Southern Alps, *Haast*. *β*. **Lord Auckland's** group and **Campbell's** Island, *J. D. H.* Var. *β* is certainly only an Antarctic form, with a more succulent, stouter habit, 1-flowered scapes, shorter petals, and larger heads of carpels.

5. **R. nivicola,** *Hook.*;—*Fl. N. Z.* i. 8. Erect, paniculately branched, 2–3 ft. high, hirsute with lax, soft, white hairs, or nearly glabrous. Leaves long-petioled, rounded cordate or reniform, 3–5 in. diam., deeply 3–7-lobed; lobes broadly cuneiform, inciso-crenate; cauline laciniate; petioles 8–12 in. Flowers numerous, large, bright yellow, 1½ in. diam. Sepals 5, linear-oblong, hirsute. Petals 10–15, narrow cuneate, notched, with one gland at the base. Achenes forming a small broadly ovoid or rounded head, glabrous; style straight, hooked at the tip.—Hook. Ic. Pl. t. 571–2.

Northern Island: near the perpetual snow on Mount Egmont, *Dieffenbach*.

6. **R. geraniifolius,** *Hook. Fl. N. Z.* i. 9. *t.* 3. Tall, slender, sparingly branched, 1–2 ft., glabrous or villous in parts, especially near the root, with white, long, silky hairs. Radical leaves broadly reniform, 2–3 in. diam., deeply 3–5-lobed; lobes cuneate, crenate-lobed; cauline sessile, cut and lobed; petioles slender, 6–8 in. Peduncles slender, glabrous or villous. Flowers ½–1½ in. diam. Sepals oblong, glabrous. Petals 10–12, golden, twice as long as broad, blunt, gland depressed, close to the base. Achenes turgid, with short flexuose styles, glabrous, collected into a small globose head.

Northern Island: snow rills on the Ruahine range, *Colenso*. **Middle** Island: top of Gordon's Nob, *Munro*. Closely allied to *R. nivicola*, but smaller in all its parts; the rhizome may be creeping, but I think not.

7. **R. Buchanani,** *Hook. f.*, *n. sp.* Stout, erect, a span and more high, silky or glabrous. Root-stalk as thick as the thumb, full of milky viscid fluid; rootlets thick, fibrous. Leaves radical on long thick petioles 1–4 in. long, reniform in outline, ternatisect, 2–6 in. broad; divisions petioled, more or less cut into linear or cuneate, 3–5-fid lobes; cauline similar, sessile. Flowers solitary, large, 2½ in. broad, white. Sepals 5, reflexed, villous. Petals 15–20, linear-oblong. Achenes turgid, pilose; styles subulate, ⅙–¼ in. long, collected into a globose head.

Middle Island: Otago, Lake district, in large patches, alt. 5–6000 ft., *Hector and Buchanan*; ? Macaulay river and Waimakeriri country, alt. 2–5000 ft., *Haast* (both without flowers). I have a fine coloured drawing of this plant from Mr. Buchanan; it is a very beautiful species, described as having the leaves at times almost entire.

8. **R. Haastii,** *Hook. f., n. sp.* Very stout, fleshy, glabrous, a span high. Radical leaves broadly reniform, 3 in. diam., palmately cut to the base into 5–7 deeply and irregularly laciniate, fleshy, blunt lobes; petiole tapering downwards, 3–4 in. long. Scape as thick as the finger, naked below, with a crowded mass of sessile laciniate cauline leaves, forming a sort of leafy involucre to the numerous 1-flowered naked peduncles. Flowers absent. Achenes forming a globose head as large as a nut, fleshy, very numerous, large, $\frac{1}{4}$ in. long, on a globose swollen torus, glabrous, turgid; style long, flattened, subulate.

Middle Island: shingle beds on Mount Torlesse and the Ribbon-wood range, alt. 4500 to 6000 ft., *Haast.* A most remarkable plant, of which I have but two imperfect specimens; it is probably very variable, and other specimens may deviate much from the above description.

9. **R. crithmifolius,** *Hook. f., n. sp.* Small, perfectly glabrous, very fleshy, glaucous, stemless; rootstock short, stout, horizontal, with thick fleshy fibres. Leaves all radical, on recurved petioles 1–2 in. long, blade broad, $\frac{1}{2}$–1 in. broad, reniform in outline, biternately multifid; segments short, linear, $\frac{1}{10}$ in. long, obtuse. Scape stout, fleshy, erect, shorter than the leaves, single-flowered. Flowers small. Sepals linear-oblong. Petals not seen. Achenes in a globose head, $\frac{1}{3}$ in. diam., turgid, keeled; style sharp, straight, subulate.

Middle Island: Wairau Gorge, on shingle slips, alt. 6000 feet, *Travers.* A very singular plant, easily recognized by its glaucous, fleshy habit, finely divided leaves, and single-flowered, short scapes.

10. **R. sericophyllus,** *Hook. f., n. sp.* Short, stout, erect, very silky, scapigerous, 1-flowered; root fibrous. Leaves all radical, 1–2 in. long, petiole stout, blade broadly ovate in outline, tripinnatisect, membranous; segments small, linear-oblong, subacute, with a pencil of silky hairs at the tip. Scape stout, erect, very silky, longer than the leaves, 1-flowered. Flower 1–1$\frac{1}{2}$ in. diam. Sepals oblong, membranous, spreading, almost as long as the petals. Petals 8–10, obovate-cuneate, rounded at the tip, bright yellow; glands 3, naked. Achenes not seen.

Middle Island: snow holes on Mount Brewster, and Hopkins river, amongst grass, alt. 5–6000 feet, *Haast.* A beautiful little plant, allied to the Tasmanian *R. Gunnii*, and like it, with 3 glands on the petals, but much more robust and silky, with large golden-yellow flowers, membranous leaves, with much smaller, shorter, ultimate segments.

11. **R. Sinclairii,** *Hook. f., n. sp.* Small, 2 to 8 in. high, almost glabrous, with a few thin long silky hairs at the base of the petioles, and sometimes on the petioles and scapes; rootstock stout, prostrate, with stout fibres. Leaves tufted, 1–4 in. long, ovate-oblong in outline, bipinnatisect or multifid; segments narrow-linear, short, spreading, primary divisions 2 to 4 pairs, opposite, ovate in outline. Scape slender, leafless, 1-flowered. Flower $\frac{1}{2}$ in. diam., golden-yellow. Sepals 5. Petals 5, linear-obovate, with a deep oblong gland below the middle. Achenes with subulate, short, straight styles.

Middle Island: mountains above Tarndale, alt 5000 ft., *Sinclair;* Wairau Gorge, alt. 4–5000 ft. *Travers;* Otago, Lake district, 6000 ft., *Hector and Buchanan.* A very peculiar little species, quite unlike any other, easily recognized by its small size and finely-cut leaves. The rootstock is prostrate, but I do not think the plant belongs to the creeping section of the genus, which is almost wholly glabrous. It resembles the Andean *R. dichotomus* more than any other. A Ruahine-range plant, of Colenso, with long hairs on scapes and petioles, and less divided leaves, but without flower, may be a form of this.

12. **R. plebeius,** *Br ;—Fl. N. Z.* i. 9.—*R. hirtus*, Banks and Sol. ; Fl. N. Z. i. 9. Short or tall, more or less branched, and covered with long spreading or appressed hairs. Leaves mostly radical, long-petioled, 3-foliolate or pinnately biternate ; leaflets usually broadly ovate, lobed and toothed. Scapes or branches numerous, ascending, slender, leafy, branched, 10–24 in. long, with few sessile or petioled leaves. Peduncles slender, glabrous or covered with appressed or patent, rigid or soft hairs. Flowers ½–¾ in. diam. Sepals oblong, reflexed. Petals twice as long, obovate, rounded at the tip, with a small depressed gland at the base. Achenes glabrous, compressed, forming a rounded head ; margin thickened ; style short, hooked ; receptacle elongate-ovoid, pilose.—*R. acris*, A. Rich. Flora, not Linn.

Abundant throughout the islands, *Banks and Solander*, etc. A common Australian plant, and probably a form of a South African and European one.

13. **R. lappaceus,** *Sm.*, var. **multiscapus.**—*R. multiscapus*, Hook. f. Fl. N. Z. i. 9. t. 5. Much smaller than *R. plebeius*, 1–10 in. high, more or less hairy or hirsute, and differing in the entire or 3-lobed leaves, slender single-flowered scapes, and spreading sepals. Leaves ¼–1 in. long, cuneate or ovate or ovate-rotundate, coarsely crenate, entire or 3-lobed, or 3-partite. Scapes usually longer than the leaves, covered with spreading or appressed hairs. Petals often large and bright yellow.

Northern and **Middle** Islands : common in many situations, especially subalpine ones, ascending to 3000 feet in Otago. A most abundant Tasmanian and Australian plant.

14. **R. subscaposus,** *Hook. f. Fl. Antarct.* i. 5. Erect or decumbent, very hairy, almost hispid, 6–10 in. high root fibrous. Leaves on petioles 4–8 in. long, covered with spreading or appressed hairs ; blade 1–1½ in. long, hairy on both surfaces, either broadly triangular-ovate and 3-lobed to the base, the lobes cuneate and incised, or entire with a cuneate base, and the margin above the middle deeply lobed or cut. Scape or stem shorter than the leaves, hispid or villous, sometimes running. Flowers small, ⅓–½ in. broad. Sepals 5, spreading, membranous. Petals 5, bright yellow. Ripe achenes not seen.

Middle Island: Hopkins river, shady gorges near snow, alt. 4–5000 ft., *Haast*. **Campbell's** Island, *Lyall*. This seems an alpine ally or a form of *R. plebeius*, remarkable for the stout hispid or hairy scapes or flowering branches being shorter than the leaves.

15. **R. macropus,** *Hook. f. Fl. N. Z.* i. 10. Perfectly glabrous. Stems slender, 2 ft. long and more, fistular prostrate and rooting. Leaves 1–3 in. diam., semicircular in outline, cut to the base into 3–5 leaflets ; leaflets broad- or narrow-cuneate, irregularly cut and lobed at the apex, lobes obtuse ; petiole 4–12 in. long, weak. Peduncles solitary, axillary, those near the ends of the branches short, the rest very long. Flower ½–¾ in. diam. Sepals 5. Petals 5–10, longer or shorter than the sepals, narrow obovate-oblong ; gland depressed, basal. Achenes turgid, glabrous, smooth, collected in a small globose head ; style subulate ; receptacle oblong, glabrous.—Hook. Ic. Pl. t. 634.

Common in pools and marshes, apparently throughout the the islands, from Poverty Bay, *Colenso*, to Otago, *Lindsay*. This differs from *R. rivularis* chiefly in its very great size, more lax habit, very long petioles, and the much broader segments of the leaves.

16. **R. rivularis,** *Banks and Sol.;—Fl. N. Z.* i. 11. Creeping, perfectly glabrous, slender. Stems tufted, or sending out creeping stolons, or prostrate and branching at the nodes, or floating and branching irregularly. Leaves broadly ovate, reniform or semicircular, $\frac{1}{2}$–1 in. diam., cut into 3–7 leaflets, which are linear or narrow-cuneate, deeply lobed and cut at the apex or to the middle, sometimes ternatisect; petioles 1–3 in. long. Scapes or peduncles slender, longer than the leaves, 1-flowered. Flower $\frac{1}{4}$–$\frac{3}{4}$ in. diam. Sepals 5. Petals 5–8, linear-oblong; gland depressed, placed just below the middle. Achenes as in *R. macropus.*

Var. *α. major*, Benth. Fl. Austral. i. 14. Suberect; leaves tufted.—*R. incisus*, Hook. f. Fl. N. Z. i. 10, t. iv.

Var. *β. subfluitans*, Benth. l. c. Floating, or prostrate in wet swamps. Leaves alternate. —*R. rivularis*, Fl. N. Z. i. 11; *R. inundatus*, Banks and Sol.; Fl. Tasm. i. 8.

Abundant in watery places throughout the islands, *Banks and Solander*, etc. Also abundant in Australia.

17. **R. acaulis,** *Banks and Sol.;—Fl. N. Z.* i. 11. Small, perfectly glabrous, rather succulent. Stems with creeping stolons, 4–6 in. long. Leaves tufted, $\frac{1}{2}$–$\frac{3}{4}$ in. broad, cut into 3, obovate, entire or 2–3-lobed, coriaceous, broad, obtuse leaflets; petioles 1–3 in. long. Scapes 1-flowered, usually shorter than the petioles. Flower $\frac{1}{4}$–$\frac{1}{3}$ in. diam. Sepals 5. Petals 5–8, spathulate; gland depressed, near the middle of the petal. Achenes as in *R. macropus.*

Northern and **Middle** Islands: sandy and gravelly places, by rills of water. Bay of Islands, not rare, but much more so than *R. inundatus*; **Lord Auckland's** group, *J. D. H.* Much smaller than either of the foregoing, and at once recognized by the nearly entire broad leaflets. The Valdivian *R. stenopetalus*, Hook. Ic. Pl. t. 667, is the same species, I think.

18. **R. gracilipes,** *Hook. f., n. sp.* Small, perfectly glabrous, slender. Stems creeping? Leaves all radical, variable in size; petioles slender, 1–4 in. long; blade pinnately divided, either pinnate, with 2–3 pairs of rounded, 3–4-lobed, sessile leaflets, $\frac{1}{4}$–$\frac{1}{3}$ in. long, or biternately divided with the leaflets petioled, wedge-shaped, deeply lobed, 1 in. long. Scapes 3–6 in. long, very slender, 1-flowered. Flowers $\frac{1}{2}$–$\frac{3}{4}$ in. diam., golden-yellow. Sepals glabrous, oblong. Petals 8–10, twice as long, obovate-cuneate, retuse; depressed gland small, near the very base.

Middle Island: banks of Lake Okau, *Haast.* A very distinct but variable species, with the habit and appearance of *R. rivularis*, but the leaves are pinnately divided. My specimens are very imperfect, and I place it in the creeping section from its resemblance to the preceding.

19. **R. pachyrrhizus,** *Hook. f., n. sp.* Small, densely matted, very succulent. Scapes and petioles with long weak hairs. Stem prostrate, cylindrical, creeping, as thick as a goose-quill. Leaves very short, small, all radical; petioles stout, fleshy, $\frac{1}{4}$–$\frac{1}{2}$ in. long; blade cuneate or obovate-cuneate, $\frac{1}{6}$–$\frac{1}{4}$ in. long, acutely lobed or cut. Scape short, stout, 1-flowered. Flowers $\frac{2}{3}$–1 in. diam. Sepals linear-oblong, membranous. Petals 10–15, obovate-spathulate, bright yellow; depressed gland near the base. Achenes not seen.

Middle Island: Otago, Lake district, alt. 6–8000 ft., covering large tracts in low matted patches, *Hector and Buchanan.* A most curious little species, best known by its habit and habitat.

20. **R. parviflorus,** *Linn.*, var. **australis,** *Benth. Fl. Austral.* i. 15.

—*R. sessiliflorus*, Br. ;—Fl. N. Z. i. 11. Slender, hairy, annual. Stem prostrate or rarely erect, branching, 1–12 in. long. Leaves few radical and many cauline, alternate, small, $\frac{1}{4}$–$\frac{1}{2}$ in. diam., orbicular, 3–5-lobed, the lobes entire or variously cut, on short slender petioles. Flowers very small, on the branches opposite the leaves, solitary, almost or quite sessile. Sepals fugacious. Petals 5, about as long as the sepals. Achenes few, in a small globular head, compressed with thin edges, pilose, the sides covered with minute tubercles or hooks; style short, hooked.

Northern Island, *Colenso* (perhaps introduced). Common in temperate Australia. The typical *R. parviflorus* is a S. European plant.

4. CALTHA, Linn.

Glabrous, tufted herbs, with most or all of the leaves radical, and 1-flowered scapes.—Sepals 5 or more, petaloid, imbricate. Petals 0. Stamens numerous. Carpels several, with many ovules in two rows on the ventral suture. Follicles splitting along the inner face, several-seeded.

A small genus, found in the temperate and cold regions of both hemispheres.

1. **C. novæ-Zelandiæ,** *Hook. f. Fl. N. Z.* i. 12. *t.* 6. Short, stout, tufted, glabrous, perennial, with a thick rootstock, numerous radical leaves, and short, thick, 1-flowered scape. Leaves spreading; blade ovate-oblong, notched at the apex, deeply cordate and auricled at the base, with the obtuse auricles turned up and appressed to the surface of the leaf; petioles 1–6 in., dilated at the base into large membranous sheaths. Flowers 1–2 in. diam. Sepals 5–7, linear-subulate. Stamens short, very numerous. Carpels 5–7, broadly ovate, gibbous; style short, hooked.

Northern Island: top of the Ruahine mountains, *Colenso*. **Middle** Island; Mount Brewster and Hopkins river, alt. 5–6000 ft., *Haast*; Otago, alpine districts, alt. 4–6000 ft., *Hector and Buchanan*. The *C. introloba*, F. Muell., of Tasmania and Victoria (a plant not discovered when this was published) is very closely allied to this, differing chiefly in the recurved styles.

ORDER II. MAGNOLIACEÆ.

Tribe WINTEREÆ.

Aromatic shrubs or trees, with alternate, exstipulate leaves.—Sepals and petals imbricated in 2, 3, or many series, very deciduous. Stamens numerous; hypogynous. Filaments often thick or dilated; anthers adnate. Carpels few, in 1 series, with 2 or more ovules attached to the ventral suture. Stigma sessile and terminal or decurrent along the suture. Ripe carpels of free, small drupes follicles or berries. Seeds few; testa shining; albumen copious, fleshy; embryo small.

This description refers only to the tribe *Wintereæ*, which alone is represented in N. Zealand. This Order contains many genera, in some of which the carpels are very numerous and spiked, in others combined, and in some the flowers are unisexual; it abounds in the southern United States, and the mountainous regions of India and Eastern Asia; its qualities are aromatic. The genus *Drimys* contains one S. American, two Australian, one alpine Bornean, and probably several New Caledonian species.

1. DRIMYS, Forst.

Sepals 2 or 3, membranous, combined into an irregularly lobed calyx. Petals 6 or more, in 2 or more series. Filaments thickened upwards; anther-cells diverging. Carpels few.

1. **D. axillaris,** *Forst.;—Fl. N. Z.* i. 12. A small, slender, evergreen tree, 10–30 ft. high, with black bark, aromatic and pungent in all its parts. Leaves 1–6 in. long, elliptical-ovate, blunt, shortly petioled, quite entire, bright-green above, glaucous below, pellucid-dotted, midrib hairy beneath. Flowers small, axillary or from scars of fallen leaves, solitary or few together; pedicels slender. Petals unequal, linear. Stamens 8–10, in several series. Berries about 3, size of a peppercorn. Seeds several, angled.—*D. colorata*, Raoul, Choix, t. 23; *Wintera axillaris*, Forst. Prodr.

Abundant in forests throughout the islands, *Banks and Solander*, etc. "Pepper-tree" of the colonists. Wood makes pretty veneers, *Buchanan.*

ORDER III. CRUCIFERÆ.

Herbs, usually with small racemose flowers.—Sepals 4, free. Petals 4, free, placed crosswise. Stamens 6 (rarely 1, 2, or 4), hypogynous, 2 longer than the others. Ovary 2-celled, with 2 or more ovules. Capsule 2- (rarely 1-) celled, bursting longitudinally by 2 valves, which fall away from the seed-bearing placentas. Seeds exalbuminous, with the radicle turned up toward the edges (accumbent) or back of the cotyledons (incumbent).

A large Order, abounding in all temperate countries, especially of Europe and Asia. Most of the New Zealand genera are British. Properties stimulant and antiscorbutic. The Mustard, Shepherd's-purse, Radish, Turnip, Cabbage, etc., all belong to this Order, and are found as escapes from cultivation.

* *Pod long and narrow (often short in* Nasturtium).

Pod terete, stout, curved. Seeds in 2 rows. Flowers yellow . . . 1. NASTURTIUM.
Pod somewhat 4-gonous. Seeds in 1 row. Flowers yellow . . . . 2. BARBAREA.
Pod terete, slender. Seeds in 1 row . . . . 3. SISYMBRIUM.
Pod flat. Valves elastic. Seeds in 1 row. Flowers white . . . . 4. CARDAMINE.

** *Pod short and broad.*

Pod with convex or keeled valves . . . . 5. BRAYA.
Pod with much-flattened, often winged, 1-seeded valves . . . . . 6. LEPIDIUM.
Pod with much-flattened, often winged, many-seeded valves . . . . 7. NOTOTHLASPI.

1. NASTURTIUM, Br.

Branching herbs, with usually yellow flowers and cut leaves.—Sepals spreading. Petals with short claws, yellow, sometimes 0. Stamens 6 or fewer. Pod subcylindric, usually curved; valves membranous, concave, many-seeded. Seeds in 2 series in each valve, minute, turgid; cotyledons accumbent.

A large British genus, of which the Watercress (*N. officinale*) is a white-flowered species, abundantly naturalized in rivers, etc.

1. **N. palustre,** *DC.—N. terrestre,* Br.; Fl. N. Z. i. 14. A suberect, glabrous or pilose, branching herb, with entire or pinnatifid leaves, auricled at the base, the lobes sinuate-toothed. Flowers on slender pedicels, small.

Petals hardly longer than the calyx. Pods turgid, oblong, as long as or shorter than their pedicels, curved.

Northern and **Middle** Islands, not uncommon in moist places. I have adopted the name *palustre* for this plant, because it is that used in most Continental works, and in Bentham's Australian Flora, but that of *terrestre* has equal claims to be retained. A very widely distributed plant in both the Old and New World; a state with almost entire leaves, *N. semipinnatifidum*, Hook., sometimes occurs.

2. BARBAREA, Br.

Stout or slender, erect, leafy, glabrous herbs, usually with angled stems, and pinnate or pinnatifid leaves.—Sepals suberect. Petals clawed, yellow. Pods erect, elongate, compressed, 4-gonous, with keeled or costate, straight, coriaceous, many-seeded valves. Seeds oblong, in one series; cotyledons accumbent.

A common European genus, of which one species was cultivated in former times in Britain as a pot-herb. The New Zealand species, which is also Australian, seems to be quite the same as the British, which is very variable.

1. **B. vulgaris,** *Linn.*—*B. australis*, Hook. f. Fl. N. Z. i. 14. Erect, rather rigid, stout, leafy, 1–2 ft. high, with green, furrowed stems. Lower leaves lyrate-pinnatifid; lobes obovate-oblong, terminal ovate and sinuate. Upper leaves entire, sinuate or pinnatifid. Flowers rather large. Pods stout, $1\frac{1}{2}$ in. long, $\frac{1}{10}$–$\frac{1}{8}$ in. broad, erecto-patent, broader than their terete pedicels; valves veined; style short, straight.

Northern Island, *Colenso*. This, the "Toi" of the natives, was formerly used by them as food.

3. SISYMBRIUM, Linn.

Herbs, usually leafy with slender stems and small white or yellow flowers.—Sepals suberect or spreading. Petals clawed. Pod slender, terete or slightly compressed; valves concave, many-seeded. Seeds in one series in each cell, oblong.

A British genus, abundant in the north temperate zone, rare in the south.

1. **S. novæ-Zelandiæ,** *Hook. f., n. sp.* Tall, very slender, 1–2 ft. high, glabrous or covered with minute stellate pubescence. Leaves chiefly radical, spreading, 1–2 in. long, few or many and crowded, narrow-obovate or linear-oblong, sinuate-pinnatifid; lobes blunt. Flowering stems very slender, sparingly branched, with few entire or toothed linear leaves. Flowers small, white. Petals narrow. Sepals erect. Pods $\frac{1}{2}$–2 in. long, $\frac{1}{15}$ in. broad, very narrow, linear, obtuse, glabrous, on slender pedicels; valves convex, 1-nerved. Seeds small. Cotyledons obliquely incumbent.

Middle Island: mountains of Nelson, *Rough;* Shingle slips, Wairau Gorge, alt. 4500 ft., *Travers.*

4. CARDAMINE, Linn.

Generally slender or small herbs, with entire or pinnate leaves, and small white flowers.—Sepals erect or spreading. Petals clawed or spathulate. Pod long, linear, compressed; valves flat, usually separating elastically and curving backwards. Seeds numerous, forming one series in each cell, flattened; cotyledons accumbent.

A very extensive genus, especially in temperate regions. Common in England.

Stems slender. Leaves pinnate . . . . 1. *C. hirsuta.*
Stem 0. Leaves spathulate . . . . 2. *C. depressa.*
Stem stout, tall. Leaves sinuate-lobed . . . . 3. *C. stylosa.*
Stem very short and stout. Leaves long, deeply toothed . . . . . . 4. *C. fastigiata.*

1. **C. hirsuta,** *Linn.*;—*Fl. N. Z.* i. 13. A very variable, slender, branched, rarely simple, glabrous or slightly hairy annual, 12–18 in. high, erect or decumbent, sometimes assuming a perennial rootstock, especially near the sea. Leaves pinnate; leaflets few, opposite or alternate, entire or lobed, orbicular oblong ovate or cordate, usually on slender petioles, sometimes reduced to one. Flowering branches sometimes reduced to capillary 1-flowered scapes. Flowers small, white (sometimes 4-androus in Europe). Pods ¾–1½ in. long, slender, on slender pedicels, obtuse or produced into acuminate styles. Seeds small, pale yellow-red.

Var. *α. debilis.* Erect or generally decumbent, much branched. Leaflets in several pairs, rounded or cordate. Pods very slender, with long slender styles.—*C. debilis,* Banks; *Sisymbrium heterophyllum,* Forst.

Var. *β. corymbosa.* Smaller in all its parts, with few-flowered corymbs. Leaflets 2 pairs. Pods with short styles.—*C. corymbosa,* Fl. Antarct. i. 6; Hook. Ic. Pl. t. 686.

Var. *γ. subcarnosa,* Fl. Antarct. i. 5. Erect, rather fleshy, with stout branches and petioles. Leaflets in several pairs, obovate or oblong. Flowers numerous, larger, in a dense corymb. Styles very short, stout.

Var. *δ. uniflora.* Very small, the leaves reduced to 1 pinnule, and the stem to a 1-flowered scape.

Abundant throughout the islands, in all situations, especially moist or shady, *Banks and Solander,* etc. Var. *α*, the most frequent; var. *β*, in woods; var. *γ*, a southern succulent form, found in **Campbell's** Island, and probably also in the Southern Island, and elsewhere; var. *δ*, rather a reduced state than a distinct race. A very common plant in all temperate and cold, and many warm parts of the world; the succulent forms are an excellent cress. In Britain and elsewhere in the northern hemisphere, this plant is an annual, in the more equable climate of the southern usually a perennial.

2. **C. depressa,** *Hook. f. Fl. Antarct.* i. 6. A glabrous or pilose stemless perennial. Leaves crowded, rosulate, 1–2 in. long, spathulate, entire or crenate, sinuate or lobulate, obtuse or retuse, narrowed into long or short petioles. Flowering stems ¼–4 in. high, erect or ascending, few-flowered. Flowers small, white. Pods ½–1½ in. long, erect, rather stout; styles short, stout.

Var. *α. depressa.* Glabrous, larger. Leaves entire or lobulate.—*C. depressa,* Fl. Antarct. i. 6. t. 3 and 4 B.

Var. *β. stellata.* Pilose, smaller. Leaves nearly quite entire.—*C. stellata,* Hook. f.; Fl. Antarct. i. 7. t. 4 A.

Middle Island: var. *α*, Lake Tennyson and Wairau mountains, alt. 4–5000 ft., *Travers;* Hopkins river, Lake Okau, etc., *Haast;* Otago wet places in the Lake district, *Hector and Buchanan.* **Lord Auckland's** group and **Campbell's** Islands, *α* and *β*, abundant, *J. D. H.* I suspect that this will prove a reduced form of *C. hirsuta,* a Tasmanian variety of which plant approaches it in foliage.

3. **C. stylosa,** *DC.*—*C. divaricata,* Fl. N. Z. i. 19. Perennial?, tall, 2 to 3 ft high, glabrous, stout, branched, leafy. Leaves 3 to 5 in. long, linear-spathulate or oblong, sagittate at the base, quite entire or toothed, or sinuate or lobed or almost pinnatifid at the base. Racemes elongated. Flowers

rather small, white. Pods stout, 1–1½ in. long, on horizontally spreading stout pedicels; valves concave; style stout, ⅛ in. long. Seeds brown, with impressed dots.—*Arabis gigantea*, Hook. Ic. Pl. t. 259.

Northern Island: Bay of Islands, *R. Cunningham;* near Auckland, *Lyall, Sinclair.* Seeds rather smaller and paler than in the Australian and Tasmanian plant; leaves in one specimen almost pinnatifid; but clearly, I think, the same species.

4. **C. fastigiata,** *Hook. f.—Arabis fastigiata*, Fl. N. Z. ii. 324. A glabrous perennial; rootstock a span long, perpendicular, tapering, fusiform, as thick as the little finger, densely clothed towards the apex with the recurved bases of the old leaves; branches ascending from the top of the rootstock amongst the leaves, 6–18 in. long, rather stout, leafy, simple or branched. Leaves densely rosulate at the apex of the rootstock, 2–3 in. long, narrow, lanceolate-spathulate, acute, coarsely inciso-serrate, very coriaceous; cauline less spathulate, narrower, with narrower serratures. Flowers very numerous, white, about ⅓ in. diam., on slender pedicels. Petals with narrow claws. Pods suberect, curved, narrow-linear, 1¼–2 in. long, $\frac{1}{10}$ in. broad, with acute ends and very short styles. Seeds (unripe) oblong, compressed, red-brown.

Middle Island: highest part of Macrae's run, *Munro;* river-bed of the Macaulay, alt. 3500 ft., *Haast.* In the New Zealand Flora, I referred this to *Arabis*, and it has equal claims to this genus and *Cardamine;* but its close affinity with *C. radicata*, of Tasmania, determines me to transfer it here.

5. BRAYA, Sternberg.

Alpine, densely tufted, perennial herbs, with long tap-roots, rosulate radical leaves, and scapes bearing short few-flowered racemes or corymbs.—Flowers white pink or purplish. Sepals short, equal. Petals obovate. Stamens 6. Pod short, thick, ovate or oblong; valves convex, with a stout costa, or keeled; septum entire or open; style very short; stigma capitate. Seeds in 1 or 2 series; funicles very short; cotyledons incumbent.

An Arctic genus, also found, but rarely, in the loftiest alps of Europe, N. Asia, and N. and S. America.

1. **B. novæ-Zelandiæ,** *Hook. f., n. sp.* A very short, depressed, alpine herb, covered with stellate pubescence; root long, tap-shaped, as thick as the finger, bearing one or several equally thick, erect or ascending cylindric branches, covered with scars of old leaves and surmounted by a head of small imbricating leaves that spread out horizontally. Leaves ⅛–¼ in. long, oblong, pinnatifidly lobed, narrowed into flat short petioles; those on the scapes with longer petioles, and a minute obovate blade, which is digitately lobed at the top. Scapes or peduncles very numerous, rising from the root below the leaves, shorter than these, and spreading horizontally, 3–5-flowered. Flowers not seen. Pods ⅛–⅙ in. long, about half as broad, laterally compressed; septum incomplete. Seeds 3–5 in each valve, obovoid.

Middle Island: Lake district, débris of schist on Mount Alta, alt. 5000 ft., *Hector and Buchanan.* A most remarkable plant; the pod is rather too much compressed for *Braya*, and approaches that of *Lepidium.*

6. LEPIDIUM, Linn.

Herbs, sometimes with an almost woody stem, toothed or pinnatifid leaves, and white, sometimes unisexual flowers.—Stamens 4 or 6. Pods broad, much flattened laterally, obtuse, winged or keeled at the back; cells 1-seeded. Cotyledons incumbent.

A large genus, commou in Englaud, and the N. and S. temperate zones generally. To this genus the garden "Cress" belougs, also *L. ruderale*, a common slender Australian and European much-branched annual, with linear leaves, which will probably soon be introduced into New Zealand.

Erect. Leaves more or less toothed or serrate . . . . 1. *L. oleraceum.*
Erect. Leaves pinnatifid . . . . 2. *L. sisymbrioides.*
Procumbent. Leaves piunatifid . . . . 3. *L. incisum.*

1. **L. oleraceum,** *Forst.;—Fl. N.Z.* i. 15. Suberect, perennial, glabrous, 10–18 in. high; stem stout, woody, scarred, branched, smelling disagreeably when bruised. Leaves obovate-cuneate or oblong-spathulate, 1–3 in. long, lower serrate, upper more entire. Flowers numerous, small, 4-androus. Pods on slender spreading pedicels, ovate, subacute, $\frac{1}{6}$ in. long, not winged at the back.—A. Rich. Flora, t. 35.

Abundant ou the shores throughout the islands, *Banks and Solander;* Otago, Lake district, Waitaki valley, *Hector and Buchanan*, a small-leaved form. **Lord Auckland's** group, *Bolton.* Not found in other countries.

2. **L. sisymbrioides,** *Hook. f., n. sp.* Erect, glabrous, slender, a span high; root woody, spindle-shaped, branching out into several heads at the top. Stems very numerous, slender, flexuose, sparingly branched, leafy. Leaves small, $\frac{1}{3}$–$\frac{2}{3}$ in. long, linear, pinnatifid; segments small, short, $\frac{1}{12}$ in. long, entire toothed or lobulate. Flowers small, in terminal racemes; petals white; pedicels slender. Pods on spreading curved slender pedicels, $\frac{1}{6}$ in. long, broadly subquadrate-ovate, acute at both ends, not winged at the back, notched at the apex.

Middle Island: Dry Grass flats, Lake Okau, alt. 2000 ft., *Haast;* Otago, grassy plains, Waitaki valley, *Hector and Buchanan;* possibly a form of *L. incisum.*

3. **L. incisum,** *Banks and Sol.;—Fl. N. Z.* i. 15. Glabrous or pilose, much branched, prostrate; root stout, perennial, woody; branches sparingly leafy, a span long, ascending at the tips. Lower leaves on long petioles, 2–3 in. long, pinnatifid, with 4–6 pairs of spreading or recurved bluntly-toothed lobes, the upper entire or toothed at the tip, broadly cuneate. Flowers in small, axillary or terminal, few-flowered racemes. Petals 0 in my specimens. Stamens 4, glands of the disk 6 or 8, elongated. Pods ovate-cordate, notched at the apex, $\frac{1}{8}$ in. long, half as long as the slender pedicels.

Northern Island. Opuraga, on the beach, rare, *Banks and Solander;* Port Nicholson, on rocks near the sea, *Colenso.* **Middle** Island. Limestone rocks in the subalpine region of Waimakeriri, alt. 2000 ft., *Haast.* I have described the habit, foliage, inflorescence, etc., of this plant, from Banks and Solander's specimen and drawing; Mr. Haast's being very young and apparently dwarfed.

7. NOTOTHLASPI, Hook. f.

Herbs, with numerous spathulate, thick, radical leaves.—Flowers rather large, white. Sepals erect. Pods very much flattened; valves winged; cells very many-seeded. Seeds on very slender funicles; radicle incumbent, sometimes very long.

A genus confined to New Zealand.

Scape very stout. Style very short . . . . 1. *N. rosulatum.*
Stem usually much branched at the base. Style long . . . . 2. *N. australe.*

1. **N. rosulatum,** *Hook. f., n. sp.* A very stout, erect, densely leafy, pyramidal, fleshy herb. Stem 0, or very short. Leaves very numerous and most densely crowded, imbricated, forming a rosette, spathulate, petioled, crenate, when young covered with weak cellular hairs, glabrous when old. Scape often thicker than the finger, a span high, bearing a profusion of white sweet-scented flowers. Pods ½–1 in. long, obovate, with a very short style. Seeds rather large, with a thin pitted testa; radicle very long, often twice folded, first upwards, then downwards and backwards over the back of the cotyledons.

Middle Island: alt. 3500 to 6500 ft.; shingle beds on the Ribbon range, Mount Torlesse and Waimakeriri valley, *Sinclair and Haast;* Wairau valley, *Maling, Travers;* Lake Tennyson, *Rough.* A most singular plant.

2. **N. australe,** *Hook. f.—Thlaspi (?) australe,* Fl. N. Z. ii. 325. A small, perennial, densely tufted, much branched (rarely simple), glabrous plant, with short leafy branches, and very numerous white corymbose flowers; roots slender, fusiform, descending deeply; branches 1–2 in. long. Leaves cauline and radical, petiolate, ½–1½ in. long, spathulate-oblong or linear-spathulate, subacute, entire or crenate, coriaceous or fleshy, often recurved. Flowers white, scentless, in many-flowered corymbs, almost involucrate by the numerous cauline leaves, about ¼ in. diam., on pedicels ⅓–1 in. long. Young pods obovate, retuse, with long stout styles; valves winged. Seeds excessively numerous.

Middle Island: top of Gordon's Nob, Upper Wairau river, and elsewhere in the Nelson Province, alt. 4–5000 ft., *Munro, Sinclair, Travers, Haast.*

The most frequent naturalized *Cruciferæ* known to me in New Zealand are all common British weeds, except *Alyssum maritimum,* mentioned by A. Cunningham, but which is not fully naturalized.

Capsella Bursa-pastoris, Linn. (Shepherd's-purse). An annual, with spreading rosulate pinnatilobed radical leaves, simple or branched scapes, many small white flowers, an obcuneate or obcordate retuse flat pod, the valves not winged, and cells many-seeded.

Senebiera pinnatifida, DC. A much branched, prostrate, glabrous, leafy annual, with bipinnatifid irregularly-cut leaves, short leaf-opposed racemes of very small, white flowers, succeeded by very small, didymous, wrinkled, indehiscent, 2-celled, 2-seeded pods. *S. Coronopus,* Poir., with subacute, crested pods, is also found.

Watercress, *Nasturtium officinale,* Linn., is a pest in the rivers about Canterbury, attaining a size never seen in Europe, and is found also abundantly elsewhere.

The Cabbage, *Brassica oleracea,* Linn., Turnip, *B. campestris,* Linn., Cress, *Lepidium sativum,* Linn., Charlock, *B. Sinapistrum,* Boiss., Horseradish, *Cochlearia Armoracia,* Linn., Radish, *Raphanus sativus,* and probably various others, occur as escapes from fields and gardens.

I have an indifferent specimen of a Cruciferous (?) plant, gathered by Haast, on terraces near Lake Okau, which I am unable to refer to any genus; it is a slender branched herb, pubescent, with simple hairs; small pinnate leaves; leaflets in 1 to 2 pairs, rounded, entire or lobed; and minute white flowers, that seem to be in an imperfect condition.

ORDER IV. VIOLARIEÆ.

Herbs or shrubs, with alternate stipulate leaves. Flowers regular or irregular.—Sepals 5, imbricate. Petals 5, imbricate. Stamens 5, hypogynous; anthers sessile or on short filaments, often united, the connective usually expanded upwards or provided with an appendage at the back, or both. Ovary with 2 to 5, parietal placentas, and one style. Fruit a capsule or berry. Embryo axile in fleshy albumen.

A large Order, widely distributed through tropical and temperate regions.

Petals spreading. Anthers united. Capsule 3-valved . . . . . 1. VIOLA.
Petals small. Anthers free. Berry with 3 or 6 placentas . . . 2. MELICYTUS.
Petals small. Anthers connate. Berry with 2 placentas . . . . 3. HYMENANTHERA.

1. VIOLA, Linn.

Herbs, with trailing stems or short woody stocks. Leaves alternate, petioled, stipulate. Flowers irregular.—Sepals 5, produced at the base. Petals unequal, spreading, lowermost often larger, spurred or gibbous at the base. Anthers 5, connective flat, produced into a thin membrane, the lower often spurred. Style capitate. Capsule 3-valved, with a parietal placenta on each valve.

A large British and widely-diffused genus in all temperate climates, of which several species produce two forms of flowers; the larger peduncled, with large petals, that often ripen few seeds; and minute ones lower down, apparently imperfect, with reduced petals or 0, that ripen abundance of seed.

Stems slender. Leaves cordate. Stipules and bracts lacerate . . . 1. *V. filicaulis*.
Stems slender Leaves cordate. Bracts entire . . . . 2. *V. Lyallii*.
Stems short. Leaves ovate. Bracts entire . . . . 3. *V. Cunninghamii*.

1. **V. filicaulis,** *Hook. f. Fl. N. Z.* i. 16. Very slender, perfectly glabrous. Stems filiform, prostrate or creeping. Leaves alternate, orbicular-cordate or broadly ovate-cordate, obtuse or acute, obtusely crenate; petioles 1–3 in. long; stipules lacerate, the teeth filiform, tipped with a gland. Peduncles very slender, 1-flowered; bracts subulate, more or less lacerate like the stipules. Flowers very pale blue, $\frac{1}{4}$–$\frac{1}{3}$ in. diam. Sepals linear-lanceolate, acuminate. Spur very short.

Northern and **Middle** Islands: as far south as Otago, in various localities, abundant. Very near the *V. serpens* of India.

2. **V. Lyallii,** *Hook. f., n. sp.*—*V. Cunninghamii*, var. γ, Fl. N. Z. i. 16. Entirely similar in most respects to *V filicaulis*, but the stipules and bracts are generally more green, and always entire, usually obtuse, and the flowers are smaller.

Northern Island: in various places, *Sinclair, Colenso*. **Middle** Island: Nelson, *Travers*; Canterbury, *Lyall*. In the N. Z. Flora I regarded this as a cordate-leaved variety of *V. Cunninghamii*, but more specimens have convinced me of its distinctness. Closely allied to the Australian *V. Caleyana*, but smaller in all its parts, and with less deeply cordate leaves.

3. **V. Cunninghamii,** *Hook. f. Fl. N. Z.* i. 16. Very variable in size, glabrous. Stem short, much branched, often thickened into a short woody stock. Leaves tufted on the top of the root or stem, or on very short branches from it, ovate ovate-oblong or triangular-ovate, narrowed into the petiole, obtuse, obscurely crenate; petioles often 4 in. long; stipules broadly adnate to the base of the petiole, slightly lacerate or entire. Peduncles slender; bracts linear, obtuse, quite entire. Flowers $\frac{1}{3}$–$\frac{2}{3}$ in. diam., pale blue. Sepals linear-oblong, obtuse.—*Erpetion spathulatum*, A. Cunn., not Don.

Very common in moist places, from the middle of the **Northern** Island southwards, ascending to 5000 ft. **Middle** Island: Hopkins river and Lake Okau, *Haast*; Wairau mountains, *Travers*; Otago, *Hector and Buchanan*. Also found in Tasmania.

2. MELICYTUS, Forst.

Shrubs, with short-petioled, toothed, minute-stipuled leaves. Flowers axillary, fascicled, small, regular, almost unisexual or polygamous. Sepals 5. Petals 5, short, spathulate, spreading. Anthers 5, free; connective produced into a membrane, and furnished with a scale at the back. Style 3–6-fid, or with a discoid stigma. Berry with few or many angled seeds, on 3 to 6 placentas.

This genus is confined to New Zealand and Norfolk Island.

Leaves oblong or oblong-lanceolate, serrate . . . . 1. *M. ramiflorus.*
Leaves large, obovate, sinuate-serrate . . . . 2. *M. macrophyllus.*
Leaves long, linear-lanceolate, sharply serrate . . . . 3. *M. lanceolatus.*
Leaves small, orbicular-ovate, sinuate . . . . 4. *M. micranthus.*

1. **M. ramiflorus,** *Forst.;—Fl. N. Z.* i. 18. A glabrous, white-barked, small tree or large shrub, 20–30 ft. high; trunk often angular, and 7 ft. in girth (*Buchanan*); branches brittle. Leaves alternate, 4–5 in. long, oblong-lanceolate, acuminate, serrate with small obtuse teeth, sometimes obscurely so; petioles slender; stipules deciduous. Flowers small, in fascicles on the branches; peduncles slender, $\frac{1}{4}$–$\frac{1}{3}$ in. long, with 2 minute bracts. Flowers minute, $\frac{1}{8}$ in. diam. Calyx-lobes obtuse, spreading, green. Anthers obtuse; stigma almost sessile, 6-lobed. Berry small, $\frac{1}{5}$ in. diam.

Abundant throughout the islands, as far south as Otago, *Banks and Solander*, etc. Leaves eaten greedily by cattle; wood soft, useless (*Buchanan*). Also found in Norfolk Island.

2. **M. macrophyllus,** *A. Cunn.;—Fl. N. Z.* i. 19. A large glabrous bush, 4–7 ft. high; bark pale-brown. Leaves 3–4 in. long, obovate or elliptical-oblong, acute, coarsely and distantly (rarely closely) sinuate-serrate, more coriaceous and broader than in *M. ramiflorus*; stipules deciduous. Flowers twice as large as in the former, in fascicles of 4 or 6 on the branches; peduncles stout, $\frac{1}{3}$–$\frac{1}{2}$ in. long, much decurved, with small, broad, opposite bracts close to the flower. Calyx-lobes short, broad, rounded. Anthers apiculate. Stigma broad, nearly sessile, discoid, lobed. Berries $\frac{1}{4}$ in. diam.

Northern Island: about the Bay of Islands only, so far as is hitherto known, *A. Cunningham*, etc. Easily distinguished from *M. ramiflorus* by the coarser habit, broader, more obovate, coriaceous leaves, fewer stout, decurved pedicels, with bracts at the apex, and larger flowers and berries.

3. **M. lanceolatus,** *Hook. f. Fl. N. Z.* i. 18. *t.* 8. A slender shrub or small tree, 10–12 ft. high; branches brittle; bark dark-brown. Leaves 4–7 in. long, narrow linear-lanceolate, acuminate, sharply erose-dentate, rather membranous. Flowers small, 2 or 3 together; pedicels short, decurved, bracteate above the middle. Calyx-lobes oblong. Petals erect, with spreading limb. Connective produced into a subulate point. Style long, 3-fid. Berry oblong, $\frac{1}{6}$ in. diam.

Northern Island: forests at Patea on the east coast, *Colenso.*

4. **M. micranthus,** *Hook. f. Fl. N. Z.* i. 18. A small, rigid shrub; branches tortuous, covered with grey or brown bark, the youngest pubescent at the tips. Leaves small and scattered, $\frac{1}{3}$–$\frac{1}{2}$ in. long, orbicular-obovate, obtusely sinuate, the youngest often obovate-oblong and pinnatilobed; petioles very short, puberulous. Flowers unisexual, very minute, axillary; pedicels

very short, curved, bracteate at the base. Calyx obtusely and shortly 4-lobed; lobes broad, ciliated. Petals small, orbicular. Anthers rounded, didymous, sessile. Ovary flagon-shaped; style short; stigma 4-lobed. Berry minute, as small as a mustard-seed, about 3-seeded.—*Elæodendron micranthus*, Hook. f. in Lond. Journ. Bot. iii. 228. t. 8.

Northern Island: east coast and interior, *Colenso*. **Middle** Island: Nelson, *Bidwill*. A plant of very different habit and appearance from the preceding species.

3. HYMENANTHERA, Br.

Woody shrubs, with alternate or fascicled, entire or toothed, minutely stipuled leaves, and small axillary, solitary, fascicled, sometimes unisexual flowers.—Sepals 5. Petals 5, short. Anthers sessile, connate; connective produced into a membrane, and furnished with a scale at the back. Style short, with a 2-lobed stigma. Berry small, with 2 or few globose seeds on 2 placentas.

A small genus, native also of Norfolk Island, Tasmania, and South-eastern Australia, the species are very variable in foliage.

1. **H. crassifolia,** *Hook. f. Fl. N. Z.* i. 17. *t.* 7. A small shrub, 2–4 ft. high, with rigid, stout, tortuous branches, rarely sending out straight shoots; branchlets pubescent; bark white. Leaves very variable, thickly coriaceous, common form linear-spathulate, 2–3 in. long, but on young shoots often larger, broader, sinuate or toothed, and in older shorter, obtuse, veined when dry; petioles short; stipules very minute and deciduous. Flowers very small, solitary or few together, axillary; pedicels short, stout, curved, with one concave appressed bract. Sepals orbicular, erose or ciliolate. Petals linear-oblong. Anthers recurved, connate into a membranous lobed tube, the lobes ciliolate. Berries $\frac{1}{4}$ in. diam., blue-purple.

Northern Island: maritime rocks opposite the Cavallos Islands, *A. Cunningham*; Cape Palliser, *Colenso*. **Middle** Island: Nelson, *Bidwill*, *Travers*.

ORDER V. PITTOSPOREÆ.

Shrubs or trees, with alternate or whorled, exstipulate, evergreen leaves and regular flowers.—Sepals 5, imbricate. Petals 5, with long, erect claws and spreading limbs, imbricate. Stamens 5, free, hypogynous, erect, with oblong or sagittate anthers. Ovary 1 or 2, rarely 3–5-celled (often imperfectly), with a short or long style; ovules many, placentas attached to the septa. Capsule usually bursting by woody valves, which bear the placentas on the middle. Seeds with a minute embryo in hard albumen.

Rather a small Order, abounding in Australia; represented only by *Pittosporum* in India and its islands, the Pacific, and Africa: absent in America.

1. PITTOSPORUM, Linn.

Flowers often polygamous. Sepals free. Petals usually recurved. Filaments subulate. Ovary perfectly or imperfectly 2–5-celled. Capsule woody or coriaceous. Seeds immersed in a transparent gluten.

Leaves alternate. Flowers solitary or 2-nate, axillary.

Leaves 1–2 in., oblong, entire. Peduncles as long as calyx, pubescent . . . . . 1. *P. tenuifolium.*
Leaves 1–2 in., oblong, entire. Peduncles very short, glabrous . 2. *P. Colensoi.*
Leaves 1–1½ in., linear-obovate, serrate, blunt, very coriaceous . . 3. *P. patulum.*
Leaves 1 in., narrow, linear-lanceolate, entire . . . . 4. *P. reflexum.*
Leaves ½ in., linear- or obovate-oblong, very coriaceous . . . . . 5. *P. rigidum.*
Leaves ⅓ in., broadly rounded-obovate or obcordate . . . . . . 6. *P. obcordatum.*

Leaves alternate. Flowers fascicled or corymbose or panicled.

Leaves glabrous. Flowers in axillary and terminal dense fascicles . 7. *P. fasciculatum.*
Leaves tomentose below. Flowers in terminal umbels . . . . . . 8. *P. crassifolium.*
Leaves glabrous. Flowers in terminal umbels . . . . 9. *P. umbellatum.*
Leaves glabrous. Flowers in terminal branched panicles . . . . 10. *P. eugenioides.*

Leaves whorled.

Leaves 1–2 in., elliptic-lanceolate . . . . 11. *P. cornifolium.*
Leaves ½–1 in., linear-oblong . . . . 12. *P. pimeleoides.*

1. **P. tenuifolium,** *Banks and Sol.*;—*Fl. N. Z.* i. 21. A bush or small tree, 20–40 ft. high, with slender trunk; young shoots and leaves often pubescent. Leaves 1–2 in. long, broadly oblong or elliptic-obovate, obtuse, acute or acuminate, quite entire, undulate, rather membranous, glabrous or pubescent on the midrib; petiole short. Flowers axillary, solitary, on curved pubescent peduncles as long or longer than the calyx, variable in size, ¼–⅔ in. long. Sepals very variable in form and shape, from broadly ovate to linear-oblong, silky or glabrous. Petals dark-purple. Ovary pubescent. Capsule size of a small nut, usually 3-valved, broadly obovoid, downy when young, glabrous and rugose when old.—*Trichilia monophylla*, A. Rich. Flora, t. 34 bis.

Abundant throughout the east coasts of the islands, as far south as Otago, *Banks and Solander*, etc. "Wood worthless for any purpose," *Buchanan.*

2. **P. Colensoi,** *Hook. f. Fl. N. Z.* i. 22. A small tree, very closely allied to *P. tenuifolium*, if not a variety of it, but the leaves are smaller, more acute and coriaceous, and not undulated; the peduncles shorter than the sepals, both of which are always glabrous; and the scarious bracts at the base of the peduncle are very persistent. The fruit also is smaller and rounder.

Northern Island: east coast and interior, *Dieffenbach, Colenso.* More specimens are much wanted to clear up this species, of which some specimens have almost the undulate, obtuse leaves of *P. tenuifolium.* Dr. Lyall's Chalky Bay specimens, in fruit only, referred here in the Fl. N. Z., I now think are more probably *P. fasciculatum.*

3. **P. patulum,** *Hook. f., n. sp.* Branches stout, glabrous; branchlets puberulous. Leaves patent or recurved, 1–1½ in. long, ⅓ in. broad, very narrow, linear-oblong, narrowed at the base, obtuse, crenate-serrate, very coriaceous and shining Fruit globose or broader than long, woody, ⅓ in. diam., compressed, on a short, stout, axillary peduncle.

Middle Island: Wairau mountains, alt. 5000 ft., *Sinclair.* I have but one, and that a fruiting specimen, of this most distinct-looking species.

4. **P. reflexum,** *R. Cunn.*—*P. pimelioides* γ, Fl. N. Z. i. 25. A small, slender, much-branched shrub, 2–3 ft. high, with almost filiform, silky-pubescent twigs. Leaves numerous, patent or recurved, very slender, linear-lanceolate, acuminate, membranous, quite entire, ¾–1¼ in. long, ⅛ in. broad.

Flowers not seen. Peduncles solitary, terminal, short, curved, pilose, 1–2-flowered. Ovary hirsute. Capsule ovoid, acuminate, $\frac{1}{3}$ in. long, compressed, 2-valved; valves with the tips recurved.—*P. radicans*, R. Cunn.

Northern Island: thickets at Wangaroa, *R. Cunningham*; Bay of Islands, *J. D. H.*; east coast ?, *Edgerley*. In the N. Z. Flora I regarded this as a variety of *P. pimeleoides*, from which it differs in the much narrower, acuminate, not whorled leaves; but more specimens are requisite to describe it fully.

5. **P. rigidum,** *Hook. f. Fl. N. Z.* i. 22. *t.* 10. A rigid, much-branched shrub; branches tortuous, woody, stout, spreading. Leaves small, shining, in young branches sinuate-dentate or pinnatilobed, in the older narrow-obovate, cuneate, elliptical or oblong, $\frac{1}{2}$ in. long, shortly petioled, very thick and coriaceous, margin recurved. Flowers axillary, solitary; pedicels short, downy. Calyx? Petals dingy-purple, nearly as long as the leaves. Capsules small, $\frac{1}{4}$–$\frac{1}{3}$ in. long, broadly ovoid, acute, compressed, pilose, 2-valved, many-seeded.

Northern Island: mountains, near the Waikare lake and Ruahine mountains, *Colenso*. **Middle** Island: mountains of Nelson, *Bidwill*. The calyx is absent in all my specimens.

6. **P. obcordatum,** *Raoul, Choix*, 25. *t.* 26;—*Fl. N. Z.* i. 23. A shrub or small tree, glabrous, with divaricating, rather slender branches; bark pale. Leaves small, $\frac{1}{3}$ in. long, remote, or 2 or 3 together, rounded or obcordate, sinuate, crenate or quite entire, suddenly contracted into a very short petiole, rather coriaceous; nerves obscure. Flowers $\frac{1}{4}$ in. long, solitary or two together, on short puberulous peduncles, white. Sepals very slender, subulate. Petals narrow-linear. Ovary pubescent. Fruit unknown.

Middle Island: shady woods, near Akaroa, *Raoul*.

7. **P. fasciculatum,** *Hook. f. Fl. N. Z.* i. 24. A branching bush, with glabrous leaves and branches, and inflorescence tomentose. Leaves alternate, coriaceous, obovate-oblong or oblong-lanceolate, acute, quite entire, pale beneath; petioles $\frac{1}{3}$ in. long. Flowers densely fascicled, axillary and terminal, sometimes collected at the base into a short cyme, together with the linear bracts and calyx densely tomentose; peduncles $\frac{1}{4}$–$\frac{1}{3}$ in. long, Sepals ovate-lanceolate. Petals linear-oblong, deep purple, $\frac{1}{3}$ in. long. Capsule on a curved pedicel, $\frac{1}{2}$–$\frac{3}{4}$ in. long, like that of *P. tenuifolium*, 2–3-valved.

Northern Island: Lake Taupo, *Colenso*. **Middle** Island: Chalky Bay, *Lyall*; Otago, Lake district, not common, *Hector and Buchanan*. Very closely allied to *P. tenuifolium*, but the flowers are densely fascicled, almost villous, and the leaves rather longer and more like those of *P. Colensoi*. Dr. Lyall's and Hector's specimens, which are in fruit only, have the foliage more like *P. tenuifolium*.

8. **P. crassifolium,** *Banks and Sol.*;—*Fl. N. Z.* i. 23. A shrub or small tree; branches erect; twigs, leaves below, petioles and inflorescence densely clothed with a thick white or buff tomentum. Leaves alternate, narrow-obovate or linear-obovate or oblong, obtuse, quite entire, 2–3 in. long, very coriaceous, margins recurved. Inflorescence terminal, usually a peduncled, simple umbel, sometimes reduced to a fascicle or a single flower; bracts broadly ovate, ciliate, imbricate. Flower nearly $\frac{1}{2}$ in. long. Sepals linear-oblong, with white tomentum. Petals narrow, deep-purple. Capsule very variable in size, nearly globose, 2–4-lobed and -valved, downy.

Northern Island: not uncommon, *Banks and Solander*, etc.

9. **P. umbellatum,** *Banks and Sol.;—Fl. N. Z.* i. 24. A small tree, 20–30 ft. high, everywhere glabrous, except the under sides and petioles of the young leaves, peduncles, and calyces, which are covered with silky fulvous hairs; branches whorled. Leaves alternate, coriaceous, bright green, 2–3 in. long, obovate- or lanceolate-oblong, obtuse or acute, quite entire, narrowed into petioles $\frac{1}{3}$–$\frac{1}{2}$ in. long. Flowers numerous, rather large, nearly $\frac{1}{2}$ in. long, in terminal umbels or corymbs. Peduncles slender, 1 in. long. Sepals ovate-lanceolate. Petals linear-oblong, obtuse. Ovary pubescent. Capsule rounded, 4-lobed, size of a hazel-nut, 2-valved; valves woody, granulated on the surface.

Northern Island: common about the Bay of Islands and elsewhere, *Banks and Solander*, etc.

10. **P. eugenioides,** *A. Cunn.;—Fl. N. Z.* i. 23. A small, branching tree, 20–30 feet high, everywhere quite glabrous, except the inflorescence and at times the youngest leaves, which may have a few scattered silky hairs; branches often whorled. Leaves 2–4 in. long, usually elliptical, acute, narrowed into long petioles, rarely broader and obovate, quite entire, undulated or crisped, rather coriaceous, with numerous fine veins. Flowers $\frac{1}{4}$–$\frac{1}{3}$ in. diam., fragrant, diœcious (more or less), males with large anthers and longer filaments, collected in branched, many-flowered corymbs, with diverging, slender peduncles and pedicels; bracteoles setaceous. Sepals very variable, ovate, acuminate, glabrous. Petals narrow and spreading, recurved. Capsules numerous, small, $\frac{1}{4}$ in. long, ovoid, acute, glabrous, 2–3-valved.—*P. elegans*, Raoul, Choix, 25; *P. microcarpum*, Putterlich.

Common on the east coast throughout the islands, as far south as Otago, *Banks and Solander*, etc. "Bark white, resinous; wood white, soft, worthless, even for firewood," *Buchanan.*

11. **P. cornifolium,** *A. Cunn.;—Fl. N. Z.* i. 23. A small slender shrub, 2–4 ft. high, with forked or whorled branches, everywhere glabrous, except the young shoots and inflorescence, which present a few long silky hairs. Leaves 1–2 in. long, whorled, obovate or elliptic-lanceolate, shortly petioled, quite entire and glabrous, coriaceous. Flowers polygamous, on very slender, terminal, 1–2-flowered peduncles, dingy red, $\frac{1}{3}$ in. long; peduncles pilose, of the males $\frac{1}{2}$–1 in., female $\frac{1}{4}$–$\frac{1}{2}$ in. long. Sepals very narrow, subulate. Petals as narrow, with slender tips. Capsule $\frac{1}{2}$ in. long, broadly oblong or obcordate, compressed; valves yellow inside, coriaceous. Seeds large.—Bot. Mag. t. 3161.

Northern Island: eastern and southern coasts, common, *Banks and Solander*, etc.: always (?) growing epiphytically on trunks of forest-trees.

12. **P. pimeleoides,** *R. Cunn.;—Fl. N. Z.* i. 25. A very slender, branched shrub, 3–5 ft. high; branches, young leaves, and inflorescence loosely pilose Leaves spreading, whorled, 1 in. long, $\frac{1}{3}$–$\frac{1}{4}$ in. wide, rather membranous, linear-oblong, obtuse or acute, quite entire. Flowers small, terminal, solitary or few; peduncles 1-flowered, very slender, $\frac{1}{4}$–1 in. long. Sepals subulate. Petals very slender, yellow-red. Capsule small, quite like that of *P. reflexum*. —*P. crenulatum*, Putterlich, Synops. Pittosp. 15.

Northern Island: on dry hills at the Bay of Islands. I have restored the var. *reflexum* of Fl. N. Z., as a different species, but with some doubts; better specimens are wanted of both.

ORDER VI. CARYOPHYLLEÆ.

Herbs, with opposite, quite entire or minutely serrulate leaves. Flowers hermaphrodite. Sepals 4 or 5, free or connate, imbricate. Petals 4 or 5 or 0, hypogynous or perigynous. Stamens 4, 5, 8, or 10, inserted with the petals, sometimes seated on or between the lobes of an annular disk. Ovary 1-celled, bearing many (rarely few) ovules on a free central or basal placenta; styles 2–5, free or connate, stigmatose at the apex or inner face. Capsule many-seeded, splitting into as many, or twice as many, valves as styles. Seeds with farinaceous albumen, and a usually curved terete embryo.

A very large Order, abounding in temperate and cold climates, of which a few foreign species are shrubby.

TRIBE I. **Sileneæ.**—*Sepals connate into a tubular calyx.*

Calyx turbinate or campanulate. Stamens 10. Styles 2 . . . . . . 1. GYPSOPHILA.

TRIBE II. **Alsineæ.**—*Sepals free.*

Petals bifid. Styles 3. Stipules 0 . . . . 2. STELLARIA.
Petals 0. Styles 4 to 5. Stipules 0 . . . . 3. COLOBANTHUS.
Petals entire. Styles 3. Stipules scarious . . . . 4. SPERGULARIA.

1. GYPSOPHILA, Linn.

Annual (or perennial) herbs, with small paniculate flowers.—Calyx more or less campanulate, 5-fid, usually 5-nerved. Petals 5, with a narrow claw, and entire or emarginate blade. Stamens 10, at the base of a small torus. Ovary many-ovuled; styles 2. Capsules ovoid or globose, 4–5-valved. Seeds laterally attached.

A large S. European genus, of which the following is the only representative in the southern hemisphere.

1. **G. tubulosa,** *Boiss.;—Fl. N. Z.* ii. 325. A small, much dichotomously-branched, glandular-pubescent annual, 4–5 in. high; stems slender, terete, erect. Leaves subulate, hardly acute, rigid. Peduncles slender, axillary, 1-flowered, $\frac{1}{3}$–$\frac{2}{3}$ in. long, diverging in fruit. Flowers small, $\frac{1}{6}$ in. long. Calyx tubular-campanulate, with 5 green ribs, 5-toothed. Petals narrow, linear, retuse or bifid, longer than the calyx. Capsule 5-valved at the tip, exserted. Seeds transversely rugose, with deep impressions.

Northern Island: east coast, Ahuriri, Raukawa, Cape Palliser, and Hawke's Bay, *Colenso.* **Middle** Island: Rangitata valley, *Sinclair and Haast.* Tarndale plain, 4000 ft., *Travers.* Moraines round Lake Okau, amongst grass, *Haast;* Otago, abundant everywhere in grassy plains, *Hector and Buchanan.* Not uncommon in South Australia, where it was discovered by Brown. Originally described from Asia Minor, and found nowhere but in that country, Australia, and New Zealand; it is worth observing whether it is rapidly increasing, for if so it is probably an importation.

2. STELLARIA, Linn.

Erect or decumbent herbs, annual (or perennial), with flat or acerose leaves, and axillary or fascicled white flowers. Sepals 5, spreading. Petals 5, bifid, or 0. Stamens 10 or fewer, hypogynous, or seated on an annular disk. Styles

3; ovules few or many. Capsule globose oblong or ovoid, splitting into 3 bifid or 6 valves. Seeds often muricate.

A very large genus in Britain, and the temperate and cold regions of both hemispheres generally.

Creeping. Leaves suborbicular. Sepals subulate . . . . 1. *S. parviflora.*
Minute, creeping or ascending. Leaves oblong or linear . . . . . 2. *S. elatinoides.*
Decumbent. Leaves ovate or obovate. Sepals blunt . . . . . . 3. *S. decipiens.*
Erect. Sepals very large. Leaves linear . . . . 4. *S. Roughii.*
Erect, rigid. Leaves acerose, with recurved margins . . . . 5. *S. gracilenta.*

1. **S. parviflora,** *Banks and Sol.;—Fl. N. Z.* i. 25. A very slender, pale green, flaccid herb, with prostrate, wiry, creeping stems and branches, a span long and upwards, wholly glabrous except a few hairs on the petioles. Leaves $\frac{1}{4}$–$\frac{1}{2}$ in. long, nearly orbicular, acute, rarely cordate at the base, longer than the petioles. Peduncles axillary, shorter than the leaves, 1–2-flowered, 2-bracteolate about the middle. Flowers minute, $\frac{1}{12}$ in. diam. Sepals 5, subulate-lanceolate or oblong, acuminate, with white scarious margins. Petals 0 or 5, shorter than the petals. Stamens 5 or 10. Capsule as long or longer than the sepals, 6-valved to the middle. Seeds about 8, pale brown, deeply pitted and reticulated.

Northern and **Middle** Islands: not uncommon in woods, *Banks and Solander*, etc.; and ascending to 5000 ft. on the Wairau mountains, *Travers.* Allied to the Tasmanian *S. flaccida.*

2. **S. elatinoides,** *Hook. f. Fl. N. Z.* i. 25. A very small, glabrous, tufted, pale-green herb. Stems $\frac{1}{2}$–1 in. long, erect or creeping, very slender or rather stout. Leaves $\frac{1}{12}$–$\frac{1}{5}$ in. long, oblong or linear, subacute, narrowed into a short petiole. Flowers $\frac{1}{10}$ in. diam., axillary, solitary, sessile or on short peduncles, large for the size of the plant. Sepals ovate-lanceolate or subulate, acuminate, with white scarious margins. Petals 0. Stamens 5 or 10. Capsule globose, as long as the sepals, 6-valved to the middle. Seeds 2–6, large, grossly tubercled.

Northern Island: grassy banks on the east coast, Cape Kidnapper, Hawke's Bay, Lake Rotoatara, etc., *Colenso.* Closely allied to the Tasmanian *S. multiflora*, Hook., but much smaller in all its parts, and with flowers sessile or almost so.

3. **S. decipiens,** *Hook. f. Fl. N. Z.* i. 27. A glabrous, laxly tufted, weak, decumbent, pale green, branching herb. Leaves $\frac{1}{4}$–$\frac{1}{2}$ in. long, ovate or obovate, acuminate, narrowed into a broad, somewhat ciliate petiole. Peduncles axillary, 1–3-flowered, 2-bracteolate in the middle, shorter or longer than the leaves. Flowers larger than in *S. parviflora* and *elatinoides.* Sepals oblong-ovate, obtuse or subacute, quite glabrous. Petals 0 or 5, small. Stamens variable.

Lord Auckland's group and **Campbell's** Island. The Northern Island plant without flower, referred here in Fl. N. Z., is more probably a form of *S. parviflora.* Closely allied to the *S. media* of Europe, which is naturalized abundantly in some parts of New Zealand, but differing in the axillary 1-flowered peduncles, and absence of a pubescent line on the branches. Also allied to the Tasmanian *S. flaccida*, but the flowers are much smaller, peduncles short, and sepals glabrous and not acuminate.

4. **S. Roughii,** *Hook. f., n. sp.* A short, much-branched, glabrous, succulent, erect or straggling green herb, 2–4 in. high. Leaves $\frac{1}{2}$–$\frac{2}{3}$ in. long, fleshy, linear, acuminate, 1-nerved. Flowers large, $\frac{1}{2}$–$\frac{3}{4}$ in. diam., terminal,

solitary, on short, stout peduncles. Sepals very large, lanceolate, acuminate, with 3 thick nerves. Petals much smaller than the sepals. Capsule subglobose, much shorter than the sepals, 6-valved to the base. Seeds about 6, very large, pale brown, covered densely with long papillæ.

Middle Island: Nelson mountains, *Rough;* Mount Torlesse, on shingle beds, alt. 4–6000 ft., *Haast;* shingle slips, Wairau Gorge, alt. 4–6500 ft., *Travers.* A very singular species, easily recognized by its large green flowers, quite unlike any other in habit.

5. **S. gracilenta,** *Hook. f. Fl. N. Z.* ii. 326. A wiry, rigid, loosely-tufted species, with suberect, nodose, scabrid stems, 2–4 in. high, and very long peduncles. Leaves opposite, each node bearing also a small fascicle of leaves, $\frac{1}{4}$ in. long, subulate or linear, curved, obtuse or acute, glabrous, shining, the margins revolute to the midrib, leaving a deep furrow between them. Peduncles axillary in the upper leaves, solitary, 1-flowered, 1–4 in. long, erect, wiry, 2-bracteolate in the middle. Flower $\frac{1}{3}$ in. diam. Sepals linear-oblong, obtuse, with white, very broad, scarious margins. Petals 5. Capsule cylindric-oblong, much longer than the sepals, 6-valved to the base. Seeds 8 to 10, yellow-brown, densely covered with long papillose hairs.

Middle Island: Nelson, *Bidwill;* Tarndale, *Sinclair, Travers;* Kowai valley, *Haast;* Manuka Island, *Munro;* Lake Tennyson, alt. 4400 ft., *Travers;* Otago, alt. 2000 ft., *Hector and Buchanan.* A very peculiar species, at once recognized by its rigid, wiry habit, narrow leaves, and very long, erect, slender peduncles.

S. media, Sm., the common English "Chickweed," is naturalized in many parts of New Zealand, and in Lord Auckland's Island. It may be distinguished from its very near congener, *S. decipiens,* by the pubescent line on the stem and more panicled inflorescence. The seeds are black and tubercled.

3. COLOBANTHUS, Bartling.

Usually densely-tufted, rigid, green herbs, with subulate, opposite leaves, and solitary green flowers on short or long terminal scapes or peduncles.—Sepals 4 or 5, coriaceous, erect. Petals 0. Stamens 4 or 5, perigynous, alternate with the sepals. Styles 4 or 5, opposite the sepals. Capsule ovoid, many-seeded.

A southern genus, found only in Australia, New Zealand, the Andes of South America, and Antarctic regions.

Leaves hardly rigid, $\frac{1}{4}$–$\frac{2}{3}$ in. with acute tips. Flower 4-merous . . 1. *C. quitensis.*
Leaves rigid, long, $\frac{1}{2}$–$1\frac{1}{2}$ in. with acicular tips. Flower 5-merous . . 2. *C. Billardieri.*
Leaves very short, $\frac{1}{6}$ in., subulate, with acicular tips . . . . 3. *C. subulatus.*
Leaves $\frac{1}{4}$ in., polished, subulate, with very long acicular points . . . 4. *C. acicularis.*
Leaves very short, $\frac{1}{6}$ in., linear, with obtuse tips . . . . 5. *C. muscoides.*

1. **C. quitensis,** *Bartl.* A small, glabrous, perennial, much-branched, tufted, green herb, 1–2 in. high. Leaves chiefly radical, $\frac{1}{4}$–$\frac{2}{3}$ in. long, subulate, acute but not acicular (or very rarely so) at the tips, concave above, convex on the back, quite entire. Scapes longer or shorter than the leaves, usually very short. Flowers 4-merous, $\frac{1}{3}$ in. long. Sepals ovate, blunt, rarely acicular at the tip.

Middle Island: Nelson mountains, *Travers;* clefts of rocks on the Kowai river, alt. 1500 ft., *Haast.* Except in the rather smaller leaves, I can find no distinction between this and the S. American plant, which is found along the Andes from Mexico to Cape Horn, and also in Amsterdam Island in the S. Indian Ocean. It is very nearly allied to *C. Bil-*

lardieri, but that is a more rigid plant, with 5-merous flowers and acicular tips to the leaves and sepals. I have an alpine Tasmanian state of *C. Billardieri*, with foliage and sepals of *C. quitensis*, but pentamerous flowers. I suspect all are varieties of one plant.

2. **C. Billardieri,** *Fenzl;—Fl. N. Z.* i. 27. A small, quite glabrous, often tufted plant, with numerous subulate, spreading or recurved radical leaves and long peduncles; rarely low, densely tufted, with shorter leaves and scapes. Leaves usually $\frac{1}{2}$–$1\frac{1}{2}$ in. long, like those of *C. quitensis*, but with acicular tips. Peduncles often 2 in. long. Flowers $\frac{1}{8}$–$\frac{1}{4}$ in. long, green, usually 5-merous. Sepals ovate, with acicular apices, longer or shorter than the capsule.—*Spergula affinis*, Hook. Ic. Pl. t. 266; *S. apetala*, Labill. Fl. Nov. Holl. t. 142.

Common throughout the islands, especially in alpine or hilly districts, *Colenso*, etc. **Campbell's** Island, *J. D. H.* Small varieties closely resemble the preceding. Abundant in Victoria and Tasmania.

3. **C. subulatus,** *Hook. f. Fl. Antarct.* i. 13 *and* 247. *t.* 93. A small, moss-like, densely cæspitose, perfectly glabrous plant, with subulate, rigid, shining leaves, forming tufts about 1 in. high. Leaves densely imbricated, about $\frac{1}{6}$ in. long, with acicular points, grooved above, convex on the back. Peduncles very short. Flowers hidden amongst the leaves. Sepals 4 or 5, ovate-subulate, with acicular apices, rigid.

Middle Island: Awatere valley, and rocky places, Sinclair range, alt. 4000 ft., *Sinclair and Haast*; Otago, Lake district, *Hector and Buchanan*; **Campbell's** Island, *J. D. H.* Also found in the alps of Victoria, and abundantly in Antarctic America.

4. **C. acicularis,** *Hook. f., n. sp.* A small, moss-like, densely cæspitose, shining, rigid, glabrous plant, forming tufts 1–2 in. high. Leaves densely imbricated all round, $\frac{1}{4}$ in. long, like those of *C. subulatus*, but paler, more shining, much larger, and with much longer acicular points. Flowers almost sessile, shorter than the leaves. Sepals 5, like the leaves, longer than the capsule.

Middle Island: dry rocky places, Wairau Gorge, alt. 4–5500 ft, *Travers*; Otago, Lake district, abundant, *Hector and Buchanan*.

5. **C. muscoides,** *Hook. f. Fl. Antarct.* i. 14. A perfectly glabrous, moss-like, densely tufted, bright green plant, soft and rather flaccid in texture, forming large patches. Stems most densely matted. Leaves densely imbricated, patent, about $\frac{1}{4}$ in. long, linear, obtuse, rather dilated at the base, green, soft. Peduncles extremely short. Flowers minute, sunk amongst the leaves, 4-merous. Sepals ovate-lanceolate, obtuse, concave, green.

Lord Auckland's group and **Campbell's** Island: rocks near the sea, *J. D. H.*

4. SPERGULARIA, Persoon.

Spreading, dichotomously-branched, perennial herbs, with linear leaves, scarious stipules, and white or rose-coloured peduncled flowers.—Sepals 5, spreading. Petals 5, obtuse, or 0. Stamens 5 or 10, hypogynous. Ovary subglobose; styles 3. Capsule 3-valved. Seeds compressed, often winged.

A small British genus, of which species are scattered over many temperate and warm parts of the globe, especially near the sea, and in waste places.

1. **S. rubra,** *Pers.*, var. **marina.**—*Arenaria media*, Linn.; Fl. N. Z. i, 26. A rather succulent, excessively-branched, prostrate or suberect, annual

herb, with perennial root, more or less pubescent with viscid down. Stems and branches terete, 1 in. to a span long. Leaves linear, $\frac{1}{3}$–1 in. long, quite entire. Flowers numerous, axillary and terminal, on long, slender peduncles, $\frac{1}{2}$–1 in. long, which are often patent or reflexed. Flowers very variable in size, $\frac{1}{4}$–$\frac{1}{3}$ in. long. Sepals green with a white membranous border. Petals shorter than the sepals, sessile, often very pale. Stamens 5. Seeds with a broad membranous wing.

Common on the shore throughout the islands, *Banks and Solander*, etc. Also found under various forms in all temperate and many tropical localities.

There are several Caryophylleous weeds introduced into New Zealand, especially from Europe, besides the *Stellaria media*, mentioned under that genus; the chief of these are:—

Polycarpon tetraphyllum, Linn. A small, tufted annual, with opposite and 4-nate obovate or oblong leaves, small green flowers, 5 sepals, small entire petals, 3 stamens, a short 3-fid style, and 3-valved capsule.—Bay of Islands, Auckland, etc., introduced probably from Australia, where it is common, as it is in many other parts of the world.

Cerastium. A genus of very common, erect or straggling, hairy European weeds, having small, ovate, oblong, or lanceolate leaves and white flowers, 5 sepals, 5 notched or 2-fid petals, 10 stamens, 5 styles, and a tubular membranous capsule, opening at the top by 10 teeth; two species *C. viscosum*, Linn., and *C. glomeratum*, are abundant weeds of cultivation.

Spergula arvensis, Linn. An annual weed, with slender stems, bearing whorls of linear leaves and cymes of white flowers, 5 sepals petals and styles, 10 stamens and a 5-valved capsule.

ORDER VII. PORTULACEÆ.

Herbs, usually glabrous, often succulent with opposite (rarely alternate) exstipulate (rarely stipulate) leaves and hermaphrodite flowers.—Sepals 2 or 3, imbricate. Petals 4 or 5, free or connate at the base, hypogynous, imbricated. Stamens 5 or fewer, often adnate to the base of the petals. Ovary 1-celled; style more or less deeply 2–3-fid, the arms stigmatose on the inner face; ovules several or many, in the base of the cell. Capsule membranous, with as many valves as styles, and one or many seeds. Seeds with a crustaceous testa, farinaceous albumen, and terete curved embryo.

A considerable Order, most abundant in America. The typical genus *Portulaca* (cultivated for salad) has a half-inferior ovary.

Leaves opposite or in pairs. Stamens 5, opposite the petals. Capsule 3- or many-seeded . . . . 1. CLAYTONIA.
Leaves opposite. Stamens 3–5, opposite the petals. Capsule 1–3-seeded . . . . 2. MONTIA.
Leaves densely imbricate. Stamens 5, alternate with petals . . . . . 3. HECTORELLA.

1. CLAYTONIA, Linn.

Herbs with opposite alternate or fasciculate leaves, and racemose or solitary flowers.—Sepals 2, herbaceous. Petals 5, united at the very base, or free. Stamens 5, adnate to and opposite to the petals. Ovary many-ovuled. Capsule 3-valved, 3- or more seeded.

A large North American genus, of which the following is the only Old World or south hemisphere species, abounding in temperate Australia.

1. **C. australasica,** *Hook. f.;—Fl. N. Z.* i. 73. A glabrous, slender,

creeping, rather succulent, tender herb, extremely variable in size. Leaves solitary or in distant pairs, narrow-linear or linear-spathulate, $\frac{1}{3}$–2 in. long, quite entire, pale green, obtuse, nerveless, the petioles dilating into membranous stipules at the base. Scapes axillary, solitary, 1-flowered, erect, usually shorter than the leaves. Flowers pure white, very variable in size, $\frac{1}{4}$–$\frac{3}{4}$ in. diam.—Hook. Ic. Pl. t. 293.

Middle Island: boggy places, probably common; Milford Sound, *Lyall;* Otago, *Lindsay;* near Lake Tennyson, *Maling;* Southern Alps, *Haast;* Mount Alta, ascending to 5000 feet, *Hector and Buchanan.*

2. MONTIA, Linn.

A small, annual, branched or simple, usually tufted, glabrous herb, with opposite, rather fleshy, spathulate leaves, and small, white, axillary, peduncled flowers.—Sepals 2 (rarely 3). Petals 5, connate at the base. Stamens usually 3 or 5, opposite the petals and adnate to them. Ovary and capsule as in *Claytonia*, but only 1–3-seeded.

A British genus, found in many parts of the north and south temperate zone.

1. **M. fontana,** *Linn.;—Fl. N. Z.* i. 74. A glabrous, slender, branching, weak, bright green marsh or water plant, very variable in size. Stems 1–12 in. high. Leaves $\frac{1}{4}$–1 in. long, from elliptical ovate to linear-lanceolate, subacute, quite entire. Flowers about $\frac{1}{8}$ in. broad.

Common in watery places throughout the alpine parts of the **Northern** and the **Middle** Islands, and in **Campbell's** Island; also found in Tasmania, Kerguelen's Land, and throughout temperate Western North America and South America, etc.; Europe, Labrador, and Greenland, but not in eastern temperate North America, central Asia, nor in the Himalaya.

3. HECTORELLA, Hook. f., *n. g.*

A small, densely tufted, glabrous, moss-like plant. Leaves most densely imbricated all round the stem, coriaceous, entire. Flowers almost sessile amongst the uppermost leaves, white.—Sepals 2, short, concave, truncate, continuous with the very short, broad, flat pedicel. Petals 5, united at the very base, erect, veined, obtuse, thickened below the tip. Stamens 5, inserted on the tube of the corolla, alternate with its petals; filaments as long as the petals; anthers linear-oblong, 2-celled. Ovary ovoid, membranous, veined, narrowed into an erect style; stigmas 1–3, linear, thickened, papillose internally; ovules 4–5, erect from the base of the cell, amphitropous, funicle slender. Fruit unknown.

A remarkable genus, allied to no other, but approaching in habit *Lyallia* of Kerguelen's Land. Named in compliment to Dr. Hector, F.G.S., during whose adventurous expedition to the Otago alps it was discovered.

1. **H. cæspitosa,** *Hook. f.* Stems 1–1$\frac{1}{2}$ in. high, most densely tufted, and with the leaves on them nearly as thick as the little finger. Leaves excessively numerous and closely imbricate, spreading, broadly triangular-ovate, $\frac{1}{6}$ in. long, much dilated and membranous below the middle, coriaceous above with thickened margins and keel; veins reticulated. Flowers several from amongst the leaves at the tips of the stems, nearly $\frac{1}{4}$ in. long, white.

Middle Island: Otago, Lake district, alt. 4–6000 feet, *Hector and Buchanan.*

ORDER VIII. ELATINEÆ.

Small water-plants, rarely shrubs, mostly prostrate, with usually opposite stipulate leaves. Flowers small or minute, hermaphrodite, regular.—Sepals 2–5, free, imbricate. Petals 2–5, free, hypogynous, imbricate. Disk 0. Stamens as many as the petals or twice as many, hypogynous, free. Ovary free, 2–5-celled; styles 2–5, stigmas capitate; ovules numerous in the angles of the cells. Capsule septicidal, the valves falling away and leaving the seeds attached to a central column. Seeds straight or curved, with terete embryo and no albumen.

A small Order, scattered over various parts of the world, as is the only New Zealand genus.

1. ELATINE, Linn.

Aquatic, creeping, glabrous, small herbs, with pellucid-dotted leaves, and minute, axillary, solitary flowers.—Sepals membranous, obtuse. Ovary globose. Capsule membranous, the septa either disappearing or persistent on the axis. Seeds oblong or cylindrical, longitudinally striated and transversely wrinkled.

The species are found in ponds, lakes, etc., in various parts of the world, including Britain.

1. **E. americana,** *Arnott;—Fl. N. Z.* i. 27. A minute, glabrous, prostrate, matted aquatic. Stem rather succulent, creeping, sometimes elongated, and 1–8 in. long. Leaves small, $\frac{1}{8}$–$\frac{1}{2}$ in. long, obovate-oblong, obtuse, quite entire, shortly petioled. Flowers minute, sessile, 3-merous. Calyx persistent. Capsule with the septa disappearing. Seeds slightly curved. —*E. gratioloides*, A. Cunn.

Northern Island, probably common: bogs at Hokianga, *R. Cunningham*; also found in Tasmania and Australia, the Feejee Islands and North and South America.

ORDER IX. HYPERICINEÆ.

Herbs shrubs or trees, with opposite, simple, quite entire or glandular-toothed, often pellucid-dotted, exstipulate leaves. Flowers regular, hermaphrodite.—Sepals 5, imbricate. Petals 5, hypogynous, imbricate. Disk 0. Stamens indefinite, hypogynous, free or polyadelphous, filaments filiform. Ovary 3–5-celled, or 1-celled with inflexed margins to the carpels; styles as many as carpels, filiform, free or connate, stigmas terminal; ovules numerous, in 2 series in the axis of the cells, or on the inflexed edges of the carpels. Fruit usually capsular. Seeds exalbuminous.

A large Order, widely distributed, as is the only New Zealand genus, which is also a British one.

1. HYPERICUM, Linn.

Sepals and petals 5. Stamens very numerous, all free or connected into bundles. Ovary 1- or 3–5-celled. Capsule septicidal. Seeds not winged, with a straight embryo.

Erect. Leaves subcordate, with revolute margins . . . . 1. *H. gramineum*.
Procumbent. Leaves oblong or obovate . . . . 2. *H. Japonicum*.

1. **H. gramineum,** *Forst.;—Fl. N. Z.* i. 36. An erect or ascending, rather wiry, quite glabrous, perennial-rooted herb, branching from the root, with 4-angled branches, 6 to 12 in. high, slender and sparingly leafy. Leaves small, ½ to 1 in. long, sessile, cordate, oblong, obtuse, quite entire, black-dotted, margins usually revolute. Flowers subsolitary or in 3-chotomous terminal cymes, ¼ to ½ in. across, on rather stout, erect, rigid peduncles. Sepals oblong or ovate, obtuse or acute, black-dotted, quite entire. Petals longer than the calyx, golden-yellow, curling inwards as they wither. Stamens nearly free. Capsule ovoid, acute, longer or shorter than the sepals.—Labill. Fl. Austr. Caled. t. 53.—*Brathys Forsteri*, Spach.

Common in grassy places throughout the islands, *Banks and Solander*, etc.; also frequent in temperate Australia, New Caledonia, and the hilly country of India.

2. **H. japonicum,** *Thunb.;—Fl. N. Z.* i. 37. A much smaller plant than *H. gramineum*, with prostrate branches, broader, flat leaves, smaller, often sessile flowers, having broader, more obtuse sepals, and shorter, rounder capsules; but apparently intermediate forms occur both in New Zealand, Australia, and India, in all which countries both occur, and I suspect they are the extreme forms of one variable species.—*H. pusillum*, Choisy; A. Cunn. Prodr.; *Ascyrum humifusum*, Labill. Fl. Nov. Holl. ii. t. 175.

Abundant in moist, grassy places, etc., throughout the islands, *Banks and Solander*, etc. Also found in many parts of temperate and subtropical Asia. A very similar and perhaps identical plant, inhabits the west coasts of temperate North and South America.

Order X. MALVACEÆ.

Herbs shrubs or trees, with (usually) tough fibrous inner bark, alternate stipulate leaves, and stellate hairs. Flowers usually hermaphrodite, regular and large.—Calyx 5-lobed, lobes valvate. Petals 5, hypogynous, usually connate at the base, adnate to the staminal tube, imbricate. Disk 0 or a small torus. Stamens very numerous, their filaments united into a tube; anthers often reniform, 1-celled. Ovary of 1 or more free or connate 1- or many-ovuled carpels, whorled round and adnate with the torus; styles as many as carpels, connate below, filiform above. Fruit of one or more indehiscent or 2-valved cocci, or capsular. Seeds often hairy; albumen little or none; cotyledons large, folded.

A very large Order, abundant both in the tropics and temperate zones, to which the *Mallow, Lavatera, Hollyhock, Cotton*, and many other well-known cultivated New Zealand garden plants belong.

Bracts 0 or small. Stigmas longitudinal. Ovules solitary . . . . . 1. Plagianthus.
Bracts 0 or small. Stigmas capitate. Ovules solitary . . . . . . 2. Hoheria.
Bracts large. Stigmas capitate. Ovules 2 to many . . . . 3. Hibiscus.

1. PLAGIANTHUS, Forst.

Shrubs or small trees, with very tough inner bark. Flowers uni- or bisexual.—Bracts 0, or small and distant from the calyx. Calyx 5-toothed or 5-fid. Staminal tube divided above into many short or long filaments. Ovary of 1 free, or many more or less united, 1-ovuled carpels; styles filiform or club-shaped, combined below, stigmatiferous towards the apex along the inner face. Fruit of 1 indehiscent or irregularly bursting carpel, or of many whorled round an axis. Seed pendulous.

A genus confined to Australia and New Zealand.

Leaves small, linear. Peduncles 1-flowered. Carpels 1 or 2 . . . 1. *P. divaricatus.*
Leaves ovate, serrate. Panicles many-flowered. Carpel solitary . . 2. *P. betulinus.*
Leaves ovate-cordate, serrate. Peduncles 1-flowered. Carpels many . 3. *P. Lyallii.*

1. **P. divaricatus,** *Forst.;—Fl. N. Z.* i. 29. A rigid, glabrous, much-branched shrub, with slender spreading tough branches, small fascicled leaves, and minute white flowers, succeeded by small globose capsules. Leaves $\frac{1}{3}$–$\frac{3}{4}$ in. long, narrow-linear or subcuneate, obtuse, quite entire, 1-nerved. Flowers in axillary fascicles or 1-flowered peduncles, shorter than the leaves, minutely bracteolate near the base. Calyx hemispherical, glabrous. Petals concave, oblong, small. Staminal tube with 6–10 large sessile anthers. Capsules size of a peppercorn, globose, rarely didymous, oblique, downy, bursting irregularly.—Hook. Bot. Mag. t. 3271.

Abundant in salt marshes throughout the islands as far south as Akaroa, *Banks and Solander*, etc.

2. **P. betulinus,** *A. Cunn.;—Fl. N. Z.* i. 29. A lofty tree, attaining 40–70 ft., when young a straggling bush with variable leaves. Leaves of young plants $\frac{1}{4}$–$\frac{1}{2}$ in. long, ovate-rounded, variously crenate and lobed, in full-grown 1–2 in. long, ovate or ovate-lanceolate, acuminate, rounded or cuneate at the base, coarsely crenate-serrate, or obtusely doubly serrate, membranous, covered on both surfaces with small stellate hairs and reticulate venation; petiole slender. Panicles terminal, much branched, very many flowered, stellate-tomentose. Flowers small, $\frac{1}{4}$ in. broad, white, on slender ebracteolate pedicels. Calyx campanulate. Petals linear-oblong, narrower in the male flowers. Staminal tube long, slender, exserted in the male, bearing many shortly-pedicelled anthers. Carpel 1. Capsule small, ovoid, acuminate, splitting down one side, 1-seeded.—*P. betulinus* and *urticinus*, A. Cunn.; *Philippodendron regium*, Poit. in Ann. Sc. Nat. ser. ii. 8. t. 3.

Abundant in forests throughout the islands, *Banks and Solander*, etc., as far south as Otago. "Ribbon-tree of Otago, wood worthless," *Buchanan.*

3. **P. Lyallii,** *Hook. f.—Hoheria Lyallii*, Fl. N. Z. i. 31. t. 11. A small branching tree, 20–30 ft. high, with the young branches, inflorescence, and leaves below covered with white stellate down. Leaves 2–4 in. long, ovate-cordate, acuminate, deeply doubly crenate, glabrous above; petioles $\frac{1}{2}$–$1\frac{1}{2}$ in. Flowers large, $\frac{3}{4}$ in. broad, white, axillary; peduncles 1-flowered, solitary or fascicled, ebracteolate, about as long as the petioles. Calyx broadly campanulate. Petals obliquely obovate-cuneate, obscurely notched on one side towards the apex. Staminal tube short, with many long filiform filaments. Ovary about 10-celled; style slender, divided into as many filiform branches, stigmatose on the inner surface towards the apex. Fruit a depressed sphere, breaking up into 10 compressed reniform membranous carpels. Seed much compressed.

In mountain districts throughout the **Middle** Island, from Nelson to Milford Sound; western districts of Otago, fringing the *Fagus* forest, *Hector and Buchanan.* Mr. Haast informs me that this forms a deciduous tree at and above 3000 ft., but is evergreen below that level; in autumn its naked branches and yellow foliage give a peculiar colour to the landscape at the higher elevation.

2. HOHERIA, A. Cunn.

Small trees, with alternate, petioled, excessively variable, pellucid-dotted foliage, and fascicled, axillary, white, hermaphrodite flowers. Peduncles jointed in the middle. Bracts 0.—Calyx with a broad cup-shaped tube and 5 teeth. Petals linear-oblong, obtuse, oblique. Stamens 5-adelphous. Ovary 5-celled; styles 5, filiform, stigmas capitate. Carpels (unknown in *H. Sinclairii*) laterally quite flat, whorled round a central axis, from which they fall away when ripe, indehiscent, crested at the back with a membranous wing. Seed pendulous.

Leaves ovate-lanceolate or linear-oblong, sharply toothed . . . . . 1. *H. populnea.*
Leaves broadly ovate, bluntly serrate . . . . 2. *H. Sinclairii.*

1. **H. populnea,** *A. Cunn.;—Fl. N. Z.* i. 30. Tree 10–30 ft. high, branches hoary. Leaves glabrous, excessively variable in size, shape and toothing, 2–5 in. long, ovate or lanceolate, generally sharply or coarsely double-toothed or serrate. Flowers abundantly produced, snow-white, glabrous or hoary, $\frac{1}{4}$–$\frac{3}{4}$ in. diam. Carpels produced backwards and upwards into a wing.—Hook. Ic. Pl. t. 565, 566.

Var. *α. vulgaris.* Leaves ovate, with large sharp teeth.

Var. *β. lanceolata.* Leaves linear or oblong-lanceolate, toothed or serrate.

Var. *γ. angustifolia.* Leaves small, linear-oblong, spinulose-toothed. Flowers small. *H. angustifolia,* Raoul, Choix, 48. t. 26.

Var. *δ. cratægifolia.* Leaves ovate, variously lobed and toothed.

Abundant throughout the islands, *Banks and Solander,* etc. The bark affords a demulcent drink, and is also used for cordage, etc.

2. **H. Sinclairii,** *Hook. f., n. sp.* Larger in all its parts than *H. populnea,* and readily distinguished by its broadly ovate, acute, obtusely serrate, coriaceous, glabrous leaves. The peduncles are usually binate, and shorter than the petioles. It much resembles *Plagianthus Lyallii,* but the leaves are not cordate, more serrate, the peduncles jointed in the middle, stigmas capitate, and carpels 5.

Northern Island: near Auckland?, *Sinclair.* I find this fine species amongst some Auckland plants, sent without localities by Dr. Sinclair.

3. HIBISCUS, Linn.

Usually erect herbs, with often lobed leaves and handsome hermaphrodite flowers. Bracts numerous (or few) below the calyx, free or connate.—Calyx 5-fid or 5-toothed. Petals usually cuneate, oblique, soon withering. Staminal tube long, 5-toothed at the mouth, below which the filaments are inserted. Ovary 5-celled, with 5 spreading styles and terminal stigmas; cells 3- or many-ovuled. Capsule 5-valved, loculicidal. Seeds glabrous or woolly.

A very large tropical genus, containing many very handsome plants, some, as the Ochra (*H. esculentus*), yielding an esculent fruit, others (*H. cannabinus*), cordage; a few are trees, some climb.

1. **H. Trionum,** *Linn.;—Fl. N. Z.* i. 28. A hispid annual, often branched; stem almost woody below, erect or with spreading branches, 1–2 ft. high.

Leaves petioled, cordate, palmately 3–5-lobed, lobes linear, often serrate or sinuate, the middle one longest. Bracts numerous, setaceous. Flowers ½–1 in. diam., yellow with a purple eye. Calyx membranous, hispid, veined. Stamens few or many Seeds dark-brown, wrinkled, glabrous. Capsule hispid.—Bot. Mag. t. 209; *H. vesicarius*, Cav.; A. Cunn. Prodr.

Scattered over the islands, and possibly introduced (*A. Cunningham*). Most common in the northern parts of the **Northern** Island, and certainly indigenous (*Colenso*). **Middle** Island: South Wanganui, *Lyall*. A very common Australian, Asiatic, and S. African plant, also found in S. Europe and elsewhere in the Old World.

ORDER XI. TILIACEÆ.

(*Including* ELÆOCARPEÆ.)

Trees or shrubs (rarely herbs) with often tough bark, alternate or opposite, often stipulate leaves. Flowers regular, hermaphrodite, rarely unisexual.—Sepals 4 or 5, free or connate, usually valvate. Petals 4 or 5, free, entire, lobed or cut. Torus generally conspicuous. Stamens usually numerous, free, inserted on the torus; filaments filiform; anthers 2-celled, often opening by terminal pores. Ovary sessile on the torus, 2–10-celled; style simple, usually divided at the apex into as many divisions as cells; ovules few or many, attached to the axis of the cells. Fruit very variable. Seeds generally with fleshy albumen, and broad, flat, thin cotyledons.

A very large tropical and subtropical Order of plants, to which the English Lime-tree (*Tilia*) belongs, together with the Indian Jute (*Corchorus*), valued for its fibre.

Leaves alternate. Petals entire. Capsule echinate . . . . 1. ENTELEA.
Leaves opposite. Petals crenate or lacerate. Berry 2–4-celled . . . 2. ARISTOTELIA.
Leaves alternate. Petals lacerate. Drupe 1-celled . . . . 3. ELÆOCARPUS.

1. ENTELEA, Br.

A small branching light-wooded tree, covered with stellate down, having large, alternate, 5–7-nerved, cordate, toothed stipulate leaves, and umbellate cymes of white flowers.—Sepals 4 or 5, free. Petals 4 or 5, undulate. Stamens very numerous, free, on a low torus, with filiform filaments and versatile anthers. Ovary 4–6-celled; style simple, stigmatiferous at the toothed apex; cells many-ovuled. Capsule globose, echinate with long rigid bristles, 4–6-valved, loculicidal.

1. **E. arborescens,** *Br.;—Fl. N. Z.* i. 33. Leaves 4–8 in. long, on long petioles, oblique, often lobed irregularly and acutely, doubly or trebly crenate or serrate; stipules persistent. Flowers white, abundant, in erect cymes, bracteate at the axils, ¾–1 in. diam., drooping. Sepals acuminate. Ovary hispid. Capsule the size of a hazel-nut, spines nearly 1 in. long. Seeds in two rows, albumen oily.—Bot. Mag. t. 2480; *Apeiba australis*, A. Rich. Flor. t. 34.

Not rare throughout the **Northern** Island, *Banks and Solander*, etc. The genus is confined to New Zealand.

2. ARISTOTELIA, L'Héritier.

Shrubs or trees, with opposite or subopposite, exstipulate leaves. Flowers usually unisexual.—Sepals 4 or 5, valvate or subimbricate. Petals 4 or 5, lobed or crenate, rarely entire, sometimes minute. Stamens 4 or 5 or numerous, inserted on the glandular torus; filaments short; anthers with short terminal slits. Ovary 2–4-celled; style subulate, entire; cells 2-ovuled. Berry 2–4-celled, few- or many-seeded. Seeds often fleshy on the outside of the hard testa.

A small genus, with two Australian and Tasmanian, one Chilian, and the three following New Zealand species.

Leaves large, membranous, pubescent. Racemes many-flowered . . . 1. *A. racemosa.*
Leaves large, membranous, glabrous. Racemes many-flowered . . . 2. *A. Colensoi.*
Leaves small, coriaceous Flowers few . . . . 3. *A. fruticosa.*

1. **A. racemosa,** *Hook. f. Fl. N. Z.* i. 33. A small handsome tree, 6–20 ft. high, with blackish bark and pubescent twigs. Leaves on long petioles, membranous, pubescent, variable in form, 3–5 in. long, ovate cordate or oblong-lanceolate, acuminate, deeply irregularly and acutely serrate, often red or purple beneath. Racemes panicled, axillary, many-flowered, peduncles and pedicels slender. Flowers diœcious, small, the males largest, $\frac{1}{6}$–$\frac{1}{3}$ in. diam., nodding. Petals 4, 3-lobed, rosy, of the female flower very small. Stamens numerous, yellow, minutely hairy; anthers longer than the filaments. Ovary usually 4-celled. Berry size of a pea, eaten by the natives. —*Friesia racemosa*, A. Cunn. Prodr.; Hook. Ic. Pl. t. 601.

Abundant throughout the islands, *Banks and Solander*, etc. "Wood white, very light, makes veneers," *Buchanan*.

2. **A. Colensoi,** *Hook. f., n. sp.* Very similar in most respects to *A. racemosa*, but differing in the much narrower, perfectly glabrous leaves, which are ovate-lanceolate, narrowed into a long acuminate point, deeply irregularly serrate; and the small fruit, which is no bigger than a peppercorn. The seeds are as in *A. racemosa*. I have seen no flowers.

Northern Island: woods in the Wairarapa valley, *Colenso*.

3. **A. fruticosa,** *Hook. f. Fl. N. Z.* i. 34. A small, rigid, erect or decumbent shrub, with woody tortuous branches, and erect or spreading downy shoots, with red-brown bark. Leaves very variable, on short, downy or glabrous petioles, coriaceous, $\frac{1}{4}$–1 in. long, ovate obovate or linear-oblong, obtuse, entire crenate toothed serrate or lobed. Flowers minute, usually solitary and axillary, rarely racemose; peduncles usually short. Petals 4, very short or as long as the calyx, entire or lobed, pink, shorter in the female. Stamens 4–6; filaments short; anthers downy. Ovary 2–4-celled. Berry globose, small, 4–6-seeded. Seed with a bony, rugged testa, covered with a thin pulp.

Common in mountain districts, alt. 2–4000 ft., throughout the islands, and varying greatly, *Colenso*, etc. I have made four varieties in the New Zealand flora, but they seem to be states determined by age and exposure, rather than hereditary races; of these the var. β is possibly a very small form of *A. racemosa*.

3. ELÆOCARPUS, Linn.

Trees, generally hard-wooded, with the branches leafy at the extremity.

Leaves generally alternate, exstipulate.—Flowers racemose, usually hermaphrodite, pendulous. Sepals 4 or 5, valvate. Petals 4 or 5, laciniate, induplicate-valvate. Stamens numerous, seated on a glandular torus; filaments short, anthers long, awned, pubescent, opening by a short, terminal slit. Ovary 2–5-celled; style subulate, stigma simple; ovules 2 or more, pendulous. Drupe with one bony, 1- or several-celled nut, which is often tubercled or wrinkled. Seed pendulous.

A very large tropical Asiatic, Australian, and Polynesian genus.

Twigs silky. Leaves with recurved margins . . . . 1. *E. dentatus.*
Twigs glabrous. Leaves flat . . . . 2. *E. Hookerianus.*

1. **E. dentatus,** *Vahl;—E. Hinau*, A. Cunn.; Fl. N. Z. i. 32. A small tree, with brown bark, which yields a permanent dye; branches fastigiate at the top of the naked trunk, silky when young. Leaves erect, petioled, 2–3 in. long, very coriaceous and variable, linear-oblong obovate or lanceolate, obtuse or acuminate, margins recurved, sinuate-serrate, below often white with silky down, and with hollows where the veins meet the midrib. Racemes glabrous or silky, of many white pendulous flowers $\frac{1}{3}$ in. diam. Petals lobed or lacerate. Anthers with a flat recurved tip. Drupe $\frac{1}{3}$–$\frac{1}{2}$ in. long, ovoid, pulp astringent but eatable; stone deeply furrowed.—Hook. Ic. Pl. t. 602; *E. Cunninghamii*, Raoul, Choix, 25; *Dicera dentata*, Forst.; *Eriostemon dentatum*, Colla, Hort. Rip. lii. t. 30.

Common throughout the islands, *Banks and Solander*, etc.

2. **E. Hookerianus,** *Raoul, Choix, t.* xxv.;—*Fl. N. Z.* i. 32. A small, quite glabrous tree, 30–40 ft. high, like *E. dentatus* in habit, but smaller in all its parts. Leaves coriaceous, elliptical or linear-oblong, obtuse, 1$\frac{1}{2}$–2 in. long, margins flat, crenate or sinuate-serrate, those of young plants linear and pinnatifid; petioles $\frac{1}{4}$–$\frac{1}{2}$ in. long. Racemes erect, shorter than the leaves, with small, drooping, whitish flowers. Sepals lanceolate. Petals rather longer, unequally cleft into obtuse lobes. Anthers obtuser than in *E. dentatus.* Drupe small, blue, $\frac{1}{3}$ in., with a furrowed rugose nut.

Hilly and other parts of the **Northern** Island; and common on the **Middle** Island, *Colenso, Raoul*, etc.

ORDER XII. LINEÆ.

Herbs or undershrubs, usually with entire, alternate leaves, and subracemose, handsome, hermaphrodite, regular flowers.—Sepals 5, free, imbricate. Petals 5, free, fugacious, contorted. Stamens 5, hypogynous; filaments united below into a cup, which has usually 5 minute glands at its base; anthers versatile. Ovary 3–5-celled; styles 3–5, stigmas terminal; cells 1–2-ovuled. Capsule splitting septicidally into indehiscent or dehiscent 1–2-seeded cocci. Seeds with scanty albumen or 0.

A rather large Order, in temperate and tropical countries, of which the tribe *Eulineæ* is chiefly temperate. The Flax, *L. usitatissimum*, belongs to the only N. Z. genus, which is a large European one.

1. LINUM, Linn.

Glabrous herbs, with narrow, quite entire leaves, fibrous bark, and usually large flowers.—Sepals 5, entire. Stamens alternating with 5 setiform processes of the staminal tube. Ovary 5-celled; cells imperfectly divided by a longitudinal septum. Cocci 5, septate, 2-seeded, or 10, by each splitting along the septum.

1. **L. monogynum,** *Forst.;—Fl. N. Z.* i. 28. A perennial, variable herb, sometimes woody at the base, simple or branched, 6–12 in. high, erect or decumbent. Leaves ¼–1 in. long, oblong linear or linear-subulate, 1–3-nerved. Flowers numerous, white or pale-blue, very variable in size. Sepals ovate-lanceolate, acute. Styles united, their free tips recurved. Capsule globose, of 10 cocci.—Bot. Mag. t. 3574.

Var. α. *grandiflorum.* Erect, branched; flowers numerous ½ to 1 in.

Var. β. *diffusum.* Decumbent; flowers few, ¼ to ½ in.

Common, especially on rocky coasts throughout the islands, *Banks and Solander*, etc. ascends 2000 ft. in the mountains of Canterbury, *Haast.* Chatham Island, var. α, *Dieffenbach.*

ORDER XIII. GERANIACEÆ.

(*Including* OXALIDEÆ.)

Herbs (the New Zealand species), with alternate, stipulate or exstipulate leaves, and regular or irregular hermaphrodite flowers.—Sepals 5, free, imbricate, one sometimes spurred. Petals 5, rarely fewer, imbricated. Disk inconspicuous or glandular. Stamens 10, hypogynous, the alternate ones often smaller, or imperfect, or without anthers; filaments often connate below; anthers versatile. Ovary 3–5-lobed, or of 3–5 carpels combined in the axis, produced into as many free or connate styles, with capitate or longitudinal stigmas; cells 1- or more ovuled. Fruit capsular, 3–5-lobed, 3–5-valved, variously dehiscing. Seeds with little or no albumen.

A very large Order, containing many genera absent in New Zealand, differing a good deal in structure, to some of which the above character does not altogether apply.

Flowers regular. Styles combined. Carpels caudate . . . . . . 1. GERANIUM.
Flowers irregular. Calyx with a spur adnate to the pedicel . . . . 2. PELARGONIUM.
Flowers regular. Styles free. Leaves 3-foliolate . . . . 3. OXALIS.

1. GERANIUM, Linn.

Stemless or branching herbs, with stipulate leaves, and axillary, 1–2-flowered, 2-bracteolate peduncles.—Flowers regular. Torus with 5 glands alternating with the petals. Stamens 10, all perfect (rarely 5 imperfect), free or united at the base. Ovary 5-celled, beaked; beak terminated by 5 short styles, which are longitudinally stigmatose; cells 2-ovuled. Capsule of 5, tailed, 1-seeded cocci, elastically curling up and separating from the axis, to which their tails remain attached. Cotyledons plicate or convolute.

A large genus in Britain, and all temperate regions of the world. The roots are astringent. All the New Zealand species are biennial- or perennial-rooted. The species are most puzzling to discriminate.

Peduncles 2-flowered. Sepals awned. Carpels hairy. Seeds pitted 1. *G. dissectum.*
Peduncles 1-flowered. Sepals hardly awned. Carpels hairy. Seeds scarcely dotted . . . . 2. *G. microphyllum.*

Peduncles 2-flowered. Sepals awned. Carpels hairy. Seeds quite even. (Stemless.) . . . . 3. *G. sessiliflorum.*
Peduncles 2-flowered. Sepals not awned. Carpels glabrous. Seeds smooth . . . . 4. *G. molle.*

1. **G. dissectum,** *Linn.*, var. **carolinianum,** *Fl. N. Z.* i. 39. Stem 1–2 ft. high, stout, erect or decumbent, branched, more or less covered with spreading, usually retrorse hairs. Leaves 1–2 in. diam., on long petioles, orbicular, deeply cut into few or many, broad or narrow, obtuse or acute lobes; stipules broad, scarious. Peduncles slender, 2-flowered, with ovate, subulate bracts. Flowers very variable in size, $\frac{1}{4}$–$\frac{3}{4}$ in. diam. Sepals hairy, awned. Petals notched or retuse, pink. Carpels and their beaks hairy, even. Seeds deeply and coarsely pitted.

Var. α. *pilosum.* Covered with spreading hairs. Petals often rather large.—*G. pilosum,* Forst.; *G. patagonicum,* Hook. f., Fl. Antarct. ii. 252.

Var. β. *patulum.* Leaves covered with spreading and retrorse hairs. Petals usually small.—*G. retrorsum,* DC. Prodr.; *G. patulum,* Forst. Prodr.

Var. γ. *glabratum.* More glabrous, the leaves 3–5-lobed, with the lobes broader.

Abundant throughout the islands, *Banks and Solander,* etc. A most puzzling plant, occurring in S. America from Canada to Cape Horn. It is described in the N. American Floras as annual or biennial; the roots seem annual in Canadian and United States specimens, but perennial in West N. America, Rocky Mountain, and Chilian ones; the petals also, which are as short as the sepals in the United States form, become larger in the West American. The New Zealand form differs from the European *G. dissectum,* chiefly in the petals being often large, always less deeply notched, and the root always more than annual; but I do not find these distinctions to be constant. The amount and depth of pitting of the seed varies much in different specimens. The species ranges through temperate Australia, and North and South America, in which latter country it has several names.

2. **G. microphyllum,** *Hook. f. Fl. Antarct.* i. 8. *t.* v.—*G. potentilloides,* L'Héritier; Fl. N. Z. i. 40. A very slender, prostrate, straggling, branched plant, 12–15 in. long, more or less covered with silky white, appressed or spreading hairs, often retrorse on the peduncles and pedicels. Leaves orbicular, $\frac{3}{4}$–1 in. broad, usually cut to or below the middle into 5–7 broad, obcuneate, toothed, obtuse lobes; petioles slender; stipules small. Peduncles rarely 2-flowered. Flowers smaller than in *G. dissectum* β, excepting in fine specimens, the sepals with very short awns, and the petals larger in proportion, white or pale-pink, retuse. Carpels even, with short hairs. Seeds minutely dotted.

Mountainous or hilly situations in all the islands, *Banks and Solander,* etc. **Lord Auckland's** group, *J. D. H.* In the N. Z. Flora I referred this to the Tasmanian *G. potentilloides,* which is however referred by Bentham, Mueller, and Archer to *G. pilosum,* Forst. (*dissectum,* Linn.), from which this differs in the much more slender habit, smaller less-lobed leaves, smaller less-pointed sepals, and very obscurely pitted seeds. The varieties *debile* and *microphyllum* of Fl. N. Z. are only forms, connected by too many intermediates to be retained as varieties.

3. **G. sessiliflorum,** *Cav.*;—*G. brevicaule,* Hook.; Fl. N. Z. i. 40. Stemless, or nearly so; root stout; branches, if present, very short, and leaves more or less covered with silky hairs, which are spreading or retrorse on the petioles, pedicels, and sepals. Leaves mostly radical, very numerous, on long, slender petioles, 3–6 in. long, orbicular, cut to or below the middle into cuneate 5–7-fid lobes; stipules broad, membranous. Peduncles short, from the root, or from branches which are rarely longer than the leaves. Flowers between those of *G. dissectum* β, and *microphyllum.* Sepals awned, very hairy.

Petals retuse, longer than the sepals. Carpels even, pilose. Seeds quite even or minutely punctulate, not pitted.—Fl. Antarct. ii. 252.

Mountain regions in the **Middle** Island. Terraces on the Kowai river, *Sinclair and Haast*; Chalky Bay, *Lyall*; Otago, *Hector and Buchanan*. Also found in Tasmania, on the Australian alps, in Fuegia, and South Chili; it scarcely differs from *G. dissectum a*, except in habit and the even seeds. Colenso's Ruahine mountain specimens are, I think, referable to *G. microphyllum*.

4. **G. molle,** *Linn.*;—*Fl. N. Z.* i. 40. A procumbent, much-branched, slender plant, covered everywhere with soft spreading hairs. Leaves 1 in. broad, orbicular or reniform, more or less 5–9-lobed, the segments 3–5-fid; petioles long; stipules very broad and membranous. Peduncles 2-flowered, with broad membranous bracts. Flowers $\frac{1}{4}$–$\frac{1}{3}$ in. broad. Sepals broadly ovate, not awned. Petals bifid, longer than the sepals, pink. Carpels wrinkled, quite glabrous. Seeds even, not dotted.

Var. *β*? Carpels not at all wrinkled.

Northern and **Middle** Islands, *Lyall* (without habitat); *β*, Hawke's Bay, *Colenso*. I am much puzzled with this plant, which is not a native of any part of Australia or of America, though common in Europe, and extending east to the Himalaya. Lyall's specimens are identical with the European, and quite characteristic of the commonest state of *G. molle* in every particular; but all Colenso's have quite even carpels, as in the European *G. pusillum*, Linn., and *G. rotundifolium*, Linn., from both of which they differ in their glabrous carpels, and from the latter also in the bifid petals.

2. PELARGONIUM, L'Héritier.

Herbs (rarely shrubs), with opposite, simple lobed or dissected, stipulate leaves, and 1- or many-flowered axillary peduncles.—Flowers slightly irregular. Sepals 5, the upper produced into a spur which is adnate with the pedicel. Petals 5 or fewer, the upper often dissimilar. Torus with 5 glands. Stamens 10, of which 7 only or fewer are antheriferous. Ovary and fruit as in *Geranium*.

A very large South African genus, rare elsewhere, and almost absent in Europe, Asia, and America; a few are Australian.

1. **P. australe,** *Willd.*, var. **clandestinum**;—*P. clandestinum*, L'Hér.; Fl. N. Z. i. 41. A more or less hairy, perennial herb; stem erect, simple or branched, 4 in.–2 feet high. Leaves orbicular or ovate, deeply cordate at the base, 3–5-lobed, lobes coarsely or finely toothed or serrate; petioles slender, 2–6 in. long. Peduncles axillary, longer than the leaves, pubescent, many-flowered. Flowers small, $\frac{1}{2}$ in. broad, in 10–12-flowered umbels, with whorled, ovate, acuminate bracts at the base; pedicels $\frac{1}{6}$–$\frac{1}{2}$ in. long, pubescent with scattered white hairs. Sepals unequal, ovate, acuminate; spur short, gibbous, or 0. Petals unequal, $\frac{1}{12}$–$\frac{1}{6}$ in. long, deep pink, longer than the sepals, spathulate, notched. Stamens about 5 fertile, the rest reduced to membranous scales. Carpels very hairy, their tails lined on the inner face with white silky hairs. Seeds minutely dotted.—*P. acugnaticum*, Thouars; *P. grossularioides*, Aiton; Harv. and Sond. Flor Cap. i. 289.

Northern and **Middle** Islands: abundant, especially near the sea, *Banks and Solander*, etc. A lotion of bruised leaves of this is applied by the natives to scalds and burns. Also found in Tristan d'Acunha, South Africa and Australia.

3. OXALIS, Linn.

Stemless or branching, usually slender, perennial, acid herbs, with alternate or tufted, usually stipulate, 3-foliolate or pinnate leaves, and obcordate leaflets.—Flowers regular. Sepals 5, imbricate. Petals 5, contorted. Disk and glands 0. Stamens 10, all fertile. Ovary 5-lobed, 5-celled; styles 5; cells 1- or more ovuled. Capsule loculicidal; valves more or less cohering together and by the septa to the axis. Seeds with an arilliform, fleshy coat, that bursts elastically; albumen fleshy.

A very large genus, abounding in South Africa, of which a few species are British, and very widely diffused in both tropical and temperate countries. The foliage affords a grateful acid.

Stems elongate, branching. Stipules 0 or small. Flowers yellow . . 1. *O. corniculata.*
Stems short or 0. Stipules large. Flowers white . . . . 2. *O. magellanica.*

1. **O. corniculata,** *Linn.;—Fl. N. Z.* i. 43. An erect or decumbent, branched, slender, glabrous or pubescent herb, usually with a perennial root; branches 1–10 in. long, erect, ascending, or more usually prostrate, sometimes matted together. Leaves 3-foliolate; leaflets deeply obcordate, $\frac{1}{8}$–1 in. long, glaucous below; stipules small or 0; petioles slender, very variable. Peduncles axillary, 1–6-flowered, variable in length. Flowers most variable in size. Petals yellow, notched. Capsules oblong or linear.

Var. *α*. Decumbent. Leaves stipulate.

Var. *β. stricta.* Erect or suberect. Leaves exstipulate.—*O. stricta,* Linn.; *O. Urvillei, lacicola, propinqua,* and *divergens,* A. Cunn. Prodr.; *O. ambigua,* A. Rich. Flor.

Var. *γ. microphylla.* Stems procumbent. Branches creeping. Leaflets very minute. Capsule oblong.—*O. microphylla* and *O. exilis,* A. Cunn. Prodr.

Var. *δ. ciliifera.* Stems procumbent, filiform. Leaflets membranous, ciliated.—*O. ciliifera* and *tenuicaulis,* A. Cunn. Prodr.

Var. *ε. crassifolia.* Stems rigid, matted together. Leaflets small, thick, pilose.—*O. crassifolia,* A. Cunn. Prodr.

Abundant, especially on the east coasts, throughout the islands, *Banks and Solander,* etc. One of the commonest and most variable weeds of warm climates.

2. **O. magellanica,** *Forst.;—Fl. N. Z.* i. 43. *t.* 13. A small, glabrous or pubescent, stemless, rather succulent species, 3–4 in. high. Rhizome creeping, perennial, covered with imbricate scales (old stipules). Leaves 3-foliolate; leaflets broadly obcordate, glabrous, glaucous below. Petioles usually hairy. Peduncles radical, 1-flowered, 2-bracteolate, often longer than the leaves. Sepals ovate, obtuse. Petals pure white, $\frac{1}{4}$–$\frac{1}{3}$ in. long, oblong-obovate or obcordate, often oblique and ciliated. Capsules globose.—*O. Cataractæ,* A. Cunn. Prodr.; Hook. Ic. Pl. t. 418.

Throughout the islands, in damp, shady, alpine, and subalpine regions. Common in similar situations in Tasmania, Fuegia, and S. Chili, very nearly allied to the European Wood Sorrel (*O. Acetosella,* Linn.).

Erodium cicutarium, Linn., a common European hairy pinnate-leaved geranium-like plant, seems now to be naturalized throughout the islands (as in Australia); it has regular flowers, 5 anthers, and the tails of the carpels bearded inside.

Order XIV. RUTACEÆ.

Shrubs and trees. Leaves opposite or alternate, exstipulate, pellucid-dotted, usually fragrant, simple or compound. Flowers regular, hermaphrodite.—Sepals and petals 4 or 5, spreading, imbricate or valvate. Stamens usually 8 or 10, free, inserted at the base of a tumid disk; anthers versatile. Ovary of 4 or 5, more or less united carpels; styles 4 or 5, wholly combined, or by a capitate stigma only; cells 2-ovuled. Fruit capsular, of 4 or 5 coriaceous, 2-valved, 1-seeded cocci; outer coat separating from the inner, which is chartaceous, dry, and elastic. Seeds oblong, testa crustaceous; albumen copious or 0.

A very large Natural Order, now including the Rue, Orange, Cape Diosmas, Australian Boronias, and many other plants differing in certain points from the above ordinal character, abounding in the temperate regions of the southern hemispheres, especially in South Africa and Australia, rarer elsewhere.

Flowers pentamerous. Leaves simple . . . . 1. PHEBALIUM
Flowers tetramerous. Leaves compound . . . . 2. MELICOPE.

1. PHEBALIUM, Ventenat.

Shrubs. Leaves alternate, pellucid-dotted. Flowers in axillary or terminal corymbs, white.—Calyx small, 4 or 5-lobed or -parted. Petals 4 or 5, imbricate or valvate. Stamens 8 or 10; filaments filiform, glabrous. Ovary 2–5-parted, almost to the base; style simple, rising from between the lobes, stigma capitate; cells 2-ovuled. Cocci 2–5, truncate or rostrate; endocarp separating. Testa smooth, black, shining.

A very large Australian genus, not found elsewhere.

1. **P. nudum,** *Hook. Ic. Pl. t.* 568;—*Fl. N. Z.* i. 44. A shrub or small tree, 12–15 ft. high, everywhere glabrous; branches very slender. Leaves alternate, 1–1½ in. long, spreading, linear-oblong or narrow oblong-lanceolate, obtuse, obscurely crenate, narrowed below into very short petioles, coriaceous, paler and dotted below. Flowers in terminal many-flowered corymbs, whitish, ¼–⅓ in. diam., on pedicels ⅙–¼ in. long. Calyx very small, 5-lobed. Petals 5, linear, obtuse, with narrowly overlapping margins. Cocci ⅛ in. long, obtusely rhomboidal, compressed, wrinkled, often only one ripens, splitting down the front and back into 2 valves.

Northern Island: Bay of Islands, *A. Cunningham;* east coast, *Edgerley, Colenso;* Auckland, *Sinclair.* Exceedingly closely allied to the Queensland *P elatius*, F. Muell., but the flowers are larger, the petals longer, the corymbs more flattened, and the leaves taper less to the base; they may prove to be local forms of one species.

2. MELICOPE, Forst.

Glabrous shrubs. Leaves opposite or alternate, simple or compound, pellucid-dotted. Flowers in axillary many- or few-flowered cymes, more or less unisexual.—Sepals 4, deciduous or persistent. Petals 4, sessile, spreading, valvate or imbricate, with inflexed tips. Stamens 8; filaments subulate. Ovary 4-lobed, 4-celled; style 1, or 4 coalescing into 1, from between the lobes, stigma capitate; cells 2-ovuled. Cocci 1–4, spreading, free; endocarp separating. Testa shining; albumen abundant; embryo slightly curved.

A Pacific Island genus, very variable in habit, which does not extend westward into Australia. The two New Zealand species are extremely dissimilar in habit.

Leaves opposite, 3-foliolate . . . . 1. *M ternata.*
Leaves alternate or fascicled, 1-foliolate . . . . 2. *M. simplex.*

1. **M. ternata,** *Forst.*;—*Fl. N. Z.* i. 43. A glabrous, small tree, 12–15 ft. high; branches rather stout. Leaves opposite, 3-foliolate; leaflets 3–4 in. long, ovate or linear-oblong, acute, quite entire, longer than the petioles. Flowers $\frac{1}{3}$ in. diam., greenish-white, in peduncled, trichotomous, axillary cymes; bracts deciduous. Petals ovate, longer than the stamens, margins imbricate. Ovary quite glabrous; style short. Carpels 4, spreading, coriaceous, strongly wrinkled; seed small, black, projecting from the fissure, attached by a slender funicle.—Hook. Ic. Pl. t. 603.

Northern and **Middle** Islands, *Banks and Solander*, etc.: not uncommon as far south as the Nelson Province.

2. **M. simplex,** *A. Cunn.*;—*Fl. N. Z.* i. 43. A glabrous shrub, 6–8 ft. high; branches slender. Leaves alternate, scattered or fascicled, small, of 1 (very rarely 3, of which the lateral are minute), orbicular obovate or ovate, obtuse, doubly crenate leaflet, $\frac{1}{2}$–$\frac{3}{4}$ in. long, jointed on the top of a flattened almost winged petiole, $\frac{1}{4}$–$\frac{1}{2}$ in. long, which is broader towards the leaflet and channelled above. Pedicels several together, axillary, slender, longer than the petioles, 1–4-flowered, bracteolate at the forks. Flowers very small, greenish-white. Petals linear-oblong, shorter in the male flowers than the stamens, valvate, or with the edges a little overlapping. Ovary oblong, hirsute; style in the female flowers slender; stigma capitate. Fruit as in *M. ternata*, but much smaller.—Hook. Ic. Pl. t. 585.

Northern and **Middle** Islands, *Banks and Solander*, etc.: not uncommon along the whole coasts to Otago. A very different-looking plant from *M. ternata*.

ORDER XV. MELIACEÆ.

Trees (rarely shrubs). Leaves usually pinnate, alternate, exstipulate. Flowers regular, generally hermaphrodite.—Calyx small, 4 or 5-lobed or -parted, imbricate. Petals 4 or 5, contorted valvate or imbricate, often long, sometimes united to the staminal tube. Disk free or wanting, sometimes tubular within the stamens. Stamens usually 8 or 10, more or less united into a tube, bearing the sessile anthers within its mouth, which is often toothed or split. Ovary free, 3–5-celled; style single, terminal, stigma capitate; cells 2-ovuled. Fruit a drupe berry or capsule, usually the latter, coriaceous, 3-celled, loculicidally 3-valved. Seeds generally solitary in the cells, mostly arillate and exalbuminous.

A very large Order of tropical forest-trees, of which various genera present characters of the flower and fruit at variance with the above description. The Mahogany (*Swietenia*), Satin-wood (*Chloroxylon*), and Pride-of-India (*Melia*), all belong to it.

1. DYSOXYLUM, Blume.

Large forest-trees, often fetid or with a garlicky smell. Leaves alternate, pinnate. Flowers rather small, in axillary panicles.—Calyx short, 4 or 5-lobed -toothed or -parted, imbricate. Petals 4 or 5, linear-oblong, valvate, sometimes united at the base and with the staminal tube, which is cylindric, 8 or 10-toothed; anthers 8 or 10, included. Disk tubular, sheathing the 3–5-celled ovary. Capsule coriaceous, globose or pyriform, 1–5-celled, 2–5-valved, splitting down the middle of the cells; valves with the septa on their faces. Seeds large, arillate or naked, oblong; hilum broad, ventral; testa brown, shining; cotyledons very large; plumule included or exserted.

A large tropical Asiatic, Australian, and Pacific Island genus of timber and forest trees.

1. **D. spectabile,** *Hook.f.;—Hartighsia spectabilis,* A. Juss.;—Fl. N. Z. i. 39. A large tree, 40–50 ft. high. Leaves 1 ft. or more long, pinnate; leaflets about 4 pairs, alternate, petioled, oblong-obovate, acute, 3-6 in. long, quite glabrous, entire, oblique at the base, narrowed into the terete petiole. Panicles 8–12 in. long, usually growing from the trunk, sparingly branched, ebracteate. Flowers shortly pedicelled, $\frac{1}{2}$ in. broad. Calyx lobes very small, ciliate. Petals linear, patent, obtuse. Staminal tube cylindric, fleshy, crenate; anthers quite included, sessile on thickened prominences. Style very slender; stigma disciform. Capsule obovate, pendulous, 1 in. long. Aril scarlet.—Hook. Ic. Pl. t. 616 and 617.

Northern Island: Bay of Islands, *A. Cunningham,* etc.; east coast, *Banks and Solander.* **Middle** Island, *Forster.* Mr. Bidwill observes that the leaves are bitter, and used for hops, and to make a stomachic infusion.

Order XVI. OLACINEÆ.

Shrubs or trees. Leaves simple, alternate, rarely opposite, exstipulate. Flowers regular, axillary or terminal, hermaphrodite or unisexual.—Calyx small, 4 or 5-toothed -lobed or -parted. Petals 4 or 5, free or coalescing into a tube, valvate. Stamens 4 or 5, hypogynous (or at the base of a disk). Ovary free, 1- or imperfectly 3–5-celled; style long or short, stigma often lobed; ovules 1–3 collaterally pendulous from below the top of the cell, or from the top of a central erect free placenta. Fruit usually a 1-celled, 1-seeded, dry or fleshy drupe. Seed pendulous, testa very thin; albumen copious, fleshy; embryo minute, terete.

A large tropical Order, containing many genera of very various characters not included in the above description, often extremely difficult of determination, on account of their minute flowers, and almost microscopic ovaries and ovules. The only New Zealand genus is also a native of Norfolk Island.

1. PENNANTIA, Forst.

Trees. Leaves alternate, entire or toothed. Cymes many-flowered. Flowers polygamous.—Calyx minute, 5-toothed. Petals 5, valvate. Stamens 5, hypogynous; filaments filiform, free, flattened at the base; anthers versatile. Ovary oblong, obscurely trigonous, 1-celled; stigma almost sessile, discoid, 3-lobed; ovule 1, pendulous below the top of the cell. Drupe small, fleshy; stone crustaceous, obtusely 3-gonous, grooved on one face, and perforated on that face below the apex; a flattened cord passes up the groove, enters the cell by the foramen, and bears the pendulous seed at its tip.

1. **P. corymbosa,** *Forst.;—Fl. N. Z.* i. 35. *t.* 12. A small, very graceful tree, covered with white sweet-smelling flowers, 20–30 ft. high; bark whitish; wood brittle; twigs and young cymes pubescent. Leaves on short petioles, 1–3 in. long, ovate obovate or oblong, obtuse, sinuate or toothed,

rarely entire, often turning black in drying. Male flowers largest, filaments longer than the petals; ovary reduced to a papilla; pedicels jointed below the calyx. Berries ovoid, black, fleshy, with purple juice, $\frac{1}{3}$ in. long.

Northern Island: chiefly in mountain woods; more abundant throughout the **Middle** Island, *Banks and Solander*, etc. Wood used by the natives for kindling fires by friction. Development of the fruit very curious, and well worth an attentive study. The only other species (*P. Endlicheri*) is a native of Norfolk Island.

ORDER XVII. STACKHOUSIEÆ.

Herbs, perennial-rooted. Leaves narrow, alternate, almost exstipulate. Flowers in terminal spikes or racemes, greenish white or yellow, hermaphrodite, regular.—Calyx with a small hemispherical tube, and 5 small imbricate lobes. Petals 5, inserted at the edge of a disk which lines the calyx tube, erect, linear or spathulate, free or united by their edges above the base only, their tips imbricate, reflexed. Stamens 5, free, erect; filaments slender, 2 shorter than the others. Ovary sessile, free, subglobose, 2–5-celled and -lobed or -parted; styles 2–5, connate or free, stigma simple or 2–5-lobed; ovules solitary and erect in the cells. Fruit of 2-5 indehiscent, globose, angled or winged cocci, attached to a central column. Testa membranous; albumen fleshy; embryo straight.

The only genus is abundant in Australia, and contains also a Philippine Island species.

1. STACKHOUSIA, Smith.

1. **S. minima,** *Hook. f. Fl. N. Z.* i. 47. A minute, slender, glabrous herb, with slender, running rhizomes, sending up erect, leafy branches, 1–2 inches high. Leaves $\frac{1}{6}$–$\frac{1}{4}$ in. long, scattered, linear or obovate, acute, fleshy, quite entire. Flowers very minute, in few-flowered spikes. Calyx lobes 5, spreading. Petals united at the middle. Anthers pubescent. Ovary 3-lobed; style 3-cleft. Cocci, usually 1 only ripens.

Northern Island: open downs on the east coast, *Colenso*. **Middle** Island: clefts of rocks, Ribbon Wood range, *Haast*.

ORDER XVIII. RHAMNEÆ.

Trees or shrubs. Leaves alternate, rarely opposite, stipulate. Flowers regular, hermaphrodite.—Calyx superior or inferior, 4 or 5-toothed or -lobed; lobes triangular, valvate, often having a raised ridge down the centre, and an incurved thickened tip. Petals 0, or 4 or 5, minute, scale-like, very concave, placed between the teeth of the calyx, and often smaller than them. Stamens 4 or 5, very small, inserted with the petals, opposite to, and often hooded by them. Disk hypogynous or epigynous. Ovary superior or inferior, 3-celled; style 1, stigma capitate or 3-fid; ovule solitary and erect in each cell. Fruit of 3 cocci, either free and subtended by the calyx, or more or less immersed in or adnate to the calyx; cocci often crustaceous, lenticular, dehiscing down the inner face. Seed erect; albumen fleshy; embryo large, cotyledons orbicular, radicle straight, terete.

A large temperate and tropical Order, of which the New Zealand genera are Australian.

Ovary inferior. Tomentose shrubs. Leaves alternate . . . . 1. POMADERRIS.
Ovary superior. Spiny, glabrous bush. Leaves 0 or opposite . . . . . 2. DISCARIA.

1. POMADERRIS, Labill.

Shrubs, more or less covered with stellate down. Leaves persistent, alternate. Flowers small, in umbellate cymes, usually pedicelled; bracts deciduous.—Calyx tube adnate with the ovary, limb 5-lobed, deciduous or reflexed. Petals 5 or 0. Stamens 5, filaments longer than the petals; anthers free. Disk epigynous. Ovary more or less inferior; style short, trifid. Capsule small, its 3-valved tip free, containing 3 plano-convex cocci, which split down the face, or open by the falling away of an oblong lid.

A considerable genus in Australia, but not found elsewhere, except in New Zealand.

Leaves 2–3 in., elliptic. Cymes many-flowered . . . . 1. *P. elliptica.*
Leaves ¾ in., oblong. Raceme many-flowered . . . . 2. *P. Edgerleyi.*
Leaves ¼–⅓ in., linear. Cymes few-flowered . . . . 3. *P. phylicifolia.*

1. **P. elliptica,** *Labill.;—Fl. N. Z.* i. 46. A branching shrub or small tree, covered, except on the upper surface of the leaf, with stellate, white or grey down. Leaves 2–3 in. long, elliptic-oblong, obtuse at both ends, or acute at the tip, quite entire, white below; petioles ⅓ in. long. Cymes terminal, much branched, very many-flowered, 2–6 in. diam. Flowers fragrant. Calyx tomentose, and covered with silky hairs. Petals with crisped margins.—*P. Kumeraho,* A. Cunn. Prodr.

Northern Island: abundant on dry hills, *Banks and Solander,* etc. The same plant is very common in temperate eastern Australia and Tasmania.

2. **P. Edgerleyi,** *Hook. f., n. sp.* A small bush, more or less covered with yellow stellate pubescence, and ferruginous flocculent tomentum, especially on the young branches. Leaves ¾ in. long, shortly petioled, oblong, obtuse at both ends, above smooth or scabrid, and covered with deeply impressed veins, below with soft, white or red tomentum, and very prominent veins. Racemes lax, few-flowered, flowers imperfect in my specimens.—*Pomaderris,* n. sp.?, Fl. N. Z. i. 46.

Northern Island: hills south of Wangarei harbour, and Coromandel harbour, *Edgerley, Sinclair, Jolliffe.* Apparently a very rare and local plant, extremely nearly allied to the Australian *P. betulina.*

3. **P. phylicifolia,** *Lodd.;—P. ericifolia,* Hook.; Fl. N. Z. i. 46. A small, erect, fastigiately-branched, heath-like downy shrub, the branches villous. Leaves small, ¼–⅓ in. long, spreading, linear-oblong, coriaceous, the margins rolled back to the midrib, obtuse, grooved down the middle above, rather scabrid above with short white hairs. Flowers numerous, in axillary few-flowered cymes, scarcely longer than the leaves, apetalous.—Lodd. Bot. Cab. t. 120.

Northern Island: abundant on dry hills, *Banks and Solander,* etc. A common plant in Tasmania.

2. DISCARIA, Hook.

Very much branched, glabrous, usually rigid, tortuous, spiny shrubs or small trees, often leafless. Branches decussately opposite, terete, green, transversely grooved or articulate at the nodes. Flowers axillary, pedicelled, nodding, small, green.—Calyx membranous, inferior or adnate at the base with the ovary, with obconic or campanulate limb and 4 or 5 recurved lobes. Petals 0 or 4 or 5. Stamens 4 or 5. Disk adnate to the base of the calyx, annular, entire or lobed. Ovary free or sunk in the base of the calyx, subglobose, 3-lobed; style slender, stigma 3-lobed. Drupe dry, coriaceous, of 3 cocci, capsular when ripe, the cocci separating and splitting down their faces.

A considerable genus in South America, of which one species is a native of Australia, and another of New Zealand.

1. **D. Toumatou,** *Raoul, Choix*, 29. *t.* 29;—*D. australis*, Hook., var. *apetala*, Fl. N. Z. i. 47. A thorny bush in dry places, becoming a small tree in damper localities, with spreading branches and branchlets reduced to spines 1–2 in. long. Leaves small, $\frac{1}{2}$–$\frac{3}{8}$ in. long, fascicled in the axils of the spines, absent in old plants, linear- or obovate-oblong, obtuse or retuse, smooth or pubescent, quite entire or serrate. Flower $\frac{1}{8}$ in. diam., apetalous, white; pedicels and calyx minutely downy. Calyx-tube short, obscure; lobes 4 or 5, broadly ovate. Disk broad, with a narrow upturned edge. Capsule size of a peppercorn.—*Notophæna Toumatou,* Miers, Contrib. 272.

East coast and interior of the southern part of the **Northern,** and throughout the **Middle** Island. I have retained this as a distinct species from the Australian plant, relying on the absence of petals and the minute pubescence on the pedicels and flowers. The spines were used in tatooing (*Raoul*). "Wild Irishman" of settlers.

ORDER XIX. SAPINDACEÆ.

Trees, rarely shrubs. Leaves alternate, rarely opposite, simple or compound, exstipulate. Flowers regular or irregular, uni- or bi-sexual.—Sepals 3–5, imbricate or valvate. Petals 0 or 3–5, generally small, often with a scale on their inner face. Disk 0, or complete or incomplete. Stamens 5–8, hypogynous or inserted within the disk. Ovary entire or lobed, usually 2–3-celled; style simple, stigma 2–3-lobed; ovules 1–2 in each cell, fixed to its axis. Fruit very various. Seeds usually exalbuminous, with large, solid, or spirally-twisted cotyledons and incurved radicle.

A very large and complicated Order of tropical and temperate plants, of which some genera present characters not noticed in the above description.

Leaves simple (in the New Zealand species). Disk 0 . . . . 1. DODONÆA.
Leaves pinnate. Disk 8-lobed . . . . 2. ALECTRYON.

1. DODONÆA, Linn.

Shrubs or trees, often covered with a viscid exudation. Leaves simple (or pinnate), exstipulate. Flowers unisexual or polygamous, apetalous, regular. —Sepals 3–5, imbricate or valvate. Male fl.: Disk 0. Stamens 5–8; filaments very short; anthers linear-oblong, 4-angled. Female fl.: Ovary sessile, 3–6-angled; cells 2-ovuled. Capsule membranous or coriaceous, septicidally 3–6-valved; valves broadly winged at the back. Cotyledons spiral.

A very large Australian genus, of which a very few species (including the New Zealand one) are scattered widely over the warmer regions of the globe.

1. **D. viscosa,** *Forst.*;—*Fl. N. Z.* i. 38. A small glabrous tree, 6–12 ft. high, with very hard wood, variegated black and white, and compressed, viscid young shoots. Leaves 2–3 in. long, on short petioles, linear-obovate, obtuse acute or retuse, quite entire, membranous, veined. Flowers small, in terminal, erect, few-flowered panicles. Sepals ovate, subacute. Anthers large, almost sessile. Fruit $\frac{1}{3}$ in. long, orbicular, 2-lobed at both ends, on slender pedicels, 2–3-valved; valves with broad, oblong, membranous veined wings. Seeds with dark red-brown testa.—*D. spathulata*, Smith.

Abundant in dry woods, etc., *Banks and Solander*, etc. Wood used for native clubs; also a native of Australia, the Pacific, and the tropics of the Old and New World.

2. ALECTRYON, Gærtner.

A lofty tree, with tomentose branchlets. Leaves alternate, unequally pinnate, exstipulate. Panicles branched, axillary and terminal, many-flowered. Flowers small, almost regular, unisexual.—Calyx 4 or 5-lobed; lobes villous within, rather unequal, imbricate. Petals 0. Disk small, 8-lobed. Stamens 5–8, inserted between the lobes of the disk; anthers large, almost sessile. Ovary obliquely obcordate, compressed, 1-celled; style short, stigma simple or 2–3-fid; cell 1-ovuled. Fruit gibbous, pubescent, tumid or globose, with a compressed prominence on one side, rather woody, indehiscent. Seed subglobose, arillate; cotyledons spirally twisted.

1. **A. excelsum,** *DC.*;—*Fl. N. Z.* i. 37. Leaves 4–10 in. long; leaflets alternate, the young ones lobed and cut, serrate in young plants, petiolate, 2–3 in. long, obliquely ovate-lanceolate, acuminate, obscurely crenate, tomentose below, as are the petioles. Panicles 6–8 in. long, much-branched, branches stout and spreading, densely tomentose, as are the flowers. Calyx pilose. Anthers deep-red. Ovary hairy, hidden by the copious hairs at the base of the calyx. Capsule $\frac{1}{3}$ in. long. Seed globose, in a large scarlet aril.—Hook. Ic. Pl. t. 570.

Northern and **Middle** Islands: in forests not uncommon, *Banks and Solander*, etc. The only species of the genus, and confined to New Zealand. The oil of the seeds was used for anointing the person, according to Cunningham.

ORDER XX. ANACARDIACEÆ.

Trees. Leaves alternate, simple or compound, exstipulate. Flowers regular, usually small, unisexual or hermaphrodite.—Calyx 3–7-fid or -partite. Petals 3–7 or 0. Disk usually annular. Stamens as many as the petals, alternating with staminodia, or twice as many, inserted on or at the base of the disk; filaments free; anthers versatile. Ovary free, in the female flower 1- or 2–5-celled; in the male often of 4 imperfect carpels; styles 1–3; ovule solitary, pendulous from a basal funicle, or from the wall or top of the cell. Fruit superior, usually 1–5-celled; drupe with hard putamen. Seed exalbuminous; cotyledons thick, fleshy; radicle short.

A very large tropical Order, rarer in temperate climates. The only New Zealand genus is endemic, and allied to the Mango.

1. CORYNOCARPUS, Forst.

A small, perfectly glabrous tree. Leaves alternate, broad, bright-green, entire, coriaceous. Flowers in branched terminal panicles, small, green.—Calyx 5-lobed; lobes rounded, imbricate. Petals 5, perigynous, rounded, concave, erect, imbricate, jagged. Disk fleshy, 5-lobed. Stamens 5, inserted between the lobes of the disk, alternating with 5 petaloid jagged scales. Ovary sessile, ovoid, 1-celled, narrowed into an erect style, stigma capitate; ovule pendulous below the top of the cell. Drupe obovoid, fleshy; endocarp coriaceous and fibrous. Seed pendulous; testa membranous, adherent to the cell; cotyledons plano-convex, radicle minute.

1. **C. lævigata,** *Forst.;—Fl. N. Z.* i. 48. A leafy tree, 40 ft. high. Leaves 4–7 in. long, oblong or lanceolate, subacute, shining, on short, stout petioles. Panicles erect, thyrsoid, spreading, 4 in. long. Flowers small, globose, $\frac{1}{6}$ in. diam., greenish-white, inodorous, on short, stout pedicels. Petals as long as the calyx-lobes, concave. Filaments stout, subulate. Ovary small, glabrous. Drupe 1 in. long, oblong.—Bot. Mag. t. 4379.

Northern and northern part of the **Middle** Islands: abundant near the sea, *Banks and Solander*, etc.; **Chatham** Island, *Dieffenbach*. Pulp of the drupe eatable; embryo considered poisonous till steeped in salt water.

ORDER XXI. CORIARIEÆ.

Sarmentose, glabrous, leafy undershrubs, with angular branches. Leaves opposite or rarely 3-nate, exstipulate, entire. Flowers hermaphrodite, solitary or racemed, axillary, regular.—Sepals 5, triangular-ovate, imbricate, persistent. Petals 5, hypogynous, triangular, shorter than the sepals, after flowering becoming fleshy and closely appressed to the carpels. Stamens 10, hypogynous, all free, or 5 of them adnate to the petals; filaments short; anthers large, rough. Carpels 5 or 10, whorled round and adnate to a fleshy torus, 1-celled; styles 5 or 10, free, flexuose, stigmatiferous all over; ovules solitary, pendulous. Fruit of 5–8, small, indehiscent, compressed, oblong, crustaceous achenes, keeled on the back and sides, enclosed in the fleshy petals. Testa membranous; albumen thin; cotyledons plano-convex; radicle very short, superior.

The only genus (of dubious affinity) is a native of southern Europe, the Himalaya, and the South American Andes; the New Zealand species and their forms are all similar to what are found in the latter country. The root of the European species is extensively used for tanning leather in Russia.

1. CORIARIA, Linn.

Stems perennial, shrubby. Leaves 1–3 in. long . . . . 1. *C. ruscifolia.*
Stems annual? Leaves ovate or lanceolate, $\frac{1}{4}$–1 in. long . . . . 2. *C. thymifolia.*
Stems annual. Leaves narrow-linear, lanceolate, $\frac{1}{4}$ in. . . . . . . 3. *C. angustissima.*

1. **C. ruscifolia,** *Linn.;—Fl. N. Z.* i. 45. A perennial shrub 10–18 ft. high; trunk 6–8 in. diam.; branches often long and flexuose. Leaves oblong or obovate, acuminate or acute, 3–5-nerved, sessile or shortly petioled. Racemes 8–12 in. long, drooping, many-flowered, pubescent; pedicels $\frac{1}{3}$ in.,

slender, bracteolate at the base. Flowers $\frac{1}{6}$–$\frac{1}{8}$ in. diam., green; anthers in some flowers imperfect. Petals full of purple juice when the fruit is ripe.—*C. sarmentosa*, Forst.; Hook. Bot. Mag. t. 2470.

Abundant throughout the islands, *Banks and Solander*, etc. **Kermadec** Island, *M'Gillivray*. The so-called berries (fleshy petals) of this and the following species vary much in succulence, the less juicy bearing seeds which, according to Colenso, are not poisonous. The juice is purple, and affords a grateful beverage to the natives; and a wine, like elderberry wine, has been made from it. The seeds alone are said by some (the whole plant by others) to produce convulsions, delirium, and death, like those of the European *C. myrtifolia*. This species is abundant in Chili. The Tua-tutu of Otago.

2. **C. thymifolia,** *Humb.;—Fl. N. Z.* i. 45. A much smaller, annual?, usually more pubescent plant than *C. ruscifolia*, sometimes only a foot or so high, with small, ovate-lanceolate leaves, shorter racemes, and smaller flowers; large forms of it however seem to be connected with smaller ones of the former, both in New Zealand and America.

Common in various dry places throughout the islands, ascending to 5000 ft: **Kermadec** Island, *M'Gillivray*. In America it ranges from Mexico to Peru, at elevations of 4–12,000 ft. The small ground Tutu of Otago.

3. **C. angustissima,** *Hook. f. n. sp.* A small bright green annual species, 6–18 in. high, with the habit of *C. thymifolia*, but the branches are glabrous, very slender, more dense, and the leaves are very narrow linear-lanceolate or subulate, $\frac{1}{4}$ in. long.

Northern Island: Mount Egmont, *Dieffenbach;* top of the Ruahine range, *Colenso*. **Middle** Island: abundant in subalpine localities from Nelson, *Sinclair*, to Otago, *Hector and Buchanan*. I think this certainly passes into *C. thymifolia*, but all my New Zealand correspondents regard it as a distinct species. The annual herbaceous Tutu of Otago.

ORDER XXII. LEGUMINOSÆ.

Tribe PAPILIONACEÆ.

Herbs, shrubs, or trees. Leaves alternate (rarely opposite), stipulate, mostly compound; leaflets usually entire. Flowers irregular, hermaphrodite.—Calyx 5-toothed or -cleft, or 4-toothed by the union of the upper lobes. Corolla papilionaceous, *i. e.* of 5 petals, the upper (standard) broadest, outside in bud, often reflexed; the 2 next lateral (wings), vertical and parallel; the 2 lowest also vertical and parallel within the wings (forming the keel), often combined by their lower edges. Stamens 10, 9 usually united into a membranous tube sheathing the ovary, the upper free, rarely all free. Ovary sessile or stipitate, generally long, flattened, 1-celled, tapering into a straight or upturned style; stigma small, simple, lateral or capitate. Fruit a 1- or more seeded *Legume*, splitting into 2 valves in all the N. Z genera, except *Carmichælia*. Seeds exalbuminous; cotyledons thick, plano-convex; radicle short.

A very large tribe, belonging to one of the largest Natural Orders on the globe, which includes the Pea, Bean, and all Legumes proper. The Order abounds in Australia, forming there a great proportion of the indigenous vegetation, herbaceous shrubby and arboreous, but is less developed in New Zealand than in any other part of the world, temperate or tropical.

Shrubs. Leafless, or leaflets in 1–3 pairs. Flowers small . . . 1. CARMICHÆLIA.
Tree, with pendulous chord-like branchlets, leafless. Flowers pink . 2. NOTOSPARTIUM.

Herb, with leaflets in several pairs. Flowers blue . . . . . 3. SWAINSONIA.
Herbaceous. Leaflets in many pairs. Flowers large, scarlet . . . 4. CLIANTHUS.
Shrubs or trees. Leaflets in many pairs. Flowers large, yellow . 5. EDWARDSIA.

1. CARMICHÆLIA, Br.

Shrubs or small trees, usually quite leafless, or leafy in a young state only; branches terete or more often flattened, grooved or striate, green. Leaflets in 1–3 pairs, obcordate. Flowers small, pink bluish or white, in small lateral fascicles or racemes, inserted in notches on the edges of the branchlets. —Calyx short, cup-shaped or campanulate, truncate, 5-toothed. Standard orbicular, usually reflexed. Wings more or less falcate, obtuse, auricled at the base. Keel oblong, incurved, obtuse. Upper stamen more or less free from the others. Ovary narrowed into a slender curved style, beardless, stigmatiferous at the very tip; ovules numerous, 2-seriate. Pod small, oblong or orbicular, straight or oblique, with a short or long rigid subulate beak, thick, very coriaceous; valves opening at the tip only, or with the centre falling away from their persistent consolidated edges. Seeds 1–3, oblong or reniform; funicle not thickened, radicle with a double flexure.

A most singular genus, confined to New Zealand, unique as regards the structure of the pod, in which the edges of the valves become consolidated, and their faces either open towards the tip as valves, or fall altogether away. Besides the species here described (which are very difficult of discrimination), I have apparently several others, especially from the Middle Island, which may be new, but not having seen fruit, I hesitate to describe them. Intermediates will probably be found between several of the following.

* *Flowers* $\frac{1}{5}$ to $\frac{1}{3}$ *in. long.*

Branches cylindric, grooved. Flowers subcapitate. Calyx woolly . . 1. *C. crassicaulis.*
Branches compressed. Flowers in 3–5-flowered, lax, pubescent racemes 2. *C. Munroi.*
Dwarf; branches flat. Flowers in lax, 1–3-flowered, nearly glabrous racemes, with long peduncles . . . . 3. *C. nana.*
Branchlets flat or terete, grooved. Flowers 6–8, subcapitate, glabrous. 4. *C. grandiflora.*

** *Flowers less than* $\frac{1}{5}$ *in. long.*

† *Ovary pubescent.*

Branchlets flat, grooved. Flowers 8–12, in erect, silky racemes . . 5. *C. pilosa.*

†† *Ovary quite glabrous.*

Branchlets flat, striate. Flowers in lax, 5–8-flowered, glabrous or slightly pilose racemes. Pod $\frac{1}{3}$ in., with usually a very short beak. 6. *C. australis.*
Branchlets flat or terete. Flowers in erect, 8–12-flowered, very pubescent racemes. Pod $\frac{1}{3}$ in., with a long, subulate beak . . . . 7. *C. odorata.*
Branchlets very slender, flat. Flowers in lax, 3–6-flowered, pilose racemes or fascicles, on slender pedicels. Pods $\frac{1}{4}$ in., with long straight beak 8. *C. flagelliformis.*
Branchlets very slender, terete or flat. Flowers in small, 2–6-flowered, pilose spreading racemes. Pod minute, $\frac{1}{10}$–$\frac{1}{8}$ in. . . . . . . 9. *C. juncea.*

1. **C. crassicaulis,** *Hook. f., n. sp.* Branchlets very robust, cylindric, $\frac{1}{3}$ in. diam., with many deep, parallel, tomentose grooves. Leaves not seen. Flowers in rounded fascicles of 6–12, $\frac{1}{4}$–$\frac{1}{3}$ in. long, shortly pedicelled. Calyx and pedicels densely woolly, lobes of the former ovate obtuse; bracteoles at its base minute. Standard large, reflexed. Ovary densely villous, with white silky hairs. Pod unknown.

Middle Island: Mount Torlesse range, alt. 3500 to 5000 ft., and old moraines round Lake Okau, *Haast;* Otago, Lindis Pass, *Hector and Buchanan.* A very fine species, at once known by its stout deeply-grooved branches, capitate inflorescence, large flowers, woolly calyx with large lobes, and villous ovary.

2. **C. Munroi,** *Hook. f., n. sp.* Branchlets stout, slightly or much compressed, $\frac{1}{6}$ in. diam., striated, not grooved, quite glabrous. Leaves not seen. Flowers $\frac{1}{4}$–$\frac{1}{3}$ in. long, in few lax-flowered racemes. Pedicels very slender, as long as or longer than the calyx, hoary. Calyx hoary, with rather large blunt teeth; bracteoles at the base minute. Standard shorter than, or as long as the wings. Ovary glabrous. Pod unknown.

Middle Island: from halfway up to the summit of Macrae's Run, *Munro.* The very stout habit, and lax, large-flowered racemes, best distinguish this.

3. **C. nana,** *Col.;—C. australis*, var. *β. nana*, Benth. in Fl. N. Z. i. 50. A very dwarf, glabrous, rigid shrub, 2 to 4 in. high, with fascicled, leafless, much compressed, minutely striated, erect branchlets, $\frac{1}{12}$–$\frac{1}{8}$ in. diam. Flowers $\frac{1}{4}$–$\frac{1}{3}$ in. long; peduncles long, slender, glabrous, 1–5-flowered, sometimes 1 in. long. Calyx shorter than the slender pedicels, glabrous or sparsely pilose, lobes rather large and obtuse; bracteoles at the base minute. Standard about as long as the wings. Ovary quite glabrous. Pod linear-oblong, $\frac{1}{3}$–$\frac{1}{2}$ in. long, with a short straight beak like the pod of *C. australis.*

Northern Island: dry mountainous country at the base of Tongariro, *Colenso;* **Middle** Island; Upper Motucka valley, *Munro;* Southern Alps, *Sinclair and Haast;* Otago, Lindis Pass and Waitaki river, *Hector and Buchanan.* Though very unlike *C. grandiflora*, I should not be surprised if this were proved to be a state of that plant. It keeps its characters in both islands.

4. **C. grandiflora,** *Hook. f.—C. australis*, var. *γ. grandiflora*, Benth. in Fl. N. Z. i. 50. A much branched, glabrous shrub; branchlets terete or compressed, $\frac{1}{12}$–$\frac{1}{6}$ in. broad, deeply grooved, often leafy. Leaflets 3, narrowly or broadly obcordate-cuneate, glabrous. Flowers $\frac{1}{4}$ in. long, in peduncled, lax, broad, obtuse, loose, 6–8-flowered, glabrous racemes, on slender pedicels half as long as the calyx. Calyx glabrous, with large, obtuse, ciliolate lobes, and minute bracteoles at its base. Standard much larger than the wings. Ovary glabrous. Pods in nodding racemes, narrow oblong, $\frac{1}{4}$ in. long, gradually narrowed into a subulate beak $\frac{1}{8}$ in. long. Seeds 2, pale brown.

Middle Island: Milford Sound, *Lyall;* river beds, Mount Cook, and elsewhere in the Southern Alps, alt. 2500 to 4500 ft., *Haast;* Otago, lake district, alt. 1500 ft., *Hector and Buchanan.* A very distinct looking species, best known by the large flowers, glabrous racemes, deeply grooved branches, 3-foliolate leaves, large calyx-lobes, and small pod with a long beak.

5. **C. pilosa,** *Col. in Fl. N. Z.* i. 50. A much distichously branched shrub; branchlets more or less silky-pubescent at the tips, notched alternately on either side, much compressed, $\frac{1}{12}$–$\frac{1}{8}$ in. broad, flexuose, deeply grooved. Leaves not seen. Flowers minute, $\frac{1}{8}$ in. long, in small, erect, dense, 10–15-flowered, silky racemes, suberect; pedicels very short. Calyx nearly glabrous; teeth very short; bracteoles minute, at its base or on the pedicel. Standard much larger than the upturned wings. Ovary silky. Pods small, in pendulous racemes, $\frac{1}{4}$ in. long, tapering into a subulate straight beak $\frac{1}{10}$ in. long. Seeds not seen.

Northern Island: east coast, *Colenso.* This has the grooved branchlets of *C. grandi-*

flora and others, and the racemes of *C. odorata*, but the silky ovary distinguishes it from these and all its congeners, except *C. crassicaulis*.

6. **C. australis,** *Br.;—Fl. N. Z.* i. 50, *excl. var.* An erect, much branched shrub or small tree; branchlets elongated, quite flat, straight, $\frac{1}{14}$–$\frac{1}{4}$ in. broad, finely striated, with distant alternate notches. Leaflets 1 or 2 pairs, membranous, broadly or narrowly obovate-cordate or obcuneate, quite glabrous. Flowers small, $\frac{1}{8}$ in. long, in small, 5–8-flowered, erect or spreading, glabrous or slightly pilose racemes; pedicels about as long as the calyx. Calyx nearly glabrous, with very small obtuse teeth. Standard broad, longer than the wings. Ovary quite glabrous. Pod oblong, spreading, $\frac{1}{3}$ in. long, suddenly narrowed to a short or rather long acute beak; valves not wrinkled. Seeds 1–4, dull red.—Bot. Reg. t. 912; *C. Cunninghamii*, Raoul, Choix, t. 28 B; *C. stricta*, Lehm.; *Bossiæa Scolopendra*, A. Rich.; *Lotus (?) arboreus*, Forst.

Common along the east coasts and interior of the **Northern** and **Middle** Islands, *Banks and Solander*, etc. The branchlets are sometimes upwards of $\frac{1}{2}$ in. broad, but this appears to be rather an anomalous than a normal state of the plant. The pod has rarely a subulate beak, as in *C. flagelliformis;* it is so in authentic specimens of Lehmann's *C. stricta*, however.

7. **C. odorata,** *Col. in Fl. N. Z.* i. 50. A much-branched shrub; branchlets distichous, terete compressed or plano-convex, $\frac{1}{12}$–$\frac{1}{10}$ in. broad, deeply grooved, pubescent towards the tips. Leaves small, silky-pubescent on both sides; leaflets in 2 pairs, very small, $\frac{1}{6}$–$\frac{1}{8}$ in. long, narrow oblong, obcuneate, 2-lobed at the tip. Flowers minute, $\frac{1}{10}$–$\frac{1}{8}$ in. long, in numerous, small, erect, many-flowered, pubescent racemes, very shortly pedicelled; bracteoles minute, on the pedicel. Calyx-teeth rather long, acute. Pods in pendulous racemes, exactly like those of *C. pilosa*.

Northern and **Middle** Islands: east coast, *Colenso;* Nelson, *Travers, Munro;* Otago, lake district, *Hector and Buchanan*. Very similar indeed to *C. pilosa*, from which the glabrous ovary at once distinguishes it.

8. **C. flagelliformis,** *Col. in Fl. N. Z.* i. 51. A much branched shrub, with almost fastigiate, numerous, very slender, compressed, rarely plano-convex grooved branchlets, $\frac{1}{16}$–$\frac{1}{10}$ in. broad. Leaves not seen. Flowers minute, $\frac{1}{10}$–$\frac{1}{8}$ in. long, in pubescent, lax, 3–6-flowered fascicles or open racemes; pedicels very slender; bracteoles above the middle, or below the calyx. Pods oblong or obliquely orbicular, about $\frac{1}{3}$ in. long, with a stout, subulate, straight beak $\frac{1}{12}$ in. long. Seeds mottled with yellow or red, brown and black.—*C. australis*, Raoul, Choix, t. 28 A.

Northern and **Middle** Islands: east coast, *Colenso;* Milford Sound, *Lyall;* Nelson, *Bidwill;* Otago, *Lindsay;* Akaroa, *Raoul*. This is possibly the true *Lotus (?) arboreus*, Forst., but it differs from *C. australis*, Br. in the narrow grooved branchlets, pubescent racemes, and in the pod.

9. **C. juncea,** *Col. in Fl. N. Z.* i. 51. A small, slender shrub, a foot or less high, with very slender, compressed, often curved, grooved branches and branchlets, less than $\frac{1}{12}$ in. broad. Leaves not seen. Flowers minute, $\frac{1}{10}$–$\frac{1}{8}$ in. long, in loose, small, nearly glabrous or puberulous 4–8-flowered fascicles; pedicels curved, about as long as the calyx or longer, bracteolate below it. Calyx small, rather membranous; lobes small, rather acute. Ovary

quite glabrous; style hooked. Pod very minute, $\frac{1}{10}$–$\frac{1}{12}$ in. long, turgid, ovoid-oblong, with a slender, curved, subulate beak. Seeds not seen.

Northern Island: East Cape, *Sinclair;* east coast, Hawke's Bay and Taupo, *Colenso.* **Middle** Island: Akaroa, *Raoul;* Canterbury plains, *Travers.*

2. NOTOSPARTIUM, Hook. f.

A shrub or small tree, with slender branches and pendulous branchlets like whipcord. Leaves not seen. Flowers rather small.—Calyx campanulate, truncate; teeth 5, short. Standard obovate-obcordate, not auricled at the base. Wings oblong, with an incurved auricle at the base, shorter than the hatchet-shaped keel. Stamens diadelphous. Ovary nearly sessile, linear, tapering into a curved style, which is ciliated on the inner margin; ovules 8–10. Pod shortly stipitate, linear-elongate, with a slender style, curved, torulose, compressed, membranous, indehiscent, many-jointed, many-seeded. Seeds solitary in the cells, oblong, with a doubly-bent and twisted club-shaped radicle.

A most curious genus, allied in habit, and many other respects, to *Carmichælia,* but widely differing from it, and from all others known to me.

1. **N. Carmichæliæ,** *Hook. f. in Kew Journ. Bot.* ix. 176. *t.* 3. A small tree, with weeping branches and pink flowers. Branchlets 1 ft. or more long, compressed, grooved, $\frac{1}{20}$ in. broad, remotely toothed, giving off at the teeth many-flowered racemes 1–1$\frac{1}{2}$ in. long. Peduncle and short pedicels pubescent. Flowers $\frac{1}{4}$–$\frac{1}{3}$ in. long. Pods 1 in. long, $\frac{1}{12}$ broad.

Middle Island: Canterbury province, *Waitts;* sandy and rocky places on the Waihopai river, *Munro;* Upper Awatere, *Sinclair,* "Pink Broom."

3. SWAINSONIA, Salisbury.

Herbs; stems prostrate erect or climbing, sometimes woody at the base. Leaves unequally-pinnate, stipulate; leaflets many. Flowers in axillary peduncled racemes.—Calyx campanulate, 5-toothed. Standard broad, open or reflexed, orbicular. Wings narrow, auricled at the base, as long as or shorter than the obtuse keel. Style slender, bearded on the upper or inner edge; ovules many. Pod ovate oblong or terete, inflated or turgid, acute. Seeds rather small.

A large Australian genus, representing *Astragalus* of other parts of the world, from which it is distinguished chiefly by the broader, more expanded standard.

1. **S. novæ-Zelandiæ,** *Hook. f., n. sp.* A small, low, sparingly-branched herb, covered with minute silky pubescence; branches 2–4 in. long. Leaves 1–2 in. long; leaflets opposite, $\frac{1}{4}$ in. long, sessile, obovate-oblong, obtuse or retuse; stipules ovate, obtuse. Peduncle 1 in. long, bearing 5–8 racemed flowers. Flowers purple, $\frac{1}{3}$ in. long; pedicels rather shorter than the calyx, bracteate at the base. Calyx with rather long teeth, villous within and on the edges, 2-bracteolate. Standard without callosities; keel and wings nearly equal, straight. Pod large, nearly 1 in. long, $\frac{1}{3}$ broad, acute at both ends, puberulous; valves thin, coriaceous. Seeds small.

Middle Island: sources of the Kowai, in shingly river-beds, alt. 2000–2500 ft., *Haast.* Very nearly allied to the Tasmanian *S. lessertiæfolia,* which is a larger plant, with many-flowered racemes, more obtuse shorter calyx-teeth, less villous within, usually more abundant blackish hairs on the inflorescence, and more curved keel.

4. CLIANTHUS, Solander.

Herbs; stems woody below, branches often trailing. Leaves pinnate, stipuled; leaflets many pairs. Flowers large, red, in pendulous racemes. —Calyx campanulate, 5-toothed. Standard ovate, reflexed, about as long as the keel. Wings oblong or lanceolate, auricled at the base, shorter than the boat-shaped keel. Ovary stipitate; style ciliated below the apex; ovules numerous. Pod stipitate, terete, narrow-oblong, turgid, rostrate, many-seeded.

A genus of most beautiful plants, consisting of an Australian (Sturt's Pea) and a New Zealand species.

1. **C. puniceus,** *Banks and Sol.;—Fl. N. Z.* i. 49. A branching, herbaceous undershrub, with prostrate or reclining branches, more or less covered with silky appressed hairs. Leaves 4–6 in. long, unequally pinnate. Leaflets in 10–14 pairs, alternate, sessile, ½–1 in. long, linear-oblong, obtuse; stipules triangular. Flowers 6–15 in a raceme, scarlet, pendulous, 2 in. long. Calyx ⅙–⅓ in. long. Standard ovate, acuminate, reflexed when fully expanded. Wings falcate, acute or obtuse, half as long as the standard. Keel very large, boat-shaped, falcate, narrowed into a long beak.—Bot. Reg. t. 1775.

Northern Island: east coast, *Banks and Solander*, etc., and various other localities, especially near native dwellings, etc. One of the most beautiful plants known; often cultivated in European greenhouses; variable in depth of colour of flower.

5. SOPHORA, Linn.

Small trees or twiggy shrubs. Leaflets in many pairs. Flowers pendulous, large, yellow in the N. Z. species.—Calyx rather inflated, urceolate hemispherical or campanulate; mouth oblique, obscurely 5-toothed. Standard obovate, very broad, shortly clawed. Wings oblong, stipitate, shorter than the straight obtuse keel. Stamens 10, all free. Ovary stipitate, linear; style slender, slightly curved, glabrous, stigma minute; ovules numerous. Pod stipitate, elongate, moniliform terete angled or 4-winged, indehiscent or 2-valved, few- or many-seeded. Seeds oblong; funicle not thickened.

A very large genus of tropical, subtropical, and Central Asiatic trees, of which the species with 4-winged pods were separated as *Edwardsia*, before it was discovered that the genus presents pods of all intermediate forms between terete, 4-angled, and 4-winged.

1. **S. tetraptera,** *Aiton;—Edwardsia grandiflora,* Salisb.; Fl. N. Z. i. 52. A small or middling-sized tree, variable in habit, foliage, and size of flower; branches in young plants slender, flexuose, in old straight, densely covered with fulvous silky tomentum. Leaves exstipulate, 1–6 in. long, petiole slender or stout, covered with silky or ferruginous hairs; leaflets 6–40 pairs, very variable, from broadly obcordate to linear-oblong, ¼–¾ in. long, rounded retuse or 2-lobed at the tip, on young plants smaller broader glabrous and membranous, in old silky or densely villous on one or both surfaces. Flowers 1–2 in. long, yellow, in axillary, pendulous, 4–8-flowered racemes; peduncles short, pedicels ½–1½ in. long, flexuose, and calyx densely

silky. Calyx rather gibbous, $\frac{1}{4}$–$\frac{1}{2}$ in. long, hemispherical or urceolate, mouth very oblique. Standard hardly reflected, always short obtuse. Wings linear-oblong. Keel nearly straight. Pods 1–5 in. long, the joints oblong, 4-angled, with 4 membranous wings, the outer walls separating from the coriaceous inner; valves hardly dehiscent. Seeds oblong, pale yellow-brown; cotyledons almost consolidated.

Var. *α. grandiflora.* Larger and more robust. Trunk sometimes 1–3 feet diam. Leaflets in 10–30 pairs, usually narrow. Flowers 2 in. long, narrower. Standard $\frac{1}{4}$ shorter than the wings.—*S. tetraptera*, Bot. Mag. t. 167; *Edwardsia grandiflora*, Salisb.

Var. *β. microphylla.* Smaller; young branches very slender and flexuose, with few obcordate membranous leaflets. Leaflets of old plants in 30–40 pairs, oblong-obcordate. Flowers 1–1$\frac{1}{2}$ in. long, broader. Standard little shorter than the wings.—*S. microphylla*, Jacq. Hort. Schœnb. t. 269; *Edwardsia microphylla*, and *Macnabiana*, Bot. Mag. t. 1442 and 3735.

Abundant throughout the islands, *Banks and Solander*, etc. I am quite unable to find permanent characters whereby to distinguish the two varieties, the wild specimens presenting intermediate ones; the extreme forms are very distinct, and remain so under cultivation. The var. *β*, and forms approaching *α*, are both common in South Chili and Juan Fernandez. Wood of var. *α* valuable for fencing and veneers; heartwood red, *Buchanan.*

Amongst the many cultivated *Leguminosæ*, several have run wild in New Zealand, and the number of colonists will annually increase; the only one known to me, however, as being generally diffused, is the common White Clover (*Trifolium repens*, Linn.).

Eutaxia Strangeana, Turczaninow (Bull. Soc. Imp. Mosc. 1853, vol. ii. p. 270), is stated by its author to have been discovered in New Zealand, no doubt through some blunder; the genus is exclusively a West Australian one.

Guilandina Bonduc, Linn., a tropical, prickly, climbing, pinnate-leaved shrub, is stated by Forster to be a native of New Zealand, but erroneously.

ORDER XXIII. **ROSACEÆ.**

Herbs shrubs or trees, erect or climbing; stems sometimes prickly or spinose. Leaves alternate, simple or compound, stipulate. Flowers regular, usually hermaphrodite.—Calyx-tube short and open, or urceolate and enclosing the carpels; lobes 4 or 5 or 0, often valvate. Petals 4 or 5, rarely 0, spreading, imbricate. Stamens numerous, few in *Acæna*, free, inserted on a perigynous ring or on the mouth of the calyx-tube; filaments subulate; anthers short, didymous. Carpels 1 or more, 1-celled, small, enclosed in the calyx-tube, or free and clustered on a torus; style lateral basal or terminal, stigma capitate; ovules 1 or 2. Fruit of 1 or more small achenes, free or enclosed in the calyx-tube, or of many small free succulent drupes. Seeds exalbuminous; testa membranous; cotyledons plano-convex, radicle short.

The above description applies chiefly to the New Zealand genera of this very extensive and universally-distributed Order, which embraces several tribes (including those of the Apple, Rose, Peach, Cherry, Strawberry, etc., in some of which the calyx is apparently inferior) that do not occur in these islands, and differ in some respects from the character above given.

Shrub, with climbing prickly stems. Fruit of many small drupes . . 1. RUBUS.
Herb, with pinnate leaves. Achenes many, with short styles . . . . 2. POTENTILLA.
Herbs, with simple or pinnate leaves. Achenes many, with elongated bent styles . . . . 3. GEUM.
Herbs, with pinnate leaves. Achenes 1 or 2, enclosed in the calyx-tube . 4. ACÆNA.

1. RUBUS, Linn.

Climbing shrubs (rarely erect or herbaceous); stems usually prickly. Leaves simple or compound. Flowers panicled (rarely solitary).—Calyx-tube open; lobes 5, large, persistent. Petals 5, rarely more, orbicular or obovate. Stamens numerous. Carpels numerous, small, crowded on a small torus; style lateral or subterminal, short, stigma capitate; ovules 2, pendulous. Fruit of small, globose, fleshy drupes.

A very large genus (to which the Bramble, Raspberry, and Dewberry belong), especially in the west of Europe and Himalaya mountains, found also in almost all other parts of the world, except the hottest and driest.

1. **R. australis,** *Forst.;—Fl. N. Z.* i. 53. *t.* 14. A lofty climber, armed with scattered recurved prickles; branches very slender, pendulous. Leaves coriaceous, 3–5-nate, or pinnate with 2 pairs of leaflets and a terminal one; leaflets on long petioles, very variable in shape, ovate cordate oblong lanceolate or reduced to prickly midribs, 2–5 in. long, coarsely serrate or toothed, shining above, glabrous pubescent or tomentose beneath. Flowers very numerous, in branched, prickly, downy panicles, ⅛–½ in. diam., fragrant. Calyx-lobes ovate, obtuse. Petals linear-oblong, pink or whitish. Stamens numerous. Female flowers rarer. Drupes numerous, with short, subterminal styles, when ripe yellowish, juicy, austere.

Var. *α. glaber*, Fl. N. Z. l. c. Leaflets 3–5-nate or pinnate, ovate-cordate, glabrous.

Var. *β. schmidelioides*, Fl. N. Z. l. c. Leaflets 3–5-nate or pinnate, ovate or cordate, pubescent or tomentose below.—*R. schmidelioides*, A. Cunn.

Var. *γ. cissoides*, Fl. N. Z. l. c. Leaflets 3–5-nate, quite glabrous, linear or elliptical-lanceolate, sometimes reduced to midribs.—*R. cissoides*, A. Cunn.

Abundant in the skirts of woods throughout the islands, *Banks and Solander;* and all the varieties found as far south as Otago. Probably the same as an Australian species of which I have seen incomplete specimens only. These varieties, though united by every intermediate form, keep their characters under cultivation.

2. POTENTILLA, Linn.

Herbs (or rarely small shrubs), often silky woolly or pubescent, with compound (rarely simple) leaves, and 1- or more flowered leafy scapes or peduncles. Flowers hermaphrodite.—Calyx-tube open; lobes 4 or 5, alternating with 4 or 5 smaller outer lobes or bracteoles, persistent, valvate. Petals 4 or 5, orbicular or obcordate. Stamens numerous. Ovaries numerous, small, crowded on a small torus; styles lateral or basal, stigmas capitate; ovule 1, pendulous. Achenes numerous, dry, with small styles.

A very large genus in the northern temperate and arctic zones, very rare in the southern.

1. **P. anserina,** *Linn.;—Fl. N. Z.* i. 54. A small stemless plant, everywhere covered with silvery hairs; root emitting slender runners. Leaves 3–6 in. long, unequally pinnate; leaflets 5–20 pairs, the alternate ones sometimes very small, ovate oblong or rounded, deeply incised. Scapes erect, villous, 1-flowered, as long as the leaves. Flowers ½–1 in. diam. Calyx very silky and villous. Petals obovate, golden-yellow. Achenes and receptacle villous.

Var. *β.* Leaflets small, rounded, sessile or petioled.—*P. anserinoides*, Raoul.

Abundant on the east side of all the islands. Var. β, most common, *Banks and Solander*, etc. This is a cosmopolite plant, being found almost throughout temperate and arctic Europe, Asia and America, also in Tasmania, but the var. β is confined to New Zealand.

3. GEUM, Linn.

Herbs with perennial rootstocks, more or less pilose; stems erect, leafy, 1- or more flowered. Radical leaves pinnate, often reduced to one large lobed leaflet.—Calyx-tube open; lobes 5, alternating with 5 smaller outer lobes or bracteoles. Petals 5, orbicular or obovate. Stamens numerous. Ovaries numerous, small, crowded on a small torus; style terminal, elongating after flowering, and then jointed bent or twisted at the middle; ovule solitary, ascending. Achenes dry, hairy, with very long, slender, hairy styles. Seed ascending.

A considerable genus, common in temperate regions, both of the north and south hemisphere. The root of one species is an astringent, and used in medicine.

Two to three feet high. Leaves interruptedly pinnate . . . . 1. *G. urbanum.*
Six inches to one foot high. Leaves with a large terminal leaflet . . 2. *G. parviflorum.*

1. **G. urbanum,** *Linn.*, var. **strictum;**—*G. magellanicum,* Commerson; Fl. N. Z. i. 55. A strict, rigid, leafy, erect, hairy silky or villous herb, 2–3 ft. high, with woody, stout rhizome. Leaves alternate, 3–6 in. long, pinnate; leaflets 3–6 pairs, 1–5 in. long, subsessile, very variable, ovate obovate or obcuneate, crenate toothed lobed or pinnatifid, alternate pairs often smaller, terminal much larger, lobed and cut; stipules large and leafy. Flowers $\frac{1}{3}$–1 in. across. Calyx-lobes triangular-ovate, and peduncles pubescent. Petals obovate, bright yellow, longer than the calyx. Carpels very numerous, densely villous, collected in a globose or oblong head; styles slender, deflexed, abruptly bent or twisted once towards the nearly glabrous apex, the portion beyond the bend is often broken off, when the style appears hooked at the top.

Common in the central and southern subalpine parts of the **Northern** Island, and throughout the **Middle** Island, *Banks and Solander,* etc. Also found in Tasmania, South and North temperate America, along the Andes to Fuegia, and in central and eastern Asia. It differs from the European *G. urbanum* chiefly in the stronger habit and larger petals.

2. **G. parviflorum,** *Commerson;—Fl. N. Z.* i. 56.—A small, silky or villous herb, 4–18 in. high, rhizome stout woody. Leaves 3–5 in. long, pinnate; leaflets 6–8 pairs, all very small, sessile, lobed and toothed; terminal very large, rounded-reniform, 1–2 in. broad, crenate. Flowers rather crowded towards the ends of the stems, about $\frac{1}{2}$ in. across. Calyx-lobes broadly ovate, obtuse. Petals yellow (sometimes white?), longer than the calyx. Styles subulate, tips hooked, villous with long hairs.—*Sieversia albiflora,* Fl. Antarct. i. 9. t. 7.

Northern Island: Ruahine Mountains, *Colenso.* **Middle** Island: common in subalpine localities, 2–5000 ft. alt., from Nelson, *Sinclair,* to Otago, *Hector.* **Lord Auckland's** group, *J. D. H.* Also found in S. Chili and Fuegia.

4. ACÆNA, Vahl.

Perennial herbs, stemless, or with prostrate, branching stems. Leaves radical, numerous, pinnate. Scapes slender, bearing dense, globose heads or spikes of minute hermaphrodite flowers.—Calyx-tube urceolate or obconic, with very contracted mouth, terete compressed or 4-angled, smooth or hispid with simple or barbed bristles, angles often bearing stiff spines barbed at the tip; limb 0. Petals 4 or 5 small, deciduous, sometimes connate at the base. Stamens 1–5; filaments short, subulate. Ovaries 1 or 2, wholly immersed in the calyx-tube; style subterminal, short, exserted, stigma capitate, plumose or fimbriate; ovule pendulous. Achene enclosed in the indurated, often armed calyx-tube; pericarp membranous, bony or coriaceous.

A remarkable genus, almost confined to the southern temperate and antarctic regions, the exceptions being a few Polynesian and Mexican species, and one Chilian, which is also found in California. The barbed calyces of various species form burrs, which in some pastures in Australia adhere to the sheep in such quantities as seriously to injure the fleece.

Fruiting calyx with 4 barbed spines. Leaves very silky . . . . 1. *A. Sanguisorbæ.*
Fruiting calyx with 4 barbed spines. Leaves glabrous . . . . . 2. *A. adscendens.*
Fruiting calyx glabrescent, with 4 usually simple red spines . . . 3. *A. microphylla.*
Fruiting calyx villous, with 4 usually simple yellow spines . . . . 4. *A. Buchanani.*
Fruiting calyx without spines. Leaflets minute . . . . 5. *A. inermis.*

1. **A. Sanguisorbæ,** *Vahl;—Fl. N. Z.* i. 54. A much branched, prostrate herb; branches rather woody at the base, tips ascending, leafy, more or less silky, especially on the leaves below, variable in size. Leaves 2–6 in. long; leaflets 8–10 pairs, very variable in shape, oblong orbicular or obovate, membranous, coarsely serrate, $\frac{1}{4}$–$\frac{2}{3}$ in. long, the upper pairs rather larger than the lower. Scapes slender, 1–2-leaved. Heads globose, $\frac{1}{3}$–$\frac{1}{2}$ in. diam. Calyx 4-angled; angles produced into purple bristles $\frac{1}{2}$ in. long, barbed at the tip. Petals 4, united at the base. Stamens 2. Stigma dilated, feathery. —*A. anserinæfolium,* Forst. Prodr.; *Ancistrum diandrum,* Forst. Char. Gen.

Abundant throughout the islands, *Banks and Solander,* etc., from **Kermadec** Islands, as far south as **Lord Auckland's** group, **Campbell's** and **Macquarrie** Islands. Leaves used medicinally and as tea by the natives of the Middle Island, *Lyall.* Also a native of Tristan d'Acunha, Australia, and Tasmania, where the heads called *burrs* are pests to the flocks, and the plant itself is a troublesome weed in gardens.

2. **A. adscendens,** *Vahl;—Fl. Antarct.* i. 10; ii. 268. *t.* 96. A prostrate, glabrous or slightly hairy herb, with rather stout, leafy stems. Leaves 2–4 in. long; leaflets 4–6 pairs, coriaceous, $\frac{1}{4}$–$\frac{1}{2}$ in. long, coarsely deeply cut towards the obtuse tip, cuneate at the base, teeth obtuse, often tipped with silky hairs. Scape almost leafless, glabrous. Heads $\frac{1}{2}$ in. across. Flowers greenish, calyx-tube silky or glabrous, 4-angled, the angles produced into purple bristles barbed at the tip. Petals 4, broad, greenish. Stamens 1–4; filaments long. Stigma linear.

Middle Island: Mountain districts, alt. 4–7000 ft., Nelson, *Munro and Sinclair;* Canterbury, *Haast and Travers;* Otago, lake district, *Haast and Buchanan.* **Macquarrie** Island, *Fraser,* in Herb. Hook. Also found in Fuegia and the Falkland Islands.

3. **A. microphylla,** *Hook. f. Fl. N. Z.* i. 55. Small, tufted, procumbent, much branched, glabrous or nearly so. Leaves 1 in. long; leaflets 2–6 pairs, suborbicular, inciso-serrate or crenate, $\frac{1}{8}$–$\frac{1}{4}$ in. long. Scapes short, 1–3 in. long, leafless. Heads large for the size of the plant, globose. Calyx 4-angled, the angles thickened or produced into rigid, divaricating red spines, that are rarely hispid at the apex. Petals green, united at the base, nearly

glabrous. Stamens 1 or 2. Achenes usually two, bony, with subclavate, fimbriate stigmas.

Northern Island: Tongariro, *Bidwill, Colenso.* **Middle** Island; Ashburton range, alt. 2000 ft., *Sinclair;* Otago, Lake district, *Hector and Buchanan.*

4. **A. Buchanani,** *Hook. f., n. sp.* Habit and size of *A. microphylla*, but the young branches petioles and calyx are villous with silky hairs; flowers much fewer in a head, spines of the calyx curved, yellow, usually with a few bristles at the tip. Stigma elongate, dilated upwards, fimbriate. Stamen 1. Achenes bony.

Middle Island: Otago, lake district, *Hector and Buchanan.* Foliage very pale. As in *A. microphylla,* the fruiting calyx is sometimes not spinescent.

5. **A. inermis,** *Hook. f. Fl. N. Z.* i. 54. Small, prostrate, slightly silky, tufted, much-branched below. Leaves 1–4 in. long; leaflets 4–6 pairs, orbicular, very small, $\frac{1}{6}$–$\frac{1}{3}$ in. long, deeply crenate, rather coriaceous. Scapes leafless, or with a small bract or solitary leaf, 2–3 in. long, silky or glabrous. Heads $\frac{1}{3}$ in. diam., globose. Calyx silky or pubescent, 4-angled, the angles thickened in fruit, not spinescent. Petals united at the base, green. Stamens 2; filaments short. Achenes 2; stigma dilated, fimbriate on one side.

Middle Island: mountainous districts, Nelson, *Bidwill;* Lake Rotuiti, etc., *Munro;* Kowai river and eastern hills, alt. 1500–3000 ft., *Haast.*

Several European *Rosaceæ* have been introduced into New Zealand, of which some, as the Briar-rose, are said to be naturalized.

Alchemilla (Aphanes) arvensis, Linn., a procumbent, much branched annual, with small cut leaves, and very minute green flowers, is sent by Mr. Travers, from Tarndale plain, alt. 4000 ft.; though stated by Brown to be indigenous to Australia, I think there can be no doubt of its being introduced there, and into New Zealand also.

Staphylorhodos Cotoneaster, Turczaninow, in Bull. Soc. Imp. Mosc. 1862, ii. 321,—a plant referred to *Rosaceæ,* but if correctly, certainly a very anomalous member, is stated by its author to be a native of New Zealand, sent by Sir Everard Home, R.N., to R. Brown (n. 563 and 579), at the sale of some of whose collections it was purchased. There is, however, no such plant nor number in the original collection of Sir E. Home, preserved in the British Museum, nor any New Zealand plant at all agreeing with the characters of this.

Order XXIV. SAXIFRAGEÆ.

Herbs, shrubs, or trees. Leaves opposite or alternate, simple or compound, stipulate or exstipulate. Flowers usually hermaphrodite and regular.—Calyx free or adnate to the ovary, lobes 4 or 5, valvate or imbricate. Petals 4 or 5, valvate or imbricate, usually small and coriaceous. Disk perigynous or epigynous. Stamens usually as many or twice as many as the petals, in *Donatia* fewer. Ovary 2–5-celled; styles 2–5, stigmas usually capitate; ovules numerous, on axile or rarely parietal placentas. Fruit generally a capsule with many seeds, almost a berry in *Carpodetus*, unknown in *Donatia*. Seeds usually small; albumen copious; embryo terete (large, with scanty albumen, in *Ixerba*).

A large Order, containing very many different forms, so that it is impossible to define it tersely. The immense herbaceous tribe of Saxifrages proper is represented by *Donatia* alone in New Zealand.

Tufted, alpine, moss-like herb, with terminal, sessile, white flowers . 1. DONATIA.
Shrubs or trees. Leaves alternate. Petals imbricate. Ovary inferior 2. QUINTINIA.
Shrub or tree. Leaves alternate and opposite, exstipulate. Ovary superior . . . . 3. IXERBA.
Shrub or tree. Leaves alternate. Petals valvate. Ovary inferior . 4. CARPODETUS.
Tree. Leaves opposite, pinnate. Calyx valvate. Ovary superior . 5. ACKAMA.
Trees or shrubs. Leaves opposite; leaflets 1–3 or pinnate. Calyx imbricate. Ovary superior . . . . 6. WEINMANNIA.

1. DONATIA, Forst.

Densely tufted, moss-like, alpine herbs, with very numerous leaves, densely imbricating all round the branches. Flowers terminal, sessile, solitary, white. —Calyx-tube obconic, adnate with the ovary; lobes 4 or 5, equal or unequal. Petals 5–10, linear or ovate. Stamens 2 or 3, inserted on an epigynous disk; filaments cohering at their bases with the bases of the styles; anthers didymous, bursting outwardly. Ovary inferior, 2 or 3-celled; styles 2 or 3, subulate, recurved, stigmas simple or capitate; ovules numerous, on placentas which are pendulous from the top of the cells. Fruit unknown.

A genus of but two species, one Fuegian, the other a native of the summits of the Tasmanian and New Zealand mountains.

1. **D. novæ-Zelandiæ,** *Hook. f. Fl. N. Z.* i. 81. *t.* 20. Stems very short, including the leaves as thick as the little finger, densely clothed with leaves, of which the upper are bright green, the lower brown. Leaves erect, appressed, $\frac{1}{3}$ in. long, coriaceous, linear-subulate, obtuse, villous at the base, nerveless, punctate. Calyx-lobes and petals 5, the latter thick, ovate-oblong, blunt, $\frac{1}{10}$ in. long. Stamens 2, filaments short. Styles 2.

Middle Island: mountains near Port Preservation, *Lyall*.

2. QUINTINIA, A. DC.

Glabrous shrubs or trees. Leaves alternate, persistent, exstipulate. Flowers small, white, in axillary or terminal many-flowered racemes.—Calyx-tube obconic, adnate to the ovary; teeth 5, persistent. Petals 5, oblong, obtuse, deciduous, imbricate. Stamens 5; filaments short, subulate. Ovary inferior, 3–5-celled, its free conical top narrowed into a conical, persistent, furrowed style; stigma capitate, 3–5-lobed; ovules very numerous. Capsule small, inferior or half superior, coriaceous, obovoid, 3–5-ribbed, 1-celled, many-seeded, dehiscing at the tip between the styles. Seeds elongate, imbricating; testa winged.

A small Australian and New Zealand genus. The New Zealand species are covered with lepidote scales, beneath which a viscid exudation is secreted.

Leaves 3–6 in., remotely serrate . . . . 1. *Q. serrata.*
Leaves 2 in., obscurely sinuate-serrate . . . . 2. *Q. elliptica.*

1. **Q. serrata,** *A. Cunn.*;—*Fl. N. Z.* i. 78. A small, erect tree; branchlets, leaves, and racemes covered with lepidote scales. Young parts viscid. Leaves petioled, 3–6 in. long, narrow linear-lanceolate or -oblong, rather obtuse, remotely irregularly and obtusely serrate, margins undulate, when dry

coriaceous, yellow-brown above, red-brown below. Racemes shorter than the leaves. Flowers $\frac{1}{4}$ in. diam. Capsules $\frac{1}{5}$ in. long.—Hook. Ic. Pl. t. 558.

Var. β. Leaves broader, very viscid, oblong-lanceolate, 3 in. long, 1 broad.

Northern Island: not uncommon in forests. Var. β. **Middle** Island: Aorere valley, alt. 1400 ft., *Travers.*

2. **Q. elliptica,** *Hook. f. Fl. N. Z.* i. 78. Very similar to *Q. serrata*, but smaller. Leaves 2 in. long, on rather longer petioles, very obtuse, broader, and very obscurely sinuate-serrate.

Northern Island: east coast, *Colenso.* Probably a variety of *Q. serrata.*

3. IXERBA, A. Cunn.

A small, evergreen, glabrous shrub or tree. Leaves exstipulate, opposite alternate or whorled, glandular-serrated. Flowers large, white, panicled.—Calyx-tube adnate with the base of the ovary; lobes 5, deciduous, imbricate. Petals 5, inserted beneath a 5-lobed disk, obovate, acute, clawed, coriaceous, imbricate. Stamens 5; filaments filiform; anthers linear-oblong. Ovary superior, conical, 5-lobed, 5-celled, narrowed into a subulate, acute, 5-furrowed, twisted style, stigmas acute; ovules 2 in each cell, collaterally suspended. Capsule very coriaceous, shortly ovoid, 5-celled, dehiscing through the style loculicidally; valves cohering below, 2-partite above. Seeds oblong, compressed, with a thick funicle and shining clouded or black testa, large embryo, and little albumen.

1. **I. brexioides,** *A. Cunn.;—Fl. N. Z.* i. 82. A small tree, 20 ft. high. Leaves coriaceous, linear or linear-lanceolate, 4 in. long by $\frac{1}{3}$ broad, but very variable. Flowers $1\frac{1}{3}$–$1\frac{3}{4}$ in. diam. Capsule $\frac{3}{4}$ in. broad. Seed $\frac{1}{5}$–$\frac{1}{4}$ in. long.—Hook. Ic. Pl. t. 577–588.

Northern Island: in woods, from the Bay of Islands to Wellington, but not very common.

4. CARPODETUS, Forst.

A shrub or tree. Leaves alternate, exstipulate, evergreen. Flowers panicled.—Calyx-tube turbinate, connate with the ovary; lobes 5 or 6, short, deciduous. Petals 5 or 6, inserted under an epigynous disk, spreading, valvate. Stamens 5 or 6; filaments short, subulate; anthers oblong. Ovary inferior, with a swollen top, 3–5-celled; style slender, stigma capitate; ovules very numerous. Fruit almost fleshy, globose, indehiscent, girt at the middle by the calyx-limb, 3–5-celled, many-seeded. Seeds small, pendulous, ovoid; testa coriaceous, pitted; embryo minute, in the base of fleshy albumen.

1. **C. serratus,** *Forst.;—Fl. N. Z.* i. 78. A shrub or small tree, 10–30 ft. high; trunk usually slender; branches spreading in a fan-shaped manner; all parts more or less pilose. Leaves petioled, ovate-oblong, obtuse, 1–$1\frac{1}{2}$ in., much smaller on young plants, rather membranous, acutely serrate. Panicle cymose, shorter than the leaves. Flowers $\frac{1}{5}$ in. diam., white. Fruit size of a peppercorn.—Hook. Ic. Pl. t. 564.

Northern and **Middle** Islands: frequent by the banks of rivers, *Banks and Solander*, etc. Wood soft, tough, used for axe-handles, *Hector.*

5. ACKAMA, A. Cunn.

A small tree. Leaves opposite, pinnate, stipulate. Flowers small, panicled, unisexual.—Calyx inferior, 5-partite; lobes valvate. Petals 5, linear-spathulate, inserted beneath a 10-lobed disk, about as long as the calyx. Stamens 10; filaments filiform, the alternate longer; anthers didymous. Ovary free, hirsute, 2-celled; styles 2, filiform, persistent; ovules very numerous, on parietal placentas. Capsule small, coriaceous, turgid, 2-celled, septicidally 2-valved, valves gaping inwardly. Seeds turgid, ovoid, apiculate; testa smooth, lax, pilose; embryo cylindric, albumen scanty.

1. **A. rosæfolia,** *A. Cunn.;—Fl. N. Z.* i. 79. ii. 329. Tree 30–40 ft. high. Leaves pubescent, imparipinnate, 3–10 in. long. Leaflets in 3–6 pairs, subsessile, linear-oblong, acute, acutely serrate, 1–2 in. long, rather membranous, the upper larger; stipules deciduous, leafy, toothed. Flowers small, $\frac{1}{12}$ in. diam., in spreading, branched panicles, sessile on the branchlets. Capsule very small.

Northern parts of the **Northern** Island. Woods at the Bay of Islands, not uncommon, *A. Cunningham*, etc. A. Gray (Bot. U. S. Expl. Exped. i. 672) regards this as a subgenus of *Weinmannia*, but I think erroneously.

6. WEINMANNIA, Linn.

Shrubs or trees. Leaves opposite, stipulate, 1–3-foliolate or pinnate. Flowers racemose, small.—Calyx inferior, tube short; limb 4–5-partite, imbricate. Petals 4 or 5, inserted under the edges of a lobed disk, imbricate. Stamens 8 or 10; filaments filiform; anthers didymous. Ovary free, ovoid or conical, 2-celled, narrowed into two beaks; styles filiform; ovules few or many, placentas inserted above the middle in the axes of the cells. Capsule small, coriaceous, 2-celled, septicidally 2-valved, valves gaping inwardly. Seeds oblong reniform or subglobose; testa membranous, often pilose; embryo terete; albumen fleshy.

A large Indian, Pacific Island, Australian, and S. American genus.

Branchlets, peduncles, and petioles pubescent. Leaves 3-foliolate or pinnate 1. *W. silvicola.*
Branchlets, peduncles, and petioles glabrous. Leaves 1-foliolate . . . 2. *W. racemosa.*

1. **W. silvicola,** *Banks and Sol.;—Fl. N. Z.* i. 79. A small tree, 20-30 ft. high, with dark bark, the young leaves, branchlets, inflorescence, and costa of leaf below all pubescent or glabrescent. Leaves opposite, 1–3-foliolate or imparipinnate (on the same specimen), 1–3 in. long; leaflets variable, 1–2 in. long, obovate-oblong or lanceolate, acute or acuminate, coarsely serrate, coriaceous, punctate beneath, young ones membranous, glabrous or pubescent below, usually narrowed into the petiole; stipules leafy, oblong, entire or toothed. Racemes 2–4 in., slender, many-flowered. Flowers $\frac{1}{10}$ in. diam., pedicelled, white. Capsule glabrous or minutely hairy, $\frac{1}{5}$–$\frac{1}{6}$ in. long.—*W. fuchsioides* and *W. betulina*, A. Cunn.

Abundant in woods throughout the **Northern** Island, and in the northern part of the **Middle** Island, *Banks and Solander*, etc. An exceedingly variable plant, of which I made three varieties in the New Zealand Flora, but find that these do not hold good: the most marked forms are a larger, often pinnated-leaved one, with almost tomentose branchlets, petioles, and peduncle, and a much smaller, nearly glabrous one, with more coriaceous, usually 3-foliolate leaves. The species is apparently identical with a New Caledonian one.

2. **W. racemosa,** *Forst.;—Fl. N. Z.* i. 80. A larger leaved, more coriaceous plant than the preceding, glabrous, except the raceme; flowers large; leaves broader, 1-foliolate. Leaflets 1–3 in., oblong-ovate or -lanceolate or orbicular-ovate, obtuse, sinuate-serrate, very coriaceous, punctate beneath. Racemes numerous, stout, erect, very many-flowered. Flowers $\frac{1}{6}$ in. diam. Capsules $\frac{1}{4}$ in. long, sometimes 3-valved.—*Leiospermum racemosum,* Don.

Middle and southern parts of the **Northern** Island, and throughout the **Middle** Island, *Banks and Solander,* etc. Very nearly allied to the preceding, and I once supposed a variety of it, but I now think them distinct.

Order XXV. CRASSULACEÆ.

Succulent plants; the New Zealand species all very small, inconspicuous herbs. Leaves usually opposite, exstipulate.—Sepals 3 or more. Petals as many as the sepals. Stamens perigynous or almost hypogynous, as many as the petals or a multiple of them. Carpels as many as the sepals, free, each often with a scale at its outer base, 1-celled; style very short, or stigma sessile; ovules few or many, attached to the suture of the carpel. Seeds minute, albuminous; embryo terete.

A very large Order, to which *Sedum* and *Crassula* belong, especially abounding in South Africa and the southern parts of Europe.

1. TILLÆA, Micheli.

Small or minute, tufted, erect or procumbent herbs. Leaves small, opposite. Flowers axillary.—Sepals, petals, stamens, and carpels 3–5, the latter with or without scales at the base. Fruit of 3–5 follicles, with few or many seeds.

An inconsiderable genus, scattered over Europe and various parts of the world.

* *A scale at the base of each carpel.*

Leaves $\frac{1}{6}$–$\frac{1}{3}$ in. Flowers $\frac{1}{6}$–$\frac{1}{5}$ in. diam. . . . . 1. *T. moschata.*
Leaves $\frac{1}{12}$–$\frac{1}{8}$ in. Flowers $\frac{1}{16}$ in. diam. . . . . 2. *T. Sinclairii.*

** *No scales.*

Stems erect. Flowers very numerous. Seeds 1 or 2 in each carpel . 3. *T. verticillaris.*
Stems prostrate. Flowers few. Seeds 1 or 2 in each carpel . . . . 4. *T. debilis.*
Stems erect or prostrate. Flowers few. Seeds numerous . . . . 5. *T. purpurata.*

1. **T. moschata,** *DC.;—Fl. N. Z.* i. 76. A tufted, rather succulent, red-brown herb; 3–5 in. high, sparingly branched, rooting at the axils of the leaves. Leaves small, $\frac{1}{6}$–$\frac{1}{3}$ in. long, oblong spathulate or obovate-oblong, obtuse, thick, quite entire. Flowers $\frac{1}{6}$–$\frac{1}{5}$ in. diam., white, axillary, shortly peduncled. Sepals 4, obtuse, half as long as the oblong obtuse petals. Stamens 4. Scales linear-cuneate. Carpels turgid, with short recurved styles, many-seeded.—Hook. Ic. Pl. t. 535 (scales omitted); *Crassula moschata,* Forst.

Moist rocks, etc., most frequent near the sea. East coast of the **Northern** Island, *Co-*

lenso. **Middle** Island more common, *Forst.*; Otago, *Lyall*, etc. **Lord Auckland's** group and **Campbell's** Island, abundant, *J. D. H.* Also a native of S. Chili, Fuegia, the Falkland Islands, and Kerguelen's Land. One of the largest species of the genus.

2. **T. Sinclairii,** *Hook. f.*, *n. sp.* A small, delicate, tufted, pale-green plant, 1–2 in. high, with very slender stems. Leaves minute, oblong, $\frac{1}{12}$ in. long. Flowers axillary, solitary, shortly pedicelled, white, $\frac{1}{16}$ in. diam. Sepals 4, ovate-oblong, obtuse. Petals 4, twice as long as the sepals, obtuse. Scales and carpels as in *T. moschata.*

Middle Island: Rangitata river, alt. 2000 ft., *Sinclair* and *Haast.*

3. **T. verticillaris,** *DC.*;—*Fl. N. Z.* i. 75. A small, erect, pale red-brown, succulent herb; stems 2–4 in. long, simple or branched from the base. Leaves very small, $\frac{1}{12}$–$\frac{1}{8}$ in., linear- or ovate-oblong. Flowers very numerous, densely crowded in the axils of the leaves, very minute, some sessile, some on slender peduncles. Sepals 4, ovate-subulate. Petals subulate, smaller than the sepals. Scales 0. Carpels 1–2-seeded, lanceolate; styles rather slender.—*T. muscosa*, Forst.

Common on dry, rocky, sunny places, *Banks and Solander*, etc. Common in Australia and Tasmania, also found in Chili (*T. minima*, Miers). A. Gray mentions this as gathered at the Auckland Islands by Wilkes' Expedition, "except the tickets be misplaced."

4. **T. debilis,** *Col. in Fl. N. Z.* i. 75. A very small, delicate species; stems intricate, filiform or capillary, prostrate, 2–3 in. long. Leaves in scattered pairs, minute, $\frac{1}{16}$–$\frac{1}{12}$ in. long, ovate-oblong or linear-oblong. Flowers minute, 1 or 2 in the axils of the leaves, sessile or on slender peduncles. Sepals 4, oblong, subacute. Petals ovate-acuminate, shorter than the sepals. Scales 0. Carpels ovate-lanceolate, 1- or 2-seeded.

Northern Island: east coast, *Colenso.*

5. **T. purpurata,** *Hook. f. Fl. N. Z.* i. 75. A small, delicate, tufted, pale red-brown species, with slender, prostrate or erect stems, sparingly branched, 1–3 in. long. Leaves linear, acuminate, $\frac{1}{8}$–$\frac{1}{6}$ in. long. Flowers solitary, axillary, on capillary peduncles much longer than the leaves. Flowers $\frac{1}{10}$ in. diam. Sepals 4, ovate, subacute. Petals smaller than the sepals, acuminate. Scales 0. Carpels broadly oblong, turgid, many-seeded.

Northern Island: east coast, Cape Palliser, etc., *Colenso.* Also a native of Tasmania and Southern Australia.

Order XXVI. DROSERACEÆ.

All the New Zealand species but one are scapigerous herbs. Leaves covered on the upper surface or margins, or both, with copious glandular hairs. Flowers regular, hermaphrodite.—Calyx 4- or 5-partite, imbricate, persistent. Petals 4 or 5, hypogynous, rarely perigynous, generally persistent. Disk 0. Stamens 4 or 5, usually hypogynous, filaments subulate or filiform; anthers short, often bursting outwardly. Ovary superior, ovoid or globose, 1-celled; styles 3–5, filiform or clavate, simple bipartite or multifid; ovules numerous, on parietal placentas. Capsule membranous, 1-celled, loculicidally 3–5-valved, many-seeded. Seeds small; testa lax, reticulate; albumen fleshy, copious; embryo rather large or minute, cylindrical.

A remarkable but small Order, of which the prevalent characters alone are given above, and these apply to all the New Zealand species; a few northern and Cape genera present some remarkable exceptions, as shrubby habit, petals connate at the base, epipetalous stamens, 3-celled few-ovuled ovary, and minute embryo.

1. DROSERA, Linn.

Stamens 4–8. Styles 2 or 3, bipartite or multifid.

An extensive genus (Sundew) found in all parts of the world, but most abundant in Australia.

Scape 1-flowered. Calyx campanulate, 5-lobed. Styles 3, multifid . 1. *D. stenopetala.*
Scape 1-flowered. Leaves linear. Styles 3 or 4, stigmas capitate . . 2. *D. Arcturi.*
Scape 1-flowered. Leaves orbicular. Styles 4, stigmas subclavate . . 3. *D. pygmæa.*
Scape many-flowered. Leaves spathulate. Styles 3, bipartite . . . 4. *D. spathulata.*
Scape many-flowered. Leaves bipartite . . . . 5. *D. binata.*
Scape many-flowered. Leaves lunate. Styles 3, penicillate . . . . 6. *D. auriculata.*

1. **D. stenopetala,** *Hook. f. Fl. N. Z.* i. 19. *t.* ix. Stemless. Leaves with the slender glabrous petiole 1–4 in. long; blade spathulate, ½–⅓ in. long, densely clothed with long, glandular hairs. Scape slender, longer than the leaves, 1-flowered. Flowers ⅓ in. diam. Calyx campanulate, 5-lobed, glabrous. Petals very narrow, linear-spathulate. Styles 3, multifid to the base.

Middle Island: Port Preservation, *Lyall.* **Lord Auckland's** group, *J. D. H.*, *Bolton.* Very nearly indeed allied to the Fuegian *D. uniflora.*

2. **D. Arcturi,** *Hook. Ic. Pl. t.* 56;—*Fl. N. Z.* i. 20. Stemless; rhizome long, slender. Leaves 2–3 in. long, linear-ligulate or linear-spathulate, covered with long glandular hairs; petiole almost as broad as the blade. Scape about as long as the leaves, 1-flowered. Flowers nearly ½ in. diam. Petals obovate, as long as the calyx. Styles 3 or 4, short; stigmas capitate.

Northern Island: Ruahine mountains, near the snow, *Colenso.* **Middle** Island: common in wet places on the mountains, alt. 3500–5000 ft. Also a native of the Tasmanian and the Australian alps.

3. **D. pygmæa,** *DC.*;—*Fl. N. Z.* i. 20. A very small, stemless species. Leaves densely crowded, rosulate, shortly petioled, orbicular, concave, subpeltate, ⅙ in. diam.; stipules very large, scarious, forming a beautiful silvery star round the base of the scape. Scape filiform, 1 in. high. Flower solitary, minute. Sepals 4, oblong, obtuse, glabrous, shorter than the white petals. Styles 4, filiform, subclavate.

Northern Island: in marshes at Cape Maria Van Diemen, *Colenso.* Also a native of the south coast of Australia and Tasmania.

4. **D. spathulata,** *Labill.*;—*Fl. N. Z.* i. 20. Stemless. Leaves rosulate, densely crowded, ½–¾ in. long, spathulate, covered with glandular hairs. Scapes 1 or several, slender, simple or bifid, 8–15- (rarely 1–2-) flowered. Flowers shortly pedicelled, secund, ⅕ in. broad. Sepals united at the base. Petals spathulate, pink red or white, twice as long as the calyx. Styles 3, 2-partite.—Bot. Mag. t. 5240; *D. propinqua*, A. Cunn.

Wet places throughout the islands, but not very common, *Banks and Solander*, etc.; ascending to 5000 ft. on the shingly banks of Lake Wanaka and Hawea, *Haast.* A common Australian plant, also found in the Philippine Islands.

5. **D. binata,** *Labill.;—Fl. N. Z.* i. 20. Stemless. Leaves on slender petioles 2–4 in. long; blade divided into two linear, strap-shaped, simple or 2-fid, divaricating lobes 2–4 in. long, covered with glandular hairs. Scapes 8–12 in. high, very slender, glabrous, bearing a branched, 6–8-flowered cyme. Flowers white, $\frac{1}{4}$–$\frac{1}{2}$ in. diam., on slender pedicels. Petals membranous, obcordate, twice as long as the glabrous or ciliate sepals. Styles 3, penicillate. —Bot. Mag. t. 3082; *D. intermedia*, A. Cunn.; *D. Cunninghamii*, Walpers.

Common in clay bogs, etc., throughout the islands, *Banks and Solander*, etc. Abundant in Tasmania and Australia. A very handsome plant.

6. **D. auriculata,** *Backhouse;—Fl. N. Z.* i. 21. Stem erect, simple or sparingly branched, 6–18 in. high, flexuose, glabrous, leafy, arising from a small coated tuber. Leaves peltate, lowest rosulate, orbicular, $\frac{1}{4}$ in. diam., on short flattened petioles; upper alternate, rather larger, lunate, on filiform petioles $\frac{1}{2}$–1 in. long; all copiously studded towards the margin with very long glandular hairs. Flowers 3–8, on slender pedicels, white or lilac, $\frac{1}{4}$–$\frac{1}{3}$ in. diam. Sepals oblong, obtuse, often jagged. Petals three times longer than the sepals, obovate or obcordate. Styles 3, penicillate below the middle.

Northern and **Middle** Islands: not uncommon, often climbing by its glandular leaves amongst grass, *Banks and Solander*, etc. A common Australian and Tasmanian plant; also found in Norfolk Island.

ORDER XXVII. HALORAGEÆ.

Herbaceous, terrestrial or aquatic plants. Leaves opposite alternate or whorled, exstipulate. Flowers small, often very inconspicuous, variously arranged, hermaphrodite or unisexual, usually regular.—Calyx-tube adnate with the ovary; limb 0 or 2–4-lobed or -toothed. Stamens 2–4 (rarely 1), epigynous. Ovary inferior, usually 4-(rarely 1-)celled, with 2–4 (rarely 1) short conical styles, or as many sessile, plumose or papillose stigmas, or with 1–4 long subulate stigmatiferous styles; ovules solitary, pendulous in each cell. Fruit small, indehiscent, 4-celled and -seeded, or breaking up into 2–4 carpels, or a 1-celled small drupe. Seeds pendulous; testa membranous; albumen fleshy; embryo minute, ovoid, or long and terete.

A small Order of dissimilar plants, scattered over the world. The genus *Callitriche*, included here, is a very doubtful member. *Balanophoreæ*, an Order closely allied to *Halorageæ*, will be found amongst *Incompletæ*.

Terrestrial. Calyx-lobes 4, evident. Fruit 4–8-gonous . . . . 1. HALORAGIS.
Aquatic. Calyx-lobes 0 or inconspicuous. Fruit 4-lobed. Stigmas 4, very short . . . . 2. MYRIOPHYLLUM.
Marsh. Calyx-lobes 2 or 3. Petals 2 or 0. Fruit drupaceous, 1-celled 3. GUNNERA.
Marsh and aquatic. Calyx-lobes 0. Fruit 4-lobed. Styles 2, very long . . . . 4. CALLITRICHE.

1. HALORAGIS, Forst.

Erect procumbent or creeping, herbaceous plants. Leaves opposite or rarely alternate, toothed, rarely lobed or pinnatifid. Flowers axillary, minute, hermaphrodite or unisexual, sometimes racemose at the ends of the branches, pendulous.—Calyx-tube terete, 4–8-angled or winged; lobes 4, erect, acute, persistent, valvate. Petals 4, small, concave, deciduous, coriaceous. Stamens 4 or 8, usually with short filaments and large linear or oblong anthers, bursting laterally. Ovary 2 or 4-celled; stigmas 2 or 4, sessile, plumose or simple. Fruit a dry, coriaceous or somewhat drupaceous, 4–8-angled or -winged, 2- or 4-celled nut. Seeds pendulous; embryo terete.

A genus almost confined to the south temperate zone, most frequent in Australia, found also in subtropical India, China, and Japan. One New Zealand species is also found in Juan Fernandez.

Glabrous. Leaves ½–1½ in. Fruit with 4 narrow wings . . . . 1. *H. alata.*
Scabrid. Leaves ¼–½ in. Flowers axillary. Fruit 8-costate, muricate. 2. *H. tetragyna.*
Slightly scabrid. Leaves ¼–⅓ in. Flowers axillary. Fruit 4–8-costate 3. *H. depressa.*
Very slender, quite glabrous. Leaves ¼–⅓ in. Flowers in naked racemes.
Fruit 8-costate . . . . 4. *H. micrantha.*

1. **H. alata,** *Jacq.;—Fl. N. Z.* i. 63. Stem herbaceous, erect or procumbent at the base, branched, 4-angled, 1–3 ft. high; branches suberect, slightly scabrid. Leaves opposite, sessile or petioled, ovate-lanceolate, acute or obtuse, very coarsely deeply serrate, coriaceous, slightly rough to the touch, ½–1½ long. Flowers minute, green, in leafy slender terminal racemes, solitary or whorled, on short curved pedicels, drooping, $\frac{1}{12}$ in. long. Calyx 4-angled, with 4 small lobes. Petals twice as long as the calyx-lobes. Nut small, with 4 narrow wings, green; sides smooth or wrinkled. Stigmas conical, not plumose.—*Cercodia erecta*, Murr.; *C. alternifolia*, A. Cunn.

Abundant on dry hills in various localities throughout the islands, *Banks and Solander*, etc.; ascending to 2000 ft. Also found in Juan Fernandez.

2. **H. tetragyna,** *Labill.;—Fl. N. Z.* i. 62. A suberect or prostrate, slender, branched herb, everywhere scabrid with short, hispid pubescence; branches slender, 2–8 in. long. Leaves all opposite, shortly petioled ¼–½ in. long, ovate or lanceolate, acute, sharply serrate, rigidly coriaceous, scabrid on both surfaces. Flowers minute, sessile, solitary in the axils of the upper leaves. Nut ovoid, $\frac{1}{12}$ in. long, 8-angled, muricate. Stigmas plumose.—*Goniocarpus tetragyna*, Labill. Fl. Nov. Holl. i. t. 53; *Cercodia incana*, A. Cunn.

Var. *α.* Branches suberect. Leaves ½–¾ in. long, many-toothed.

Var. *β. diffusa.* Branches diffuse, prostrate. Leaves ¼–½ in. long, broader, obtuser, with fewer teeth.

Northern Island: on dry hills; *α*, Bay of Islands; *β*, common there and at Auckland. I have seen no Middle Island specimens. Var. *α* is a native of Tasmania and Southern Australia, also of China (*G. scaber*, Kœnig), Borneo, and the mountains of East Bengal.

3. **H. depressa,** *Hook. f. Fl. N. Z.* i. 63. A very slender, prostrate, branched herb, slightly scabrid with short white hairs; branches 1–12 in. long, wiry, almost filiform. Leaves sessile, broadly ovate, subacute, ¼–⅓ in. long, with 3–5 deep close serratures on each side, slightly scabrid on both surfaces, coriaceous. Flowers solitary, axillary or subterminal, sessile. Nut ⅛ in. long, 4-angled, or obscurely 4–8-costate, smooth or wrinkled, but not muricate. Stigmas plumose.—*Goniocarpus depressus*, A. Cunn.

Common on dry hills, throughout the islands. Readily distinguished from *G. tetragyna* by being almost glabrous and very slender. In the New Zealand Flora I considered this as the same with the Tasmanian *H. (Goniocarpus) serpyllifolius*, Hook. Ic. Pl. t. 290, and it is certainly most closely allied to it; but the fruit is very much larger in the New Zealand plant.

4. **H. micrantha,** *Br.;—H. tenella*, Brongn.; Fl. N. Z. i. 63. Small, slender, tufted, perfectly glabrous; stems and branches filiform, ascending or erect, 2–6 in. high, running out into leafless racemes. Leaves few, $\frac{1}{6}$–$\frac{1}{3}$ in. long, all opposite, subsessile, broadly ovate, subacute, with very few shallow crenatures on each side, glabrous. Flowers few, very minute, scattered along the naked filiform tips of the branches; pedicels very short. Nut minute, oblong, 8-ribbed. Stigmas plumose.—*Goniocarpus citriodorus*, A. Cunn.—*G. tenellus*, Brongn.

Common in rather moist and dry places in various parts of the **Northern** Island. **Middle** Island: Aglionby plains, *Munro*. Common in Tasmania, Southern Australia, the Eastern Himalaya, hilly parts of East Bengal, and in Japan. This is the *Goniocarpus citriodorus* of A. Cunningham, but I have never perceived any smell of lemons; possibly he dried it with *Micromeria*.

2. MYRIOPHYLLUM, Vaillant.

Water or marsh plants, with slender, terete, sparingly branched stems. Leaves usually whorled, entire toothed or pinnatifid, the submerged often capillaceo-multifid, the uppermost more entire, often opposite or alternate. Flowers small, in the axils of the upper leaves, almost sessile, uni- or bisexual. —Calyx-tube in the male 0, in the female ovoid or 4-gonous, limb 4-toothed or truncate. Petals 0 in the female flower, in the male 4, membranous, deciduous. Stamens 4 or 8, filaments short; anthers linear. Ovary 4-celled; styles 4, very short, or 0; stigmas often plumose. Fruit usually separating into 4 small, hard, indehiscent, 1-seeded nuts; testa membranous; albumen fleshy; embryo terete.

A very widely-diffused genus in both tropical and temperate regions. The species have wide ranges in their several zones, but those of the southern regions are hardly the same with the northern. The foliage is very variable; the upper leaves in the two first New Zea- and species are sometimes as much divided as the lower.

Upper leaves whorled, short and broad, serrate . . . . 1. *M. elatinoides.*
Upper leaves whorled, narrow-linear, nearly entire . . . . . . 2. *M. variæfolium.*
All the leaves pectinate-pinnatifid. Fruit large . . . . . . 3. *M. robustum.*
All the leaves opposite, linear, narrow, entire . . . . 4. *M. pedunculatum.*

1. **M. elatinoides,** *Gaudichaud;—Fl. N. Z.* i. 63. Stems 6 in.–3 ft. long, depending on the depth of the water. Leaves whorled, usually 4-nate, the lower pectinately pinnatifid, 1–2 in. long, the upper much smaller, sessile, oblong, obtuse, serrate. Flowers small, white; female with two linear entire or serrate bracts. Stigmas plumose. Fruit of 4 minute, oblong, terete, smooth nuts.—*M. propinquum*, A. Cunn.

Still waters throughout the islands, ascending to 3000 ft. Abundant in Tasmania, Australia, and the lakes of the Cordillera from the equator to Chili and the Falkland Islands.

2. **M. variæfolium,** *Hook. f. in Ic. Pl. t.* 289;—*Fl. N. Z.* i. 65. Habit of the last, but leaves in whorls of 5–7, the lower multifid or pectinate-pinnatifid, the upper very narrow linear, $\frac{1}{2}$–$1\frac{1}{2}$ in. long, quite entire or serrate. Flowers and fruit very small, like those of *M. elatinoides,* but the bracts are much shorter than the nuts.

Common in still water throughout the islands, as far south as Canterbury. Also found in Tasmania, Australia, and South America.

3. **M. robustum,** *Hook. f.*, *n. sp.*;—*M. variæfolium*, β, Fl. N. Z. i. 64. Stems robust. Leaves 1–2 in. long, 5–7 in a whorl, all pectinately-pinnatifid. Flowers and nuts much larger than in the preceding species, the latter nearly ⅛ in. long, tumid, short, rounded at the back, smooth or obscurely tubercled, bracts very short.

Northern Island, *Colenso*. A robust species, of which I have imperfect specimens, full of fruit, which is as large as in the Indian *M. tuberculatum*, but the nuts are more rounded, and the foliage is course and more crowded. It much resembles some states of the European *M. verticillatum*. I do not recognize it amongst the Australian and American species.

4. **M. pedunculatum,** *Hook. f. Fl. Tasm.* i. 123. *t.* 23 *B.* Small, tufted, erect, slender, simple or sparingly branched, 2–4 in. high. Leaves opposite, narrow linear, quite entire or obscurely toothed, ⅛–⅓ in. long. Flowers minute, sessile or shortly pedicelled. Stamens 8. Styles recurved; stigmas plumose. Nuts ovoid, somewhat wrinkled.

Middle Island: moist places, Canterbury plains, *Sinclair and Haast*. Found also in Tasmania and Victoria, where the leaves sometimes become broad and flat.

3. GUNNERA, Linn.

Stemless herbs, usually growing in swamps or boggy places, with stout or slender creeping or subterranean rhizomes. Leaves petioled, radical. Scapes bearing small, racemose or panicled, unisexual, rarely bisexual, flowers.—Male fl.: Sepals 2 or 3. Petals 2 or 3 or 0. Stamens 2 or 3, opposite the petals, filaments slender; anthers large, often tetragonous. Female fl.: Calyx-tube ovoid, lobes 2 or 3, often toothed. Petals 2 or 3 or 0. Stamens 0. Ovary 1-celled; styles usually 2, linear subulate or filiform, covered with stigmatic papillæ; ovule solitary, pendulous. Fruit a small, rather fleshy drupe. Seed filling the cavity of the drupe, to the walls of which the membranous testa often adheres; albumen fleshy and oily; embryo very minute, broadly pyriform or cordate.

An almost exclusively southern genus, scattered over the Australian, Malayan, New Zealand, Pacific, and South American islands, also found in South Africa and in the Andes, as far north as the Gulf of Mexico. The Chilian species is noted for its immense leaves, sometimes 6 ft. across. The New Zealand species have long filiform styles, and no petals.

Leaves orbicular-reniform. Flowers lax, subpanicled. Drupe very small, ovoid . . . . 1. *G. monoica.*
Leaves reniform or cordate. Flowers in a dense spike. Drupe oblong. 2. *G. densiflora.*
Leaves oblong- or ovate-oblong. Flowers spicate. Drupe much larger, obovoid . . . . 3. *G. prorepens.*

1. **G. monoica,** *Raoul, Choix* xv. *t.* 8;—*Fl. N. Z.* i. 65. Glabrous or minutely hairy; rhizome creeping, tufted, forming matted patches. Leaves reniform or orbicular, ⅓–⅔ in. diam., lobed and crenate, or crenate only, both surfaces glabrous or covered with scattered small hairs, petiole ½–2 in. long, pilose, rarely glabrous. Panicle longer or shorter than the leaves, very slender, branched, rarely simple. Flowers sparse, males with long filaments; bracts

and calyx-lobes ciliate, rarely entire. Drupes very small, ovoid, about $\frac{1}{12}$ in. ong.

Moist shady places throughout the islands, *Banks and Solander*, etc.

2. **G. densiflora,** *Hook. f.*, *n. sp.* Rhizome creeping, tufted. Leaves orbicular cordate or broadly ovate-cordate, glabrous, minutely toothed, $\frac{1}{2}$–1 in. diam.; petioles 1–1$\frac{1}{2}$ in., villous or glabrescent. Spike on a stout, erect, villous or glabrescent peduncle. Flowers crowded, males with short filaments; bracts and calyx-teeth small, subulate. Drupes crowded, small, pendulous, $\frac{1}{10}$ in. long, oblong.

Middle Island: Acheron and Clarence rivers, alt. 4000 ft., *Travers*. Similar to *G. monoica*, but the leaves are never lobed, though minutely and sharply toothed, the peduncles stout, flowers spiked, crowded, filaments short, and drupes larger, oblong, and pendulous.

3. **G. prorepens,** *Hook. f. Fl. N. Z.* i. 66. Usually much larger than the two preceding species, often 1 ft. high, but sometimes small. Leaves ovate or oblong, rounded at the top, rounded or cordate at the base, crenulate, $\frac{1}{2}$–2 in. long; petiole 1–8 in. long, glabrous or more or less hairy. Peduncle stout, as long as the petiole. Flowers spicate. Drupes sessile, large, $\frac{1}{8}$–$\frac{1}{6}$ in. long, obovoid.

Var.? *β.* Smaller, leaves ovate-oblong, narrowed into the petiole, deeply toothed; probably a distinct species.

Northern Island: in subalpine wet localities, *Colenso*. **Middle** Island: west coast, *Lyall*. Var.? *β.* **Northern** Island, *Colenso*. **Middle** Island: mountains of Nelson, *Rough*; Clarence valley, 3000 ft., *Travers*; alps of Canterbury, *Haast*; and of Otago, *Hector and Buchanan*.

I have what appears to be a fourth species of Gunnera, from the base of Tongariro, *Colenso*, etc.; and the Middle Island, *Lyall*; with the leaves of *G. densiflora*, and fruit of *G. monoica*, but too imperfect for description.

4. CALLITRICHE, Linn.

Small, branched, glabrous, delicate aquatic or marsh plants. Leaves opposite, linear or spathulate, membranous, quite entire.—Flowers minute, solitary, rarely 2 together, axillary, sessile, or very shortly peduncled. Male fl.: Bracts 0 or 2, deciduous, narrow-oblong, membranous, curved. Perianth 0. Stamen 1; filament capillary; anther 2-celled, bursting by lateral slits, which becoming confluent above, assume a horse-shoe form. Female fl.: Bracts as in the male or 0. Perianth (if any) adherent to the ovary, which consists of 4 laterally compressed 1-celled carpels; styles 2, filiform, papillose all over; ovule 1 in each cell, pendulous. Fruit 4 minute, flattened, keeled or winged carpels. Seed oblong or cylindric; testa membranous; albumen fleshy; embryo small, terete.

A small genus of doubtful affinity, found in all parts of the temperate world.

1. **C. verna,** *Linn.;—Fl. N. Z.* i. 64. Stems 5–10 in. long in water, shorter when growing on damp soil. Leaves $\frac{1}{6}$–$\frac{2}{3}$ in. long, obtuse, 3-nerved, orbicular spathulate or linear-obovate. Male and female flowers sometimes collateral on the same pedicel. Carpels narrowly winged at the back.

Common in wet soil and watery places throughout the islands, *Banks and Solander*, etc. **Lord Auckland's** group, and **Campbell's** Island, *J. D. H.* Abundant in the antarctic and northern and southern temperate regions.

Order XXVIII. MYRTACEÆ.

Shrubs or trees, sometimes subscandent. Leaves opposite or alternate, exstipulate, quite entire, furnished with pellucid glands full of volatile fragrant oil. Flowers hermaphrodite, regular, solitary or in axillary or terminal racemes or cymose panicles, often showy.—Calyx-tube adnate with the ovary, ovoid oblong or obconic, sometimes produced beyond the ovary, limb 4- or 5-cleft, deciduous or persistent. Petals 4 or 5, orbicular, concave, sessile, imbricate. Stamens very numerous in the New Zealand genera, filaments short or long, incurved in bud; anthers small, didymous. Ovary inferior, usually 4–5-celled; style slender or short, simple, stigma simple or capitate; ovules numerous, on projecting placentas. Fruit dry or succulent, indehiscent or capsular, 1- or more celled, 1- or more seeded. Seeds very various, exalbuminous; embryo straight or curved.

A very large Order, especially in the tropics, rare in the north temperate zone, represented in Europe by the Myrtle alone, but common in the south temperate zone, where one species advances to Cape Horn, and another to Campbell's Island.

Leaves alternate. Flowers solitary or fascicled . . . . 1. Leptospermum.
Leaves opposite.
Flowers cymose or racemed. Capsule many-seeded. Seeds linear . 2. Metrosideros.
Flowers usually solitary. Berry many- or few-seeded. Seeds with bony testa. . . . . 3. Myrtus.
Flowers cymose. Berry 1- or few-seeded. Seeds large, angular . 4. Eugenia.

1. LEPTOSPERMUM, Forst.

Shrubs or trees. Leaves alternate, small, evergreen, coriaceous. Flowers solitary or fascicled, shortly peduncled, white or pink.—Calyx-tube turbinate, lobes 5, often persistent, valvate. Petals 5, rounded, concave. Stamens numerous, filaments short. Ovary 4- or 5-celled; style straight; ovules very numerous. Capsule coriaceous or woody, broadly hemispherical, bursting within the calyx-border by 4 or 5 valves. Seeds very numerous, small, linear, pendulous from the inner upper angle of the cell; testa membranous.

A very large subtropical and temperate Australian genus, extending northwards to the Malayan islands.

Leaves pungent. Flowers solitary, sessile. Calyx-lobes deciduous . . 1. *L. scoparium.*
Leaves not pungent. Flowers fascicled, peduncled. Calyx-lobes persistent . . . . 2. *L. ericoides.*

1. **L. scoparium,** *Forst.;—Fl. N. Z.* i. 70. A large bush or small tree; trunk erect; branches fastigiate, prostrate in alpine localities; young branches and leaves silky. Leaves very variable, $\frac{1}{6}$–$\frac{3}{4}$ in. long, sessile, from linear-lanceolate to orbicular, acuminate, rigid, pungent, concave, veinless, dotted, patent or recurved. Flowers sessile, solitary, axillary or terminal, $\frac{1}{4}$–$\frac{3}{4}$ in. diam. Calyx-tube broadly turbinate, lobes orbicular, deciduous. Petals orbicular, clawed, crumpled. Capsule woody, turbinate, half sunk in the calyx-tube; the free portion 5-valved.

Var. *α.* Fl. N. Z. l. c. Erect, leaves lanceolate.
Var. *β. linifolium,* l. c. Erect, leaves narrow, linear-lanceolate.
Var. *γ. myrtifolium,* l. c. Erect, leaves ovate, spreading or recurved.

Var. δ. *prostratum*, l. c. Prostrate, branches ascending. Leaves ovate or orbicular, recurved.

Abundant throughout the islands, *Banks and Solander*, etc. Wood hard. Leaves used as tea in Tasmania and Australia, where the plant is equally abundant and variable.

2. **L. ericoides,** *A. Rich.*;—*Fl. N. Z.* i. 70. A large, erect shrub or tree, 10–40 ft. high, similar to the *L. scoparium*, but with more slender, glabrous branches; narrower, less coriaceous, acute, not pungent leaves, which are fascicled and erect or recurved; pedicelled, smaller flowers, and the smaller capsule immersed in the calyx-tube.

Abundant throughout the islands, *Banks and Solander*, etc.

2. METROSIDEROS, Br.

Erect or subscandent trees or large shrubs, often embracing the trunks of forest trees and strangling them. Leaves opposite, sometimes distichous, coriaceous. Flowers in terminal or axillary racemes or cymes, white pink or scarlet.—Calyx-tube globose oblong or turbinate, sometimes produced into a tubular or campanulate limb, lobes 5, concave, imbricate. Petals 5. Stamens very numerous; filaments filiform. Ovary 3-celled; style slender; ovules very numerous, linear, radiating outwards from tumid axile placentas. Capsule coriaceous, sunk deeply in the calyx, or its tip free, 3-celled, 3-valved, bursting usually loculicidally, rarely irregularly. Seeds very numerous, densely packed, linear; testa membranous.

A comparatively small genus, consisting of several Pacific Island species, 10 New Zealand, an Indian, and a South African one. Its total absence in Australia is a most remarkable fact.

Capsule coriaceous or woody, wholly enclosed in the calyx-tube, which is produced beyond it, not dehiscent to the base; bursting irregularly, or by 3 valves within the calyx.

Scandent. Leaves obtuse. Calyx glabrous . . . . 1. *M. florida.*
Erect. Leaves acuminate. Calyx silky . . . . 2. *M. lucida.*

Capsule rather membranous, wholly enclosed in the calyx-tube, which is produced far beyond it, dehiscent to the base.

Leaves decussate, very acuminate. Petals white . . . . 3. *M. albiflora.*
Leaves decussate, ¾–1¼ in., oblong, obtuse. Petals scarlet . . . 4. *M. diffusa.*
Leaves distichous, ⅓–⅔ in., oblong-ovate, subacute. Branches glabrescent 5. *M. hypericifolia.*
Leaves distichous, ovate-lanceolate, acuminate. Branches tomentose 6. *M. Colensoi.*

Capsule girt below the middle by the calyx-tube, the upper half exposed, 3-*valved.*

Erect. Leaves 1–1½ in., decussate, glabrous, obtuse . . . . 7. *M. robusta.*
Erect. Leaves 1–3 in., decussate, downy below . . . . 8. *M. tomentosa.*
Erect. Leaves ¾–1¼ in., decussate, obovate, downy below . . . . 9. *M. polymorpha.*
Scandent. Leaves ⅓–½ in., distichous, glabrous, 3-nerved . . . . 10. *M. scandens.*

1. **M. florida,** *Sm.*;—*Fl. N. Z.* i. 67. *t.* 15. A lofty climber, often clothing the tall forest-trees. Trunk very stout; branches and branchlets terete, glabrous or puberulous. Leaves 1–3 in., on short petioles, elliptic-oblong, obtuse, coriaceous, glabrous; midrib stout; nerves numerous, slender, parallel, nearly horizontal. Flowers with the stamens 1 in. long, collected into terminal, branching, few- or many-flowered cymes; peduncles and pedicels short and thick. Calyx obconic or turbinate, costate, glabrous, produced beyond the ovary into an obconic cup. Petals orbicular, pale pink. Stamens scarlet, numerous, and style stout. Ovary wholly adnate with the base of the

calyx-tube. Capsule thick, almost woody, urceolate, with 5 stout ribs, the lobes finally deciduous.—*Melaleuca florida*, Forst. Prod.; *Leptospermum scandens*, Forst. Gen.

Common in forests throughout the islands, *Banks and Solander*, etc. **Lord Auckland's** group, *Bolton.*

2. **M. lucida,** *Menzies;—Fl. N. Z.* i. 67. A small, erect, branching evergreen tree; bark pale, papery; branches obscurely tetragonous; twigs and very young leaves silky. Leaves 1–3½ in. long, elliptic-lanceolate, acuminate at both ends, shortly petioled, very coriaceous, shining above, quite glabrous, midrib stout, lateral nerves faint, oblique; under surface with large glands. Flowers 3 or more together, sessile or shortly pedicelled at the ends of the branches. Calyx broadly obconic, ⅓ in. long; tube silky; lobes ovate, obtuse, persistent. Petals small, oblong or linear, scarlet. Stamens scarlet, very numerous and style very stout, nearly 1 in. long. Ovary free above the middle. Capsule broadly urceolate with 5 stout ribs, the upper free 3-valved part enclosed in the coriaceous calyx-limb.—*M. umbellata*, Cavanilles; *Melaleuca lucida*, Forst.; *Agalmanthus umbellatus*, Homb. and Jacq.

Mountainous districts of the **Northern** Island, *Colenso.* Abundant in the **Middle** Island, and forming the greater part of the wood on **Lord Auckland's** group.

3. **M. albiflora,** *Banks and Sol.;—Fl. N. Z.* i. 67. A lofty climber; trunk stout; branches perfectly glabrous; branchlets terete. Leaves decussate, 1½–3 in. long, elliptic-lanceolate, acuminate at both ends, shining above, very coriaceous, shortly petioled; margins recurved when dry; midrib stout, lateral veins numerous, delicate, leaving the costa at an acute angle. Flowers in terminal, leafless, many-flowered, branched, glabrous cymes; peduncles and pedicels rather slender. Calyx scarcely ¼ in. long, campanulate; lobes shortly ovate, obtuse, persistent. Petals small, white, orbicular. Stamens and style very slender. Ovary wholly adnate to the base of the calyx-tube. Capsule ⅓ in. long, urceolate, turgid, 3-lobed, bursting loculicidally, crowned by the much narrower tubular calyx, which is terminated by the reflexed persistent lobes.—Hook. Ic. Pl. t. 569; *M. diffusa*, A. Cunn., not Smith.

Northern Island: dense forests in the Bay of Islands, and east coast, *Banks and Solander*, etc.

4. **M. diffusa,** *Smith;—Fl. N. Z.* i. 67. A large, scandent, glabrous shrub, with pale, ragged bark, and slender, obscurely 4-gonous branches. Leaves somewhat distichous, ¾–1¼ in., shortly petioled, elliptic-oblong, obtuse; midrib and oblique lateral veins prominent. Flowers in terminal and lateral many-flowered, puberulous, often leafy cymes. Calyx ⅓ in. long, narrow-oblong, suddenly expanding into a cup-shaped limb, with orbicular deciduous lobes. Petals orbicular, scarlet, shortly clawed, jagged. Stamens and style ⅔ in. long, slender. Ovary wholly adherent to the lower part of the calyx-tube. Fruit shortly pyriform, turgid, 5-ribbed, rather membranous, bursting loculicidally, crowned with the short calyx-limb.—*M. myrtifolia*, Gærtner; *M. lucida*, Linn. f.; *M. Homeana*, Turczaninow.

Northern Island: frequent in forests, *Banks and Solander*, etc.

5. **M. hypericifolia,** *A. Cunn.;—Fl. N. Z.* i. 67. *t.* 16. A large, rambling, climbing shrub, with ragged bark, and slender, divaricating, 4-angled

branches, glabrous or minutely pubescent. Leaves distichous, sessile, $\frac{1}{3}$–$\frac{2}{3}$ in., ovate-oblong or oblong-lanceolate, acute or apiculate, glabrous on both surfaces, rather membranous; midrib and lateral nerves delicate. Flowers small, with the stamens $\frac{1}{3}$–$\frac{1}{2}$ in. long, in short, lateral, few-flowered, glabrous or pubescent racemes; peduncles and pedicels slender. Calyx-tube pyriform, suddenly dilating into a short campanulate or cup-shaped spreading limb; lobes triangular-ovate. Petals orbicular, scarlet, shortly clawed. Stamens and style slender. Ovary wholly adherent with the base of the calyx-tube. Capsule small, short, $\frac{1}{8}$ in. long, globose or turgid, 3-lobed, bursting loculicidally, crowned with narrower short tubular calyx-limb,

Humid forests throughout the islands, *Banks and Solander*, from Bay of Islands, *A. Cunningham*, to Chalky Bay, *Lyall.*

6. **M. Colensoi,** *Hook. f. Fl. N. Z.* i. 68. A slender, twiggy, climbing shrub; branches very slender (and young leaves tomentose), terete, tetragonous at the tips. Leaves distichous, scarcely petioled, $\frac{1}{3}$–$\frac{2}{3}$ in., ovate or ovate-lanceolate, acuminate, rather membranous when old, glabrous on both surfaces; midrib and veins delicate. Flowers small, in axillary, rarely terminal, branched, many-flowered cymes, 1–2 in. long; peduncles and pedicels slender. Flowers with the stamens $\frac{1}{2}$ in. long, puberulous. Calyx narrow funnel-shaped, slightly swollen at the base, produced far beyond the ovary; lobes small, broadly ovate, obtuse. Petals orbicular, scarlet. Stamens and style very slender. Ovary wholly adherent to the base of the calyx-tube. Capsule $\frac{1}{8}$ in. long, globular, 3-lobed, bursting loculicidally, crowned by the longer narrower tubular calyx limb.

Northern Island: forests of the Bay of Islands and east coast, *Edgerley, Colenso*, etc. Closely allied to *M. hypericifolia*, but at once distinguished by the slender habit, glabrous, not divaricating, branches, and acuminate leaves.

7. **M. robusta,** *A. Cunn.;—Fl. N. Z.* i. 68. *t.* 17. A tall, erect (never? climbing), glabrous tree, 50–60 ft. high; bark smooth, pale; branches robust, terete, the ultimate 4-gonous. Leaves decussate, petioled, 1–1$\frac{1}{2}$ in. long, elliptic- or ovate-oblong, obtuse or retuse at the tip, very coriaceous, glabrous on both surfaces, reticulated with numerous very fine close-set veins. Flowers with the stamens 1 in. long, in dense, many-flowered, terminal, pubescent cymes; peduncles and pedicels short and thick. Calyx-tube short, obconic, scarcely produced beyond the ovary; lobes triangular, short. Petals orbicular, scarlet, shortly clawed. Stamens and style rather stout. Ovary adnate below the middle with the calyx-tube. Capsule urceolate, small, $\frac{1}{4}$–$\frac{1}{3}$ in., girt round the middle by the calyx-tube, the upper part free and loculicidally 3-valved.—*M. florida*, Hook. Bot. Mag. t. 4471, not Smith.

Northern Island: abundant in forests, *Banks and Solander*, etc. Wood hard, close-grained, good for naval purposes.

8. **M. tomentosa,** *A. Cunn.;—Fl. N. Z.* i. 68. A short, stout, much-branched tree, 30–40 ft. high; branches very stout, terete, densely tomentose. Leaves decussate, petioled, 1–3 in. long, variable in shape, from linear-lanceolate to oblong or orbicular-oblong, rounded or narrowed at the tip, very coriaceous, margins recurved, glabrous, finely reticulated above, densely or laxly clothed with snow-white tomentum below, rarely glabrous. Flowers large, with the stamens 1$\frac{1}{2}$ in. long, in terminal many-flowered lax or dense cymes,

densely clothed with white tomentum; peduncles and pedicels very stout. Calyx-tube obconic; lobes short, triangular-ovate. Petals orbicular, scarlet, clawed, tomentose on the back. Stamens and style very long and stout. Ovary adnate below the middle with the base of the calyx-tube. Capsule $\frac{1}{2}$ in. long, woody, girt round the middle by the calyx-tube, bursting loculicidally.—A. Rich. Flor. t. 37.

Northern Island: common on the rocky shores, *Banks and Solander*, etc.

9. **M. polymorpha,** *Forst.* A small, branching tree, excessively variable in foliage and in being woolly downy or glabrous; the variety from Kermadec Island has downy branches, coriaceous, petioled, decussate leaves, $\frac{3}{4}$–$1\frac{1}{4}$ in. long, obovate, rounded at the tip; veins finely reticulated; margins recurved. Flowers in terminal, many-flowered cymes; peduncles and calyx white, woolly. Calyx $\frac{1}{4}$ in. long, hemispherical; lobes triangular, persistent, often tipped at the back with a black gland. Petals white or scarlet. Stamens $\frac{1}{2}$–$\frac{3}{8}$ in. long. Ovary woolly at the top. Capsule 3-lobed, 3-valved, sessile in the broad calyx-tube, which is attached to its middle.

Kermadec Islands, where it is the prevailing tree, and reaches 60 ft. high, *M'Gillivray*. One of the most variable trees in the world, abundant throughout all the Pacific islands and New Caledonia, but not found in Australia.

10. **M. scandens,** *Banks and Sol.;—Fl. N. Z.* i. 69. A large climbing shrub or small tree, with spreading, terete, tomentose branches. Leaves distichous, $\frac{1}{3}$–$\frac{1}{2}$ in., sessile, broadly ovate oblong or orbicular, obtuse, glabrous, the younger pubescent, very coriaceous, below covered with large glandular dots, 3-nerved; margins recurved. Flowers with the stamens $\frac{1}{2}$ in. long, in short, few-flowered, subglobose, terminal, rarely lateral, pubescent cymes; peduncles and pedicels rather slender. Calyx-tube very broadly obconic; lobes short, obtuse, deciduous. Petals orbicular, whitish. Stamens and style rather slender. Ovary adnate below its middle with the base of the calyx-tube. Capsule small, globose, $\frac{1}{6}$ in. diam., girt at the middle by the calyx-tube, the free portion loculicidally 3-valved.—*M. perforata*, A. Rich.; *M. buxifolia*, A. Cunn.; *Melaleuca perforata*, Forst. Prod.; *Leptospermum perforatum*, Forst. Gen.

Common in forests of the **Northern** and **Middle** Islands, *Banks and Solander*, etc. **Lord Auckland's** group, *Bolton*. Wood hard and heavy.

3. MYRTUS, Linn.

Trees or shrubs. Leaves evergreen, opposite, often coriaceous, pellucid-dotted. Flowers axillary, solitary or in few-flowered cymes.—Calyx-tube globose; limb 4- or 5-lobed, imbricate. Petals 4 or 5. Stamens very numerous, longer than the petals. Berry globose, crowned with the persistent calyx-lobes, 2- or 3-celled; cells many- or few-seeded. Seeds reniform or angular; testa bony; embryo terete curved, cotyledons small.

A genus scattered over the temperate and warmer countries of the globe, of which the common Myrtle is the only European example, but of which many species inhabit the south temperate zone, one extending to Cape Horn, and two to Akaroa. The genus is absent in Tasmania.

Leaves 1–2 in., tumid between the veins . . . . 1. *M. bullata.*
Leaves $\frac{3}{8}$–1 in., flat . . . . 2. *M. Ralphii.*

Leaves obcordate, small . . . . 3. *M. obcordata.*
Leaves obovate . . . . 4. *M. pedunculata.*

1. **M. bullata,** *Banks and Sol.;—Fl. N. Z.* i. 70. An erect shrub or small tree, 10–15 ft. high; branches terete, tomentose, as are the young leaves, peduncles, and calyx. Leaves shortly petioled, 1–2 in. long, broadly elliptic-ovate or orbicular-ovate, obtuse acute or apiculate, coriaceous, often red-brown when living, the surface tumid between the veins. Peduncles axillary, ½–1½ in., 1-flowered. Calyx-tube turbinate, 2-bracteolate at the base; lobes 4, broadly ovate, obtuse. Petals 4, orbicular, pale pink. Stamens numerous, white; filaments and style slender. Berry urceolate, ⅓ in. long, turgid, black, 2-celled, with several flat reniform seeds in each cell; testa very thick and bony; embryo very slender, curved.—Hook. Ic. Pl. t. 557.

Northern Island: frequent in forests, *Banks and Solander*, etc.

2. **M. Ralphii,** *Hook. f. Fl. N. Z.* ii. 329. Very similar in habit, form of the foliage, and flowers to *M. bullata*, but the leaves are smaller and quite plane, the branchlets more slender, scarcely tomentose.

Northern Island: forests near Wellington, *Ralph*; east coast, *Colenso*. This may be a variety of *M. bullata*. I have seen no fruit.

3. **M. obcordata,** *Hook. f. Fl. N. Z.* i. 71. A small, much-branched, erect or decumbent shrub, with slender pubescent branches. Leaves small, ¼–½ in. long, obcordate, narrowed into a short petiole, coriaceous, glabrous on both surfaces. Peduncles solitary, slender, about as long as the leaves, 1-flowered, pubescent. Flowers small, about ¼ in. long. Fruit oblong, ¼ in. long, black or violet, with 1 or 2 bony reniform seeds.—*Eugenia obcordata*, Raoul, in Ann. Sc. Nat. ser. iii. p. 123.

Northern Island: mountainous localities in the east coast and interior, *Colenso*. **Middle** Island: Nelson, *Bidwill*: Lake Wainaka, *Haast*; Akaroa, *Raoul*.

4. **M. pedunculata,** *Hook. f. in Ic. Pl. t.* 629;—*Fl. N. Z.* i. 71. A straggling shrub, 10–12 ft. high, glabrous, much resembling *M. obcordata*, but the branchlets are 4-gonous and perfectly glabrous, the leaves obovate or oblong-obovate, and rounded at the tip, rarely acute, and the peduncles and calyx perfectly glabrous. Berry small, very few-seeded, orange-yellow.—*Eugenia Vitis-Idæa*, Raoul, in Ann. Sc. Nat. ser. iii. 2. p. 122.

Northern Island: hilly districts on the east coast and interior, *Colenso*. **Middle** Island: Nelson, *Bidwill*; Akaroa, *Raoul*.

4. EUGENIA, Linn.

Characters almost identical with those of *Myrtus*, except that the seeds are few, large, with a thick embryo, often presenting no traces of cotyledons. The New Zealand species of *Myrtus* further differ from *Eugenia Maire*, in wanting the corymbose inflorescence.

A tropical genus, abounding in Asia and America, and extending into New Zealand and Chili; it is absent in Tasmania.

1. **E. Maire,** *A. Cunn.;—Fl. N. Z.* i. 71. A small tree, with the habit of a *Metrosideros*, 30–50 ft. high, everywhere glabrous; bark white; branches slender; twigs 4-angled. Leaves petioled, 1–2 in. long, oblong-lanceolate, acuminate, membranous, with delicate veins. Flowers almost unisexual,

in terminal, branched, many-flowered corymbs 2–3 in. diam., white; peduncles and pedicels very slender, the latter often in threes. Calyx-tube broadly obconic; lobes very short, broad, deciduous. Petals orbicular, often falling before expanding. Stamens and style very slender, $\frac{3}{4}$ in. long. Ovary wholly adnate with the base of the calyx-tube, 2-celled, with many ovules on the septum of each cell. Berry $\frac{1}{3}$ in. long, urceolate or broadly pyriform, crowned with the narrower cup-shaped calyx-limb, 1-celled, 1-seeded. Seed large; testa coriaceous; cotyledons large, amygdaloid.

Northern parts of the **Northern** Island: in swampy ground, *Cunningham*, etc.

Order XXIX. ONAGRARIEÆ.

Herbs shrubs or rarely trees. Leaves opposite or alternate, simple, exstipulate. Flowers usually regular, hermaphrodite.—Calyx-tube often elongate, adherent to the ovary, and sometimes produced beyond it; limb 4- (rarely 2–5-) lobed, valvate. Petals as many as lobes of calyx, or 0, inserted on the calyx, spreading or convolute. Stamens as many or twice as many as the petals, inserted with them, free. Ovary inferior, 2–4-celled; style straight, stigma capitate or clavate; ovules usually very numerous in the angles of the cells. Fruit a capsule or berry. Seeds small, exalbuminous, sometimes provided with a tuft of hairs; embryo straight.

A large Order in all temperate and many tropical parts of the world, to which the *Œnothera*, *Clarkia*, and various other garden plants now cultivated in New Zealand belong.

Ovary oblong, terete. Fruit a berry . . . . 1. Fuchsia.
Ovary linear, 4-angled. Fruit a capsule . . . . 2. Epilobium.

1. FUCHSIA, Linn.

Shrubs or small trees. Leaves alternate or opposite or whorled. Flowers pendulous, long-pedicelled.—Calyx-tube ovoid, produced into a tubular or campanulate, deciduous, 4-lobed limb, the lobes large, spreading, acuminate. Petals 0 or 4, concave or convolute. Stamens 8, much longer than the petals. Ovary crowned with an urceolate disk, 4-celled; style slender, stigma capitate or clavate; ovules very numerous. Berry ovoid, fleshy, 4-celled, many-seeded.

A very large South American, and especially Andean genus of most beautiful plants, found nowhere in the Old World but in New Zealand.

Stem erect, subarboreous. Leaves 2–3 in. Petals small . . . . 1. *F. excorticata.*
Stem prostrate. Leaves slender. Petals 0, or small . . . . . . 2. *F. procumbens.*

1. **F. excorticata,** *Linn. f.;—Fl. N. Z.* i. 56. A large bush or small tree, 10–30 ft. high, with the trunk sometimes 3 ft. diam., covered with ragged, papery bark; branches brittle. Leaves alternate, twice as long as their slender petioles, very variable in length, $1\frac{1}{2}$–$3\frac{1}{2}$ in. long, lanceolate or ovate-lanceolate, acuminate, obscurely toothed, membranous, green above, silvery below. Flowers axillary, solitary, on long, slender, drooping peduncles, shorter than the petioles, $\frac{3}{4}$–1 in. long. Calyx dingy-purple, globular above the ovary, then contracted and lengthened into a funnel-shaped 4-cleft tube; lobes lanceolate, very acuminate. Petals much smaller than the calyx-lobes, lanceolate, red-purple. Stamens exserted; anthers oblong. Ovary linear-oblong;

stigma globose. Berry oblong, fleshy.—Bot. Reg. t. 857.; *Skinnera excorticata*, Forst.

Damp woods throughout the islands, abundant, *Banks and Solander*, etc. The wood, which is soft and useless, contains much tannic and gallic acids, *Buchanan*. A small form, found at Otago by Lindsay and Hector, has much smaller and broader leaves, approaching *F. procumbens*.

2. **F. procumbens,** *R. Cunn.;—Fl. N. Z.* i. 57. A procumbent, scarcely shrubby plant, with prostrate, slender branches. Leaves small, $\frac{1}{2}$–1 in. long, as long as or shorter than the very slender petioles, broadly ovate or orbicular, cordate or rounded at the base, acute, very obscurely toothed. Flowers as in *F. excorticata*, but usually smaller.—Hook. Ic. Pl t. 421.

Damp forests in the **Northern** Island, from the east coast southward. **Middle** Island: Canterbury plains, *Travers*. The extreme forms of this and the former differ greatly, but I have apparently intermediate states.

2. EPILOBIUM, Linn.

Perennial herbs. Stem sometimes woody below, creeping prostrate or erect from a decumbent base. Leaves opposite or alternate, exstipulate, usually toothed. Flowers often drooping in bud, then erect, white pink rosy or purplish, axillary or subterminal.—Calyx-tube very long, adnate with the tetragonous ovary; limb of 4 deciduous lobes. Petals 4, spreading, obovate or obcordate. Stamens 8, perigynous. Ovary 4-celled; style short, stigma obliquely clavate (4-cleft in some European species); ovules very numerous. Capsule very long, linear, splitting into 4 recurved, slender valves, very many-seeded. Seeds small, with a pencil of white hairs at the tip.

A very large genus in almost all temperate climates, rare in the tropics, especially abounding in New Zealand. The species are, without exception of locality or country, extremely variable, and probably hybridize. I have repeatedly studied the New Zealand ones, many of which completely puzzle me. The following descriptions represent in many cases perhaps prevalent forms rather than species; and the student will certainly find intermediates between most of them. It is useless attempting to name many species, till copious suites of specimens are collected, the characters being to a great extent comparative.

* *Small, glabrous species, 1–6 in. long. Branches prostrate and creeping, their tips only ascending. Leaves all opposite, uniform in size. Flowers few, chiefly from the axils of leaves remote from the ends of the branches.*

Leaves $\frac{1}{6}$–$\frac{1}{3}$ in., subsessile, scarcely toothed . . . . 1. *E. nummularifolium.*
Leaves $\frac{1}{3}$ in., oblong or orbicular, coriaceous, purple below . . 2. *E. purpuratum.*
Leaves $\frac{1}{2}$ in., petioled, orbicular, sharply toothed . . . . . 3. *E. linnæoides.*
Leaves $\frac{1}{2}$–$\frac{2}{3}$ in., oblong, obscurely toothed . . . . 4. *E. macropus.*

** *Small alpine and subalpine, glabrous, glossy species. Branches creeping, ascending at the tips, rarely suberect. Leaves all opposite, nearly uniform, obtuse, coriaceous, nerveless. Flowers few, usually small, from towards the tips of the branches.*

Leaves $\frac{1}{3}$–$\frac{1}{2}$ in., oblong, sinuate-toothed. Flowers small . . 5. *E. confertifolium.*
Leaves 1 in., subsessile, spathulate-oblong. Flowers rather large 6. *E. crassum.*
Leaves 1 in., petioled, narrow oblong, subacute . . . . . 7. *E. brevipes.*

*** *Small, slender, usually prostrate, lowland species. Branches ascending or suberect. Leaves opposite or upper alternate (not glossy), $\frac{1}{6}$–1 in. long. Flowers towards the ends of the branches, usually numerous and small.*

Leaves $\frac{1}{6}$–$\frac{1}{3}$ in., petioled. Flowers very small. Peduncles of fruit long . . . . 8. *E. alsinoides.*

Stem rigid, black. Leaves $\frac{1}{8}$–$\frac{1}{6}$ in., sessile. Flowers very small. Peduncles of fruit short . . . . 9. *E. microphyllum.*

Leaves $\frac{1}{4}$–$\frac{2}{3}$ in., broad, sharply toothed. Flowers larger. Peduncles of fruit long . . . . 10. *E. rotundifolium.*

**** *Stem erect or decumbent at the very base, 6–24 in. high. Leaves opposite or alternate. Flowers subterminal.*

† *Flowers usually less than $\frac{1}{3}$ in. diam.*

Glabrous. Leaves opposite, narrowed at base, in scattered pairs, coriaceous . . . . 11. *E. glabellum.*

Glabrous. Stems rigid, black. Leaves opposite, narrowed at base, crowded, linear . . . . 12. *E. melanocaulon.*

Glabrous. Leaves opposite, broad at base, sessile or $\frac{1}{2}$ amplexicaul . . . . 13. *E. tetragonum.*

Pubescent. Leaves opposite and alternate, sessile . . . . . . 14. *E. junceum.*

†† *Flowers usually $\frac{1}{3}$–1 in. diam.*

Pubescent. Leaves most alternate, all petioled, broad. Flowers numerous . . . . 15. *E. pubens.*

Glabrous. Leaves mostly opposite, sessile, narrow at base, linear-oblong . . . . 16. *E. Billardierianum.*

Tall, puberulous above. Leaves opposite, sessile, 2–3 in., linear 17. *E. pallidiflorum.*

1. **E. nummularifolium,** *A. Cunn.;—Fl. N. Z.* i. 57. Small, prostrate, branched, slender; branches 2–6 in. long, glabrous or bifariously pubescent. Leaves $\frac{1}{6}$–$\frac{1}{3}$ in. long, sessile or petioled, numerous, opposite, rather crowded, uniform in size, orbicular or oblong, obtuse, entire or waved at the margin, coriaceous or thin, flat or convex, usually with the under surface lying flat on the soil. Peduncles axillary, slender, $\frac{1}{4}$–4 in. long. Flowers very small, pink or whitish. Ovary and capsule generally hoary.

Var. α. *pedunculare,* Fl. N. Z. i. 58. Leaves membranous; peduncles long and slender; capsule glabrous.—*E. pedunculare,* A. Cunn.; *E. pendulum,* Banks and Sol.

Var. β. *nerteroides,* Fl. N. Z. l. c. Leaves thick, with recurved margins; peduncles long; capsule glabrous.—*E. nerteroides,* A. Cunn.

Var. γ. *brevipes,* Fl. N. Z. l. c. Leaves thick, coriaceous, with recurved margins; peduncles very short; capsule puberulous.

Abundant in wet places, var. α and β, throughout the islands, *Banks and Solander,* var. β extending to **Lord Auckland's** group; var. γ, dry places, Cape Palliser, *Colenso.* A very common and variable plant, best known by its small size, creeping stems, small, crowded, rounded, and scarcely toothed leaves, which are uniform in size throughout the plants, and flowers axillary and distant from the ends of the branches. States with ascending branches approach very closely *E. alsinoides.*

2. **E. purpuratum,** *Hook. f., n. sp.* Small, glabrous, creeping; branches short, black, leafy, 2–3 in. long. Leaves spreading horizontally, shortly petioled, $\frac{1}{3}$ in. long, thickly coriaceous, orbicular or ovate-oblong, obtuse, almost entire, under surface and edges dark-purple; veins obscure. Peduncles from the axils distant from the ends of the branches, stout, black, 1–2 in. long. Capsules $1\frac{1}{2}$–2 in. long, black, quite glabrous.

Middle Island: alps of Otago, alt. 6000 ft., *Hector.* Closely allied to *E. nummularifolium,* but much larger, with stout black stems and peduncles and long capsules.

3. **E. linnæoides,** *Hook. f. Fl. Antarct.* i. 9. *t.* 6;—*Fl. N. Z.* i. 58. Slender, prostrate, branched, creeping, everywhere quite glabrous. Leaves all opposite, petioled, orbicular, sharply toothed, uniform in size, very membranous, $\frac{1}{2}$ in. long. Peduncles very slender, fruiting ones 3–5 in. long.

Flowers pale rose-coloured, petals cleft nearly to the middle. Capsules 2–3 in. long, glabrous.

Mountainous districts throughout the islands, in wet places, *Banks and Solander*, etc.; most abundant in **Lord Auckland's** group and **Campbell's** Island, *J.D.H.* The slender creeping habit, uniform, membranous, orbicular, sharply toothed, opposite leaves, and few axillary flowers, well distinguish this species, which is one of the best characterized of the genus. It most nearly approaches *E. rotundifolium*, but differs in all the leaves being opposite, and branches creeping.

4. **E. macropus,** *Hook. Ic. Pl. t.* 812;—*Fl. N. Z.* i. 58. Glabrous throughout, branching from the base; branches slender, creeping below, then ascending or erect, 4–8 in. long, often purplish-green or blackish. Leaves almost all opposite, petioled, oblong or ovate-oblong, obtuse, very obscurely toothed, rather thick, $\frac{1}{3}$–$\frac{2}{3}$ in. long, uniform in size. Flowers few, distant from the ends of the branches, almost $\frac{1}{3}$ in. diam., on slender long peduncles. Capsules quite smooth.

Middle Island: marshy places in the mountains from Nelson to Otago, alt. 2500–5000 ft. The narrower, uniformly opposite, almost entire leaves, and large flowers distinguish this at once from *E. rotundifolium*; the less crowded leaves and large flowers from *E. confertifolium*; and the habit, and fewer flowers usually distant from the ends of the branches from various species of the following sections, especially *E. Billardierianum*.

5. **E. confertifolium,** *Hook. f. Fl. Antarct.* i. *p.* 10. Small, almost glabrous, sparingly branched, prostrate; branches stout, 1–4 in. long, densely leafy, their tops erect or almost wholly erect. Leaves all opposite, $\frac{1}{8}$–$\frac{1}{2}$ in. long, oblong linear-oblong or obovate-oblong, obtuse, coriaceous, shining, obscurely toothed, nearly uniform, or the uppermost rather smaller. Flowers in the axils of the upper leaves, on rather short peduncles, small, $\frac{1}{6}$ in. diam. Capsules glabrous or puberulous.—Hook. Ic. Pl. t. 685.

Var. *α*. Stems short. Leaves crowded, scarcely toothed. Peduncles very short.

Var. *β*. *tenuipes*. Leaves linear-oblong, more conspicuously toothed. Peduncles in fruit very slender.—*E. tenuipes*, H. f. Fl. N. Z. i. 59.

Mountain swamps, var. *α*, in the **Middle** Island; Canterbury and Otago mountains, alt. 3–5000 ft.; abundant in **Lord Auckland's** group and **Campbell's** Islands. Var. *β*. **Northern** Island: head of the Wairarapa valley, *Colenso*. Also found in the Tasmanian alps, and a very similar plant occurs in the Chilian Andes (*E. nivale*, Meyen). A very variable species, differing chiefly in habit from *E. Billardierianum*.

6. **E. crassum,** *Hook. f. Fl. N. Z.* ii. 328. Glabrous, prostrate, creeping, glossy; branches short, thick, red, densely leafy, 2–4 in. long. Leaves ascending, opposite, $\frac{3}{4}$–$1\frac{1}{4}$ in. long, obovate- or spathulate-oblong, obtuse, very obscurely toothed, narrowed into a broad red petiole, very coriaceous, bright and glossy. Flowers few, from towards the ends of the branches. Peduncles short and stout, lengthening in fruit. Flowers large, nearly $\frac{1}{2}$ in. across. Capsules glabrous, stout, $1\frac{1}{2}$ in. long.

Middle Island: in alpine localities. Top of Macrae's Run, alt. 4000 feet, *Munro*; Wairau mountains, alt. 4–5000 ft., *Travers*. A very fine species, allied to the following, and not readily distinguished except by the more creeping habit, and leaves broadest towards the tips, larger flowers, and long pedicels of the capsules.

7. **E. brevipes,** *Hook. f. Fl. N. Z.* ii. 328. Glabrous, stout, branching, coriaceous, glossy; branches all prostrate, red, almost woody below, leafy, 4–6 in. long. Leaves all opposite, spreading, $\frac{3}{4}$–1 in. long, narrow oblong, subacute, minutely remotely toothed, very coriaceous, glossy, the lower red-

dish, nerveless, petioles slender red. Flowers numerous towards the ends of the branches, nearly $\frac{1}{3}$ in. across, on very short peduncles that do not elongate after flowering. Capsules slender, glabrous, about 1 in. long.

Middle Island: Macrae's Run, alt. 4000 ft., *Munro;* Kaikora mountains, *M'Donald.* A remarkable species, of which I have fine specimens, but from Dr. Munro only, by whose shepherd it was collected: it may be a luxuriant form of *E. crassum.*

8. **E. alsinoides,** *A. Cunn.;—Fl. N. Z.* i. 59. Very slender, branched, prostrate, pubescent, small-leaved; branches green or reddish, 2–8 in. long. Leaves all opposite, in scattered or crowded pairs, $\frac{1}{6}$–$\frac{1}{3}$ in. long, oblong-orbicular or ovate- or linear-oblong, obtuse, obscurely toothed, sessile or shortly petioled. Flowers few, very small, towards the very ends of the branches, on short peduncles which lengthen much after flowering. Capsule very slender, pubescent.—*E. thymifolium* and *atriplicifolium,* A. Cunn.

Abundant in lowland marshes and wet ground throughout the islands, *Banks and Solander,* etc. The very slender habit, less coriaceous leaves, and pubescence distinguish this at once from *E. confertifolium,* the terminal inflorescence from *E. nummularifolium,* the small sessile leaves from *E. rotundifolium,* and the habit and pale stems from *E. microphyllum.*

9. **E. microphyllum,** *A. Rich. Flor.* 325. *t.* 36;—*Fl. N. Z.* i. 59. Rigid, strict, erect, wiry, generally much branched at the base; branches stout, simple, black, 4–6 in. long, glabrous or bifariously pubescent. Leaves all opposite, very uniform, small, $\frac{1}{8}$–$\frac{1}{6}$ in. long, sessile or very shortly petioled, ovate oblong or orbicular, obtuse, obscurely toothed, coriaceous. Flowers few, towards the ends of the branches, very small, on short peduncles which elongate but little in fruit. Capsules glabrous or pubescent, 1 in. long.

Northern and **Middle** Islands: not uncommon in moist places, *Banks and Solander,* etc.

10. **E. rotundifolium,** *Forst.;—Fl. N. Z.* i. 58. Erect or decumbent; stems weak, terete, 6–12 in. high, pubescent or glabrous. Leaves uniform throughout the plant, all opposite or more often the upper alternate, $\frac{1}{4}$–$\frac{2}{3}$ in., shortly petioled, orbicular or oblong, unequally toothed, rarely ovate or oblong and entire, glabrous or puberulous. Flowers in the axils of the uppermost leaves, small, pink, $\frac{1}{6}$–$\frac{1}{4}$ in. diam.; peduncles very short, in fruit long and slender. Capsule glabrous or pubescent, $1\frac{1}{2}$ in. long, very slender.

Abundant in moist places throughout the islands, *Banks and Solander,* etc. Very distinct in its usual form; distinguished from *E. alsinoides* by its greater size, from *E. linnæoides,* by the more or less erect habit, subterminal inflorescence, and smaller flowers.

11. **E. glabellum,** *Forst.;—Fl. N. Z.* i. 59. Stem erect or decumbent at the base, often much-branched below, terete, 6–12 in. high, glabrous or bifariously pubescent. Leaves in rather scattered pairs, sessile or shortly petioled, $\frac{1}{3}$–$\frac{2}{3}$ in. long, oblong ovate- or lanceolate-oblong, obtuse, obtusely sinuate-toothed, rather coriaceous, sometimes shining. Flowers in the axils of the upper leaves, variable in size, pink, $\frac{1}{6}$–$\frac{1}{3}$ in. diam.; peduncles very short, when fruiting rarely longer than the leaves. Capsules 1–$1\frac{1}{2}$ in. long, slender, glabrous or puberulous.

Abundant in upland districts in the **Northern** Island and throughout the **Middle** Island, *Banks and Solander,* etc. Abundant in temperate South America and Tasmania, where it becomes very large and the leaves very membranous. Nearly allied to *E. Billardieria-*

num, which differs only in its large flowers; extremely near *E. confertifolium*, which creeps and has crowded leaves. Small states pass into *E. alsinoides*, and others into *E. tetragonum*.

12. **E. melanocaulon,** *Hook. Ic. Pl. t.* 813;—*Fl. N. Z.* i. 60. Erect, 6–12 in., rigid; branches simple from a decumbent woody stem, terete, shining, black or dark-red, glabrous. Leaves crowded, mostly opposite, $\frac{1}{3}$–$\frac{2}{3}$ in. long, subpetioled, linear-oblong, obtuse, deeply lobed or sinuate-toothed, coriaceous, glabrous, often shining, yellow-red when dry. Flowers small, $\frac{1}{8}$–$\frac{1}{4}$ in. diam., pink, sessile in the upper leaves; ovaries short; peduncles very short, rarely exceeding the leaves when fruiting. Capsules $\frac{1}{2}$–1 in. long.

Common in mountainous districts of the **Northern** and throughout the **Middle** Island: alt. 2–5000 ft. A very well-marked species, readily known, in its usual form, by its erect, rigid habit, dark stem and branches, very numerous, crowded, small, coriaceous leaves and short, sessile ovaries. States of it are difficult to separate from *E. confertifolium, glabellum*, and *brevipes*.

13. **E. tetragonum,** *Linn.*;—*Fl. N. Z.* i. 60. Stems robust, erect or with a very short decumbent base, 12–18 in. high, stout, terete or 4-gonous, glabrous or puberulous. Leaves opposite, 1–2 in. long, sessile or semiamplexicaul, broadly oblong or obovate-oblong, obtuse, irregularly toothed, glabrous, rather membranous, never shining, green. Flowers rather small, $\frac{1}{3}$ in. diam., pink, very numerous towards the ends of the branches; peduncles shorter than the leaves, scarcely lengthening in fruit. Capsules 1–2 in. long, hoary.

Northern and **Middle** Islands: not uncommon in moist places. A common plant in the temperate regions of both hemispheres, very variable, distinguished by the small flowers and sessile or semiamplexicaul leaves, but states of it are difficult to distinguish from *E. Billardierianum, junceum*, and *pubens*.

14. **E. junceum,** *Forst.*;—*Fl. N. Z* i. 60. Erect, very pubescent or tomentose, rather rigid, leafy; stem stout, decumbent and rather woody at the base; branches stout or slender, 4–24 in. high, terete, often proliferous at the axils of the leaves. Leaves narrow, very variable, 1–3 in. long, alternate and opposite, sessile, the upper gradually smaller, linear-oblong, sinuate-toothed or nearly entire, narrowed or not at the base, usually pubescent on both surfaces. Flowers copiously produced towards the ends of the branches, rather small, $\frac{1}{6}$–$\frac{1}{4}$ in. diam., purple; peduncles generally shorter than the leaves, lengthening in fruit. Capsules pubescent, 1–3 in. long.—*E. cinereum*, A. Rich.; *E. virgatum, confertum, hirtigerum*, and *incanum*, A. Cunn.

Abundant throughout the islands, *Banks and Solander*, etc. Abundant in Tasmania, Australia, and South Chili, passing into *E. tetragonum* and *glabellum*.

15. **E. pubens,** *A. Rich. Flor.* 329. *t.* 36;—*Fl. N. Z.* i. 61. Stems erect from a woody decumbent base, 6–24 in. high, simple or branched; branches terete, pubescent, leafy, often proliferous at the axils. Leaves almost always alternate, long-petioled, ovate-oblong, obtuse, rather sharply toothed, membranous, pubescent on both surfaces, the upper gradually smaller. Flowers numerous towards the ends of the branches, $\frac{1}{4}$–$\frac{1}{3}$ in. diam., white pink or purplish; peduncles of both flower and fruit always shorter than the leaves. Capsules pubescent, 1–2 in. long.—*E. haloragifolium*, A. Cunn.

Abundant on dry hills, etc., throughout the islands, *Banks and Solander*, etc. One of the most distinct species of the genus, at once known by its erect stems, slender petioles, and usually alternate, ovate-oblong, obtuse leaves, etc.; it is allied to the European *E. roseum*.

16. **E. Billardierianum,** *Seringe;—Fl. N. Z.* i. 61. Slender, simple or sparingly branched, glabrous or puberulous; stems decumbent at the base; branches 6–18 in. high, terete, sparingly leafy. Leaves opposite and alternate, ½–1 in. long, sessile or very shortly petioled, linear-oblong or narrow ovate-oblong, obtuse, toothed, membranous, the upper gradually smaller. Flowers few towards the ends of the branches, rather large, ½–¾ in. diam., white or rosy; peduncles usually longer than the leaves. Capsules 1½–3 in. long, pubescent.

Mountainous parts of the **Northern** Island, *Colenso.* More common in the **Middle** Island, *Lyall,* etc. A most puzzling plant, and one of which I am almost inclined to refer the various forms to large-flowered states of *tetragonum, glabellum, alsinoides, junceum,* and perhaps others. None of the New Zealand specimens have leaves so sessile and semi-amplexicaul as those figured in the Tasmanian Flora: nor am I certain that either is the true plant of Seringe, which is very imperfectly described.

17. **E. pallidiflorum,** *Sol.;—Fl. N. Z.* i. 61. The largest New Zealand species. Stems erect, 1–3 ft. high, terete, as thick as a goose-quill, leafy, simple or branched, glabrous or puberulous. Leaves opposite, 2–4 in. long, sessile, semiamplexicaul, linear-lanceolate or -oblong, acute, jagged or toothed, glabrous. Flowers large, ¾–1 in. diam., white or pale-pink, numerous towards the ends of the branches; peduncles always shorter than the leaves. Capsule 2–3 in., hoary.—*E. macranthum,* Hook. f. Ic. Pl. t. 297.

Abundant in wet places throughout the islands, *Banks and Solander,* etc. A very distinct species, also common in Tasmania, at once recognized by its great size, long, sessile, narrow leaves and large flowers.

ORDER XXX. PASSIFLOREÆ.

Climbers (rarely erect shrubs), with lateral tendrils or 0. Leaves alternate, stipuled, entire or lobed. Flowers axillary, solitary or rarely racemose, usually hermaphrodite, often showy.—Calyx free; tube short; lobes 4 or 5 (rarely 3) long, imbricate. Petals 3–5, inserted at the base of the calyx, usually persistent (rarely 0), sometimes with a crown of filaments at their base. Stamens generally as many as the petals; filaments often adnate to the stalk of the ovary. Ovary often stalked, 1-celled; styles 3–5; ovules many on 3–5-parietal placentas. Fruit coriaceous or succulent, indehiscent or splitting between the placentas. Seeds often arilled, exalbuminous; cotyledons flat.

A large tropical family, abounding in America, rarer in Asia.

1. PASSIFLORA, Linn.

Flowers unisexual or hermaphrodite.—Calyx-lobes 4 or 5. Petals as many, with a ring of filaments at their base. Stamens 4 or 5; filaments quite adnate with the stalk of the ovary; anthers versatile, those of the female flower, imperfect. Styles 2 or 3, with capitate stigmas. Fruit an indehiscent berry.

A very large American genus, of which the few Old-World representatives are more or less diœcious and tetramerous.

1. **P. tetrandra,** *Banks and Sol.;—Fl. N. Z.* i. 73. A lofty, glabrous climber, with slender, terete branches. Leaves 1½–4 in., petioled, lanceolate, acuminate, quite entire, eglandular; tendrils long and slender. Flowers ½ in.

diam., green, in 2–4-flowered cymes; peduncles and pedicels slender, the latter jointed below the middle, ebracteate. Sepals 4, oblong, obtuse. Petals like the sepals. Filaments of corona slender, shorter than the petals. Stamens with very slender filaments. Fruit nearly globular, orange-coloured, 1–1½ in. diam., many-seeded. Seeds flat; testa wrinkled.—*Tetrapathæa australis*, Raoul, Choix, t. 27.

Northern and **Middle** Islands: not uncommon in skirts of woods, as far south as Akaroa, *Banks and Solander*, etc.

ORDER XXXI. CUCURBITACEÆ.

Climbing or prostrate, weak, straggling herbs, rarely stout and woody, with tendrils from the sides of the stem near the petioles. Leaves alternate, exstipulate, usually palmately veined and lobed. Flowers mostly unisexual.—Calyx superior (or adherent at the base) and produced above the ovary, usually campanulate, 5-toothed or -lobed. Petals 5, free or united into a 5-lobed corolla, often continuous and confluent below with the calyx-tube. Male fl.: Stamens 3 or 5, free or more or less combined; anthers often confluent into a waved or curved mass. Female fl.: Ovary inferior, usually 1-celled when very young, placentæ becoming thick and meeting together in the axis of the ovary, which they divide into 3–6 cells; styles 3-fid or 3-partite, stigmas entire or lobed; cells 1- or more ovuled. Fruit succulent or coriaceous, indehiscent or bursting irregularly, few- or many-seeded. Seeds usually flat; testa coriaceous or bony; albumen 0; cotyledons large, flattish.

A large Order, abundant in the tropics, rare in the temperate zone, to which the Cucumber, Melon, and whole family of Gourds, Pumpkins, etc., belong. All are bitter and poisonous in a wild state.

1. SICYOS, Linn.

Prostrate or climbing herbs.—Male flowers racemed. Calyx campanulate, 5-toothed. Corolla continuous with the calyx, 5-lobed. Anthers 3–5, on the summit of a short dilated column; cells confluent. Female flowers capitate on an axillary peduncle, solitary or umbelled. Calyx-tube adnate with the ovary; limb campanulate, free, 5-toothed. Ovary 1-celled, 1-ovuled, with a trifid style. Fruit a small coriaceous spinous nut.

A small, not uncommon, tropical and subtropical genus.

1. **S. angulatus,** *Linn.;—Fl. N. Z.* i. 72; ii. 329. Stems climbing and trailing, often several feet long, glabrous or nearly so. Leaves petioled, 2–5 in. diam., broadly reniform with a deep sinus, palmately 5–7-lobed, toothed, glabrous or scabrid above and pubescent beneath. Tendrils palmately divided at the apex of a long petiole. Male and female inflorescences often from the same axil; male peduncle often several in. long. Flowers green, ⅓ in. diam., pedicelled. Corolla-lobes obtuse. Anthers 3, convolute on the circumference of a very short, broad, dilated column. Female flower on shorter peduncles. Ovary hispid with stiff bristles. Nuts 5–7 together, ovate, sessile, compressed, 1-seeded, densely clothed with barbed spines.

Northern and northern parts of the **Middle** Islands: on the coast abundant in many places, *Banks and Solander*, etc. **Kermadec** group, *M'Gillivray*. Also a native of Australia, Tasmania, and both shores of America. A. Gray (Bot. U. S. Expl. Exped. i. 648) keeps this distinct from *S. angulatus*, as *S. australis*, Endl., but I find no characters.

ORDER XXXII. FICOIDEÆ.

The New Zealand species are succulent herbs, with opposite or alternate exstipulate leaves, but the habit of the Order is very various. Flowers uni- or bi-sexual, regular.—Calyx superior or inferior, 4- or many-cleft, persistent. Petals perigynous or epigynous, 4 or 5 or very numerous. Stamens 4 or 5 or very numerous; filaments free; anthers oblong. Ovary inferior or superior, 2- or more celled; styles as many as the ovary cells; ovules solitary or numerous in the cells. Fruit very various, capsular drupaceous or baccate. Seeds usually with farinaceous albumen and terete curved or annular embryo.

A considerable Order, scattered over the temperate and tropical regions of the globe.

Petals numerous. Capsule dehiscent at the top . . . . 1. MESEMBRYANTHEMUM.
Petals 0. Drupes 3–8-celled . . . . 2 TETRAGONIA.

1. MESEMBRYANTHEMUM, Linn.

Succulent, usually creeping plants.—Calyx-tube adnate to the ovary, limb 5-parted. Petals very numerous, linear. Stamens very numerous. Ovary inferior, 4- or more celled; styles free or connate; ovules very numerous, attached to the bases of the cells by long funicles. Capsule bursting by valves at the exposed top, the rest included in the calyx-tube. Seeds numerous, subglobose or pyriform.

An immense South African genus, rare elsewhere.

1. **M. australe,** *Soland.*;—*Fl. N. Z.* i. 76. Stems woody, terete, creeping, 1–2 ft. long, rooting at the nodes. Leaves opposite, 1–3 in. long, fleshy, linear, triquetrous, acute, glaucous, punctate, variable in thickness. Peduncles axillary, solitary, very thick, shorter or longer than the leaves. Flowers $\frac{2}{3}$–1 in. diam., pink or white, unisexual. Calyx-tube fleshy, obconic, lobes 5, unequal, the inner smaller. Petals 50–60, linear. Capsule 5-celled, 5-valved. Seeds minute, with a dark-brown, smooth testa.

Common on rocky and sandy shores throughout the islands, *Banks and Solander*, etc. Fruit full of red colouring-matter, *Buchanan*. Also abundant in Australia and Tasmania, and is, I suspect, not very different from the Cape of Good Hope *M. crassifolium*, Linn.

2. TETRAGONIA, Linn.

Scandent or trailing (or erect), succulent herbs. Leaves alternate, petioled. Flowers axillary, peduncled.—Calyx-tube adnate with the ovary, limb 3- or 4-cleft. Petals 0. Stamens 1 or more, perigynous, filaments filiform. Ovary inferior, 3–8-celled; styles 3–8; ovules solitary. Fruit a drupe or bony nut, globose or obconic and 4-angled with the angles produced and sometimes proliferous. Testa membranous.

A large South African genus, containing also a few Japan, Australian, New Zealand, and American species.

Leaves 1–3 in. long. Fruit 4-sided, turbinate . . . . 1. *T. expansa*.
Leaves 1 in. long. Fruit globose, red . . . . 2. *T. trigyna*.

1. **T. expansa,** *Murray;—Fl. N. Z.* i. 77; ii. 329. A trailing or subscandent herbaceous plant, glabrous or sparsely puberulous, glistening with cellular papillæ. Leaves petioled, 1–3 in. long, ovate or rhomboid, obtuse, entire or sinuate. Flowers on short peduncles, $\frac{1}{4}$ in. long. Calyx urceolate, 4-lobed. Stamens 4–8, irregularly inserted. Styles 3–8, recurved. Fruit about $\frac{1}{3}$ in. long, turbinate or obpyramidal with the angles produced, 3–8-celled.—Bot. Mag. t. 2362; *T. halimifolia*, Forst.

Abundant near the sea, in many parts of both the islands, *Banks and Solander*, etc. Has been cultivated in Europe as New Zealand Spinach. Also a native of Japan, Australia, Tasmania, and South America.

2. **T. trigyna,** *Banks and Sol.* Smaller than *T. expansa;* leaves broader. Flowers smaller; styles 3. Drupe small, globose, fleshy, bright-red, $\frac{1}{6}$ in. diam.

Northern Island: east coast, *Banks and Solander*. Sand-hills, Cape Turnagain, *Colenso*. Auckland, *Sinclair*: In the dried state it is difficult to distinguish this from *T. expansa,* but the figure in the Banksian collection represents a very different plant, of smaller size, with rounder leaves, and with unarmed and fleshy drupes.

ORDER XXXIII. UMBELLIFERÆ.

Herbs, rarely shrubs, often aromatic or rank-smelling when bruised. Leaves chiefly radical, usually much divided; petioles sheathing at the base. Flowers small, in lateral or terminal umbels (capitate in *Eryngium* and sometimes in *Hydrocotyle*), which are simple or compound; bracts when present forming involucres.—Calyx-tube adnate with the ovary, limb truncate or 5-toothed. Petals 5, inserted round an epigynous disk, their tips often inflexed, valvate or imbricate. Stamens 5; filaments incurved. Ovary inferior, 2-celled; styles 2, stigmas terminal; ovules 1 in each cell, pendulous. Fruit separating into 2 dry carpels, each often with prominent ribs or wings, the pericarp often traversed with longitudinal oil-canals (vittæ). Seed linear; embryo minute; albumen horny.

A very large Order, especially in the northern temperate zone, rarer in the southern, to which the Celery, Carrot, Parsnip, etc., belong. Distinguished from *Araliaceæ* chiefly by habit.

Stems rhizomes or scions creeping. Umbels simple or capitate. Leaves never pinnate.
Stems usually slender and creeping. Leaves broad, simple or 3-foliolate, not spiny. Flowers capitate.
Fruit flat, much laterally compressed . . . . 1. HYDROCOTYLE.
Fruit 4-gonous, dorsally rounded or compressed . . . . . 2. POZOA.
Stems creeping. Leaves terete, fistular, septate. Umbel simple . 3. CRANTZIA.
Stems creeping. Leaves spinescent. Umbels crowded into a very dense ovoid head . . . . 4. ERYNGIUM.
Stems erect, rarely scandent, without creeping rhizomes or scions.
Umbels simple. Leaves pinnate or 2–3-pinnate . . . . . . 5. OREOMYRRHIS.
Umbels compound.
Stems prostrate. Umbels lateral. Involucre 0 . . . . . . 6. APIUM.
Stems rigid, spinescent. Umbels spiked or panicled . . . . 7. ACIPHYLLA.

Aromatic herbs. Umbels terminal. Carpels with 3–5 equal, narrow wings . . . . 8. LIGUSTICUM.
Herbs or climbing shrubs. Leaves simple or compound. Carpels with 2 broad, lateral wings . . . . 9. ANGELICA.
Herbs. Leaves compound. Carpels prickly . . . . . . 10. DAUCUS.

1. HYDROCOTYLE, Linn.

Herbs, with long, slender, creeping, often tufted stems. Leaves alternate, rarely fascicled, reniform or orbicular, petioles slender, dilated and membranous at the base. Umbels or heads small, simple. Involucral leaves few or 0. Flowers very minute.—Calyx-limb obsolete. Petals valvate, flat or incurved, but tips not inflexed. Fruit orbicular, much laterally flattened. Carpels placed edge to edge, each with one or few (sometimes obsolete) ribs on each face.

A rather extensive genus, found in all moist temperate regions. Some of the species are very widely diffused; the New Zealand ones numbered 8, 9, 10, are not satisfactorily distinguished.

Umbels on long peduncles. Fruits on very slender pedicels . . 1. *H. elongata.*
Umbels sessile in the axils of the leaves. Fruits sessile . . . 2. *H. americana.*
Umbels more or less peduncled. Fruits nearly sessile, capitate.
Leaves tufted. Involucre 2-leaved. Fruits 2 or 3, large . . 3. *H. asiatica.*
Leaves 3–6-partite. Peduncles slender. Umbels 3–6-flowered 4. *H. muscosa.*
Leaves 5–7-partite. Peduncles slender. Umbels 40–50-flowered 5. *H. dissecta.*
Leaves more or less deeply 5-7-lobed. Peduncles short, slender. Umbels 3–5-flowered. Fruit hispid . . . . 6. *H. heteromeria.*
Peduncles long, slender. Umbels 3–5-flowered. Carpels very flat, broadly winged . . . . 7. *H. pterocarpa.*
Lobes very obscure, obscurely crenate. Peduncles short, slender. Umbels loosely 5–10-flowered . . . . 8. *H. novæ-Zelandiæ.*
Hairy or tomentose. Lobes distinct, toothed. Umbels densely many-flowered . . . . 9. *H. moschata.*
Nearly glabrous, small. Peduncles slender. Umbels many-flowered, very small . . . . 10. *H. microphylla.*

1. **H. elongata,** *A. Cunn.;—Fl. N. Z.* i. 84. Slender, very pilose or almost glabrous; stems 8–10 in. long. Leaves $\frac{1}{2}$–1 in. diam., orbicular-reniform, deeply 5–7-lobed, lobes sharply toothed; petioles 1–2 in. long, slender; stipules small. Peduncles very slender, longer than the leaves. Umbels 20–50-flowered. Flowers very minute, on slender strict pedicels, $\frac{1}{12}$–$\frac{1}{4}$ in. long. Fruit very small. Carpels with 1 rib on each face, brown.

Wet places in the **Northern** and **Middle** Islands, not uncommon. Very near indeed to the Chilian *H. geranioides*, but that is a much more robust plant.

2. **H. americana,** *Linn.;—Fl. N. Z.* i. 82. Perfectly glabrous, slender, creeping, shining. Leaves $\frac{1}{4}$–$\frac{1}{2}$ in. diam., orbicular-reniform, 5–7-lobed, lobes shallow crenate; stipules small. Umbels very small, 3–5-flowered, sessile in the axils of the leaves. Flowers sessile. Fruit minute, pale-yellowish. Carpels with 1 rib on each face, glabrous, one often seedless and hispid.

Northern Island: marshy ground, not unfrequent. A Canadian, United States, and Brazilian plant, not hitherto found elsewhere in the Old World. The American specimens have sometimes pedicelled flowers.

3. **H. asiatica,** *Linn.;—Fl. N. Z.* i. 82. Stems creeping, rather robust. Leaves fascicled, cordate-oblong or oblong-reniform, obtuse, sinuate-toothed or entire, usually glabrous, $\frac{1}{4}$–$1\frac{1}{2}$ in. long; stipules membranous. Umbels 2–3-flowered, on short peduncles. Involucral leaves 2 or 3, large, pilose. Flowers shortly pedicelled. Fruit large, often pilose. Carpels with 2 or 3 stout ribs on each face, reticulated.—*H. cordifolia,* Hook. Ic. Pl. t. 303.

Common in marshy places throughout the islands, *Banks and Solander*, etc. Found also in Australia, Tasmania, and in the warmer parts of Asia, Africa, and America.

4. **H. muscosa,** *Br.;—H. tripartita,* Br. ?, Fl. N. Z. i. 83. Stems glabrous or pilose, tufted. Leaves $\frac{1}{4}$ in. diam., 3-5-partite to the base, the lobes obcuneate, 2–3-toothed at the tip, coriaceous, stipules rather large. Peduncles shorter than the leaves. Umbels 3–6-flowered. Flowers almost sessile. Fruit glabrous. Carpels rather turgid, with 1 slender rib on each face.

Northern Island: swamps at the foot of Tongariro, *Colenso.* **Middle** Island: abundant, *Hector and Buchanan*, etc. Also a native of Tasmania and Southern Australia.

5. **H. dissecta,** *Hook. f. Fl. N. Z.* i. 84. Small, slender, hispid-pilose. Leaves alternate and fascicled, 1 in. diam., cut nearly to the base into 5–7 obovate-cuneate, acute, deeply toothed or laciniate segments, deep green, hairy on both surfaces, teeth rather diverging. Peduncles slender. Umbels very many-flowered. Flowers densely capitate. Fruit small, glabrous. Carpels rather turgid, with 1 obscure rib on each face.

Northern Island, *Colenso.* A most distinct species, of which I have very imperfect specimens.

6. **H. heteromeria,** *DC.;—Fl. N. Z.* i. 82. Glabrous or sparsely pilose. Stems very slender. Leaves $\frac{1}{2}$–$\frac{3}{4}$ in. diam., reniform with a deep sinus, obtusely crenate and obscurely lobed, shining, membranous; petioles very slender; stipules membranous. Peduncles slender, half the length of the petioles, or shorter. Umbels 3–5-flowered. Flowers very shortly pedicelled. Fruit very small, pale-coloured. Carpels with 1 evident rib on each face, one often hispid.

Northern Island: in marshes, Bay of Islands, Auckland, etc. Very near *H. americana*, and perhaps a peduncled form of that plant.

7. **H. pterocarpa,** *F. Muell.* Perfectly glabrous; stems very long, slender. Leaves 1–2 in. diam., orbicular-reniform with a narrow deep sinus, obscurely broadly crenate-lobed, very membranous; petioles a span long; stipules small, membranous. Peduncles long, slender, shorter than the leaves. Umbels 3–5-flowered. Flowers small, shortly pedicelled. Fruit larger than in any other New Zealand species, very flat. Carpels with 1 rib on each face, glabrous, mottled with brown.—Fl. Tasm. i. 153. t. 33.

Northern Island, *Colenso.* A native of Tasmania and Victoria, where the peduncles are shorter.

8. **H. novæ-Zelandiæ,** *DC.;—Fl. N. Z.* i. 83. Glabrous or pilose; stems very slender, 5–10 in. long. Leaves $\frac{1}{2}$–$\frac{3}{4}$ in. diam., reniform, with usually an open sinus, very obscurely 5–7-lobed, lobes obscurely and obtusely crenate, membranous; petioles slender; stipules rather large. Peduncles slender, much shorter than the leaves. Umbels 5–10-flowered. Flowers very

shortly pedicelled. Fruit glabrous, $\frac{1}{12}$ in. diam., rather turgid, pale brown. Carpels rounded on the back, with 1 obscure rib on each face.

Abundant throughout the islands in wet places, *Banks and Solander*, etc. Very near the Indian *H. rotundifolia*, Roxb., the Australian *H. vagans*, H. f., and the American *H. Bonplandi*, Rich., but the leaves are less deeply lobed.

9. **H. moschata,** *Forst.*;—*Fl. N. Z.* i. 83. Hispidly pilose tomentose or glabrescent; stems rather slender, tufted, 2–6 in. long. Leaves $\frac{1}{4}$–$\frac{1}{2}$ in. diam., reniform or orbicular with usually an open sinus, 5–7-lobed, generally very distinctly, but never to the middle, lobes acutely toothed, rather coriaceous, and generally hispid on both surfaces; petioles short, $\frac{1}{2}$–1 in. long; stipules large, membranous. Peduncles very variable, as long as the petioles or shorter or 0. Umbels few- or many-flowered. Fruit minute, $\frac{1}{20}$ in. diam., generally crowded, glabrous, turgid. Carpels with 1 rib on each face, acute at the back.

Abundant in moist places throughout the islands, *Banks and Solander*, etc. I suspect the Tasmanian *H. hirta*, Br., to be one form of this, the Tristan d'Acunha *H. capitata*, Thouars, and the Chilian *H. Bonplandi*, Rich., others. In some New Zealand specimens the leaves are scarcely lobed, and the teeth obtuse.

10. **H. microphylla,** *A. Cunn.*;—*Fl. N. Z.* i. 84. Glabrous or sparingly hairy; stems almost filiform, often stout at the very base, 1–4 in. long. Leaves $\frac{1}{8}$–$\frac{1}{3}$ in. diam., orbicular-reniform, with a closed sinus, 5–7-lobed, lobes shallow, obtusely 3–4-crenate; petioles $\frac{1}{4}$–$\frac{1}{2}$ in. long; stipules rather large. Peduncles as long as or shorter than the petioles, slender. Umbels very small, 3–5-flowered. Flowers almost sessile. Fruit very minute, brown, glabrous, $\frac{1}{14}$ in. diam. Carpels rather turgid, with an indistinct rib on each face, obtuse at the back.

Not uncommon in moist places, from the Bay of Islands to Nelson, *Travers*; but easily overlooked, from its small size.

2. POZOA, Lagasca.

Small herbs, with radical leaves and scapes, or, like *Hydrocotyle*, with creeping stems and fascicled simple or 3–5-foliolate leaves. Umbels simple. Involucral leaves free or connate, forming a toothed or lobed cup. Flowers small, sometimes unisexual.—Calyx 5-toothed. Petals valvate, without inflexed tips. Fruit almost tetragonous, the carpels dorsally compressed rounded flat or concave, more or less evidently 5-ribbed.

A small temperate South American, Tasmanian, and New Zealand genus; the species have often a rank odour when bruised.

Very minute, stemless. Leaves nearly orbicular, obscurely lobed . 1. *P. exigua.*
Rhizome stout. Leaves orbicular. Stipules ciliate. Fruit shorter than its pedicel . . . . 2. *P. Haastii.*
Rhizome creeping. Leaves orbicular. Stipules entire. Fruit longer than its pedicel . . . . 3. *P. reniformis.*
Stems creeping, slender. Leaves 3-foliolate. Leaflets petioled . 4. *P. trifoliolata.*
Stems creeping, stout. Leaves 3–5-foliolate. Leaflets sessile, coriaceous . . . . 5. *P. hydrocotyloides.*
Rhizome stout. Leaves 3–5-foliolate. Leaflets sessile, coriaceous . 6. *P. Roughii.*

1. **P. exigua,** *Hook. f., n. sp.* A very minute plant, $\frac{1}{2}$–1 in. high. Rhizome stout; branches 0. Leaves numerous from rhizome, orbicular-ovate

cordate at the base, $\frac{1}{6}$ in. diam., obscurely 3–5-lobed, coriaceous, glabrous; stipules obsolete; petiole very stout. Scape short, stout; involucral leaves linear-oblong, connate at the base. Calyx-teeth very unequal. Fruit not seen.

Middle Island: Otago, lake district, alpine, *Hector and Buchanan.*

2. **P. Haastii,** *Hook. f., n. sp.* Very variable in size. Rhizome stout or slender, sometimes as thick as the little finger, often sending out creeping suckers, or prostrate leafing and flowering branches 6–10 in. long. Leaves 1–1$\frac{1}{2}$ in. broad, coriaceous, bright-green and glossy, reniform or orbicular, rounded or deeply lobed at the base; margin almost cartilaginous, crenate-lobed; lobes very broad, shallow; petioles stout, 1–3 in. long; stipules ciliate or laciniate. Umbels many-flowered, rising from the suckers in the axils of shortly petioled, 3–4-lobed leaves, shortly peduncled. Involucral leaves broadly linear, obtuse, longer than the pedicels. Pedicels $\frac{1}{4}$ in. long, generally 4 times as long as the shortly oblong fruit.

Middle Island: Hopkins river, 4000 ft.; Mount Darwin, Waitaki river, and Hunter river, on shingle, alt. 3500–4500 ft., *Haast.* The leaves vary in being sometimes deeply lobed, at others rounded and not lobed at all at the base.

3. **P. reniformis,** *Hook. f. Fl. Antarct.* i. 15. *t.* 11. Small, perfectly glabrous, shining, bright green. Rhizome slender, creeping or sending out stolons. Leaves $\frac{1}{2}$–$\frac{2}{3}$ in. diam., coriaceous, reniform or orbicular with a rather open sinus, crenate-lobate; petioles stout, 1–2 in. long; stipules acuminate, quite entire. Peduncles shorter than the leaves. Involucral leaves linear, obtuse, membranous, with a green nerve. Umbels 6–10-flowered. Flowers very small, shortly pedicelled. Fruit glabrous, linear-oblong, longer than its pedicel, $\frac{1}{8}$ in. long.

Lord Auckland's group: in clefts of rocks, *J. D. H.* Very near indeed to ***P. Fragosia,*** F. Muell., of the Victoria Alps, which has ciliate stipules.

4. **P. trifoliolata,** *Hook. f. Fl. N. Z.* i. 85. *t.* 18. Very slender, glabrous. Stem 2–6 in. long, filiform, creeping. Leaves 3-foliolate; leaflets with the slender petioles $\frac{1}{4}$–$\frac{1}{3}$ in. long, obovate-cuneate, unequally 3–4-lobed; lobes obtuse, apiculate, narrowed into a petiole 1–3 in. long; stipules very obscure, ciliate. Peduncles shorter than the petioles, slender. Umbels 3–6-flowered; involucral leaves subulate, small, 2–5-fid. Flowers minute, almost sessile; petals acute. Fruit very small, $\frac{1}{20}$ in. long, dark-brown or pale-yellow. Carpels almost terete at the back.

Var. *β. tripartita;* very small, with a few bristly hairs here and there; leaflets sessile, more coriaceous, $\frac{1}{4}$ in. long.

Northern Island: Totara-nui, *Banks and Solander;* stony and rocky places on the east coast and Ruahine range, *Colenso.* Var. *β.* Crags on Titiokura, *Colenso.*

5. **P. hydrocotyloides,** *Hook. f., n. sp.* Glabrous, stout, tufted. Stems creeping and rooting by stout runners 3–5 in. long. Leaves $\frac{1}{3}$–$\frac{2}{3}$ in. diam., orbicular-reniform, coriaceous, 3–5-foliolate; leaflets sessile, broadly obovate-cuneate, 3–5-lobed or -crenate at the tip; petioles $\frac{1}{2}$–1 in. long, stout; stipules subulate, entire or ciliate. Peduncles as long as the leaves, solitary from the nodes of the stem, or several at the apex of a leaf-bearing scion. Umbels as in *P. Roughii,* but much smaller.

Middle Island: grassy plains, Rangitata river, *Sinclair*; moist ground, near the sources of the Kowai, alt. 2–3000 ft., *Haast*. This approaches the var. β of *P. trifoliolata*, but the whole plant is much more robust and the umbels are many-flowered; it differs from *P. Roughii* in its small size, creeping stems, and habit, but considering how variable alpine plants are, it may prove a var. of it. It is nearly allied to the Falkland Island *P. Ranunculus*, Hook. f. (*Azorella Ranunculus*, D'Urv.).

6. **P. Roughii,** *Hook. f., n. sp.* Glabrous, coriaceous, shining. Rhizome often as stout as the little finger, crowned with numerous leaves, and with prostrate slender leafy scions bearing umbels and leaves. Leaves $\frac{3}{4}$ in. diam., orbicular, very coriaceous, 5-foliolate; leaflets sessile, broadly obcuneate, 5-lobed at the end; lobes rounded, obtuse; petioles 1–2 in. long, very stout; stipules laciniate or toothed. Flowering-scions longer than the leaves, bearing 1 sessile and 2 or more peduncled umbels, rising from the axils of small, simple or lobed leaves. Peduncles $\frac{1}{2}$–$\frac{3}{4}$ in. long. Umbels many-flowered. Involucres linear, obtuse. Flowers small. Fruit small, $\frac{1}{12}$ in., twice as long as the slender pedicels. Carpels rounded at the back.

Middle Island: mountains, near Nelson, *Rough*. The habit of this is that of *P. Haastii*.

3. CRANTZIA, Nuttall.

A small, glabrous, creeping herb. Leaves fascicled, terete, fistular, internally septate, rarely flattened and linear towards the tip. Umbels simple, shortly peduncled. Involucral leaves few, very short. Flowers very minute, pedicelled.—Calyx minutely 5-toothed. Petals not inflexed at the tip. Fruit ovoid-globose. Fruit terete; carpels not contracted at their opposed faces, spongy, each with 5 thick ribs separated by slender furrows.

1. **C. lineata,** *Nutt.*;—*Fl. N. Z.* i. 87; *Fl. Antarct.* ii. 287. *t.* 100. Rhizome rank-smelling, as thick as a crowquill, 2–6 in. long. Leaves very variable, $\frac{1}{2}$–4 in. long, the tip when expanded to a lamina $\frac{1}{14}$–$\frac{1}{6}$ in. broad. Peduncles much shorter than the leaves. Flowers few in an umbel, spreading.

Northern and **Middle** Islands: not uncommon in wet, sandy, gravelly, and boggy places. East coast, *Colenso*; Nelson, *Bidwill*. The only species is a native of the United States and Mexico, the Andes of New Granada and Peru, Chili and the Falkland Islands, Tasmania, and Victoria.

4. APIUM, Linn.

Erect or prostrate, rank-smelling, glabrous herbs. Leaves alternate or radical and fascicled, 2-pinnate or 3-foliolate. Umbels shortly peduncled or subsessile, simple or compound. Involucres 0.—Calyx-limb 0. Petals broad, with incurved tips. Fruit subglobose. Carpels with 5, slender or thick, spongy ribs, and 1–3 vittæ between each rib, contracted at their contiguous faces.

The genus to which the cultivated Celery belongs. The New Zealand species very closely resemble the European, but differ in the thick, spongy ribs of the carpels; one of them, which is also a native of Fuegia, in the latter country forms an excellent salad and potherb in its wild state.

Branches stout. Leaves pinnate or 2–3-pinnate . . . . 1. *A. australe.*
Branches slender. Leaves 3-foliolate . . . . 2. *A. filiforme.*

1. **A. australe,** *Thouars;—Fl. N. Z.* i. 86. Root stout. Stems prostrate, ascending, rarely erect, branched from the base, grooved, thickness of a sparrow- to a goose-quill, leafy, 6 in.–2 ft. long. Leaves 3–8 in. long, pinnate or 2–3-pinnate; leaflets petioled or sessile, very variable in size, linear obovate or obcuneate, variously lobed or cut, membranous or coriaceous. Umbels usually axillary, compound and sessile, hence appearing like many simple umbels; rays many, 1–2 in. long; partial ones $\frac{1}{4}$ in. long. Flowers small, white. Fruit ovoid, glabrous, $\frac{1}{12}$–$\frac{1}{10}$ in. long.

Var. *α.* Leaflets variously cut, broadly obovate or obcuneate.

Var. *β.* Leaflets cut into narrow linear or lanceolate lobes or leaflets.

Abundant on the shores of all the islands, *Banks and Solander*, etc. A most variable plant, found also in Australia, Tasmania, temperate and cold South America, the Cape of Good Hope, Tristan d'Acunha and St. Paul's Island.

2. **A. filiforme,** *Hook.;—Fl. N. Z.* i. 87. Stems prostrate, slender, 6–12 in. long. Leaves 3-foliolate, rarely pinnate; leaflets petioled, obovate or orbicular-cuneate, 3-lobed or variously lobed or cut; petioles 1–2 in. long. Umbels as in *A. australe*, but smaller, sometimes peduncled, sometimes sessile and reduced to one primary ray. Fruit short, with thick spongy ribs, very variable in size.—Hook. Ic. Pl. t. 819.

Northern and **Middle** Islands: generally on rocky shores, not uncommon. This is I expect, a variety of the preceding: also found in the Isle of Pines.

5. ERYNGIUM, Linn.

Harsh, rigid, generally spinous herbs or shrubs. Umbels small, collected into dense, spiny, involucrate, globose or oblong heads. Involucral leaves spiny, involucels scattered amongst the flowers. Flowers very minute, sessile. —Calyx-limb of 5, erect lobes; tube covered with scales. Petals deeply lobed, with an inflexed lacinia from between the lobes. Fruit obovoid or subterete, often scaly or tubercled. Fruits nearly terete, without ribs or vittæ.

A large European, Oriental, and South American genus.

1. **E. vesiculosum,** *Labill. Fl. Nov. Holl.* i. 73. *t.* 98;—*Fl. N. Z.* i. 85. A small, rigid, spinous herb, 2–10 in. high. Root stout, crowned with a tuft of leaves and throwing out stout suckers 4–8 in. long, which bear small cuneate toothed leaves at the tip. Leaves tufted, oblong-lanceolate or linear-lanceolate, 3–6 in. long, narrowed into long flat petioles, rarely 1 in. broad, deeply toothed or pinnatifid, the lobes spinescent. Peduncles radical, bearing one small, broadly ovoid head. Involucral leaves 8–10, stellate, rigid, subulate, pungent, $\frac{1}{4}$–$\frac{1}{3}$ in. long. Flowers very inconspicuous, mixed with the pungent projecting scales of the involucels. Calyx densely clothed with imbricating, chaffy, convex scales.

East coast of the **Northern** and **Middle** Islands, from Auckland to Otago. A native of Tasmania and South Australia.

6. OREOMYRRHIS, Endlicher.

Silky hairy or nearly glabrous, scapigerous or branched herbs. Leaves pinnate or decompound. Umbels simple, solitary on the scapes, with broad, linear involucral leaves.—Calyx-limb obsolete. Petals with an incurved tip. Fruit linear or ovate-oblong, somewhat laterally compressed; carpels with 5 rather obtuse ridges, 3 dorsal and 2 lateral.

A small genus, confined to Tasmania, Southern Australia, New Zealand, and the Andes from Mexico to Chili.

Scapes long, radical. Fruit glabrous . . . . 1. *O. Colensoi.*
Scapes long, radical. Fruit tomentose . . . . 2. *O. Haastii.*
Stem branched. Umbels axillary. Fruit glabrous . . . . 3. *O. ramosa.*

1. **O. Colensoi,** *Hook. f. Fl. N. Z.* i. 92. Very variable in stature, from 2–10 in. high, glabrous or pilose. Leaves numerous, linear-oblong, pinnate or 2-pinnate; leaflets opposite, sessile or petioled, $\frac{1}{6}$–$\frac{1}{4}$ in. broad, broadly oblong, inciso-serrate or inciso-pinnatifid, turning brown when dry; petioles wiry; sheaths short, membranous. Scapes several, pubescent or woolly, especially upwards, where the hairs point downwards. Involucre of 6–8 ovate leaflets, $\frac{1}{6}$–$\frac{1}{4}$ in. long. Flowers small, white, almost sessile in the involucres; pedicels very short, elongating in fruit to 1 in. or less. Fruit quite glabrous.

Northern Island: mountainous districts of the east coast and interior, abundant in grassy plains, etc. **Middle** Island: Waimakeriri county, 2500 ft., and Kowai river, *Haast.*

2. **O. Haastii,** *Hook. f., n. sp.* Very similar to *O. Colensoi*, but more flaccid. Leaves pinnate, with the leaflets more membranous, petioled, broadly ovate, variously lobed or 3-foliolate. Fruit linear, densely tomentose; pedicels usually shorter than the involucral leaves.

Middle Island: Mount Darwin, alt. 3200 ft., *Haast.*

3. **O. ramosa,** *Hook. f., n. sp.* Stem 6–24 in. high, much-branched from the base, and branches very slender, glabrescent or pilose. Leaves pinnate or 2-pinnate, ovate-oblong in outline; leaflets or primary divisions in few pairs, opposite, long petioled, $\frac{1}{4}$–$\frac{1}{2}$ in. long, ovate, deeply 3–5-lobed or partite or again pinnate, the lobes obtuse, membranous pilose; petioles general and partial very slender. Umbels 6–8-flowered; peduncles axillary, usually shorter than the leaves; involucral leaves small. Flowers nearly sessile. Fruits unequally pedicelled. $\frac{1}{6}$–$\frac{1}{5}$ in. long, very narrow, glabrous.

Middle Island: Otago, river flats in the lake district, *Hector and Buchanan.*

7. ACIPHYLLA, Forst.

Erect, simple or branched, glabrous, rigid, spinescent, diœcious or polygamous herbs, sometimes tremendously armed throughout with long, rigid, skewer- or sword-like, spinose leaves. Leaves very thick and coriaceous, with sheathing bases, pinnate or 2–3-pinnate, the rhachis jointed at the insertion of the leaflets. Umbels densely fascicled or spicate or panicled; males panicled, on spreading, slender peduncles, oblique; females usually shortly peduncled, in the sheaths of bracteal leaves. Involucres spinescent. Flowers unisexual; males with imperfect ovaries and large depressed stylopodia; females with longer ovaries and erect, tumid stylopodia.—Calyx-tube ovoid or oblong; limb 5-toothed, often unequally, or obsolete. Petals incurved, without an inflexed tip. Fruit linear or oblong; carpels usually each with 3 dorsal and 2 marginal narrow wings, sometimes 1 carpel is 5- and the other 3-winged.

A very remarkable genus, confined to New Zealand and the Australian Alps, only distinguished from *Ligusticum* by its curious habit and spinescent character.

Stem 2–9 ft. Female inflorescence much contracted.
Stem 5–9 ft. Spinous leaflets of bracts slender, the middle one refracted 1. *A. squarrosa.*
Stem 6–9 ft. Spinous leaflets of bracts lanceolate, middle one patent or suberect . . . . 2. *A. Colensoi.*
Stem 2 ft. Spinous leaflets of bracts erect. . . . . 3. *A. Lyallii.*
Stem 8–12 in. high. Female inflorescence an open panicle or in globose heads.
Umbels with slender rays. Involucral leaves linear . . . . 4. *A. Munroi.*
Umbels densely capitate. Involucral leaves very small . . . . . 5. *A. Dobsoni.*

1. **A. squarrosa,** *Forst.*;—*Fl. N. Z.* i. 87. Tall, unbranched, stout, aromatic, 5–9 ft. high, covered with the very long, spreading, spinous leaflets. Root very stout. Stem 2–4 in. thick below, grooved. Radical leaves 1 ft. or more, pinnate or 2-pinnate; leaflets crowded, a span or more long, very narrow, $\frac{1}{10}$–$\frac{1}{5}$ in. across, strict, rigid, pungent, striated, rough to the touch; sheath flattened, 2–3 in. long, terminating in long spines. Inflorescence a dense, oblong, cylindrical panicle, consisting of numerous floral leaves, with small umbels in their axils. Bracts with 3 spinous leaflets, the middle one 4–6 in. long, refracted when the plant is in fruit; the lateral shorter, erect, Umbels shortly peduncled, few-flowered; involucre 0. Flowers diœcious. Calyx-limb obsolete. Fruit $\frac{1}{3}$ in. long; one carpel 3-, the other 5-winged. with about 3 vittæ in the interstices.—Forst. Gen. t. 38; Hook. Ic. Pl. t. 607, 608.

Middle and southern parts of the **Northern** Island, and throughout the **Middle** Island, generally below 2000 ft. elevation. The "Wild Spaniard" of settlers, forming a thicket impenetrable to man: it exudes an aromatic gum-resin, and the roots are devoured by pigs.

2. **A. Colensoi,** *Hook. f., n. sp.* Stem 6–9 ft. high, and leaves much larger, broader, and more robust than in *A. squarrosa.* Leaves forming a circle 5–6 ft. diam. of bayonet-like spikes, 2 ft. long, pinnate, or 2-pinnate at the base with only 1 or 2 leaflets to each pinna; leaflets 8–10 in. long, $\frac{1}{2}$ in. broad, very thick and coriaceous, narrow-linear, acuminate, striate; margins rough or serrulate, cartilaginous; sheath 3 fingers broad, very thick and leathery, terminated on each side above by a short, simple or 3-foliolate leaf. Inflorescence much more lax than in *A. squarrosa,* the bracts broader and shorter; the middle lobe not refracted. Male umbels on branching peduncles 2–4 in. long, which often exceed the bract-sheaths, many-flowered. Flowers and fruit as in *A. squarrosa,* but the wings of the carpels broader.

Northern Island: top of Ruahine mountains, *Colenso.* **Middle** Island: Nelson mountains, above 2000 ft., *Bidwill, Munro,* etc.; Canterbury, ascending to 5500 ft., *Raoul, Haast,* etc.; Otago, *Lindsay.* There are apparently two varieties, one with the leaflets grooved and their edges serrulate or rough, the other with thicker, scarcely striated leaflets, having smooth margins; both are called "Spear-grass" and "Wild Spaniard." Munro states that it forms a thicket impenetrable to men and horses. Sinclair suspected that this was only a form of *A squarrosa,* but the wings of the fruit are broader.

3. **A. Lyallii,** *Hook. f., n. sp.* About 2 ft. high, similar in habit and general appearance to *A. Colensoi,* but perfectly smooth, polished, and much smaller. Stem deeply grooved. Leaves simply pinnate or 3-foliolate; leaf-

lets narrow, sword-shaped, rigid, pungent, striate, 4–6 in. long, $\frac{1}{8}$–$\frac{1}{3}$ broad, quite smooth. Inflorescence long and contracted, female umbels almost concealed in the tumid sheaths of the bracts; male umbels on spreading, branched, slender peduncles, with subulate involucral leaves; bracts with 3 suberect leaflets, the middle one sometimes spreading. Fruit small, $\frac{1}{6}$ in. long, both carpels 5-winged.

Middle Island: Dusky Bay, *Lyall;* Rangitata range, Ashburnham glacier, etc., alt. 3–5000 ft., *Sinclair and Haast;* Otago, lake district, alpine, *Hector and Buchanan.*

4. **A. Munroi,** *Hook. f. Fl. N. Z.* ii. 330. Small, rarely 1 ft. high, more flaccid than the preceding, perfectly smooth and shining. Leaves numerous, 3–5 in. long, pinnate, rarely 2-pinnate at the base; leaflets 1–2 in. long, $\frac{1}{6}$ in. broad, linear, pungent, midrib obscure; sheath linear, with one subulate leaflet on the top on each side. Scape rather slender. Umbels in an open branched panicle, compound. Bracts spreading, with linear sheaths and 1–3 spreading subulate leaflets. Fruit small, $\frac{1}{6}$ in. long, narrow; carpels some 5-winged, a few 3-winged.

Middle Island: Macrae's Run, and other exposed alpine places, alt. 4500 ft., *Munro;* Awaterc, *Sinclair;* Discovery Peak, 5800 ft., *Travers;* alps of Canterbury, not uncommon near glaciers, alt. 3000–6500 ft., *Haast;* Otago, lake district, subalpine, *Hector and Buchanan.*

5. **A. Dobsoni,** *Hook. f., n. sp.* Very robust, a span high, perfectly smooth and glabrous, yellow-brown when dry. Leaves (radical) very numerous, consisting of a broad sheath, $1\frac{1}{2}$–2 in. long and $\frac{3}{4}$ in. broad, bearing 3 erect, rigid, very thick, subulate or dagger-shaped, rather concave, jointed leaflets, 1 in. long, and $\frac{1}{3}$ in. broad at the base, pungent and keeled at the back towards the apex. Flowering-stem as thick as the little finger, terete, striate, bearing at the top 2 small leaves like the radical, and 5 peduncled, densely capitate, globular umbels of fruit. Peduncles unequal, $\frac{3}{4}$–$1\frac{1}{2}$ in. long, stout, grooved. Umbels (or heads) 1 in. diam., compound, both peduncles and pedicels very short and thick. Fruits densely packed, mixed with short, subulate, involucral leaves, linear-oblong, $\frac{1}{8}$ in. long. Calyx-teeth rather large, unequal. Carpels usually with 5 narrow wings.

Middle Island: summit of Mount Dobson, near Lake Tekapo, alt. 7500 ft., *Dobson and Haast;* amongst shingle on the summits of the ranges near Lake Hawea, alt. 6000 ft., *Haast.* A most remarkable plant, of which more specimens are required to make a good specific character.

8. LIGUSTICUM, Linn.

Glabrous, perennial, erect, often large and robust, aromatic or rank-smelling herbs. Leaves pinnate or decompound, the rhachis jointed at the insertion of the leaflets. Umbels compound, panicled, polygamous; involucral leaves few or numerous. Flowers often unisexual, white or pink, females on very short pedicels.—Calyx-teeth oblong; limb obsolete, or of 5 short, often unequal teeth. Petals with an incurved tip. Styles long or short. Fruit elliptic-oblong or linear-oblong; carpels dorsally compressed or rounded, with 5 nearly equal winged ridges, or one with 3 the other with 5 ridges, rarely each with 3 wings.

A genus of temperate Europe, Asia, and America, not hitherto found in South America nor in Australia, except *Aciphylla* be joined with it, as perhaps it should be. The New Zealand species have ranked under *Anisotome*, in the New Zealand and Antarctic Floras. *Angelica montana* has quite the habit of some of the species, but the broad lateral wings of the carpels keep it distint. Some forms of *L. aromaticum*, however, approach so near *A. decipiens*, that it is very difficult to distinguish them except by the carpels; and some of the *Ligustica* here described without fruit may be *Angelicæ*.

Leaves 2–3-pinnate or decompound.

Very tall and robust. Leaflets decurrent, their lobes pungent . . 1. *L. latifolium*.
Robust, 1 ft. high. Leaflets contracted at base, lobes obtuse . . 2. *L. intermedium*.
Very tall and robust. Leaves decompound; leaflets subulate . . 3. *L. antipodum*.
Tall, 1½–2 ft. Leaves decompound; leaflets cut into narrow blunt lobes . . . . 4. *L. Lyallii*.
Tall, 1–2 ft. Leaves decompound; leaflets cut into narrow piliferous lobes. Styles slender . . . . 5. *L. Haastii*.
Slender, 8–12 in. Leaves 2-pinnate; leaflets cut into narrow piliferous lobes. Styles very minute . . . . 6. *L. brevistyle*.
Small and very slender. Leaflets few, flaccid, filiform . . . . . 7. *L.? filifolium*.
Short, stout, fleshy, deformed. Leaves crowded; leaflets subulate 8. *L.? carnosulum*.

Leaves pinnate or 3-foliolate (*triternate? in* 12).

Tall, 1–2 ft., robust. Leaflets broad, lobed or partite, with piliferous lobules . . . . 9. *L. piliferum*.
Small, 2–10 in. Leaflets sessile, short, broad, variously toothed or incised . . . . 10. *L. aromaticum*.
Very short, densely tufted, much branched. Leaves imbricate. Umbels almost sessile . . . . 11. *L. imbricatum*.
Small, 4-8 in. Leaflets few, petiolate, rhombeo-orbicular . . . 12. *L.? trifoliolatum*.

1. **L. latifolium,** *Hook. f.*—*Anisotome*, Fl. Antarct. i. 16. t. 8. Tall, 3–4 ft. high, very robust and coriaceous. Stem as thick as the wrist at the base. Radical leaves 1–2 ft. long; petiole as thick as the finger; lamina ovate, 2-pinnate; primary divisions linear-oblong; leaflets 2 in. long, 1–1½ broad, obliquely cuneate-oblong, with broad decurrent bases, unequally 3–5-lobed and lobulate; lobules acuminate, with needle-like pungent points, margin thickened, nerves all reticulate. Bracts with very large concave bases, 2–3 in. diam. Umbels very numerous and large, 2–3 in. diam.; involucral leaves of male flower as long as the rays, linear, membranous, 3–4-nerved. Flowers pale-pink. Fruit ⅙ in. long, on pedicels as long, ovoid; carpels with 5 wings, rarely 4 or 3.—*Calosciadium latifolium*, Endl. Gen. Pl. Suppl.

Lord Auckland's group and **Campbell's** Island: in moist places, abundant, *J.D.H.*

2. **L. intermedium,** *Hook. f.*—*Anisotome*, Fl. N. Z. i. 89. Rather stout, 6–12 in. high. Leaves 4–10 in. long, ovate-oblong or linear-oblong, 2-pinnate; primary divisions 6–8 pairs; leaflets ½–1 in. long, short, sessile, cuneate-ovate or triangular-ovate, deeply unequally cut to the middle or below it into broad linear obtuse lobes, coriaceous, veins reticulate; petiole 1–3 in. long, stout, sheath membranous, narrow. Umbels few, subterminal, males 1½–2 in. diam.; rays many; involucral leaves linear-lanceolate. Fruit as in *L. latifolium*, but rather longer, ⅕ in. long.

Middle Island: Port Preservation, *Lyall;* on trap cliffs, Shaw's Bay, Otago, *Lindsay*. This approaches *L. Haastii*, but the leaves are less compound, sometimes almost simply pinnate, and the leaflets are cut into broader, blunt lobes. The Otago specimens are less coriaceous and larger than those of Port Preservation, but many more specimens are wanted.

3. **L. antipodum,** *Homb. and Jacq Voy. t.* 3.—*Anisotome*, Fl. Antarct.

i. 17. t. 9 and 10. Stem very robust, 3–4 ft. high, 2–4 in. thick at the base, deeply furrowed. Leaves 1–2 ft. long, oblong, 2–3-pinnate; leaflets excessively numerous, all narrow, linear-subulate, rigid, pungent, 1 in. long, $\frac{1}{14}$ broad, 1-nerved, all pointing upwards and forwards; rhachis and petiole very stout, as thick as the thumb, with a membranous sheathing base. Male umbels numerous, 2 in. broad, compound, dense; involucral leaves very narrow. Flowers pink. Fruit $\frac{1}{2}$ in. long, very narrow-oblong; carpels one 3- the other 5-winged. —*Calosciadium antipodum*, Endl. Gen. Pl. Suppl.

Lord Auckland's group and **Campbell's** Island, in marshy places, *J. D. H.*

4. **L. Lyallii,** *Hook. f.—Anisotome*, Fl. N. Z. i. 88. Very stout, 1$\frac{1}{2}$–2 ft. high, purplish, obscurely grooved. Stem 1–2 in. thick. Leaves linear-oblong, 2–3-pinnate; leaflets 8–10 pairs, linear-oblong; pinnules crowded, 1 in. long, obovate-cuneate, cut to the base into linear, obtuse lobes, $\frac{1}{12}$ in. broad and 1-nerved; petiole as thick as the little finger, with a narrow sheathing base. Fruiting umbels only seen, compound, very many-flowered. Fruit $\frac{1}{4}$ in. long, longer than its pedicel, linear-oblong.

Middle Island: Port Preservation, *Lyall.* A smaller plant than *L. antipodum*, with more flaccid leaves, much shorter and broader obtuse leaflets, and smaller fruit; it looks like a large state of the following.

5. **L. Haastii,** *F. Muell.* Stem rather stout, 1–2 ft. high, $\frac{1}{2}$ in. thick, grooved, purplish. Leaves $\frac{1}{2}$–1$\frac{1}{2}$ ft. long, oblong or ovate-oblong, 2–3-pinnate; leaflets $\frac{1}{4}$–$\frac{3}{4}$ in. long, crowded, flaccid, narrow-cuneate, obovate, deeply cut into narrow rather membranous linear lobes, $\frac{1}{4}$–$\frac{1}{2}$ in. long, $\frac{1}{24}$ in. broad, with hair-like points. Umbels numerous, panicled, fruiting 2–3 in. diam.; involucral leaves filiform; primary rays much larger than the bracts. Fruit ovoid-oblong, $\frac{1}{6}$ in. long; carpels usually 5-winged; styles slender, spreading.

Middle Island: Black Hill, Rangitata range, Ashburnham glacier, and other places in the Southern Alps, alt. 3–5000 ft., *Sinclair and Haast;* Otago, lake district, alpine, *Hector and Buchanan.* Very similar to *L. Lyallii,* but the foliage is much more flaccid, the leaflets cut into much narrower lobes, which are all hair-pointed.

6. **L. brevistyle,** *Hook. f., n. sp.* Stem 8–12 in. high, slender, striate. Leaves 4–6 in. long, linear-oblong in outline, pinnate or 2-pinnate; leaflets $\frac{1}{4}$–$\frac{1}{2}$ in. long, flaccid, rather sparse, broadly ovate or oblong in outline, pinnatifid or cleft to the base into few very narrow-linear acuminate acicular-pointed spreading lobes, $\frac{1}{4}$–$\frac{1}{2}$ in. long. Umbels few, loosely panicled, fruiting 1$\frac{1}{2}$–2 in. diam.; involucral leaves filiform, short; primary rays 1 in. long; secondary shorter than the fruit. Fruit oblong, with narrow ridges, $\frac{1}{10}$ in. long; calyx-lobes obsolete; style very short.

Middle Island: gullies and hillsides, Lake Hawea and Waitaki, alt. 1000 ft. *Haast;* Otago, lake district, *Hector and Buchanan.* Very similar indeed to *L. Haastii*, but much smaller, more slender, and at once distinguished by the small fruit, with narrower ridges and very minute style. Perhaps only a slender, lax form of *L. aromaticum.*

7. **L.(?) filifolium,** *Hook. f., n. sp.* Small, slender, grass-green, 4–12 in. high. Stem slender, as thick as a sparrow's quill, grooved, often much branched above. Leaves on very slender petioles, broad in outline, of a few opposite, very narrow, almost filiform, flaccid flat acute leaflets, $\frac{1}{10}$–$\frac{1}{20}$ in. diam., $\frac{1}{2}$–1$\frac{1}{2}$ in. long, as narrow as the flattened rhachis. Umbels small,

on long slender peduncles; rays slender, very unequal, $\frac{1}{4}$–2 in. long; involucral leaves very short, subulate. Flowers very small, pink. Fruit immature.

Middle Island: rocky places, Nelson mountains, *Munro*; Dun mountain, 2400–4000 ft., *Travers*. Various places in the Southern Alps, alt. 2–3000 ft., *Sinclair and Haast.*

8. **L. (?) carnosulum,** *Hook. f., n. sp.* Small, stout, dark-green, fleshy, 4–6 in. high. Stem $\frac{1}{4}$ in. thick, tortuous amongst shingle. Leaves crowded or whorled about the upper part of the stem, very numerous, often exceeding the umbels, 2-3-ternately divided; leaflets as narrow as the rhachis, $\frac{1}{14}$–$\frac{1}{12}$ in. broad, $\frac{1}{4}$–$\frac{3}{8}$ in. long, curved, subacute, fleshy, indistinctly articulate with the rhachis; petiole 1–2 in. long, stout and fleshy, scarcely sheathing at the base. Umbels small, $\frac{1}{4}$–$\frac{1}{2}$ in. diam., compound, on very thick peduncles which are crowded amongst the leaves, involucral leaves like the cauline, exceeding the peduncles. Flowers very small, pinkish (males only seen). Fruit unknown.

Middle Island: shingly places on Mount Torlesse, alt. 6–7000 ft., *Haast*. A most singular plant, quite unlike any other in New Zealand: its tortuous stem, leafy at the top only, and fleshy habit are both indicative of its habitat amongst loose dry shingle.

9. **L. piliferum,** *Hook. f., n. sp.* Erect, robust, very coriaceous. Stem 12–20 in. high, $\frac{1}{2}$ in. thick, branched above, striated, red-purple. Leaves a span to a foot long, linear, pinnate; leaflets 10–12 pairs, $\frac{1}{2}$–1 in. long, broadly deltoid, ovate or orbicular-ovate, sessile, often imbricating, very coriaceous, 2–3-lobed to the base or subpinnatifid, the margin lobulate; lobules ending in a bristle; petiole and rhachis very stout, as thick as a goose-quill; sheath narrow, membranous. Umbels on stout branches; rays 1–1$\frac{1}{2}$ in. long; involucral leaves very numerous, linear or linear-lanceolate, membranous. Flowers white; styles slender. Fruit $\frac{1}{6}$ in. long; carpels 3-winged, polished.

Var. *α*. Leaflets very broad and coriaceous, 3-lobed to the base or ternate, closely imbricating.

Var. *β*. Leaflets longer, pinnatifidly cut into narrower lobes.

Middle Island: var. *α*. Hurumiri mountains. alt. 4500–6000 ft., *Travers;* Mount Torlesse, 4700 ft., and Hopkins river by running water, 3–5000 ft., *Haast;* Otago, lake district, alpine, *Hector and Buchanan.* Var. *β*. Great Clyde glacier, 3800–4000 ft., Mount Darwin and Lake Tekapo, *Haast.*

10. **L. aromaticum,** *Banks and Sol.—Anisotome,* Fl. N. Z. i. 99. Small, 2–10 in. high, sometimes in alpine places tufted matted and depressed, bright green, shining, very aromatic. Roots often large and stout; stem sparingly branched above, $\frac{1}{8}$ in. thick. Leaves numerous, spreading, 1–6 in. long, linear, pinnate; leaflets 8–12 pairs, $\frac{1}{6}$–$\frac{1}{3}$ in. long, sessile, deltoid-ovate or orbicular, more or less toothed and cut, sometimes to the base into separate leaflets, or rarely pinnate, the teeth and lobes usually piliferous, very coriaceous, reticulated closely with veins; petiole very stout, with a rather broad, short, membranous sheath. Umbels small, 1–1$\frac{1}{2}$ in. diam.; rays slender; involucral leaves few, linear-subulate. Flowers white; styles slender. Fruit linear-oblong, $\frac{1}{6}$ in. long; carpels equally 5-winged.

Mountainous regions in the **Northern** and **Middle** Islands, alt. 4–6500 ft., very common, *Banks and Solander*, etc. Foliage very variable indeed, the leaflets being entire and toothed, or more or less deeply cut into lobes till they are almost again pinnate, with narrow segments.

11. **L. imbricatum,** *Hook. f., n. sp.* Stems most densely tufted, forming large, flat, depressed patches, very much branched, as thick as the little finger, densely covered with closely imbricating, persistent, coriaceous, shining, green leaves. Leaves with the sheaths ½ in. long; petiole and rhachis very stout; leaflets 4–6 pairs, very small, closely imbricate, $\frac{1}{12}$–⅛ in. long and broad, palmately 3–5-lobed; lobes terminated by a stout bristle; sheaths very large, broad, membranous, produced upwards into a large hood. Flowering-stem sunk amongst the leaves. Umbel compound, male only seen. Involucral leaves few, oblong, obtuse. Calyx-teeth large, acute.

Middle Island: Otago, dry débris on the alps of the lake district, alt. 5–6000 ft. A highly curious plant, very unlike an Umbelliferous one; but I suspect only a form or state of *L. aromaticum*.

12. **L. (?) trifoliolatum,** *Hook. f., n. sp.* Small, glabrous. Stem slender, 6 in. high, sparingly divided above. Leaves 3-foliolate or pinnate; leaflets in 1 or 2 distant pairs, 1⅓ in. long, on slender petioles, rhombeo-orbicular, cuneate at the base, the rounded tip crenate, glaucous below, reticulated with veins, lowermost sometimes lobed or 3-fid; petioles slender, sheaths short, broad. Umbels small, few-flowered; rays short or long, unequal, slender; involucral leaves very short. Flowers white; styles slender. Fruit unknown.

Middle Island: watercourses by the Kowai river, alt. 2–3000 ft., *Haast*. A curiou little species, at once known by the few petioled leaflets; it is probably 2-pinnate, or 2 ternately pinnate. I have only two specimens, and, in the absence of fruit, am not certain of its genus.

I have another small, pinnate-leaved species of this genus?, somewhat similar to *L. aromaticum*, gathered in the **Southern** Island by Dr. Lyall, but not in flower or fruit, with small orbicular crenate leaflets, and the upper part of the rhizome sheathed with the rigid bases of the old petioles.

9. ANGELICA, Linn.

Erect herbs or subscandent undershrubs. Leaves pinnate or 2- or 3-pinnate. Umbels compound, often polygamous.—Calyx-limb 5-toothed, often unequally, or obsolete. Petals with an incurved or rarely inflexed tip. Styles generally long and slender. Fruit oblong, more or less cordate at the base; carpels much dorsally compressed, with 3 slender filiform dorsal ribs, and 2 broad lateral membranous wings.

A small genus, of temperate Europe, Asia, and America, hitherto found nowhere in the southern hemisphere, except in New Zealand, where the two subscandent species are anomalous in the Order, and the erect ones are in many characters more closely allied to *Ligusticum* than to *Angelica*.

Herbaceous, stem erect. Leaves radical, pinnate; leaflets crenate . . 1. *A. Gingidium.*
Herbaceous, stem erect. Leaves radical, pinnate; leaflets laciniate . . 2. *A. decipiens.*
Suffruticose, subscandent. Leaves on the branches, pinnate . . . . 3. *A. rosæfolia.*
Suffruticose, subscandent. Leaves on the branches, 1-3-foliolate . . 4. *A. geniculata.*

1. **A. Gingidium,** *Hook. f.;—Anisotome,* Fl. N. Z. i. 89. Stem stout, erect, 1–1½ ft. high, striate, branched sparingly above. Leaves 4–10 in.

long, pinnate; leaflets 4–8 pairs, sessile, 1–2 in. long, broadly obliquely ovate-oblong, obtuse, crenate, sometimes obscurely lobed on one side, closely reticulated, glaucous below; petioles stout, rhachis jointed at the insertion of the leaflets; sheath narrow. Umbels compound, 1–3 in. diam., polygamous; rays slender; involucral leaves very variable, long or short. Flowers white. Fruit $\frac{1}{6}$ in. long, ovate-oblong.

Var. β. Leaflets more membranous, deeply inciso-serrate.

Northern Island: moist grassy plains, Wairarapa valley, *Colenso.* More frequent in the Nelson and Canterbury provinces of the **Middle** Island. Otago, lake district, *Hector and Buchanan.*

2. **A. decipiens,** *Hook. f., n. sp.* Root stout, sometimes very thick. Stem 6–8 in. high. Leaves numerous, most or all radical, spreading, 4–6 in. long, linear, pinnate; rhachis stout, jointed at the leaflets; leaflets 6–10 pairs, ovate-oblong, sessile, $\frac{1}{4}$–$\frac{1}{2}$ in. long and broad, flaccid, irregularly pinnatifid; lobes linear, acute, not awned, 1-nerved. Stems or peduncles not longer than the leaves, rather slender, striate, usually unbranched. Umbels 1–1$\frac{1}{2}$ in. broad; involucral leaves rather membranous; primary rays $\frac{1}{2}$–1 in. long, secondary $\frac{1}{8}$ in. Flowers minute. Fruit ovate-oblong, $\frac{1}{10}$ in. long, rounded or cordate at the base, lateral wings coriaceous, calyx-lobes distinct; styles long, recurved.

Middle Island: Terraces near Lake Okau, Mount Cook, Rangitata range, and Black Birch Creek, alt. 2–5000 ft., *Sinclair, Haast.*

3. **A. rosæfolia,** *Hook.—Anisotome,* Fl. N. Z. i. 90. Stem branched, subscandent, or trailing over rocks, etc., several feet long, woody, as thick as a goose-quill, branches leafy at the top, below often covered with persistent leaf-sheaths. Leaves 2–4 in. long, pinnate; leaflets 1–1$\frac{1}{2}$ in. long, sessile, obliquely ovate-oblong or lanceolate, acute, serrate, coriaceous, with evident midrib and reticulating veins; petiole slender, rigid; sheath bilobed at the top. Umbels terminal, peduncled, very compound; involucral leaves linear-subulate, shorter than the slender rays. Flowers small, white. Fruit $\frac{1}{10}$ in. long, ovate-cordate, with broad white lateral wings.—Hook. Ic. Pl. t. 581.

Northern Island: East Cape, *Sinclair, Colenso;* Great Barrier Island, *Lyall;* Tongariro, *Bidwill.* **Middle** Island: Akaroa, *Raoul.*

4. **A. geniculata,** *Hook. f.—Anisotome,* Fl. N. Z. i. 90. *t.* 19. Weak, suffruticose, much branched; branches slender, flexuose, divaricating, forming tangled masses over rocks and shrubs, internodes 2–3 in. long. Leaves small, alternate, 1-foliolate (young 3-foliolate); leaflets $\frac{1}{4}$–$\frac{1}{3}$ in. diam., broadly orbicular-ovate, or transversely oblong, rarely oblong or reniform, obscurely crenate, glaucous below, with radiating finely reticulate veins; petiole slender, $\frac{1}{4}$ in. long, with a very small bilobed sheath. Umbels small, terminal, shortly peduncled; rays few, slender, longer than the short involucral leaves. Flowers small, white; petals with an inflexed tip. Fruit $\frac{1}{5}$ in. long, broadly ovoid, cordate at the base; lateral wings broad, white, membranous.

Northern Island: east coast, *Colenso.* **Middle** Island: common in the Canterbury and Otago provinces.

10. DAUCUS, Linn.

Erect, branching, often hispid herbs. Leaves decompound. Umbels compound.—Calyx-limb obscurely 5-toothed. Petals with an inflexed tip. Fruit oblong, dorsally rather compressed; carpels each with 5 primary and 4 secondary ridges, both covered with hispid short hairs, or with stout bristles.

A considerable genus in temperate Europe and Asia, rarer in America and in the southern hemisphere. The Carrot, *D. Carota*, has been gathered at the Bay of Islands, doubtless introduced (*A. Gray*).

1. **D. brachiatus,** *Sieber;—Fl. N. Z.* i. 91. Pilose or glabrate, 6–12 in. high. Leaves 2-pinnatisect; leaflets incised, the ultimate divisions linear. Umbels axillary and terminal, of 8–10 unequal rays; involucral leaves simple or compound. Flowers small, red. Fruit $\frac{1}{12}$ in. long, larger ridges with a row of stiff barbed bristles, intermediate ones much smaller, with a double row of bristles pointing right and left.—*Scandix glochidiata*, Labill. Fl. N. Holl. t. 102.

Northern and **Middle** Islands: not uncommon in dry grassy pastures from Auckland to Otago, *Banks and Solander*, etc. Abundant in Australia, Tasmania, and Western America, from Mexico to Chili.

ORDER XXXIV. ARALIACEÆ.

Trees, rarely herbs. Leaves alternate (rarely opposite), simple or compound, usually evergreen, glossy, stipulate or exstipulate. Flowers in umbels, less commonly in panicles racemes or heads, often unisexual.—Calyx-tube adnate to the ovary, limb truncate or 5-, rarely 3- or many-toothed or lobed; lobes persistent in fruit. Petals usually 5, rarely 0, coriaceous, valvate, rarely imbricate, deciduous. Stamens inserted round an epigynous disk, usually 5; filaments incurved, subulate; anthers oblong. Ovary 2- or more celled; styles short, subulate, recurved, stigmatiferous on their inner faces; ovule pendulous in each cell. Fruit succulent or coriaceous, flattened angled or terete, 2- or many-celled, cells cartilaginous, 1-seeded. Seed flattened, testa very thin; albumen copious, fleshy or horny; embryo small.

A larpe tropical Natural Order, to which the Ivy belongs, but rare in the north temperate zone, chiefly distinguished from *Umbelliferæ* by the arboreous habit, evergreen foliage, often many-celled ovary, and fruit never splitting into its component carpels. Some tropical species have anomalous characters, not introduced into the above ordinal character, as 1-celled ovaries and coherent petals.

Herbaceous. Pedicel jointed with flower. Petals imbricate . . . 1. STILBOCARPA.
Trees or shrubs. Pedicel jointed with flower. Petals valvate . . . 2. PANAX.
Tree. Pedicel not jointed with flower. Petals valvate . . . . . 3. SCHEFFLERA.
Tree. Pedicel not jointed with flower. Petals 0 . . . . 4. MERYTA.

1. STILBOCARPA, Decaisne and Planchon.

A large, much branched, stout, fleshy herb. Stem fistulose, usually beset with large soft bristles or stout hairs. Leaves orbicular, entire, with foliaceous stipulary sheaths. Umbels polygamous, compound, subglobose, with foliaceous involucres.—Flowers jointed on the top of the pedicel. Calyx-tube 3- or 4-grooved, limb entire. Petals 5, obovate-oblong, imbricate. Male fl.: stamens 5; style 0; lobes of disk flat. Female: stamens as in the male or 0; lobes of disk 3 or 4, subreniform, surrounding a cavity in the axis of the ovary; styles 3 or 4, subulate, recurved. Ovary broadly turbinate, 3- or 4-celled. Fruit globose, axis hollow, 3- or 4-furrowed, corky, covered with a brilliant black shining epidermis, containing 3 or 4 horny nuts.

The only species of the genus, and a most remarkable plant, allied to the Chinese and Himalayan herbaceous Aralias.

1. **S. polaris,** *Dcne. and Planch.;—Aralia polaris*, Fl. N. Z. i. 95. Stems 2–4 ft. high, from a long, annulate, thick, prostrate rhizome, much branched, 1–2 in. diam., grooved, succulent, of a rank odour when bruised. Leaves 1–1½ ft. broad, almost fleshy, orbicular-reniform, many-lobed and -toothed, bristly on both surfaces, veins flabellate; petiole erect, semiterete, 2 ft. long; sheaths semiamplexicaul, produced upwards into a foliaceous, truncate, laciniate ligule. Umbels terminal and axillary, compound, as large as the human head, composed of myriads of yellowish waxy flowers, with a purple disk; involucral leaves various, lower foliaceous. Flowers $\frac{1}{6}$–$\frac{1}{4}$ in. diam., on short clavate pedicels. Fruit size of a small peppercorn, black, brilliant.—*Aralia polaris*, Fl. Antarct. i. 21; Hook. Ic. Pl. t. 747.

Southern Island, *Lyall*. Abundant in **Lord Auckland's** group and **Campbell** Island, *J. D. H.* Covering large tracks of ground with huge orbicular masses, very conspicuous from the yellowish waxy flowers and black shining fruit. Lyall's Southern Island specimen wants the long bristles, and may belong to another species.

2. PANAX, Linn.

Trees or shrubs. Leaves evergreen, 1–7-foliolate, rarely simple; stipules 0, or sheathing or subulate. Umbels simple or compound.—Flowers polygamous, jointed on the summit of the pedicel. Calyx-limb toothed sinuate or entire. Petals 5, valvate. Stamens 5. Ovary 2–5-celled, with 2–5 short recurved styles, sometimes connate at the base. Fruit coriaceous or fleshy, 2–5-celled.

I have adopted the character of *Panax* from the observations of Decaisne and Planchon, in their classification of the Order in the 'Revue Horticole,' 1854, p. 105. The genus, as thus defined, is a very considerable one, and found in many parts of the world.

Leaves in old plants simple or 1-foliolate, in young usually 3–5-foliolate.
- Stipules 0. Leaflet lanceolate, serrate. Styles 2 . . . . . . 1. *P. simplex.*
- Stipules 0. Leaflet oblong, quite entire. Styles 3 or 4 . . . . . 2. *P. Edgerleyi.*
- Stipules minute. Leaflet small, rounded. Umbels minute. Styles 2. 3. *P. anomalum.*
- Stipules subulate. Leaves 2 in., very coriaceous, linear. Styles 3 or 4 4. *P. lineare.*
- Stipules 0. Leaves linear, 3–10 in., very coriaceous, green, toothed. Styles 5 . . . . 5. *P. crassifolium.*
- Stipules 0. Leaves 4–12 in., very narrow, purple below, always simple, toothed . . . . 6. *P. longissimum.*

Leaves in old plants 3–5-foliolate.
- Stipules 0. Leaflets sessile. Styles 5 . . . . 7. *P. Lessonii.*
- Stipules sheathing. Leaflets sessile. Styles 2 . . . . 8. *P. Colensoi.*
- Stipules sheathing. Leaflets petiolate. Styles 2 . . . . . . 9. *P. arboreum.*
- Stipules 0. Leaflets sessile. Styles 2 . . . . 10. *P. Sinclairii.*

1. **P. simplex,** *Forst.;—Fl. N. Z.* i. 93. A small evergreen glossy tree, 12–20 ft. high. Leaves with slender petioles 1–3 in. long, coriaceous, very glossy, young 3–5-foliolate, older 1-foliolate; leaflets 2–4 in. long, oblong- or obovate-lanceolate, subacute or acuminate, coarsely serrate, rarely entire, in young plants deeply sinuate-pinnatifid; stipules 0. Umbels small, axillary and terminal, on peduncles shorter than the leaves, partial 10–15-

flowered, rays $\frac{1}{4}$–$\frac{1}{3}$ in. Flower small, greenish-white. Ovary compressed, 2-celled. Styles small, free, recurved. Fruit $\frac{1}{8}$ in. diam., broadly urceolate.—Fl. Antarct. i. 18. t. 12.

Northern Island: mountains of the interior; Tongariro, *Colenso;* abundant throughout the **Middle** Island at 2–4000 ft. elevation, and in **Lord Auckland's** group, *J. D. H.* A very variable plant in the size and serration of the leaflets, which are obtuse acute or acuminate and sometimes opposite. In **Lord Auckland's** group I collected young plants 4 in. high, with unifoliolate, coriaceous leaves, but both Bidwill and Hector sent as the foliage of young plants 5-foliolate leaves, the leaflets membranous and sinuate pinnatifid.

2. **P. Edgerleyi,** *Hook. f. Fl. N. Z.* i. 94. A small tree, 20–40 ft. Leaves on long (2–3 in.) slender petioles, rather membranous, very glossy; leaflet 2–9 in. long, obovate- or oblong- or linear-lanceolate, acuminate, quite entire, in young plants 3-foliolate, leaflets deeply pinnatifid; stipules 0. Umbels as in *P. simplex,* but ovary 3- or 4-celled, and the styles connate at the base.

Northern Island: mountainous regions of the interior, *Edgerley, Colenso.* **Middle** Island: Nelson mountains, *Bidwill;* Otago, *Hector.*

3. **P. anomalum,** *Hook. Lond. Journ. Bot.* ii. 422. *t.* 12;—*Fl. N. Z.* i. 93. A small tree or shrub, with divaricating woody branches, rough with minute hispid hairs. Leaves very shortly petioled, 1-foliolate; leaflet oblong orbicular or obovate, rounded at the tip, obscurely crenate, rather membranous, not glossy; petiole flattened, $\frac{1}{12}$ in. long, pubescent, with minute subulate stipules at its base, and stipes at its tip. Umbels minute, axillary, simple, very shortly peduncled, 1–4-flowered. Flowers very minute, green, shortly pedicelled. Fruit $\frac{1}{6}$ in. diam., 2-celled, with free recurved styles, white mottled with brown.

Northern Island: pine swamps, Bay of Islands, and elsewhere not unfrequent. **Middle** Island: Nelson, *Bidwill.* A very anomalous species, in habit resembling *Melicope simplex* and *Elæodendron micranthum.*

4. **P. lineare,** *Hook. f. Fl. N. Z.* i. 93. A small tree (?); branches stout, woody. Leaves mixed with trifid coriaceous scales, patent, simple, not jointed on the extremely short petiole, 2–3 in. long, $\frac{1}{3}$–$\frac{1}{2}$ wide, minutely serrate, obtuse or acute, excessively coriaceous, with thickened margins, not glossy; petiole $\frac{1}{8}$ in. long, stout; stipule small, subulate. Umbels almost sessile amongst the upper leaves, of 4 or 5 short stout bracteolate rays. Fruit ovoid, 3- or 4-celled; styles connate into a cone, their summits free, recurved.

Middle Island: Chalky Bay, *Lyall.* I have only two small specimens of this curious species.

5. **P. crassifolium,** *Dcne. and Planch.*—*Aralia crassifolia,* Banks and Sol.; Fl. N. Z. i. 97. A slender, sparingly branched, glabrous tree, bark green. Leaves extremely variable, in young plants 3-foliolate; leaflets very narrow-linear, $\frac{1}{3}$–$\frac{2}{3}$ in. broad, of the upper leaves broader, $\frac{2}{3}$–1 in., deeply remotely sinuate-serrate, serratures cuspidate; in older plants, simple, narrow, linear-obovate or linear, quite entire or sinuate or coarsely acutely serrate at the broader part beyond the middle, extremely coriaceous, narrowed into a short, stout, exstipulate petiole of variable length; veins numerous, divergent. Umbels unisexual, terminal, compound, of several very spreading rays; peduncles and rays 1–3 in. long; pedicels short, of the male flowers subrace-

mose. Flowers rather large. Fruit subglobose, as large as a pea, 5-celled, the 5 styles connate into a cone, with their summits free and recurved.—Hook. Ic. Pl. t. 583–4.

Common in forests throughout the islands, *Banks and Solander*, etc. **Chatham** Islands, *Dieffenbach*. A very puzzling plant; the lower and younger leaves are always 3-foliolate, the upper 1-foliolate or simple, all excessively coriaceous. For notes on some supposed varieties of this, see end of genus.

6. **P. longissimum,** *Hook. f.*, *n. sp.* Stem forming a straight, slender, erect, simple rod, 4–10 ft. high; bark dark-green, striped with brown. Leaves all uniform, quite simple, 6–18 in. long, ¼–¾ in. broad, spreading or drooping, excessively stiff, coriaceous, coarsely regularly or irregularly toothed, dirty-purple below, very deep-green above, with often a light-green spot at the base of each tooth; petiole very short, exstipulate, not jointed with the blade; midrib very thick; veins 0.

Throughout the islands, *Banks and Solander*, etc., from the Bay of Islands to Otago. This remarkable plant has been for fifteen years in cultivation at Kew (as a state of *P. crassifolium*), and never changes its habit; it has been collected by Banks, and many succeeding travellers, but no one has identified any flower or fruit with it. I have, however, in the Herbarium male flowers and leaves of a *Panax* closely resembling *P. crassifolium*, differing chiefly from that plant in the less coriaceous, more acute leaves, with more regular serratures, that may prove to belong to this.

7. **P. Lessonii,** *DC.—Aralia Lessonii*, Fl. N. Z. i. 96. A small, glabrous tree, branches very stout. Leaves in old plants 3–5-foliolate; leaflets 1–4 in. long, sessile, oblong- or obovate-lanceolate, subacute, more or less sinuate-serrated or quite entire, very thick and coriaceous, not shining; nerves very indistinct; petiole 4–8 in. long, stout or slender, neither sheathing nor stipulate at the base. Umbels branched, on long stout peduncles; pedicels racemose, unisexual. Flowers rather large. Fruit ovoid, ¼ in. long, 5-celled, with 5 styles connate into a cone, and very short, recurved tips.

Northern Island: east coast, *Cunningham;* Auckland, *Sinclair.* **Middle** Island: Bream Bay, *D'Urville.*

8. **P. Colensoi,** *Hook. f. Fl. N. Z.* i. 94. *t.* 21. A small tree; branches stout. Leaves glossy, on long petioles, 3–5-foliolate; leaflets 4–6 in. long, very coriaceous, sessile or very shortly petioled, obovate- or oblong-lanceolate, coarsely serrated, veins very indistinct; petiole stout, 6–10 in. long, with a short 2-lobed coriaceous sheath. Umbels compound, terminal; primary rays long, stout, divaricating, in the largest specimens 2–3 in.; secondary rays 1 in., pedicels short. Calyx-limb truncate or sinuate. Fruit flattened, nearly orbicular, 2-celled, with 2 divergent styles.

Northern Island: Ruahine Mountains and Tararua, *Colenso.* **Middle** Island: Otago, *Lindsay, Hector.* **Southern** Island: *Herb. A. Richard.* Ivy-tree of Otago. Colenso sends as a young state of this, a plant with pinnatifid leaflets, which perhaps belongs to *P. Sinclairii*, for both Lindsay and Hector observe that the leaves are 3–5-foliolate in all stages, and never sinuate or lobed. Wood useless, trunk exuding large quantities of gum, *Buchanan*. Hector observes that it is the *Aralia trifolia* of Sydney Garden.

9. **P. arboreum,** *Forst.;—Fl. N. Z.* i. 94. A robust, leafy, glossy tree, 12–20 ft. high; branches stout. Leaves on long petioles, 5–7-foliolate; leaflets with petioles ½–1 in. long, very coriaceous, 4–6 in. long, broadly or narrowly oblong or obovate-oblong, subacute, sinuate-serrate; veins distinct;

petiole 2–8 in. long, robust, with a short, broad, coriaceous, 2-lobed sheath. Umbels unisexual, terminal, very numerous, large; peduncles 2–3 in. long, spreading; rays very numerous, 1 in. long; pedicels ¼ in. Flowers large. Calyx-margin sinuate. Fruit nearly orbicular, flattened, grooved on the faces, with 2 short recurved styles, connate to the middle.—Hook. Lond. Journ. Bot. ii. 421. t. 11.

Common in forests throughout the islands, *Banks and Solander*, etc. **Kermadec** Islands, *M'Gillivray*.

10. **P. Sinclairii,** *Hook. f., n. sp.* A small tree. Leaves on rather short petioles, 3–5-foliolate; leaflets sessile, 1–2 in. long, very coriaceous, not glossy, obovate- or oblong-lanceolate, acute, sharply serrate, veins obscure; petiole 1–1½ in. long, neither stipulate nor sheathing. Umbels small, unisexual, on short, terminal peduncles ½–1 in. long or less; rays few; pedicels very short. Fruit nearly orbicular, compressed, 2-celled, with 2 short recurved styles.

Northern Island: Ruahine Mountains, etc., *Colenso*. Auckland?, *Herb. Sinclair*. Sinclair sends both old specimens with normal 5–7-foliolate serrate leaflets, and young ones with pinnatifid leaflets; Colenso sends the pinnatifid-leaved state as the young of *P. Colensoi*.

There are two species of *Panax* mentioned as natives of New Zealand, in Planchon's 'Hortus Donatensis' (p. 10), and which, being founded on garden specimens that have never flowered, are very doubtful: these are *P. pentadactylum*, Dcne. and Pl., from New Zealand? (*A. pentaphylla*, Hort., *A. quinquevulnera*, Makoy), which Planchon suspects is a variety of *P. crassifolium*, with green not brown leaves, 5- rarely 3-foliolate, and the leaflets elegantly curved instead of being stiff and straight as in *P. crassifolium* and *tridactylum*. The other is *P. tridactylum*, Dcne. and Pl. (*A. triphylla* and *A. trifoliata*, Hort.), of which he says that this also is perhaps a variety of *P. crassifolium*, sometimes bearing only one leaflet, whilst the leaflets of *P. crassifolium* become ternate, especially towards the upper part of large trees. Professor Planchon is, however, in some misconception here, for the true *P. crassifolium* is green and not brown, both in a native state and cultivated, and has 3-foliolate leaves, except only at the tops of the older branches, where they become 1-foliolate and (losing their joint) simple.

3. SCHEFFLERA, Forst.

Trees. Leaves digitately 7-foliolate. Umbels racemed.—Flowers polygamous, not jointed on the top of the pedicel. Calyx-tube minutely 5-toothed. Petals 5, valvate. Stamens 5. Ovary 10-celled, with 10 short styles united into a cone to above the middle, their tips free. Fruit rather fleshy, 10-celled, 10-ribbed.

I do not find that this old genus of Forster's is retained in the rearrangement of the Order by Decaisne and Planchon, who, however, I cannot suppose would leave it in *Panax*, from which it differs in habit, inflorescence, the 10-celled ovary, and the absence of any evident joint at the summit of the pedicel with the flower, whilst it differs still further from *Aralia* in the valvate petals, digitate leaves with regular serratures, and 10 styles.

1. **S. digitata,** *Forst.*—*Aralia Schefflera*, Spr.;—Fl. N. Z. i. 95. A small umbrageous tree or large shrub, with stout branches. Leaves on long petioles, digitately 7–11-foliolate; leaflets petiolate, 4–7 in. long, oblong-lanceolate, acuminate, membranous, sharply finely serrate; veins diverging, delicate; petiole 3–7 in. long, terete, with a short 2-lobed sheath at the base. Umbels racemose on the branches of a very large, unisexual, compound, axillary panicle, small, ¼–½ in. diam., many-flowered; rhachis of panicle 1 ft. and

more, branches diverging, a span long, peduncles of umbels $\frac{1}{2}$ in., pedicels $\frac{1}{4}$ in. Flowers rather small. Fruit nearly globose, $\frac{1}{10}$ in. diam., black, pulpy, deeply furrowed when dry.

Common throughout the island in forest regions, *Banks and Solander*, etc.

4. **MERYTA,** Forst.

Small trees; trunk slender, erect, sparingly branched; branches terminated by a crown of very large leaves.—Flowers in terminal, involucrate, panicled heads, polygamous. *Male:* Calyx 3–5-lobed, lobes valvate?. Petals 0. Stamens 3–5, inserted beneath a glandular disk. Ovary 0. *Female:* Calyx-limb 5–9-partite, lobes valvate. Petals 0. Ovary 5–9-celled, with as many short stigmas united below, free and recurved above. Berry ovoid or oblong, 5–9-celled. Embryo terete, curved, with small, flattened cotyledons.

A small genus of singular-looking trees, natives of the Pacific Islands, one of which Dr. Seemann (in Herb. Hook.) has rightly referred to the imperfectly-known genus *Meryta* of Forster.

1. **M. Sinclairii,** *Hook. f.—Botryodendrum*, Fl. N. Z. i. 97. Leaf on a long petiole, very coriaceous, shining, 5–20 in. long, oblong or obovate-oblong, sometimes contracted below the middle, cordate at the base, margin quite entire or waved, strongly veined; petiole 3–10 in. long, stout, very obscurely if at all jointed with the blade. Panicles terminal, branched, erect, very stout, 6–8 in. long. Flowers not seen. Fruit oblong, $\frac{1}{2}$ in. long, 5-celled.

Northern Island: east coast, between Capes Rodney and Brett, *Sinclair and Colenso*. The fruits are quite separate, and do not appear as if they had belonged to a capitate inflorescence, but my specimens are very imperfect. I described the petiole as jointed with the blade, in Fl. N. Z., I believe erroneously.

ORDER XXXV. **CORNEÆ.**

Shrubs or trees, rarely herbs. Leaves alternate, entire, exstipulate. Flowers regular, hermaphrodite or unisexual, usually small and green.—Calyx-tube adnate to the ovary; limb 4- or 5-toothed. Petals 4 or 5, rarely 0, small, often triangular, valvate. Stamens inserted at the base of an epigynous disk; filaments subulate anthers didymous or linear. Ovary inferior, 1–3-celled; style 0 or very short, stigma capitate, or 2 or 3 short recurved stigmas; ovules solitary, pendulous in each cell (if more than one) or 3 pendulous from the top of a column in 1-celled ovaries. Fruit usually a drupe, with a bony inner wall. Testa very thin, adherent; albumen copious, fleshy; embryo very small.

A small European Order, scattered over various parts of the world; to which the Cornel and Dog-wood belong.

Leaves glabrous, broad, and shining . . . . 1. GRISELINIA.
Leaves with silky pubescence beneath . . . . 2. COROKIA.

1. **GRISELINIA,** Forst.

Shrubs, with green branches, transversely scarred at the insertion of the leaf, said to be parasitical on roots of trees. Leaves very coriaceous, glabrous, broad, green, very oblique; veins irregularly netted, the main ones very oblique, the petiole expanded into a small sheath, jointed on the stem. Flowers in axillary panicles, diœcious.—*Male:* Calyx 5-toothed. Petals 5, valvate. Stamens 5; filament very short. *Female:* Calyx-tube adherent to the ovary; limb 5-toothed. Petals 5, obscurely imbricate. Stamens 0. Ovary ovoid, 1- or 2-celled; styles 3, very short, recurved, subulate. Fruit a fleshy 1- or 2-celled, 1-seeded berry; cotyledons divaricating.

To this should be referred *Decostea*, Ruiz and Pavon, of Peru and Chili, a genus of several species. It is also very nearly allied to the Japanese *Aucuba* of our gardens. Raoul figures petals in the female flower of *G. littoralis*, but I do not find them. I am doubtful whether the following species may not be varieties of one; both are in cultivation and look remarkably distinct. In the 'New Zealand Flora' they are regarded as varieties.

Leaves 2–7 in. Veins strong beneath . . . . 1. *G. lucida.*
Leaves 1–3 in. Veins very indistinct . . . . 2. *G. littoralis.*

1. **G. lucida,** *Forst.;—Fl. N. Z.* i. 98. An erect much-branched bush, 10–12 ft. high. Leaves 2–7 in. long, very obliquely ovate obovate or oblong, quite entire, obtuse or rounded at the tip, very unequal towards the base, one side much narrower than the other, which is often gibbous, in the upper shoots narrowed at the base, bright green, polished; veins very distinct on the under surface; petioles $\frac{1}{2}$–1 in. long, rather slender. Panicles axillary, often as long as the leaves, much branched, minutely pubescent with spreading, golden (when dry) hairs. Flowers minute; pedicels jointed, very short. Drupe $\frac{1}{3}$ in. long, the fleshy part full of oil canals.

Var. *β. macrophylla*, very robust. Leaves almost orbicular, almost cordate at the base (perhaps only young shoots of *G. lucida*).

Northern Island: not unfrequent, *Banks and Solander*, etc. Paliser Bay, Cape Turakirae, on maritime rocks, and head of Ruamahanga river, *Colenso.* **Middle** Island: Dusky Bay, *Menzies;* Chalky Bay, *Lyall.* Var. *β.* **Northern** Island: Bay of Islands, *A. Cunningham.* Auckland, growing on *Metrosideros tomentosa*, Sinclair.

2. **G. littoralis,** *Raoul, Choix*, xxii. *t.* 19. A small bush (or tree, 60 ft. high, *Raoul*), with paler foliage, less glossy than the two preceding. Leaves 1–3 in., ovate or oblong, less oblique at the base, wedge-shaped or narrowed into the slender rather long petiole; veins very obscure below. Peduncles much shorter than the leaves, pubescence and flowers as in *G. lucida.*

Northern and **Middle** Islands: not uncommon. Hawkes Bay, *Colenso;* ascending to 2500 ft. in the Canterbury forests, *Haast;* Otago, *Hector and Buchanan.* In cultivation, this is a very small shrub, more yellow-green and less shining than *G. lucida.*

2. COROKIA, A. Cunn.

Shrubs or small trees; branchlets and leaves below silvery with appressed pubescence. Leaves alternate, exstipulate, evergreen, quite entire, coriaceous. Flowers axillary or terminal, solitary or in panicles.—Calyx-tube turbinate, silky; limb 5-toothed, valvate. Petals 5, small, valvate, with a small scale at their base, silky outside, yellow. Stamens 5, filaments short; anthers linear. Ovary 1- or 2-celled; style short, stigma 2-lobed. Drupe small, ovoid, 1- or 2-celled. Seeds elongated, with a small slender embryo.

A small genus, confined, in so far as is at present known, to New Zealand.

Leaves lanceolate . . . . 1. *C. buddleoides.*
Leaves spathulate . . . . 2. *C. Cotoneaster.*

1. **C. buddleoides,** *A. Cunn.*;—*Fl. N. Z.* i. 98. An erect shrub or small tree, 10–40 feet high. Leaves shortly petioled, 2–6 in. long, narrow-lanceolate or oblong-lanceolate, acuminate, shining above, densely silky-tomentose beneath. Panicles short, 6–20-flowered. Pedicels bracteolate at the base, short. Flowers ¼ in. long; petals oblong-spathulate, obtuse. Drupe red, ⅓–½ in.—Hook. Ic. Pl. t. 424.

Var. β, Fl. N. Z. l. c. Leaves oblong-lanceolate, broader, not shining above, more densely tomentose below.

Northern Islands: margins of woods, etc., from the Bay of Islands to the east coast, *Banks and Solander*, etc. Var. β. **Chatham** Islands, *Dieffenbach.*

2. **C. Cotoneaster,** *Raoul, Choix*, 22. *t.* 20;—*Fl. N. Z.* i. 98. A low, rigid, spreading, much-branched bush, with woody, black, tortuous branches. Leaves alternate or fascicled, ⅓–1 in. long, orbicular obcordate oblong ovate or obovate, suddenly contracted into a flat linear petiole; margins recurved; tip rounded or emarginate, shining above, white beneath. Flowers axillary or terminal, solitary or few together. Pedicels much shorter than the petioles, bracteolate at the middle. Flowers as in *C. buddleoides*, but petals narrower and drupe shorter.

Northern Island: east coast, at Matapouri and Manukau Bay, etc., *Colenso.* **Middle** Island: common from Nelson to Otago, on the east side.

ORDER XXXVI. LORANTHACEÆ.

Parasitical shrubs, with opposite or alternate, quite entire, coriaceous, exstipulate leaves, or jointed leafless branches. Flowers hermaphrodite or unisexual; bracts and bracteoles 1 or 3, sometimes connate or 0.—Calyx-tube adherent to the ovary; limb 0 or truncate or 4–8-toothed. Petals as many, rarely 0, sometimes united into a tubular corolla with valvate lobes. Stamens as many as the petals, opposite to and usually inserted on them. Ovary inferior, 1-celled; style long or short; stigma simple; ovule solitary, erect, adnate to the walls of the cavity. Berry 1-seeded. Seed albuminous; embryo straight; radicle superior.

A large tropical and subtropical Order of parasites, often very handsome, to which the common Mistletoe belongs. It is not Tasmanian, though extending to the south island of New Zealand.

Leaves opposite. Flowers hermaphrodite . . . . 1. LORANTHUS.
Leaves opposite and alternate. Flowers diœcious . . . . 2. TUPEIA.
Leaves 0. Branches jointed, flattened . . . . 3. VISCUM.

1. LORANTHUS, Linn.

Parasitic shrubs. Leaves opposite or subopposite in the New Zealand species, jointed to the stem. Flowers axillary, solitary racemed or panicled, jointed to the pedicels and these to the peduncles.—Calyx-tube ovoid; limb short, truncate, cup-shaped. Corolla tubular, the petals 4, usually narrow, free or united into a tube below, upper part spreading or recurved. Stamens 4, filaments filiform; anthers linear, 2-celled. Style slender, jointed to the top of the ovary, deciduous, stigma capitate or clavate.

A very large tropical genus, of which many species differ from the character given above, in their alternate leaves and other characters.

Flowers 1½–2 in., spicate, in opposite pairs. Petals free. . . . . . 1. *L. Colensoi.*
Flowers 1 in., subsolitary, axillary. Petals free . . . . 2. *L. tetrapetalus.*
Flowers 1 in., panicled. Petals united below . . . . 3. *L. tenuiflorus.*
Flowers ½ in., racemed. Petals united below . . . . 4. *L. flavidus.*
Flowers ⅛ in., panicled. Petals free . . . . 5. *L. micranthus.*

1. **L. Colensoi,** *Hook. f. Fl. N. Z.* i. 99. A large glabrous shrub. Branches woody, terete. Leaves 1½–2 in., broadly oblong orbicular or obovate-oblong, narrowed into a short stout petiole, ⅓–½ in. long, rounded at the tip, very coriaceous, quite nerveless or nerves very obscure and oblique. Flowers 1½–2 in. long, scarlet, sessile, in pairs on a short, stout, 3–9-flowered peduncle, ½ in. long, erect. Bracts 0. Corolla terete, swollen at the base, straight; petals free. Anthers long, linear.—Hook. Ic. Pl. t. 633.

Northern Island: parasitic on *Metrosideros tomentosa,* Lake Waikare, *Colenso.* **Middle** Island: Nelson, *Bidwill, Munro:* Otago, *Lindsay.*

2. **L. tetrapetalus,** *Forst.;—Fl. N. Z.* i. 99. A middling-sized glabrous shrub. Branches terete. Leaves ¾–1¼ in. long, linear-oblong or oblong, rounded at the tip, narrowed into very short petioles, very coriaceous, almost veinless or with 3 or 4 veins diverging from the base of the leaf. Flowers about 1 in. long, shortly pedicelled, solitary, rarely 2 together, axillary, erect. Calyx obscurely 4-toothed. Corolla terete, straight, rather swollen at the base; petals very narrow, linear, free. Anthers linear. Stigma very small.

Northern and **Middle** Islands: common on *Metrosideros* and *Vitex,* from the Bay of Islands to Canterbury, *Banks and Solander,* etc.

3. **L. tenuiflorus,** *Hook. f. Fl. N. Z.* i. 100. Glabrous. Branches slender, terete. Leaves 1–1½ in. long, obovate-oblong, rounded at the tip, narrowed at the base into a short petiole; veins slender and few. Flowers 1 in. long, in axillary trichotomous puberulous panicles; peduncles and pedicels slender, divaricating, about ¼ in. long. Corolla slender, terete, curved; petals very narrow, united about ⅔ way up, perhaps separating afterwards. Anthers oblong. Stigma simple, flexuose.

Locality unknown. I have but one bad specimen of this very distinct species.

4. **L. flavidus,** *Hook. f. Fl. N. Z.* i. 100. *t.* 27. A glabrous shrub, 1–2 ft. high. Branches terete, the ultimate compressed or angled. Leaves 1–2 in. long, linear-oblong, obtuse or apiculate, narrowed into a short petiole, coriaceous, veinless or with 3–5 slender veins diverging from their base. Racemes axillary, drooping, 10–12-flowered, 1–2 in. long; peduncle rather slender; pedicels opposite, ⅛ in. long. Flowers yellow, ½ in. long. Corolla slender, straight, slightly swollen above the base; petals combined to above the middle. Anthers oblong. Stigma capitate, oblique.

Northern Island: Totaranui, *Banks and Solander.* On *Fagus,* in the Ruahine Mountains, *Colenso.* **Middle** Island: Nelson, on *F. Solandri, Sinclair and Munro;* Otago, lake district, *Hector and Buchanan.*

5. **L. micranthus,** *Hook. f. Fl. N. Z.* i. 100. A large glabrous shrub. Branches terete or angled; twigs much compressed, 2-edged. Leaves 1½–3 in. long, oblong- ovate- or obovate-oblong, obtuse or rounded at the tip, narrowed into short petioles, with small tumid, brown, scaly buds in their axils, veinless or with few slender veins branching from the midrib. Panicles

small, $\frac{1}{2}$–$\frac{3}{4}$ in. long, glabrous, with divaricating peduncles and pedicels $\frac{1}{8}$–$\frac{1}{6}$ in. long. Flowers minute, green, $\frac{1}{8}$ in. long. Petals free, spreading, linear. Anthers oblong. Style suddenly tortuous; stigma capitate.

Common throughout the islands, from the Bay of Islands to Otago, *Banks and Solander*, etc.

2. TUPEIA, Chamisso and Schlechtendal.

A parasitical diœcious shrub. Branches terete, jointed. Leaves opposite and alternate.—Flowers panicled, not jointed to the pedicels, which are jointed to the peduncles. *Male:* Calyx 0. Petals 4, free, linear, recurved. Stamens 4; anthers broadly oblong. Ovary, style, and stigma 0. *Female:* Ovary ovoid with no calyx-limb. Petals 4, reflexed, linear, free. Stamens 0. Style short, straight, constricted below the capitate stigma, deciduous.

A genus of but one species.

1. **T. antarctica,** *Cham. and Schl.;—Fl. N. Z.* i. 101. *t.* 26. Branches terete; bark pale. Leaves $\frac{1}{2}$–$1\frac{1}{2}$ in. long, oblong obovate lanceolate or oblong-lanceolate, subacute, narrowed into short petioles, pale green, veinless or with faint veins branching from the midrib. Panicles small, axillary and terminal, 6–10-flowered. Peduncles and pedicels $\frac{1}{8}$–$\frac{1}{6}$ in. long, divaricate, very slender. Flowers ebracteate, $\frac{1}{8}$ in. diam., greenish-yellow. Filaments about as long as the oblong anthers. Berry white or spotted with pink.—*T. pubigera* and *T. Cunninghamii*, Miquel, in Linnæa, xviii. 85.

Abundant throughout the islands, *Banks and Solander*, etc.; parasitic on various bushes, sometimes on *Loranthus micranthus*. A very variable plant, in size, colour, and robustness.

3. VISCUM, Linn.

Parasitical shrubs, of which the New Zealand species are small, leafless, with flattened, jointed branches, much constricted at the joints.—Flowers very minute, spicate or solitary or whorled at the joints of the branches, diœcious or monœcious. *Male:* Perianth 3- or 4-leaved. Anthers sessile and adnate to the perianth lobes; pollen in many cavities which open by pores. *Female:* Perianth 3- or 4-lobed. Stamens 0. Style 0 or short, stigma obtuse. Berry ovoid, succulent.

A very large, chiefly tropical and subtropical genus, including the Mistletoe, of which many species have the habit of *Loranthus* and otherwise differ from the two New Zealand ones.

Joints $\frac{1}{5}$ in. broad, very flat. Flowers spiked . . . . 1. *V. Lindsayi.*
Joints $\frac{1}{24}$ in. broad, terete. Flowers on the joints . . . . 2. *V. salicornioides.*

1. **V. Lindsayi,** *Oliver, n. sp.* A small, glabrous, branching, succulent shrub, 4–6 in. high. Joints of stem obovate, very flat, rather longer than broad. Spikes often 3, starting from the upper joint, divaricating, $\frac{1}{4}$ in. long. Peduncle jointed. Flowers very minute, whorled on the joints of the peduncle. Perianth 3-lobed; lobes persistent.

Middle Island: Otago, *Lindsay*, *Hector*. Parasitic on *Coprosma*, *Metrosideros*, and *Melicope*.

2. **V. salicornioides,** *A. Cunn.;—Fl. N. Z.* i. 101. A very small, glabrous, succulent plant, much branched, 3–4 in. high. Joints of stems

$\frac{1}{2}$–$\frac{1}{3}$ in. long, terete, with a rather expanded top, not $\frac{1}{20}$ in. broad. Flowers very minute, solitary or few together in the tips of the upper joints. Perianth 3-lobed; lobes persistent.

Northern and **Middle** Islands: common on *Leptospermum*, *Dracophyllum*, and various other shrubs.

ORDER XXXVII. CAPRIFOLIACEÆ.

Shrubs or trees, rarely herbs. Leaves usually simple and exstipulate. Flowers hermaphrodite.—Calyx-tube adnate with the ovary; limb 4–5-lobed or toothed. Corolla tubular campanulate or rotate, with 4 or 5 imbricate teeth or lobes. Stamens as many as the lobes of the corolla, inserted on its tube, equal or unequal. Ovary inferior, 2–5-celled, surmounted with an epigynous disk; style often filiform, stigma simple or divided; ovules 1 or more in each cell, pendulous. Fruit generally a 1–5-celled berry. Seeds with fleshy albumen; embryo axile.

A small Order, chiefly confined to the north temperate zone, where the Honeysuckles and Viburnums form its largest known genera. It is chiefly distinguished from *Rubiaceæ* by the absence of stipules and often serrate leaves.

1. ALSEUOSMIA, A. Cunn.

Shrubs. Leaves rather membranous, alternate, polymorphous, with minute tufts of red (when dry) hairs in their axils. Flowers axillary, solitary or fascicled, deliciously scented, greenish; pedicels bracteolate.—Calyx-tube small, ovoid; limb 4- or 5-lobed, deciduous. Corolla tubular or funnel-shaped, with 4 or 5 small spreading lobes; edges inflexed, toothed or lobulate. Stamens 4 or 5; filaments very short, inserted at the mouth of the corolla; anthers oblong. Ovary 2-celled; style filiform, stigma subclavate; ovules several, in 2 series, inserted in the axis of the cells. Berry ovoid, 2-celled, several-seeded, with a broad terminal areola. Seeds angular; testa bony.

A small genus, confined to New Zealand, of which the species are excessively variable, apparently passing into one another, so that their forms are very difficult to discriminate. The alternate leaves are unusual in the Order.

Leaves 3–7 in., broad. Flowers 1 in., usually 5-merous . . . . 1. *A. macrophylla.*
Leaves 1–3 in., oblong or obovate. Flowers $\frac{1}{2}$ in., 4- or 5-merous . 2. *A. quercifolia.*
Leaves $\frac{1}{2}$–2 in., obovate or rhomboid. Flowers $\frac{1}{3}$–$\frac{1}{2}$ in., 4- or 5-merous 3. *A. Banksii.*
Leaves 1–1$\frac{1}{2}$ in., linear. Flowers $\frac{1}{3}$–$\frac{1}{2}$ in., 4- or 5-merous . . . . 4. *A. linariifolia.*

1. **A. macrophylla,** *A. Cunn.;—Fl. N. Z.* i. 102. *t.* 23. A shrub, 6–8 ft. high, perfectly glabrous; branches stout. Leaves 3–7 in. long, linear- or oblong-obovate, obtuse, entire or sparingly toothed, narrowed into short petioles, rather coriaceous. Flowers 1 in. long, fascicled or subpanicled. Calyx-lobes lanceolate, tips recurved. Corolla-lobes 3–5, with incurved, fimbriate margins. Berry about $\frac{1}{2}$ in. long.

Northern Island: from the Bay of Islands to the Thames, in woods.

2. **A. quercifolia,** *A. Cunn.;—Fl. N. Z.* i. 103. A slender twiggy shrub, 2–3 ft. high, glabrous or the branchlets minutely puberulous. Branches

slender. Leaves 1–3 in. long, most variable, lanceolate, or oblong- or obovate-lanceolate or oblong, obtuse, entire or more or less deeply sinuate-toothed or lobed, like the Oak; petioles $\frac{1}{4}$–$\frac{1}{2}$ in. long. Flowers about $\frac{1}{3}$–$\frac{1}{2}$ in. long, 4- or 5-lobed; pedicels slender. Berry $\frac{1}{3}$ in. long.—*A. Ilex*, A. Cunn. Prodr.

Northern Island: Bay of Islands, *A. Cunningham*, to Te Hawera, *Colenso*.

3. **A. Banksii,** *A. Cunn.;—Fl. N. Z.* i. 102. *t.* 24. A small straggling bush, with spreading branches, which are slender and puberulous. Leaves about 1 in. long, very variable, usually broadly-oblong obovate or rhomboid, with waved toothed or lobed margins. Fowers $\frac{1}{3}$–$\frac{1}{2}$ in. long, generally solitary. Berry $\frac{1}{6}$ in. long.—*A. atriplicifolia* and *palæiformis*, A. Cunn.

Northern Island, *Banks and Solander;* Bay of Islands, in woods, common, *A. Cunningham*, etc.

4. **A. linariifolia,** *A. Cunn.;—Fl. N. Z.* i. 102. *t.* 25. A small erect shrub, 1–3 ft. high. Branches slender, glabrous or with minute spreading pubescence. Leaves 1–1$\frac{1}{2}$ in. long, linear or linear-lanceolate, entire or sinuate-toothed, generally on very short petioles. Flowers $\frac{1}{3}$–$\frac{1}{2}$ in. long, on slender pedicels, solitary or 2 or 3 together. Lobes of corolla obscurely toothed. Berry small, less than $\frac{1}{4}$ in. long.—*A. ligustrifolia*, A. Cunn.

Northern Island: Bay of Islands, in woods, *A. Cunningham*, etc.

ORDER XXXVIII. RUBIACEÆ.

Trees shrubs or herbs. Leaves opposite with interpetiolar stipules, or whorled, simple, quite entire, usually turning black when dried. Flowers regular, hermaphrodite or unisexual.—Calyx-tube adherent to the ovary, limb 4- or 5-toothed or -lobed or 0. Corolla rotate tubular funnel-shaped or campanulate, 4–9-lobed -toothed or -parted, valvate or imbricate. Stamens inserted on the corolla, as many as its lobes. Ovary usually 2- or 3-celled; styles 1 or 2, long, stigma 2- or 3-lobed or capitate; ovule 1 in the New Zealand species. Fruit various, in the New Zealand species dry or fleshy, with 2–6 cells or nuts and 1 seed in each cell. Testa very thin, albumen horny; embryo rather small, with flat cotyledons.

One of the largest Orders of flowering plants, of which the shrubby and arboreous species are very rare in the northern temperate zone, but frequent in the southern.

Shrubs or trees or woody-stemmed herbs. Leaves opposite. Flowers unisexual . . . . 1. COPROSMA.
Herbs, creeping. Leaves opposite. Flowers hermaphrodite . . . . . 2. NERTERA.
Herbs. Leaves whorled. Corolla rotate . . . . 3. GALIUM.
Herbs. Leaves whorled. Corolla funnel- or bell-shaped . . . . 4. ASPERULA.

1. COPROSMA, Forst.

Trees or shrubs, sometimes small and creeping, often fetid when bruised and drying. Leaves evergreen, opposite, with interpetiolar stipules. Flowers unisexual, often diœcious, often minute green and inconspicuous, solitary and sessile, or 2 or more clustered on the branches of a peduncle.—Calyx: *male*, cupular, equally or unequally 2–5-lobed or toothed, or truncate and erose, *female*, tube ovoid; limb 4–5-toothed or -lobed. Corolla tubular or funnel-

or bell-shaped; tube straight or curved; lobes 4 or 5, rarely 6–9, valvate. Stamens 4 or 5, rarely 6–9, inserted on the base of the corolla; filaments filiform, exserted; anthers linear, apiculate, pendulous. Ovary 2- rarely 3- or 4-celled; styles 2 (3 or 4), filiform, exserted, pubescent; ovule 1, erect in each cell. Drupe ovoid or round, with 2 (rarely 4) plano-convex 1-seeded cocci.

A large and extremely variable New Zealand genus, of which a few species inhabit the Pacific islands, Tasmania, temperate Australia, and the lofty mountain of Kini Balou, in Borneo. The New Zealand species are most difficult of discrimination, owing to their extreme variability, their being diœcious, and their very small flowers. I distinguish two chief groups, one with solitary female flowers, the other with these fascicled; then again some of the former have the flower obviously terminal on the shoot, between the uppermost pair of leaves; others have it apparently lateral, springing from the side of the shoot or branch below the leaves,—in other words, the flower is still terminal, but in shoots that are habitually leafless. In all cases connate stipules are abundantly developed below the flowers, forming one or more series of cup-shaped involucels to the calyx. In some cases the calyx of the male flower is reduced or absent, and the upper series of connate stipules may be taken for it. It is not always easy to say whether this is so or no; and the structure of the calyx wants working out from fresh specimens. The following analysis of the supposed species is very imperfect; it has cost me several weeks' assiduous study of many specimens; many more are, however, wanted to establish the constancy or even prevalence of the characters employed, especially of the Northern Island species. The descriptions of the ten species originally published by A. Cunningham (Ann. Nat. Hist. (1839), ii. 206) are very faulty and incomplete, and the specimens in his Herbarium were much intermixed. Those in my 'Flora of New Zealand,' also, are very defective in regard to the alpine and Middle Island species, of which I have now better materials. As with so many other genera, it is vain to expect to name single specimens of single species; the characters are in most cases comparative, and until some general notion of the prevalent forms of the genus is obtained, the distinctive characters of the species, such as they are, cannot be appreciated, nor in some cases understood.

I. *Shrubs or trees, erect, rarely prostrate, never creeping and rooting.*

A. *Female flowers aggregated (rarely solitary). Leaves glabrous, rarely fascicled.*

α. *Leaves* 3–8 *in., membranous, greenish when dry. Peduncles* 1–2 *in.*

Leaves 3–6 in. Corolla ⅛ in., tubular . . . . 1. *C. lucida.*
Leaves 4–8 in. Corolla ¼–⅓ in., bell- or funnel-shaped . . . . 2. *C. grandifolia.*

β. *Leaves* 1–3 *in., black or brown when dry. Peduncle* 0 *or* ⅛–1 *in. long.*

Robust, maritime; twigs glabrous. Leaves obovate, margins recurved . . . . 3. *C. Baueriana.*
Slender, maritime; twigs pubescent. Leaves 1–2 in., obovate . 4. *C. petiolata.*
Robust. Leaves 2–3 in., oblong, coriaceous, acute . . . . . 5. *C. robusta.*
Leaves 1–2 in., linear or linear-oblong, subacute . . . . . . 6. *C. Cunninghamii.*
Leaves 1–2 in., membranous, ovate-oblong, acute . . . . . . 7. *C. acutifolia.*
Slender. Leaves orbicular or spathulate, petioles broad, flat . . 8. *C. spathulata.*

B. *Female flowers solitary, males solitary or fascicled.*

α. *Twigs hirsute or densely pubescent. Leaves more or less pubescent, especially the young.*

Leaves ¼–¾ in. orbicular, membranous, cuspidate. Drupe didymous . . . . 9. *C. rotundifolia.*
Leaves ¼–⅔ in., oblong or obovate, blunt or subacute . . . . . 10. *C. ciliata.*

β. *Twigs glabrous or pubescent. Leaves glabrous* (*or puberulous in* C. divaricata *and* tenuicaulis).

Flowers lateral. Leaves broad, small, ¼–¾ in., spathulate oblong or orbicular.
Leaves puberulous below, veins reticulate. . . . . 11. *C. tenuicaulis.*
Twigs very pubescent. Female corolla long, tubular, 4- or 5-fid 12. *C. rhamnoides.*
Twigs almost glabrous. Female corolla 4- or 5-partite . . . 13. *C. divaricata.*

Twigs pubescent. Leaves very coriaceous. Female corolla 4- or 5-partite . . . . 14. *C. parviflora*.
Flowers lateral. Leaves linear . . . . 15. *C. propinqua*.
Flowers terminal. Leaves oblong, or obovate- or cuneate-oblong.
Leaves $\frac{1}{2}$–2 in. Flowers $\frac{1}{4}$–$\frac{1}{3}$ in. long . . . . 16. *C. fœtidissima*.
Leaves $\frac{1}{4}$–$\frac{3}{4}$ in., straight, oblong or oblong-lanceolate, retuse . . 17. *C. Colensoi*.
Leaves $\frac{1}{4}$–$\frac{3}{4}$ in. recurved, coriaceous, cuneate-oblong . . . . . 18. *C. cuneata*.
Flowers terminal. Leaves very narrow, linear or linear-oblong.
Leaves geminate, in opposite pairs, linear, acerose, $\frac{1}{20}$ in. broad . 19. *C. acerosa*.
Leaves in opposite fascicles, linear-lanceolate, rigid, concave above, $\frac{1}{16}$ in. broad . . . . 20. *C. depressa*.
Leaves linear or lanceolate, flat, $\frac{1}{4}$–$\frac{1}{3}$ in. long, $\frac{1}{12}$–$\frac{1}{15}$ broad . . 21. *C. microcarpa*.
Leaves $\frac{1}{2}$–1 in. long, $\frac{1}{10}$–$\frac{1}{6}$ in. broad, linear or lanceolate, concave; upper stipules tubular . . . . 22. *C. linariifolia*.

II. *Stems prostrate and rooting. Small-leaved alpine species.*

Leaves ovate-oblong or obovate. Corolla tubular, curved . . . 23. *C. repens*.
Leaves oblong or linear-oblong. Male corolla $\frac{1}{3}$–$\frac{1}{2}$ in. long . . . 24. *C. pumila*.

1. **C. lucida,** *Forst.;—Fl. N. Z.* i. 104. A small diœcious or monœcious, perfectly glabrous, leafy tree; bark pale. Leaves 3–6 in. long, oblong-obovate or -lanceolate, acute or acuminate, narrowed into a petiole $\frac{1}{4}$–$\frac{1}{2}$ in. long, membranous, bright green when dry, shining above, paler with veins finely reticulated below. Peduncles 1–2 in. long, 3-chotomously branched, bracteate at the axils. Flowers often crowded in threes at the ends of the pedicels, 5-merous;—*male:* calyx hemispherical, limb toothed; corolla tubular, about $\frac{1}{8}$ in. long, 5-cleft to the middle, lobes erect; anthers linear, as long as the corolla;—*fem.:* calyx-limb truncate; corolla-lobes linear, reflexed; stamens minute, included; styles filiform, 3 or 4 times longer than the corolla. Drupe $\frac{1}{3}$ in. long, pink.

Northern Island, abundant. **Middle** Island, *Banks and Solander*, etc.; Otago, lake district, and south-east coast, *Hector and Buchanan*. Fruit edible.

2. **C. grandifolia,** *Hook. f. Fl. N. Z.* i. 104. A small, perfectly glabrous tree, very similar to *C. lucida*, but with larger, more membranous, less shining leaves, and longer corollas. Bark pale. Leaves 4–8 in. long, elliptic- or obovate- or lanceolate-oblong, acute or cuspidate, green when dry but hardly shining, pale below, veins obscurely reticulated; petioles $\frac{1}{3}$–1 in. long. Peduncles 1–3 in. 2-bracteate, 3-chotomous; male pedicels bearing a dense head, females 3-flowered;—*male:* calyx minute, urceolate, 4- or 5-lobed; corolla $\frac{1}{4}$–$\frac{1}{3}$ in. long, funnel- or bell-shaped, shortly 4- or 5-lobed; anthers much shorter than the corolla;—*fem.:* calyx-limb 4-toothed; corolla very slender, styles $\frac{1}{2}$ in. Berry as in *C. lucida*, but seed usually longer.—*Ronabea australis*, A. Rich.

Northern Island: Bay of Islands to Wellington, common, *Banks and Solander*, etc. The corolla varies a good deal in shape and breadth.

3. **C. Baueriana,** *Endl.;—Fl. N. Z.* i. 105. A small, perfectly glabrous, robust, monœcious, branching tree or large shrub, with almost fleshy bright-green foliage, black when dry. Branches stout, terete or 4-angled, perfectly glabrous; bark pale, wrinkled. Leaves $\frac{1}{2}$–$1\frac{1}{2}$ in. long, broadly oblong-obovate, obtuse rounded or retuse at the tip, rarely smaller, oblong-lanceolate and subacute, narrowed into short petioles, margins recurved, pale below with few veins. Peduncles as long as the petioles or shorter, stout,

rarely-branched, 3–many-flowered. Flowers $\frac{1}{8}$–$\frac{1}{6}$ in. long, capitate;—*male:* calyx minute, shortly 4-lobed; corolla bell-shaped; anthers broad;—*fem.:* calyx-limb short, truncate or 4-lobed; corolla-tube short, terete, lobes 4, short, obtuse, spreading; styles very stout, about twice as long as the corolla. Drupe broadly obovoid, $\frac{1}{4}$–$\frac{1}{3}$ in. long.—Endl. Icon. Pl. iii.; *C. retusa,* Banks and Sol.; Hook. f. in Journ. Bot. iii. 415.

Northern Island: common on maritime rocks. **Middle** Island: Massacre Bay, *Lyall.* This precisely accords with Endlicher's figure, drawn from a Norfolk Island specimen, and differs from the following in the perfectly glabrous branchlets and young leaves.

4. **C. petiolata,** *Hook. f. in Journ. Linn. Soc. Bot.* i. 128 (1857). A maritime shrub, very similar indeed to *C. Baueriana,* but the leaves are less fleshy, and the young foliage and branches covered with a minute pubescence. Leaves 1–2$\frac{1}{2}$ in. long, oblong-obovate, obtuse rounded or retuse, narrowed into rather slender petioles. Peduncles and flowers as in *C. Baueriana,* but the latter rather smaller.

Northern Island: maritime rocks south of Castle Point, *Colenso.* **Kermadec** Islands, *M'Gillivray.* A variable plant. Colenso's specimens are very small, from a small dense rigid shrub; the Kermadec Island ones have small leaves with margins hardly recurved. The same plant in Norfolk Island and Lord Howe's Island has much larger, more obovate leaves, rather green when dry. It may prove to be a variety of *C. Baueriana,* but the pubescent branches at once distinguish it.

5. **C. robusta,** *Raoul;—Fl. N. Z.* i. 105. A large erect, stout, perfectly glabrous, leafy, glossy green shrub. Branches glabrous; bark pale-brown, wrinkled, shining. Leaves numerous and close-set, 2–3 in. long, $\frac{1}{2}$–1$\frac{3}{4}$ broad, elliptic-oblong or lanceolate, acute or acuminate, narrowed into short stout petioles, when dry brown, paler below, margins slightly recurved. Flowers very densely capitate, sessile or on short stout peduncles, heads nearly $\frac{1}{2}$ in. diam.;—*male:* calyx minute, cupular, 4- or 5-toothed; corolla $\frac{1}{6}$ in. long, bell-shaped, 4- or 5-lobed or -parted; anthers linear-oblong;—*fem.:* calyx-tube tubular, truncate, entire or with 5–8 glandular points; corolla tubular, unequally 2–5-toothed; styles very stout. Drupe small, less than $\frac{1}{4}$ in. long.—Raoul, Choix, xxiii. t. 21.

Abundant in forests throughout the islands, *Banks and Solander,* etc. Raoul's figure is not characteristic of the excellent specimens he gave me. I suspect that this species may vary into the following.

6. **C. Cunninghamii,** *Hook. f., n. sp.* An erect twiggy bush, with slender or stout, erect, rarely divaricating, glabrous branches and twigs, and pale-brown bark. Leaves erect, often crowded, $\frac{1}{2}$–2 in. long, $\frac{1}{8}$–$\frac{1}{4}$ broad, linear or linear-lanceolate or -oblong, obtuse or acute, dark-brown when dry, paler below, coriaceous, flat; veins few, dark; stipules short, rather membranous. Flowers sessile, glomerate on lateral branchlets, $\frac{1}{8}$ in. long;—*male:* calyx-limb cupular, short, 4- or 5-toothed; corolla bell-shaped, 4- or 5-partite;—*fem.:* calyx-limb tubular, 5–8-toothed; corolla shortly bell-shaped, 4–5-lobed to the middle or lower. Drupe ovoid, $\frac{1}{4}$ in. long, crowned with the tubular calyx-limb.—*C. fœtidissima,* A. Cunn. in part, not Forst.

Northern Island: common towards the seacoast. **Chatham** Island, *Dieffenbach.* **Middle** Island: Akaroa, *Raoul.* An extremely variable plant, perhaps a form of *C. robusta,* but very different-looking, with much smaller narrower leaves, sessile flowers, and smaller fruit. In the New Zealand Flora' I had confounded it with *C. propinqua*, to which it is very near, but differs in the fascicled female flowers, larger leaves, and branches rarely divaricating at right angles. In Cunningham's Herbarium it is mixed with *propinqua*, under the name of *fœtidissima*, Forst. From *C. linariifolia* it differs in the fascicled flowers, calyx and stipules not long and sheathing. I have gathered at the Bay of Islands what appears to be an extremely slender and narrow-leaved form of this, with almost white bark, and female flowers sometimes solitary.

7. **C. acutifolia,** *Hook. f. in Journ. Linn. Soc. Bot.* i. 128 (1857). A small tree, quite glabrous. Branches slender; bark pale. Leaves 1–2½ in. long, ⅓–1 broad, ovate or oblong-ovate, acute, membranous, dark-brown when dry; veins spreading, veinlets finely reticulated below; petioles slender, ¼ in. long; stipules broad, very deciduous. Peduncles as long as the petioles, slender, simple or 3-chotomously branched; branches 3-flowered;—*male:* ¼ in. long; calyx minute, cupular; corolla between bell- and funnel-shaped, 4- or 5-lobed;—*fem.:* unknown.

Kermadec group, from the coast to the mountain-tops, *M'Gillivray*. A most distinct species. I have small specimens of a very similar and perhaps identical plant, with more coriaceous leaves and sessile drupes, gathered by Colenso in the Ruahine district, where it forms a tree 12–18 ft. high.

8. **C. spathulata,** *A. Cunn.;—Fl. N.Z.* i. 106. A slender, glabrous, shining shrub, 3–5 ft. high and more, sparingly leafy. Branches spreading, twigs obscurely puberulous; bark pale, ashy or grey. Leaves ½–2 in. long, orbicular or broadly ovate, rounded retuse or 2-lobed at the tip, suddenly contracted into a flat winged petiole longer or shorter than the blade, rather coriaceous, glossy above, paler below, margins recurved, veins few, diverging, faint, not reticulated, brown when dry; stipules cuspidate, deciduous. Flowers sessile, axillary, small, ⅛ in. long;—*male:* solitary geminate or fascicled; calyx cupular, 4- or 5-lobed to the base, lobes linear, obtuse; corolla bell-shaped, 4- or 5-partite;—*fem.:* fascicled; calyx-tube ovoid, glabrous, limb 4- or 5-partite, lobes linear, erect, ciliate; corolla 4- or 5-partite to the base; styles twice as long as the corolla. Drupe ovoid, ¼–⅓ in.

Northern Island: Bay of Islands to Auckland, not uncommon in forests, etc., *Banks and Solander*, etc. A well-marked species, from the linear broad petioles and orbicular blade of the leaf, the long lobes of the calyx of both sexes, and deeply divided corolla of the female.

9. **C. rotundifolia,** *A. Cunn.;—Fl. N. Z.* i. 109. A large shrub, much and diffusely branched, 3–6 ft. high. Branches lax, long, very slender, divaricating at right angles; twigs pubescent, ultimate villous with spreading or appressed hairs; bark grey or pale-brown. Leaves in rather remote pairs, ¼–¾ in. long, orbicular or broadly oblong, cuspidate or abruptly acute, suddenly narrowed into slender ciliate petioles ⅛–⅓ in. long, very membranous, more or less ciliate and hairy, especially on the under surface; veins finely reticulate beneath; stipules small, membranous. Flowers very minute, $\frac{1}{12}$–$\frac{1}{10}$ in. long;—*male:* solitary or 2 or 3, sessile; calyx cup-shaped, membranous, 4-toothed, teeth equal or 2 longer; corolla short, deeply 4-cleft;—*fem.:* calyx-tube hairy, limb shortly tubular, 4- or 5-toothed; corolla shortly funnel-shaped, unequally 4- or 5-cleft to the middle. Drupe very small, didymous, much broader than long, ⅙ in. broad.

Northern Island: marshy ground, river banks, etc., Bay of Islands, *A. Cunningham*, to Wellington, *Ralphs*. **Middle** Island: Otago, *Hector and Buchanan* (a scrap only). I have only male flowers of Ralphs' specimen, which has smaller, oblong acuminate leaves. A most distinct form of species, from the slender habit, very membranous, ciliate (at length glabrous) cuspidate leaves, minute flowers, and didymous drupe.

10. **C. ciliata,** *Hook. f. Fl. Antarct.* i. 22. A lax or densely-branched shrub, 4–10 ft. high. Branches stout or slender, ultimate villous with rather rigid hairs; bark very pale, nearly white. Leaves tufted on short lateral branchlets, $\frac{1}{4}$–$\frac{2}{3}$ in. long, oblong, rounded or subacute at the tip, flat, rather membranous, very shortly petioled, petiole margin and under surface more or less ciliated, dusky brown when dry; veins few, nearly parallel to margin, not reticulated; stipules broad, villous, cuspidate. Flower and fruit unknown.

Lord Auckland's group and **Campbell's** Island, abundant, *J. D. H.* I have seen no New Zealand specimens of this most distinct species, which a good deal resembles *C. pilosa*, Endl., of Norfolk Island.

11. **C. tenuicaulis,** *Hook. f. Fl. N. Z.* i. 106. An erect shrub, 4–6 ft. high. Branches slender, divaricating, but not at right angles, puberulous, bark pale-grey or brown. Leaves generally in distant pairs, $\frac{1}{4}$–$\frac{1}{3}$ in., orbicular-spathulate, or broadly ovate-spathulate, obtuse or subacute, flat, puberulous on the under surface, veins few, reticulated in large areoles, suddenly narrowed into short petioles, not coriaceous; stipules pubescent, not ciliated. Flowers axillary;—*male* solitary or fascicled; calyx cupular, minute, 4-toothed; corolla $\frac{1}{8}$–$\frac{1}{5}$ in. diam., between funnel- and bell-shaped, 4-lobed nearly to the base; anthers nearly as large as the lobes; *fem.:* unknown. Drupes very small, globose, $\frac{1}{12}$–$\frac{1}{8}$ in. diam.

Northern Island: Bay of Islands, *Colenso*, etc. What I have described in Fl. N. Z. as the female flower of this, I now refer without hesitation to *C. divaricata*, which this species closely approaches, but differs in the bark, in the less divaricating branches, leaves in series of pairs on the twigs (not in terminal pairs on lateral branchlets), pubescent, reticulated veined below, and in the globose drupe.

12. **C. rhamnoides,** *A. Cunn.;—Fl. N. Z.* i. 107. A densely branched bush, 2–4 ft. high or more. Branches slender, much divaricating, divided often flexuous and interlaced, pubescent; bark pale, not smooth and even. Leaves $\frac{1}{4}$–$\frac{1}{2}$ in. long, $\frac{1}{6}$–$\frac{1}{2}$ broad, orbicular or oblong, rounded retuse or acute, narrowed suddenly into a short petiole, flat, greenish-brown, quite glabrous; veins very indistinct, not reticulated. Flowers minute, solitary, axillary, $\frac{1}{10}$ in. long, on short curved bracteolate peduncles;—*male:* calyx cupular, 4-lobed, membranous; corolla bell-shaped, 4-lobed;—*fem.:* calyx-tube ovoid, glabrous, limb shortly tubular, 4- or 5-lobed; corolla tubular, rather slender, curved, mouth shortly 4- or 5-cleft; styles filiform. Drupe very small, $\frac{1}{10}$–$\frac{1}{6}$ in. diam., globose or oblong.—? *C. gracilis*, A. Cunn.

Northern and **Middle** Islands: not uncommon from the Bay of Islands to Otago. The divaricating slender pubescent branches, small broad glabrous leaves, minute axillary flowers, and tubular female corolla, are the best marks of this species, which in Cunningham's Herbarium is mixed with the following.

13. **C. divaricata,** *A. Cunn.* A laxly branched shrub, 8–10 ft. high; branches very slender, widely divaricating, glabrous or minutely pubescent; bark dark red-brown, quite even. Leaves in pairs on very short lateral shoots, $\frac{1}{3}$–$\frac{3}{4}$ in. long, orbicular or obovate-spathulate, shortly petioled, in small varieties

coriaceous, in larger membranous and puberulous when young, contracted into petioles of variable length; veins not reticulated. Flowers;—*male:* not seen; *fem.:* axillary, solitary, minute, $\frac{1}{12}$ in. long; calyx-tube glabrous, limb short, tubular, unequally and often irregularly 4- or 5-lobed; corolla 4- or 5-partite, almost to the base; styles slender. Drupe small, obovoid, $\frac{1}{6}$–$\frac{1}{5}$ in. long, variable in size and colour.

Common in the **Northern** and probably throughout the islands, Otago, *Lyall.* I am greatly puzzled with this plant, which is the *C. divaricata*, δ. *latifolia*, and ε. *coriacea*, of my N. Z. Flora, and in Cunningham's Herbarium is mixed with *C. rhamnoides*, and is with difficulty distinguished from it; its prominent characters are, the very slender branches divaricating at right angles, the smooth brown bark, almost glabrous twigs, and female corolla not tubular and cleft at the mouth, but spreading and divided to the base. The large-leaved form (*latifolia*) has larger membranous rounded leaves, somewhat pubescent below, the smaller (*coriacea*) has more coriaceous spathulate leaves. Of the variety γ. *pallida* of Fl. N. Z., which has young male flowers only, I can make out nothing more; it has pale bark, and spathulate leaves, opposite throughout the twigs, it was collected by Colenso in the Wairarapa valley, where it forms a small tree 5–7 ft. high, and where also the true *divaricata* grows. Cunningham's *C. gracilis* (*divaricata*, β. *gracilis*, Fl. N. Z.) has neither flower nor fruit, and is, I now think, referable to *C. rhamnoides*, having more pubescent twigs than *C. divaricata*.

14. **C. parviflora,** *Hook. f. Fl. N. Z.* i. 107. A large dense rigid branching leafy bush. Branches slender, pubescent, divaricating more or less; bark pale-grey. Leaves small, rather close-set, fascicled on lateral branchlets, very uniform, $\frac{1}{4}$–$\frac{1}{3}$ in. long, obovate or linear-oblong, rounded at tip, rarely subacute, very shortly petioled, rigid, flat, not shining, dark-brown when dry, very obscurely veined; stipules pubescent, ciliate. Flowers minute, lateral, solitary, subsessile;—*male:* calyx cupular, 4-toothed; corolla $\frac{1}{10}$ in. long, broadly bell-shaped, 4-partite;—*fem.:* calyx-limb glabrous, tube very short, jagged, and ciliolate; corolla $\frac{1}{12}$ in., 4- or 5-cleft $\frac{3}{4}$ way down; style $\frac{1}{6}$ in. Drupe ovoid, $\frac{1}{6}$–$\frac{1}{4}$ in.—*C. myrtillifolia*, *a*, Fl. Antarct. i. 21; Fl. N. Z. i. 108.

Northern and **Middle** Islands: low grounds; common in woods as far south as Otago, also on the mountains, Tongariro and Ruahine range, *Colenso and Bidwill.* **Auckland's** Island, *J.D.H.* I think that my *C. myrtillifolia* of Lord Auckland's Island is referable to this very common species, and as it was described without flower or fruit, that name had better be abandoned. Its axillary minute flowers distinguish it well from *C. Colensoi*, and its very coriaceous uniform leaves from *C. divaricata*, *rhamnoides*, and *tenuicaulis*.

15. **C. propinqua,** *A. Cunn.;—Fl. N. Z.* i. 109. Tall, erect, glabrous, woody, sparingly leafy, 4–8 ft. high. Branches divaricating, glabrous, scarcely puberulous at the tips; bark brown. Leaves usually in pairs on short axillary branchlets, $\frac{1}{3}$–$\frac{1}{2}$ in. long, $\frac{1}{12}$–$\frac{1}{10}$ broad, narrow linear-oblong, obtuse, narrowed at the base, nearly flat, coriaceous, veins indistinct; stipules obtuse, glabrous. Flowers lateral;—*male:* fascicled, $\frac{1}{10}$ in. long; calyx cupular, 4- or 5-toothed; corolla bell-shaped, 4- or 5-cleft nearly to the base;—*fem.:* calyx glabrous, limb tubular, 4- or 5-toothed; corolla not seen. Drupe ovoid, $\frac{1}{6}$–$\frac{1}{4}$ in. long.

Northern Island: common from the Bay of Islands, *A. Cunningham*, to Hawke's Bay, *Colenso*, in wet places. This is the *C. propinqua*, α and β, Fl. N. Z.;—the γ. *linariifolia* is certainly a different species, of which I have since procured fine specimens.

16. **C. fœtidissima,** *Forst.;—Fl. N. Z.* i. 105, *and* ii. 331. A large glabrous shrub or small tree 8–20 ft. high, intensely fetid after being bruised,

sometimes smaller and procumbent. Branches slender, not divaricate, the ultimate minutely puberulous, bark very pale, white or grey. Leaves 1–2½ in. long, ⅓–¾ broad, oblong or linear- or obovate-oblong, narrowed into a rather slender petiole ¼–¾ in. long, obtuse or subacute, flat, rather membranous, pale or dark-brown when dry, not shining; veins 0, or diverging and indistinct, not reticulated; stipules glabrous or puberulous and minutely ciliate, cuspidate, the uppermost sheathing. Flowers terminal on the branchlets, solitary, sessile, pendulous, rather large, ¼–⅓ in. long;—*male:* calyx cupular, 4- or 5-toothed; corolla broadly campanulate, 4- or 8-lobed to the middle; anthers large, linear-oblong, as long as the corolla;—*fem.:* calyx-tube narrow-ovoid, limb short, tubular, toothed; corolla tubular, curved, 4- or 5-cleft ⅓–½ way down; styles very stout and long, ⅓–¾ in. Drupe generally large, ovoid, very fleshy, ¼–⅓ in. long.—Fl. Antarct. i. 20. t. 13; *C. affinis*, Fl. Antarct. l. c.; *C. repens*, A. Rich.; *C. pusilla*, Forst.?

Mountainous districts of the **Northern** and **Middle** Islands: abundant. **Lord Auckland's** group and **Campbell's** Island: abundant. The fetid odour, oblong leaves, and large solitary terminal flowers, well characterize this most distinct species; the leaves are very variable.

17. **C. Colensoi,** *Hook. f., n. sp.*—*C. myrtillifolia, γ. linearis*, Fl. N. Z. i. 108. A small, subalpine, slender, perfectly glabrous shrub, 3–4 ft. high. Branches slender, spreading, puberulous; bark white or grey. Leaves usually fascicled on short lateral twigs, ¼–¾ in. long, ⅙–⅓ broad, very variable in shape, narrow-oblong lanceolate- or obovate-oblong, more or less obtusely truncate and emarginate, narrowed into short slender petioles, margins sub-recurved, pale yellow-brown when dry, coriaceous, almost veinless; stipules glabrous, not ciliated. Flowers solitary, terminal, on very short curved pedicels;—*male:* ⅙ in. long; calyx cupular, 4-lobed, 2 lateral lobes much longer, patent; corolla bell-shaped, 4-lobed, but not to the middle;—*fem.:* calyx-tube cupular, unequally jagged and ciliate; corolla $\frac{1}{12}$ in. long, 5-partite, lobes spreading. Drupe small, ⅛ in. long, ovoid.

Mountainous regions in the **Northern** Island, near Cook's Straits, *Colenso*. A very distinct species, allied to *C. parviflora*, in the retuse apices of the coriaceous leaves, the white bark, and olive-brown hue when dry; but differing in the slender habit, lax foliage, leaves not recurved, and stipules not ciliated. When out of flower with difficulty to be distinguished by descriptions from *divaricata, tenuicaulis*, and other lowland species, with which it has no real affinity however.

18. **C. cuneata,** *Hook. f. Fl. Antarct.* i. 21. *t.* 15; *Fl. N. Z.* i. 110. A rigid, woody, much branched, erect or prostrate, very leafy shrub, 1–7 ft. high. Branches stout, glabrous or puberulous; bark white brown or black. Leaves fascicled, close-set, ¼–½ in. long, ⅙–¼ in. broad, cuneate-oblong or obovate-lanceolate or linear-obovate, retuse obtuse or subacute, often recurved, very rigid coriaceous and shining, a deeply sunk midrib above, almost veinless below; stipules (young) with fimbriate or densely ciliate margins. Flowers terminal;—*male:* calyx cupular, jagged and fimbriate at the mouth; corolla ¼ in. long, nearly ⅓ diam., with short funnel-shaped limb, and 4 or 5 long spreading lobes; anthers ⅕ in. long;—*fem:* calyx glabrous, with a 4- or 5-lobed limb; corolla not seen. Drupe globose or oblong, rather small, ⅙–¼ in. long.

Mountains of the **Northern** and **Middle** Islands: Ruahine and Lake Taupo ranges, and Mount Hikurangi, *Colenso;* Mount Egmont, *Dieffenbach;* Otago, lake district, *Hector and Buchanan.* Common in **Lord Auckland's** group and **Campbell's** Island, *J. D. H.* I believe this to be one of the most distinct forms of the genus, but almost impossible to distinguish by foliage from some states of *C. parviflora* and others: its prevalent characters are the very stout leafy habit, small recurved, very rigid and coriaceous, more or less cuneate leaves, fimbriate or densely ciliate young stipules, and terminal, solitary, rather large flowers; but flat-leaved forms approach *C. parviflora,* narrow-leaved ones *C. propinqua,* lax-leaved ones *C. Colensoi,* and the stipules are sometimes glabrous.

19. **C. acerosa,** *A. Cunn.;—Fl. N. Z.* i. 109. A low, excessively branched, spreading shrub, 2–5 ft. high. Branches divaricating, flexuous and tortuous, puberulous; bark yellow-brown. Leaves uniform, in opposite pairs or opposite fascicles on the twigs, $\frac{1}{4}$–$\frac{1}{3}$ in. long, $\frac{1}{20}$ broad, linear, subacute, erecto-patent, veinless; stipules short, puberulous, not ciliate. Flowers;—*male:* calyx cup-shaped, 4-lobed, 2 lobes short acute, 2 much longer oblong obtuse; corolla $\frac{1}{6}$–$\frac{1}{4}$ in. diam., broadly bell-shaped, 4-cleft to or below the middle;—*fem.:* minute, calyx-limb very short, 4-toothed; corolla $\frac{1}{12}$ in., 4-cleft to the middle; styles as long as the leaves. Drupe ovoid, nearly $\frac{1}{4}$–$\frac{1}{2}$ in. long.

Northern, Middle, and **Southern** Islands: not uncommon in rocky places, sand-banks, forests, etc.; Hokianga, etc., in salt marshes, *A. Cunningham.* Colenso remarks that at Ahuriri the berries are sky-blue, transparent, as large as sloes, and eaten by the natives. A most distinct plant. I have a prostrate form from the mountains of Canterbury, alt. 4000 ft., and the lake district, Otago.

20. **C. depressa,** *Col.;—Fl. N. Z.* i. 110. A small, dense, prostrate, much branched bush, 1–4 ft. high. Branches pubescent; bark grey. Leaves in opposite fascicles, $\frac{1}{6}$–$\frac{1}{4}$ in. long, $\frac{1}{16}$ wide, spreading, rather recurved, linear-lanceolate, acute or obtuse, rigid, coriaceous, nerveless, rather concave, yellow-green when dry; stipules pubescent and ciliate. Flowers;—*male:* calyx minute, cupshaped, equally 4 toothed; corolla $\frac{1}{10}$ in. long, bell-shaped, 4- or 5-cleft;—*fem.:* calyx-limb short, 4-toothed; corolla not seen. Drupe orange-yellow, sweet, about $\frac{1}{6}$ in. long.

Northern Island: in mountainous localities, Lake Taupo, top of Ruahine and Hawke's Bay ranges, *Colenso.* Perhaps a form of *C. cuneata,* but with much smaller narrower leaves. Some forms approach *C. acerosa,* but the leaves are always broader and rather concave when dry.

21. **C. microcarpa,** *Hook. f. Fl. N. Z.* i. 110, *and* ii. 331. A leafy shrub, 1–10 ft. high. Branches slender, close-set, divaricating, pubescent, leafy; bark grey. Leaves in pairs on short slender lateral branchlets, $\frac{1}{4}$–$\frac{1}{3}$ in. long, $\frac{1}{15}$–$\frac{1}{12}$ in. broad, spreading, linear or linear-lanceolate, acute, flat, veinless, dark-brown when dry, not coriaceous; stipules short, ciliate. Flowers minute;—*male:* calyx cup-shaped, 4-toothed; corolla broadly bell-shaped, $\frac{1}{5}$ in. diam., 4-partite, lobes narrow acuminate, long;—*fem.:* calyx-limb short, tubular, 4-toothed; corolla $\frac{1}{12}$ in., tubular or funnel-shaped, 4-cleft $\frac{1}{4}$ way down. Drupe very small, globose, $\frac{1}{10}$ in. diam.

Northern Island: tops of the Ruahine mountains, *Colenso.* Perhaps a variety of *C. cuneata.*

22. **C. linariifolia,** *Hook. f.;—C. propinqua, γ,* Fl. N. Z. i. 109. A tall, erect, branching shrub. Branches spreading, slender or rather stout, twigs puberulous; bark grey or black. Leaves all opposite (rarely fascicled), $\frac{1}{2}$–1 in. long, $\frac{1}{10}$–$\frac{1}{6}$ in. broad, linear linear-lanceolate or oblong-lanceolate,

acute or acuminate, flat, rather coriaceous, black when dry; stipules puberulous, the upper produced into long sheaths, acute, ciliate. Flowers;—*male:* not seen;—*fem.:* calyx-limb of 4 long, linear, erect, obtuse teeth; corolla 4-partite, $\frac{1}{10}$ in. long, lobes narrow acuminate; styles as long as the leaves. Drupe ovoid, $\frac{1}{4}$ in. long, generally crowned with the calyx-limb.

Northern and **Middle** Island: in mountain localities; Ruahine mountains and woods at Manawarakua, etc., *Colenso;* banks of the Waihopai, *Munro;* Lake Tekapo, *Haast;* Otago, *Lindsay.* A very well marked species by the foliage, long calyx-lobes and styles of the female flower, and long sheathing stipules; in many respects it closely resembles states of the *C. Cunninghamii* and *C. propinqua.*

23. **C. repens,** *Hook. f.;—Fl. Antarct.* i. 23. *t.* 16 A (*not A. Rich.*); *Fl. N. Z.* i. 110. A small alpine matted species, with long prostrate, creeping, rather stout flaccid stems and branches, densely leafy. Branches glabrous, bark pale. Leaves close-set and fascicled, $\frac{1}{6}$–$\frac{1}{4}$ in. long, spreading or recurved, ovate-oblong or obovate, rounded at the tip, concave, very coriaceous, rather shining, black when dry, with thickened margins; veins 0; stipules short, obtuse, glabrous. Flowers:—*male:* not seen;—*fem.:* minute, sessile; calyx shortly 4-toothed; corolla $\frac{1}{11}$ in. long, tubular, curved, 4-cleft at the mouth; styles sometimes 3 or 4. Drupe large, very succulent, sometimes nearly $\frac{1}{3}$ in. long, orange-yellow, with 2–4 nuts.

Northern Island: open grounds near Lake Taupo, *Colenso.* **Lord Auckland's** group and **Campbell's** Island: common on the hills, *J. D. H.* Probably common in the Southern Alps, but I have seen no specimens.

24. **C. pumila,** *Hook. f. Fl. Antarct.* i. 22. *t.* 16 B; *Fl. N. Z.* i. 3. A small, alpine, creeping, tufted species. Branches densely leafy, glabrous; bark pale. Leaves imbricate, patent or suberect, $\frac{1}{4}$–$\frac{1}{3}$ in. long, oblong linear-oblong or -obovate, obtuse or acute, rigid, very coriaceous, shining and rather concave when dry, veinless; stipules glabrous, obtuse. Flowers sessile, erect;—*male:* very large, $\frac{1}{3}$–$1\frac{1}{2}$ in. long; calyx cupular, 4-lobed; corolla tubular, curved, shortly 4- or 5-lobed; anthers very large;—*fem.:* calyx 5-toothed; corolla like the male, but half the size; styles 2. Drupe $\frac{1}{4}$ in. long, orange-yellow.

Mountains of the **Northern** Island: summit of Ruahine mountains, *Colenso.* **Middle** Island: subalpine regions near the Great Godley glacier, *Haast;* Otago, lake district, *Hector and Buchanan.*

2. NERTERA, Banks and Solander.

Small creeping herbs. Leaves petioled, ovate, stipulate. Flowers axillary, solitary, sessile, hermaphrodite.—Calyx-tube ovoid; limb truncate or obscurely 4-toothed. Corolla tubular or funnel-shaped, 4- or 5-lobed. Stamens 4 or 5, inserted at the base of the corolla; filaments exserted; anthers large, pendulous. Ovary 2-celled; styles 2, filiform, very long, hirsute; cells 1-ovuled. Drupe red, fleshy, containing 2 plano-convex, coriaceous, 1-seeded nuts.

A small genus, of which one species is widely diffused in the southern hemisphere, the others are natives of New Zealand. It scarcely differs from *Coprosma,* except in the herbaceous habit.

Perfectly glabrous. Leaves broadly-ovate . . . . 1. *N. depressa.*
Perfectly glabrous. Leaves narrow-ovate . . . . 2. *N. Cunninghamii.*

Hairy or villous. Leaves cordate-ovate, long petioled . . . . 3. *N. dichondræfolia.*
Hispid or glabrate. Leaves ovate or oblong, short petioled . . . 4. *N. setulosa.*

1. **N. depressa,** *Banks and Sol.;—Fl. N. Z.* i. 112. Perfectly glabrous. Stems creeping and rooting, tufted, 6–10 in. long, 4-gonous. Leaves $\frac{1}{4}$–$\frac{1}{2}$ in. long, broadly ovate, acute or obtuse, coriaceous or almost fleshy; petioles as long as the blade or shorter. Stipules very small. Flowers minute, sessile, $\frac{1}{10}$ in. long. Calyx-limb 4-toothed. Corolla funnel-shaped.

Middle Island: Wai-au-au mountains, alt. 3000 ft., *Travers.* **Southern** Island, *Lyall* **Lord Auckland's** group, *J. D. H.* Also found in Tasmania, Tristan d'Acunha, and the Andes of South America, from Mexico to Fuegia.

2. **N. Cunninghamii,** *Hook. f. Fl. N. Z.* i. 112. Altogether very similar to *N. depressa*, but smaller, more slender, with more acute leaves. Calyx-limb obscurely 4-lobed. Corolla shortly funnel-shaped. Stamens erect.

Northern Island: Bay of Islands, common in moist places; Taupo plains, *Colenso.* This appears also to be a Philippine Island plant.

3. **N. dichondræfolia,** *Hook. f. Fl. N. Z.* i. 112. *t.* 28 A. Stems slender, creeping, hairy or villous, a span to 2 feet long. Leaves $\frac{1}{4}$–$\frac{2}{3}$ in. long, broadly ovate-cordate, acute or apiculate, membranous, more or less hairy above, glabrous below; petiole slender, as long as the blade. Flowers very small and fruit as in *N. depressa.*—*N. gracilis*, Raoul, in Ann. Sc. Nat.; *Geophila (?) dichondræfolia*, A. Cunn.

Northern and **Middle** Islands: common in damp places. Very variable in size and hairiness.

4. **N. setulosa,** *Hook. f. Fl. N. Z.* i. 112. *t.* 28 B. Small, hispid or glabrate; stem rigid, wiry, slender, 2–3 in. long; branches erect or ascending. Leaves $\frac{1}{4}$–$\frac{1}{2}$ in. long, broadly ovate or oblong, obtuse, with stiff white hairs on the upper surface, glabrous below; petiole short; stipules small, notched. Flowers white, inconspicuous, as long as the leaves. Calyx-tube hispid. Corolla very slender, 4- or 5-toothed. Filaments long, exserted; anthers pendulous, linear-oblong, 2-lobed at the base. Styles very long. Drupe small, hispid. Embryo very minute.

Northern Island: Ahuriri and head of Wairarapa valley, *Colenso.*

3. GALIUM, Linn.

Slender, weak, prostrate, erect or climbing, often hispid herbs; stems 4-angled. Leaves whorled, entire or ciliated. Flowers minute, white or yellow, on axillary, simple or branched peduncles.—Calyx-tube globose; limb 0. Corolla rotate, 3- or 4-partite; lobes valvate. Stamens 3 or 4; filaments short; anthers didymous. Ovary 2-celled; styles 2, short, with simple or capitate stigmas; ovules 1 in each cell. Fruit minute, of 2 dry, 1-seeded carpels.

A very large European and Oriental genus, found also in most other parts of the world.

Leaves 4-nate, linear-lanceolate . . . . 1. *G. tenuicaule.*
Leaves 4-nate, oblong, mucronate . . . . 2. *G. umbrosum.*

1. **G. tenuicaule,** *A. Cunn.;—Fl. N. Z.* i. 113. Stem straggling; branches 4 in.–3 ft. long, glabrous or scabrid. Leaves whorled in fours, $\frac{1}{4}$–$\frac{3}{4}$ in. long, oblong- or linear-lanceolate, awned or acuminate, scabrid on

the edges and midrib below. Peduncles 1–3-flowered, spreading, longer or shorter than the leaves, decurved in fruit. Flowers white, $\frac{1}{12}$ in. diam. Fruit of 2 minute, globose, glabrous carpels.

Northern and **Middle** Islands: abundant in grassy situations, ascending to 4000 ft., *Banks and Solander*, etc.

2. **G. umbrosum,** *Forst.*—*G. propinquum*, A. Cunn.;—Fl. N. Z. i. 113. Very variable, annual, usually erect and rather stiff, but often weak and straggling, 1–10 in. long, glabrous or ciliated on the stem and leaves. Leaves in whorls of 4, $\frac{1}{10}$–$\frac{1}{3}$ in. long, broadly oblong, acuminate or awned, marked with pellucid dots when seen between the eye and light. Peduncles 1–3-flowered, longer or shorter than the leaves. Flowers very minute, white. Fruit of 2 globose, smooth, minute carpels.

Abundant throughout the islands, *Banks and Solander*, etc. I suspect that the Tasmanian *G. ciliare*, nob., does not differ from this.

4. ASPERULA, Linn.

Characters of *Galium*, but corolla more or less bell- or funnel-shaped.

A less frequent genus than *Galium*, but very large and with about the same distribution.

1. **A. perpusilla,** *Hook. f. Fl. N. Z.* i. 114. A small, slender, decumbent, inconspicuous annual, everywhere perfectly glabrous. Stems very short, 1–2 in., filiform, branched. Leaves 4 in a whorl, $\frac{1}{10}$–$\frac{1}{12}$ in. long, lanceolate, acuminate, awned, often curving to one side. Flowers solitary, sessile, white. Calyx-tube glabrous. Corolla funnel-shaped, $\frac{1}{12}$ in. diam., 4-partite; lobes linear. Styles united, their tips free, divergent.

Northern and **Middle** Islands: not uncommon in dry and sandy places; base of Tongariro and of the Tararua range, *Colenso*; upper Motueka alps, 2000 ft., *Haast, Munro*; Canterbury, *Travers*; Otago, *Hector and Buchanan*. The smallest known species of the genus, and the smallest flowering plant, except *Tillæa* and *Lemna* in the islands.

Two species of the Australian genus *Opercularia* are described (erroneously) by Gærtner as having been found in New Zealand by Banks and Solander; there are no specimens of them in Banks's Herbarium. The flowers are in involucrate heads, and the capsules open by transverse lids.

Order XXXIX. COMPOSITÆ.

Herbs shrubs or trees. Leaves alternate, rarely opposite, simple in most of the New Zealand species, exstipulate. Flowers minute (florets), sessile, densely crowded on flat or conical receptacles, forming heads surrounded by an involucre consisting of 1 or more series of linear, erect bracts. Receptacle naked or bearing scales, smooth, pitted or papillose.—Calyx-tube adnate with the ovary; limb 0 or represented by bristles scales or hairs (pappus). Corolla of two forms, tubular and 4- or 5-cleft with valvate lobes, or tubular below with a long linear limb; usually both kinds occur in each head, the outer ligulate (ray-florets) forming a ray of 1 or 2 series round the inner which are tubular (disk flowers); the ray-flowers are female or hermaphrodite, the disk-flowers male or hermaphrodite. Stamens 5, inserted on the tube of the corolla; anthers usually united in a tube which sheaths the style; cells often produced downward (tailed). Ovary inferior, 1-celled; style filiform, with 2 stigmatic branches; ovule 1, erect. Fruit a dry nut or achene. Seed erect; testa membranous; albumen 0; cotyledons oblong, plano-convex, radicle short, inferior.

One of the largest Orders of flowering-plants, found in almost every part of the globe to which flowering plants extend. The New Zealand genera are with few exceptions Australian also.

Subord. 1. **Corymbiferæ.**—*Florets all tubular, or of the ray ligulate. Herbs shrubs or trees, without milky juice.*

(1. Asteroideæ.)

α. *Ligulate florets, when present, never yellow. Pappus rigid, of long unequal scabrid hairs.*

Shrubs or trees. Peduncles 1- or many-flowered. Leaves whitish below. Veins branching . . . . 1. Olearia.
Herbs. Peduncles many-flowered. Leaves large, with many parallel ribs . . . . 2. Pleurophyllum.
Herbs. Leaves all radical. Scape 1-flowered . . . . 3. Celmisia.
Branched herb, leafy, branches 1-flowered . . . . 4. Vittadinia.

β. *Ligulate florets white or purplish, revolute. Pappus 0, or of small scales or short hairs.*

Scapigerous herbs. Achene beaked. Pappus 0 . . . . 5. Lagenophora.
Scapigerous herbs. Achene not beaked. Pappus very short or 0 6. Brachycome.

(2. Senecionideæ.)

γ. *Ligulate florets white or yellow. Pappus of 2 bristles* . . . 7. Bidens.

δ. *Florets all tubular. Pappus 0.*

Tufted moss-like herbs. Florets purplish . . . . 8. Abrotanella.
Heads yellow, unisexual or with ♀ and ♂ florets; outer fl. ♀ in 1 or many series . . . . 9. Cotula.

ε. *Florets all tubular, yellow. Heads all collected into one dense globose ball. Pappus soft and plumose.*

Scapigerous herbs, with globose inflorescence . . . . 10. Craspedia.

ζ. *Ligulate florets, if present, yellow (inner scales of involucre sometimes white, spreading, and resembling a ray). Pappus of soft, white hairs.*

† *Shrubs. Leaves all small. Involucral scales in many series. Ligulate fl.* 0.

Receptacle narrow, with scales amongst the flowers . . . . . 11. Cassinia.
Receptacle narrow, not without scales amongst the flowers . . . 12. Ozothamnus.

†† *Herbs. Involucral scales in 2 or several series. Ligulate fl.* 0.

Small, creeping, tufted, alpine herbs. Heads small, sessile . . . 13. Raoulia.
Erect or decumbent herbs. Scales of involucre with or without white rays . . . . 14. Gnaphalium.
Tufted, woolly, alpine herbs, with very broad sessile heads . . . 15. Haastia.

††† *Herbs shrubs or trees. Involucral scales in 1 series. Ligulate fl. 0 or yellow.*

Herbs. Outer florets tubular, excessively slender, female . . . 16. Erechtites.
Outer florets ligulate or similar to interior. Pappus soft, white . 17. Senecio.
Outer florets small, irregularly cleft . . . . 18. Brachyglottis.
Ligulate fl. 0. Pappus very rigid, yellowish . . . . 19. Traversia.

(3. CICHORACEÆ.)

Subord. 2. **Liguliflorae.**—*Florets all ligulate. Herbs with milky juice.*

Herbs with radical leaves and leafless 1-flowered scapes.

Achene not beaked. Pappus flattened, subulate . . . . 20. MICROSERIS.
Achene not beaked. Pappus of soft, simple hairs . . . . 21. CREPIS.
Achene muricate upwards, beaked. Pappus plumose . . . . . . 22. TARAXACUM.

Branched leafy herbs.

Achene ribbed, compressed. Pappus plumose . . . . 23. PICRIS.
Achene flat or winged. Pappus of simple, very soft hairs . . . 24. SONCHUS.

1. OLEARIA, Mœnch.

(Eurybia, *Cass.*)

Shrubs or trees. Leaves alternate or fascicled, very rigid, coriaceous, with white or buff tomentum below.—Heads large or small, rayed, corymbose or solitary, peduncled or sessile. Involucre of imbricate, rigid scales. Receptacle convex, alveolate or almost even. Florets 20–100 (rarely 1–4); outer female, in one row, ligulate, white, the rest tubular, hermaphrodite; anthers with very short tails. Pappus of one or more rows of long, rigid, usually unequal, scabrid hairs, acute or thickened at the points. Achenes ribbed, terete, not flattened.

A very large New Zealand and Australian genus, which will possibly have to be united with some others of the Old or New World; most were included, by De Candolle, under *Eurybia*, Cass., a genus established on the single pappus, but which cannot be separated from *Olearia*. The small N. Zealand section, having opposite pairs or fascicles of leaves, might (on account of its habit) be separated generically, were it not for intermediate Australian forms. The terete (not compressed) achene and short tails to the anther-cells, distinguish the genus from the Northern shrubby *Asters;* and the branched, shrubby or arboreous habit alone from *Celmisia* and *Pleurophyllum*. The three first species, or two of them, may prove varieties of one, as may the fourth and fifth of another.

I. *Leaves alternate, not fascicled.*

A. *Head solitary on a bracteate peduncle,* 1–1½ *in. diam.; florets* 50–100. *Rays large, long. Leaves very coriaceous,* 2–5 *in. long.*

Leaves 2–4 in., closely obtusely toothed. Bracts imbricate . . 1. *O. operina.*
Leaves 2–4 in., closely obtusely toothed. Bracts few, leafy . . 2. *O. angustifolia.*
Leaves 1–2½ in., toothed towards the tip . . . . 3. *O. semidentata.*

B. *Heads panicled,* 1 *in. diam.; florets* 50–80, *rays very short. Leaves broad,* 2–5 *in. long, very coriaceous, toothed.*

Leaves acutely irregularly toothed . . . . 4. *O. Colensoi.*
Leaves crenulate . . . . 5. *O. Lyallii.*

C. *Head very large, solitary on a long naked peduncle; florets very numerous, rays short. Leaves* 3–6 *in., entire* . . . 6. *O. insignis.*

D. *Heads panicled, small,* ⅛–⅓ *in. diam.; florets* 8–20. *Leaves* 1½–4 *in. long, oblong or ovate, coriaceous, entire, waved or toothed.*

Leaves rather shining below, oblong, obtuse. Heads ⅓ in. long . 7. *O. furfuracea.*
Leaves silvery below, oblong or ovate, acute, toothed. Heads ⅙–¼ in. long . . . . 8. *O. nitida.*
Leaves silvery below, oblong, acute at both ends, toothed; veins diverging . . . . 9. *O. dentata.*

Leaves shining below, linear-oblong, spinous, truncate at the base 10. *O. ilicifolia*.
Leaves with soft white wool below, acute, toothed. Achenes glabrous . . . . 11. *O. Cunninghamii*.

E. *Heads solitary or corymbose, small*, $\frac{1}{6}$–$\frac{1}{3}$ in. *diam.; florets* 10–20. *Leaves less than* 1 *in. long, entire.*

Leaves 1 in., flat. Heads corymbose; involucre closely imbricate 12. *O. Haastii*.
Leaves $\frac{1}{3}$–$\frac{2}{3}$ in., flat. Heads corymbose; involucre lax . . . . . 13. *O. moschata*.
Leaves $\frac{1}{4}$–$\frac{1}{3}$ in., convex. Head solitary . . . . 14. *O. nummularifolia*.

F. *Heads corymbose, very narrow; florets* 1–4. *Leaves* 1–3 *in. long.*

Leaves waved, oblong, obtuse, reticulate on both sides . . . . . 15. *O. Forsteri*.
Leaves lanceolate-oblong, acute, flat, reticulate on both sides . . 16. *O. avicenniæfolia*.
Leaves oblong, obtuse, waved, not evidently reticulate . . . . . 17. *O. albida*.

II. *Leaves in opposite pairs or opposite fascicles*, $\frac{1}{6}$–1 *in. long.*

Leaves $\frac{1}{4}$–$\frac{2}{3}$ in., pale below. Heads pedicelled . . . . 18. *O. virgata*.
Leaves 1–2 in., obovate or oblong, grey and silvery below . . . 19. *O. Hectori*.
Leaves $\frac{1}{6}$ in., linear, yellow below. Heads sessile . . . . . . 20. *O. Solandri*.

1. **O. operina,** *Hook. f. Fl. N. Z.* i. 114. A stout branching shrub. Branches thick; leaves below and inflorescence covered with white tomentum. Leaves crowded, 2–4 in. long, $\frac{1}{3}$–$\frac{2}{3}$ broad, rigidly coriaceous, narrowly obovate-lanceolate, acuminate; teeth with obtuse, callous tips, glabrous above; veins almost obliterated beneath. Peduncles 2 in. long, stout, thickly clothed with imbricating, ovate-lanceolate bracts $\frac{1}{3}$–$\frac{1}{2}$ in. long; heads 1–1$\frac{1}{2}$ in. diam. Corolla-tube of ray glabrous, of disk puberulous. Pappus hairs very unequal. Achene $\frac{1}{4}$ in. long, ribbed, silky.—*Arnica operina*, Forst.

α. Branches as thick as a goose-quill, glabrous below. Flower of ray $\frac{3}{4}$ in. long.
β. Branches as thick as the finger, tomentose. Flower of ray very short.
Middle Island: *α*, Dusky Bay, *Forster*, *Menzies*; *β*, Port Preservation, *Lyall*.

2. **O. angustifolia,** *Hook. f. Fl. N. Z.* i. 115. Very similar to *O. operina*, *β*, but the leaves much more attenuated, longer, with 3–5 distinct parallel ribs below, the teeth closer, and the peduncles bearing fewer leaf-like bracts, of which the upper are 1–2 in. long. The specimens are past flower.

Stewart's Island, *Lyall*.

3. **O. semidentata,** *Decaisne*;—*Fl. N. Z.* i. 115. A much more slender species than the preceding. Branches as thick as a crow-quill. Leaves 1$\frac{1}{2}$–2$\frac{1}{2}$ in. long, narrow oblong- or linear-lanceolate, acute, serrate towards the tip, white below. Peduncles $\frac{1}{2}$–1$\frac{1}{2}$ in. long, with few bracts; head 1$\frac{1}{4}$ in. diam., purple? Achene $\frac{1}{8}$ in. long, ribbed, almost glabrous.

Chatham Island, *Dieffenbach*. A beautiful plate of this plant was prepared by M. Decaisne, in 1845, for a work (hitherto unpublished) on the botany of the French expedition in the 'Venus.' The leaves are broader in the figure than in my specimens.

4. **O. Colensoi,** *Hook. f. Fl. N. Z.* i. 115. *t.* 29. A stout branching shrub. Leaves very thick and rigid, 1$\frac{1}{2}$–4 in. long, broadly oblong- or lanceolate-obovate, acute, acutely irregularly toothed, shining above, with thick buff or white tomentum beneath; ribs evident or obliterated below. Panicles 6–10-headed, bracteate. Heads $\frac{1}{3}$ in. diam.; involucral scales linear, in 1 or 2 series,

scarious, villous at the tips. Ray florets very short. Pappus white, unequal, in several rows. Achene silky.

α. Leaves without obvious petioles, shorter than the panicles.
β. Leaves in distinct petioles, longer than the panicles.

Northern and **Middle** Islands: var. α, Mount Hikurangi, *Colenso;* mountains of Canterbury, *Haast;* var. β, Dusky Bay, *Lyall.*

5. **O. Lyallii,** *Hook. f. Fl. N. Z.* i. 116. A small, very robust tree, much resembling *O Colensoi* and similarly clothed, but far more robust, the leaves very broadly elliptical or oblong, with diverging veins, obtusely crenulate, the panicles much stouter; involucral bracts in more series. Petioles very broad and thick.

Lord Auckland's Islands, *Lyall, Bolton.* A noble plant. I have the foliage of what is possibly the same, gathered by Lyall in Milford Sound, but the branch and petioles are more slender.

6. **O. insignis,** *Hook. f. Fl. N. Z.* ii. 331. A very robust and remarkable species. Branches as thick as the finger, very densely clothed with thick fulvous or reddish tomentum, as are the young leaves, old leaves below, petioles, peduncles, and involucres. Leaves 3–5 in. long, 2–3 broad, oblong, obtuse, quite entire, excessively thick and hard, unequal at the base, shining above; petiole $\frac{1}{3}$–1$\frac{1}{2}$ in. long. Peduncle as thick as a goose-quill, 6–10 in. long. Head hemispherical. Involucral scales in very many imbricating series, subulate-lanceolate, acuminate, with rigid, recurved, needle-like points, the outer obtuse. Florets excessively numerous, of ray with filiform pilose tubes, $\frac{1}{2}$ in. long, and narrow short rays. Pappus rufous, of one series of equal scabrid hairs, thickened at their tips. Achene very slender, densely silky.

Middle Island: banks of the Waihopai, on the driest rocks, *Munro;* Awatere valley, *Sinclair.* A most magnificent plant, almost generically distinct from *Eurybia* in the simple pappus of equal hairs, thickened at the tip, but it approaches some Australian species in habit, size, and other respects.

7. **O. furfuracea,** *Hook. f.—Eurybia,* DC.; Fl. N. Z. i. 117. A small tree, 10–15 ft. high. Branches terete, velvety; twigs deeply grooved, and inflorescence and leaves below covered with closely-appressed matted hairs, producing a shining surface. Leaves 1$\frac{1}{2}$–2$\frac{1}{2}$ in. broad, ovate-oblong, obtuse, waved, rarely sinuate-toothed, rounded and unequal at base, reticulated above; petiole $\frac{1}{2}$–1 in. long. Corymbs large, branched, loose, spreading. Heads very numerous, $\frac{1}{3}$ in. long, $\frac{1}{6}$ diam.; involucral scales imbricate, oblong, obtuse. Florets longer than involucre, of ray 3 or 4 with short broad rays, of disk 6–8. Pappus white, outer hairs short. Achene slender, angled, puberulous, and glandular.—*Haxtonia furfuracea,* A. Cunn.; *Aster,* A. Rich.; *Shawia,* Raoul.

Northern Island: Bay of Islands to the east coast, *Banks and Solander.*

8. **O. nitida,** *Hook. f.—Eurybia,* Fl. N. Z. i. 117. A small tree, resembling *O. furfuracea,* but the tomentum is more silvery and shining, and leaves more ovate, less coriaceous, sinuate-toothed, and acute or acuminate. Heads in close, many-flowered, rounded corymbs, on silky pedicels, $\frac{1}{6}$–$\frac{1}{4}$ in. long, about as broad; involucral scales few, pilose, or laxly villous; florets 15–20,

rays short. Pappus unequal. Achene silky.—*Eurybia alpina*, Lindl. and Paxt. Mag.; *Solidago arborescens*, Forst., not A. Cunn.

Mountainous regions of the **Northern** Island, *Banks and Solander*, etc.: Mount Egmont, alt. 4000 ft., *Dieffenbach;* Tongariro and Ruahine range, *Colenso.* Abundant throughout the **Middle** Island.

9. **O. dentata,** *Hook. f.—Eurybia dentata, a,* Fl. N. Z. i. 118. A shrub or small straggling tree, 20 ft. high, smelling of musk. Branches, panicles, and leaves below with very appressed down. Leaves oblong or linear-oblong, acute or acuminate, young pubescent above, 2–4 in. long, toothed and rather waved along the margins, not truncate at the base; veins at an obtuse angle to the midrib. Heads white, in large rounded or flattened, many-headed corymbs, $\frac{1}{4}$–$\frac{1}{3}$ in. long, broadly campanulate; involucral scales few, pubescent and pilose; florets 10–12, of ray short; pappus white or red, in one series, hairs thickened at tips. Achene short, $\frac{1}{12}$ in. long, pilose and ribbed.

Northern Island: Ruahine range, *Colenso.* **Middle** Island: Otago, *Hector.* Wood streaked with yellow, makes fair veneers, *Buchanan.*

10. **O. ilicifolia,** *Hook. f.—Eurybia dentata, β,* Fl. N. Z. i. 118. Very similar to *O. dentata*, and also smelling of musk, but branches and twigs nearly glabrous. Leaves longer and narrower, usually truncate at the base, much and deeply waved at the margin, with spinous teeth, yellowish below when dry, and veins at right angles to the midrib. Heads as in *O. dentata.*

Northern Island: Tongariro, *Bidwill;* Ruahine range, *Colenso.* **Middle** Island: Haast river, Canterbury, *Haast;* Otago, *Lyall;* lake district, *Hector and Buchanan.* Certainly distinct from *O. dentata*, and remarkable for its yellow hue when dry, waved spinous leaves, and horizontal veins.

11. **O. Cunninghamii,** *Hook. f.—Eurybia,* Fl. N. Z. i. 117. *t.* 30. A small tree, 12–20 ft. high. Branches, panicles, petioles, and under surface of leaves covered with soft white or brown tomentum. Leaves 2–5 in. long, oblong ovate-oblong or lanceolate-oblong, acute, more or less toothed, finely reticulated above; veins diverging, but not at right angles. Heads very numerous, in large broad panicles with diverging branches, broadly campanulate, $\frac{1}{4}$ in. diam.; involucre shortly turbinate, scales very woolly or nearly glabrous, obtuse or acute, much shorter than the florets; florets 12–20; rays few, broad; pappus white or reddish, unequal, rather thickened at the tips. Achene quite glabrous.—*Brachyglottis Rani*, A. Cunn.

Northern Island: not uncommon in woods, *Banks and Solander*, etc. **Middle** Island: Nelson, common, *Travers;* Ship Cove, *Lyall.* Very variable in size of foliage, in its toothing, and in the involucre sometimes of very tomentose obtuse scales, at others of nearly glabrous acute ones.

12. **O. Haastii,** *Hook. f., n. sp.* A small shrubby tree. Branches stout and corymbs hoary. Leaves $\frac{2}{3}$–$1\frac{1}{3}$ in. long, oblong or ovate- or linear-oblong, obtuse, very coriaceous, quite entire, reticulate and shining above, below white and smooth, with close appressed white down, not shining, flat; veins very obscure, at right angles to midrib. Heads in rather loose peduncled corymbs, $\frac{1}{3}$ in. long, $\frac{1}{3}$ in. diam., 10–15-flowered; involucre cylindric, scales closely imbricate, oblong, obtuse, glabrous, inner nearly as long as the florets; ray florets few, broad; pappus white, unequal. Achene loosely pilose.

Middle Island: near the glacier of Lake Okau, alt. 4–4500 ft., *Haast*. I received this plant first from the late Mr. Veitch, of Exeter, who cultivated it from seeds brought from New Zealand, bearing the MS. name *Eurybia parvifolia*.

13. **O. moschata,** *Hook. f., n. sp.* A much-branched shrub, smelling strongly of musk. Branches, peduncles, and leaves below covered with densely appressed, white or yellowish tomentum. Leaves $\frac{1}{3}$–$\frac{2}{3}$ in. long, narrow obovate-oblong, obtuse, quite entire, flat, narrowed into a very short petiole, hardly reticulated above; veins quite obsolete below. Corymbs axillary, long-peduncled. Heads few, long-pedicelled, campanulate, $\frac{1}{4}$ in. diam.; involucral scales in few series, outer short, obtuse, white, tomentose, inner brown; florets 12–20; pappus whitish, unequal. Achene silky.

Middle Island: Mount Cook, 2500–3500 ft., and banks of Haast and Hopkins rivers, *Haast;* Otago, lake district, alt. 2000 ft., *Hector and Buchanan*.

14. **O. nummularifolia,** *Hook. f.—Eurybia*, Fl. N. Z. i. 118. A rigid, erect shrub, 1–10 ft. high, more or less viscid. Branches stout, erect, often glutinous. Leaves close-set, erect spreading or reflexed, $\frac{1}{4}$–$\frac{1}{3}$ in. long, orbicular oblong or obovate, obtuse, quite entire, margins recurved, very hard and coriaceous, reticulate and shining above, below white or yellow, covered with appressed down. Heads solitary, on peduncles longer or shorter than the leaves, $\frac{1}{3}$–$\frac{1}{2}$ in. long, $\frac{1}{3}$ broad; involucre turbinate, scales appressed, imbricate, broad, obtuse, nearly glabrous or the outer pubescent; florets 8–10, rays rather broad; pappus white, unequal. Achene pubescent.

Alps of the **Northern** and **Middle** Islands, alt. 4000 ft: Tongariro, *Bidwill;* Mount Hikurangi, *Colenso;* Nelson, *Bidwill;* Southern Alps, in various places, *Sinclair, Haast, Travers;* Otago, lake district, *Hector and Buchanan*. Travers sends a variety from the Wairau gorge, alt. 4500 ft., with the margins of leaves so recurved that the latter are cymbiform, and the outer scales of the involucre are tomentose.

15. **O. Forsteri,** *Hook. f.—Eurybia*, Fl. N. Z. i. 119. A small tree, closely resembling *O. furfuracea* in habit and foliage. Leaves 2–3 in. long, oblong, obtuse, margins undulate, both surfaces finely reticulate, lower white with densely appressed down; petioles $\frac{1}{2}$–1 in. long. Corymbs shorter than the leaves, peduncled, spreading, many-headed. Heads fascicled and sessile on the branches of the panicle, $\frac{1}{5}$–$\frac{1}{6}$ in. long; involucre narrow, scales few, lax, coriaceous, shining, white, obtuse, nearly glabrous; florets 1 or 2, one often ligulate; pappus white, unequal. Achene pubescent.—*Shawia paniculata*, Forst.

Northern and **Middle** Islands, *Banks and Solander*, etc.; head of Ruamahanga and banks of the Pahawa river, *Colenso;* Akaroa, *Raoul*.

16. **O. avicenniæfolia,** *Hook. f.—Eurybia*, Fl. N. Z. i. 120. A small tree. Branches hoary. Leaves $1\frac{1}{2}$–$2\frac{3}{4}$ in. long, elliptic- or lanceolate-oblong, narrowed at both ends, subacute, margins flat, reticulate on both surfaces, white or rufous with closely appressed tomentum below; petioles $\frac{1}{3}$–$\frac{1}{2}$ in. long. Corymbs peduncled, densely very many-headed. Heads shortly pedicelled, $\frac{1}{6}$ in. long, narrow; involucre cylindric, scales few, imbricate, coriaceous, glabrous, ciliate, not shining; florets 3 or 4, one with a broad ray; pappus white. Achene silky.—*Shawia avicenniæfolia*, Raoul, t. 13.

Middle Island: Akaroa, *Raoul*; Nelson, *Bidwill*; Southern Alps, *Haast*; Otago, lake district, *Hector and Buchanan*.

17. **O. albida,** *Hook. f.—Eurybia*, Fl. N. Z. i. 118. A small tree, very like *O. Forsteri* in habit and foliage, but the leaves are not evidently reticulate on either surface, the under surface is rather softer and very white, the panicles larger, more effuse. Heads quite different, pedicelled, $\frac{1}{5}$ in. long, subcylindric; involucral scales imbricate, short, obtuse, pubescent; florets 2 or 3, one often ligulate; pappus white, unequal. Achene pubescent.

Northern Island, *Colenso*; Auckland, *Sinclair*.

18. **O. virgata,** *Hook. f.—Eurybia*, Fl. N. Z. i. 119. An erect, twiggy shrub, 3–8 ft. high. Branches opposite, angled or grooved, glabrous or pubescent. Leaves $\frac{1}{4}$–$\frac{2}{3}$ in. long, linear-oblong obovate or spathulate, obtuse, rarely quite linear, with white or fulvous tomentum beneath, veinless, margins flat or recurved. Heads solitary or fascicled, on short or long slender pedicels, rarely sessile, broadly campanulate, $\frac{1}{4}$ in. diam.; involucral scales woolly, few, short; ray florets few; pappus white or reddish. Achene pubescent.

Var. *α*. Leaves $\frac{1}{4}$–1 in., narrow linear-obovate, flat.

Var. *β*. Leaves $\frac{1}{2}$ in., linear, margins recurved.

Var. *γ*. Leaves 1–1$\frac{1}{2}$ in., excessively narrow linear, margins recurved.

Northern Island, var. *α*, boggy places, Wairarapa, *Colenso*. **Middle** Island, var. *α* and *β*, Wairau Pass, by watercourses, *Bidwill*; Rangitata river, *Sinclair*; Lake Tikapo and Wainaka, *Haast*; Otago, *Lindsay*. Var. *γ*, Otago, *Lindsay*; lake district, *Hector and Buchanan*. Very variable in size, breadth, and colour of foliage.

19. **O. Hectori,** *Hook. f., n. sp.* Habit of *O. virgata*, but leaves 1–2 in. long, obovate, on slender petioles, membranous, covered below with pale-grey silvery tomentum, the veins distinct.

Middle Island: Otago, lake district, *Hector and Buchanan*, "a beautiful shrubby tree." Possibly an extreme form of *O. virgata*, but a most distinct-looking plant.

20. **O. Solandri,** *Hook. f.—Eurybia*, Fl. N. Z. i. 119. An erect shrub, 5 ft. high. Branches stout, upper opposite, angled, often viscid. Leaves in opposite pairs or fascicles, $\frac{1}{6}$ in. long, uniform, narrow linear or linear-obovate, obtuse, margins recurved, covered with yellow tomentum below. Heads solitary, sessile, $\frac{1}{4}$ in. long; involucre turbinate, scales numerous, short, imbricated, obtuse, fulvous; florets 8–10, ray short; pappus red, equal. Achene pubescent.

Northern Island: sandy shores of the east coast, *Banks and Solander*, *Cunningham*, etc. Habit of *Cassinia fulvida*.

2. PLEUROPHYLLUM, Hook. f.

Tall, succulent, robust, leafy, silky or woolly herbs. Leaves radical and cauline, with numerous stout parallel ribs.—Heads racemed. Involucre hemispherical, scales in 2 or 3 series, herbaceous. Receptacle flat, alveolate, toothed. Florets very numerous, of the ray in 1–3 series, ligulate, female; of the disk tubular, 4- or 5-toothed, hermaphrodite. Pappus of 2 or 3 series, rigid, scabrid, unequal. Anthers shortly tailed. Achene angled, densely setose.

The only species known are confined to the islets south of New Zealand, and are noble plants. The genus is best distinguished from such Olearias as *O. Colensoi* by its herbaceous habit and ribbed leaves.

Heads with long, purple, 3-toothed rays . . . . 1. *P. speciosum.*
Heads with very short 2–3-partite rays . . . . 2. *P. criniferum.*

1. **P. speciosum,** *Hook. f. Fl. Antarct.* i. 31. *t.* 22, 23. Stem 2–3 ft. high; root thick, fleshy. Leaves villous with bristles intermingled, radical oblong, 1 ft. long, 6–8 in. broad, quite entire, ribs 18–20, cauline smaller. Raceme leafy, sometimes compound below Heads 2 in. broad, very handsome, purple.

Lord Auckland's group and **Campbell's** Island, in wet places, *J. D. H.*

2. **P. criniferum,** *Hook. f. Fl. Antarct.* i. 32. *t.* 24, 25. Stem 2–6 ft. high, sometimes branched at the base, covered at the base with the curled hair-like dry ribs of the old leaves. Radical and lower leaves silky-villous, 2 ft. long, 1 broad, amplexicaul at the base, broadly oblong, quite entire, ribs about 20–40, upper narrow, densely silky. Raceme simple. Heads subglobose; involucral scales ciliate-dentate, glabrous or tomentose.

Lord Auckland's group and **Campbell's** Island, in swampy places, *J. D. H.* **M'Quarrie's** Island, *Fraser.*

3. CELMISIA, Cass.

Perennial Aster-like herbs, with fusiform roots, or creeping branched rhizomes. Leaves all radical, rosulate, simple, entire or toothed, most frequently covered with more or less appressed white or buff tomentum. Scapes with linear bracts, 1-headed.—Heads large, rayed. Involucral scales in few or many series, linear or subulate, usually pubescent or cottony or glandular, often recurved. Receptacle plane or convex, even or deeply alveolate. Florets of ray *female*, in one series, with long spreading or revolute, white or purplish ligules; arms of style with linear obtuse even arms, thickened at the edges: of disk *hermaphrodite*, tubular, 5-toothed, tube often thickened below; anthers with very short tails; arms of style shorter, tipped with long or short glandular cones. Pappus of about 2 series of rather few, unequal, rigid, scabrid, white or reddish bristles, the outer shorter. Achene linear, often as long as the pappus, angled or terete, rarely compressed, glabrous or silky, the hairs usually bifid at the very tip.

A most beautiful genus, abundant in New Zealand, and, as in all the other large genera of these islands, the species are very variable, difficult to discriminate, and intermediate forms may be expected between those here described. It is very closely allied in characters but not in habit to the large northern genus *Erigeron*, and the minute obscure tails to the anthers is the only diagnostic mark; three South American Erigerons, indeed, have all the habits of *Celmisia*, they have, however, the a nthers of *Erigeron*, in which genus Weddell has placed them. From *Aster*, the same characters of the anthers, and the rarely flattened achenes distinguish them. From *Olearia* and *Pleurophyllum*, they differ only in habit. Two of the New Zealand species, and another are Australian; the genus is unknown elsewhere.

A. *Leaves more or less toothed or serrate (often obscurely), white or buff beneath (glabrous in a var. of* Sinclairii). *Involucres generally viscid.*

Leaves 6–9 × 1½–2 in., acutely serrate, lanceolate . . . . 1. *C. holosericea.*
Leaves 3–6 × ⅔–1⅓ in., crenate, narrow linear-oblong . . . . . . 2. *C. densiflora.*

Leaves ½–2⅓ × ¼–⅓ in., subserrate, linear or spathulate, coriaceous, white below . . . . 3. *C. discolor.*
Leaves 1½–2 × ½–¾ in., crenate-serrate, obovate-oblong, buff below. Achene silky . . . . 4. *C. hieracifolia.*
Leaves 1–2 × ⅓–½ in., obovate-oblong or spathulate, white below, margins recurved. Bracts many. Achene glabrous . . . . 5. *C. Haastii.*
Leaves 1–2 × ½–¾ in., serrulate, coriaceous, obovate, with white, soft tomentum below . . . . 6. *C. incana.*
Leaves 2–2½ × ½ in., scarcely serrate, linear-oblong. Rays distant, corolla-tube rigid . . . . 7. *C. Lindsayi.*
Leaves 1½–2 × ⅓–⅔ in., scarcely serrate, oblong or spathulate, membranous . . . . 8. *C. Sinclairii.*

B *Leaves* 3–18 × ¼–3 *in., quite entire* (*obscurely toothed or rough in* C. verbascifolia, spectabilis, *and* petiolata), *silvery silky or woolly below* (*glabrate in* C. Mackaui).

Leaves 4–8 × ½–3 in., oblong or lanceolate, not silvery below. Achene glabrous . . . . 9. *C. verbascifolia.*
Leaves 10–18 × ½–2½ in., lanceolate, silky below. Achene pilose . . 10. *C. coriacea.*
Leaves 10 × 1½ in., narrow lanceolate, glabrate on both sides . . . 11. *C. Mackaui.*
Leaves 3–5 × ⅓–⅔ in., linear, silvery above, shining below . . . . 12. *C. Munroi.*
Leaves 8–18 × ¼–¾ in., ensiform, even and glabrous above, silvery below 13. *C. Lyallii.*
Leaves 3 × ¼ in., linear, obtuse, viscous above, grooved below . . . 14. *C. viscosa.*
Leaves 6–12 × ¾–1¾ in., linear-oblong, with long purple sheaths . . 15. *C. petiolata.*
Leaves 4–6 × ½–1 in., linear-lanceolate, with opaque, thick, buff, loose tomentum below . . . . 16 *C. spectabilis.*
Leaves 9 × 1½ in., oblong, and scape and head, with dense velvety, rusty tomentum below . . . . 17. *C. Traversii.*

C. *Leaves* 1–12 × $\frac{1}{16}$–¼ *in., very narrow-linear or linear-oblong, white below, or on both sides.*

Leaves 2–18 × $\frac{1}{16}$–¼ in., narrow, margins recurved, cottony and silvery 18. *C. longifolia.*
Leaves ½–1 × $\frac{1}{16}$ in., acerose, pungent, silvery . . . . 19. *C. laricifolia.*
Leaves ½–¾ × ⅙–¼ in., linear-spathulate, silky on both sides . . . . . 20. *C. Hectori.*
Leaves ½–¾ × $\frac{1}{12}$ in., shortly silky on both sides. Head sessile . . 21. *C. sessiliflora.*

D. *Leaves perfectly glabrous on both surfaces, or with minute glandular pubescence only.* (*See* 8, Sinclairii, *in* A, *and* 11, Mackaui, *in* B.)

Leaves ½ × ⅛ in., obtuse, green, quite entire . . . . 22. *C. bellidioides.*
Leaves ¾–1 × ⅓ in., spathulate, acute, serrate . . . . 23. *C. glandulosa.*
Leaves 1–4 × ⅙–¼ in., linear, curved, coriaceous, shining . . . . . . 24. *C. vernicosa.*

1. **C. holosericea,** *Hook. f. Fl. N. Z.* i. 121. *t.* 31. Leaves 6–9 in. long, 1½–2 broad, flat, leathery, lanceolate, acute, acutely serrate, quite glabrous above, with fine, nearly parallel veins, covered below with a thin layer of white appressed tomentum, except the midrib; petiolar part broadly sheathing, 1 in. long and broad, furrowed, quite glabrous, shining, margins membranous. Scape 1–1½ ft. long, slender, glabrous, with linear bracts 1 in. long. Head 2–3½ in. broad; involucral scales in many series, subulate, outermost leafy, the rest glabrous or viscid, with recurved tips; florets glabrous; ray narrow, ¾–1 in. long; pappus ⅙ in. long, as long as the pilose achene.—*Aster holosericeus*, Forst.

Middle Island: Dusky Bay, *Forster;* Port Preservation, *Lyall.* A magnificent plant.

2. **C. densiflora,** *Hook. f., n. sp.* Leaves 3–6 in. long, ⅜–1⅓ broad, narrow linear-oblong, obtuse, flat, leathery, crenate-dentate, quite glabrous but viscid above, below covered (except the midrib) with appressed white

tomentum; veins 0; petiolar part 1–1½ long, ¾ broad, sheathing, glabrous, grooved, shining. Scape 8–12 in. long, slender, with few linear bracts 1 in. long. Head 1–2 in. broad; involucral scales excessively numerous, subulate, recurved, pubescent and viscid; florets densely crowded, glabrous; rays twisted, numerous, ⅙–½ in. long, narrow; pappus ⅙ in. long, as long as the slender rather silky achene.

Middle Island: mountains near Lake Okau, Dobson's and Hopkin's rivers, alt. 2–5000 ft., *Haast;* Otago, lake district, subalpine, *Hector and Buchanan.* The excessively numerous florets and involucral scales (with recurved tips), short rays, and obtuse crenate leaves, well distinguish this.

3. **C. discolor,** *Hook. f. Fl. N. Z.* i. 123. A rather small species, 4–12 in. high. Leaves imbricating upon short branches, from oblong-spathulate to linear, the former 1–1½ in. long by ⅓ broad, the latter 2 in. long by ¼ broad, sometimes with slender petioles, obtuse or subacute, serrulate to the touch or quite entire, opaque, glabrous or hoary above, white below, coriaceous. Scape slender, viscid; bracts many, linear. Head ¾ in. diam.; involucral scales in few series, linear-subulate, recurved, viscid, pubescent or tomentose; rays narrow; pappus ⅕ in. long. Achene silky.

Middle Island: abundant on the mountains between 2000 and 6000 ft., from Nelson to Milford Sound. The var. β of Fl. N. Z. is clearly a large form. Some states approach *C. incana* in the softness and laxity of the wool.

4. **C. hieracifolia,** *Hook. f. Fl. N. Z.* i. *t.* 34 *B.* Closely allied to *C. discolor*, of the same size, but the branches are much shorter, the leaves 1½–2 in. long, ½–¾ in. broad, obovate-oblong, sessile, obtuse, obtusely serrate, less coriaceous, with appressed buff (when dry) tomentum beneath. Scape stout, very pubescent and viscid, with many rather leafy, erect or recurved bracts not white below. Head as in *discolor*, but rather larger.

Middle Island: Nelson mountains, *Bidwill, Munro.* Very near *C. discolor*, but I think distinct.

5. **C. Haastii,** *Hook. f., n. sp.* Size and habit of *C. hieracifolia*, the leaves similar in dimensions form and texture, pale yellow-green above when dry, white with closely appressed tomentum below, margins narrowly recurved, with a few distant obscure teeth or stout bristles. Scape stout, cottony, covered with linear obtuse or acute cottony or glabrate bracts. Head ⅔–1 in. diam. Involucral scales linear, obtuse, hoary, membranous, erect, appressed, hardly glandular; rays ⅓ in. long. Achene glabrous.

Middle Island: Hopkins River and Mount Darwin, alt. 4500–6000 ft., *Haast;* Otago, lake district, alpine, *Hector and Buchanan.* Near *C. hieracifolia* and *incana*, but differing wholly in the stout cottony scape, numerous bracts, linear membranous not recurved involucre-scales, and glabrous achene.

6. **C. incana,** *Hook. f. Fl. N. Z.* i. 123. *t.* 34 A. Robust, 4–10 in. high, branches from the rooting rhizome short, stout, densely clothed with fibrous leaf-remains. Leaves crowded, 1–2 in. long, ½–¾ broad, oblong-spathulate, obtuse or acute, densely coriaceous, serrulate, thickly covered below or on both surfaces with snow-white soft, not appressed tomentum. Scapes stout, numerous, linear; bracts woolly. Heads 1 in. diam.; involucral scales linear or subulate-lanceolate, recurved, pubescent and viscid; rays numerous, narrow; pappus ⅕ in. long. Achene silky.

Northern Island: Mount Hikurangi and Ruahine range, *Colenso.* **Middle** Island: Upper Wairau, *Sinclair;* top of Gordon's Nob and Macrae's Run, alt. 4–5000 ft., *Munro;* Otago, lake district, *Hector and Buchanan.* Generally a very distinct species, at once distinguished by the size, soft, loose, snow-white tomentum, and very coriaceous serrate leaves, but I fear it may pass into *C. discolor.*

7. **C. Lindsayi,** *Hook. f., n. sp.* Rhizome creeping, as thick as a quill, clothed with coriaceous, glabrous, short leaf-sheaths. Leaves 2–2½ in. long, ½ in. broad, linear-oblong, subacute, with a few small serratures, snow-white below with appressed shining tomentum, midrib black, cottony towards the short sheath, opaque and glabrous above. Scape slender, flexuous, nearly glabrous. Head 1½ in. diam.; involucral scales subulate, imbricate, glabrous, short; rays patent, few, distant, not recurved, ½ in. long; tube of corolla much thickened, rigid; pappus red, ¼ in. long. Achene silky.

Middle Island: Otago, Trap Cliffs at Shaw's Bay, the Nuggets, mouth of the Clutha river, *Lindsay.* A very distinct species, remarkable for the slender, flexuous, glabrous scape, small imbricate involucre, few straight distant rays, and much thickened, rigid corolla-tubes.

8. **C. Sinclairii,** *Hook. f., n. sp.* Rhizome long, as thick as a quill. Leaves 1½–2 in. long, ⅓–⅔ in. broad, linear-oblong or obovate-spathulate, obtuse or acute, obscurely toothed, glabrous on both sides, or white below, membranous, contracted into membranous, glabrous or hoary sheaths 1 in. long. Scape slender, nearly glabrous. Head 1–1¼ in. diam.; involucral scales subulate, squarrose, recurved, pilose; rays numerous, ⅓ in. long; pappus whitish, ⅙ in. Achene silky.

Middle Island: Dun mountains, *Sinclair* (leaves white below); Tarndale, *Sinclair* (leaves glabrous on both surfaces). Leaves more membranous than usual in the genus. I have only three specimens.

9. **C. verbascifolia,** *Hook. f. Fl. N. Z.* i. 121. Very large. Root thicker than the thumb, spindle-shaped. Leaves oblong or lanceolate, narrowed into very broad, purple, glabrous, sheathing petioles, with woolly edges, 4–8 in. long, 1½–3 broad, coriaceous, scarcely crenulate, opaque, glabrous above, below thickly clothed with pale, buff, loose, soft, thick tomentum; veins diverging. Scapes as long as the leaves, and heads densely woolly; bracts long, linear. Head 2 in. diam.; involucral scales erect, linear-subulate; tube of corolla much thickened below; rays very slender, ¾ in. long; pappus ¼ in. long. Achene glabrous, ripe not seen.

Middle Island: Milford Sound and Port Preservation, *Lyall.* A splendid species.

10. **C. coriacea,** *Hook. f. Fl. N. Z.* i. 121. Leaves 10–18 in. long, ½–2½ broad, lanceolate, coriaceous, narrowed into broad woolly sheaths, covered above with a silvery pellicle of matted cottony hairs, below with dense white silvery tomentum. Scapes very stout, very cobwebby and cottony, with many linear bracts. Head 1½–3 in. diam.; involucral scales very numerous, linear-subulate, cottony or glabrous; rays excessively numerous, 1½ in. long; tube of corolla pubescent. Achene linear, compressed, ¼ in. long, pilose.—*Aster coriaceus,* Forst.

Middle Island: abundant in the mountains, from Nelson to Dusky Bay, *Forster,* etc. Not hitherto found in the Northern Island, the plant alluded to in the Fl. N. Z., as from the

Ruahine range, proves to have been found in the Middle Island. "Leather plant" of colonists. A magnificent species, which Mueller appears to have found in the Australian alps, but his specimens not being in flower, I cannot well determine the point.

11. **C. Mackaui,** *Raoul, Choix,* xix. *t.* 18;—*Fl. N. Z.* i. 122. Altogether similar to *C. coriacea*, but the mature leaves are glabrate on both sides, rather glaucous below, sparingly cottony at the base only; the scape and involucre also glabrate, the achene quite glabrous, and the tube of the corolla thickened at the base as in *C. verbascifolia*.

Middle Island: rocky places at Akaroa, *Raoul.*

12. **C. Munroi,** *Hook. f., n. sp.* Leaves 3–5 in. long, $\frac{1}{3}$–$\frac{2}{3}$ broad, narrow linear-oblong, strict, subacute, very coriaceous, with a silvery pellicle of matted cottony hairs above, below with dense white silvery tomentum, wrinkled in parallel lines when dry; margins slightly recurved; sheaths densely woolly and snow-white. Scape longer than the leaves, stout, very woolly and cottony; bracts numerous, linear. Head 1–1$\frac{1}{2}$ in. diam.; involucral scales numerous, linear-subulate, erect or recurved, woolly or glabrate; rays very numerous, $\frac{3}{4}$ in. long; pappus $\frac{1}{3}$ in. long, Achene glabrous or hispidulous.

Middle Island: Upton Downs, Awatare, elev. 1500 ft., *Munro;* Hopkins River, Mount Cook, and elsewhere in the alps of Canterbury, 3–4500 ft., *Travers and Haast.* Best distinguished from *C. coriacea* by the narrow, smaller, linear, longitudinally wrinkled leaves. Munro's specimens have quite glabrous achenes, Haast's are hispidulous.

13. **C. Lyallii,** *Hook. f., n. sp.* Leaves 8–18 in. long, $\frac{1}{4}$–$\frac{3}{4}$ broad, narrow ensiform, gradually tapering from a silky sheathing grooved base to the tip, very rigidly coriaceous, quite smooth glabrous and even above, below with dense appressed white tomentum, even or grooved, tip tapering, rigid, pale yellow when dry. Scape densely cottony; bracts rather rigid, linear. Head 1$\frac{1}{4}$–1$\frac{1}{3}$ in. diam.; involucral scales subulate-lanceolate, rigid, rather recurved, glabrous or cottony at the margins; rays twisted, $\frac{1}{2}$ in. long; pappus $\frac{1}{4}$ in. long. Achene pubescent or silky.

Middle Island: Dusky Bay, *Lyall;* Hurrumui valleys, 1200–1600 ft., *Travers.* Abundant in the Rangitata, Two-thumb, and Malvern ranges, alt. 3–5000 ft., and alps of Hopkins River, alt. 2500 ft., *Sinclair and Haast;* Otago, lake district and Lindis Pass, displacing the tussock grass at 4000 ft., *Hector and Buchanan.* A well-marked species, best distinguished by the very long, narrow, rigid leaves, which gradually taper from base to tip, and are usually grooved with parallel ribs beneath.

14. **C. viscosa,** *Hook. f., n. sp.* Rhizome very stout, together with leaf-sheaths, 1 in. diam. Leaves crowded, 3 in. long, $\frac{1}{4}$ broad, exactly linear, obtuse, sessile on the broader, glabrous, grooved, brown, short sheaths, very rigid, even or grooved above, white with appressed down beneath and ridged with parallel ribs. Scapes much longer than leaves, stout; bracts numerous, pubescent and viscid. Head 1$\frac{1}{2}$ in. diam.; involucral scales numerous, subulate, woolly and viscid; rays short, revolute; pappus $\frac{1}{4}$ in. long. Achene silky.

Middle Island: summit of Mount Torlesse, alt. 5–6000 ft., *Haast;* Otago, lake district, subalpine, *Sinclair and Haast.* The short, linear, rigid, entire, viscid leaves, which are white and ribbed below, and viscid scape, are the best marks of this species; it approaches states of *C. discolor*, but is much larger.

15. **C. petiolata,** *Hook. f., n. sp.* Leaves with the long purple hairy grooved petiole 6–12 in. long, ¾–1¾ broad; blade linear-oblong or oblong-lanceolate, subacute, hairy or loosely villous above, greenish-white, with villous appressed tomentum below, membranous; midrib purple below; veins slender, diverging; margins entire or denticulate. Scapes longer than the leaves, hairy or villous, often purple; bracts slender. Head 1½–2½ in. diam.; involucral scales narrow, subulate, tomentose or glabrate; rays ½ in. long; pappus ¼ in. long. Achene glabrous or silky.

Middle Island: Hurrumui mountains, in moist valleys, alt. 3000 ft., *Travers;* Hopkin's River and Great Clyde glacier, alt. 2–6000 ft., *Haast;* Rangitata river, alt. 1800 ft., *Sinclair;* Mount Brewster, alt. 4–4500 ft., *Hector* The long purple petioles and purple midrib of the rather membranous leaf, are constant characters in all my specimens.

16. **C. spectabilis,** *Hook. f. Fl. N. Z.* i. 122. *t.* 33. Rhizome very stout, together with leaf-sheaths 1–1½ in. diam. Leaves 4–6 in. long, ½–1 broad, narrow linear-oblong, obtuse or acute, sessile on the long, excessively silky and shaggy sheaths, glabrous and even, young silvery above, densely clothed with soft, loose, buff tomentum below, very thick and coriaceous; margins recurved, sometimes obscurely toothed. Scape stout, much longer than leaves, densely cottony; bracts linear. Head 1–1½ in. broad; involucral scales subulate, recurved, densely woolly; rays short, narrow, very numerous; tube of corolla thickened below; pappus nearly ⅓ in. long. Achene quite glabrous.

Northern Island: Tongariro, *Bidwill;* Mount Hikurangi and Ruahine range, *Colenso.* Abundant throughout Nelson and Canterbury provinces of the **Middle** Island, alt. 2–5000 ft. The rather short, rigid leaves, covered below with loose, not appressed, buff, woolly tomentum, and glabrous achenes, are the best characters of this species, which varies extremely in stature; very dwarf varieties are common.

17. **C. Traversii,** *Hook. f., n. sp.* Leaves beneath margins and scape most thickly clothed with soft, velvety, rusty-brown (when dry) tomentum. Leaves with petiole 9 in. long, 1½ broad, oblong-lanceolate, obtuse, coriaceous, glabrous and opaque above, except the silky midrib, below purple, dilating into the broad petiolar sheath, which is densely clothed with snow-white cotton. Scape very stout; bracts linear. Head 1½ in diam.; involucral scales very numerous, linear, acute, densely clothed with brown velvety tomentum; rays narrow, rather short, ½ in. long; pappus ¼ in. Achene glabrous.

Middle Island: Discovery Peaks, Nelson, alt. 5500 ft, *Travers.* I have but one scape and two leaves of this very remarkable and handsome plant, the dense, rusty, velvety tomentum of scape and leaf below, contrasting with the snow-white cotton of the sheaths, are most singular and beautiful features.

18. **C. longifolia,** *Cass.;—C. gracilenta,* Fl. N. Z. i. 122. Leaves usually very numerous, 1–18 in. long, $\frac{1}{10}$–¼ in. broad, narrow-linear or narrow-lanceolate, acute or acuminate, not pungent, upper surface glabrous or silvery, under silky-tomentose, margins often much recurved or revolute, rather soft and flexuose, never rigidly coriaceous. Scapes slender, cottony; bracts linear. Head ½–1¼ in. broad; involucral scales subulate, glabrous or cottony, acute or obtuse, never recurved in flower, black at tip; rays few or many, long or short, broad or narrow; pappus reddish or white. Achene glabrous or silky.

Abundant throughout the islands, ascending to 5500 ft., and varying excessively in stature, robustness, length, and breadth of leaves, one or many (8–12) slender or robust scapes, and large or small, cottony or glabrate heads. This is the commonest *Celmisia* in the islands, and is also found on the Australian and Tasmanian alps and Blue Mountains. Some states almost pass into *C. Munroi.*

19. **C. laricifolia,** *Hook. f. Fl. N. Z.* ii. 331. A small species, with prostrate rhizomes, densely leafy for 2–3 inches. Leaves densely crowded, $\frac{1}{2}$–1 in. long, by $\frac{1}{16}$ broad, acerose, pungent, somewhat recurved, glabrous above, silvery or cottony below; margins recurved. Scapes very slender, 2–4 in. long, glabrate; bracts few, setiform. Head $\frac{1}{3}$ in. diam.; involucral scales few, linear-subulate, erect, cottony; florets few. Achene hispid.

Middle Island: Gordon's Nob and other Nelson mountains, alt. 4000 ft., *Munro, Sinclair;* summit of Mount Torlesse, alt. 4500–6000 ft., *Haast.*

20. **C. Hectori,** *Hook. f., n. sp.* A densely tufted species, forming extensive patches. Leaves densely imbricate, spreading, $\frac{1}{2}$–$\frac{3}{4}$ in. long, $\frac{1}{6}$–$\frac{1}{4}$ broad, linear- or spathulate-oblong, obtuse, silky on both surfaces; margins recurved; sheaths short, coriaceous, shining, glabrous, cottony at the margins. Scape very robust, 4–5 in. long, villous and silky; bracts many, linear. Head large for the plant, 1 in. diam.; involucral scales linear-subulate, few, obtuse, pubescent; rays $\frac{1}{3}$ in., long, rather broad; pappus $\frac{1}{6}$ in. long. Achene silky.

Middle Island: Otago, Mount Brewster, alt. 5–6000 ft., *Hector and Buchanan.* A very pretty and distinct little species, remarkable for the small foliage and large head.

21. **C. sessiliflora,** *Hook. f., n. sp.* A small, robust, densely tufted, greenish-grey species, 2–3 in. high. Leaves most numerous and densely tufted, $\frac{1}{2}$–$\frac{3}{4}$ in. long, $\frac{1}{12}$ in. broad, strict, narrow-linear, obtuse, thick, convex at the back, equally hoary or silky on both surfaces; sheath as long and rather broader than the blade, membranous, with villous margins. Head $\frac{2}{3}$ in. diam., sunk amongst the leaves; involucral scales few, slender, subulate; florets few; rays $\frac{1}{3}$ in. long; pappus white or reddish, $\frac{1}{4}$ in. long. Achene silky.

Middle Island: Upper Wairau, etc., 4000 ft., *Sinclair;* Discovery Peaks, 5800 ft., *Travers;* grassy flats, Mount Cook and Macaulay river, 4000 ft., *Haast;* Mount Brewster, and elsewhere in the lake district, 3–5000 ft., *Hector and Buchanan.*

22. **C. bellidioides,** *Hook. f., n. sp.* A small glabrous species. Rhizome creeping, branched, often stout and woody. Leaves spreading, linear-oblong, $\frac{1}{3}$–$\frac{2}{3}$ in. long, $\frac{1}{8}$–$\frac{1}{6}$ broad, linear-oblong or linear-spathulate, rounded at the tip, green on both surfaces; margin sometimes obscurely toothed, rather coriaceous, veinless, narrowed into short rather cottony petioles. Scape slender, 1–2 in. high; bracts many, leafy, narrow. Head $\frac{3}{4}$ in. diam.; involucral scales few, green, often purple-margined, broadly linear; rays many, $\frac{1}{3}$ in. long; pappus $\frac{1}{4}$ in. long; corolla-tube pilose. Achene villous or silky.

Middle Island: Tarndale mountains, in loose shingle, alt. 5000 ft., *Sinclair;* fissures of rocks, Mount Torlesse, Macaulay and Hopkins rivers, Lake Hawai, etc., alt. 3–6000 ft., *Haast.*

23. **C. glandulosa,** *Hook. f. Fl. N. Z.* i. 124. A small glabrous plant, 1–3 in. high, covered with minute and glandular pubescence. Leaves $\frac{3}{4}$–1 in. long, $\frac{1}{3}$ in. broad, petioled, ovate- or oblong-spathulate, acute, acutely serrate

or toothed; petioles dilating into very short sheaths, rigid but thin. Scape slender; bracts few, linear, acuminate. Head $\frac{3}{4}$ in. diam.; involucral scales linear-subulate, erect or recurved, pubescent and ciliate; receptacle deeply alveolate; rays few, $\frac{1}{4}$ in. long; pappus $\frac{1}{6}$ in. long. Achene silky.

Northern Island: base of Tongariro, *Colenso*. **Middle** Island: Mount Cook, alt. 5-6000 ft., *Haast*.

24. **C. vernicosa,** *Hook. f.; Fl. Antarct.* i. 34. *t.* 26 *and* 27. A densely tufted, very leafy, perfectly glabrous species, everywhere polished and shining. Leaves excessively densely crowded, rosulate, straight or curved, 1–4 in. long, $\frac{1}{6}$–$\frac{1}{4}$ broad, linear, obtuse, extremely coriaceous, convex above, sometimes obscurely toothed; margins often recurved. Scapes solitary or numerous, very stout, flexuose, 1–8 in. high, covered with leafy coriaceous bracts. Head 1–1$\frac{1}{2}$ in. diam.; involucral scales linear, erect; rays numerous, $\frac{1}{3}$–$\frac{1}{2}$ in. long; disk purple; corolla-tube pilose; pappus $\frac{1}{6}$ in. long. Achene hispid.

Lord Auckland's group and **Campbell's** Island: common from the sea to 1000 ft., *J. D. H.* A most beautiful plant, quite unlike any hitherto found in New Zealand proper.

4. VITTADINIA, A. Rich.

Herbs or undershrubs, branched, leafy. Heads on the ends of the branches, their characters those of *Celmisia*, but involucre obconic; achene compressed, very slender; rays short, and anthers not tailed.

A small genus, confined to Australia, Tasmania, New Zealand, and the Sandwich Islands.

1. **V. australis,** *A. Rich.*;—*Eurybiopsis australis*, DC.;—Fl. N. Z. i. 125. A small, much branched, more or less pubescent, hispid and glandular herb, 6-12 in. high, with a short, woody rhizome, and erect, leafy, slender branches. Leaves $\frac{1}{4}$–$\frac{1}{2}$ in. long, cuneate or linear-spathulate, 3–5-lobed or -toothed at the tip. Heads solitary, short-peduncled; involucral scales few, in two series, linear, rigid, erect; rays one series, white tipped with pink; pappus white or reddish, $\frac{1}{4}$ in. long. Achene not so long as pappus, linear, compressed, hispid, hairs forked at the tip.

Common in dry places throughout the islands, from Auckland southwards, Wairau gorge, alt. 4–5000 ft., *Travers* (a very tomentose form); in Otago it ranges from the sea-level to the snow-ranges, *Hector and Buchanan*.

5. LAGENOPHORA, Cass.

Small perennial herbs. Leaves oblong or spathulate, usually membranous. Scapes single-headed.—Involucre of two series of narrow, appressed, acute scales. Receptacle rather broad, naked. Florets many, of ray in one series, female not yellow, with revolute ligule, and compressed, beaked, often glandular achene: of disk tubular, 5-toothed; achene often imperfect; pappus 0.

A small genus, found in Australia, Tasmania, India, China, and extratropical S. America.

Petioles slender. Leaves hairy or glabrate. Heads $\frac{1}{3}$–$\frac{1}{2}$ in. . . . . . 1. *L. Forsteri.*
Petioles slender. Leaves hairy or glabrate. Heads $\frac{1}{6}$–$\frac{1}{4}$ in.. . . . . 2. *L. petiolata.*
Petioles short. Leaves hirsute. Achenes glabrous . . . . . . 3. *L. lanata.*
Petioles broad. Leaves hirsute, pinnatifid. Achenes viscid . . . . . 4. *L. pinnatifida.*

1. **L. Forsteri,** *DC.*;—*Fl. N. Z.* i. 125. A small, glabrous pilose or hirsute, daisy-like, slender plant. Leaves with the slender petiole $\frac{1}{2}$–$1\frac{1}{2}$ in. long, blade $\frac{1}{2}$–1 in., obovate- or orbicular-oblong, obtuse, crenate toothed or lobed towards the base. Scape slender, 2–6 in. long, with few minute linear bracts, or 0. Head $\frac{1}{8}$–$\frac{1}{2}$ in. diam.; involucral scales linear, acute, appressed, with hyaline margins, outer sometimes recurved; rays very numerous, short, revolute. Achene $\frac{1}{14}$ in. long, much compressed, edges thick, narrowed suddenly into a short, hardly viscid beak.—*Calendula pumila*, Forster; *Microcacalia australis*, A. Rich.

Abundant throughout the islands, *Banks and Solander*, etc.

2. **L. petiolata,** *Hook. f. Fl. N. Z.* i. 125. In all respects like *L. Forsteri*, but the heads are very much smaller, only $\frac{1}{6}$–$\frac{1}{4}$ in. diam., with much fewer florets, smaller rays, and more viscid achenes.

Common throughout the islands, ascending to 4000 ft. **Kermadec** Islands, *M'Gillivray*. **Auckland** Island, *Bolton*. A variety, I suspect, of *L. Forsteri*.

3. **L. lanata,** *A. Cunn.*;—*Fl. N. Z.* i. 126. Hirsute and tomentose. Leaves obovate or obovate-spathulate, narrowed into very short petioles, $\frac{1}{2}$–1 in. long, obtusely coarsely crenate-dentate. Scapes very slender, glabrous, 2–5 in. long, naked. Heads as in *L. petiolata*, but achenes glabrous.

Northern Island: Bay of Islands, etc., common on dry grassy hills. Very similar to *L. hirsuta*, Pœpp., of Chili.

4. **L. pinnatifida,** *Hook. f. Fl. N. Z.* i. 126. Hirsute. Leaves 2–4 in. long, narrow obovate-oblong or spathulate, narrowed into long broad petioles, deeply obtusely crenate-lobate or pinnatifid. Scapes 4–10 in. high, slender, pilose. Heads $\frac{1}{4}$–$\frac{1}{2}$ in. diam.; involucral scales as in *L. petiolata*. Achene viscid.

Northern Island: east coast, *Colenso*; Auckland, *Sinclair*. **Middle** Island: Canterbury, *Sinclair and Haast*; Otago, *Lindsay*. Probably a form of the Australian *L. Billardieri*.

6. BRACHYCOME, Cass.

Scapigerous, perennial small herbs. Leaves chiefly radical, spreading.—Heads with one series of rayed female, and numerous tubular hermaphrodite florets. Receptacle convex, naked, papillose. Involucral scales in 1 series, appressed, oblong-linear. Corolla of ray white blue or purplish, revolute; of disk 5-toothed. Pappus of very short bristles or 0. Achene compressed, without beak.

A large Australian and Tasmanian genus, closely allied to the European Daisy.

Leaves all radical, entire or lobed. Scapes with 1 head . . . . . . 1. *B. Sinclairii.*
Leaves radical and cauline, lobulate. Heads terminating branches . . 2. *B. odorata.*
Leaves all radical, pinnatifid. Scapes with 1 head . . . . 3. *B. pinnata.*

1. **B. Sinclairii,** *Hook. f.*, *n. sp.* More or less glandular or glabrate. Rhizome short, simple or branched, as thick as a crow-quill. Leaves all radical, 1–2 in. long, petioled, spathulate, round at the tip, entire lobed or subpinnatifid, coriaceous, nerveless. Scape solitary or several, strict, 2–10 in. high, glandular; bracts 0 or 1 or 2 Head $\frac{1}{4}$–1 in. diam., yellow with white ray; involucral scales with or without a purple membranous jagged

border; pappus minute, bristly or 0. Achene much compressed, narrow linear-obovoid, glandular or eglandular, margins thickened.

Var. *α*. Leaves obtusely lobulate. Involucral scales with white or pale-purple edges. Achene glandular; pappus 0.

Var. *β*. Leaves lobulate or entire. Involucral scales as in *α*. Achene glabrous; pappus evident.

Var. *γ*. Smaller, alpine. Leaves quite entire. Scape short, stout, very glandular. Involucral scales with broad purple edges. Achene glabrous; pappus 0. Ray sometimes very long.

Northern Island: not common; var. *α*, grassy places, east coast, *Colenso*. **Middle** Island: var. *α* and *β*, common in subalpine localities, Gordon's Nob, and top of Macrae's Run, *Munro;* shingle beds, Ahuriri, Mount Misery, alt. 2-4000 ft., and elsewhere in Southern Alps, *Sinclair, Travers, Hector, and Haast;* var. *γ*, Otago alps, alt. 6-7000 ft., *Hector*. A common and variable plant in the Middle Island; the only representative of the Daisy in New Zealand.

2. **B. odorata,** *Hook. f.*, *n. sp.* A small species, 2–4 in. high, branched from the base, pubescent and subglandular. Rhizome woody, as thick as a crow-quill, perpendicular, with ascending leafy branches at the tip, terminating in stout 1- or 2-bracteate scapes. Leaves few, ½–1 in. long; petiole slender; blade spathulate, deeply unequally 3–8-lobed. Scape 2–3 in. long. Head ¼ in. diam.; involucral scales short, not purple at the tips; ray very short; pappus very short, obscure. Achene linear-clavate, densely glandular.

Northern Island: Patea village, *Colenso*, who observes that it is a favourite plant with the natives, is called "Ronia," and worn round the neck for its scent. The *B. radicata* of Fl. N. Z. was founded partly on this, and partly on specimens of a *Brachycome* in A. Cunningham's herbarium, which I am now convinced were introduced there by accident, and belong to an Australian species. The name of *B. radicata* had therefore better be abandoned.

3. **B. pinnata,** *Hook. f.*, *n. sp.*—*B. radicata*, var. *β*, Fl. N. Z. i. 127. Glabrous or minutely pubescent. Rhizome stout, ascending. Leaves all radical, ½–1 in. long, ⅙ in. broad, narrow linear-oblong, gradually dilated upwards, pinnatifid, the lobes uniform, orbicular, close-set, quite entire, coriaceous, concave beneath. Scape slender, ebracteate, 3 in. long. Head ⅓ in. diam.; involucral scales and florets as in *B. Sinclairii*.

Southern Island, *Lyall*. I have but one specimen of this pretty little plant.

7. BIDENS, Linn.

Erect herbs. Leaves opposite, simple or compound. Heads erect, usually terminal and peduncled.—Involucral scales few, erect, narrow. Receptacle chaffy. Florets of ray ligulate, neuter, or 0; of disk hermaphrodite, tubular, 5-toothed. Styles bifid, arms with subulate points. Pappus of 2–4 barbed hispid rigid awns. Achene narrow, 4-angled, or flattened, sometimes beaked.

A large genus, found in all warm and most temperate parts of the world.

1. **B. pilosa,** *Linn.* An erect, glabrous or slightly hairy annual, 1–2 ft. high, with angular branches. Leaves membranous, lower generally simple, upper pinnately divided, segments 3–5, stalked, ovate or ovate-lanceolate, 1–2 in. long, serrate, rarely lobed. Heads few, terminal on slender pedun-

cles; involucral scales $\frac{1}{4}$–$\frac{1}{3}$ in. long; ray fl. few, white or 0. Achenes slender, the inner longer, exceeding the involucre, 4-angled.

Northern Island: a weed of cultivation, not indigenous. **Kermadec** Island: on the margin of the crater of Sunday Island, *M'Gillivray*. A common cosmopolitan weed of warm countries.

8. ABROTANELLA, Cassini.

Including CERATELLA, *Hook. f., and* TRINEURON, *Hook. f.*

Small, glabrous, alpine, tufted herbs. Leaves lax or imbricate, long or short.—Heads subsessile, small, inconspicuous, 4–15-flowered; outer fl. female; inner male or hermaphrodite. Involucre cylindric; scales few, erect, in 1 or 2 series, coriaceous. Receptacle narrow, flat, papillose. *Female* fl.: pappus 0 or very small; corolla tubular, unequally 3–5-cleft at the mouth, inflated; style bifid, inserted on a spherical disk; achene 3- or 4-angled, or costate, obovoid or obconic. *Hermaphrodite* fl.: corolla narrow campanulate, 4-cleft; stamens almost free; style truncate.

A small genus, of which the species included under *Ceratella* and *Trineuron* (genera founded by myself on Auckland Island plants) differ much in lax habit and foliage from the original Fuegian *Abrotanella emarginata*, which has closely imbricate, minute leaves; my Tasmanian genus *Scleroleima* again, has the habit of the Fuegian one. I follow A. Gray (Proceedings Amer. Acad.) in reducing all to *Abrotanella*. The section *Trineuron* is found on the Australian alps and New Zealand; and one species is Fuegian.

Leaves $1\frac{1}{2}$–1 in., narrow linear-spathulate . . . . 1. *A. spathulata.*
Leaves $\frac{1}{4}$–$\frac{1}{3}$ in., ovate-lanceolate, acute recurved . . . . 2. *A. rosulata.*
Leaves $\frac{1}{6}$–$\frac{1}{3}$ in., very narrow linear, recurved . . . . 3. *A. pusilla.*
Leaves $\frac{1}{4}$ in., narrow linear-oblong, straight . . . . 4. *A. inconspicua.*

1. **A. spathulata,** *Hook. f.;—Trineuron spathulatum*, Fl. Antarct. i. 23. t. 17. Stems short, tufted, 1–2 in. high, leafy. Leaves spreading, $\frac{1}{2}$–1 in. long, $\frac{1}{10}$–$\frac{1}{12}$ broad, narrow linear-spathulate, acute or obtuse, coriaceous, nerveless. Heads crowded amongst the upper leaves, which are sometimes raised above the others on a short stem, $\frac{1}{8}$ in. long; involucral scales 8, oblong-lanceolate, with 3 pellucid nerves. *Male* fl.: corolla 4-angled, angles pellucid;—*fem.* fl.. corolla 4-toothed. Achene obovoid, flattened, with 3 cellular ribs.

Lord Auckland's group and **Campbell's** Island: in peaty soil, *J. D. H.*

2. **A. rosulata,** *Hook. f.;—Ceratella rosulata*, Fl. Antarct. i. 25. t. 18 A small, densely tufted, moss-like herb; stems 1–$1\frac{1}{2}$ in. high. Leaves imbricating, patent and recurved, rigid, coriaceous, $\frac{1}{4}$–$\frac{1}{3}$ in. long, narrow ovate or lanceolate, acute, concave above. Heads aggregated amongst the upper leaves, $\frac{1}{10}$ in. long; involucral scales 8–10, linear-oblong, coriaceous, with pellucid veins. *Male* fl.: corolla 4-angled, angles pellucid;—*fem.* fl.: corolla tubular, 4-toothed. Achene 4-angled, the angles produced into short horns.

Campbell's Island: in crevices of rocks, *J. D. H.*

3. **A. pusilla,** *Hook. f.;—Trineuron pusillum*, Fl. N. Z. i. 130. Stems slender, 1 in. high, slightly puberulous. Leaves $\frac{1}{6}$–$\frac{1}{3}$ in. long, narrow linear,

recurved, sometimes secund. Head solitary, terminal, $\frac{1}{12}$ in. long; involucral scales linear, obtuse, ribbed. Achene linear-clavate, 4-angled.

Northern Island: snowy places amongst the Ruahine mountains, *Colenso.*

4. **A. inconspicua,** *Hook. f. n. sp.* A small, glabrous, tufted, very inconspicuous, moss-like herb. Stems $\frac{1}{2}$ in. high, densely leafy. Leaves spreading, $\frac{1}{4}$ in. long, $\frac{1}{14}$ in. broad, linear, or very narrow linear-oblong, subacute, flat, coriaceous, rather rigid when dry. Head solitary, sunk in the upper leaves; involucral scales linear-oblong, dilated upwards, obtuse, nerved; florets about 16, outer with very slender corollas.

Middle Island: forming soft patches on Mount Alta, elev. 6000 ft., *Hector and Buchanan.*

9. COTULA, Linn.

Including LEPTINELLA, *Cass., and* MYRIOGYNE, *Less.*

Herbs, usually perennial, flaccid or succulent, full of minute oil-glands, very aromatic when bruised. Stems creeping or prostrate, terete. Leaves rarely entire, usually pinnatifid. Scapes short or long, ebracteate or bracteate.—Heads small, yellow, subglobular, unisexual, or the outer fl. female, the inner male. Involucre of 1 or more series of few or many scales with scarious margins. Receptacle hemispherical or conical, papillose, the exterior papillæ generally elevated into pedicels for the florets. Florets glandular or eglandular; pappus 0. *Male* fl.: tubular or funnel-shaped, with minute imperfect achenes; 4 or 5 loosely cohering stamens, and stout exserted style with discoid or bifid stigma. *Female* fl.: corolla 0, or short or inflated, contracted at the unequally 3–5-toothed mouth; style exserted, 2-lobed, seated on a spherical disk. Achene compressed, turgid or winged, obtuse or 2-lobed, usually thick and spongy.

A large genus, abounding in many tropical and extratropical countries, especially of the southern hemisphere. I have reduced here *Leptinella* and *Myriogyne* to it. The species are extremely difficult to make out, and better specimens are much wanted.

1. **Cotula.** *Heads bisexual, outer florets in few series, female, with flat winged achenes. Corolla* 0, *or a very minute deformed one.*

Stout, glabrous. Leaves $\frac{1}{2}$–2 in., variously cut . . . . 1. *C. coronopifolia.*
Slender, much branched. Leaves $\frac{1}{2}$–1 in. Heads $\frac{1}{10}$ in. . . . . 2. *C. australis.*

2. **Leptinella.** *Heads uni- or bi-sexual, female with inflated corolla. Achenes not winged.*

* *Flower-heads bisexual, outer florets female, in* 1 *or few series.*

Scapes very leafy. Leaves much cut. Heads blackish . . . . . 3. *C. atrata.*
Scapes ebracteate.
Stem stout, softly woolly. Leaves 2-pinnatifid, 2–4 in. Florets eglandular . . . . 4. *C. plumosa.*
Stem stout, woolly. Leaves pinnate, 2–3 in. Florets glandular . 5. *C. lanata.*
Stem slender. Leaves 1–1$\frac{1}{2}$ in, pinnatifid, lobes short, distant . . 6. *C. minor.*
Stem wiry, slender. Leaves $\frac{1}{4}$ in., pinnatifid. Heads $\frac{1}{10}$ in. . . . 7. *C. filiformis.*
Stem rigid. Leaves 1–1$\frac{1}{2}$ in., pectinate, lobes few, entire . . . . 8. *C. pectinata.*

** *Flower-heads unisexual.*

Stout, glabrous or glabrate. Leaves pinnatifid. Scape bracteate. Heads $\frac{1}{3}$ in. . . . . 9. *C. pyrethrifolia.*
Slender, silky with white hairs. Leaves pinnatifid. Scape short, ebracteate. Heads $\frac{1}{8}$–$\frac{1}{6}$ in. . . . . 10. *C. perpusilla.*

Stout, glabrate. Leaves serrate or lobulate. Scape ebracteate. Heads $\frac{1}{3}$ in. . . . . 11. *C. dioica.*
Slender, sub-woolly. Leaves pinnatifid, lobes cut. Heads $\frac{1}{8}$–$\frac{1}{3}$ in. . 12. *C. squalida.*

3. **Myriogyne.** *Heads bisexual: female florets in very many series; males very few, in the centre. Heads sessile* . . . . . . 13. *C. minuta.*

1. **C. coronopifolia,** *Linn.;—Fl. N. Z.* i. 127. Perfectly glabrous; stems succulent, creeping, branching, ascending, 2–10 in. long. Leaves scattered, $\frac{1}{2}$–2 in. long, lanceolate or oblong, variously toothed lobed or pinnatifid; petiole dilated into a broad toothed or lobed sheath. Heads $\frac{1}{3}$–$\frac{1}{2}$ in. diam. on the scape-like ends of the branches; involucral scales in 2 or 3 series, linear-oblong, obtuse, membranous: ray fl. 1 series, on slender pedicels; corolla 0; achene flat, broadly winged, wings lobed at top and enclosing the 2-fid style, glandular on the inner face: disk fl. shortly pedicelled; corolla subcylindric, 4-toothed.

Northern and **Middle** Islands: marshy spots in various localities from the Bay of Islands to Otago, *Banks and Solander*, etc. A widely-spread plant, found in Australia, S. America, S. Africa, N. and S. Europe, and N. Africa.

2. **C. australis,** *Hook. f. Fl. N. Z.* i. 128. A very slender, much-branched, flaccid herb, glabrous hairy or woolly at the nodes, 2–4 in. high. Leaves $\frac{1}{2}$–1 in. long, deeply pinnatifid or 2-pinnatifid; lobes linear, entire. Heads minute, $\frac{1}{10}$ in. broad, on long slender peduncles; involucral scales in 2 series, membranous, linear-oblong: ray fl. in 3 series, pedicelled, without corolla; achene obovate, broadly winged, wing 2-fid at the top, glandular on the inner face: disk fl. tubular, subcylindric, 4-toothed, teeth glandular.—Fl. Tasm. t. 50; *Soliva tenella*, A. Cunn.

Northern and **Middle** Islands: not rare in waste places; also found in Australia, Tasmania, S. Africa, and Tristan d'Acunha.

3. **C. plumosa,** *Hook. f.—Leptinella*, Fl. Antarct. i. 24. t. 20. A large, tufted, creeping, aromatic, feathery species, more or less covered with soft, matted, villous hairs. Stems creeping, as thick as a goose-quill. Leaves long-petioled, membranous, 2–4 in. long, oblong in outline, pinnate; leaflets close-set, slender, linear, recurved, pinnatifid to the base along the upper side only, ultimate divisions $\frac{1}{6}$ in. long, linear, toothed on one side. Scapes woolly, shorter than the leaves. Heads $\frac{1}{3}$ in. diam.; involucral scales 20–30, broad-oblong, woolly, with broad black margins; receptacle conic: ray fl. 2 or 3 series, shortly pedicelled; corolla inflated, much compressed, mouth contracted, unequally 4-toothed, cordate at base; achene obovoid: disk fl. funnel-shaped, 5-toothed.

Lord Auckland's group, **Campbell's** Island, *J. D. H.* **M'Quarrie's** Islands, *Fraser.* Also found in Kerguelen's Land. The largest and most compound-leaved N. Z. species.

4. **C. lanata,** *Hook. f.—Leptinella*, Fl. Antarct. i. 25. t. 19. A smaller plant than *C. plumosa;* stems creeping, 5–12 in. long, robust, densely woolly or glabrate. Leaves 1–3 in. long, rather thick, with broad petioles, pinnate; pinnules close-set, curved, 3–5-lobed or pinnatifid along the upper edge, minutely glandular. Scapes short, stout, woolly. Heads as in *C. plumosa*, but smaller; involucral scales not purple-edged. Floret of ray narrower, and all florets covered with minute conglobate glands.

Lord Auckland's group and **Campbell's** Island, *J. D. H.* The *Leptinella propinqua*, Hook. f., of Campbell's Island, is a broader-leaved variety of this species.

5. **C. atrata,** *Hook. f., n. sp.* Robust, pubescent; stems shortly creeping, ascending or erect, stout, 2–4 in. high, very leafy. Leaves 1 in. long, linear-oblong, erect, pinnatifid; lobes close-set, crenate or toothed, thick and fleshy. Scapes stout, pubescent, clothed with pinnatifid bracts. Heads large, ⅓–½ in. diam., subglobose; involucral scales in 2 or 3 series, oblong, obtuse, entire or pectinate or pinnatifid; receptacle conic: florets excessively numerous, black when dry; outer female in several series; tube of corolla cylindric, rugose, 3- or 4-toothed; achene linear-oblong, rugose: disk fl. funnel-shaped, with very long rugose tube, 4-toothed.

Middle Island: shingle heaps on the alps, alt. 2–6500 ft.; Tarndale, *Sinclair*; Wairau gorge, *Travers*; Ashburton glacier, Mount Torlesse, and Macaulay river, *Haast*. A very singular plant. Haast observes that the heads, when fresh, present a dark-yellow eye with a brown rim. The involucral scales are sometimes quite entire, at others all pinnatifid.

6. **C. minor,** *Hook. f.*—*Leptinella*, Fl. N. Z. i. 129. A small, creeping, pubescent or glabrate species, 1–15 in. long. Leaves either rosulate or alternate on long creeping runners, narrow linear-oblong in outline, pinnatifid almost to the base; leaflets rather distant, short, recurved, obovoid oblong or linear, entire or cut on the upper edge only. Scapes slender, naked. Heads small, ¼ in. diam.; involucral scales 8–20, orbicular, with purple edges: florets yellow, glabrous or glandular, of ray in several series; corolla flattened, ovoid, inflated, with narrow 2- or 3-toothed mouth; achene obcuneate: disk fl. funnel-shaped, with 4 large teeth.

Northern and **Middle** Islands: east coast, *Colenso*; Foxhill, Nelson, *Munro*; Canterbury plains, *Lyall*, *Travers*.

7. **C. filiformis,** *Hook. f., n. sp.* A very slender, rigid, creeping plant, glabrous or pilose. Leaves minute, ¼ in. long, oblong, pinnatifid, segments subulate. Scapes filiform, 1 in. long, naked. Heads minute, $\frac{1}{10}$ in. diam.; involucral scales 6–8, orbicular, with purple edges; receptacle conical; ray-fl. about 20; corolla short, compressed, inflated, very broad oblong, 2-lobed above; achene obconic, glandular; disk fl. funnel-shaped, 4-lobed; lobes glandular.

Middle Island: Canterbury plains, amongst grass, *Haast*.

8. **C. pectinata,** *Hook. f., n. sp.* Stems short, tufted, 1–3 in. long, silky-pilose, at length glabrate. Leaves 1–1½ in. long, rigid, narrow linear-oblong, pectinately pinnatifid, the lobes short, subulate, entire, pointing upwards. Scapes slender, naked, 1–1½ in. long. Heads ¼ in. diam.; involucral scales about 15, pubescent, orbicular or oblong, with purple toothed edges; receptacle conic: ray fl. in many series; corolla oblong, compressed, 2-lobed above; achene cuneate, compressed, glandular: fl. of disk funnel-shaped; lobes 4, glandular.

Middle Island: Canterbury plains, and crevices of rocks on Mount Torlesse, *Haast*; Otago, grass land in the lake district, *Hector and Buchanan*. I have also a specimen without habitat from Sinclair's Herbarium. The Mount Torlesse specimens have woolly stems, but the structure of the flower seems the same as the plains ones.

9. **C. pyrethrifolia,** *Hook. f. n. sp.* A small, robust species, glabrous

or sparingly pilose; rhizome branching, tortuous, 1–2 in. long. Leaves $\frac{1}{2}$–1 in. long, thick and coriaceous, petiolar part longest; blade pinnatifidly cut into 5–8 alternate linear-oblong or obovate, short, thick, entire lobes, $\frac{1}{12}$–$\frac{1}{6}$ in. long. Scapes 1–2 in. long, with one or more linear bracts. Heads unisexual, $\frac{1}{3}$ in. diam.; involucral scales linear, in several series, herbaceous, with broad, membranous, purple, jagged tips; receptacle hemispheric: florets covered with globose glands;—*fem.:* corolla oblong, ovoid, inflated, with broad truncate base and contracted 4-lobed mouth; achene oblong-cuneate;—*male:* funnel-shaped, 4-lobed.

Middle Island: rocks on the Kowai river, and elsewhere in the alps of Canterbury, *Sinclair* and *Haast; Tarndale plains, alt. 4000 ft., *Travers;* Otago, *Hector and Buchanan.* A very distinct species, small but very robust; the leaves have few lobes, and the scapes usually many bracts. It is very odorous when bruised.

10. **C. perpusilla,** *Hook. f.—Leptinella pusilla,* Fl. N. Z. i. 129. A small silky species, with wiry runners 2–5 in. long. Leaves tufted, $\frac{1}{2}$–$\frac{3}{4}$ in. long, $\frac{1}{6}$ in. broad, sessile, narrow linear-oblong in outline, pinnatifid to the base; leaflets close-set, recurved, serrate along the upper edge, silky on both surfaces. Scapes short, $\frac{1}{4}$ in. long, stout, silky, quite naked. Heads unisexual;—*fem.* $\frac{1}{6}$ in. diam.; involucral scales 3 or 4 series, longer than the florets, incurved, orbicular-oblong, silky, coriaceous, with broad membranous toothed purple edges; receptacle conic; florets eglandular; corolla as in *C. pyrethrifolia,* obovoid;—*male* heads much smaller; scales in 1 row; florets fewer, funnel-shaped, glabrous.

Northern Island: east coast, *Colenso.* **Middle** Island: grassy places, Kowai river, *Haast;* Tarndale plains, 4000 ft., *Travers.*

11. **C. dioica,** *Hook. f.—Leptinella dioica,* Fl. N. Z. i. 129. Glabrous or slightly hairy. Stems creeping, rather robust, short, 1–3 in. long. Leaves petioled, 1–2 in. long, $\frac{1}{4}$–$\frac{1}{3}$ broad, linear obovate or spathulate, obtuse, crenate-serrate lobulate or semipinnatifid; lobes entire or serrate on the upper edge, glabrous. Scapes longer or shorter than the leaves, without bracts. Heads $\frac{1}{6}$–$\frac{1}{3}$ in., male and female similar; involucral scales 2 or 3 series, oblong-orbicular, hairy, with broad, purple, toothed margins; receptacle conical; florets as in *C. perpusilla,* but eglandular.—*Soliva tenella,* A. Cunn.

Northern Island: Cape Turnagain, east coast, *Colenso.* **Middle** Island: Canterbury plains and Acheron valley, alt. 4000 ft., *Travers;* Akaroa, *Raoul;* grassy terraces, Kowai river, *Haast;* Otago, sands and swamps near the sea, and in the interior, *Lindsay, Hector and Buchanan.* Very near and perhaps not different from the Chilian and Fuegian *Leptinella scariosa,* Cass. (*Soliva dioica,* Schultz, *L. ancistroides,* H. and A.), but that has deeply pinnatifid leaves and rather glandular flowers. The scarcely pinnatifid leaves of this species are its most prominent characters.

12. **C. squalida,** *Hook. f.—Leptinella,* Fl. N. Z. i. 129. A slender species, with long, weak, creeping, woolly stems, and long soft hairs on the leaves and scapes. Leaves 1–2 in. long, $\frac{1}{4}$–$\frac{1}{2}$ in. broad, long linear-obovate, flaccid, petioled, pinnatifid; leaflets rather lax, recurved, incised along the upper margin. Scapes slender, ebracteate, tomentose, longer than the leaves. Heads unisexual;—*fem.:* $\frac{1}{3}$ in. diam.; involucral scales numerous, in many series, longer than the florets, incurved, orbicular, with erose purple margins, silky; receptacle conical; florets as in *C. perpusilla,* eglandular;—*male* heads much smaller, $\frac{1}{6}$ in. diam.; involucral scales very few, and florets few.

Northern Island: Hawke's Bay, etc., east coast, *Colenso.* **Middle** Island: Akaroa, *Raoul;* Canterbury, *Travers.* Also very nearly allied to the South American *Leptinella scariosa*, but the large female heads, with large incurved involucral scales concealing the florets, as in *L. perpusilla*, are very different.

13. **C. minuta,** *Forst.—Myriogyne minuta*, Less.—Fl. N. Z. i. 130. A glabrous, prostrate, excessively branched, annual herb. Leaves alternate, sessile on the branches, $\frac{1}{4}$–$\frac{3}{8}$ in. long, lanceolate or oblong-lanceolate, unequally sparingly toothed. Heads $\frac{1}{6}$–$\frac{1}{4}$ in. diam., axillary, sessile, depressed, spherical; involucral scales in 2 series, linear, obtuse;—*fem.* fl. very numerous, densely packed; corolla short, 4-cleft; achene linear, angled, pilose;—*male* fl. very few, central; corolla broadly campanulate.

Northern and **Middle** Islands, in waste places near settlements, *Forster*, etc. Very fragrant when bruised. Also a native of India, China, Japan, Australia, the Pacific islands, and Chili (*M. elatinoides*, Less.).

10. CRASPEDIA, Forster.

Perennial, erect, silky, simple, leafy herbs, bearing one globose, terminal, dense, involucrate, compound head, formed of numerous slender flower-heads, aggregated on a small receptacle. Leaves radical and alternate, simple, long. —Heads narrow, of 5–8 small, yellow, tubular, 5-toothed florets. Involucral scales long, membranous, hyaline, linear. Receptacle very narrow, bearing hyaline scales amongst the florets. Pappus of 1 row of very soft feathery hairs. Anthers with 2 slender tails. Styles included. Achene silky, oblong, narrow.

A small genus of Tasmanian, Australian, and New Zealand plants, excessively variable.

Pubescent or tomentose . . . . 1. *C. fimbriata.*
Lanate with white wool . . . . 2. *C. alpina.*

1. **C. fimbriata,** *DC.;—Fl. N. Z.* i. 131. Slender or robust, 4–15 in. high. Leaves usually all radical, with only bracts on the scape; radical petiolate, 1–8 in. long, spathulate, obtuse, quite entire, usually fringed with white tomentum, glabrate pubescent or woolly. Compound head $\frac{1}{4}$–2 in. diam., soft, white dotted with yellow florets.—*C. uniflora*, Forst.

Northern and **Middle** Islands: abundant, from the East Cape southwards, *Banks and Solander*, etc., ascending to 5500 ft. on the Discovery Peaks. I cannot distinguish this satisfactorily from the Australian and Tasmanian *C. Richea.*

2. **C. alpina,** *Backhouse, in Fl. Tasm.* i. 198.—*C. fimbriata*, ε. *lanata*, Fl. N. Z. i. 132. Very near *C. fimbriata*, and perhaps only a variety, but at once distinguished by the clear white cottony wool on the lower or on both surfaces of the leaf and scape.

Middle Island: Nelson mountains, *Bidwill;* Upper Waihopai and Wairau, *Munro;* grassy places on the Rangitati, Kowai and Godley rivers, alt. 4–6000 ft. *Haast*, Otago, lake district, *Hector and Buchanan.* Intermediates between this and the preceding may be found. It is a native of the Tasmanian and Victorian alps.

11. CASSINIA, Br.

Shrubs. Leaves alternate, small, persistent, simple, often white or rusty below. Heads in terminal panicles or corymbs, small, white.—Involucre cylindric turbinate or campanulate, of many or few, short, obtuse, imbricating scales, the innermost with short, white, dilated rays. Receptacle contracted, covered with slender scales like the inner involucral, except *C. fulvida.* Florets tubular, all similar and hermaphrodite, or the outer very slender and female. Corolla 4- or 5-toothed. Anthers 2-tailed. Arms of the style long, truncate, glandular. Pappus of 1–4 series of soft slender hairs, rather thickened at the tip. Achene small, oblong or obovate.

A small Australian, New Zealand, and New Caledonian genus, distinguished from *Ozothamnus* by the linear scales amongst the florets. I fear that the first three species may prove forms of one. *C. pinifolia* is a native of New Caledonia, not of New Zealand.

Leaves oblong or obovate, whitish below. Heads few . . . . . 1. *C. retorta.*
Leaves linear, narrow, whitish below, not glutinous. Heads numerous 2. *C. leptophylla.*
Leaves linear, narrow, fulvous below, glutinous. Heads numerous . 3. *C. fulvida.*
Leaves obovate or oblong, fulvous below, glutinous. Heads numerous 4. *C. Vauvilliersii.*

1. **C. retorta,** *A. Cunn.;—Fl. N. Z.* i. 132. A shrub, 10–15 ft. high. Branches and leaves below covered with white tomentum, not glutinous. Leaves close-set, $\frac{1}{6}$ in. long, spreading or recurved, linear-obovate or linear-oblong, obtuse, with recurved margins, opaque above. Heads 1–8 together, 6–8-flowered, turbinate, shortly pedicelled, $\frac{1}{4}$ in. long; involucral scales with white tomentum.

Northern Island: not rare, especially near the coasts.

2. **C. leptophylla,** *Br.;—Fl. N. Z.* i. 133. A shrub like *C. retorta* in habit and pubescence, also glutinous, but more slender. Leaves erect or spreading, rarely recurved, $\frac{1}{12}$–$\frac{1}{10}$ in. long, $\frac{1}{20}$ broad, narrow linear, with recurved margins, glabrous above and more or less shining. Heads numerous, in terminal hemispherical corymbs, $\frac{1}{6}$–$\frac{1}{4}$ in. long, 8–10-flowered, narrow turbinate or tubular; involucral scales few, scarious, glabrate, shining.—*Calea leptophylla,* Forst.

Northern and **Middle** Islands, *Banks and Solander;* Cape Palliser and east coast, *Colenso;* East Cape, *Sinclair.* **Middle** Island: Port Underwood, *Lyall.*

3. **C. fulvida,** *Hook. f.—C. leptophylla,* γ, Fl. N. Z. i. 133. A shrub with the habit, etc., of *C. leptophylla*, but glutinous, with foliage larger and tomentum fulvous. Branches covered with subviscid tomentum. Leaves $\frac{1}{6}$–$\frac{1}{4}$ in. long, spreading, linear, obtuse, margins recurved, fulvous below, more or less shining and glutinous above. Heads very numerous, in terminal corymbs, $\frac{1}{6}$–$\frac{1}{4}$ in. long, 4- or 5-flowered; involucral scales pubescent or glabrate; scales amongst the florets few or 0.

Northern Island: Cape Palliser, *Colenso.* **Middle** Island, *Lyall;* mountains of Nelson, *Munro;* Look-out Point, Dunedin, *Lindsay;* river-beds in the Kowai, alt. 2–4000 ft., *Haast;* Otago, common, *Hector and Buchanan.* I am still in some doubt as to the validity of this species, of which I had but one scrap when the Fl. N. Z. was prepared; the various specimens received since all agree with the original, and differ from *C. leptophylla* (to which, however, it is very nearly allied), by the strong fulvous colour and glutinous foliage. I find few or no scales amongst the florets, so, by right, the plant should perhaps be referred to *Ozothamnus.*

4. **C. Vauvilliersii,** *Hook. f. Fl. N. Z.* i. 133. An erect, dense, fastigiately-branched shrub, 2–10 ft. high. Branches and leaves below covered with fulvous tomentum. Leaves erect or patent, $\frac{1}{4}$–$\frac{1}{3}$ in. long, linear-

oblong or obovate, obtuse, margins recurved, costate below, opaque or shining and generally glutinous above. Heads numerous, in terminal globose corymbs, $\frac{1}{5}$ in. long, turbinate, on tomentose, very short pedicels; involucral scales few, scarious, woolly; scales amongst the florets numerous.—*Ozothamnus Vauvilliersii,* Homb. and Jacq. Voy. au Pôl. Sud, Bot. t. 5.

Northern and **Middle** Islands: common on the mountains, *Bidwill*, etc. Scarcely distinguishable from a true *Ozothamnus* of Tasmania (*O. cuneifolius*, A. C.).

12. OZOTHAMNUS, Br.

Characters of *Cassinia*, but without any scales amongst the florets; inner scales of the involucre without white radiating tips in the New Zealand species, and hairs of pappus not always thickened at the tip.

A large Australian and Tasmanian genus, very variable in habit.

Heads corymbose. Slender shrub. Leaves orbicular, petiolate, lax . 1. *O. glomeratus.*
Heads solitary, terminal, sessile.
Leaves sub-4-farious, keeled and polished at back . . . . 2. *O. microphyllus.*
Leaves sub-6-farious, linear, hoary or silky . . . . 3. *O. depressus.*
Leaves imbricated in very many series, polished and convex on back 4. *O. coralloides.*
Leaves imbricated in 6–8 series, polished and keeled on back . . 5. *O. Selago.*

1. **O. glomeratus,** *Hook. f. Fl. N. Z.* i. 133. A spreading bush; branches slender, flexuous, tomentose at the tips. Leaves scattered, $\frac{1}{4}$–1 in. broad, orbicular or broadly ovate or spathulate, quite entire, often apiculate, margins recurved, white and cottony beneath, suddenly contracted into slender petioles. Heads in small, lateral, subglobose, sessile or peduncled corymbs, small, $\frac{1}{10}$ in. long, pedicelled or sessile; involucral scales scarious, woolly at the base, a few outer florets slender, female. Pappus-hairs thickened at the tip. Achene puberulous.—*Swammerdamia glomerata*, Raoul, Choix, 20. t. 16.

Northern and **Middle** Islands: dry hills from the Bay of Islands to Otago, *Banks and Solander*, etc.

2. **O. microphyllus,** *Hook. f. Fl. N. Z.* i. 134. *t.* 35 A. A depressed woody shrub. Branches tomentose, ascending, crowded, densely covered with imbricating leaves, $\frac{1}{12}$ in. diam. Leaves minute, $\frac{1}{16}$–$\frac{1}{12}$ in. long, closely imbricating, almost quadrifariously, appressed to the stem, triangular, ovate, thick, obtuse, obtusely keeled, woolly next the stem, green and polished at the back, with sometimes an oblong spot below the tip. Heads solitary, terminal, sessile, turbinate, $\frac{1}{6}$–$\frac{1}{4}$ in. long; involucral scales scarious, glabrous or pubescent, inner somewhat dilated at the tip. Pappus not thickened at the tip. Achene pubescent.

Middle Island: Wairau Pass, 4000 ft., *Bidwill, Sinclair;* Southern Alps, clefts of perpendicular rocks, *Sinclair and Haast.* The Nelson specimens have more slender branches than the more southern.

3. **O. depressus,** *Hook. f. Fl. N. Z.* i. 134. A prostrate, silver-grey, woody shrub, sometimes 5 ft. high, with straggling, divaricating, rigid, slender branchlets, hoary or with appressed cottony down. Leaves minute, $\frac{1}{12}$ in. long, closely appressed to the branchlets and imbricating towards their tips, linear, obtuse, silky, woolly above. Heads $\frac{1}{5}$–$\frac{1}{4}$ in. long, solitary, sessile on the tips of the branchlets; involucral scales very narrow, acuminate, recurved, cottony at the base. Pappus very slender. Achene glabrous.

Middle Island: Nelson mountains, banks of streams, Clarence and Wairau valleys, alt. 3–5000 ft., *Bidwill, Sinclair, Travers.* Great Tasman glacier and elsewhere, on shingle beds in the Southern Alps, *Haast*, and Otago mountains, *Hector and Buchanan.*

4. **O. coralloides,** *Hook. f. Fl. N. Z.* ii. 332. A very remarkable, woody, short, stout, branched shrub. Branches (with the leaves on) cylindric, as thick as the finger, the leaves resembling tubercles on their surface. Leaves in very numerous series, closely and densely imbricating, oblong, obtuse, $\frac{1}{8}$–$\frac{1}{4}$ in. long, upper part very thick, convex and shining, lower part membranous; surface next to the stem densely clothed with white cotton. Head small, solitary, hidden amongst the uppermost leaves at the tip of the branch; involucral scales linear, recurved, with membranous tips.

Middle Island: Kaikora mountains, *M'Donald; Upper* Awatare valley, *Sinclair.*

5. **O. Selago,** *Hook. f. Fl. N. Z.* i. 332. A good deal similar to *O. coralloides*, also very stout and woody, but intermediate between it and *O microphyllus.* Branchlets with leaves on $\frac{1}{6}$ in. diam. Leaves imbricating in 5 or 6 series, oblong-ovate, obtuse, or subacute, trigonous, the exposed part of back keeled, shining, surface next the stem densely cottony. Heads terminal, solitary, sessile, exposed; involucral scales linear-oblong, obtuse, lower half coriaceous, upper scarious.

Middle Island: Kaikora mountains, *M'Donald, Sinclair.*

13. **RAOULIA,** Hook. f.

Very small, generally tufted (often most densely), alpine and subalpine, slender, or rigid and stout herbs. Leaves minute, usually silky woolly or cottony, often most densely imbricated. Heads small, terminal, sessile.—Involucre oblong; scales scarious, in 2 or 3 series, the inner often white and radiating. Receptacle very narrow, papillose or fimbrillate. Florets of circumference in 1 or 2 series, female; corolla filiform, 3- or 4-toothed; arm of style exserted. Florets of disk numerous, hermaphrodite; corolla funnel-shaped above; anthers with slender tails; arms of style shorter. Pappus of 1 row of slender or stout scabrid hairs, sometimes thickened at the tips. Achene small, oblong.

A genus founded on habit more than on any good characters that can separate it from *Gnaphalium*, section *Helichrysum;* its herbaceous habit distinguishes it from *Ozothamnus.* It contains two natural and most distinct sections, of which one, containing *R. subulata, eximia, grandiflora, mammillaris*, and *bryoides* has a convex, often hispid receptacle; achenes with very long, silky hairs, a thickened areole at their base; and stout, rigid, opaque pappus hairs, thickened at the tip; these probably constitute a good genus, to which the name *Raoulia* may be retained: the others may perhaps fall into *Gnaphalium* or *Helichrysum*, but until all the Gnaphalioid *Compositæ* are worked up, it is impossible to settle the limits of the genera. I at one time suspected that the white radiating involucres indicated sexual differences, but I have failed to prove this. The style has often 3 arms in this genus. *Helichrysum Youngii*, has much the habit of a *Raoulia*, but more that of the genus I have placed it in.

1. *Involucral scales without white or radiating tips. Pappus-hairs numerous, slender, not thickened at the tip.*

Leaves $\frac{1}{12}$ in., erect or recurved, spathulate, obtuse . . . . 1. *R. australis.*
Leaves $\frac{1}{12}$ in., recurved, narrow oblong or spathulate, acute or mucronate 2. *R. tenuicaulis.*

Leaves $\frac{1}{14}$ in., erecto patent, broadly ovate, glabrate . . . . 3. *R. Haastii.*
Leaves $\frac{1}{8}$ in., patent, recurved, linear, obtuse, silky . . . . 4. *R. Munroi.*

2. *Involucre as* 1. *Pappus-hairs few, rigid, thickened at the tip. Achenes with long hairs and a thickened areole at the base.*

Leaves glabrous, subulate, rigid . . . . 5. *R. subulata.*
Leaves most densely imbricate, hidden in silky wool . . . . . . 6. *R. eximia.*
Leaves ovate, obtuse, silvery, grooved when dry . . . . 7. *R. Hectori.*

3. *Involucre with the inner scales white-tipped, and radiating. Pappus as in* 1.

Stems long. Leaves loosely imbricate, linear-oblong, glabrate . . . 8. *R. glabra.*
Stems short. Leaves densely imbricate, linear-oblong, glabrate or silky 9. *R. subsericea.*

4. *Involucre as in* 3. *Achenes and pappus as in* 2.

Leaves closely imbricate, ovate-subulate, silvery . . . . 10. *R. grandiflora.*
Leaves most densely imbricate, with a velvety silky tuft above middle . 11. *R. mammillaris.*
Leaves most densely imbricate, tips covered with appressed wool . . . 12. *R. bryoides.*

1. **R. australis,** *Hook. f. Fl. N. Z.* i. 135. A small, moss-like, densely-tufted plant; stems 1–2 in. high; branches slender, erect or prostrate. Leaves minute, laxly or densely imbricate, $\frac{1}{16}$–$\frac{1}{12}$ in. long, spathulate, erect or recurved, rounded at the tip, covered with silky appressed wool. Heads $\frac{1}{12}$–$\frac{1}{6}$ in. long; outermost scales spathulate, inner linear, scarious, shining, yellow or pale brown, not dark at tips, nor white and radiating; florets about 12, outer few. Pappus hairs excessively slender, subpilose, not thickened at the tips. Achene glabrous or puberulous.—Raoul, Choix, t. 15.

Northern Island: Tongariro and Waikato, *Bidwill;* and elsewhere, in lofty, rocky hills, *Colenso.* **Middle** Island: Nelson mountains, *Bidwill;* rocky hills at Akaroa, *Raoul, Lyall;* Clarence, Tarndale plain, and Wairau valleys, alt. 3500 ft., *Travers;* Southern Alps, *Haast;* Otago, *Hector and Buchanan.* A very variable plant.

2. **R. tenuicaulis,** *Hook. f. Fl. N. Z.* i. 135. *t.* 36 A. Stems generally slender, loosely tufted, prostrate, creeping, 1–10 in. long, with ascending branches. Leaves loosely imbricating, spreading and recurved, $\frac{1}{12}$ in. long, linear-oblong or spathulate-lanceolate, apiculate or acuminate, rarely broadly spathulate, grey with appressed silvery tomentum, rarely glabrate. Heads as in *R. australis*, but involucral scales with brown acute tips.

Northern Island: gravelly beds of rivers, Wairarapa, Ruanahanga, Palliser Bay, etc., *Colenso.* **Middle** Island: Kowai river, alt. 1–2000 ft., *Haast.*

3. **R. Haastii,** *Hook. f., n. sp.* A small, densely tufted, nearly glabrous species; stems rather stout, prostrate; branches 1 in. high. Leaves densely imbricate, erecto-patent, $\frac{1}{16}$ in. long, broadly sheathing, broadly ovate-subulate, obtuse, coriaceous, obscurely woolly or silky. Heads as in *R. australis*, but narrower, with 6–8 florets; involucral scales obtuse, not brown nor with a white radiating tip.

Middle Island: gravelly terraces, Kowai river, *Haast;* Waiauna valley, alt. 3000 ft., *Sinclair, Travers.* I was at first disposed to regard this as a form of *R. tenuicaulis*, but the leaves are very different in shape and in their broad bases, and the involucral scales are not brown at the tips.

4. **R. Munroi,** *Hook. f., n. sp.* Stems slender, creeping, with very long, wiry, filiform rootlets. Branches slender, ascending, 1–2 in. high. Leaves laxly imbricate, patent and recurved, $\frac{1}{8}$–$\frac{1}{6}$ in. long, linear, obtuse, uniformly clothed with grey silky tomentum. Heads narrow, $\frac{1}{6}$ in. long; involucral

scales glabrous, linear, green, with rather dilated scarious brown tips; florets about 12; pappus as in *R. australis.*

Middle Island: Waihopai valley, *Munro;* Canterbury plains, *Travers.* The wiry stems and very long filiform rootlets, are prominent characters, as are the uniformly grey, silky, linear leaves, and narrow heads with brown-tipped involucral scales.

5. **R. subulata,** *Hook. f., n. sp.* A small, very densely tufted, rigid, moss-like species, quite glabrous throughout, blackish when dry. Stems stoutish, branches ½ in. high. Leaves most densely imbricate, patent or suberect, rigid, subulate, acuminate. Heads large for the size of the plant, $\frac{1}{6}$ in. diam.; involucral scales linear-oblong, scarious, shorter than the leaves; receptacle convex, hispid; florets of circumference in several rows. Pappus of rigid, scabrid hairs, rather thickened at the tips. Achene silky.

Middle Island: Nelson mountains, *Sinclair:* Otago mountains, alt. 5–6000 ft., *Hector and Buchanan.* A remarkable and very small species, differing much from the foregoing in the pappus, hispid receptacle, and foliage.

6. **R. eximia,** *Hook. f., n. sp.* A small, most densely tufted, hard little plant, forming large woolly balls on the mountains, enveloped in soft, velvety, white tomentum. Branches very short, with the leaves forming cylindric or mammilliform knobs, ¼ in. diam. Leaves most densely compacted, wholly hidden amongst woolly hairs, imbricated all round in many series, ⅛ in. long, membranous, broadly linear- or obovate-oblong, rounded at the tip, bearing at the back above the middle a dense thick pencil of white velvety hairs, these bundles of hairs, meeting beyond the leaves, envelope the whole. Heads minute, sunk amongst the upper leaves; involucral scales about 10, linear, with subulate or obtuse tips, and a tuft of hairs on the back above the middle; receptacle convex, naked; florets about 10. Pappus of few rigid hairs, thickened upwards. Achene silky, with very long hairs.

Middle Islands: Riband-wood rage, Mount Arrowsmith and Dobson, alt. 5500–6000 ft., *Sinclair, Haast.* A most singular plant, forming hemispherical cushions on the mountains, 2 ft. high and 3 in. diam., called "Vegetable Sheep." Very near allied to *R. mammillaris.*

7. **R. Hectori,** *Hook. f., n. sp.* Most densely tufted, 1–2 in. high; branchlets erect, densely leafy, silvery at the tips. Leaves closely imbricate, erecto-patent, $\frac{1}{12}$ in. long, broadly ovate, obtuse, coriaceous, more membranous below the middle, upper half covered with appressed silvery shining tomentum, back grooved longitudinally when dry. Heads small, sunk amongst the uppermost leaves; involucral scales about 10–12, scarious, linear-oblong, obtuse or subacute, yellowish, glabrous; receptacle conical, pilose; florets about 20. Pappus of few, rigid, scabrous hairs, thickened upwards. Achene silky.

Middle Island: Otago, lake district, in dry, subalpine places, *Hector and Buchanan.* A very distinct species, resembling in habit some states of *R. australis.*

8. **R. glabra,** *Hook. f. Fl. N. Z.* i. 135. Stems elongate, slender, prostrate, branching, 2–10 in. long; branches ascending. Leaves laxly imbricate, spreading, hardly ever recurved, $\frac{1}{6}$ in. long, linear or linear-oblong, acute or obtuse, glabrous or nearly so, rarely silky, 1-nerved, green. Heads rather large, ¼–⅓ in. diam.; outer involucral scales leaf-like, but with broader bases; inner linear, with short, white, radiating tips; florets numerous, outer in 2 series,

Pappus of numerous soft, white, slender hairs as in *R. australis*. Achene puberulous.

Middle Island: Nelson mountains, *Bidwill;* Milford Sound, *Lyall;* shingle beds, Rangitata river, and Mount Cook, alt. 1800–3200 ft., *Sinclair, Haast;* Otago mountains, alt. 3–4000 ft., *Hector and Buchanan*. Haast sends apparently a silky variety, with rather broader leaves, from mountains near Lake Hawea, alt. 4000–5000 ft.

9. **R. subsericea,** *Hook. f. Fl. N. Z.* i. 136. Very similar in most characters to *R. glabra*, and perhaps an alpine variety of that, but a much more densely tufted plant. with very short stems and branches, closely imbricated, linear-oblong leaves, glabrous or covered loosely with silvery tomentum, green or silvery-white. Heads similar, but larger, $\frac{1}{3}$ in. diam.

Middle Island: Wairau mountains, Clarence valley, Aglionby plains, alt. 3000–4000 ft., and elsewhere in Nelson, *Munro, Sinclair, Travers;* Port Cooper, *Lyall;* Godley rivulet and Mount Darwin, alt. 3–5000 ft., *Haast;* Otago mountains, alt. 4000 ft., *Hector and Buchanan.*

10. **R. grandiflora,** *Hook.f. Fl. N. Z.* i. 136. A very short, erect, densely tufted species, with very long, wiry, thread-like roots. Stems 1 in. high, densely leafy, with the leaves on as thick as the little finger. Leaves imbricating all round the stem, $\frac{1}{6}$–$\frac{1}{4}$ in. long, erecto-patent, ovate-subulate, rigid, shining with white silky hairs, cottony at the base, striate. Heads large, $\frac{1}{3}$–$\frac{2}{3}$ in. diam. involucral scales 1 or 2 series, long, white, linear, spreading, $\frac{1}{4}$ in. long; receptacle convex, hispid. Pappus hairs few, rigid, swollen towards the tip. Achene silky.

Northern Island: summits of the Ruahine range, *Colenso.* **Middle** Island: top of Gordon's Nob, *Munro;* Upper Wairau, *Sinclair;* top of Big Ben, Mounts Cook, Darwin and Torlesse, alt. 5–7000 ft., *Haast;* Mount Brewster, alt. 5–6000 ft., forming carpets, *Hector and Buchanan.* Allied in many respects to *R. subulata*, and especially in the hispid receptacle.

11. **R. mammillaris,** *Hook. f., n. sp.* Like *R. eximia*, forming large, hard, hemispherical balls and patches on the ground, sometimes 8 ft. long and 3 high. Branches very short, thick, with the leaves on forming cylindric or mammillary knobs, $\frac{1}{4}$ in. diam. Leaves most densely compacted, imbricated in many series, spreading, $\frac{1}{10}$–$\frac{1}{12}$ in. long, obovate cuneate or spathulate, obtuse, membranous, cottony below, with a dense brush of velvety hairs on both surfaces beyond the middle, which does not exceed the tip of the leaf. Heads very small, $\frac{1}{6}$ in. diam., about 10-flowered; inner involucral scales with short, white, acute, radiating tips; receptacle convex, naked. Pappus of few rigid hairs thickened at the tips. Achene with a swollen areole at the base and long white silky hairs.

Middle Island: Mount Torlesse, on hard soil and rocky places, alt. 3–5000 ft., *Haast.* Very similar in many respects to *R. eximia*, and closely allied to it, but the leaves are smaller, with the velvety hairs not so long as to hide them, more cottony and obovate, and the inner involucral scales are distinctly rayed.

12. **R. bryoides,** *Hook. f. Fl. N. Z.* ii. 322. Forming hard, dense, convex, hoary patches, with an even surface. Branches $\frac{1}{2}$–$1\frac{1}{2}$ in. long, densely compacted, with the leaves on cylindric, $\frac{1}{10}$–$\frac{1}{6}$ in. diam. Leaves most densely imbricate all round the branches, erecto-patent, $\frac{1}{12}$–$\frac{1}{10}$ in. long, broad, linear,

rather dilated at the obtuse tip, membranous, coriaceous; margins cottony, glabrous below the upper $\frac{1}{3}$, above that covered with appressed silky wool, 1-nerved. Heads $\frac{1}{4}$ in diam., about 12-flowered; involucral scales with white, subacute, radiating tips; receptacle tumid. Pappus-hairs few, rigid, with thickened tips. Achene with very long white hairs and a thickened areole at the base.

Middle Island: top of Gordon's Nob, *Munro;* Clarence and Wairau valleys, alt. 3000–4000 ft., *Sinclair, Travers.*

14. GNAPHALIUM, Linn.

Herbs of very various habit, annual or perennial, the New Zealand species all more or less densely covered on the leaves below, or all over, with white cottony wool. Heads solitary or corymbose or fascicled.—Involucre campanulate hemispherical or turbinate; scales narrow, in several series, all similar, scarious and shining, or the inner produced into white spreading rays. Receptacle flat or conic, papillose or alveolate. Florets of ray female, in 1 or more series, very slender, tubular, 3–5-toothed; of disk hermaphrodite, funnel-shaped above. Anthers with slender tails. Pappus hairs in one series, slender or stout and thickened at the tip, slightly cohering at the base. Achene small, linear-oblong, usually pubescent.

I have in vain sought to arrange the New Zealand *Gnaphalia* and *Helichrysa* under these genera as defined by De Candolle and other authors, any separation of them into these involves bringing together plants most different, and separating most closely allied ones. Thus, those with white radiating involucral scales form, I think, a most natural genus or group; but I cannot identify them as a group either with *Helichrysum, Antennaria,* or *Anaphalis,* of authors, to many species of which they seem naturally allied. The bracteate species again so closely resemble the European and Himalayan *Leontopodia,* that they seem naturally congeneric, but they differ in several very important floral characters. I have not neglected to examine Weddell's character of the pappus hairs cohering or free at the base, but cannot apply it to the New Zealand species, in all of which the hairs very partially cohere, but are so readily separated that the character is valueless. The thickness of the pappus hairs, and their thickening upwards, constitutes an excellent character, and generally goes with that of the outer florets being in one series; these together should perhaps distinguish the New Zealand *Helichrysa* or *Antennariæ* from *Gnaphalium,* and I have indicated them accordingly under the former name.

1. *Heads solitary, inner involucral scales white, radiating. Florets of circumference in several series. Pappus hairs very slender.*

Stem 8–18 in. Leaves $\frac{1}{4}$ in., apiculate. Head sessile . . . . . 1. *G. prostratum.*
Stem 2–10 in. Leaves $\frac{1}{4}$ in., apiculate. Head on long peduncle . 2. *G. bellidioides.*
Stem very short, densely tufted. Leaves obtuse. Head sessile . . 3. *G. Youngii.*

2. *Heads corymbose, not bracteate. Inner involucral scales white, radiating. Florets of circumference in many series* (*except in* 7). *Pappus hairs very slender* (*except* 7).

Stem robust. Leaves 2 × $\frac{1}{4}$–$\frac{1}{3}$ in. Corymbs dense. Heads $\frac{1}{2}$ in. diam. 4. *G. Lyallii.*
Stems slender. Leaves $\frac{1}{2}$–1 in. Heads few, $\frac{3}{8}$ in. diam. . . . . 5. *G. trinerve.*
Stem slender. Leaves $\frac{1}{2}$–2 in. Heads $\frac{1}{3}$ in. diam. . . . . 6. *G. Keriense.*
Stem 2–4 in. Leaves tomentose on both sides. Heads numerous, $\frac{1}{4}$ in. diam. . . . . 7. *G. Sinclairii.*

3. *Heads solitary. Inner involucral scales not white and radiating.*

Stem filiform, erect. Leaves scattered, cottony below . . . . . 8. *G. filicaule.*
Leaves radical, densely cottony on both sides. Scape cottony . . 9. *G. Traversii.*

Stems short, tufted. Leaves silky on both surfaces . . . . 10. *G. nitidulum.*

4. *Heads corymbose, not bracteate. Inner involucral scales not white nor radiating. Pappus hairs very slender* . . . . . 11. *G. luteo-album.*

5. *Heads collected into a dense, bracteate globe. Inner involucral scales not white nor radiating. Bracts broad, densely woolly. Female florets in* 1 *series. Pappus-hairs stout, rigid, thickened upwards.*

Leaves $\frac{1}{3}$–$\frac{1}{2}$ in. long, rosulate. Inflorescence on a stout, terminal peduncle . . . . 12. *G. Colensoi.*
Leaves $\frac{1}{4}$ in. long, densely imbricate, recurved. Inflorescence sessile 13. *G. grandiceps.*

6. *Heads collected into a dense, bracteate globe. Bracts linear. Inner involucral scales not white nor radiating. Female florets in many series. Pappus hairs very slender.*

Stem erect, branched, leafy. Leaves linear, glabrous above . . . 14. *G. involucratum.*
Leaves radical, spreading, and scape densely cottony . . . . 15. *G. collinum.*

1. **G. prostratum,** *Hook. f. Fl. Antarct.* 30, *t.* 21;—*Fl. N. Z.* i. 137. Stems rather slender, prostrate, woody at the base, 8–18 in. long, with many ascending leafy branches. Leaves uniform, loosely imbricating, $\frac{1}{4}$ in. long, spreading or recurved, flat, obovate-spathulate, covered below or on both surfaces with dense, white, appressed wool, 1-nerved. Heads sessile at the tips of the branches, $\frac{1}{3}$ in. diam.; involucral scales in many series, white, radiating, $\frac{1}{4}$ in. long, with white cottony claws; receptacle conical; female florets in many series; pappus hairs very slender. Achene glabrous.

Northern Island: Mount Egmont, alt. 4000 ft., *Dieffenbach;* top of Titiokura, *Colenso.* Abundant in **Lord Auckland's** group and **Campbell's** Island, *J. D. H.*

2. **G. bellidioides,** *Hook. f. Fl. N. Z.* i. 137. Stems and leaves and flower-heads as in *G. prostratum,* of which it is probably a variety, but the ends of the branches are produced into slender bracteate peduncles, 2–5 in. long. Receptacle conical, or plane with a conical mammilla.—*Xeranthemum,* Forst.

Abundant in alpine districts in the **Northern** and **Middle** Islands, ascending to 5000 ft., *Banks and Solander,* etc.

3. **G. (Helichrysum) Youngii,** *Hook. f., n. sp.* A small, prostrate, densely tufted species, 1–2 in. high. Leaves imbricating on the short branches, erecto-patent, $\frac{1}{6}$–$\frac{1}{4}$ in. long, obovate-spathulate, obtuse, densely clothed on both surfaces with snow-white cottony wool. Head sessile amongst the leaves, $\frac{1}{3}$ in. diam.; involucral scales in 2 or 3 series, the inner white, radiating; receptacle very small, narrow; florets about 12, female in 1 series; pappus hairs thickened upwards, rather stout. Achene pubescent.

Middle Island: mountains above Lake Hawea, alt. 6000 ft., forming patches; summit of Mount Torlesse and Mount Cook, on shingle, alt. 6500–7000 ft., *Haast;* Otago, lake district, *Hector and Buchanan.* Named in honour of Mr. William Young, Mr. Haast's fellow-traveller and able assistant, both as a surveyor and botanical collector. A beautiful little snow-white plant, intermediate in habit between *Raoulia* and the two preceding *Gnaphalia.*

4. **G. Lyallii,** *Hook. f. Fl. N. Z.* i. 137. Stem very stout, almost woody, prostrate, branched, as thick as a swan's quill; branches stout, erect, leafy, 5–10 in. high, cottony above. Leaves close-set, spreading, 2 in. long, $\frac{1}{4}$–$\frac{1}{3}$ broad, narrow oblong-lanceolate, broader upwards, acute, 3-nerved, glabrous above, appressed tomentose beneath. Heads $\frac{1}{2}$ in. diam., forming dense corymbs, 2–4 in. across; pedicels cottony; involucral rays very many, $\frac{1}{4}$ in.

long, white, radiating, with short cottony claws; receptacle plane; florets numerous; female in 2 or 3 series; pappus hairs few, slender. Achene glabrous.

Middle Island: Massacre Bay, *Lyall.* A very handsome species, at once known by its robust habit and large leaves, but probably a variety of the following.

5. **G. trinerve,** *Forst.;—Fl. N. Z.* i. 138. Stem rather slender and branches 6–24 in. long, ascending, glabrous or cobwebby, produced into bracteate peduncles. Leaves lax, uniform, flat, spreading or recurved, $\frac{1}{2}$–1 in. long, obovate- or spathulate-lanceolate, acute or apiculate, faintly 3-nerved, glabrous above, white with appressed wool below. Heads 3–5, corymbose at the ends of the produced, slender, peduncle-like branches, $\frac{3}{8}$ in. diam.; pedicels slender or short and stout; inner involucral scales numerous, white, radiating, clawed, $\frac{1}{4}$ in. long. Achene and pappus as in *G. Lyallii.*

Middle Island: Dusky Bay, *Forster;* Milford Sound, *Lyall;* Dunedin, on sand dunes, abundant, *Lindsay*, etc. The Ruahine mountain plant referred to this in Fl. N. Z., is, I am now sure, a var. of *G. Keriense.*

6. **G. Keriense,** *A. Cunn.;—Fl. N. Z.* i. 138. A very variable plant, smaller than either of the above. Stems prostrate, with slender or stout, ascending, leafy branches, 2–6 in. high, produced into bracteate peduncles. Leaves spreading, very variable, $\frac{1}{2}$–2 in. long, $\frac{1}{10}$–$\frac{1}{3}$ in. broad, from narrow-linear to oblong-spathulate or -lanceolate, acute, 1- rarely 3-nerved. Heads numerous, corymbose, on cottony pedicels, like those of *G. trinerve*, but much smaller, $\frac{1}{3}$ in. diam.—*Helichrysum micranthum*, A. Cunn. in DC. Prodr.; *G. dealbatum*, Forst. Prodr.?

Var. *β. linifolia.* Stems erect. Leaves excessively narrow; pedicels capillary.

Northern Island: very abundant in moist places, falls of the Keri-Keri river, and elsewhere. **Middle** Island: near Nelson, *Travers;* Dusky Bay, *Lyall;* var. *β*, banks of the Manawatu and Ruahine range, *Colenso.*

7. **G. (Helichrysum) Sinclairii,** *Hook. f., n. sp.* A small, subalpine species; stems and branches ascending, leafy, 2–4 in. high. Leaves close-set, spreading, $\frac{1}{4}$–$\frac{1}{3}$ in. long, $\frac{1}{6}$ broad, linear-oblong or obovate-spathulate, obtuse, densely covered with pale, cottony tomentum on both surfaces. Heads $\frac{1}{4}$ in. diam., in numerous, rounded, terminal, dense corymbs, $\frac{1}{2}$–1 in. across; peduncles and pedicels short, densely cottony; outer scales of involucre cottony, inner shortly radiating; female florets in 1 series; pappus of few stout hairs, thickened towards the tip. Achene glabrous.

Middle Island: Upper Awatere valley, *Sinclair.* Very closely allied to the *Raoulia catipes* of Tasmania, but the leaves are much smaller, and the heads not half the size, and much more numerous.

8. **G. filicaule,** *Hook. f.;—Helichrysum,* Fl. N. Z. i. 140. t. 36 B. Stems usually simple, very slender, flexuose, cottony, 6–10 in. high, terminating in a filiform peduncle. Leaves scattered, $\frac{1}{4}$–$\frac{1}{3}$ in. long, obovate-oblong, obtuse or apiculate, glabrous above, white and cottony beneath. Heads solitary, $\frac{1}{3}$ in. diam.; involucral scales numerous, linear, scarious; outer cottony at the base; receptacle small, convex; pappus of very slender filiform hairs. Achene puberulous.

Northern Island: dry hills towards the east coast; Wairarapa valley, Cape Kidnapper, and Puehutai, *Colenso.* **Middle** Island?, *Banks and Solander;* Canterbury, *Haast;* Otago, grass flats in the lake district, *Hector and Buchanan.*

9. **G. Traversii,** *Hook. f., n. sp.* A small, slender, erect, almost simple plant, 2–3 in. high; stem and leaves on both sides covered loosely with snow-white cottony wool. Leaves radical, petiolate, spreading, $\frac{1}{3}$–$\frac{1}{2}$ in. long, spathulate-obovate. Head solitary, $\frac{1}{3}$ in. diam., on a scapiform slender stem, with 2 or 3 linear bracts; involucral scales numerous, linear, scarious, hyaline, shining, outer cottony at the base, inner with erect paler tips; receptacle flat; pappus hairs excessively fine. Achene puberulous.

Middle Island: Wairau mountains, alt. 3–4000 ft., *Travers;* alps of Canterbury, *Haast.* Mueller has sent this same plant from the Victorian alps, as *G. involucratum,* var. *monocephalum,* but besides the totally different habit it differs from that plant in the heads not being bracteate and twice as large, and in the looser cottony tomentum.

10. **G. nitidulum,** *Hook. f., n. sp.* A small, densely tufted species, covered with appressed, silky, shining, yellowish tomentum. Leaves closely imbricated at their bases, above spreading, flat, $\frac{1}{3}$ in. long, linear, obtuse; lower $\frac{1}{3}$, membranous, glabrous, upper $\frac{2}{3}$ densely silky. Heads terminal, solitary, large, $\frac{1}{2}$ in. broad, on very short, slender peduncles; involucral scales in 2 series, erect, linear, hyaline, shining, with pale erect tips; florets not seen.

Middle Island: Nelson mountains, *Sinclair;* Clarence and Wairau valleys, alt. 3500 ft., *Travers.*

11. **G. luteo-album,** *Linn.;—Fl. N. Z.* i. 139. Stems simple or branched from the base, erect or ascending, 6–18 in. high, and leaves densely covered with cottony tomentum. Leaves scattered, $\frac{2}{3}$–2 in. long, narrow-linear or linear-spathulate. Heads fascicled, $\frac{1}{6}$ in. long, fascicles collected in corymbs, dusky-yellow, cottony at the base; involucral scales erect, tips incurved, numerous, linear-oblong, scarious, hyaline, shining.

Northern and **Kermadec** Islands; very common in some places, rarer in the **Middle** Island; Kowai valley and Rangitata ranges, *Haast;* sand dunes, Dunedin, *Lindsay;* Otago, lake district, *Hector and Buchanan.* **Lord Auckland's** group, *Lyall.* A very abundant tropical weed.

12. **G. (Helichrysum) Colensoi,** *Hook. f.;—Helichrysum Leontopodium,* Fl. N. Z. i. 141. t. 37 B. A tufted, alpine, very silky species. Stems ascending, 1–4 in. high, terminating in stout woolly bracteate or leafy peduncles 1–2 in. long. Leaves densely imbricate, rosulate, patent or reflexed, flat, $\frac{1}{3}$–$\frac{1}{2}$ in. long, linear-oblong, subacute, uniformly clothed with silky shining hairs, striate when dry; those on the peduncle smaller and shorter. Heads densely crowded into a bracteate capitulum $\frac{1}{3}$ in. diam.; bracts many, spreading, $\frac{1}{4}$–$\frac{1}{3}$ in. long, ovate-oblong, obtuse, most densely woolly; each head $\frac{1}{6}$ in. long; involucral scales linear-lanceolate, erect, scarious, shining, woolly at the back; pappus hairs few, stout, scabrid, thickened upwards. Achene silky.

Northern Island: Ruahine range and Mount Hikurangi, *Colenso.* **Middle** Island: Tarndale mountains, in shingle, alt. 5000 ft., *Sinclair.*

13. **G. (Helichrysum) grandiceps,** *Hook. f., n. sp.* Densely tufted. Stems ascending, and branches 2–3 in. long, uniformly and densely clothed to the tips with leaves, and terminated by the sessile bracteate heads. Leaves

closely imbricating, small, recurved, ¼ in. long, concave, oblong-spathulate, obtuse, recurved, densely covered with thick white silvery tomentum. Heads and florets in *G. Colensoi*, but bracts shorter.

Middle Island: Mount Sinclair; moraines of the Great Clyde glacier; mountains near Lake Hawea, Mountains Cook and Torlesse, alt. 4000–6000 ft., *Sinclair, Haast;* Otago, lake district, subalpine, *Hector and Buchanan*. Allied to *G. Colensoi*, but very different in the small, broad, recurved, closely imbricated, uniform, less silvery leaves, and inflorescence not peduncled.

14. **G. involucratum,** *Forst.;—Fl. N. Z* i. 139. An erect annual. Stems branched at the base, 1–2 ft. high, branches often proliferous, stiff, cottony. Leaves scattered, spreading, often fascicled, 1–3 in. long, narrow-linear or lanceolate, acute, glabrous above, cottony beneath; margins often recurved and wrinkled. Heads small, ⅙–¼ in. long, collected into dense axillary and terminal globular balls ¼–1 in. diam., which are subtended by numerous spreading or reflexed linear foliaceous bracts 1 in. long and of variable breadth; involucral scales erect, hyaline, linear, acute; female florets numerous, hermaphrodite very few; pappus hairs very slender. Achene glabrous.—*G. virgatum*, Banks and Sol.;—Fl. N. Z. i. 139; *G. lanatum*, Forst.; *G. Cunninghamii*, DC.

Abundant throughout the islands, in waste places, *Banks and Solander*, etc.. The *G. virgatum*, B. and S., not to be distinguished even as a variety I fear. Abundant in Australia and Tasmania.

15. **G. collinum,** *Labill. Fl. Nov. Holl. t.* 189;—*Fl. N. Z.* i. 139. Perennial, roots with runners. Stems scapiform. Leaves chiefly radical, petiolate, 2–4 in. long, lanceolate-spathulate, acuminate, covered with cottony wool on both surfaces. Scape 4–8 in. high, slender, white and cottony; bracts 2 or 3, foliaceous. Heads as in *G. involucratum*, but collected into a smaller fewer-headed subglobose capitulum, with only 1 or 2 linear bracts at their base.—*G. simplex*, Forst.?

Northern Island: dry hills, Bay of Islands, east coast, etc. **Middle** Island: not uncommon in the Canterbury district; Otago, lake district, *Hector and Buchanan*. Though extremely dissimilar in its typical state from *G. involucratum*, I find now so many almost intermediate forms, that I suspect their permanent difference, and am disposed to refer the vars. β and γ of the N. Z. Flora to that plant.

15. HAASTIA, Hook. f., nov. gen.

Very densely tufted, low, woolly herbs, forming balls or cushions on the lofty mountains. Leaves crowded, broad. Flower-heads large, solitary, terminal, sunk amongst the uppermost leaves.—Involucral scales in 2 series, very numerous, narrow, herbaceous with scarious tips, free or connate at the base, acuminate. Receptacle narrow, flat, papillose. Florets of circumference in 2 or more series, female: corolla very short, slender, tubular, with a crenulate mouth; styles with very long exserted arms, papillose at the tips. Florets of disk very numerous, funnel-shaped, hermaphrodite; arms of style shorter; anthers without tails. Pappus of 1 series of rather rigid, white, slender, scabrid hairs, thickened towards the tip. Achene compressed, linear or oblong, even or grooved.

A very singular and distinct genus, differing from the other Gnaphalioid *Compositæ* in the tailless anthers.

Leaves most densely imbricate, 3-nerved, crenulate. Pappus hairs free. 1. *H. pulvinaris.*
Leaves laxly imbricate, recurved. Pappus hairs united below . . . . 2. *H. recurva.*
Leaves laxly imbricate, suberect. Pappus hairs free . . . . 3. *H. Sinclairii.*

1. **H. pulvinaris,** *Hook. f.* Plants forming dense hemispheres or cushions, 3 ft. across, covered with fulvous wool; branches with the leaves on as thick as the thumb. Leaves patent, ½ in. long, crenulate, most densely imbricate, broadly obcuneate, with dilated rounded tips, margins recurved towards the tip, membranous, 3-nerved when the wool is removed. Heads ⅓ in. broad. Pappus hairs free to the base. Achene glabrous.

Middle Island; Kaikora mountains, and Mowatt's Mountain, alt. 5000 ft., *Sinclair;* Discovery Peaks, alt. 5800 ft., *Travers.* One of the most extraordinary plants in the islands. Sinclair says the patches are so dense, that the finger cannot be thrust between the branches.

2. **H. recurva,** *Hook. f.* More laxly tufted, as densely covered with wool, which is more rufous when dry. Leaves loosely imbricating, ½ in. long, obovate-spathulate, recurved. Heads ½ in. diam. Pappus hairs paleaceous and united at the base. Achene glabrous.

Middle Island: shingle-beds above 5000 ft., Tarndale, *Sinclair;* Discovery Peaks, alt. 5800 ft., *Travers;* Mount Torlesse, alt. 6000 ft., *Haast.*

3. **H. Sinclairii,** *Hook. f.* Loosely tufted, branches ascending, erect, covered with paler cottony wool. Leaves erect, imbricating, ½–¾ in. long, oblong-obovate or rounded-obovate, obtuse, not recurved, 5–7-nerved. Head ⅓–1 in. diam.; involucral scales broader than in the preceding species. Pappus hairs free to the base. Achene glabrous.

Middle Island: shingle beds, alt. 4–6000 ft., Wairau and Awatere mountains, and at the Wairau pass, *Sinclair;* Mounts Darwin and Cook, *Haast;* Mount Brewster, dry débris, alt. 6000 ft., *Hector and Buchanan.*

16. ERECHTITES, Rafinesque.

Tall, perennial, glabrous or cottony herbs. Leaves alternate, simple or runcinate-pinnatifid.—Heads corymbose, bracteolate, very narrow, cylindric or bell-shaped; involucral scales in 1 series, herbaceous, narrow linear, appressed. Receptacle papillose. Florets of circumference in 2 or more series, female, excessively slender, tubular, 2–4-toothed; of the disk hermaphrodite, campanulate above; anthers without tails; arms of style with short terminal cones. Pappus in many series of excessively slender, soft, roughened hairs. Achene oblong, striated, glabrate or hispid, obtuse or narrowed at the tip, terminated by a disk-like thickened top.

A small genus, the species are natives chiefly of Australia and Tasmania; a few others are American and Indian.

Glabrous. Leaves toothed or pinnatifid. Achene slender . . . . 1. *E. prenanthoides.*
More or less cottony. Leaves pinnatifid. Achene short . . . . 2. *E. arguta.*
Hispid, except the heads and pedicels . . . . 3. *E. scaberula.*
Cottony and white. Leaves linear, long, with revolute margins . 4. *E. quadridentata.*

1. **E. prenanthoides,** *DC.;—Fl. N. Z.* i. 141. A tall, glabrous, simple or branched herb, 1–3 ft. high, rarely slightly hairy. Leaves 3–6 in. long, linear-oblong or lanceolate, lower petiolate, upper sessile with auricled bases, toothed lobed or pinnatifid. Corymb lax; heads quite glabrous, nu-

merous, on very slender pedicels, each $\frac{1}{4}$–$\frac{1}{3}$ in. long; involucral scales green with white margins. Achenes slender, grooved.—*Senecio*, A. Rich.

Not uncommon in moist woods, etc., throughout the islands, *Banks and Solander*, etc. A common S. Australian and Tasmanian plant.

2. **E. arguta,** *DC.;—Fl. N. Z.* i. 142. A stout, erect herb, 1–2 ft. high, more or less cottony, especially on the leaves below. Leaves 2–4 in. long, linear-oblong or lanceolate, irregularly pinnatifid and toothed, sessile and auricled at the base. Heads corymbose, numerous, small; florets $\frac{1}{4}$ in. long; involucral scales cottony below, shorter than the florets. Achene short, grooved, hispidulous.—*Senecio argutus*, A. Rich.

Common throughout the islands, *Banks and Solander*, etc. Abundant in Tasmania and Southern Australia.

3. **E. scaberula,** *Hook. f., n. sp.;—E. hispidula*, Fl. N. Z. i. 142, not DC. A slender, erect herb, 1–1$\frac{1}{2}$ ft. high, hispid all over, except the pedicels and involucre. Leaves 1–3 in. long, linear-oblong or lanceolate, toothed or pinnatifid, sessile. Heads $\frac{1}{4}$ in. long, laxly corymbose or divaricating. Achene linear, indistinctly grooved, pubescent.

Northern and **Middle** Islands: in various places, but so common as *E. arguta* and *quadridentata*. This differs from the Tasmanian *E. hispidula*, in the achene only $\frac{1}{12}$ in. long, and not attenuate at the top, also in the smaller glabrous heads, and short involucral scales.

4. **E. quadridentata,** *DC.;—Fl. N. Z.* i. 142. A stout species, 1–3 ft. high, sometimes woody below, more or less covered with cottony wool, often snow-white all over. Leaves 2–6 in. long, very narrow linear, margins revolute. Heads corymbose, numerous, very narrow, $\frac{1}{3}$ in. long; involucral scales glabrous or cottony. Achene grooved and angled, hispid, $\frac{1}{12}$ in. long. —*Senecio quadridentatus*, Lab. Fl. Nov. Holl. t. 194.

Abundant throughout the islands, *Banks and Solander*, etc. Also frequent in Southern Australia and Tasmania.

I have a specimen of another *Erechtites* from the bed of the Godley river, alt. 3000 ft., *Haast*, apparently differing from any of the above, but too imperfect to determine; it is nearly glabrous, with narrow, linear-oblong, slightly toothed, coriaceous leaves, 1–2 in. long, sparingly cottony below, not auricled at base: heads glabrous, immature.

17. SENECIO, Linn.

Herbs shrubs or trees, of very various habit. Leaves entire toothed lobed or pinnatifid. Flower-heads bracteate at the base, corymbose, rarely solitary.—Involucral scales in 1 or 2 rows, linear, erect, herbaceous or coriaceous, rarely scarious, often brown at the tip. Receptacle plane or tumid, papillose or alveolate. Florets yellow, all similar and hermaphrodite, or the outer series female and ligulate; the inner tubular, campanulate above, hermaphrodite. Anthers with very short tails or 0. Pappus of 1 or several rows of soft or stiff, smooth or scabrid, equal or unequal, slender or stout hairs, sometimes thickened at the tips. Achene linear-oblong, never beaked, usually with a thickened disk-like top, terete or grooved.

One of the largest genera in the vegetable kingdom, found in all parts of the world, and containing a multitude of dissimilar, always most variable species. I have classed the New

Zealand ones into natural groups, which are, in point of habit and many other characters, the equivalents of genera in other parts of the vegetable kingdom, but which cannot be made genera of, because of the numberless connecting forms found in other countries. *S. Forsteri*, of Fl. N. Z., differs from all others in the scarious involucral scales, and hence I have restored it to the genus *Brachyglottis*, of Forster. The shrubby tomentose species, with very rigid coriaceous, almost woody involucres, also form a peculiar set, for the most part; but *S Greyii* and *Munroi*, which should belong to the same group, have membranous scales. The New Zealand species are excessively variable, the first four may prove one, the remainder seem distinct enough

1. *Scapigerous herbs. Leaves all broad, radical. Scapes 1- or many-headed. Outer florets rayed. Involucral scales in about 2 series. Achenes glabrous. Pappus hairs white, slender.*

Leaves 1–4 x 1–3 in. rugose and hispid above, woolly below. Scape glandular . . . . 1. *S. Lagopus.*
Leaves ½–4 x ½–1½ in., rugose and hispid above, glabrate below. Scape cottony . . . . 2. *S. bellidioides.*
Leaves 3–6 x 1–3 in., woolly above and below. Scape glandular . 3. *S. saxifragoides.*
Leaves 2–5 x 1–3 in., and whole plant cottony, snow-white . . . 4. *S. Haastii.*

2. *Branched herbs, usually glabrous. Leaves all or upper sessile and auricled at the base. Heads corymbose. Outer florets rayed (rarely tubular in* S. lautus). *Involucral scales linear, in 1 series, herbaceous, connate into an obconic base. Achenes ribbed. Pappus of white slender hairs.*

Tall. glabrous. Leaves membranous, ovate, petioled, lobed, and toothed . . . . 5. *S. latifolius.*
Stout, short, glabrous or glabrate. Leaves 1–2 in., rather fleshy . 6. *S. lautus.*
Stout. Leaves cottony, white, lobed and toothed . . . . 7. *S. Colensoi.*
Tall, robust. Leaves leathery, glabrous, toothed, and veined . . 8. *S. odoratus.*

3. *An erect, simple, leafy herb, corymbose above. Heads very large. Involucral scales 2 series, outer florets' rays very long. Pappus hairs rigid, scabrid* 9. *S. Lyallii.*

4. *Shrubby, glabrous. Leaves crowded, linear, margins revolute to the midrib. Head solitary; outer florets with broad rays. Pappus of soft hairs* . . 10. *S. bifistulosus.*

5. *Shrubs with glabrous or woolly or tomentose or cottony stems, leaves, and involucres. Corymbs branched (heads solitary in* S. cassinioides). *Involucral scales of 1 series. Receptacle pitted. Outer florets rayed or not. Pappus of rigid, often unequal, scabrid hairs.*

α. *Glabrous or nearly so. Heads corymbose.*

Leaves 2–4 in., linear or oblong-lanceolate. Rays very long . . 11. *S. glastifolius.*
Leaves 1–2 in., orbicular, toothed. Rays very short . . . . . . 12. *S. sciadophilus.*
Leaves 1–1½ in., oblong, crenate-toothed. Rays short . . . . . 13. *S. perdicioides.*

β. *Leaves quite entire, very coriaceous indeed, covered on both surfaces or below with tomentum. Heads panicled or corymbose.*

Leaves 2–4 in., oblong, softly cottony below. Rays long . . . 14. *S. Greyii.*
Leaves 2–4 in., oblong, with appressed brown tomentum below. Ray 0. 15. *S. elæagnifolius.*
Leaves 2–6 in., orbicular, with brownish tomentum below . . . . 16. *S. rotundifolius.*
Leaves ⅓–1 in., oblong, with whitish tomentum below . . . . 17. *S. Bidwillii.*

γ. *Leaves crenate, undulate, white and cottony below* . . . 18. *S. Munroi.*
δ. *Leaves ⅛ in. Heads solitary* . . . . 19. *S. cassinioides.*

1. **S. Lagopus,** *Raoul, Choix*, 21. *t.* 17;—*Fl. N. Z.* i. 143. Root of stout fibres; crown loaded with long, woolly, matted hairs. Leaves 1–4 in. long, on short, stout, villous petioles, broadly cordate-oblong, rounded at the tip, even or crenulate on the margin, densely clothed below with white wool, rugose above, and covered with short scattered bristles. Scape 2–6 in. high, branched above, bracteate, 2–8-headed, and involucre pubescent and pilose with black glandular hairs. Heads on slender pedicels, ½–1 in. diam., broadly

obconic · involucral scales tomentose; rays $\frac{1}{4}$–$\frac{1}{3}$ in. long; pappus hairs white, slender, unequal. Achene linear, slender.

Northern and **Middle** Islands: subalpine pastures and rocky places, Ruahine mountains, *Colenso;* Nelson, *Bidwill, Sinclair, Munro;* Canterbury, *Raoul, Lyall.* This and the two following, though most dissimilar in their usual states, appear to me to be united by intermediate forms.

2. **S. bellidioides,** *Hook. f. Fl. N. Z.* i. 145. Smaller than *S. Lagopus,* quite similar in habit and woolly crown, but the petioles often more slender. Leaf $\frac{1}{2}$–4 in. long, broadly oblong or linear-oblong or cordate, obtuse at both ends or cordate or acute at the base, entire or crenate, margins often edged with white wool, above rugose and rough with hispid scattered hairs, below glabrate, with raised veins; petioles more or less densely clothed with soft wool. Scapes slender, cottony and glandular, 1–4-flowered. Heads, etc., as in *S. Lagopus.*

α. Leaves broad-oblong, glabrous below. Scapes cottony.
β. Leaves broad-oblong, densely woolly below. Scapes tomentose and glandular.
γ. Leaves linear-oblong, glabrate or villous below. Scapes glabrate.

Middle Island, abundant in hilly pastures, etc., from Nelson to Otago, *Munro,* etc., ascending to 7000 ft. in the Otago alps, *Hector and Buchanan.*

3. **S. saxifragoides,** *Hook. f. Fl. N. Z.* i. 144. Much larger than the two preceding, top of root sometimes 1 in. thick, covered with long villous wool. Leaves 3–6 in. long, with short, very stout, woolly petioles, broadly oblong or orbicular, cordate, crenate, villous and silky above, densely woolly below. Scape 2–10 in. high, stout, bracteate, cottony or tomentose, and covered with black, spreading, glandular hairs, 4–10-flowered. Heads $\frac{1}{2}$–$1\frac{1}{2}$ in. diam., as in *S. Lagopus.*

Middle Island: Port Cooper, *Lyall.*

4. **S. Haastii,** *Hook. f., n. sp.* Wholly covered with white, lax or appressed, cottony wool; rootstock woody, covered with soft wool or naked. Leaves with long, slender petioles, 2–8 in. long, blade 2–5 in. long, 1–3 broad, oblong or broad cordate-oblong, obtuse, obscurely crenulate. Scape slender, 8–14 in. high, and peduncles very long, cottony, and slightly glandular. Heads as in *S. Lagopus.*

Middle Island: Mount Cook, alt. 2700–4000 ft., shores of Lake Okau, Lake Hawea, and mountains near the sources of the Ahuriri, *Haast;* Otago, lake district, alt. 2–3000 ft., *Hector and Buchanan.* The Lake Okau specimen is more buff-coloured, less cottony; its tomentum is more glandular all over, and shows a tendency to pass into *S. bellidioides.*

5. **S. latifolius,** *Banks and Sol.;—Fl. N. Z.* i. 145. A tall branching glabrous herb, 2–3 ft. high. Stems flexuose, furrowed. Leaves membranous, 4–8 in. long, very variable, lower on long winged petioles with toothed auricles, oblong, lobulate and toothed; upper more fiddle-shaped or ovate-oblong, acute, contracted below the middle and again expanding into toothed auricles, 1 in. broad; in young specimens the leaves are scaberulous, lanceolate, and toothed. Heads $\frac{1}{2}$ in. diam., in large corymbs with spreading branches; involucral scales short, in 1 series, linear, acuminate; rays slender, $\frac{1}{4}$ in. long; pappus very soft, white, slender. Achene strongly ribbed, hispidulous.

Northern Island: in wooded districts, *Banks and Solander, Dieffenbach, Bidwill;* Wairarapa, Hawke's Bay, and Ruahine range, *Colenso.*

6. **S. lautus,** *Forst.;—Fl. N. Z.* i. 145. An excessively variable, perfectly glabrous or slightly cottony, branched herb, 6 in. to 2 ft. high. Stem stout, grooved, green, flexuose. Leaves 1–2 in. long, sessile with amplexicaul auricles, or petioled, rather fleshy, ovate-oblong or linear, entire toothed lobed or pinnatifid; lobes long or short, broad or narrow. Heads in few-flowered corymbs, $\frac{1}{4}$–$\frac{1}{3}$ in. long, $\frac{1}{4}$–$\frac{1}{2}$ in. diam., broad-campanulate; involucral scales linear, acuminate; outer florets with short revolute rays, rarely 0; pappus fine, soft, white. Achene glabrous, ribbed or puberulous.—*S. neglectus* and *S. rupicola*, A. Rich. Fl. t. 37.

A most abundant plant, especially on maritime rocks and sands, ascending the mountains to 6000 ft., *Banks and Solander.* Some Otago mountain specimens are slightly cottony, but none others that I have seen. Equally abundant in Tasmania and South Australia.

7. **S. Colensoi,** *Hook. f. Fl. N. Z.* i. 147. An erect, much-branched herb, covered with white cottony or cobwebby tomentum, except peduncles and heads. Stem flexuose, grooved. Leaves 1–4 in. long, coriaceous, linear-oblong, lyrate or contracted in the middle, sessile with broad auricled bases, deeply irregularly toothed or lobulate, obtuse or acute, tomentose on both surfaces. Heads few, corymbose, $\frac{1}{4}$–$\frac{1}{3}$ in. long, broadly campanulate; rays short, revolute. Achene small, grooved, silky.

Northern Island: cliffs near the sea, Bay of Islands, East Cape, and Cape Kidnapper, *Colenso.* This has the cottony tomentum and much the appearance of *Erechtites arguta.* I have immature specimens from Colenso of a variety or an allied species, covered with loose coarse white tomentum.

8. **S. odoratus,** *Hornemann.—S. Banksii,* Fl. N. Z. i. 146. A perfectly glabrous or sparsely pilose, tall, slender or very robust plant, 2–5 ft. high. Stem flexuose, grooved. Leaves coriaceous, sessile, with auricled amplexicaul bases, 2–4 in. long, $\frac{1}{2}$–2 in. broad, broadly oblong to lanceolate, acute or acuminate, strongly veined. Heads small, $\frac{1}{4}$ in. long, corymbose on slender pedicels, broadly campanulate; rays short, revolute. Achene slender, grooved, pubescent.

Northern Island: east coast, *Banks and Solander;* on maritime cliffs, *Colenso.* Also found in South Australia and Tasmania, where the leaves are sometimes cottony.

9. **S. Lyallii,** *Hook. f.—S. Lyallii* and *S. (?) scorzoneroides,* Fl. N. Z. i. 146. Glabrous, or glandular-pubescent. Rootstock very thick, crowned with long silky hairs. Stem stout, simple, erect, 1–2 ft. high, leafy, ending in a branched corymb of many flower-heads. Leaves all quite entire; radical petiolate, oblong-lanceolate or very narrow linear, 2–10 in. long, $\frac{1}{12}$–$\frac{2}{3}$ broad, subacute, 1–5-nerved, cauline numerous, sessile, stem-clasping, gradually narrowed from the base to the obtuse tip. Branches of corymb (peduncles) simple, bracteate. Heads large, 1–2 in. diam.; involucre broadly turbinate, scales in 1 series, linear, pubescent; rays $\frac{1}{2}$–1 in. long; pappus of rigid, dirty-white, unequal scabrid hairs. Achene narrow, silky, ribbed.

Middle Island: west coast, Milford Sound and Dusky Bay, *Lyall;* abundant on the alps, ascending to 5000 ft., *Munro, Sinclair, Travers, Haast,* etc. A truly magnificent, but excessively variable plant. Buchanan observes that the flowers vary in colour from salmon-colour to bright-yellow. *S. scorzonerioides* is a very broad-leaved, glandular state.

10. **S. bifistulosus,** *Hook. f. Fl. N. Z.* i. 145. Glabrous? Stem woody; branches scarred like a pine-branch. Leaves densely crowded, spreading, 1 in. long, $\frac{1}{12}$ broad, narrow-linear, margins revolute to midrib and united to it by their woolly borders, appearing crenate from the impressed veins. Flowering branch or scape from below the leaves, 4 in. long, with foliaceous bracts. Head $1\frac{1}{4}$ in. diam.; involucral scales few, broad, cottony at back; rays few, broad; pappus very soft. Achene short, glabrous, ribbed.

Middle Island: Dusky Bay, *Lyall.* A very curious plant, allied to no other in New Zealand, but apparently to the Tasmanian *S. pectinatus.* I have but one specimen.

11. **S. glastifolius,** *Hook. f. Fl. N. Z.* i. 147. *t.* 39. A perfectly glabrous, branched, woolly shrub, 6–10 ft. high. Branches brittle, stout or slender; bark pale-brown. Leaves shortly petioled, 2–4 in. long, $\frac{1}{3}$–1 broad, oblong or lanceolate, entire or sinuate-toothed, acute or obtuse; nerves very faint, brown when dry. Corymbs lax, much spreading; bracts leafy. Heads large, $1\frac{1}{2}$–2 in. diam.; involucre broad-campanulate, scales broadly linear, membranous; rays few, very long, $\frac{1}{2}$–$\frac{3}{4}$ in.; anthers with short tails; pappus hairs rigid, unequal, scabrid, dirty-white. Achene glabrous, terete,. expanded and thickened at the tip.—*Solidago arborescens*, A. Cunn., not Forst.

Northern Island: in woods, from the Thames river northward, *Banks and Solander*, etc. A curious plant, in many respects allied to the Tasmanian *Centropappus Brunonis.*

12. **S. sciadophilus,** *Raoul, Choix*, 21. *t.* 18;—*Fl. N. Z.* i. 153. A climbing shrub, perfectly glabrous or sparingly pubescent, with slender flexuose branches. Leaves 1–2 in. long, petiolate, orbicular, coarsely toothed, membranous, brown when dry. Heads in few-flowered corymbs, on slender axillary peduncles, sparingly divided, broadly campanulate, $\frac{1}{4}$ in. diam.; involucral scales few (8–10) in 1 series, broadly linear, subacute; rays $\frac{1}{6}$ in. long, revolute; pappus hairs numerous, in several series, rigid, white, scabrid. Achene nearly glabrous, grooved.

Middle Island: Akaroa, in woods, *Raoul.*

13. **S. perdicioides,** *Hook. f. Fl. N. Z.* i. 149. A shrub with slender, pubescent, striated, scarred branches. Leaves on slender petioles, quite glabrous, 1–$1\frac{1}{2}$ in. long, ovate-oblong, obtuse, crenate and toothed, finely reticulated on the under surface. Heads obconic, $\frac{1}{4}$ in. long, 8-flowered, in axillary and terminal corymbs; peduncles pubescent; involucral scales few, broad, obtuse; rays few, broad; pappus of 2 series of scabrid white hairs. Achene oblong, glabrous, deeply grooved.

Northern Island: Tolaga, in woods, *Banks and Solander.* This plant has not been found since Cook's voyage, a century ago.

14. **S. Greyii,** *Hook. f. Fl. N. Z.* i. 148. *t.* 38. A shrub, about 5 ft. high; branches terete, woody, and petioles and leaves below covered with white appressed tomentum. Leaves with petioles $\frac{1}{2}$–$1\frac{1}{2}$ in. long, blade 2–4 long, oblong, obtuse, coriaceous, quite entire, unequal at the base, under side thickly softly cottony, upper glabrate, nerveless, edged with white. Corymbs large, terminal, panicled, drooping; bracts large, foliaceous, sessile, obovate-oblong, cottony or glabrate; peduncles either white and cottony or not cottony and densely glandular-pubescent. Heads campanulate, $\frac{1}{2}$ in. long;

involucral scales thick, linear, acute; rays $\frac{1}{4}$–$\frac{1}{2}$ in. long; anthers with short tails; pappus of many series of rigid scabrid white hairs. Achene silky.

Northern Island: from Pawahati, Cape Palliser, to Pahawa, in rocky and stony places, *Colenso*. A very beautiful plant.

15. **S. elæagnifolius,** *Hook. f. Fl. N. Z.* i. 150. *t.* 41. A small, robust shrub, 6–8 ft. high; branches stout and petioles and leaves beneath densely covered with an appressed smooth layer of dirty-white or buff tomentum. Leaves with stout petioles, $\frac{1}{2}$–1 in. long, obovate or lanceolate-oblong, obtuse, coriaceous, quite entire, glabrous and obscurely veined above. Panicle terminal, stout, branched, and involucres densely covered with thick buff wool. Heads $\frac{1}{3}$ in. long, campanulate; involucral scales very coriaceous and woolly; ray 0; anthers tailed; pappus rigid, scabrid, white. Achene grooved, pubescent.

Northern Island: woods on the Ruahine mountains, *Colenso*: Mount Egmont, alt. 6000 ft., *Dieffenbach*. **Middle** Island: Rangitata range and Mount Cook, alt. 3–4000 ft., *Sinclair, Haast*; Otago, in the bush, rare, *Hector and Buchanan*.

16. **S. rotundifolius,** *Hook. f. Fl. N. Z.* i. 149. A small tree with very robust branches, covered as are the petioles, leaves below, and inflorescence with buff tomentum. Leaves with stout petioles 1–3 in. long, blade 3–6 in. diameter, orbicular, unequal or rounded or cordate at the base, very coriaceous, shining above with reticulate veins, margin tomentose. Corymbs terminal, short, close, much branched. Heads very numerous, cylindric-campanulate, $\frac{1}{3}$–$\frac{1}{2}$ in. long; involucral scales very thick, woolly; rays extremely short; anthers without tails; pappus hairs rigid, scabrid, white. Achene glabrous, grooved.—*Cineraria rotundifolia*, Forst.

Middle Island: Dusky Bay, *Forster*; Milford Sound, *Lyall*. Closely allied to the last, though differing in the large orbicular leaves and the glabrous achene. The Mount Egmont plant referred to this in Fl. N. Z. is certainly *S. elæagnifolius*.

17. **S. Bidwillii,** *Hook. f. Fl. N. Z.* i. 150. A small, very robust, alpine shrub. Branches, petioles, and leaves below densely covered with appressed, whitish tomentum. Leaves shortly petioled, $\frac{1}{3}$–1 in. long, oblong or obovate-oblong, extremely thick and coriaceous, round at the tip, glabrous above, nerveless or reticulated; petioles $\frac{1}{6}$–$\frac{1}{3}$ in. jointed on to the branch. Corymbs with many long-peduncled heads, sometimes abbreviated; peduncles and heads with softer, more woolly tomentum than the leaves below. Heads campanulate, $\frac{1}{4}$ in. long; involucral scales very thick; rays few, very short; anthers shortly tailed; pappus hairs white, rigid, scabrid. Achene grooved, glabrous.

Northern Island: Mount Hikurangi, and Ruahine range, *Colenso*. **Middle** Island: mountains of Nelson, alt. 6000 ft., *Bidwill, Rough*; Discovery Peaks, alt. 5800 ft., *Travers*.

18. **S. Munroi,** *Hook. f. Fl. N. Z.* ii. 333. A woody shrub, or small tree; branches, petioles, and leaves below covered with white appressed tomentum. Leaves petioled, $\frac{1}{3}$–1 in. long, narrow oblong, obtuse, glabrous or viscid above, coriaceous, margin wrinkled and crenate; petiole $\frac{1}{4}$–$\frac{1}{3}$ in. long. Corymbs lax, terminal, leafy; peduncles and pedicels tomentose, the latter and involucre glandular and pubescent. Head broadly turbinate, $\frac{1}{2}$–$\frac{2}{3}$ in. diam.; involucral scales spreading, few, short, membranous, obtuse; rays

nearly $\frac{1}{4}$ in. long, revolute; anthers very shortly tailed; pappus hairs white, rather slender, scabrid. Achene grooved, pubescent.

Middle Island: Nelson mountains, *Munro;* Upper Awatere valley, *Sinclair.*

19. **S. cassinioides,** *Hook. f., n. sp.* A small, woody, very robust, small-leaved shrub; branches stout, covered with deciduous bark, marked, like a pine-tree, with scars of fallen leaves. Leaves imbricate, $\frac{1}{6}$–$\frac{1}{5}$ in. long, sessile, linear-oblong, obtuse, coriaceous, covered below, like the branchlets and involucres, with yellowish very appressed tomentum. Heads solitary, sessile, terminal, $\frac{1}{3}$ in. long; involucral scales in 1 series, broadly linear, obtuse, coriaceous, with membranous margins, much shorter than the florets; rays very short, revolute; anthers scarcely tailed; pappus of rigid, scabrid white hairs. Achene grooved, glabrous.

Middle Island: Wairau Pass, alt. 5000 ft., and Rangitata river, alt. 3000 ft., *Sinclair;* Godley river, alt. 4–5000 ft., *Haast.* A remarkable species, resembling *Cassinia Vauvilliersii.*

18. BRACHYGLOTTIS, Forst.

A tree. Leaves very large, tomentose below, as are the branches.—Heads small, excessively numerous, disposed in very large branching panicles, campanulate, 10–12-flowered.—Involucral scales in 1 series, linear, scarious, obtuse, shining, with subulate bracts at the base. Receptacle very narrow, alveolate. Florets of circumference female, irregularly lobed or 2-lipped, outer lip very short broad recurved, inner narrow revolute; of disk tubular, campanulate above. Anthers with short tails. Arms of style truncate, papillose at the tip. Pappus of 1 series of white, rather stout, scabrid hairs. Achene short, terete, papillose.

I have restored this genus of Forster's, because of its very different habit from the other species of *Senecio,* its scarious shining involucral scales and two-lipped ray-florets; the latter, however, vary much in form.

1. **B. repanda,** *Forst. Char. Gen.* 46.—*Senecio Forsteri,* Fl. N. Z. i. 148. A small branching tree, 10–20 ft. high. Branches, petioles, leaves below, and inflorescence covered densely with soft, white, cottony tomentum. Leaves very large, 6–12 in. long, very broadly ovate-oblong or cordate-oblong, irregularly lobed or waved along the margin, membranous, glabrous above; petioles 1–3 in. long. Panicles larger than the leaves, spreading, drooping or erect; branches slender, flexuose. Heads excessively numerous, sessile or pedicelled, minute, $\frac{1}{6}$ in. long.—*Cineraria repanda,* Forst. Prod.

Abundant in forests throughout the islands, *Banks and Solander,* etc.

19. TRAVERSIA, Hook. f., n. g.

Leaves alternate, sessile, serrate, with reticulate venation.—Heads in corymbs, broadly campanulate, 10–12-flowered, bracteolate. Involucre of 1 series of 6–8 broad, oblong, obtuse, erect, spreading, very rigid and coriaceous scales, shorter than the florets. Receptacle sinuous, alveolate. Florets all tubular, campanulate above, with 5 revolute long lobes. Anthers obtuse, without tails, exserted. Arms of style truncate, papillose. Pappus of about 2 series of very rigid, unequal, scabrid, dirty-white hairs. Achene short, glabrous.

A remarkable plant, allied to *Senecio*, but differing remarkably in the rigid pappus, very coriaceous involucral scales, and venation of the leaves, which recalls that of the Juan Fernandez genera, *Balbisia* and *Robinsonia*.

1. **T. baccharoides,** *Hook. f.* Apparently a small shrub, perfectly glabrous, somewhat glutinous on the leaves and base of the involucres. Branches slender, angled, with raised lines that are decurrent from the leaves. Leaves 1½–2 in. long, coriaceous, obovate-lanceolate, serrate, acute, the nerves running nearly parallel to the midrib, and anastomosing. Heads on slender, rigid pedicels, 1–2 in. long, about ⅓ in. diam.; involucral scales green, shorter than the florets.

Middle Island: Upper Awatere and Wairau valley, alt. 5000 ft., *Sinclair;* Discovery Peaks, 5500 ft., *Travers.*

20. MICROSERIS, Don.

Glabrous herbs, with milky juice, perennial roots, linear leaves, and naked scapes.—Involucre narrow, scales in 1 series, slender, with a few short small ones at the base. Receptacle naked, pitted. Florets all ligulate, yellow. Pappus of 1 series of rather rigid, flat, brown hairs, broadest at the base, produced upwards into fine scabrid points. Achene slender, striate, not beaked.

A genus of only two species, a South American one and the present.

1. **M. Forsteri,** *Hook. f. Fl. N. Z.* i. 151. Very variable in size, from 2–12 in. Leaves narrow, flaccid, quite entire toothed pinnatifid or irregularly cut. Scapes glabrous or downy above. Heads ⅓–⅔ in. long; involucral scales rather fleshy; borders membranous. Pappus pale yellow-brown.—Fl. Tasman. i. 226. t. 66; *M. pygmæa,* Raoul, not Hook.; *Scorzonera scapigera,* Forst.

Northern and **Middle** Islands: common from the Thames river, southward, ascending to 3000 ft. on the mountains. Also found in Tasmania and Southern Australia. This differs from the Chilian species only in the narrower pappus-scales.

21. CREPIS, Linn.

Branching or (the New Zealand species) scapigerous herbs, with milky juice.—Heads campanulate, bracteate at the base. Involucral scales in 2 or 3 series, herbaceous, green, outer short, inner longer, often tipped with black. Receptacle naked. Florets all ligulate. Pappus of several series of excessively fine, silky, more or less pilose hairs. Achene linear, not beaked.

A large genus of the north temperate zone, not found in Tasmania or Australia.

1. **novæ-Zelandiæ,** *Hook. f., n. sp.* A scapigerous herb, very variable in stature, 2–12 in. high, glabrous all over, or with the leaves, or scape, or involucre, or all, white and tomentose. Root stout, fleshy. Leaves spreading, gradually widened upwards into a large, entire, obtuse lobe, or linear and pinnatifid to the base, or petioled and lobed along the margin, lobes toothed; sometimes the leaf is pinnatifid to the midrib, with all the lobes toothed, and the alternate smaller. Scape slender, naked, glabrous or tomentose, and studded with long, black, glandular hairs. Heads ⅓–1⅓ in.

broad, like those of *Taraxacum;* involucral scales linear, broader at the base, with obtuse black tips, glabrous or cottony and covered with black glandular bristles. Pappus hairs almost simple. Achene short, glabrous, compressed, ribbed.

Middle Island: Totara-nui, or Queen Charlotte's Sound, *Banks and Solander;* open hillsides in the Southern Alps, Canterbury plains, ascending to 3000 ft., *Sinclair and Haast;* Tuapeka ranges, Otago, *Lindsay;* lake district, *Hector and Buchanan.* A very curious plant, not well according with any described genus, alluded to as *Hieracium fragile,* Banks and Solander, in the N. Z. Flora, i. 153.

22. TARAXACUM, Juss.

Herbs with milky juice and perennial roots. Leaves all radical, spreading, pinnatifid or lobed. Scapes single-headed, hollow.—Involucre of long, erect, herbaceous scales, the outer shorter and usually reflexed. Receptacle naked. Florets all ligulate. Pappus soft, white, of many series of extremely fine unequal hairs. Achene long, ribbed, tapering into a slender filiform beak; the ribs muricate or toothed.

A genus of probably only one species (the "Dandelion"), which varies very greatly, and is found in all parts of the temperate world.

1. **T. Dens-leonis,** *Desf.—Fl. N. Z.* i. 152. An extremely variable weed. Root stout, dark-coloured, bitter. Leaves toothed, sinuate, pinnatifid or runcinate, rarely entire, narrow, linear-obovate, 2–5 in. long, round at the tip. Scapes 2–8 in., leafless, glabrous or pubescent, or woolly above. Heads 1–1½ in. broad, yellow. Involucre green; scales with membranous edges, often thickened at the tip, reflexed after flowering. Achene spreading, the pappus hairs diverging and forming a circle round the top of its beak.

Northern and **Middle** Islands: in various places, ascending the mountains to 5000 ft., *Banks and Solander,* etc. Very variable in foliage and stature, the larger forms are probably introduced with cultivation; the smaller are certainly indigenous.

23 PICRIS, Linn.

Erect, hispid, leafy herbs, with milky juice.—Heads corymbose. Involucre campanulate; scales in 2 or 3 series, outer often recurved. Florets all ligulate, yellow. Receptacle naked. Pappus of 1 series of soft, white, plumose hairs. Achene turgid below, narrowed above, with tubercled ridges.

A small European genus, of which one species abounds in many parts of the world.

1. **P. hieracioides,** *Linn.;—Fl. N. Z.* i. 151. Hispid or pilose. Stem 2–4 ft. high. Root-leaves petioled, linear-oblong, obtuse, toothed; cauline smaller, sessile, linear, acuminate. Peduncles slender, pedicels bracteate. Heads ⅓–½ in. long; involucral scales hispid and pubescent.

Northern Island. dry hills, Bay of Islands, etc. A very common plant in Europe, some parts of Tasmania, Australia, and various other parts of the world.

24. SONCHUS, Linn.

Tall leafy herbs, usually with hollow stems and milky juice. Leaves alternate.—Heads corymbose, cylindric or ovoid. Involucral scales imbricate, green, herbaceous, connivent after flowering. Receptacle naked. Florets all ligulate. Pappus soft, white; hairs simple, excessively fine and silky. Achene flat, not beaked, ribbed or striate.

A genus of many species, natives of temperate countries, and several (Sowthistles) are weeds of cultivation.

1. **S. oleraceus,** *Linn.*;—*Fl. N. Z.* i. 153. An erect, glabrous annual, 2–4 ft. high. Leaves undivided or pinnatifid, toothed and prickly at the margins, with a broad, cordate or triangular, terminal lobe, upper narrow, 2-lobed, and stem-clasping at the base. Heads pale-yellow, $\frac{2}{3}$–1 in. diam., corymbose panicled or umbellate.

Var. *α*. Achene glabrous; ribs muricate.

Var. *β*. Achene oblong, broad, winged; ribs smooth.—*S. aspera*, Vill.

Common throughout the islands; var. *α* perhaps only introduced, being a weed of cultivation throughout the world. Var. *β*. Certainly indigenous, being found by *Banks and Solander* and *Forster*, and at Chalky Bay by *Lyall*, and in the interior of the Northern Island by *Colenso*, who says that it was formerly eaten by the natives, but that the introduced variety is preferred from being less bitter. The var. *β* is a European and Australian plant.

ORDER XL. STYLIDIEÆ.

Herbs. Leaves alternate, exstipulate. Inflorescence various.—Calyx-tube adnate with the ovary; lobes 5–9. Stamens 2, filaments connate with the style, forming a column; anthers large, transverse. Ovary inferior, crowned with epigynous glands, 1- or 2-celled; style slender, stigma 2-lobed, almost hidden between the anthers; ovules numerous, attached to the septum or to a central axis. Capsule 2-valved, dehiscing downwards, or coriaceous and indehiscent. Seeds minute; albumen fleshy; embryo most minute.

A large Australian Order, extending westwards to Bengal and northwards to China, also found in antarctic America.

Flowers 1–3, on long scapes. Column straight . . . . 1. FORSTERA.
Flowers solitary, sessile at the ends of the branches . . . . 2. HELOPHYLLUM.
Flowers 1 or racemed. Column bent towards the top . . . . . . 3. STYLIDIUM.

1. FORSTERA, Linn.

Small, alpine, glabrous, perennial herbs, with tufted or creeping stems. Leaves small, imbricating or spreading, thick. Scapes terminal, slender, 1–2-flowered. Flowers white, sometimes unisexual.—Calyx 1–3-bracteolate; tube ovoid, lobes 5 or 6. Corolla campanulate or tubular, with spreading limb, 5–9-fid; throat naked or glandular. Epigynous glands 2; style straight, erect. Capsule ovoid. Seeds unknown.

A small genus, confined to the alps of Tasmania, New Zealand, and antarctic America. It is probable that the three following species may be united by intermediates.

Leaves $\frac{1}{6}$–$\frac{1}{4}$ in., imbricate, obovate, recurved; midrib cuneate . . . . 1. *F. sedifolia.*
Leaves $\frac{1}{4}$–$\frac{1}{2}$ in., patent and recurved, linear-oblong; margins recurved . 2. *F. Bidwillii.*
Leaves $\frac{1}{4}$–$\frac{1}{3}$ in., patent, not imbricate nor shining; midrib slender . . 3. *F. tenella.*

1. **F. sedifolia,** *Linn. f.*;—*Fl. N. Z.* i. 154. Stems very stout, 3–12 in., densely covered throughout their length with leaves. Leaves imbricating, sessile, $\frac{1}{6}$–$\frac{1}{4}$ in. long, oblong- or obovate-spathulate, obtuse, recurved, very coriaceous, shining, midrib thickened, margin cartilaginous, the lower red-

brown. Scape very slender, strict, 2–4 in., 1- rarely 2-flowered. Flower very variable; bracts linear-oblong. Calyx-lobes linear, obtuse. Corolla $\frac{1}{4}$–$\frac{1}{2}$ in. diam.; tube very short; lobes linear-oblong. Stamens included.

Middle and **Southern** Islands: Dusky Bay, *Forster;* Chalky Bay, *Lyall;* Rangitata range, alt. 2500–5000 ft. *Sinclair;* Mount Dobson, 6500 ft., *Haast;* grassy banks, Otago alps, alt. 4–6000 ft., *Hector and Buchanan.*

2. **F. Bidwillii,** *Hook. f. Fl. N. Z.* i. 155. Stems 2–8 in., more slender than in *F. sedifolia.* Leaves numerous, close-set, but not imbricate, patent and recurved, $\frac{1}{4}$–$\frac{1}{2}$ in. long, linear or linear-oblong, coriaceous, not shining, green; margins recurved when dry; midrib beneath very indistinct. Flowers smaller than in *F. sedifolia.*

Northern Island: Tongariro, *Bidwill;* summit of the Ruahine range in shady places, *Colenso.*

3. **F. tenella,** *Hook. f. Fl. N. Z.* i. 154. Stem short, 1–2 in. Leaves few, lax, erect or patent, not recurved, $\frac{1}{4}$–$\frac{1}{3}$ in. long, narrow, oblong-obovate, contracted into a short petiole, subacute, margins recurved, midrib obsolete, not shining nor thickly coriaceous. Scape and flower as in *F. sedifolia.*

Middle Island: Otago, on a hill 1800 ft. high, and at Milford Sound, *Lyall;* Southern Alps, *Haast.* Intermediate in some respects between *F. Bidwillii* and *F. sedifolia,* but has shorter stems and less coriaceous foliage than either; the leaves are as short as in *sedifolia,* but neither imbricate, recurved, coriaceous, shining, nor furnished with a thick cuneiform midrib. At the same time I can hardly doubt these three forms having been very recent offshoots from one.

2. HELOPHYLLUM, Hook. f.

Densely tufted, moss-like plants. Leaves most closely imbricating, terminated by a globose knob. Flowers solitary, sessile in the uppermost leaves, 2- or 3-bracteate, white or pink, polygamo-diœcious.—Calyx-tube obconic; lobes 5–7, obtuse. Corolla-tube short; limb spreading, unequally 4–9-partite, often with thickened glands at the base of the lobes. Ovary incompletely 2-celled. Epigynous glands 2, semilunar. Fruit small, turbinate, coriaceous, 1-celled, dehiscent only by the falling away of the summit. Seeds numerous, obovoid; testa brown, coriaceous; albumen very fleshy; embryo not seen, probably most minute.

In the 'Flora Antarctica,' I regarded the first-discovered species of this genus as a section (*Helophyllum*) of *Forstera.* I did not then know the fruit, which being turbinate and indehiscent, together with the most peculiar habit of the three species now known, establishes an excellent genus. The three species here described appear very distinct in the drawings sent me by Mr. Buchanan, but I suspect they may prove forms of one very variable plant. There is much in the habit and characters of *Donatia* that approaches this genus, and I think that *Stylidieæ* are more nearly allied to *Saxifrageæ* than to any other Order.

Leaves linear, broad at the base. Column scarcely exserted. Flowers white . . . . 1. *H. clavigerum.*
Leaves broadly ovate at the base. Column much exserted. Flowers white . . . . 2. *H. Colensoi.*
Leaves linear, not broader at the base. Column included. Flowers red 3. *H. rubrum.*

1. **H. clavigerum,** *Hook. f.;—Forstera clavigera,* Fl. Ant. i. 38. t. 16. Stems 1–2 in. long, with the leaves $\frac{1}{4}$ in. diam. Leaves erect, densely imbricated all round the stem, linear-oblong, broad at the base, $\frac{1}{6}$ in. long,

thickly coriaceous; tips globose, shining, bright green, concave above, convex below, with an obscure gland on the back below the tip. Flower $\frac{1}{8}$–$\frac{1}{4}$ in. diam., white. Corolla 5–7-cleft. Column very shortly exserted. Stigmas of female flower uncinate, plumose; of male obtuse, 4-lobed.

Middle Island: alps of Otago, alt. 6000 ft., *Hector and Buchanan.* **Lord Auckland's** group and **Campbell's** Island: abundant on the hills, *J. D. H.*

2. **H. Colensoi,** *Hook. f., n. sp.—Forstera clavigera,* Fl. N. Z. i. 155. A smaller species than *H. clavigera,* with very much broader leaves, smaller flowers, and the staminal column much exserted. Flowers minute, usually 5-cleft, sunk amongst the uppermost leaves, white.

Northern Island: summits of the Ruahine range, *Colenso.* **Middle** Island: Gordon's Nob, *Munro;* summit of Wairau range, alt. 4500 ft.; crater of a volcano near the lake above Tarndale, alt. 6000 ft., and Rangitata range, *Sinclair;* Mount Torlesse, alt. 4500–6500 ft., *Haast;* Otago alps, alt. 6000 ft., *Hector and Buchanan.*

3. **H. rubrum,** *Hook. f., n. sp.* Habit and size of *H. clavigerum,* but the leaves are narrower, more coriaceous, with larger, thicker knobs. Flowers larger, dark red. Corolla unequally 7-cleft. Column included.

Middle Island: Otago, alpine, *Hector and Buchanan.*

3. STYLIDIUM, Swartz.

Herbs, usually rigid. Leaves various, chiefly radical.—Corolla irregular, 5-lobed, one lobe smaller and deflexed, the others ascending in pairs. Upper part of the staminal column bent down, and irritable at the flexure, springing up with elastic force when touched, and discharging the pollen.

A very large Australian genus, with one or two East Indian species.

Tall. Flowers spicate . . . . 1. *S. graminifolium.*
Short. Flowers solitary . . . . 2. *S. subulatum.*

1. **S. graminifolium,** *Swartz;—Fl. N. Z.* ii. 333. Stems tufted, 8–10 in. Leaves very numerous, all radical, narrow-linear, rigid, grass-like, 2–6 in. Scape 6–18 in., rather stout, pubescent and glandular, edges serrulate to the touch. Spike 3–4 in. long. Flowers rather distant, $\frac{1}{2}$ in. long, glandular. Calyx-lobes short, obtuse. Corolla-tube shorter than the ovary.—Bot. Mag. 44. t. 1918.

Northern Island: clay-hills near Auckland, *Bolton,* December, 1851. Only one specimen found, and I suspect introduced; it is a most abundant S.E. Australian and Tasmanian plant.

2. **S. (?) subulatum,** *Hook. f., n. sp.* A small, tufted, subsquarrose plant. Stems excessively short, $\frac{1}{2}$ in. long. Leaves very closely imbricate at the base, patent and recurved, $\frac{1}{2}$ in. long, narrow subulate, rigid, pungent, concave above, convex at the back. Scape extremely short, stout, (and ovary) glandular, 1-flowered. Calyx-limb unequally 4- or 5-lobed; lobes short, obtuse. Corolla-tube very short; limb irregularly 3–5-partite; lobes linear, obtuse. Column short, stout, straight? Capsule ovoid, $\frac{1}{2}$ in. long, 1-celled by the rupture of the septum. Seeds numerous, small, obovoid; testa coriaceous, rugose, brown; albumen very fleshy; embryo minute, globular.

Middle Island: Nelson mountains, *Travers; Haast.* A very singular little plant, the flowers of which are in a very imperfect state; it resembles *Colobanthus Billardieri.*

S. spathulatum, Br., introduced into A. Cunningham's and other catalogues, is an Australian plant.

Order XLI. CAMPANULACEÆ.

(*Including* Lobeliaceæ *and* Goodeniaceæ.)

Herbs, rarely shrubs. Leaves alternate, exstipulate, entire, rarely pinnatifid. Inflorescence various.—Calyx-tube adnate with the ovary; lobes 5, rarely 2–10, persistent on the fruit. Corolla regular or irregular, usually tubular or campanulate, 5-lobed, often 2-lipped, and split to the base posteriorly; lobes valvate or induplicate. Stamens 5, epigynous, rarely epipetalous; anthers free or united. Ovary inferior, 2–5-celled; style simple, often hairy at the top, 2–5-cleft, or with 2–5 stigmas, the latter sometimes surrounded with a cup, or ring of hairs; ovules few or many. Fruit a capsule berry or drupe. Seeds few or many; albumen fleshy; embryo straight.

A very large Natural Order, found in all parts of the world, but of which the tribe *Goodeniaceæ* is chiefly confined to Australia and the islands of the tropical and southern oceans.

1. Campanulaceæ.—Corolla campanulate, regular. Stamens free . 1. Wahlenbergia.
2. Lobeliaceæ.—Corolla 2-lipped, split to the base posteriorly. Anthers united.
   Tall. Leaves large. Flower racemed. Berry indehiscent . . 2. Colensoa.
   Erect. Flowers axillary. Capsule coriaceous, 3-valved at the top 3. Lobelia.
   Creeping. Flowers axillary. Berry indehiscent . . . . . . 4. Pratia.
3. Goodeniaceæ.—Corolla 1–2-lipped, posteriorly split to the base. Anthers free.
   Creeping herb. Corolla-lobes valvate Berry many-seeded . . 5. Selliera.
   Suberect, rather shrubby. Corolla-lobes induplicate. Drupe 2-celled 6. Scævola.

1. WAHLENBERGIA, Schrader.

Erect or ascending, generally slender, simple or branched herbs; juice milky. Leaves alternate. Flowers terminal, regular or nearly so, drooping in bud, white or blue, rarely reddish.—Calyx-lobes 3–5. Corolla campanulate, 5-lobed or -partite. Stamens 5, epigynous; filaments dilated at the base; anthers free. Ovary 2–5-celled; style simple, hairy at the top; stigmas 2 or 3; ovules numerous in each cell. Capsule ovoid or turbinate, opening at the top with 2–5 valves.

A very large genus in Europe, most abundant in South Africa, also found in other parts of the world.

Annual, usually branched. Stems leafy. Corolla 5-cleft . . . . 1. *W. gracilis.*
Perennial, glabrous. Leaves all radical; scape naked, 1-flowered. Corolla 5-cleft . . . . 2. *W. saxicola.*
Short, stout. Leaves thick, with broad cartilaginous margins. Corolla 5-partite . . . . 3. *W. cartilaginea.*

1. **W. gracilis,** *A. Rich.;—Fl. N. Z.* i. 159. A very slender annual, branched, glabrous, hispid or pilose herb. Stem 1–24 in. high, often ascending, angular; branches terminating in very slender, 1-flowered peduncles. Leaves ½–2 in. long; radical spathulate, petioled, toothed; cauline sessile, linear-oblong, entire toothed or sinuate, acute or acuminate, rarely spathulate; margins cartilaginous. Flowers extremely variable in size and form. Calyx-

tube ovoid; lobes 3–5, linear, long or short. Corolla $\frac{1}{8}$–$\frac{1}{2}$ in. long, blue purplish or white, 3-5-lobed. Capsule elongate, obconic ovoid or club-shaped, $\frac{1}{8}$–$\frac{1}{2}$ in. long, ribbed.—*Campanula gracilis*, Forst.; Bot. Mag. t. 691.

Var. *α*. Stem tall, generally glabrous. Flowers large. Capsule large, elongate, obconic.

Var. *β*. *capillaris*. Stem 1–8 in., covered with spreading hairs. Flowers small or minute. Capsule small, ovoid.

Abundant in dry places throughout the islands, *Banks and Solander*, etc., ascending to 4–5000 ft. An equally abundant and variable temperate and tropical Australian and Pacific Island plant, probably not different from an Indian and South African species.

2. **W. saxicola,** *A. DC.;—Fl. N. Z.* i. 160. A small, perennial, scapigerous, perfectly glabrous herb, 2–8 in. high. Leaves all radical, petiolate, 1–2 in. long, spathulate, obovate lanceolate or narrow-linear, shining, margins often white, entire or toothed. Scape naked, 1-flowered. Calyx-tube turbinate. Corolla oblique, 5-lobed, $\frac{1}{3}$–$\frac{3}{4}$ in. diam., pale-blue. Anthers linear-oblong, 1 or 2 of them unguiculate at the tip. Ovary 2- or 3-celled. Capsule obovoid.—*W. albo-marginata*, Hook. Ic. Pl. t. 818; *Campanula saxicola*, Br. Prodr.

Northern and **Middle** Islands: not uncommon in hilly and subalpine districts, ascending to nearly 6000 ft. The Blue-bell of Otago.

3. **W. cartilaginea,** *Hook. f., n. sp.* A small, low, stout, glabrous or pubescent herb, 1–3 in. high. Leaves chiefly radical, $\frac{1}{3}$–1 in. long, broadly spathulate, obtuse; petioles broad, thickly coriaceous; margins broad, white, entire, cartilaginous. Scape short, very stout, erect, naked or 1- or 2-leaved, sometimes forked, as if the plant might have a branched stem. Flower large, $\frac{1}{3}$ in. diam. Calyx-tube subspheric or obconic, lobes large, linear-oblong; margins cartilaginous, longer than the corolla, which is short, broad, and included within the calyx-lobes, and 5-partite almost to the base. Capsule turbinate.

Middle Island: Nelson mountains, in shingle-beds, *Rough;* Tarndale, alt. 4000 ft., *Sinclair;* Clarence and Wairau valleys, alt. 4–6500 ft., *Travers*. A highly curious little species. Calyx sometimes 10-lobed. Flower very sweet-scented.

2. COLENSOA, Hook. f.

A tall, glabrous, milky herb. Leaves alternate, large, membranous, with very long petioles. Flowers large, racemose.—Calyx-tube turbinate; lobes 5, subulate, equal. Corolla very long, slightly curved, tubular, split to the base down the back, 2-lipped; upper lip of 2 linear-acute lobes, one on either side the fissure, lower of 3 oblong, acute, spreading lobes. Stamens exserted; anthers cohering, pubescent, hairy towards their tips. Ovary 2-celled; style bifid; arms spreading; ovules numerous. Berry globose, fleshy and coriaceous, 2-celled, many-seeded. Seeds small, globose, attached to broad, peltate placentæ.

This fine plant was separated from *Lobelia* chiefly on account of the baccate fruit.

1. **C. physaloides,** *Hook. f. Fl. N. Z.* i. 157. Stem flexuose, branched, 2–3 ft. high, woody at the base. Leaves with petioles 3–6 in. long, ovate, acute, doubly serrate or toothed, membranous, glabrous or pilose. Racemes terminal, shorter than the leaves, 6–12-flowered; pedicels 1 in., bracteolate at the base.

Corolla 1–2 in. long, blue, pubescent. Berry ½ in. diam.—*Lobelia physaloides*, A. C.; Hook. Ic. Pl. t. 555–6.

Northern parts of the **Northern** Island, from Wangaroa to the North Cape, *Cunningham, Dieffenbach*, etc.

3. LOBELIA, Linn.

Erect or ascending, generally glabrous, milky herbs. Leaves alternate. Flowers usually racemose or axillary.—Calyx-lobes 5. Corolla-tube split to the base down the back, rarely of 3 petals, 2-lipped, upper lip of 2 lobes (or two petals when separate), one on each side the fissure, lower with 3 spreading lobes. Anthers connate, 2 upper often pilose at the tip. Ovary 2-celled; style undivided or 2-lobed. Capsule membranous or coriaceous, usually dehiscing by 2 valves at the top.

A very large genus, found in all parts of the world. The corolla is generally split but once to the base, but in *L. Roughii*, the two dorsal petals are free to the base, and it is hence split thrice to the base.

Erect, glabrous. Stems flat or 3-gonous. Flowers small, axillary . . . 1. *L. anceps.*
Small. Stems very short. Leaves broad, with deep teeth. Flowers on stout peduncles, large . . . . 2. *L. Roughii.*

1. **L. anceps,** *Thunberg;—Fl. N. Z.* i. 158. A branched, leafy, erect or ascending, glabrous, subsucculent herb, 6–12 in. high. Stems flattened, trigonous, sometimes winged. Leaves 1–3 in. long, variable in shape, spathulate linear obovate or oblong, entire sinuate or toothed, narrowed into decurrent petioles. Flowers small, ¼ in. long, pale blue, on short, solitary, axillary peduncles. Ovary elongated. Capsule linear or clavate, often ½ in. long.

Northern and **Middle** Islands: abundant in wet places, as far south as Banks's Peninsula, *Banks and Solander.* **Kermadec** Islands: *M'Gillivray*, a large leaved form.

2. **L. Roughii,** *Hook. f., n. sp.* A very short, glabrous, scapigerous plant, full of acrid, milky fluid. Stems branched, very short, slender, tortuous amongst shingle, leafy at the top. Leaves ½ in. long, shortly petioled, obovate, very deeply toothed or lobed, the sinus round, coriaceous, nerveless. Peduncles stout, erect, axillary, 1-flowered. Calyx-tube globose; lobes linear, coriaceous, obtuse, growing out as the fruit ripens. Corolla-tube 3-partite, about as long as the calyx-lobes, its lobes obtuse, short. Anthers glabrous. Capsule ovoid, globose, leathery, ¼–⅓ in. long, dehiscing by cartilaginous valves between the calyx-lobes.

Middle Island: Nelson mountains, *Rough;* Wai-au-na valley, alt. 3000 ft., and Wairau Gorge, alt. 4–6500 ft. on shingle, *Travers.* A very curious little species, its habit is that of *Wahlenbergia cartilaginea.*

4. PRATIA, Gaudichaud.

Small, creeping, herbaceous plants, with prostrate stems. Leaves small, broad, rounded or oblong, sinuate or toothed. Peduncles axillary, often long, single-flowered.—Flowers the same as in *Lobelia.* Fruit fleshy, indehiscent.

A small genus, natives of marshy places in Australia, Tasmania, India, and temperate South America.

Stems long. Leaves $\frac{1}{8}$–$\frac{1}{2}$ in., obtusely toothed or sinuate . . . . 1. *P. angulata.*
Stems short. Calyx-lobes lanceolate or subulate. Leaves $\frac{1}{12}$ in. long, sharply deeply toothed . . . . 2. *P. (?) perpusilla.*
Stems stout. Leaves $\frac{1}{4}$–$\frac{1}{3}$ in., very broad, fleshy, deeply coarsely toothed . . . . 3. *P. (?) macrodon.*
Stem slender. Leaves $\frac{1}{6}$ in., orbicular, obtusely toothed or sinuate, purple below. Calyx-lobes $\frac{1}{16}$ in., triangular, obtuse . . . . 4. *P. (?) linnæoides.*

1. **P. angulata,** *Hook. f. Fl. Antarct.* i. 41; *Fl. N. Z.* i. 157. Glabrous, very variable. Stems slender, 6–12 in. Leaves $\frac{1}{6}$–$\frac{1}{2}$ in. long, petioled or nearly sessile, orbicular, broadly oblong or obovate-oblong, obtusely sinuate-toothed, membranous. Peduncles short or long, $\frac{1}{2}$–4 in., slender. Flowers pale-blue, $\frac{1}{4}$–$\frac{1}{2}$ in. long, sometimes $\frac{1}{2}$ in. broad. Calyx-lobes linear or triangular-ovate, erect, obtuse. Anthers glabrous. Berry globose or ovoid, sometimes $\frac{1}{2}$ in. diam.—*Lobelia angulata*, Forst.; *L. littoralis*, A. Cunn.

Var. *α.* Leaves orbicular, sinuate-toothed, shortly petioled. Peduncle long and slender. Calyx-lobes short.

Var. *β.* Leaves obovate, on slender petioles, deeply toothed, acute. Peduncle long. Calyx-lobes long, almost subulate.

Var. *γ.* Leaves as in *α*, but much larger and very obscurely toothed. Peduncles very short.—*P. arenaria*, Fl. Antarct. i. 41. t. 29.

Var. *δ.* Smaller. Leaves $\frac{1}{8}$–$\frac{1}{4}$ in. long.

Abundant in watery places, moist banks, etc., *Banks and Solander*. Ascending the Southern Alps to 5000 ft. Var. *γ*. **Lord Auckland's** group. Very nearly allied to the Tasmanian *Lobelia pedunculata*, Br., and possibly only a large form of that plant, but glabrous, the leaves more petioled, less crowded, and the aspect is different.

2. **P. (?) perpusilla,** *Hook. f.*—*Lobelia perpusilla*, Fl. N. Z. i. 158. Very minute, glabrous. Stems matted, stout for the size of the plant, 3–4 in. Leaves $\frac{1}{12}$ in. long, sessile, oblong, acute, deeply toothed. Flowers almost sessile, $\frac{1}{4}$ in. long. Calyx-tube hairy; lobes ovate-subulate, recurved. Anthers glabrous. Fruit not seen.

Northern Island: Hawke's Bay, in muddy places, *Colenso*. The fruit being unknown, I am doubtful as to the genus; but the habit is altogether that of *Pratia*. Allied to the Tasmanian *Lobelia irrigua*, but a very much smaller plant.

3. **P. (?) macrodon,** *Hook. f., n. sp.* Perfectly glabrous. Stems matted, short, rather stout. Leaves crowded, very shortly petioled, $\frac{1}{4}$–$\frac{1}{3}$ in. long, broadly obovate-orbicular, cuneate at the base, coriaceous, deeply coarsely 6–8-toothed. Peduncle short. Flower large. Calyx-lobes subulate-lanceolate. Corolla-tube cylindric, $\frac{1}{4}$–$\frac{1}{3}$ in. long, broadest at the base. Fruit not seen.

Middle Island: Southern Alps, Discovery Peaks, Acheron and Clarence rivers, alt. 5500 ft., *Travers*; summit of Mount Torlesse, alt. 4500–6000 ft., *Haast*. This again is doubtful as to genus till the fruit is known; it differs much from *P. angulata* in the more coriaceous (perhaps fleshy), deeply toothed leaves, large almost sessile flower, and long corolla-tube dilated below.

4. **P. (?) linnæoides,** *Hook. f., n. sp.* Glabrous; stem slender, creeping, 1–3 in. Leaves coriaceous, very shortly petiolate or sessile, orbicular, $\frac{1}{6}$ in. diam., coarsely obtusely sinuate-toothed, often purple below. Scape slender,

dark-coloured, $1\frac{1}{2}$–2 in. Calyx-lobes very small, broadly triangular, obtuse, $\frac{1}{16}$ in. long. Corolla $\frac{1}{3}$ in. long, as in *P. angulata*.

Middle Island: Observatory Hill, Macaulay river, alt. 4500 ft., *Haast;* Otago, subalpine, Lindis Pass, *Hector and Buchanan.* This approaches small-leaved states of *P. angulata*, but the coriaceous or somewhat fleshy leaves, and very small calyx at once distinguish it, whether as a species or variety.

5. SELLIERA, Cavanilles.

A small, glabrous, creeping, rather fleshy herb. Leaves narrow-linear or lanceolate. Peduncles axillary, 1- or 2-flowered.—Calyx-lobes 5, equal. Corolla 1-lipped, split posteriorly to the base; lobes 5, ovate, acute, valvate, not winged. Stamens 5, epigynous; anthers free. Ovary 2-celled; style simple, stigma in a 2-lipped cup; ovules many. Berry 2-celled, indehiscent. Seeds numerous, imbricating upwards, compressed.

1. **S. radicans,** *Cav.*—*Goodenia repens*, Lab.;—Fl. N. Z. i. 156. Stems succulent, 2–10 in. long. Leaves $\frac{1}{2}$–5 in. long, linear-spathulate or linear, obtuse, quite entire, nerveless, petiole half-clasping the stem. Peduncles axillary, solitary or several together, 1- or 2-flowered, with 2 subulate bracts above the middle. Flower $\frac{1}{3}$ in. long. Berry very variable in size.—Lab. Fl. Nov. Holl. i. 53. t. 76.

Abundant in salt marshes throughout the islands, *Banks and Solander*, etc.; Otago, Lower Waitaki river, *Hector and Buchanan*, apparently far from the sea. Also common in Tasmania, South Australia, and Chili.

6. SCÆVOLA, Linn.

Erect or ascending, shrubby or half-shrubby plants. Flowers axillary or spiked.—Calyx-lobes 5, equal or very unequal, sometimes obsolete. Corolla split to the base posteriorly, 1-lipped; lobes induplicate, winged. Stamens 5; anthers free. Ovary 2-(rarely 1–4)-celled; style simple; stigma in a cup; ovules erect, solitary in each cell. Drupe dry or fleshy, usually 2-celled, with 1 erect seed in each cell.

A rather extensive genus in Australia and the Pacific Islands.

1. **S. gracilis,** *Hook. f. in Journ. Linn. Soc. Bot.* i. 129. A procumbent shrub, covered with silky hairs. Leaves lanceolate or oblong-lanceolate, 1–3 in. long, acute, serrate, hairy on both surfaces. Flowers axillary, $\frac{2}{3}$ in. long, subsessile, on short branchlets, with 4 linear-lanceolate bracts at their base. Calyx silky, bracteolate; lobes 5, very unequal, 3 subulate, 2 intermediate very short. Corolla yellow, its lobes very long, linear; tube villous within.

Kermadec Islands, *M'Gillivray.*

S. Novæ-Zelandiæ, A. Cunn., is *Hymenanthera crassifolia*, Hook. f.

Order XLII. ERICEÆ.

(*Including* Epacrideæ.)

Shrubs or trees. Leaves evergreen and coriaceous in all the New Zealand species, simple, exstipulate. Inflorescence various. Flowers usually white, bracteate.—Sepals inferior, usually 5, free or united into a 5-partite calyx, imbricate. Corolla usually tubular or campanulate, glabrous or bearded on the throat or on the lobes; lobes 5, generally short, imbricate or induplicate-valvate. Stamens 5–10, hypogynous or epipetalous. Anthers 1- or 2-celled, cells awned at the back or tip or awnless, opening by slits or terminal pores. Disk 5–10-lobed, or of 5 scales. Ovary 1–12-celled; style simple, stigma capitate or truncate, simple or lobed. Ovules 1 or many in each cell. Fruit a capsule berry or drupe, free or enclosed in the fleshy calyx. Seeds minute; testa reticulate; albumen copious, fleshy; embryo small.

A large Order, found in all parts of the world.

SUBORDER I. **Ericeæ.**—*Stamens in the New Zealand genera hypogynous; anthers 2-celled, opening by pores.*

Capsule dry, 5-valved, often enclosed in the fleshy calyx . . . . . 1. GAULTHERIA.
Berry fleshy, with the small withered calyx at its base . . . . . . 2. PERNETTYA.

SUBORDER II. **Epacrideæ.**—*Stamens in the New Zealand genera epipetalous; anthers 1-celled.*

Fruit a drupe with a 1–10-celled bony nut; cells with 1 pendulous seed.
Pedicels covered with imbricating bracts. . . . . 3. CYATHODES.
Pedicels with few bracts placed close under the calyx. . . . . . 4. LEUCOPOGON.
Fruit a drupe, with 5 or more minute 1-seeded nuts . . . . . . 5. PENTACHONDRA.
Fruit a many-seeded capsule.
Leaves not amplexicaul (except *E. purpurascens*). Flowers solitary.
Pedicels covered with imbricating bracts . . . . 6. EPACRIS.
Leaves not amplexicaul. Flowers racemed. Bracts few or 0 . . 7. ARCHERIA.
Leaves with broad, sheathing bases . . . . 8. DRACOPHYLLUM.

1. GAULTHERIA, Linn.

Shrubs. Leaves coriaceous, toothed, alternate, rarely opposite, never sheathing at the base. Flowers white or pink, axillary solitary or racemose.—Calyx 5- or 6-lobed or -partite, often becoming fleshy and enclosing the capsule. Corolla ovoid or urceolate; mouth contracted; lobes 5, recurved. Stamens 10, included, hypogynous; filaments flat; anthers opening by pores, each pore with 1 or 2 awns. Disk cup-shaped, 10-lobed, or of 10 glands. Ovary 5-celled; cells many-ovuled. Capsule free or enclosed in the baccate calyx, 5-valved loculicidally; valves separating from a central axis which bears the seeds; sometimes the capsule becomes fleshy and indehiscent.

A large genus, especially in the American and Indian mountains, unknown in Europe and Africa, found also on the Australian and Tasmanian alps. The baccate calyx is a variable character in New Zealand, occurring in *G. antipoda*, sometimes on the same fruiting raceme with simple calyces, in which plant further, dry dehiscent, and baccate indehiscent capsules occur also on the same branch, thus uniting the characters of *Pernettya* and *Gaultheria*.

Leaves alternate. Flowers axillary, or racemes leafy . . . . 1. *G. antipoda.*
Flowers racemose. Leaves lanceolate oblong or rounded . . . . 2. *G. rupestris.*
Flowers racemose. Leaves ovate-oblong, cordate . . . . 3. *G. fagifolia.*
Leaves opposite, cordate. Flowers racemose . . . . 4. *G. oppositifolia.*

1. **G. antipoda,** *Forst.;—Fl. N. Z.* i. 161. A rigid bush, erect or prostrate, extremely variable in stature and habit. Branches pubescent, covered with scattered black or yellow-brown bristles. Leaves very coriaceous, veined, shortly petioled, orbicular oblong-lanceolate or linear-lanceolate, acute

obtuse or acuminate, $\frac{1}{2}$–$\frac{2}{3}$ in. long in *α* and *β*, $\frac{1}{6}$–$\frac{1}{3}$ in *δ*, serrate, the teeth sometimes ending in a bristle. Flowers small, white or pink, axillary and solitary, or crowded towards the ends of the branchlets, which form leafy racemes; peduncles curved, pubescent. Calyx rarely 6-lobed, lobes red at the tips. Corolla $\frac{1}{10}$ in. long or more. Capsule with the enlarged calyx $\frac{1}{2}$ in. diam., purple red or white; sometimes the capsule is itself baccate, and the calyx unchanged, at others neither capsule nor calyx are fleshy.

Var. *α*. Erect. Branches pubescent. Leaves broad.

Var. *β*. Erect. Leaves oblong or lanceolate. Flowers small, almost racemed, on slender, glabrate pedicels.

Var. *γ*. Depressed or prostrate. Leaves orbicular. Branchlets covered with brown, stiff hairs. Flowers axillary. Fruit very large.—Fl. Tasman. i. 241. t. 73.

Var. *δ*. Prostrate, small. Leaves ovate or lanceolate, very small.

Var. *ε*. Erect or prostrate. Leaves lanceolate, serrate, with the teeth terminating in bristles.

Throughout the islands, abundant, *Banks and Solander*, etc.; var. *β*, descending to nearly the level of the sea, as far north as the Bay of Islands; the other varieties more or less alpine. Var. *γ*, ascending to 5000 ft. Found also on the Tasmanian alps.

2. **G. rupestris,** *Br.;—Fl. N. Z.* i. 162. *t.* 42. A bush or small tree. Branches stout, glabrous or pubescent, sometimes hispid towards the tips. Leaves close-set, alternate, $\frac{1}{3}$–1 in. long, from ovate-obtuse to oblong or oblong-lanceolate and acute, crenulate or serrulate, glabrous, extremely coriaceous, reticulate on both surfaces. Racemes short or long, $\frac{1}{5}$–1$\frac{1}{2}$ in. long, terminal and axillary, sometimes fastigiate at the ends of the branchlets, few- or many-flowered; pedicels longer or shorter than the bracteoles. Calyx-lobes ovate, acute, often thickening in fruit. Capsule dry.—*Andromeda rupestris,* Forst.

Var. *α*. Leaves oblong-lanceolate, acute, serrulate. Branchlets glabrous.

Var. *β*. Leaves smaller, oblong. Branchlets glabrous.

Var. *γ*. Leaves as in *α*, but more membranous. Branchlets pubescent and setose.

Var. *δ*. Leaves $\frac{1}{4}$ in. long, nearly orbicular. Branchlets glabrous.—*G. Colensoi,* Hook. f. Fl. N. Z. i. 162.

Var. *ε*. Leaves oblong, 1 in. long, obtuse. Branchlets hispid, setose at the ends. Calyx baccate. ? A hybrid between *rupestris* and *antipoda.*

Northern and **Middle** Islands: abundant on all the mountains, ascending to 5000 ft., and excessively variable, *Banks and Solander*, etc. Var. *ε* is a singular plant; with the habit and baccate calyx of *G. antipoda*, it has the racemose inflorescence of *G. rupestris*; I have it in fruit only, and the locality is doubtful. This species covers large tracts of ground in the lake district of Otago, and the fruit, which is white or red, is eaten by the ground-parrots.

3. **G. fagifolia,** *Hook. f. Fl. N. Z.* i. 162. A shrub, 4–5 ft. high. Branchlets sparingly setulose. Leaves petiolate, $\frac{3}{4}$ in. long, ovate-oblong, cordate at the base, crenate-serrate, coriaceous, waved. Racemes $\frac{1}{2}$ in. long. Flowers $\frac{1}{8}$ in. long. Calyx-lobes ovate-acute, not enlarging after flowering.

Northern Island: Motukino, east of Lake Taupo, *Colenso.* I have but two specimens of this most distinct species.

4. **G. oppositifolia,** *Hook. f. Fl. N. Z.* i. 162. *t.* 43. A large shrub. Branchlets glabrous or setose. Leaves 1$\frac{1}{2}$–1$\frac{3}{4}$ in. long, sessile, cordate, oblong, obtuse or acute, serrulate or crenate, glabrous or setulose on both surfaces. Racemes axillary and terminal, sometimes panicled (with opposite

branches) at the ends of the branches. Flowers numerous, small, as in *G. rupestris*. Calyx-lobes not becoming fleshy. Capsule small.

Northern Island: Mount Egmont, *Bidwill;* cliffs between Hawkes' Bay and Taupo, *Colenso.*

2. PERNETTYA, Gaud.

Characters of *Gaultheria,* but the calyx is unchanged, or but slightly enlarged after flowering, and the fruit is baccate instead of capsular.

A large South American genus, and there confined to the Andes, Chili and Fuegia.

1. **P. tasmanica,** *Hook. f. Fl. Tasm.* i. *t.* 73. A very small, creeping, perfectly glabrous plant. Stems slender, 2–3 in. long, without hairs or bristles. Leaves $\frac{1}{6}$ in. long, sessile, oblong, subacute, subveined, coriaceous, very obscurely crenate. Flowers axillary, solitary, larger than the leaves. Pedicels as long as the leaves, with 3–5 obtuse, imbricate bracts at the base. Corolla campanulate.

Middle Island: amongst grass on the Hopkins river, alt. 2500 ft., *Haast.*

This is a most puzzling plant, and seems to unite the genera *Pernettya* and *Gaultheria,* as much as do the varieties of *G. antipoda,* mentioned under that plant. The Tasmanian specimens have no awns to the anther-cells, the New Zealand ones have very short awns: the plants are otherwise undistinguishable. They differ from the small states of *G. antipoda,* with baccate fruit and slightly swollen calyx, only in the minute more obtuse leaves, and total absence of black bristles on the branches; and these latter are so very rare in some states of *G. antipoda,* that I cannot doubt but that all will prove one plant, however anomalous this opinion may appear.

3. CYATHODES, Br.

Shrubs. Leaves acerose, rigid, pungent or oblong and obtuse, glaucous, parallel-veined below, and not sheathing at the base. Flowers small, white or yellowish.—Pedicels covered with bracts, which become larger upwards, and hide the calyx. Corolla funnel-shaped or urceolate; tube scarcely longer than the calyx, glabrous; lobes glabrous or bearded. Ovary with 3–10 1-ovuled cells. Drupe with a bony 3–10-celled nut.

A large Australian and Pacific island genus.

Leaves $\frac{1}{6}$–$\frac{1}{2}$ in., very narrow, with rigid, pungent points . . . . . . 1. *C. acerosa.*
Leaves $\frac{1}{2}$–$\frac{2}{3}$ in., narrow, linear-oblong, subacute . . . . 2. *C. robusta.*
Leaves $\frac{1}{8}$–$\frac{1}{6}$ in., narrow, linear, patent or recurved, obtuse . . . . . 3. *C. empetrifolia.*
Leaves $\frac{1}{8}$–$\frac{1}{6}$ in., linear or oblong, suberect, obtuse. Corolla bearded . 4. *C. Colensoi.*

1. **C. acerosa,** *Br.;—Fl. N. Z.* i. 163. A large shrub or small tree, with blackish branches. Leaves spreading, about $\frac{1}{6}$–$\frac{1}{2}$ in. long, acerose or very narrow linear or lanceolate, pungent, margins often recurved and ciliate, under side glaucous, with 3–7 parallel veins, of which the outer often branch outwards. Flowers solitary, minute. Bracts and calyx-lobes obtuse. Corolla glabrous; lobes spreading. Drupe red, $\frac{1}{8}$–$\frac{1}{3}$ in. diam.

Var. *α.* Leaves usually only $\frac{1}{4}$ in. long, linear, lateral nerves often branching outwards.—*C. acerosa,* Br.

Var. *β.* Leaves often $\frac{1}{3}$–$\frac{3}{4}$ in. long, with longer pungent points, nerves all simple.—*C. oxycedrus,* Br.;—Fl. N. Z. i. 164.

Abundant throughout the islands, *Banks and Solander,* etc.; ascending the mountains in a stunted form. The same branch bears minute, dry, and large succulent drupes. Both varieties are common New Zealand plants.

2. **C. robusta,** *Hook. f., n. sp.—C. acerosa, β, latifolia,* Fl. N. Z. i. 163. A much larger and more robust plant than *C. acerosa.* Leaves spreading, $\frac{1}{2}$–$\frac{2}{3}$ in. long, narrow linear-oblong, $\frac{1}{6}$ in. broad, subacute, not pungent, 5–11-nerved below, the outer nerves sometimes branching outward.

Chatham Island, *Dieffenbach* and *Herb. Mueller.* I was at first disposed to refer this to a form of *C. acerosa,* but more specimens, received from Dr. Mueller, quite agreeing with Dieffenbach's, and equally without the pungent tips to the leaves, seem to indicate its specific distinctness. I have seen no flowers; the drupe is $\frac{1}{3}$ in. diam.

3. **C. empetrifolia,** *Hook. f. Fl. N. Z.* i. 164. A small, procumbent, alpine, straggling plant, with slender, leafy, tomentose branches, 8–24 in. long. Leaves erect, spreading or recurved, $\frac{1}{8}$–$\frac{1}{6}$ in. long, linear, obtuse, glabrous, pilose or pubescent, convex above, glaucous below; margins recurved, ciliated; midrib stout; veins 0. Flowers minute, axillary or terminal, solitary or 2 or 3 together. Peduncle very short. Corolla-tube not longer than the calyx; lobes glabrous, acute. Drupe 3–5-celled.—*Androstoma empetrifolia,* Fl. Antarct. i. 44. t. 30.

Common on the mountains of the **Northern** and **Middle** Islands, and of **Lord Auckland's** group and **Campbell's** Island.

4. **C. Colensoi,** *Hook. f.—Leucopogon Colensoi,* Fl. N. Z. i. 165. Stems robust, prostrate, with erect or ascending, puberulous, leafy branches, 4–10 in. high. Leaves erect or suberect, $\frac{1}{8}$–$\frac{1}{6}$ in. long, linear-oblong or obovate-oblong, obtuse, glabrous, convex above, glaucous below, with 3 stout nerves, the outer branching towards the margin; margins thin, cartilaginous or membranous, especially towards the tips. Flowers in short 3–5-flowered racemes. Corolla-lobes bearded. Drupe white or red, 5-celled.

Northern Island: base of Tongariro, Lake Taupo, Ruahine range, etc., *Colenso.* **Middle** Island: Gordon's Nob and Fairfield Downs, *Munro;* Wairau mountains, alt. 4–5500 ft., *Travers;* Common on the Southern Alps, *Haast;* Otago, lake district, alt. 2000 ft., *Hector and Buchanan.* Intermediate between *Cyathodes* and *Leucopogon* in characters, but with the habit of the former genus, to which I have referred it, both on this account and because of its extremely close affinity with *C. Tamaiameiæ,* Cham., of the Sandwich Islands; it is also most closely allied to *Leucopogon suaveolens* of the Borneo alps, which may be a *Cyathodes.*

4. LEUCOPOGON, Br.

Shrubs, sometimes very small, or trees. Leaves imbricating or scattered, coriaceous, parallel-veined below, not sheathing at the base. Flowers solitary spiked or racemose, white or pink.—Pedicels with 2 or 3 bracts below the calyx. Corolla tubular, campanulate or funnel-shaped; lobes spreading, bearded. Ovary with 2–10 1-ovuled cells. Drupe with a bony 1–10-celled nut.

Chiefly an Australian genus, but also found on the mountains of the Malayan and Pacific Islands.

Leaves subwhorled, spreading. Flowers minute, spiked . . . . . 1. *L. fasciculatus.*
Leaves imbricate, with long pungent tips. Flowers solitary, large . . 2. *L. Frazeri.*

1. **L. fasciculatus,** *A. Rich.;—Fl. N. Z.* i. 164. A large shrub or small tree, with slender, spreading, puberulous branchlets. Leaves somewhat whorled, patent, $\frac{1}{3}$–1 in. long, linear-lanceolate or oblong, obtuse or acuminate, flat, glabrous, striated above, glaucous and obscurely veined below; margins denticulate or ciliolate. Flowers minute, on drooping, filiform, fascicled

spikes, crowded or distant, greenish; bracts and calyx-lobes obtuse. Drupe small, hard, fleshy, oblong, 2-celled.—*Epacris fasciculatus*, Forst.

Var. *α*. Leaves linear-lanceolate, acute or acuminate.

Var. *β*. Leaves broader, oblong, hardly acute.—*L. brevilabris*, Stsch., in Bull. Soc. Nat. Hist. Mosc. xxxii.

Abundant throughout the islands, *Banks and Solander*, etc. Var. *β*, in mountainous districts.

2. **L. Frazeri,** *A. Cunn.;—Fl. N. Z.* i. 165. A very small, erect or ascending plant, 2–4, rarely 6–8 in. high, branches often curving, densely covered with imbricating leaves. Leaves close-set, $\frac{1}{6}$–$\frac{1}{4}$ in. long, obovate-oblong or linear-oblong, with long, pungent, mucronate tips, glabrous above, margins flat or recurved, serrulate, cartilaginous, glaucous below, the nerves branching outwards. Flowers solitary, axillary, sessile, large; bracts short, broad. Calyx-lobes lanceolate, twice as long as the bracts. Corolla tubular, $\frac{1}{2}$ in. long; lobes short. Drupe 1- or more celled.—*L. nesophilus*, DC. Prodr. vii. 752; *L. Bellignianus*, Raoul, Choix, 18. t. 12.

Abundant throughout the islands in dry soil, ascending to 5000 ft., *Banks and Solander*, etc. The sweetish orange drupe is edible. Also found on the Tasmanian and Victorian alps, and very closely allied to a Bornean species.

5. PENTACHONDRA, Br.

Small, alpine, procumbent, straggling or cæspitose plants. Leaves imbricating.—Flowers solitary, axillary, sessile. Bracts 4 or more. Corolla tubular or funnel-shaped; lobes spreading, densely bearded within. Drupe with 5 or more small 1-seeded nuts.

A small genus, natives of the alps of Australia, Tasmania, and New Zealand.

1. **P. pumila,** *Br.;—Fl. N. Z.* i. 166. Stems woody, procumbent, with numerous, short, ascending, tufted, leafy branches, 1–4 in. high. Leaves close-set, suberect, imbricating, $\frac{1}{8}$ in. long, oblong, obtuse, concave, margins ciliate, glossy on both surfaces, glabrous or ciliated, striated, 3–5-nerved. Flowers twice as large as the leaves. Bracts and calyx-lobes short, ciliated. Corolla tubular; lobes 5, short, densely bearded. Berry large, often $\frac{1}{2}$ in. long, red, succulent. Nuts 5 or more, small, almost reniform.—*Epacris pumila*, Forst. Prodr.

Northern and **Middle** Islands: abundant on the mountains, *Forster;* ascending to 5500 ft. in the Nelson ranges. Also found on the Victorian and Tasmanian alps.

6. EPACRIS, Smith.

Shrubs, often small, or small trees. Leaves shortly petioled, usually closely imbricating, rarely sheathing at the base.—Flowers solitary, shortly peduncled; bracts numerous, imbricating, covering the peduncle and concealing the base of the calyx. Corolla tubular or bell-shaped; limb not bearded. Capsule dry, 5-celled, 5-valved; cells with numerous seeds attached to a central placenta.

A very large Australian and Tasmanian genus, found also in New Zealand, but nowhere else.

Leaves with long, pungent points . . . . 1. *E. purpurascens.*
  Leaves not pungent, acuminate. Bracts and sepals acute . . . 2. *E. pauciflora.*
  Leaves obtuse. Bracts and sepals acute . . . . 3. *E. Sinclairii.*
  Leaves ovate obtuse. Bracts and sepals obtuse . . . . 4. *E. alpina.*

1. **E. purpurascens,** *Br.* An erect, tall, sparingly-branched shrub, 6 ft. high; branches flexuose, ½–1 ft. long, densely clothed throughout with closely imbricating, sheathing, spreading, recurved, pungent leaves. Leaves ½ in. long, coriaceous, very convex, sheathing by their lower part, but attached by a small broad petiole, broadly ovate-cordate, suddenly contracting into the rigid patent needle-like tip, quite glabrous. Flowers very numerous, often one in the axil of every leaf for a large portion of the branches; bracts and sepals ovate-lanceolate, acuminate, pungent. Corolla ¼ in. long; lobes ovate, acute.—*E. pungens*, Bot. Mag. t. 844.

Northern Island: Papakura, 18 miles from Auckland, *Sinclair*. A New South Wales plant, and I cannot but suspect introduced (like *Stylidium graminifolium*) into New Zealand, but Dr. Sinclair, with whom I had a good deal of correspondence on the subject, regarded it as indigenous.

2. **E. pauciflora,** *A. Rich.;—Fl. N. Z.* i. 166. A glabrous, erect, twiggy shrub, 1–2 ft. high, branches puberulous, stoutish, leafy. Leaves very coriaceous, suberect, imbricating, ⅛–⅕ in. long, concave, broadly ovate or oblong-lanceolate, suddenly acuminate, obtuse at the tip, nerveless, quite smooth and glabrous on both surfaces. Flowers small, white, scarcely longer than the leaves, numerous towards the tips of the branches; bracts very numerous, ovate, acute, closely imbricating in 5 or 6 ranks. Corolla with a very short tube, and patent, broad, rounded lobes. Capsule small.

Northern and **Middle** Islands: on dry hills and in swampy grounds, *Banks and Solander*, etc. Common as far south as Nelson, very near the Tasmanian *E. virgata.*

3. **E. Sinclairii,** *Hook. f., n. sp.* A foot high, much branched; branches stout, puberulous, leafy. Leaves erect, imbricating, ⅙–¼ in. long, densely coriaceous, narrow lanceolate-oblong or oblong, obtuse, smooth and glabrous on both surfaces. Bracts, calyx, etc., as in *E. pauciflora.*

Northern Island, *Herb. Sinclair.* Numerous fine specimens of this are in Sinclair's Herbarium, but without ticket; it resembles *E. pauciflora*, but is a much less twiggy and more branched plant, with narrower, less concave not acuminate leaves.

4. **E. alpina,** *Hook. f. Fl. N. Z.* i. 167. A small, alpine, tufted or straggling shrub, 6–10 in. high, much branched, erect or decumbent; branches puberulous. Leaves spreading or suberect, very coriaceous, ⅕–⅙ in. long, broadly ovate, obtuse, glabrous and smooth on both surfaces. Flowers axillary towards the ends of the branches; bracts and calyx-lobes broadly ovate, obtuse. Corolla not seen.

Northern Island: base of Tongariro and Lake Taupo, *Bidwill, Colenso.* **Middle** Island: Southern Alps, growing with *Pentachondra, Sinclair and Haast.* A smaller and more straggling plant than either of the preceding, with obtuse bracts and calyx-lobes; very closely allied to the Tasmanian *E. serpyllifolia*, but the flowers are different, and leaves not mucronulate.

7. ARCHERIA, Hook. f.

Branched shrubs, with coriaceous, evergreen leaves. Flowers in the New Zealand species in terminal racemes. Bracts few or 0. Corolla as in *Epacris.* Capsule deeply 5-lobed. Seeds ascending from basilar or subbasilar placentæ.

This genus was established in the Tasmanian Flora, for a set of peculiar *Epacrideæ*, differing from *Epacris* in the absence of bracts, the usually more deeply lobed ovary, with basilar placentas, and was divided into two sections, one with axillary flowers and long styles, the other (to which the New Zealand species belong) with racemed flowers and short styles.

Leaves $\frac{1}{3}$–$\frac{1}{2}$ in., narrow linear or linear-lanceolate . . . . 1. *E. Traversii.*
Leaves 1 in., broadly obovate or oblong . . . . 2. *E. racemosa.*

1. **A. Traversii,** *Hook. f.*, *n. sp.* A small (?) shrub with slender spreading branches. Leaves loosely set, very spreading, $\frac{1}{3}$–$\frac{1}{2}$ in. long, narrow linear-lanceolate, acute, quite glabrous, smooth and shining on both surfaces, margin recurved, midrib very thick below. Flowers small, in very short, terminal, puberulous racemes. Bracts few, deciduous, oblong, obtuse. Sepals oblong; margins membranous. Corolla not seen. Capsule minute.

Middle Island: woods, Aorere valley, alt. 1400 ft., "not observed on the Nelson side of the valley," *Travers.*

2. **A. racemosa,** *Hook. f.;—Epacris racemosa,* Fl. N. Z. i. 167. A shrub. Branches very slender. Leaves scattered in fascicles or almost whorled, very spreading, 1 in. long, $\frac{1}{3}$ broad, elliptic- or obovate-oblong, acute, flat, nerveless. Raceme $\frac{1}{2}$ in. long, downy; bracts and calyx-lobes oblong; margins membranous. Corolla $\frac{1}{6}$ in. diam.; tube short, broad; lobes broad, ovate, obtuse.

Great Barrier Island, *Rough.*

8. DRACOPHYLLUM, Lab.

Shrubs or trees, sometimes prostrate or tufted. Leaves long, rigid or grassy, usually crowded at the ends of the branchlets, their bases broad, sheathing, suddenly contracting into a long subulate, usually concave, very narrow blade, which tapers from the base to the tip.—Flowers in axillary or terminal branched panicles racemes or spikes, rarely solitary; pedicel bracteate. Sepals 5, ovate or lanceolate, persistent, longer or shorter than the corolla. Corolla tubular or campanulate, usually white; lobes 5, spreading, ovate or lanceolate, obtuse, their tips more or less inflexed. Anthers 5, sessile at the mouth of the corolla. Disk of 5 erect scales. Ovary 5 -or 6-celled; style shortish, stout; ovules numerous in each cell, attached to a pendulous placenta. Capsule shorter than the sepals, 5- or 6-celled, 5- or 6-valved.

A large genus in New Zealand, having several representatives in Tasmania, temperate Australia and New Caledonia.

1. *Leaves patent or recurved* (*or suberect in* 3). *Flowers panicled or spiked.*

Flowers in terminal panicles, 8–16 in. long. Corolla $\frac{1}{3}$ in. . . . 1. *D. latifolium.*
Flowers in lateral panicles, 3–6 in. long. Corolla $\frac{1}{4}$ in. . . . . 2. *D. Menziesii.*
Flowers in terminal panicles, 2–4 in. long. Corolla $\frac{1}{8}$ in. . . . 3. *D. strictum.*
Flowers few, spiked. Leaves 2–5 in., pungent . . . . 4. *D. squarrosum.*
Flowers in capitate spikes. Leaves $\frac{2}{3}$–1 in., obtuse . . . . . 5. *D. recurvum.*

2. *Leaves erect, with acicular or pungent tips. Flowers spiked* (*solitary in* 9).

Leaves 4–9 in., pubescent or glabrate. Sheath $\frac{1}{3}$–$\frac{1}{2}$ in. broad . 6. *D. longifolium.*
Leaves 1–4 in., glabrous. Sheath $\frac{1}{8}$–$\frac{1}{6}$ in. broad . . . . . . 7. *D. Urvilleanum.*
Leaves $\frac{1}{4}$–$\frac{3}{4}$ in., glabrous or puberulous. Flowers 2 or 3 . . . . 8. *D. subulatum.*
Leaves $\frac{1}{2}$–$\frac{2}{3}$ in., glabrous or puberulous. Flower solitary . . . . 9. *D. uniflorum.*

3. *Leaves erect, not pungent, obtuse at the very tip. Flowers solitary or 2- or 3-spiked.*

Leaves $\frac{1}{6}$–1 in. Flowers lateral . . . . 10. *D. rosmarinifolium.*
Leaves $\frac{1}{10}$ in. Flower solitary, terminal . . . . 11. *D. muscoides.*

1. **D. latifolium,** *A. Cunn.;—Fl. N. Z.* i. 167. A small tree, 8–15 ft. high. Bark black. Leaves spreading and squarrose, 10–24 in. long, 1½ broad at the base, serrulate, gradually tapering into very long, fine points, very concave, rarely nearly flat. Panicle terminal, very large and dense, 8–16 in. long, narrow oblong, cernuous in fruit; rachis and branches stout, pubescent. Flowers innumerable, densely crowded, ⅕ in. long. Sepals broadly ovate, obtuse, a quarter the length of the shortly campanulate corolla. Style short, stout. Capsule ⅛–⅙ in. diam.

Northern Island: common in woods from the Bay of Islands to Auckland, *Banks and Solander*, etc.

2. **D. Menziesii,** *Hook. f. Fl. N. Z.* i. 168. A small tree, Leaves similar to those of *D. latifolium*, but less serrulate, only 6–8 in. long, and ⅔ broad at the base. Panicle lateral, 3–6 in. long, pubescent, cernuous in fruit, sparingly branched, not very many-flowered. Flowers on curved peduncles, ¼ in. long. Sepals broadly ovate, acute, ¼ shorter than corolla. Style long and stout. Capsule nearly ¼ in. diam.

Middle Island. Dusky Bay, *Menzies;* Port Preservation, *Lyall;* Otago, lake district, *Hector and Buchanan.*

3. **D. strictum,** *Hook. f. Fl. N. Z.* i. 168. A small (?) shrub. Leaves strict, patent or suberect, sword-shaped, flattish, 1½–3 in. long, ⅛–⅓ in. broad at the sheathing base; margin scarcely serrulate. Panicle terminal, 2–4 in. long, puberulous. Flowers rather numerous, ⅙ in. long. Sepals broadly ovate, subacute, ⅓ as long as the corolla. Style short, stout. Capsule $\frac{1}{10}$ in. diam.

Var. *α*. Larger. Leaves suberect, 2–3 in. long, ⅓ broad at the base.
Var. *β*. Smaller. Leaves patent, 2 in. long, ¼ broad at the base.—*D. affine*, Hook. f. Fl. N. Z. i. 168.

Northern Island. Var. *α*, Tongariro, *Bidwill;* var. *β*. *Dieffenbach.* **Middle** Island: Otago, lake district, *Hector and Buchanan;* var. *β*, Southern Alps, *Sinclair and Haast.*

4. **D. squarrosum,** *Hook. f. Fl. N. Z.* i. 169. A shrub, with branchlets as thick as a crow-quill. Leaves patent and recurved, 2–5 in. long, ⅙–¼ in. broad at the sheathing base, which is not auricled, and gradually attenuated, rather soft and grass-like, margins of young leaves serrulate, ciliolate towards the base. Flowers, ¼ in. long, in short, simple, 3–5-flowered, spiciform racemes ½–1 in. long. Sepals equalling the tube of the corolla, ovate-lanceolate, acuminate, ciliate; corolla-lobes ovate-lanceolate. Style rather long.

Northern Island: east coast, *Banks and Solander;* Auckland, *Sinclair;* Manakau Bay, *Colenso.*

5. **D. recurvum,** *Hook. f. Fl. N. Z.* i. 171. A small, erect, much branched shrub, 1 ft. or more high. Leaves much recurved, ⅔–1 in. long, sheathing base membranous, ⅙–¼ in. broad, suddenly contracted into a rigid, concave, subulate, recurved lamina, $\frac{1}{10}$ in. broad at the base, obtuse at the very tip, minutely serrulate, almost keeled at the back. Flowers ¼ in. long, in terminal, oblong, bracteate spikes ½ in. long; bracts foliaceous, almost hiding the flowers. Sepals ovate-lanceolate, acute, as long as the corolla-tube. Corolla-lobes ovate, acute. Capsules small, shorter than the sepals.

Northern Island: Tongariro, *Bidwill;* Mount Hikurangi and tops of the Ruahine range, *Colenso.* Probably a recurved-leaved form of *D. rosmarinifolium*, but it looks very different in the foliage and dense heads of flowers.

6. **D. longifolium,** *Br.;—Fl. N. Z.* i. 169, and *D. Lyallii*, Hook. f. l. c. A small tree, with black bark. Leaves erect, 3–9 in. long, sheath ⅓ in. broad, lamina at the base ⅛–⅙ in. broad, concave, rigid, pubescent above, rarely glabrous, ciliate at the margins. Racemes lateral 1–1½ in. long, 6–12-flowered. Flowers crowded, ¼ in. long. Sepals broadly ovate, acute, ciliated, longer than the corolla-tube.—*Epacris longifolia*, Forst.

Middle Island: Dusky Bay, *Forster;* Thomson's Sound, *Lyall;* Otago, *Lindsay;* Dunedin, ascending to 3000 ft., *Hector and Buchanan;* Southern Alps, above the *Fagus* forest, alt. 3–4000 ft., *Haast;* Hopkins river, 4000 ft., *Haast* (leaves 2–3 in. long). **Lord Auckland's** group and **Campbell's** Island: abundant near the sea, *J. D. H.* The Campbell's Island specimens have more obtuse sepals and shorter leaves (2–3 in. long). Mr. Buchanan observes that the wood is soft, makes pretty veneers, and burns well when new-cut. None of the Middle Island specimens are so large in the foliage as the Auckland Island, which also are most pubescent. The *D. Lycllii* I find to be connected by too many intermediate forms to rank as a separate variety.

7. **D. Urvilleanum,** *A. Rich.;—Fl. N. Z.* i. 170. A much branched shrub, 6–8 ft. high. Branches slender, quite black or dark chestnut. Leaves erect, very slender, flexuous, 1–4 in. long; sheath membranous, ⅛–⅙ in. broad, auricled, lamina $\frac{1}{20}$ in. broad at the base, concave, glabrous, not serrulate. Racemes short, lateral, 6–10-flowered. Flowers small, ⅛ in. long. Sepals ovate-acuminate, equalling the corolla-tube, ciliate or glabrous.

Var. *α*. Leaves 1½–2 in. long.

Var. *β*. Branches paler. Leaves 2–4 in. long, more flexuose.—*D. filifolium*, Hook. f. Fl. N. Z. i. 169, *D. setifolium*, Stsch., in Bull. Soc. Nat. Hist. Mosc. xxxii. 23.

Var. *γ*. Branches chestnut-brown. Leaves 1–1½ in. long. Sepals ovate-lanceolate, exceeding the corolla.—*D. Lessonianum*, A. Rich.—Fl. N. Z. i. 171.

Var. *δ*. More robust. Branches chestnut-brown. Leaves 1–1½ in. long, often ciliate or tomentose on the edges. Sepals ovate, acute, fimbriate, rather shorter than the corolla.—*D. scoparium*, Hook. f. Fl. Ant. 47. t. 33;—Fl. N. Z. i. 171.

Northern Island: var. *α*, Tasman's Bay, *D'Urville;* Bay of Islands, banks of the Keri-Keri river, *A. Cunningham*, etc.; var. *β*, various places from Auckland to Wellington, *Colenso*, etc.; var. *γ*, Bay of Islands to Auckland not unfrequent, on dry hills; var. *δ*, mountainous districts of the **Northern** Island: top of Ruahine mountains, *Colenso*. **Middle** Island: ascending to 4000 ft. in Nelson, *Travers;* Southern Alps, *Sinclair and Haast*. **Chatham** Island, *Dieffenbach* (edges of leaves very downy). **Campbell's** Island: common near the sea, *J. D. H.*

8. **D. subulatum,** *Hook. f. Fl. N. Z.* i. 171. An erect shrub, 2–4 ft. high, with slender, sparingly leafy branches, and red-brown bark. Leaves erect, rigid, pungent, ¼–¾ in. long, strict or flexuose, sheath $\frac{1}{12}$ in. broad, blade $\frac{1}{36}$ in. broad at the base, semiterete, concave, puberulous or glabrate above, margin most minutely ciliate. Racemes small, lateral, 2–5-flowered. Flowers small, $\frac{1}{10}$–⅛ in. long. Sepals broadly ovate, subacute, as long as the short tube of the corolla, quite glabrous.

Northern Island: barren plains of Tewahiti, base of Tongariro, Tarawera, etc., *Colenso*. A very distinct little species, and quite different from any form of *D. Urvilleanum*.

9. **D. uniflorum,** *Hook. f., n. sp.* A stout, erect shrub, with dark red-brown bark. Leaves erect, rigid, coriaceous, pungent, ½–⅔ in. long, sheath $\frac{1}{10}$–⅛ in. broad, not auricled; blade $\frac{1}{15}$–$\frac{1}{20}$ in. broad at the base, semiterete, concave and puberulous above, margin most minutely ciliate. Flower solitary, lateral, almost hidden by sheathing bracts, ¼–⅓ in. long. Sepals lanceolate, acute, as long as the corolla-tube.

Middle Island: Wairau mountains, alt. 4000 ft., *Travers;* Rangitata range, 3–5000 ft., *Sinclair and Haast;* forming much of the subalpine vegetation between 3000 and 5000 ft. in the Southern Alps, *Haast;* Otago, lake district, alpine, *Hector and Buchanan.* A very distinct species, remarkable for the short, pungent leaves, and large solitary flower.

10. **D. rosmarinifolium,** *Forst.;—Fl. N. Z.* i. 171. A very small, often prostrate, woody shrub, a few inches to 1 ft. high. Leaves erect or somewhat recurved, $\frac{1}{6}$–1 in. long, rigid, straight or curved; sheath $\frac{1}{8}$–$\frac{1}{6}$ in. broad; blade $\frac{1}{20}$ in. broad at the base, semiterete, concave, minutely ciliate, keeled towards the obtuse tip. Flowers solitary or in 2-flowered spikes, $\frac{1}{8}$–$\frac{1}{6}$ in. long. Sepals ovate, subacute, glabrous, as long as the corolla-tube.

Middle Island: Dusky Bay, *Forster, Lyall;* Nelson mountains, *Bidwill;* Wairau mountains, altitude 4–5000 ft., *Travers;* Otago alps, altitude 5–7000 ft., *Hector and Buchanan;* common in the Southern Alps, ascending to 6300 ft., *Sinclair and Haast.* Allied to *D. subulatum,* but the obtuse leaves at once distinguish it; nearer *D. recurvum.*

11. **D. muscoides,** *Hook. f., n. sp.* A most densely tufted little species, with woody subterranean stem, and compacted, short branches, densely covered with minute, imbricate leaves. Leaves $\frac{1}{10}$ in. long, ovate-subulate from a broad sheathing base, obtuse, coriaceous, semiterete, shining, most minutely ciliolate. Flower solitary, terminal, $\frac{1}{8}$ in. long. Sepals ovate, subacute, as long as the corolla-tube.

Middle Island: alps of Otago, alt. 7–8000 ft., *Hector and Buchanan.* A very singular little plant, closely allied to the *D. minimum,* F. Muell., of the Victorian alps, but differing in the longer branches, covered with shorter, more imbricating leaves.

ORDER XLIII. MYRSINEÆ.

Shrubs or trees. Leaves alternate, simple, exstipulate, full of pellucid glandular dots or lines. Flowers small, regular, or nearly so.—Calyx inferior in the New Zealand genus, 4- or 5-cleft, imbricate. Corolla, 4- or 5-cleft or 5-partite, rarely of 5 free petals. Stamens 4 or 5, opposite to and inserted on the corolla-lobes, or almost free. Ovary 1-celled; style simple, stigma simple or lobed; ovules 1 or more, inserted on a free central, often fleshy placenta. Berry indehiscent, 1-celled, 1–many-seeded. Seeds sometimes enclosed within the withered placentas, albuminous; embryo transverse, terete.

A tropical and subtropical Order, advancing much further south in the New Zealand Islands than in any other longitude.

1. MYRSINE, Linn.

(Suttonia, *Fl. N. Z.*)

Trees and shrubs, sometimes small and creeping. Flowers small, usually in lateral fascicles or umbellate, rarely axillary and solitary, hermaphrodite or polygamous.—Calyx 4- or 5-fid, inferior, rarely 2-fid or 0. Petals 4 or 5, free or tapering at the base, reflexed, deciduous. Stamens 4 or 5; filaments free or attached to the base of the petals. Ovary subglobose, 1-celled; style short or 0; stigma concave or fimbriate; ovules 1–5, sunk in the fleshy placenta. Fruit a berry or drupe with a crustaceous nut. Seeds solitary or few, usually enclosed in the papery remains of the placenta.

A large genus, found in all tropical and warm countries, rare in temperate.

Tree. Leaves 4–6 in., linear, glands oblong	1. *M. salicina.*
Shrub, erect. Leaves 1–1½ in., oblong or obovate	2. *M. Urvillei.*
Shrub, erect. Leaves ⅓ in., obovate or obcordate, retuse or 2-lobed .	3. *M. divaricata.*
Tree. Leaves ½–1 in., oblong-obovate, obtuse; cuticle loose below .	4. *M. montana.*
Small, trailing shrub. Leaves ¼–⅓ in., orbicular or obovate . . .	5. *M. nummularia.*

1. **M. salicina,** *Heward, mss.—Suttonia*, Fl. N. Z. i. 172. t. 44. A small, erect, perfectly glabrous tree. Leaves 4–6 in. long, ½–⅔ broad, narrow linear or linear-oblong, obtuse, quite entire, flat, much veined, pellucid glands oblong. Flowers in dense lateral many-flowered fascicles, hermaphrodite, $\frac{1}{12}$ in. broad; pedicels ¼ in. long, stout. Calyx oblique; lobes rounded, ciliate. Petals cohering at the base. Stamens adhering to the petals. Berry ovoid, ¼ in. long, 2-seeded.

Northern Island: in woods, from Bay of Islands to Wellington, *Cunningham, Colenso*, etc.

2. **M. Urvillei,** *A. DC.—Suttonia australis*, A. Rich. Flor. t. 38;—Fl. N. Z. i. 173. A shrub, 8–10 ft. high, perfectly glabrous; bark nearly black. Leaves 1–1½ in. long, oblong or obovate, obtuse, coriaceous, undulate, much veined, studded with rounded pellucid glands. Flowers in capitate, lateral fascicles, $\frac{1}{16}$ in. broad. Calyx 0 or 2–4-lobed. Petals revolute. Stamens adherent to the petals; anthers large; stigma sessile, capitate. Berries small, globose, ⅛–⅙ in. diam.—*M. Richardiana*, Endl.

Northern and **Middle** Islands, abundant, *Banks and Solander*, etc. Resembles *Pittosporum undulatum* a good deal in foliage.

3. **M. divaricata,** *A. Cunn.—Suttonia*, Fl. N. Z. i. 173; Fl. Antarct. i. 51. t. 34. A small, very straggling, twiggy, branched bush. Leaves alternate or fascicled, ⅓ in. long, broadly obovate or obcordate, obtuse retuse or 2-lobed, very coriaceous, reticulated, pellucid glands rounded. Flowers minute, $\frac{1}{12}$ in. diam., fascicled. Calyx 4- or 5-lobed. Petals 4 or 5, obovate. Style short; stigma cup-shaped, lacerate. Berry small, depressed, spherical.

Northern Island: moist woods, abundant, **Lord Auckland's** group, *J. D. H.* Habit of a *Coprosma*.

4. **M. montana,** *Hook. f.—Suttonia*, Fl. N. Z. ii. 334. A small tree, branches robust; bark dark red-brown. Leaves alternate, ½–1 in. long, oblong-obovate, obtuse, very coriaceous, reticulate above, cuticle beneath loose when dry, pellucid glands rounded. Flowers not seen. Fruit globose.

Northern Island: top of the Ruahine range, *Colenso*. Possibly only a variety of *M. divaricata*, as I originally supposed, but the habit is very different, branches more robust, leaves longer and more coriaceous, not fascicled, reticulate above only, and never obcordate or 2-lobed.

5. **M. nummularia,** *Hook. f.—Suttonia*, Fl. N. Z. i. 173. t. 45. A small prostrate shrub, with very slender branches, straggling, 6–18 in. long. Leaves alternate, ¼–⅓ in. long, orbicular or broadly obovate, reticulate above, wrinkled below, pellucid glands rounded. Flowers minute, scattered, solitary, axillary or lateral. Calyx very minute, 4-lobed. Petals 4, concave, ciliate. Stamens large, inserted on the petals. Stigma sessile, conical. Berry globose, $\frac{1}{10}$ in. diam.

Northern Island: not uncommon on the mountains; top of the Ruahine range, Lake Rotoatara, etc., *Colenso*. **Middle** Island: alps near Haast's Pass, *Haast*; Otago, lake district, scrambling over rocks, *Hector and Buchanan*.

Order XLIV. PRIMULACEÆ.

Characters of *Myrsineæ*, but plants herbaceous, leaves less coriaceous, without pellucid glands, and corolla usually less deeply divided.

A large Order in the northern hemisphere, especially in the cold, temperate, and mountainous regions, rare in the tropics, and still more so in the southern hemisphere.

1. SAMOLUS, Linn.

Creeping, rarely erect herbs. Leaves alternate. Flowers axillary and solitary in the New Zealand species.—Calyx half-superior, persistent, 5-cleft. Corolla campanulate, tube very short, 5-cleft. Stamens 5, inserted on the corolla, alternating with 5 staminodia. Ovary subglobose; style straight, stigma capitate: ovules numerous. Capsule half-inferior, 5-valved at the top, 1-celled, many-seeded.

A genus found in Europe, and in various temperate and subtropical parts of the world.

1. **S. littoralis,** *Br.;—Fl. N. Z.* i. 207. A small prostrate and creeping, branched, perfectly glabrous herb. Leaves fleshy, $\frac{1}{6}$–1 in. long, linear-spathulate or oblong-spathulate, often recurved. Peduncles longer than the leaves. Flowers white, $\frac{1}{4}$–$\frac{1}{3}$ in. diam. Capsule crowned with the persistent stigma.—*Sheffieldia repens*, Forst.

Marshy places near the sea throughout the islands, *Banks and Solander*, etc. Also abundant in Australia, Tasmania, the Pacific Islands, and found in South Chili.

Anagallis arvensis, Linn., the "Scarlet Pimpernel" or "Poor Man's Weather-glass," has been introduced into cultivation.

Order XLV. SAPOTEÆ.

Trees or shrubs, juice usually milky. Leaves alternate, entire, exstipulate, coriaceous. Flowers axillary, regular, solitary or fascicled.—Calyx free, 4–8-toothed or -partite. Corolla 4–8-lobed; lobes imbricate. Stamens 4–8 or 8–16, sometimes with alternating staminodia. Ovary superior, 2–12-celled; style simple, stigma simple or lobed; ovules solitary in each cell. Fruit a berry or drupe, 1–4-seeded. Seeds usually with a crustaceous, shining testa, marked with a large unpolished hilum, albuminous with foliaceous cotyledons or exalbuminous with fleshy cotyledons.

Almost exclusively a tropical Order, found in both the Old and New World.

1. SAPOTA, Linn.

Trees, with milky juice. Leaves generally fascicled at the ends of the branches. Flowers polygamous, in axillary or lateral fascicles or umbels. —Sepals 4–6, orbicular, imbricate. Corolla 4–6-lobed. Stamens 4–6, short, inserted at the base of the corolla-lobes and opposite them, alternating with as many staminodia. Ovary hirsute, 4–12-celled; style straight, stigma simple. Berry with 1 nut-like seed. Seeds elongate, compressed; testa hard, crustaceous, shining, with a long grooved opaque hilum; embryo with flat foliaceous cotyledons and a short terete radicle.

A small genus of chiefly tropical trees.

1. **S. costata,** *A. DC.;—Fl. N. Z.* i. 174. A tree, 20 ft. high, branches hoary. Leaves 2–3 in. long, coriaceous, obovate-oblong, obtuse, with numerous parallel veins diverging from the midrib. Pedicels stout, curved, $\frac{1}{4}$ in. long. Flowers globose, $\frac{1}{5}$ in. diam., usually 4-merous. Sepals very coriaceous. Corolla-lobes scarcely longer than the sepals. Filaments short, fleshy. Ovules suspended. Berry $\frac{2}{3}$–1 in. long.

Northern Island: Wangarei Bav, *Colenso;* coast opposite the Cavalhos Islands, *R. Cunningham.* The same with the Norfolk Island plant, except that the flowers are very rarely pentamerous, and the calyx smaller. There is also an allied Australian species.

ORDER XLVI. JASMINEÆ.

(OLEINEÆ, *Fl. N. Z.*)

Trees or shrubs. Leaves opposite or subopposite, exstipulate. Flowers small, in axillary or terminal clusters racemes or panicles, often unisexual. —Calyx small, 2–4-toothed, often unequally. Petals 0 in the New Zealand species. Stamens 2, epipetalous or hypogynous. Ovary 2-celled; stigma simple or 2-fid; ovules 1 or 2 in each cell. Fruit drupaceous in the New Zealand species, containing a bony 1- or 2-celled nut. Seeds with or without albumen; embryo straight.

A considerable Order of temperate and tropical plants, to which the Jasmine and Olive belong.

1. OLEA, Linn.

Shrubs or trees. Leaves opposite or subopposite, entire, coriaceous. Flowers small, unisexual, in short axillary racemes or panicles.—*Male* fl.: Calyx unequally 2–4-lobed. Petals 0 in the New Zealand species. Stamens 2, with large exserted anthers; ovary rudimentary. *Female* fl.: Calyx urceolate, unequally 4-lobed. Anthers compressed, imperfect, included. Ovary oblong, 2-celled; style short; stigmas 2. Drupe oblong, 1- or 2-celled.

A large genus, scattered over the globe; the New Zealand species belong to a peculiar small section with apetalous flowers.

Leaves 3–6 in., obtuse, veins obscure. Racemes stout. Drupe $\frac{1}{2}$ in. 1. *O. Cunninghamii.*
Leaves 2–4 in., acute, veins distinct. Racemes slender. Drupe $\frac{1}{2}$ in. 2. *O. lanceolata.*
Leaves 1–2 in., narrow linear, obtuse. Drupe $\frac{1}{4}$ in. . . . . . 3. *O. montana.*

1. **O. Cunninghamii,** *Hook. f. Fl. N. Z.* i. 175. A lofty tree, branches with white bark, young pubescent. Leaves nearly opposite, 3–6 in. long, coriaceous, narrow oblong-lanceolate or narrow linear-oblong, obtuse, very coriaceous, nerves very obscure on both surfaces. Racemes $\frac{1}{2}$–$\frac{3}{4}$ in. long, stout, erect, 10–15-flowered; bracts ovate, concave, membranous. Flowers shortly pedicelled. Male calyx of 2 very small and 2 large lobes. Drupe obliquely ovoid, $\frac{1}{2}$ in. long, 1- or 2-celled and -seeded.—*O. apetala*, A. Cunn., not Vahl.

Northern and eastern parts of the **Northern** Island, *Banks and Solander*, etc. Conounded by A. Cunningham with the Norfolk Island closely-allied plant.

2. **O. lanceolata,** *Hook. f. Fl. N. Z.* i. 176. A small tree, 30 ft. high, much less robust than *O. Cunninghamii*, branches slender; bark white. Leaves opposite, 2–4 in. long, narrow lanceolate or linear-lanceolate, acute,

with raised veins on both surfaces. Racemes very slender, 6–10-flowered, sparsely pilose. Flowers minute, much as in *O. Cunninghamii*, but smaller, pedicels slender. Berry ovoid, crimson, ½ in. long.

Northern Island: woods of the east coast and interior, *Colenso.*

3. **O. montana,** *Hook. f. Fl. N. Z.* i. 176. *t.* 46 A *and* B (*not* C). A large bushy-headed tree, 40–50 ft. high; branches slender; bark reddish. Leaves opposite, 1–2 in. (in young plants 3–4 in.) long, very coriaceous, narrow linear, ⅙–¼ in. broad, obtuse, nerveless. Racemes slender, ¼–1 in. long, puberulous, 6–8-flowered. Flowers very minute, as in *O. Cunninghamii.* Drupe narrow ovoid, ¼ in. long.

Northern Island: Bay of Islands, *Cunningham;* east coast, interior, and Wairarapa Valley, *Colenso.*

ORDER XLVII. APOCYNEÆ.

Trees or shrubs, often climbing, with milky juice. Leaves opposite exstipulate. Flowers in axillary or terminal cymes or panicles.—Calyx 5-partite or -lobed; lobes imbricate. Corolla with a short or long tube; lobes 5, contorted in bud. Stamens 5, inserted on the corolla; anthers often sagittate and adhering by their anterior face to the stigma. Ovary 2-celled (rarely 1-celled), or of 2 carpels combined by the styles or stigmas; style long or short, stigma usually angular; ovules many. Fruit of 2, slender, 1-celled capsules (rarely a berry or drupe), opening inwardly. Seeds pendulous, exalbuminous, often with a tuft of silky hairs.

A large tropical Natural Order.

1. PARSONSIA, Br.

Slender climbing plants; branches terete. Leaves excessively variable in form and size. Flowers small, panicled.—Calyx 5-partite; lobes within furnished with a small scale at the base. Corolla urceolate campanulate or shortly funnel-shaped; lobes 5, reflexed, eglandular. Stamens 5; anthers sagittate, included or exserted, adhering to the stigma, one cell without pollen. Hypogynous scales 5. Ovary 2-celled; style slender. Fruit of 2 long narrow, terete, acute, 1-celled capsules. Seeds with a fine silky tuft of hairs.

A small tropical Asiatic and Australian genus. I am convinced that there are but two New Zealand species of this genus, to which Raoul's names of *P. albiflora* and *P. rosea* had on the whole better be retained, to avoid the confused synonymy of Forster's name of *capsularis*, which has been variously applied.

Flowers ¼ in. long. Anthers included . . . . 1. *P. albiflora.*
Flowers ⅛ in. long. Anthers exserted . . . . 2. *P. rosea.*

1. **P. albiflora,** *Raoul.*—*P. heterophylla*, A. Cunn.;—Fl. N. Z. i. 181. Stems stout, glabrous or pubescent, as thick as a crow-quill. Leaves most often 1–2 in. long, coriaceous, oblong ovate or lanceolate, with transverse veins, more rarely linear or obovate, or narrow linear-lanceolate, 3–4 in. long, with lobed margins, in young plants spathulate. Panicles many-flowered. Flowers white, odorous, ¼ in. long. Corolla-lobes shorter than the tube. Anthers included. Capsule 3–4 in. long.—*P. capsularis*, Endl.; Deless. Ic. Sel. v. t. 49 (bad); *? P. variabilis*, Lindl.;—Fl. N. Z. i. 181.

Northern and **Middle** Islands: abundant from the Bay of Islands to Otago, *Banks and Solander*, etc. I know nothing of the *P. variabilis*, Lindl., but suppose it to be founded on young specimens of this.

2. **P. rosea,** *Raoul, Choix*, xvi. *t.* 12;—*Fl. N. Z.* i. 180. A more slender plant than the preceding. Leaves usually very long, linear, membranous, obtuse or acute at both ends, entire or undulate at the margins, 2–3 in. long, in young plants obovate or spathulate. Panicles with few scattered flowers. Corolla shortly campanulate $\frac{1}{10}$ in. long; lobes as long as the tube. Anthers exserted.—*P. rosea* and *P. capsularis*, Raoul, l. c.; *Periploca capsularis*, Forst.

Northern and **Middle** Islands, abundant, *Banks and Solander*, etc.

ORDER XLVIII. LOGANIACEÆ.

Shrubs or trees (rarely herbs). Leaves opposite, with interpetiolar stipules. Flowers usually in cymes or corymbs, regular, and hermaphrodite.—Calyx 4- or 5-lobed, -toothed, or -parted. Corolla 4- or 5-lobed, lobes imbricate, contorted or valvate, often hairy at the throat. Stamens 4 or 5, alternate with the corolla-lobes. Ovary free, 2–5-celled; style simple, stigma simple or lobed; ovules 1 or more in each cell. Fruit capsular in the New Zealand genera, 2-valved, many-seeded. Seeds albuminous.

A small Order, consisting of plants variously related, some to *Rubiaceæ*, others to *Scrophularineæ*, others to *Gentianeæ* and *Apocyneæ*. The leaves often turn black when dry.

Calyx 5-partite . . . . 1. LOGANIA.
Calyx 5-fid . . . . 2. GENIOSTOMA.

1. LOGANIA, Br.

Herbs or shrubs. Flowers small, axillary or solitary, diœcious?—Calyx 5-partite, imbricate. Corolla campanulate, tube bearded, limb 5-parted, lobes imbricate. Stamens 5, inserted on the corolla. Ovary 2-celled; stigma simple; ovules numerous. Capsule 2-celled, splitting into two valves, with the seeds on their margins; valves 2-fid. Seeds small; albumen fleshy.

A large New Holland genus, not extending into Tasmania.

Leaves spreading, linear, obovate or oblong. Flower minute . . . . . 1. *L. depressa.*
Leaves 4-fariously imbricate. Flower as large as leaves . . . . . . 2. *L. tetragona.*

1. **L. depressa,** *Hook. f. Fl. N. Z.* i. 177. A prostrate, rigid, woody shrub; branches densely interlaced, puberulous. Leaves $\frac{1}{6}$–$\frac{1}{4}$ in. long, coriaceous, veinless, linear-obovate or oblong, obtuse. Flowers minute, axillary, pedicelled, bracteate, solitary or in 3–5-flowered panicles, male only seen. Sepals oblong, obtuse, ciliate. Corolla scarcely longer than the calyx; lobes rounded. Filaments slender, anthers large, 2-cleft for halfway up. Ovary imperfect in my specimens (which are probably unisexual); style short, clavate; stigma oblong, thick. Fruit unknown.

Northern Island: Ruahine mountains, *Colenso*. Very closely allied to the *L. fasciculata*, Muell., of the Australian alps. Habit of an alpine *Coprosma*.

2. **L. tetragona,** *Hook. f., n. sp.* Rigidly coriaceous, decumbent or prostrate; stem woody, creeping at the base, densely tufted; branches ascending, short, densely leafy, pubescent, 1–2 in. long, with the leaves on $\frac{1}{4}$ in. diam. Leaves densely 4-fariously imbricate, spreading, oblong, obtuse, quite entire,

very coriaceous, concave, ciliated towards the base, connate in pairs at the very base, keeled margins thickly cartilaginous. Flower solitary, sessile, terminal, seen with old fruit only. Calyx about as large as the leaves; tube turbinate; lobes 4, oblong, obtuse, ciliated, rather distant. Capsule coriaceous, 4-valved.

Middle Island: Otago, lake district, alpine, *Hector and Buchanan.*

2. GENIOSTOMA, Forst.

Characters of *Logania*, but the valves of the capsule not 2-fid nor 2-partite, the æstivation of the corolla-lobes contorted, and the calyx less deeply divided.

A considerable genus of Madagascar and Bourbon, the Asiatic and Polynesian islands, not found in continental Africa, nor in Australia.

1. **G. ligustrifolium,** *A. Cunn.;—Fl. N. Z.* i. 177. A perfectly glabrous shrub or small tree; branches slender. Leaves 1½–3 in. long, ovate-oblong, acuminate, membranous. Flowers in short, axillary and lateral corymbs, ⅛ in. diam., white; peduncles and pedicels bracteolate. Corolla-tube very short, lobes reflexed. Ovary globose, with a very short style and large 2-lobed stigma. Capsules on slender divaricating pedicels, globose, mucronate, ⅛ in. diam., valves separating from the placentiferous axis.—Hook. Ic. Pl. t. 430.

Northern Island: not uncommon in woods, etc., *Banks and Solander*, etc.

ORDER XLIX. GENTIANEÆ.

Herbs, usually glabrous and bitter. Leaves opposite, exstipulate, quite entire. Flowers solitary or cymose, often handsome.—Calyx 4- or 5-lobed. Corolla 4- or 5-cleft or lobed, contorted in bud, persistent. Stamens 5, inserted on the corolla. Ovary 1-celled, with 2 parietal placentas often projecting into the cavity, and dividing it into 2–4 cells; styles 1 or 2, stigma 2-lobed or 2-capitate; ovules numerous. Capsule (rarely a berry) usually membranous, elongate, septicidally 2-valved. Seeds small; albumen fleshy.

A large tropical and temperate Order, especially abounding in mountainous regions.

Flowers large, white or yellowish or purplish. Style 1. Stigma 2-lobed . 1. GENTIANA.
Flowers small, yellow. Ovary with 2 styles . . . . 2. SEBÆA.

1. GENTIANA, Linn.

Erect or ascending, simple or branched herbs, with conspicuous flowers.—Calyx 4- or 5-cleft. Corolla 4- or 5-cleft, campanulate or rotate. Anthers not twisted, turning back, and so becoming extrorse. Ovary linear; style 1, stigma bifid. Capsule septicidal, elongated, membranous.

A large mountain genus, found in all parts of the world: The species are most variable, and the New Zealand ones especially so, insomuch that in framing characters for the species I have had to take prevalent forms and to disregard intermediate ones, which occur in abundance between all. Some Chili species are too close to the New Zealand, but I hesitate to unite them, without better materials.

Root annual (sometimes perennial in 3).
Stems very slender, 1- or few-flowered. Stem leaves few. Calyx-lobes subulate . . . . 1. *G. montana.*

Stems short, robust, very leafy, many-flowered. Calyx-lobes linear-oblong, obtuse . . . . 2. *G. concinna.*
Stems robust, many-flowered. Stem-leaves few. Flowers corymbose. Calyx-lobes oblong . . . . 3. *G. pleurogynoides.*
Root perennial.
Stems ascending, few-leaved, 1- or many-flowered. Calyx-lobes oblong or ovate . . . . 4. *G. saxosa.*
Stems prostrate, very leafy. Calyx-lobes spathulate, as long as the corolla . . . . 5. *G. cerina.*

1. **G. montana,** *Forst.;—Fl. N. Z.* i. 178. Root slender, filiform, annual. Stems generally many from the root, very slender, 4–18 in. high, usually ascending, 1- or few-flowered, sparingly leafy. Leaves: radical spathulate, with long or short petioles, $\frac{1}{2}$–2 in. long, rather membranous; cauline sessile, ovate or oblong, obtuse. Flowers on slender pedicels, $\frac{1}{3}$–$\frac{2}{3}$ in. long. Calyx deeply divided; lobes subulate-lanceolate, acuminate. Corolla-lobes oblong, subacute.—*G. Grisebachii,* Hook. f. in Hook. Ic. Pl. 636.

Northern Island: mountainous districts; Tongariro, *Bidwill.* Common throughout the **Middle** Island, *Forster,* ascending to 3000 ft. Also common in Tasmania and the alps of Victoria, in both which countries it attains a larger size, and more corymbose habit. I cannot distinguish seedling states of *G. saxosa* and *pleurogynoides* from this.

2. **G. concinna,** *Hook. f.; Fl. Antarct.* i. 53. *t.* 35. Root slender, annual. Stems numerous from the root, erect or ascending, 2–12 in. long, rather stout, leafy. Leaves all linear-oblong or spathulate-oblong, $\frac{1}{2}$–1$\frac{1}{2}$ in. long, coriaceous, obtuse, often recurved. Flowers about $\frac{1}{2}$ in. long. Calyx deeply divided; lobes often as long as corolla, linear-oblong, obtuse. Corolla lobes narrow, obovate-oblong, obtuse.

Lord Auckland's group and **Campbell's** Island: abundant on the hills, *J. D. H.* The stouter leafy habit and form of the calyx-lobes best distinguish this from *G. montana.* It is extremely variable; shoots have often leaves 3 in. long, linear-oblong, and 3-nerved.

3. **G. pleurogynoides,** *Griseb.*—*G. saxosa,* γ, Fl. N. Z. i. 178. Root slender or stout, usually annual. Stems solitary or numerous from the root, erect, rarely ascending, always stout, 4–20 in. high, sparingly leafy. Leaves: radical $\frac{1}{2}$–3 in. long, very coriaceous, rosulate, petioled, spathulate; cauline in distant pairs, oblong-ovate or ovate-cordate, short or long. Flowers yellowish, very handsome, usually large, $\frac{1}{2}$–1 in. long, in terminal umbels or corymbs. Calyx not deeply divided; lobes oblong or ovate, subacute or acute, very variable in length. Corolla-lobes usually much larger and broader than in the two preceding.

Northern Island: summit of Ruahine range, *Colenso.* **Middle** Island: abundant on all the mountains, ascending to 5000 ft., *Forster,* etc. The usually simple, stout, erect, sparingly leafy stem, and corymbose large flowers are the best characters of this beautiful form. A common Tasmanian plant.

4. **G. saxosa,** *Forst.;—Fl. N. Z.* i. 178. Root stout or slender, perennial. Stems usually numerous, erect or ascending, stout or slender, 2–6 in. high, sparingly leafy, 1- rarely many-flowered. Leaves: radical numerous, often rosulate, spathulate or oblong, $\frac{1}{2}$–3 in. long, coriaceous; cauline oblong-ovate or ovate-cordate. Flowers usually large, $\frac{1}{2}$–$\frac{2}{3}$ in. long. Calyx-lobes excessively variable in depth and form. Corolla-lobes broadly obovate-oblong, obtuse.

Var. *α*. Stems numerous, short, ascending, 1–2-flowered. Leaves rosulate, spathulate. Calyx divided ⅔ way down; lobes ovate-oblong, acute. (Like *G. montana*, but perennial.)—*G. bellidifolia*, Hook. f. in Hook. Ic. Pl. t. 635.

Var. *β*. Stems stout, branched, erect, leafy, many-flowered. Calyx divided ⅔ way down, lobes linear-oblong. (Like *G. pleurogynoides*, but perennial.)

Var. *γ*. Stems erect, very stout, simple, sparingly leafy. Leaves often large, thick, and fleshy. Flowers very numerous, large, corymbose. Calyx short, divided to the middle; lobes ovate-acute or acuminate. (Like *G. pleurogynoides*, but perennial, and calyx very peculiar, unlike any except some forms of var. *α*.)

Northern and **Middle** Islands. Var. *α*. Abundant in the mountains, *Forster*, etc., ascending to 6000 ft. Var. *β*. Sinclair range and elsewhere; Southern Alps, ascending to 6000 ft., *Sinclair and Haast*. Var. *γ*. Nelson mountains, *Bidwill* (with *G. pleurogynoides*); Port Cooper, *Lyall* (very large state); Upper Wairau, *Sinclair* (root leaves 3 in. long, linear-oblong); Mount Darwin and mountains near Lake Tekapo, alt. 3–5000 ft., *Haast* (very stout forms with very broad and fleshy leaves). The calyx of var. *γ*, in conjunction with its habit, would indicate a different species, were it not that the same calyx occurs in genuine *G. saxosa*, var. *α*.

5. **G. cerina,** *Hook. f. Fl. Antarct.* i. 55. *t.* 36. Root perennial, stems very numerous, branched, stout, trailing, very leafy, 8–16 in. long, as thick as a quill. Leaves very thick, coriaceous, obovate- or spathulate-oblong, ⅔–1½ in. long; radical and cauline similar, 3-nerved. Flowers several together, crowded towards the ends of the branches, sunk amongst the leaves, ⅓–½ in. long. Calyx deeply divided; lobes large, oblong-spathulate, often recurved, longer than the corolla-tube. Corolla-lobes broad, white, with red-purple nerves.

Lord Auckland's group: on rocky islets, etc., near the sea, abundant, *J. D. H.* A most beautiful plant; remarkable for the thick, trailing leafy stems, bright-green, shining, succulent foliage, and large calyx-lobes.

2. SEBÆA, Solander.

Erect, glabrous, annual herbs, with simple or divided, sparingly leafy stems. Flowers cymose, small.—Calyx 4- or 5-parted; lobes keeled or winged. Corolla 4- or 5-fid, persistent in fruit; tube straight; lobes twisted after flowering. Stamens 5, at the mouth of the corolla; anthers finally twisted a little. Ovary 2-celled; styles 2, straight, stigmas capitate. Capsule of 2 linear-pointed valves, separating from a seed-bearing axis.

A small, tropical, and Southern African genus, also found in Australia and Tasmania.

1. **S. ovata,** *Br*;—*Fl. N. Z.* i. 179. Stem slender, 4–10 in. high, 4-angled. Leaves 2 or 3 pairs, ¼ in. long, sessile, obtuse, very broadly ovate. Flowers ¼ in. long, yellow, 5-fid. Calyx-lobes ovate-lanceolate, keeled.—*S. gracilis*, A. Cunn. Prodr.

Northern and **Middle** Islands: bogs at Hokianga, *A. Cunningham;* grassy places, Ahuriri, *Colenso;* Port Cooper, *Lyall*. Also found in Australia and Tasmania, and very nearly related to a Madagascar species.

ORDER L. BORAGINEÆ.

Herbs, often hispid with stiff hairs. Leaves alternate, simple, quite entire, exstipulate. Flowers rarely solitary, usually in 1-sided, scorpioid racemes or spikes, often variable in colour.—Calyx 5-lobed or -partite. Corolla regular,

5-lobed, imbricate in bud, throat often closed with swellings or scales opposite the lobes. Stamens 5, inserted in the throat of the corolla, included or exserted. Anthers with 2 parallel cells. Ovary deeply 4-lobed, 4-celled; style inserted between the lobes, stigma capitellate; ovule 1 in each cell. Fruit of 4 simple, smooth crested spinous or winged nuts, often resembling naked seeds. Embryo straight; albumen little or 0.

A large Natural Order, especially in Europe and oriental regions; also found all over the globe. The above character does not include the tribe *Cordiaceæ*, of which there is no New Zealand representative, and which includes shrubs and large trees, with the leaves often serrated, the ovary not lobed, and the fruit a 4-celled drupe or nut.

Corolla salver-shaped. Stamens included. Nuts minute, polished . . 1. MYOSOTIS.
Corolla more or less campanulate. Stamens exserted. Nuts minute, polished . . . . 2. EXARRHENA.
Corolla rotate. Nuts large, with broad wings . . . . 3. MYSOTIDIUM.

1. MYOSOTIS, Linn.

Annual or perennial herbs. Leaves usually spathulate or ovate, radical petioled, cauline sessile. Flowers small, in scorpioid racemes, or solitary and axillary, or solitary sessile and terminal.—Calyx 5-lobed or -partite. Corolla slender, long or short, cylindric; limb expanded; lobes 5, patent; throat with 5 swellings. Stamens 5; anthers sessile or filaments very short, included. Nuts minute, ovoid, compressed, very shining.

A large European genus, to which the "Forget-me-not" belongs. Three New Zealand species differ from all their congeners in having solitary sessile terminal flowers.

1. *Flowers solitary, sessile, terminal. Leaves small, imbricate.*

Stems 1 in., tufted. Leaves narrow linear-oblong . . . . 1. *M. uniflora.*
Stems 1 in., tufted. Leaves broadly oblong-quadrate . . . . . . 2. *M. pulvinaris.*
Stems 1 in., tufted. Leaves broadly obovate-spathulate . . . . 3. *M. Hectori.*

2. *Flowers all solitary and in the axils of the leaves or below them.*

Leaves distant, petioled. Flowers pedicelled . . . . 4. *M. spathulata.*
Leaves close-set, sessile. Flowers minute, sessile . . . . 5. *M. antarctica.*

3. *Flowers all in terminal racemes, or the lower only axillary.*

Erect, very hispid. Flowers all racemose. Pedicels short. Nuts ovoid, black . . . . 6. *M. australis.*
Prostrate, slender. Lower flowers axillary. Petioles and pedicels slender. Nuts orbicular, pale . . . . 7. *M. Forsteri.*
Stout, erect. Petioles broad. Flowers pedicelled. Calyx-hairs appressed . . . . 8. *M. capitata.*
Stout, erect, very hispid. Flowers sessile. Calyx-hairs spreading . . 9. *M. Traversii*

1. **M. uniflora,** *Hook. f., n. sp.* A small, densely-tufted, much-branched perennial, forming rounded masses; hoary with appressed, rigid, spicular hairs, that are rough under the microscope; root woody, tortuous; branches slender, erect, fascicled, $\frac{1}{2}$–$1\frac{1}{2}$ in. high, densely leafy throughout. Leaves erect, imbricating, close-set, $\frac{1}{6}$–$\frac{1}{4}$ in. long, sessile, narrow linear-oblong, obtuse, a little dilated at the base and often above the middle, nearly glabrous on the upper surface, rather coriaceous. Flower terminal, yellow, solitary, sessile, nearly $\frac{1}{4}$ in. long. Calyx-lobes linear-oblong, covered with straight rigid hairs. Corolla with a long, rather slender tube, twice or more as long as the calyx; lobes short, rounded. Stamens included. Nuts ovoid, acute.

Middle Island: on shingle beds on the mountains, often forming rounded masses; Hopkins river, alt. 2–3500 ft., *Haast.*

2. **M. pulvinaris,** *Hook. f., n. sp.* A small, densely-tufted, much-branched perennial, forming soft, rounded, mossy cushions, hoary with soft white hairs. Branches 2–3 in. long. Leaves most densely imbricated in many series all round the branches, closely overlapping, sessile, $\frac{1}{6}$–$\frac{1}{4}$ in. long, $\frac{1}{8}$ broad, broadly obovate or oblong-quadrate, rounded or retuse at the herbaceous tip, slightly narrowed below and very membranous, 1-nerved. Flower white, terminal, solitary, sessile, nearly $\frac{1}{4}$ in. long. Calyx-lobes linear, obtuse, covered with soft white hairs. Corolla-tube funnel-shaped, twice as long as the calyx; lobes short, rounded. Stamens included. Nuts not seen.

Middle Island: alps of Otago, alt. 6000 ft., *Hector and Buchanan.* A most remarkable little plant.

3. **M. Hectori,** *Hook. f., n. sp.* Habit and appearance of *M. pulvinaris*, but less soft. Leaves broadly obovate-spathulate, contracted into a broad, coriaceous, glabrous petiole. Flowers white, shortly peduncled. Nuts narrow ovate-oblong, shining.

Middle Island: Otago, dry localities in the lake district, *Hector and Buchanan.*

4. **M. spathulata,** *Forst.;—Fl. N. Z.* i. 201. A flaccid, decumbent, pilose or slightly hispid annual. Stems branched from the base, prostrate, ascending at the tips, 3–10 in. long, leafy at intervals. Leaves with short or long petioles, $\frac{1}{10}$–$\frac{1}{2}$ in. long, blade $\frac{1}{4}$–$\frac{1}{2}$ in. long, orbicular-obovate or -oblong, apiculate, membranous, hispidulous on both surfaces. Flowers all axillary or on the stem below the leaves, solitary, on slender peduncles longer than the petiole, white with a yellow eye. Calyx-lobes linear, acuminate, shorter than the short funnel-shaped corolla-tube. Corolla $\frac{1}{3}$–$\frac{1}{4}$ in. diam.; lobes rounded. Stamens included. Nuts ovoid, pale, very shining, compressed edges thin. —*Anchusa spathulata*, Rœm. and Schultes.

Northern Island: dry stony places, not unfrequent, *Banks and Solander*, etc. **Middle** Island: Nelson, *Travers;* Wakefield, *Munro.* Both this and *M. Forsteri* are figured amongst Forster's drawings as *M. spathulata.*

5. **M. antarctica,** *Hook. f. Fl. N. Z.* i. 201. A small, very hispid, much-branched, depressed perennial. Stems many, spreading from the root, 1–4 in. long, prostrate with ascending tips, rather stout, very leafy. Leaves sessile, somewhat recurved, $\frac{1}{4}$–$\frac{1}{3}$ in. long, obovate- or spathulate-oblong, obtuse or apiculate, hoary on both surfaces with rigid white hairs. Flowers solitary, axillary, blue yellow or white, nearly sessile, $\frac{1}{10}$ in. long. Calyx-lobes short, ovate-lanceolate, subacute, hispid with appressed straight hairs. Corolla-tube cylindric, as long or twice as long as the calyx; lobes short, rounded. Stamens included. Nuts ovoid, very shining, black.—Fl. Antarct. 57. t. 38.

Northern Island: mountainous districts of the east coast and interior, *Colenso.* **Middle** Island: Upper Waihopai, *Munro;* Tarndale plains, alt. 4000 ft., *Travers;* terraces on the Hopkins, alt. 2–3000 ft., *Haast;* Otago, Torbury Heads, Dunedin, *Lindsay;* lake district, subalpine, *Hector and Buchanan.* **Campbell's** Island, *J. D. H.* This seems to be identical with a plant from the Straits of Magalhaens.

6. **M. australis,** *Br.;—Fl. N. Z.* i. 201. An erect hispid annual herb,

6–10 in. high, branched from the base; branches rather stout, ascending, sparingly leafy. Leaves, radical 1–2 in. long, narrowed into long petioles, oblong-spathulate, obtuse, hispid on both surfaces; cauline shorter, sessile, linear-oblong or spathulate, all very hispid with rigid hairs on both surfaces. Racemes terminal, elongate, many-flowered, hispid with spreading simple and hooked hairs. Flowers yellowish, shortly pedicelled. Calyx oblong, 5-lobed to near the base; lobes linear, acute, hispid with spreading hooked hairs. Corolla very variable, $\frac{1}{6}$–$\frac{1}{4}$ in. broad, tube funnel-shaped; lobes short, rounded. Stamens included. Nuts ovoid, black, very shining.

Middle Island: abundant in dry stony places, Hurumui and Wairau valleys, 1–3500 ft., *Travers;* Tarndale, alt. 4–5000 ft., *Sinclair;* terraces and moraines at Lake Okau, *Haast;* Waihopai and Aglionby plains, *Munro;* Otago, *Lyall.* Also a native of Australia and Tasmania.

7. **M. Forsteri,** *Rœm. and Sch.;—Fl. N. Z.* i. 200. A prostrate, branched, slender, subhispid and pilose annual. Stems branching from the base, ascending, flaccid, 8–16 in. long. Leaves, radical and lower cauline with long slender petioles, $\frac{1}{2}$–2 in. long; blade oblong-spathulate, obtuse, membranous, hispidulous on both surfaces; upper leaves oblong-obovate. Flowers axillary and in terminal hispid racemes, variable in size, lower on pedicels $\frac{1}{4}$–$\frac{1}{2}$ in. long. Calyx $\frac{1}{6}$–$\frac{1}{4}$ in. long, campanulate, 5-lobed to the middle; lobes linear-oblong, acute, hispid with spreading hooked hairs. Corolla-tube funnel-shaped, as short as or longer than the calyx; limb $\frac{1}{12}$–$\frac{1}{5}$ in. broad; lobes rounded. Stamens included. Nuts nearly orbicular, pale, very shining.

Northern Island: dry places, east coast, etc., *Banks and Solander, Colenso.* **Middle** Island: Nelson, *Sinclair:* Milford Sound, *Lyall;* Canterbury, *Raoul; Travers.*

8. **M. capitata,** *Hook. f. Fl. Antarct.* 56. *t.* 37; *Fl. N. Z.* i. 201. A robust perennial, covered with soft, appressed, scarcely hispid hairs; stems ascending 6–18 in. high, stout, leafy. Leaves, radical linear-obovate or lanceolate, obtuse, 2–4 in. long, narrowed into broad petioles, hispid-pilose on both faces; cauline linear-oblong or spathulate, sessile. Racemes large, simple or branched, often forming a dense very many-flowered head. Flowers very crowded, violet-blue or purple; pedicels short, stout. Calyx $\frac{1}{6}$–$\frac{1}{4}$ in. long, 5-partite; lobes hispid with appressed straight hairs. Corolla-tube cylindric, little longer than the calyx; limb $\frac{1}{4}$ in. diam.; lobes rounded. Stamens included. Nut ovoid, obtuse, polished.

Middle Island: upper part of Macrae's Run, *Munro;* Ruapuke Island and Port William, *Lyall;* trap cliffs at Shaw's Bay, Otago, *Lindsay.* **Lord Auckland's** group, *J. D. H.*

9. **M. Traversii,** *Hook. f., n. sp.* An erect, rather rigid, densely hispid perennial, 3–8 in. high; stems several from the root, stout, erect or ascending, leafy, very hispid with erect or spreading hairs. Leaves: radical narrow linear-spathulate, obtuse, 1–1$\frac{1}{2}$ in. long, narrowed into short petioles, hispid on both surfaces; cauline linear-oblong. Racemes densely hispid, capitate, many-flowered. Flowers almost sessile, lemon-coloured. Calyx $\frac{1}{6}$ in. long, 5-partite; lobes linear, hispid with rigid, spreading, simple and hooked bristles. Corolla funnel-shaped, rather longer than the calyx; lobes rounded. Stamens included. Nuts narrow ovoid, obtuse, very polished.

Middle Island: shingle beds on Tarndale, alt. 5–6000 ft., *Sinclair;* Wai-au valley and

Discovery Peaks, alt. 5500 ft., *Travers;* Waimakeriri valley, alt. 2500 ft., and Mount Darwin, alt. 4500–6500 ft., *Haast;* Otago, lake district, *Hector and Buchanan.* Near *M. capitata*, but smaller, much more hispid, with often hooked bristles on the calyx, and nuts much longer and narrower. In the form of the flower this tends towards *Exarrhena.*

2. EXARRHENA, Br.

Hispid, erect, rarely prostrate herbs. Leaves, radical petioled, cauline sessile. Flowers usually large, in scorpioid cymes.—Calyx narrow, 5-lobed or -partite. Corolla narrow, funnel-shaped or tubular, with an expanded limb; throat usually without thickenings; lobes 5, spreading. Stamens with long exserted filaments. Nuts as in *Myosotis.*

An Australian, Tasmanian, and New Zealand genus, very near to *Myosotis*, but no species of the latter genus have flowers so large and campanulate as the majority of *Exarrhenæ* have, and the length of the filaments, though a very variable character, is a very manifest one.

Slender, diffuse, prostrate. Leaves broad, on slender petioles . . . 1. *E. petiolata.*
Stout, suberect, very hispid. Calyx 5-partite. Corolla $\frac{1}{3}$–$\frac{2}{3}$ in. long. Nuts linear . . . . 2. *E. macrantha.*
Stout, suberect, sparingly hairy. Flowers pedicelled. Calyx 5-lobed. Nuts broad, short, black. . . . . 3. *E. Lyallii.*
Suberect, hispid. Flowers sessile. Calyx 5-partite. Corolla $\frac{1}{4}$–$\frac{1}{8}$ in. Nuts linear . . . . 4. *E. saxosa.*

1. **E. petiolata,** *Hook. f.—Myosotis petiolata*, Fl. N. Z. i. 202. A slender diffuse perennial?, covered with short scabrid or hispid hairs. Stems many from the root, 3–12 in. long, slender, prostrate and ascending, sparingly leafy. Leaves: radical and lower cauline with long slender petioles, $\frac{1}{2}$–2 in. long, rounded elliptic-oblong, apiculate, $\frac{2}{3}$–$1\frac{1}{2}$ in. long; cauline sessile, obovate-spathulate; all membranous, with short scattered hairs on both surfaces. Racemes slender, elongate, many-flowered, simple or forked. Flowers on slender pedicels. Calyx $\frac{1}{8}$ in. long, parted deeply; lobes linear, hairs straight, appressed. Corolla with a very short funnel-shape tube, and 5 oblong spreading lobes, $\frac{1}{4}$–$\frac{1}{2}$ in. diam. Filaments very slender, elongate; anthers shortly oblong. Nuts broadly ovoid, very shining, red-brown.

Northern Island: dry stony places and alluvial river banks, Cape Turnagain, Puehutai, Hawke's Bay, etc., *Colenso.* Habit of *Myosotis Forsteri*, but flowers totally different.

2. **E. macrantha,** *Hook. f., n. sp.* A rather stout, suberect perennial, covered with appressed or spreading, rather hispid hairs. Stems 6–12 in. high, ascending, stout, leafy. Leaves: radical 2–4 in. long, lanceolate-oblong, narrowed into broad petioles; cauline 1–2 in., linear-oblong, all softly hispid on both surfaces. Raceme short, many-flowered, simple or branched, very hispid. Flowers purple or white, close-set, large, all pedicelled. Calyx $\frac{1}{4}$–$\frac{1}{3}$ in. long, deeply 5-partite; lobes linear, hairs appressed or spreading, simple or hooked. Corolla $\frac{1}{3}$–$\frac{2}{3}$ in. long, tube much longer than the calyx, funnel- or almost bell-shaped; lobes broad, oblong. Stamens exserted, filaments slender; anthers linear. Nuts linear-oblong, $\frac{1}{6}$ in. long.

Middle Island: Dun mountain, *Sinclair;* Wairau and Wai-au, on mountains, alt. 3–5000 ft., *Travers;* Hopkins river, by waterfalls, alt. 3500 ft., *Haast;* Otago, Lindis Hills, subalpine, *Hector and Buchanan.* A good deal like *Myosotis capitata* in habit, foliage, and pubescence.

3. **E. Lyallii,** *Hook. f.—Myosotis Lyallii*, Fl. N. Z. i. 202. A rather short and stout tufted perennial, slightly hispid with appressed hairs. Stems several from the root, erect or ascending, rather stout, 2–6 in. high. Leaves: radical oblong-spathulate or obovate-lanceolate, subacute, 1–1½ in. long, narrowed into rather slender petioles; cauline narrow linear-oblong or oblong-spathulate; all slightly hispidulous on both surfaces with appressed hairs. Raceme short, simple or forked. Flowers very shortly pedicelled. Calyx ¼ in. long, hispid with appressed or patent, simple and hooked bristles. Corolla $\frac{1}{6}$–¼ in. long; tube cylindric, longer than the calyx; lobes short, rounded. Stamens with long slender filaments; anthers linear. Nuts broadly ovate or orbicular, very black and shining.

Middle Island: Milford Sound, *Lyall.* Habit of a small specimen of *Myosotis capitata*, but the flower is very different. I have but two specimens.

4. **E. saxosa,** *Hook. f.—Myosotis saxosa*, Fl. N. Z. i. 202. A small, rigid, prostrate perennial, more or less densely hispid (sometimes white) with rather spreading white hairs. Stems 2–6 in. long, rather stout, leafy, prostrate; racemiferous branches ascending, stout or slender. Leaves: radical obovate- or lanceolate-spathulate, acute, ½–¾ in. long, narrowed into a broad or narrow petiole, rather harsh and rigid, uniformly hispid-pilose on both surfaces; cauline linear-oblong, ⅓–½ in. long. Raceme small, few-flowered. Flowers nearly sessile. Calyx ⅙–¼ in. long, deeply partite; lobes linear, acute. Corolla-tube cylindric, short or rather elongate; lobes short, rounded, ⅕ in. diam. Stamens exserted; filaments slender; anthers linear-oblong. Nuts $\frac{1}{10}$ in. long, narrow linear-oblong.

Northern Island: east coast, crags at Tetiokura, *Colenso* (very white and hispid). **Middle** Island: Dun mountain, in open stony places, *Munro, Travers* (much less hispid).

3. MYOSOTIDIUM, Hook.

A succulent herb, 1–3 ft. high, perennial or biennial. Leaves large, lower petioled, ovate, obtuse, with parallel veins. Flowers pale blue, in dense branched scorpioid racemes.—Calyx 5-partite. Corolla rotate; lobes 5, rounded, expanded; throat closed with 5 protuberances. Stamens 5, inserted within the tube; anthers included. Ovary 4-lobed; style very short, stigma capitate. Fruit large, between globose and pyramidal, 4-angled, of 4 dorsally much-flattened coriaceous winged nuts, adhering to a central fleshy column.

A remarkable genus, perhaps too near to *Cynoglossum* and *Omphalodes*, differing from the former in the margined nuts, which do not bear barbed bristles, and from the latter in the wing of the nut not being inflexed; all should probably merge into one genus.

1. **M. nobile,** *Hook. Bot. Mag. t.* 5137. Stem stout, pilose. Radical leaves a span long, broadly ovate or ovate-cordate, very thick and fleshy, glabrous, shining, bright green; cauline sessile, oblong. Racemes collected into a dense large subglobose head, 2–5 in. diam. Flowers pedicelled. Calyx-lobes oblong, obtuse, hispid. Corolla ½–⅗ in. diam., deep azure in the centre with a purple eye, fading towards the ends of the lobes. Fruit as large as a hazel-nut.—*Cynoglossum nobile*, Hook. f. in Gard. Chron. 1858, p. 240.

Chatham Island, *Watson.* I have no native specimens, the above description being drawn up from cultivated ones.

I find amongst Sinclair's plants, without habitat, a small scrap of a *Cynoglossum*, from the neighbourhood of Auckland. The genus may be known from *Myosotis* by the nuts covered with barbed bristles; the species looks like the common tropical *C. micranthum*, and is probably an introduced weed.

ORDER LI. CONVOLVULACEÆ.

Climbing or trailing, rarely erect herbs or shrubs, usually with milky juice. Leaves alternate, exstipulate (0 in *Cuscuta*). Flowers regular, hermaphrodite, axillary or terminal, solitary or cymose, often large.—Sepals 5, rarely united, imbricate, persistent. Corolla bell- or funnel-shaped or rotate, limb 5-angled and plaited or 5-lobed and imbricate. Stamens 5, inserted on the corolla, alternate with its lobes, often unequal; anthers free. Ovary free, undivided or 2-lobed, 2–4-celled; style simple or 2-fid, or styles 2, stigmas various; ovules 1 or 2, erect in each cell. Fruit various. Albumen mucilaginous or 0; cotyledons usually folded; embryo curved or spiral in *Cuscuta.*

A very large tropical Order, rarer in the temperate zones, though common in Europe.

Stems leafy, prostrate or twining.
  Corolla plaited. Style 1; stigmas 2 . . . . 1. CONVOLVULUS.
  Corolla plaited. Style 1; stigma capitate, lobed . . . . . . 2. IPOMŒA.
  Corolla rotate. Styles 2; stigmas capitate . . . . 3. DICHONDRA.
Stems leafless, twining, parasitic . . . . 4. CUSCUTA.

1. CONVOLVULUS, Linn.

Climbing or prostrate herbs, with milky juice, slender stems, and usually large perennial rhizomes. Flowers large, axillary, solitary or cymose.—Corolla funnel- or bell-shaped, border 5-angled, plaited. Stamens nearly equal, included. Ovary on an annular disk, incompletely 2-celled; style slender, stigmas 2; ovules 4. Capsule 1-celled, 2–4-seeded.

A very large and widely distributed genus.

Bracts large, enclosing the calyx. Peduncles terete.
  Climbing. Leaves 2–4 in., oblong-sagittate, acuminate, deeply 2-lobed at base . . . . 1. *C. Sepium.*
  Prostrate. Leaves ½–1½ in., ovate or deltoid cordate, acute . . . 2. *C. Tuguriorum.*
  Prostrate. Leaves reniform . . . . 3. *C. Soldanella.*
Bracts enclosing the calyx. Peduncles winged . . . . 4. *C. marginata.*
Bracts small, on the peduncles . . . . 5. *C. erubescens.*

1. **C. Sepium,** *Linn.;—Calystegia Sepium,* Br.; Fl. N. Z. i. 183. Stem slender, climbing, and leaves glabrous or pubescent. Leaves large, 2–4 in. long, oblong-sagittate, acuminate, deeply lobed at the base, lobes rounded angled or truncate. Bracts enclosing the calyx and longer than it, ovate or oblong, obtuse or acute. Peduncles 1-flowered, generally twice as long as the petioles, angled or margined. Corolla 2–4 in. broad, white or rose-coloured.

Abundant throughout the islands, *Banks and Solander,* etc. The common Convolvulus or "Bindweed" of England. Rhizome eaten by the natives. Also common in Europe, Australia, and various temperate countries in both hemispheres.

2. **C. Tuguriorum,** *Forst.*;—*Calystegia Tuguriorum*, Br.; Fl. N. Z. i. 183. t. 47. Stem slender, prostrate, rarely climbing, 12–24 in. long, and leaves glabrous. Leaves ½–1½ in. long, broadly ovate-cordate or deltoid, acute or obtuse, entire lobed sinuate or angled, sinus at the base broad. Peduncles longer than the petioles, terete or margined. Bracts as long as the calyx and enclosing it, orbicular or cordate. Corolla 1–2 in. across, white or rose-coloured. Capsule ovate, acute, ¼ in. long. Seeds yellow, small.

Abundant throughout the islands, *Banks and Solander*, etc. A much smaller plant than the preceding, but large specimens are often difficult to distinguish. The same plant is found in Valdivia and Chiloe.

3. **C. Soldanella,** *Linn.*;—*Calystegia Soldanella*, Br.; Fl. N. Z. i. 183. Stems prostrate, glabrous or puberulous, 1 ft. long, stouter than in the preceding. Leaves broader than long, ½–1½ in. diam., reniform or cordate-reniform, acute or obtuse and apiculate, sometimes rather fleshy, sinus at the base broad and open, entire or rarely lobed. Peduncles terete, longer than the leaves. Bracts enclosing the calyx, broadly orbicular or cordate, obtuse or apiculate. Corolla rose-red, 1–2 in. broad. Capsule large, globose. Seeds large, black.

Northern Island: shores near Auckland, *Sinclair*. **Middle** Island: Canterbury, *Haast*; Bluff Island, *Lyall*. The southern specimens are identical with the European and Australian; the Auckland ones approach *C. Tuguriorum* in foliage, and in the absence of fruit may be referable to a state of that plant. This species is found in many temperate and tropical shores.

4. **C. marginata,** *Hook. f.*—*Calystegia marginata*, Br.; Fl. N. Z. i. 184. t. 48. Stems slender, climbing, quite glabrous. Leaves 1–2 in. long, broadly sagittate, acuminate, the basal lobes long, diverging, obtuse or acute, entire lobed or 2-fid. Peduncle shorter than the petiole, with two often crisped wings. Flowers small. Bracts cordate-ovate, obtuse, half as long as the corolla. Corolla white, ½ in. diam.

Northern Island: Wangarui and Owai, on the east coast, *Colenso*; Bay of Islands?, *Sinclair*. Also a native of eastern Australia.

5. **C. erubescens,** *Br.*;—*Fl. N. Z.* i. 185. Stems prostrate, 4–12 in. long, rarely twining, and leaves glabrous pubescent or silky. Leaves very variable, ⅙–½ in. long, oblong hastate or cordate, obtuse, quite entire or sinuate. Peduncles longer than the petioles, with two small subulate bracts above the middle. Sepals broadly oblong, rounded at the tip, coriaceous, silky. Corolla white or rose-coloured, ⅓–⅔ in. diam. Capsule globose. Seeds rugose.

Northern Island: south-west head of Palliser Bay, *Colenso*. **Middle** Island: banks of the Waihopai, *Munro*; Port Cooper, *Lyall*, *Bolton*; Otago, Lower Waitaki river, *Hector and Buchanan*. The flowers appear to be most frequently white in New Zealand. A very common Australian plant, closely allied to the European *C. arvensis*, and some others of very wide distribution.

2. IPOMŒA, Linn.

Climbing herbs (rarely erect), with milky juice, resembling *Convolvulus*, except that the stigma is capitate, 2- or 3-lobed, and the capsule 2- or 3-celled.

A very large tropical and subtropical genus, to which the Sweet-Potato, or Kumerabo, of the New Zealander belongs; this (*Batatas edulis*, Choisy; *Convolvulus chrysorhizus*, Forst.) is cultivated all over the Pacific, and was introduced by the earliest inhabitants.

1. **I. tuberculata,** *Rœm. and Sch.*;—*I. pendula*, Br.; Fl. N. Z. i. 185. Stems slender, glabrous, twining, sometimes tubercled. Leaves 5-foliolate; leaflets $\frac{1}{3}$–1$\frac{1}{2}$ in. long, sessile, lanceolate, acuminate, quite entire, the outer sometimes 2-lobed. Peduncles 1–3-flowered. Flowers drooping, large, rose-coloured. Sepals obtuse. Corolla 1–3 in. diam. Seeds silky.—*Convolvulus mucronatus*, Forst.

Northern Island: east coast, *Banks and Solander*; Bay of Islands, *Cunningham*; Cavalhos Island, *Colenso*. An Australian, Pacific island, and Indian plant, probably the same as *I. palmata*, Forst.

3. DICHONDRA, Forst.

Prostrate herb. Flower solitary, axillary, small.—Calyx 5-partite. Corolla nearly rotate, 5-lobed, imbricate. Ovary of 2 distinct carpels; styles 2, distinct, stigmas capitate; ovule 1 in each carpel, erect. Capsule membranous, indehiscent, 1-seeded.

1. **D. repens,** *Forst.*;—*Fl. N. Z.* i. 185. A small, procumbent, branched, creeping, tufted, silky herb. Leaves petiolate, reniform, $\frac{1}{2}$–1 in. broad, entire or emarginate at the tip. Flowers small, yellow. Corolla shorter than the calyx.

Abundant throughout the islands, *Banks and Solander*, etc. A very common tropical plant in both hemispheres; also found in Tasmania, Australia, and the warmer regions of America.

4. CUSCUTA, Linn.

Leafless, rootless, slender, twining, parasitical herbs, adhering by small lateral suckers to herbs or shrubs, which they derive their nutriment from, and eventually strangle. Flowers white yellow or pink, marked with transparent oil-glands, small, clustered or racemose.—Calyx 5-lobed. Corolla ovoid globular or urceolate. Stamens very short, inserted at the union of the lobes of the corolla, with as many scales below them. Ovary 2-celled; styles 2, stigmas capitate; cells 2-ovuled. Capsule membranous, 2-celled, 2-seeded, dehiscing transversely at the base. Seeds albuminous; embryo terete curved or spiral; cotyledons 0.

A considerable genus, found in Europe and many tropical and temperate parts of the globe: it resembles *Cassytha* in habit; and some species (Dodders) are pests in clover-fields and other crops in England.

1. **C. densiflora,** *Hook. f. Fl. N. Z.* i. 186. Stems slender, matted and twisting together, as thick as stout thread. Flowers crowded into very short 6–10-flowered racemes. Calyx-lobes short, oblong, obtuse. Corolla $\frac{1}{8}$ in. long, bell-shaped; lobes short, rounded, recurved. Scales oblong, fimbriated, united at their bases by a thin membrane. Filaments longer than the anthers; styles rather long.

Middle Island: Port Underwood, *Lyall*. Dr. Engelmann, who has examined all the *Cuscutæ* of the Hookerian Herbarium, observes of this that it hardly differs from the South Brazilian *C. racemosa*, Martius; but according to the descriptions of that plant, the corolla-lobes are erect and acute, and the filaments and style much shorter.

Order LII. SOLANEÆ.

Herbs shrubs or trees. Leaves alternate, exstipulate. Flowers usually in cymes, regular or irregular, hermaphrodite.—Calyx usually 5-lobed, inferior. Corolla 5-lobed, folded in bud. Stamens inserted on the corolla, alternating with the lobes; anthers free or cohering, opening by slits or pores. Ovary superior, 2-celled; style simple, stigma entire or lobed; ovules numerous, on placentas attached to the axis. Fruit a berry (rarely capsular) indehiscent, several-seeded; albumen copious, fleshy; embryo usually curved or spiral.

A very extensive Natural Order, abounding in all temperate and hot latitudes.

1. SOLANUM, Linn.

Herbs shrubs or small trees. Flowers regular.—Calyx 4- or 5-cleft. Corolla rotate or bell-shaped, 4- or 5-fid, with plaited æstivation. Stamens 4 or 5, equal or unequal, filaments short; anthers linear, free or conniving, opening by 2 terminal pores. Style simple, stigma obtuse; ovules very numerous. Berry oblong or globose, 2-celled, cells many-seeded. Seeds flattened or reniform; embryo curved.

A very large genus, found in all temperate and tropical parts of the world, absent in the coldest.

Tall and stout. Flowers $\frac{1}{4}$–$\frac{1}{2}$ in. diam., blue or purple. Anthers spreading 1. *S. aviculare.*
Slender. Flowers $\frac{1}{4}$ in. diam., whitish. Anthers connivent . . . . . 2. *S. nigrum.*

1. **S. aviculare,** *Forst.;—Fl. N. Z.* i. 182. Stem tall, herbaceous, glabrous, angled, shrubby at the base, branched, leafy. Leaves very variable, 4–10 in. long, lanceolate or oblong, entire or variously lobed or pinnatifid, membranous, glabrous, veins divaricating at right angles. Flowers in axillary or supra-axillary 3–10-flowered cymes, large, purplish or bluish. Calyx-lobes short, obtuse. Anthers spreading. Berry ovoid, edible.—*S. laciniatum,* Aiton; Bot. Mag. t. 349.

Throughout the islands, common in woods, *Banks and Solander,* etc. Also found in Norfolk Island, in Southern Australia, and Tasmania. The spreading anthers are unusual in the genus.

2. **S. nigrum,** *Linn.;—Fl. N. Z.* i. 182. Stems slender, branched, glabrous, 1–3 ft. high. Leaves petioled, 1–4 in. long, ovate, acuminate, rarely sinuate or lobed. Flowers $\frac{1}{4}$–$\frac{1}{3}$ in. diam., umbelled, umbels on long supra-axillary peduncles. Berry $\frac{1}{4}$–$\frac{1}{3}$ in. diam., globose, black or red.

Abundant in waste places near houses, etc., throughout the islands. One of the commonest weeds in the world.

The "Cape Gooseberry," *Physalis peruviana,* Linn., is naturalized in the northern districts of New Zealand. The Potato, Capsicum, and Tomato also occur as escapes from cultivation.

Order LIII. SCROPHULARINEÆ.

Herbs, shrubs, or small trees. Leaves opposite, (except *Pygmæa*) exstipulate. Flowers axillary or in terminal racemes cymes or panicles, irregular, rarely regular, hermaphrodite.—Sepals 5, free or variously cohering, inferior, very rarely half-superior, persistent. Corolla regular or 2-lipped, imbricate in bud. Stamens 2 or 4, with sometimes a rudimentary fifth between the upper corolla-lobes. Ovary 2-celled; style simple, stigma simple or 2-lobed, or of 2 plates; ovules numerous in each cell. Capsule 2-celled, many-seeded. Seeds albuminous.

One of the most extensive Natural Orders, found in all quarters of the globe. The New Zealand genus *Pygmæa*, of which I have seen no fruit, is a doubtful member of the Order; unlike its co-ordinates, the leaves appear to be imbricated all round the stem, and not opposite.

Stamens 2.

Calyx 4-partite. Corolla with 2 inflated lips. Stigma subcapitate . 1. CALCEOLARIA.
Calyx 5-partite. Corolla 2-lipped. Stigma 2-lamellate . . . . . 4. GRATIOLA.
Calyx 4- or 5-partite. Corolla 4- or 5-lobed. Stigma subcapitate. Leaves opposite . . . . 7. VERONICA.
Calyx 5- or 6-partite. Corolla 5- or 6-lobed. Stigma subcapitate. Leaves imbricate, alternate . . . . 8. PYGMÆA.

Stamens 4.

Calyx 5-toothed. Corolla 2-lipped, tumid at throat. Stigma 2-lamellate . . . . 2. MIMULUS.
Calyx 5-lobed. Corolla 2-lipped, not tumid at throat. Stigma 2-lamellate . . . . 3. MAZUS.
Calyx 3–5-lobed. Corolla minute. Anther 1-celled. Stigma spathulate . . . . 5. GLOSSOSTIGMA.
Calyx 5-toothed. Corolla minute, rotate. Anther 1-celled. Stigma clavate . . . . 6. LIMOSELLA.
Calyx 5-partite. Corolla 5-lobed. Stigma capitate . . . . . . 9. OURISIA.
Calyx 4-lobed or -toothed. Corolla 2-lipped. Stigma dilated . . . 10. EUPHRASIA.

1. CALCEOLARIA, Linn.

Herbs. Leaves radical and cauline. Flowers in axillary or terminal racemes.—Calyx 4-partite, inferior or half-superior. Corolla-tube very short; limb 2-lipped; lips nearly equal and both inflated in the New Zealand species, the upper small and lower very large and inflated in the American ones. Stamens 2; anthers 2-celled. Ovary 2-celled; style simple, stigma subcapitate; ovules numerous, placentas on the septum. Capsule ovoid, septicidal, 2-valved; valves 2-fid. Seeds numerous, striate.

A very large South American and especially Andean genus, of which the only extra American species are the New Zealand ones, which belong to a section (*Jovellana*), having the two lips of the corolla nearly equal.

Stem erect. Leaves oblong, 2–4 in. long. Calyx inferior . . . . . 1. *C. Sinclairii.*
Stem creeping. Leaves ovate, $\frac{1}{3}$–$\frac{3}{4}$ in. long. Calyx half-superior. . . 2. *C. repens.*

1. **C. Sinclairii,** *Hook. Ic. Pl. t.* 561;—*Fl. N. Z.* i. 187. An erect, slender, glandular, pubescent herb, 6–18 in. high. Leaves with slender petioles, 2–4 in. long, oblong, coarsely-toothed or lobulate, the lobes again toothed, very membranous, sometimes cordate at the base. Panicle branched. Calyx-lobes ovate, obtuse or acute. Corolla downy, $\frac{1}{3}$ in. diam., yellow? spotted with purple. Capsule $\frac{1}{6}$ in. long.

Northern Island: East Cape, *Sinclair;* Hawke's Bay, *Colenso.* Very closely allied to the Chilian *C. punctata*, but differing a little in foliage, and in the corolla apparently yellow, not purple.

2. **C. repens,** *Hook. f. Fl. N. Z.* i. 187. A very slender, branched, creeping, pubescent herb. Leaves with slender petioles, $\frac{1}{3}$–$\frac{3}{4}$ in. long, orbicular, broadly ovate-oblong or ovate-cordate, irregularly and unequally doubly-toothed or crenate, very membranous. Panicles very few-flowered. Flowers $\frac{1}{4}$ in. across. Calyx half-superior.

Northern Island: ravines and forests at the base of the Ruahine range, *Colenso.*

2. MIMULUS, Linn.

Erect or creeping herbs. Flowers solitary and axillary in the New Zealand species.—Calyx tubular or short, terete or 5-angled, 5-toothed or -lobed. Corolla campanulate, 2-lipped; upper lip erect or reflexed, 2-lobed; lower 3-lobed, usually with 2 protuberances at the throat, lobes all flat. Stamens 4; anther-cells diverging, finally confluent. Stigma 2-lamellate. Capsule loculicidal, 2-valved; valves separating from the placentas.

A considerable genus, found in various parts of the world, but not in Europe, except as an introduced plant.

Glabrous. Leaves sessile . . . . 1. *M. repens.*
More or less pilose. Leaves petiolate . . . . 2. *M. radicans.*

1. **M. repens,** *Br.;—Fl. N. Z.* i. 188. A small, creeping, succulent, perfectly glabrous herb; stems branched, 1–5 in. long. Leaves $\frac{1}{6}$–$\frac{1}{4}$ in. long, oblong or broadly ovate, sessile or stem-clasping, quite entire. Peduncle axillary, 1-flowered, longer or shorter than the leaves. Calyx variable in form, from obconic to hemispherical; lobes obscure. Corolla large, $\frac{1}{2}$ in. across, pale-blue with yellow throat.—Bot. Mag. t. 5423.

Northern and **Middle** Islands: not rare in muddy places, bogs, etc., *Banks and Solander*, etc. Also a common Tasmanian and South-Eastern Australian plant.

2. **M. radicans,** *Hook. f. Fl. N. Z.* i. 188. Stem creeping and rooting, with short, leafy branches. Leaves spreading, close together, $\frac{1}{4}$–1 in. long, petiolate, obovate, obtuse, quite entire, glabrous or pilose. Peduncle stout, erect, longer or shorter than the leaf, 1- or 2-flowered, with a subulate bract in the middle or at the fork. Calyx 5-cleft, pilose. Corolla large, $\frac{1}{2}$–$\frac{3}{4}$ in. broad.

Northern Island: Tararua mountains and Wairarapa valley, *Colenso.* **Middle** Island: common in swampy places, ascending to 1200 ft., from Nelson, *Munro*, etc., to Otago, *Lindsay.*

3. MAZUS, Loureiro.

Herbs. Leaves opposite or fascicled. Flowers in terminal, leafless racemes, or solitary.—Calyx bell-shaped, 5-fid. Corolla: upper lip 2-fid; lower larger, 3-fid, with two protuberances at the throat. Stamens 4; anther-cells diverging. Stigma equally 2-lamellate. Capsule globose or compressed, loculicidal, 2-valved; valves entire, separating from the placentas.

A small Indian, Chinese, and Australian genus.

1. **M. Pumilio,** *Br.;—Fl. N. Z.* i. 189. Stem creeping underground, sending out very short, leafy branches. Leaves fascicled, spreading, $\frac{1}{2}$–2 in. long, petioled, narrow obovate-spathulate, obtuse, nearly entire or lobulate, membranous, glabrous or sparingly pilose. Scape slender, 1–6-flowered.

Flowers on slender, curved pedicels, with a subulate bract at the base or middle. Corolla blue, $\frac{1}{4}$-$\frac{1}{2}$ in. diam.—Hook. Ic. Pl. t. 567.

Northern and **Middle** Islands, *Banks and Solander*, etc.: common as far south as Canterbury. Also common in South-Eastern Australia and Tasmania.

4. GRATIOLA, Linn.

Erect or creeping herbs. Leaves small, usually sessile. Peduncles axillary, 1-flowered.—Calyx 5-partite. Corolla: upper lip entire or shortly 2-fid; lower 3-fid; throat without protuberances. Stamens 2 fertile, and 2 reduced to filaments; anther-cells distinct, parallel. Stigma inflated or 2-lamellate. Capsule 4-valved; valves falling away from the placentas. Seeds numerous, small.

A considerable tropical and subtropical genus, scattered over the world, rare in temperate regions though found in Europe.

Leaves $\frac{1}{4}$–$\frac{3}{4}$ in. long, glabrous, toothed . . . . 1. *G. sexdentata.*
Leaves $\frac{1}{6}$–$\frac{1}{4}$ in. long, glabrous or puberulous, obtusely toothed . . . 2. *G. nana.*

1. **G. sexdentata,** *A. Cunn.*;—*Fl. N. Z.* i. 189. Stems stout, ascending or suberect, 6–18 in. long, glabrous. Leaves $\frac{1}{4}$–$\frac{3}{4}$ in. long, sessile, oblong, subacute, with scattered teeth. Peduncles very short. Flowers $\frac{1}{3}$ in. long, yellow. Anthers 2-celled, one cell sometimes empty; sterile filaments elongated.

Northern Island: marshy places, not uncommon, *Banks and Solander*, etc., probably overlooked in the Middle Island. Also found in South-Eastern Australia and Tasmania; and most closely allied to the South American *G. peruviana.*

2. **G. nana,** *Benth.*;—*Fl. N. Z.* i. 189. Stems short, matted, creeping, and as well as the leaves, glabrous or puberulous. Leaves $\frac{1}{6}$–$\frac{1}{4}$ in. long, oblong, obtuse, obtusely-toothed. Peduncles very short. Flowers as in *G. sexdentata.*

Northern Island: Bay of Islands, *R. Cunningham*, and elsewhere probably common but overlooked. **Middle** Island: Kowai valley, *Haast.* Much smaller than *G. sexdentata.* Flowers white or pinkish. Also found on the alps of Tasmania. Very like *Mimulus repens*, in general appearance.

Herpestes cuneifolia, Spr., is introduced into Raoul's catalogue of New Zealand plants, no doubt by mistake for a *Gratiola*, or for *Mimulus repens*, which it closely resembles.

5. GLOSSOSTIGMA, Arnott.

Minute, tufted, creeping herbs. Peduncles axillary, 1-flowered.—Calyx bell-shaped, 3–5-lobed. Corolla most minute, upper lip 2-lobed, lower 3-lobed. Stamens 2–4; anthers 1-celled, exserted. Stigma large, dilated, spathulate. Capsule subglobose, loculicidal, 2-valved; valves separating from the placentas.

Minute Indian, African, and Australian herbs.

1. **G. elatinoides,** *Benth.*;—*Fl. N. Z.* i. 189. Glabrous. Stems rooting at the nodes, 1–2 in. long. Leaves petioled, spathulate, $\frac{1}{6}$–$\frac{1}{4}$ in. long, quite entire, obtuse. Peduncles shorter than the leaves. Flowers $\frac{1}{12}$ in. long. Stamens 4; anthers peltate, exserted.—*Tricholoma elatinoides*, Benth.; *Lobelia submersa*, A. Cunn.

Northern Island: common in wet places, *A. Cunningham*, etc.

6. LIMOSELLA, Linn.

Small, tufted, marsh or water plants. Leaves linear or linear-spathulate. Peduncles axillary, 1-flowered. Flowers minute.—Calyx campanulate, 5-toothed. Corolla rotate; limb 5-fid; segments unequal. Stamens 4; anthers 1-celled, included. Stigma subclavate. Capsule subglobose, 2-valved; valves entire, separating from the placenta.

A small genus of plants found in all parts of the globe, probably all the species enumerated in books are varieties of one.

1. **L. aquatica,** var. **tenuifolia,** *Linn.;—Fl. N. Z.* i. 190. Leaves $\frac{1}{2}$–$1\frac{1}{2}$ in. long, obtuse. Peduncles solitary or several together. Flowers white, $\frac{1}{12}$ in. across or less.—*L. australis*, Br.; *L. tenuifolia*, Nuttall.

Throughout the islands, common in wet places. A widely distributed plant in the temperate and cold regions of both hemispheres and tropical mountains.

7. VERONICA, Linn.

Herbs or shrubs, rarely small trees. Leaves very various, opposite, often connate at the base, small and scale-like or large, sometimes minute and most densely imbricating quadrifariously. Flowers small, usually in axillary racemes, sometimes spiked corymbose or panicled, rarely solitary.—Sepals 4, rarely 5 (one being 2-fid or 2-partite). Corolla with a short or long tube, and expanded 4- rarely 5-lobed limb; lobes unequal. Stamens 2, filaments long or short, inserted at the throat of the corolla. Ovary small, compressed, 2-celled; style slender, stigma capitate; ovules numerous. Capsule 2-celled, ovoid orbicular or didymous, dorsally or laterally compressed, septicidally dehiscing; valves often splitting longitudinally, falling away from the seed-bearing septum. Seeds numerous or few.

A very large European, Oriental, and New Zealand genus, comparatively rare in other parts of the globe. In New Zealand it forms a more conspicuous feature of the vegetation than in any other country, both from the number, beauty, and ubiquity of the species, from so many forming large bushes, and from the remarkable forms the genus presents. The species are excessively difficult of discrimination, present numerous intermediate forms between many most distinct-looking ones, vary extremely in all their organs and hybridize most freely; many probably are, if not bisexual, still partially so, the two sexes presenting differences in the size of the stamens and calyx and capsules, a point worthy of the close attention of the colonist.

Between the first 19 species it is most difficult to draw any contrasting specific characters, they appear to present a graduated scale of forms. *V. elliptica* alone, I find it impossible to confound with any other which is the more remarkable, as it is the only New Zealand shrubby species that extends beyond the islands, and inhabits South Chili, etc.; yet, except for the large size of the white flower and large fruit, it is difficult to point out any character of importance to distinguish it from forms of several others.

Of the curious species of section 4 (viz. 20 to 25), all seem very distinct and well marked; though intermediates are quite conceivable, and may be found, it would be instructive to know if they will hybridize together and with the other sections.

Section 5 also presents a most remarkable form of the genus, quite new, and peculiar to New Zealand; the two species it includes seem distinct.

Of the 5 species in section 6, *V. macrantha* and *Benthami* are very distinct, and the latter a most beautiful and remarkable plant; the three others are more closely allied, but I think distinct.

In section 7, *V. linifolia* and *Anagallis* are very different from one another and any others; 34 to 37 are probably all forms of one plant, as is perhaps *V. spathulata* of *V. elongata.*

So many new species of this genus (19) have been found since the publication of the 'Flora Novæ-Zelandiæ,' that probably many more will reward the researches of collectors; on the other hand, no doubt some of the species here described will be reduced by future observers.

In the following Conspectus I have had regard to *prevalent* prominent differences only; there is not an organ that does not vary conspicuously in every species, and I regret to add that I have been obliged to neglect sundry specimens from inability to refer them exactly to any species. In such a genus as this, characters must be arbitrarily adopted, and be regarded as provisional only.

I. *Capsule dorsally compressed, ovoid, turgid; valves 2-fid at the tip.*

A. *Leaves quite entire (rarely toothed in* V. Haastii).

§ 1. *Large shrubs. Leaves oblong or obovate* 1–4 *in. long, obtuse or subacute. Racemes simple, short, stout, many-flowered.*

Leaves 2–4 × $1\frac{1}{2}$ in., obovate. Stamens stout. Capsule $\frac{1}{4}$ in. . . . 1. *V. speciosa.*
Leaves 3 × 1 in., linear-oblong. Stamens slender. Capsule $\frac{1}{4}$ in. . . 2. *V. Dieffenbachii.*
Leaves 1–2 × $\frac{1}{2}$–1 in., obovate-oblong, subacute. Stamens slender. Capsule $\frac{1}{8}$ in. . . . . 3. *V. macroura.*

§ 2. *Large shrubs. Leaves linear or lanceolate,* 2–6 *in. long. Racemes simple, long, slender, very many-flowered.*

Leaves 2–6 in. Capsule $\frac{1}{8}$ in., scarcely longer than sepals . . . 4. *V. salicifolia.*
Leaves 3–6 in. Capsule $\frac{1}{4}$ in., 2 or 3 times as long as sepals . . 5. *V. macrocarpa.*
Leaves 1–3 in. Capsule a little longer than the small, obtuse sepals 6. *V. parviflora.*
Leaves 1–3 in. Capsule $\frac{1}{4}$ in., 2 or 3 times as long as the lanceolate sepals . . . . 7. *V. ligustrifolia.*
Leaves 1–2 in., hairy on both surfaces . . . . 8. *V. pubescens.*

§ 3. *Large or small, erect shrubs. Leaves* $\frac{1}{6}$–1 *in. long, coriaceous. Racemes or spikes peduncled, usually short, simple or corymbose, or collected into heads. Bracts usually small, large in* 13 *and* 18.

α. *Leaves lax, spreading, not imbricate. Branches even.*

Leaves $\frac{3}{4}$–1 × $\frac{1}{6}$–$\frac{1}{3}$ in. Racemes long. Pedicels usually distinct . . 9. *V. Traversii.*
Leaves $\frac{1}{3}$–$\frac{1}{2}$ × $\frac{1}{4}$–$\frac{1}{3}$ in. Spikes attenuate. Pedicels 0 . . . . . 10. *V. vernicosa.*
Leaves $\frac{1}{3}$–$\frac{2}{3}$ × $\frac{1}{4}$–$\frac{1}{3}$ in., petioled and apiculate. Branches hoary. Corolla $\frac{1}{2}$–$\frac{3}{4}$ in., white . . . . 11. *V. elliptica.*
Leaves $\frac{1}{2}$–$\frac{2}{3}$ × $\frac{1}{6}$ in., keeled. Flowers corymbose. Pedicels slender . 12. *V. diosmæfolia.*
Leaves $\frac{1}{2}$–1 × $\frac{1}{4}$–$\frac{1}{3}$ in., not keeled. Spikes short. Pedicels 0. Bracts large . . . . 13. *V. Colensoi.*

β. *Leaves excessively thick, concave, more or less imbricate, often closely. Branches with close-set transverse scars.*

Leaves $\frac{1}{3}$–$\frac{1}{2}$ × $\frac{1}{4}$–$\frac{1}{3}$ in., keeled, not truncate nor cordate at base . . 14. *V. lævis.*
Leaves $\frac{1}{4}$–$\frac{1}{3}$ × $\frac{1}{5}$–$\frac{1}{4}$ in., keeled, subcordate at base . . . . 15. *V. buxifolia.*
Leaves $\frac{1}{3}$–$\frac{2}{3}$ × $\frac{1}{4}$–$\frac{1}{3}$ in., midrib obscure. Capsule ovate, acute, glabrous 16. *V. carnosula.*
Leaves $\frac{1}{6}$–$\frac{1}{2}$ × $\frac{1}{8}$–$\frac{1}{2}$ in., midrib obscure. Capsule broad, obtuse, pubescent . . . . 17. *V. pinguifolia.*
Leaves $\frac{1}{6}$–$\frac{1}{4}$ in. Spikes tomentose, distichous. Bracts large, concave 18. *V. pimeleoides.*
Leaves $\frac{1}{8}$–$\frac{1}{4}$ in., orbicular, spreading and recurved. Spike tomentose, dense. Ovary villous . . . . 19. *V. Buchanani.*

§ 4. *Erect or decumbent shrubs. Leaves very minute, thick and short,* $\frac{1}{20}$–$\frac{1}{12}$ *in. long, densely 4-fariously imbricate, or in distant pairs. Flowers fascicled or capitate at the ends of the branches.*

α. *Leaves most densely imbricate, connate in pairs.*

Branches square, $\frac{1}{12}$–$\frac{1}{8}$ in. diam. Leaves black when dry, tumid, obtuse . . . . 20. *V. tetragona.*

Branches square, $\frac{1}{8}$–$\frac{1}{10}$ in. diam. Leaves brown when dry, abruptly acuminate . . . 21. *V. lycopodioides.*
Branches square, $\frac{1}{12}$ in. diam. Leaves very broad, subacute, black when dry . . . 22. *V. tetrasticha.*
Branches terete. Leaves connate to middle, brown when dry . . . 23. *V. Hectori.*
Branches terete. Leaves truncate, yellowish when dry . . . 24. *V. salicornioides.*
β. *Leaves in distant, opposite pairs* . . . 25. *V. cupressoides.*

§ 5. *Small decumbent shrubs, with short flexuous branches, densely clothed with short, broad, rigid, densely imbricating leaves,* $\frac{1}{6}$–$\frac{1}{3}$ *in. long. Flowers in terminal, sessile, ovoid heads, continuous with the branches.*

Leaves $\frac{1}{4}$–$\frac{1}{3}$ in., not keeled. Sepals not ciliate . . . 26. *V. Haastii.*
Leaves $\frac{1}{6}$–$\frac{1}{3}$ in., recurved, keeled. Sepals ciliate . . . 27. *V. epacridea.*

B (§ 6). *Leaves more or less toothed or serrate.* (*See* Haastii, *in* § 5.)

Glabrous. Flowers racemose, $\frac{3}{4}$ in. diam. . . . 28. *V. macrantha.*
Flowers sessile on branches of a loose panicle 3–10 in. long . . . 29. *V. Hulkeana.*
Flowers sessile, in short, spreading, puberulous panicles. Bracts acute 30. *V. Lavaudiana.*
Flowers sessile, in dense oblong panicles. Bracts obtuse . . . 31. *V. Raoulii.*
Flowers indense, leafv racemes. Leaves edged with down . . . 32. *V. Benthami.*

II (§ 7). *Capsules laterally compressed, didymous.—Herbs with creeping, diffuse, slender stems (or erect in* 40). *Leaves serrate or toothed, except in* V. linifolia. *Flowers in slender, axillary, long-peduncled racemes (short in* 38 *and* 39).

Leaves $\frac{1}{3}$–1 in., linear, obtuse, quite entire . . . 33. *V. linifolia.*
Leaves $\frac{1}{4}$–$\frac{1}{2}$ in., ovate, glabrous, serrate. Raceme glandular . . . 34. *V. nivalis.*
Leaves $\frac{1}{4}$–$\frac{1}{2}$ in., ovate or oblong, serrate, glabrous. Peduncle long . 35. *V. Lyallii.*
Leaves $\frac{1}{12}$–$\frac{1}{4}$ in., ovate or oblong, glabrous, with 1 or 2 teeth on each side. Peduncle long . . . 36. *V. Bidwillii.*
Leaves 1–5 in., oblong ovate or lanceolate, glabrous, deeply serrate . 37. *V. cataractæ.*
Leaves $\frac{1}{2}$–1 in., glabrous or pubescent, broad, ovate-cordate, coarsely toothed . . . 38. *V. elongata.*
Leaves $\frac{1}{6}$–$\frac{1}{4}$ in., glandular-pubescent, ovate-spathulate. Peduncles very short . . . 39. *V. spathulata.*
Erect glabrous. Leaves oblong, 1–2 in., crenate . . . 40. *V. Anagallis.*

1. **V. speciosa,** *R. Cunn.*;—*Fl. N. Z.* i. 191. A glabrous stout shrub; branches angled, as thick as a goose-quill. Leaves sessile or with very short thick petioles, 2–4 in. long, 1–1½ broad, obovate-oblong, rounded at the tip, very coriaceous, shining, quite entire, downy on the midrib above, veins obsolete. Racemes dense-flowered, not longer than the leaves, 1 in. diam., stout, erect; pedicels short. Calyx-lobes oblong-ovate, subacute. Corolla deep blue-purple, $\frac{1}{3}$ in. diam., lobes obtuse. Stamens and style very long and stout. Capsule $\frac{1}{4}$ in. long, broadly ovate, thrice as long as the calyx.—Bot. Mag. t. 4057.

Northern Island: seacoast at Hokianga, *Cunningham*. **Middle** Island: Ship Cove and Port Nicholson, *Lyall*. I have a cultivated hybrid between this and *V. elliptica*, with leaves only 1½ in. long, and racemes 1 in. long, of blue flowers, raised by I. A. Henry, Esq., of Trinity, Edinburgh.

2. **V. Dieffenbachii,** *Benth.*;—*Fl. N. Z.* i. 191. A stout glabrous shrub; branches terete, almost as stout as a quill. Leaves sessile by a subcordate base, 3 in. long, 1 broad, linear-oblong, acute, rather coriaceous, quite entire, downy on the edges towards the base. Raceme longer than the leaves, $\frac{3}{4}$ in. diam., strict, glabrous; pedicels slender. Sepals small, ovate-

lanceolate, acute. Corolla $\frac{1}{4}$ in. diam. Stamens and style slender. Capsule $\frac{1}{4}$ in. long, broadly ovate, thrice as long as the calyx.

Chatham Island, *Dieffenbach*. I have seen but one specimen.

3. **V. macroura,** *Hook. f. Fl. N. Z.* i. 191. A glabrous shrub, 1–6 ft. high; branches terete, as stout as a crow-quill. Leaves narrowed into very short petioles, 1–2 in. long, $\frac{1}{2}$–1 broad, linear-oblong or slightly obovate, acute, glabrous. Racemes rather longer than the leaves, stout, slightly curved, dense-flowered, $\frac{1}{2}$–1 in. diam., puberulous or pubescent; pedicels slender. Sepals small, ovate, acute. Corolla $\frac{1}{4}$–$\frac{1}{8}$ in. diam.; tube rather long. Stamens and style slender. Capsules small, $\frac{1}{8}$ in. long, most densely crowded, recurved, very little longer than the sepals.

Northern Island: East Cape, Wangarei, Cook's Straits, etc., *Colenso*. **Middle** Island: Tarndale, *Sinclair* (flowers $\frac{1}{4}$ in. diam.).

4. **V. salicifolia,** *Forst.*;—*Fl. N. Z.* i. 191. A large glabrous shrub; branches terete, as thick as a crow-quill. Leaves sessile, 2–6 in. long, linear- or oblong-lanceolate, acuminate, quite entire, glabrous. Racemes much longer than the leaves, simple, very many-flowered, pubescent or glabrate; pedicels slender. Flowers extremely variable in size and length of tube of corolla, bluish-purple or white. Sepals oblong-lanceolate, obtuse, rarely acute, glabrous or pubescent. Corolla $\frac{1}{8}$–$\frac{1}{4}$ in. diam. Capsule $\frac{1}{8}$ in. long, ovate, acute, not twice as long as the calyx.—*V. Lindleyana*, Paxt. Mag. Bot.; *V. stricta*, Banks and Solander.

Abundant throughout the islands, *Banks and Solander*, etc. The *V. Foukii*, Philippi, of Chili, is (no doubt) founded on cultivated specimens of the New Zealand plant, erroneously supposed to have been sent from Guayticas Archipelago. This species passes into *V. parviflora* by becoming smaller-leaved, and into *macrocarpa* by the fruit. Many varieties of this, and hybrids between it and *V. speciosa*, *macrocarpa*, and others, are extensively cultivated in England under various names (*Kermesina*, *Lindleyana*, and *Andersoni* of gardens, *versicolor* and *linariæfolia*, Visiani, etc. etc.).

5. **V. macrocarpa,** *Vahl*;—*Fl. N. Z.* i. 192. Characters of *V. salicifolia*, but flowers usually larger, and capsule nearly $\frac{1}{4}$ in. long, three times as long as the calyx.—*V. myrtifolia*, Banks and Solander; *V. salicifolia*, A. Cunn. Herb.

Northern and **Middle** Islands: not uncommon, Bay of Islands, *A. Cunningham*, etc.; Mount Egmont, *Dieffenbach*; Cook's Straits, *D'Urville*; Bay of Islands, *Logan*; Port William, *Lyall*. I have specimens of this from Lyall, gathered at Otago and Port William, with oblong-lanceolate leaves, 1 in. long, and short broad racemes 1–2 in. long, with very large flowers upwards of $\frac{1}{3}$ in. diam. This species is too closely allied to *V. salicifolia*, and indeed differs materially in the capsule only.

6. **V. parviflora,** *Vahl*;—*Fl. N. Z.* i. 192. A glabrous shrub, usually 4–6 ft. high. Leaves erect or spreading, 1–3 in long, lanceolate oblong-lanceolate or linear-lanceolate, flat or concave and keeled, quite entire, acute or acuminate. Racemes generally strict, about twice as long as the leaves, dense-flowered, pubescent; pedicels short. Flowers small. Sepals small, ovate, obtuse, puberulous. Corolla $\frac{1}{8}$–$\frac{1}{6}$ in. diam. Capsule $\frac{1}{8}$ in., a little longer than the calyx.—*V. angustifolia*, A. Rich.; *V. stenophylla*, Steud.

Northern and **Middle** Islands: abundant. This seems to be a small-leaved form of *V. salicifolia*. Specimens with more oblong obtuser leaves almost pass into *V. macroura*.

Vahl describes the capsule as twice as long as the calyx; it is rarely so long in my specimens but it is variable in this respect, passing thus into *V. ligustrifolia*, as *V. salicifolia* does into *V. macrocarpa*, and *V. macroura* into *V. Dieffenbachii*. Colenso sends specimens from Cook's Straits with very narrow leaves, and fruit as large as in *V. ligustrifolia*.

7. **V. ligustrifolia,** *A. Cunn.*;—*Fl. N. Z.* i. 192. A large glabrous diffusely-branched shrub. Leaves 1½–3 in. long, usually very narrow linear-lanceolate, acuminate, ⅙–¼ in. broad, flat or concave and keeled at the back, quite entire, sometimes broader, ½–¾ in. and more, obtuse. Racemes about twice as long as the leaves, rather slender, lax-flowered, puberulous; pedicels slender, often ⅙–¼ in. long. Flowers rather large. Sepals lanceolate, acuminate. Corolla ¼ in. across. Capsule nearly ¼ in. long, twice or thrice as long as the calyx.—*V angustifolia*, A. Cunn. not A. Rich.; *V. acutiflora*, Benth.

Northern Island: abundant at Bay of Islands, near the falls of the Keri Keri and elsewhere. **Middle** Island: Otago and Port William, *Lyall*. **Kermadec** Island, *M'Gillivray*. This is probably a small-leaved form of *V. macrocarpa*: it differs from *V. parviflora* in the usually larger flowers, acuminate lanceolate sepals, and larger capsule.

8. **V. pubescens,** *Banks and Solander*;—*Fl. N. Z.* i. 193. A shrub, 6 ft. high, covered everywhere with (when dry) red-brown hairs. Leaves 1–2 in. long, oblong-lanceolate, entire. Racemes many-flowered. Sepals oblong-lanceolate, acuminate, pubescent. Capsule twice as long as the calyx.

Northern Island: Opuragi, in woods. A remarkable plant, of which I have seen no specimens but the Banksian.

9. **V. Traversii,** *Hook. f., n. sp.* A small glabrous shrub; branches terete. Leaves spreading, sessile, ¾–1 in. long, ⅙–⅓ broad, obovate or linear-oblong, acute or obtuse, entire, coriaceous, flat; midrib strong. Racemes longer than the leaves, subterminal, 1–2 in. long, many-flowered, puberulous; pedicels distinct, slender or reduced to 0; bracts very small. Sepals $\frac{1}{16}$ in. long, ovate, obtuse or subacute, ciliate, 2–5 times shorter than the cylindric corolla-tube. Corolla-tube long or short; lobes ¼ in. diam. Capsule ⅕ in. long, oblong, acute, 3 or 4 times longer than the calyx.

Middle Island: hills near Canterbury and abundantly in river beds, *Travers*; Southern Alps, abundant, ascending to 4000 ft., *Haast*; Otago, abundant in the lake district, *Hector and Buchanan*. Allied to *V. parviflora*, of which it has the calyx, whilst the fruit is like that of *V. ligustrifolia*. Mr. Travers sends an instructive series of specimens, showing the passage from long to no pedicels, and from long to very short corolla.

10. **V. vernicosa,** *Hook. f., n. sp.* A small stout glabrous shrub. Leaves close-set, spreading, petioled, ⅓–½ in. long, ¼–⅓ broad, obovate-oblong, obtuse or apiculate, varnished on the upper surface, entire, flat or a little concave; midrib evident. Racemes crowded at the ends of the branches, puberulous, 1–1½ in. long, often peduncled, tapering or caudate; pedicels 0 or very short; bracts very small. Sepals $\frac{1}{16}$ in. long, oblong, obtuse. Corolla-tube very short; limb ⅙–¼ in. diam. Capsules pedicelled, ovoid, ⅙ in. long, twice as long as the calyx.

Middle Island: common, Canterbury hills, alt. 1200 ft., *Travers*; Upper Wairau, *Munro*; Southern Alps, alt. 1500–3000 ft., *Haast*; Nelson mountains, *Rough* (flower very large); Otago, Dun mountains, 3900 ft., *Lindsay*. A very pretty and distinct-looking plant, remarkable for the numerous spreading, broad, short, varnished leaves, and much attenuated racemes of nearly sessile flowers.

11. **V. elliptica,** *Forst.;—Fl. N. Z.* i. 193. A bushy large shrub or small tree, 5–20 ft. high; branches stout, hoary all round or in 2 lines, Leaves close-set, spreading, uniform, petiolate, $\frac{1}{3}$–$\frac{2}{3}$ in. long, $\frac{1}{4}$–$\frac{1}{3}$ broad, linear-oblong or obovate-oblong, more or less truncate at the base, entire, flat, coriaceous, not shining; midrib strong, produced at the tip. Racemes very short, few-flowered, forming together a loose subcorymbose head at the ends of the branches, glabrous; pedicels distinct; bracts small, lanceolate. Sepals large, $\frac{1}{8}$ in. long, ovate, acuminate. Corolla large, white; tube short; limb $\frac{1}{3}$–$\frac{2}{3}$ in. broad. Capsule $\frac{1}{4}$ in. long, turgid, twice as long as the calyx. —*V. decussata,* Aiton; Bot. Mag. t. 242; *V. Menziesii,* Benth.

Middle Island: Dusky Bay, *Forster;* Canterbury, *Haast;* Otago, abundant by the sea. **Lord Auckland's** group and **Campbell's** Island, frequent, *J. D. H.* Also found in South Chili, Fuegia, and the Falkland Islands. A very handsome and most distinct species, easily recognized by its oblong, spreading, apiculate, petioled leaves, and large white subcorymbose flowers.

12. **V. diosmæfolia,** *R. Cunn.;—Fl. N. Z.* i. 193. A glabrous shrub, 3–12 ft. high; branches rather slender. Leaves petioled, close-set, spreading, rigidly coriaceous, $\frac{1}{2}$–$\frac{2}{3}$ in. long, $\frac{1}{6}$ broad, linear-oblong, acute at both ends, entire, not shining, sharply keeled by the midrib below. Corymbs terminal, depressed, many-flowered; pedicels slender; bracts lanceolate, acuminate. Sepals small, $\frac{1}{10}$–$\frac{1}{20}$ in. long, ovate, obtuse or acute, glabrous. Corolla-tube short; limb $\frac{1}{6}$–$\frac{1}{4}$ in. across. Capsule twice as long as the calyx.

Northern Island: Bay of Islands and Hokianga, *Cunningham, Edgerly, Colenso,* etc. A very distinct species, known by its fascicled branches, narrow rigid acute patent leaves, and small truly corymbose flowers.

13. **V. Colensoi,** *Hook. f.—V. Menziesii,* Benth. in part; Fl. N. Z. i. 193. A small glabrous shrub. Leaves patent or erecto-patent, almost sessile, $\frac{1}{2}$–1 in. long, $\frac{1}{4}$–$\frac{1}{3}$ in. broad, very coriaceous, linear-oblong, or narrow oblong-obovate, acute, entire, narrowed into the petiole, flat, not keeled, opaque, sometimes glaucous; midrib distinct, excurrent. Racemes subterminal, often compound, short, peduncled, hardly longer than the leaves, puberulous; pedicels very short or 0; bracts as long as the sepals, coriaceous. Sepals ovate-oblong, obtuse or subacute, $\frac{1}{12}$ in. long. Corolla white pink or bluish; tube short; limb $\frac{1}{8}$–$\frac{1}{4}$ in. diam. Capsule $\frac{1}{5}$ in., twice as long as the calyx.

Northern Island: Ruahine mountains, *Colenso* (racemes simple, leaves glaucous). **Middle** Island: Nelson mountains, *Bidwill;* Tarndale, *Sinclair;* Southern Alps, abundant, *Haast;* ascending to 4000 ft., Wairau valley, and 3–5000 ft., *Travers;* Rotaiti Lake, *Munro* (midrib nearly obsolete, and leaves most coriaceous). A very variable plant, difficult in some states to distinguish from *V. lævis* and *V. Traversii,*; best characterized by the subsessile flowers, large coriaceous bracts, and subcorymbose inflorescence.

14. **V. lævis,** *Benth.; Fl. N. Z.* i. 194. A small stout glabrous shrub, 2–4 ft. high. Leaves erect and appressed and imbricating, rarely patent, $\frac{1}{3}$–$\frac{1}{2}$ in. long, $\frac{1}{4}$–$\frac{1}{3}$ broad, broadly obovate-oblong, obtuse or acute, extremely coriaceous, entire, concave, sharply keeled by the stout prominent midrib, narrowed rather suddenly into the very stout, short petiole. Racemes short, twice as long as the leaves, usually crowded at the ends of the branches, puberulous; pedicels short; bracts short, coriaceous. Sepals $\frac{1}{16}$ in. long, oblong-ovate, obtuse. Corolla-tube short; lobes $\frac{1}{4}$ in. across. Capsule not seen.

Northern Island: Tongariro, *Bidwill;* Ruahine range, *Colenso.* **Middle** Island: Nelson mountains, *Bidwill,* alt. 2–6000 ft. In its normal state well marked by the small, concave, acute, very coriaceous, keeled leaves, not truncate at the base; but it passes into *V. buxifolia,* which has truncate bases, and into *V. carnosula.*

15. **V. buxifolia,** *Benth.;—Fl. N. Z.* i. 194. A small stout glabrous shrub, 2–3 ft. high, very closely allied to *V. lævis,* but the leaves are closely imbricated and cordate at the base. Leaves $\frac{1}{4}$–$\frac{1}{3}$ in. long, $\frac{1}{5}$–$\frac{1}{4}$ broad, broadly oblong-obovate, obtuse, suddenly truncate or cordate at the very short thick petiole, excessively thick and coriaceous, concave, keeled by the prominent midrib, usually polished. Racemes very short, dense-flowered, crowded at the ends of the branches and subcapitate, puberulous or glabrous; pedicels short; bracts as large as the sepals. Sepals oblong, obtuse. Corolla-tube short; limb $\frac{1}{4}$–$\frac{1}{3}$ in. across. Capsule $\frac{1}{8}$ in. long, twice as long as the calyx. —*V. odora,* Fl. Antarct. 62, t. 41.

Northern Island: Tongariro?, *Dieffenbach;* Ruahine range, *Colenso.* **Middle** Island: abundant on the alps, Nelson, *Bidwill;* Wairau mountains, alt. 4–5500 ft., *Travers;* Southern Alps, ascending to 4000 ft., *Sinclair* and *Haast;* Hopkins river, alt. 3000 ft., *Haast* (with spreading leaves); Otago, subalpine, in the lake district, *Hector and Buchanan.* **Lord Auckland's** Island, *J. D. H.* Usually a very distinct form, best distinguished by the closely imbricate, small, concave, keeled leaves, truncate or cordate at the base, and the crowded terminal inflorescence.

16. **V. carnosula,** *Hook. f., n. sp.* A small, often prostrate, glaucous, stout shrub; branches covered with transverse scars, pubescent towards the tips. Leaves closely imbricate, suberect, $\frac{1}{3}$–$\frac{2}{3}$ in. long, broadly obovate or oblong or orbicular, round at the tip, entire, concave, extremely thick and coriaceous, almost sessile or suddenly contracted to the extremely broad thick petiole, not keeled, midrib very obscure or 0. Spikes short, pilose and pubescent, crowded together and forming heads at the ends of the branches, very dense-flowered; pedicels 0; bracts coriaceous, as long as the calyx. Sepals ovate, obtuse, scarcely ciliate. Corolla-tube very short; limb $\frac{1}{4}$–$\frac{1}{3}$ in. diam. Capsules $\frac{1}{6}$ in., ovate, acute, glabrous, twice as long as the calyx.—*V. lævis, β. carnosula,* Fl. N. Z. i. 194.

Middle Island: Nelson, Morse's Mountain, 5000 ft., *Bidwill;* upper Wairau, *Munro.* The glaucous habit, broader, obtuse, excessively thick, concave leaves, without keel, distinguish this at once from *V. lævis;* from *V. pinguifolia* the ovary and capsules alone distinguish it.

17. **V. pinguifolia,** *Hook. f., n. sp.* A small, erect or decumbent, glaucous, robust shrub, 4 in. to 4 ft. high; branches pubescent above, covered with close-set transverse scars. Leaves sessile, erecto-patent, imbricate, $\frac{1}{6}$–$\frac{1}{2}$ in. long, $\frac{1}{8}$–$\frac{1}{2}$ broad, obovate-oblong, obtuse, entire, excessively thick and coriaceous, concave, not keeled; midrib very obscure. Spikes very short, longer than the leaves, pilose and pubescent, crowded into heads at the tops of the branches, very dense-flowered; pedicels 0; bracts coriaceous, as large as the sepals, ciliated. Sepals oblong, obtuse, ciliate and puberulous. Capsule obovate-oblong, obtuse, rounded or emarginate, pubescent, not much longer than the calyx.

Middle Island: common on the mountains of Nelson and Canterbury, Wairau gorge, alt. 3–5000 ft., *Travers;* Southern Alps, common, ascending to 5000 ft., *Haast.* The small size, robust habit, glaucous obtuse imbricate excessively thick small leaves without midrib, short dense spikes crowded into a head, distinguish this from all except *V. carnosula*, which has very different capsules.

18. **V. pimeleoides,** *Hook. f. Fl. N. Z.* i. 195. A small, suberect, branched, shrubby plant, 4–10 in. high; branches erect, as thick as a sparrow-quill, pubescent, covered with close-set transverse scars. Leaves sessile, imbricate, erecto-patent, $\frac{1}{6}$–$\frac{1}{4}$ in. long, broadly obovate-oblong, obtuse, rather concave, obtusely keeled, rather glaucous; midrib obscure. Spikes short, very pubescent or tomentose, subdistichous. Flowers opposite, in the axils of large leafy ciliated bracts. Sepals oblong, obtuse, ciliated. Corolla deep purple; tube very short; limb $\frac{1}{6}$ in. diam. Capsule ovate, acute, glabrous, twice as long as the calyx.

Middle Island: Port Cooper, *Lyall;* stony flats on the Hurumui mountains, alt. 800–1000 ft., *Travers;* Southern Alps, amongst shingle and grass, Hopkins, Godley, and Macaulay rivers, alt. 2–4000 ft., *Haast.* A very distinct little species, allied in some respects to *V. pinguifolia*, but more so to *V. Lavaudiana*, etc.

19. **V. Buchanani,** *Hook. f., n. sp.* Stems much branched from the base, 4–8 in. high; branches terete, stout, woody, tortuous, black, closely scarred, glabrous or upper pubescent. Leaves small, $\frac{1}{8}$–$\frac{1}{4}$ in. long, closely 4-fariously imbricate, spreading or recurved, nearly orbicular, sessile by a very broad base, concave, very obscurely keeled, nerveless, very thick and coriaceous, not shining. Spikes short, dense, terminal, short-peduncled, subcapitate, $\frac{1}{4}$–$\frac{1}{3}$ in. long; peduncle and rachis tomentose. Flowers sessile; bracts and sepals oblong-ovate, obtuse, ciliate. Corolla white, about $\frac{1}{5}$ in. broad. Ovary and base of style villous,

Middle Island: Otago, lake district, alpine, alt. 3–5000 ft., *Hector and Buchanan.*

20. **V. tetragona,** *Hook. Ic. Pl. t.* 580;—*Fl. N. Z.* i. 104. A small, erect or prostrate, much-branched shrub; branches (with the leaves on) obtusely 4-angled, $\frac{1}{12}$–$\frac{1}{8}$ in. diam. Leaves erect, densely imbricate, extremely coriaceous, ovate, obtuse, tumid, $\frac{1}{12}$ in. long, the upper ciliated or pubescent on the edges, opposite pairs connate at the base, brown-black when dry. Flowers 3–5, sessile amongst the uppermost leaves. Sepals and bract linear-oblong, obtuse, ciliated. Corolla-tube short; limb $\frac{1}{4}$ in. diam.—*Podocarpus Dieffenbachii*, Hook. Ic. Pl. t. 547.

Northern Island: Tongariro, *Bidwill;* Hikurangi, *Colenso.* **Middle** Island: Queen Charlotte's Sound, *Dieffenbach;* Gordon's Nob, *Munro;* Wai-au-na valley, alt. 3000 ft., *Travers.*

21. **V. lycopodioides,** *Hook. f., n. sp.* An erect?, very much branched, stout shrub; branches (with the leaves on) acutely 4-angled, $\frac{1}{8}$–$\frac{1}{10}$ in. diam., pale yellow-brown when dry. Leaves most densely and closely 4-fariously imbricate, thickly coriaceous, very broadly reniform-ovate, much broader than long, abruptly narrowed into an acute tip, about $\frac{1}{10}$ in. broad, obtusely keeled, opposite pairs connate at the base, glabrous or pubescent on the edges. Flowers sessile, in small, dense, oblong heads at the ends of the branches; rachis tomentose; bracts larger than the leaves, not so broad in proportion, grooved in parallel lines when dry, ciliated. Sepals linear-oblong, obtuse, ciliated. Corolla-tube very short; limb $\frac{1}{5}$ in. diam. Capsule broadly-oblong, obtuse, glabrous, not longer than the calyx.

Middle Island: Southern Alps, Ribbon-wood range, Big Ben, Macaulay river, etc., alt.

P 2

3–5000 ft., *Sinclair* and *Haast;* Wairau gorge, alt. 4–5000 ft., *Travers;* Otago, Lindis Pass, subalpine, *Hector and Buchanan.*

22. **V. tetrasticha,** *Hook. f., n. sp.* A small, much branched, decumbent plant, 3–6 in. high; branches (with the leaves on) acutely 4-farious, $\frac{1}{12}$ in. diam., black when dry. Leaves spreading, most densely imbricate, coriaceous, broadly ovate, subacute, opposite pairs connate at the base, not keeled, ciliate, not shining. Flowers in very small, short, terminal, 2- or 3-flowered spikes, distinct from the branches; rachis tomentose; bracts and sepals ovate-oblong, obtuse, ciliated. Corolla-tube very short; limb $\frac{1}{10}$ in. broad. Capsule not seen.

Middle Island: Southern Alps, fissures of rocks on Big Ben, and Hopkins river, alt. 4–5000 ft., *Haast;* Wairau mountains, alt. 4–5000 ft., *Travers.* Very different from *V. lycopodioides,* though not easily defined. Whole plant much smaller, black when dry. Leaves more spreading, though as densely imbricate; spikes not forming a sessile head to the branch; and flowers very small.

23. **V. Hectori,** *Hook. f., n. sp.* A robust, small, much-branched shrub, 6–24 in. high; branches (with the leaves on) terete or obscurely 4-gonous, $\frac{1}{10}$–$\frac{1}{6}$ in diam. Leaves closely but not densely imbricate, extremely thick and coriaceous, tumid, broader than long, very broadly ovate or orbicular, very obtuse, nearly $\frac{1}{8}$ in. across, opposite pairs connate to the middle, puberulous along the edges, shining, not keeled. Flowers collected into an ovate terminal head; rachis villous; bracts broader than the leaves, striated. Sepals broadly oblong, obtuse, ciliate. Corolla as in *V lycopodioides,* pink and white. Ovary glabrous. Capsule as long as the sepals.

Middle Island: Southern Alps, *Haast;* Otago, Mount Alta, alt. 7–7500 ft., *Hector and Buchanan.* Very closely allied to *V. lycopodioides,* but the branches are scarcely tetragonous and the leaves very different, being connate to the middle and very obtuse.

24. **V. salicornioides,** *Hook. f., n. sp.* A small, much branched, erect or ascending, woody species, yellow-green when dry; branches (with the leaves on) terete, $\frac{1}{16}$–$\frac{1}{12}$ in. diam. Leaves closely imbricating, and closely appressed to and adnate with the branch, extremely short, opposite pairs connate almost throughout their length, each pair forming a short narrow ring about $\frac{1}{20}$–$\frac{1}{10}$ in. deep around the branch, truncate, scarcely acute, minutely ciliate. Flowers in small, short, oblong, terminal, 3–6-flowered heads; rachis villous; bracts short and very broad. Sepals oblong, obtuse. Corolla with a very short tube; limb $\frac{1}{4}$ in. diam. Capsule small, oblong, obtuse, glabrous, a little longer than the calyx.

Middle Island: Nelson mountains, *Rough;* Wairau mountains, 4–5500 ft., *Travers;* Rangitata, and Waimakuriri mountains, alt. 3–4000 ft., *Haast.*

25. **V. cupressoides,** *Hook. f., n. sp.* A dense, excessively-branched bush, 1–6 ft. high; branches erect, fastigiate, slender, glabrous. Leaves very minute, $\frac{1}{16}$ in. long, ovate-oblong, obtuse, not broader than the branch, opposite pairs connate at the base, erect or appressed, glabrous, fleshy. Flowers very small, 3 or 4 at the ends of the slender branchlets; bracts much larger than sepals, both broadly oblong, obtuse, not ciliated. Corolla violet; tube short; limb $\frac{1}{16}$ in. diam. Capsule not seen.

Middle Island: Upper Wairau, alt. 4000 ft., and Tarndale, *Sinclair;* Wai-au-na valley,

3500 ft., *Travers;* Ashburton valley, 2–5000 ft., *Haast;* Otago, river flats of the lake district and Lindis Pass, *Hector and Buchanan.* A most curious species.

26. **V. Haastii,** *Hook. f., n. sp.* A tortuous, decumbent or ascending, woody, sparingly branched, glabrous species, 4–10 in. long; branches uniformly and densely leafy, with the leaves on $\frac{1}{2}$–$\frac{2}{3}$ in. diam., obscurely 4-angled. Leaves densely 4-fariously imbricated, spreading, sessile, $\frac{1}{4}$–$\frac{1}{3}$ in. long, broadly obovate or orbicular, rounded at the tip, very coriaceous, opposite pairs almost connate at the base, sometimes with 1 obscure tooth on each side, slightly ciliate at the very base. Flowers sessile in pairs or more amongst the uppermost (floral) leaves, being apparently in reduced spikes, together collected into an oblong head $\frac{1}{2}$–$1\frac{1}{2}$ in. long. Bracts ovate-oblong. Sepals $\frac{1}{6}$ in. long, linear-oblong, scarcely ciliate. Corolla not seen. Capsule oblong, obtuse, scarcely longer than the calyx.

Middle Island; summits of Mounts Darwin, Dobson, Torlesse, and on Mount Cook, alt. 6–7000 ft., *Haast.* This and the following are most remarkable plants, of a different habit from any hitherto described.

27. **V. epacridea,** *Hook. f., n. sp.* A small, tortuous, prostrate or ascending, glabrous, rigid, much-branched species; branches uniformly and densely leafy, with the leaves on $\frac{1}{4}$ in. diam., obscurely 4-angled. Leaves sessile, densely 4-fariously imbricate, spreading and recurved, $\frac{1}{8}$–$\frac{1}{6}$ in. long, very broadly obovate-oblong, concave, keeled, rigid, glabrous, round or subacute at the tip. Flowers collected into terminal, ovoid, leafy heads, as in *V. Haastii;* bracts broadly ovate, ciliate, shorter than the sepals, which are linear-oblong, obtuse, ciliate. Corolla with a long tube; limb $\frac{1}{12}$ in. diam. Capsule small, ovate-oblong, obtuse.

Middle Island: Tarndale, alt. 3500 ft., *Sinclair;* Discovery Peaks, alt. 5800 ft., *Travers;* Southern Alps, Ashburton valley, Godley glacier, and forming the highest vegetation on Mount Darwin, alt. 5–6500 ft., *Haast;* Wai-au-ua valley, 3500 ft., *Travers.* Nearly allied to *V. Haastii*, but quite distinct, much smaller, with recurved leaves and ciliate bracts and sepals.

28. **V. macrantha,** *Hook. f., n. sp.* A short, stout, erect, rigid shrub, perfectly glabrous, sparingly branched; branches as thick as a small quill, terete. Leaves $\frac{1}{2}$–1 in. long, obovate-lanceolate, acute, obtusely serrate, narrowed into a short thick petiole, excessively thick and coriaceous, smooth, nerveless; midrib very indistinct. Racemes axillary, 5–7-flowered; peduncle longer than the leaves. Flowers very large, on short pedicels; bracts subulate-lanceolate, rigid, and the lanceolate, attenuate, acuminate, coriaceous sepals $\frac{1}{4}$ in. long. Corolla $\frac{3}{4}$ in. broad; tube very short. Capsule not seen.

Middle Island, *Travers;* Southern Alps, grassy hillsides, 2500–4000 ft., sources of the Waitaki, etc., *Haast.* This must be a very beautiful plant when fresh, the flowers are larger than in *V. elliptica*, white according to Haast.

29. **V. Hulkeana,** *F. Muell.* A slender, erect, sparingly leafy shrub, 1–3 ft. high; stem nearly simple, terete, puberulous above. Leaves in distant pairs, 1–$1\frac{1}{2}$ in. long, oblong-ovate, obtuse or acute, obtusely or acutely coarsely serrate, rather coriaceous; petiole $\frac{1}{4}$–$\frac{3}{4}$ in. long. Spikes spreading, puberulous and glandular, arranged in long terminal opposite-branched panicles 4–10 in. long and 2–4 broad. Flowers sessile; bracts broadly ovate, obtuse, $\frac{1}{16}$ in. long, nearly as long as the similar but broader sepals.

Corolla $\frac{1}{4}$ in. across, lilac; tube very short. Stamens short. Capsule small, oblong, obtuse, twice as long as the sepals.

Middle Island: Wairau mountains, 1500–2000 ft., *Travers;* Macrae's Run, halfway up, in rocky places, *Munro;* Kaikoras mountains, *Sinclair.*

30. **V. Lavaudiana,** *Raoul, Choix, t.* 10;—*Fl. N. Z.* i. 195. A small, rather stout, puberulous species; stem decumbent at the base; branches ascending, terete, 4–8 in. high. Leaves rather crowded, short-petioled, $\frac{1}{3}$–$\frac{2}{3}$ in. long, broadly obovate-spathulate, obtuse, crenate-serrate, very coriaceous. Spikes short, $\frac{1}{4}$–$\frac{1}{2}$ in. long, crowded on a low spreading puberulous and glandular corymb 1–2 in. broad. Flowers sessile; bracts and sepals nearly equal, ovate or lanceolate, acuminate, ciliate and pubescent, $\frac{1}{10}$ in. long. Corolla $\frac{1}{3}$ in. diam., purple; tube rather long. Stamens very short. Capsule oblong, obtuse, a little longer than the calyx.

Middle Island: Akaroa, rocky mountains, *Raoul;* Port Cooper, on stony ground, *Lyall;* river bed of the Ashley, Canterbury plains, *Travers.*

31. **V. Raoulii,** *Hook. f., n. sp.* A small, robust, much-branched, erect shrub, 6–12 in. high; stem often tortuous at the base; branches terete, puberulous, erect or ascending, leafy. Leaves shortly petioled, rather crowded, suberect or spreading, $\frac{1}{2}$–$\frac{3}{4}$ in. long, oblong-spathulate, obtuse or apiculate, crenate, very thick and coriaceous, opaque or shining. Spikes very short, arranged in terminal pedicelled or subsessile crowded glabrous or puberulous corymbs $\frac{1}{2}$ in. broad, or in heads, or all congested into an oblong crowded thyrsus 1–2 in. long. Flowers sessile; bracts and sepals very broadly oblong-ovate, obtuse, nearly glabrous. Corolla $\frac{1}{4}$ in. broad; tube short. Stamens short. Capsule broadly oblong, obtuse, rather longer than the calyx.

Middle Island: Akaroa, *Raoul* (sent with *V. Lavaudiana*); Upper Wairau valley, alt. 3000 ft., *Munro, Travers;* rocks on the Kowai river, alt. 2500 ft., *Haast.*

32. **V. Benthami,** *Hook. f. Fl. Antarct.* i. 60. *t.* 39 *and* 40. An erect, much-branched shrub, 2–4 ft. high; branches very robust, closely transversely scarred, puberulous on the opposite sides. Leaves crowded towards the ends of the branches, sessile, $\frac{1}{2}$–$1\frac{1}{2}$ in. long, linear- or obovate-oblong, obtuse; margin with a few deep serratures and edged with down, very coriaceous, flat, veinless, opposite pairs connate at the very base. Racemes terminal, continuous with the ends of the branches, clothed with imbricating foliaceous obovate bracts, $\frac{1}{4}$–$\frac{1}{3}$ in long, which are edged with down. Flowers pedicelled, shorter than the bracts Sepals unequal, $\frac{1}{6}$ in. long, oblong-spathulate, edged with down. Corolla fine bright blue, nearly $\frac{1}{3}$ in. diam.; tube short. Stamens short. Capsule very broadly ovate, acute, as broad as long, about as long as the sepals. Seeds broadly winged.

Lord Auckland's group and **Campbell's** Islands: abundant in rocky places, *J. D. H.* A most beautiful plant, quite unlike any New Zealand congener. The flowers are sometimes 5- or 6-merous, with three stamens and a 3-carpelled ovary.

33. **V. linifolia,** *Hook. f., n. sp.* A small, perfectly glabrous, leafy, herbaceous, procumbent, branched species; branches slender, terete, 2–6 in. long. Leaves rather close-set, spreading, sessile or shortly petioled, $\frac{1}{2}$–1 in. long, $\frac{1}{12}$–$\frac{1}{6}$ broad, linear, obtuse, quite entire, flat, not very coriaceous,

greenish when dry. Peduncles slender, axillary, longer or shorter than the leaves, 3–5-flowered; bracts leaf-like; pedicels often very long, $\frac{3}{4}$–1 in. long, curved. Sepals $\frac{1}{6}$ in. long, linear-oblong, obtuse. Corolla $\frac{1}{2}$ in. broad; tube very short. Stamens short. Capsule broadly obcordate, shorter than the calyx.

Middle Island: Southern Alps, alt. 2500–4000 ft., Trinity Hill, Forest Creek, Ashburton and Great Clyde glaciers, clefts of rocks on Mount Darwin, etc., *Sinclair and Haast.*

34. **V. nivalis,** *Hook. f. Fl. N. Z.* i. 196. Stems long, rather stout, prostrate, sparingly branched, flexuose, puberulous, 6–10 in. long. Leaves on short petioles, $\frac{1}{4}$–$\frac{1}{2}$ in. long, ovate or broadly oblong-ovate, obtuse, deeply obtusely serrate, very coriaceous, black when dry. Peduncles 1–2 in. long, axillary, glandular-pubescent, 6–8-flowered; bracts minute; pedicels $\frac{1}{10}$–$\frac{1}{3}$ in. Sepals $\frac{1}{8}$ in. long, oblong, obtuse, glandular. Corolla $\frac{1}{3}$ in. broad, white with pink veins; lower lobe 2-fid; tube very short. Stamens short. Capsule transversely oblong, shorter than the calyx.—Hook. Ic. Pl. t. 640; *V. Hookeriana*, Walp. Rep.

Northern Island: Tongariro, *Bidwill;* summit of the Ruahine range, *Colenso.*

35. **V. Lyallii,** *Hook. f. Fl. N. Z.* i. 196. Stems slender, prostrate and rooting, diffusely branched, 5–15 in. long, glabrous or sparingly puberulous. Leaves short-petioled, $\frac{1}{4}$–$\frac{1}{2}$ in. long, ovate oblong-ovate or ovate-lanceolate, obtuse or acute, glabrous, with a few coarse serratures, coriaceous, black when dry. Peduncles axillary, slender, 3–8 in. long, glabrous, many-flowered; pedicels slender, lower $\frac{1}{2}$ in. long; bracts very variable, oblong linear or spathulate. Sepals as variable, glabrous. Corolla nearly $\frac{1}{2}$ in. diam.; tube very short. Stamens very short. Capsule transversely oblong, didymous, as long as or longer than the sepals.

Northern Island: rocky cliffs near Patea (leaves larger, acute, passing into *V. cataractæ, β*) and summit of Ruahine range (plants very small), *Colenso.* **Middle** Island: abundant in many places; Otago, *Hector*, etc.; Milford Sound, *Lyall;* rocky rivulets, Ashburton river and Rangitata range, 2–4000 ft., *Haast.* This seems to me to be a plant which, assuming widely different forms, passes into *V. nivalis*, by the raceme becoming few-flowered and glandular; into *V. Bidwillii* by the leaves becoming much smaller, etc.; into *V. cataractæ* by the leaves becoming larger and longer; and into var. *diffusa* of *V. cataractæ* by the leaves becoming larger and broader. Most of these forms, however, appear to keep their characters over large areas, where several of them occur together, and the extreme states are very widely dissimilar.

36. **V. Bidwillii,** *Hook. Ic. Pl. t.* 814;—*Fl. N. Z.* i. 196. Stems prostrate, slender, creeping at the base, 3–6 in. long, glabrous or puberulous. Leaves minute, subsessile, $\frac{1}{12}$–$\frac{1}{4}$ in. long, broadly ovate oblong or oblong-ovate, obtuse, with 1 or 2 deep notches on each side, coriaceous, black when dry. Peduncles axillary, usually very long and slender, 2–10 in. long, few- or many-flowered, glabrous. Flowers racemed or sometimes in interrupted whorls; pedicels slender, lower $\frac{1}{2}$ in. long; bracts and sepals very small, $\frac{1}{12}$–$\frac{1}{6}$ in., oblong, obtuse, puberulous. Corolla $\frac{1}{4}$ in. diam. or more, violet white or pink; tube very short. Stamens small. Capsule didymous, $\frac{1}{10}$ in. broad, longer than the calyx.

Middle Island: bed of the Wairau, alt. 2–3000 ft., *Bidwill, Travers*; Rotuite Lake, *Munro;* Southern Alps, abundant above 1500 ft., *Sinclair and Haast;* Otago, lake district and

Lindis Pass, alt. 1–3C00 ft., *Hector and Buchanan*. A much smaller plant than *V. Lyallii*, with usually smaller flowers and sepals, and leaves with only one or two serratures on each side.

37. **V. cataractæ,** *Forst.* ;—*Fl. N. Z.* i. 195. Quite glabrous. Stems suberect or prostrate at the base and ascending, 10 in. to 2 ft. long, branched, rather slender, terete. Leaves very variable, sessile or petioled, ½–5 in. long, ovate-oblong or narrow-lanceolate, acuminate, deeply acutely serrate, coriaceous, when dry black above and brown or whitish beneath. Racemes axillary, slender, 3–8 in. long, very many-flowered; pedicels very slender, ½–1 long; bracts linear-subulate. Sepals linear-oblong, acute, glabrous. Corolla ½–¾ in. diam.; tube very short. Stamens short. Capsule small, didymous, $\frac{1}{10}$–$\frac{1}{6}$ in. broad, as long as or longer than the calyx.

Var. *α*. Leaves 2–5 in. long, lanceolate, stems long.

Var. *β*. *diffusa*. Leaves ovate or oblong, ½–1½ in. long.—*V. diffusa*, Hook. Ic. Pl. t. 645; Fl. N. Z. i. 195.

Var. *γ*. *lanceolata*. Stems short. Leaves 1 in. long, $\frac{1}{12}$ broad, very narrow linear-lanceolate.—*V. lanceolata*, Benth.

Northern Island: east coast, Taupo, Tongariro, and Ruahine range, *Colenso and Bidwill*. **Middle** Island: Dusky Bay, *Menzies*, *Lyall*.

38. **V. elongata,** *Benth.* ;—*Fl. N. Z.* i. 197. A very slender, prostrate, straggling herb, glabrous or pubescent, diffusely branched; branches terete. Leaves petioled, ½–1 in. long, broadly ovate-cordate or deltoid-ovate, coarsely irregularly crenate-toothed, 3-nerved, glabrous or hairy; petioles ¼–½ long, dilated at the blade. Peduncles axillary, stout, 2–6 in. long, few- or many-flowered; bracts rather large, linear-oblong or spathulate; pedicels slender, ½ in. long. Sepals foliaceous, obovate-oblong, obtuse, growing out much after flowering. Corolla small, ⅛–⅙ in. diam.; tube very short. Stamens very short. Capsule broadly didymous, enclosed in the calyx.

Northern Island: Bay of Islands, *Cunningham*, etc. A very distinct species, allied to the Australian *V. calycina*.

39. **V. spathulata,** *Benth.* ;—*Fl. N. Z.* i. 197. Stems prostrate, short, stout, tufted, much-branched, leafy, pubescent, 2–4 in. long. Leaves narrowed into broad petioles, ⅙–¼ in. diam., broadly ovate-spathulate, obtuse, crenate, coriaceous, pubescent, black when dry. Peduncles shorter or rather longer than the leaf, stout, erect, pubescent, 2- or 3-flowered; pedicels stout, ⅛ in. long; bracts foliaceous. Sepals oblong, obtuse. Corolla ¼ in. diam.; tube very short. Stamens short. Capsule didymous, longer than the calyx.

Northern Island: Tongariro, *Bidwill*. Perhaps an alpine form of *V. elongata*. I have seen but one specimen.

40. **V. Anagallis,** *Linn.* ;—*Fl. N. Z.* i. 197. A rather stout, erect, succulent, glabrous herb; stem rooting at the base, 6–18 in. high. Leaves 1–2 in. long, oblong or linear-oblong, obtuse crenate, contracted and half stem-clasping at the base, membranous when dry. Peduncles extremely slender, axillary, many-flowered, longer than the leaves; bracts small, linear or lanceolate; pedicels very slender, ¼ in. long, spreading or reflexed. Sepals oblong, obtuse, $\frac{1}{10}$ in. long, growing out after flowering. Corolla small, ¼ in. diam., pale-blue or flesh-coloured; tube very short. Stamens short. Capsule broadly oblong, included in the calyx.

Northern Island: watery places on the east coast, *Colenso.* A very common subaquatic plant, found in very many parts of the world.

Veronica serpyllifolia and *V. arvens s* are both found as weeds of cultivation.

8. PYGMEA, Hook. f., nov. gen.

Moss-like, tufted, low herbs. Leaves densely imbricated all round the short branches, ciliated. Flowers solitary, terminal, sessile or shortly peduncled.—Sepals 5 or 6, linear-spathulate, unequal, ciliate. Corolla salver-shaped; tube longer than the calyx; limb 5- or 6-lobed; lobes linear- or oblong-spathulate. Stamens 2, inserted at the throat of the corolla; filaments variable in length; anthers large. Disk annular, rather large. Ovary broadly ovoid, compressed, 2-celled; style capillary; ovules numerous, attached to placentas on the septum. Fruit not seen.

The following are the only known species. The genus appears closely allied to *Veronica*, but differs in the 5- or 6-parted flowers, and leaves not quadrifariously arranged.

Leaves $\frac{1}{8}$ in., ciliate at the edges only . . . . 1. *P. ciliolata.*
Leaves $\frac{1}{6}$ in., hispid and ciliate above the middle . . . . 2. *P. pulvinaris.*

1. **P. ciliolata,** *Hook. f., n. sp.* Patches broad, moss-like, rather hoary from the long white hairs on the leaves; branches $\frac{1}{2}$ in. high, with the leaves on $\frac{1}{4}$ in. diam. Leaves most densely imbricate, $\frac{1}{8}$ in. long, broadly oblong-spathulate, obtuse, quite entire, ciliate from above the middle, coriaceous, glabrous. Flowers $\frac{1}{8}$ in. long, emerging from the tips of the branches. Sepals shorter than the corolla-tube.

Middle Island: Discovery Peaks, alt. 5500 ft., *Travers;* Hopkins river, *Haast.*

2. **P. pulvinaris,** *Hook. f., n. sp.* Patches broad, moss-like; habit and branches as in *P. ciliolata,* but more hoary, and almost white from the much longer white hairs. Leaves most densely imbricate, $\frac{1}{6}$ in. long, narrow linear, somewhat dilated towards the tip, upper half covered with long, white, rigid, hispid hairs on both surfaces and edges. Flowers shortly peduncled. Sepals linear, obtuse. Corolla as in *P. ciliolata,* but larger.

Middle Island: summit of Mount Torlesse, forming large, hoary, moss-like patches, alt. 5500–6500 ft., *Haast.*

9. OURISIA, Comm.

Perennial herbs. Leaves chiefly radical, petioled, crenate or entire. Scapes bracteate, 1- or many-flowered. Flowers rather large, whorled racemed corymbose or solitary.—Calyx 5-lobed or -partite. Corolla irregular, funnel-shaped, incurved or oblique; limb 5-fid. Stamens 4, included, didynamous; anther-cells divaricating, confluent. Stigma capitate. Capsule 2-valved, loculicidal. Seeds numerous; testa lax.

A beautiful genus, confined to the alps of Tasmania, New Zealand, the Andes of South America, and Fuegia.

Upper bracts whorled. Calyx $\frac{1}{4}$ in., glandular. Capsule $\frac{1}{4}$ in. long . 1. *O. macrophylla.*
Upper bracts whorled. Calyx $\frac{1}{2}$ in. glabrous. Capsule $\frac{1}{2}$ in. long . 2. *O. macrocarpa.*
Upper bracts in pairs.
Erect. Leaves crenate; petioles slender . . . . 3. *O. Colensoi.*

Erect. Leaves crenate; petioles short and broad . . . . 4. *O. sessilifolia.*
Creeping, cæspitose. Leaves entire or subcrenate. Scape and sepals glabrous . . . . 5. *O. cæspitosa.*
Creeping, cæspitose. Leaves crenate. Scape and sepals glandular and hairy . . . . 6. *O. glandulosa.*

1. **O. macrophylla,** *Hook. Ic. Pl. t.* 545, 546;—*Fl. N. Z.* i. 197. Erect, more or less pubescent or tomentose. Rhizome short, decumbent at the very base. Leaves all radical, long-petioled, 1–6 in. long, oblong or ovate-cordate, obtuse, crenate. Scape 2–30 in. high. Bracts on the scape, lower pair opposite, upper whorled, all oblong-obovate or linear, very variable in form and size, crenate. Flowers umbelled, in superimposed whorls; pedicels 1–2 in. long, very slender. Calyx 5-partite; lobes glandular hairy or glabrate, linear or lanceolate, $\frac{1}{4}$ in. long. Corolla $\frac{1}{2}$ in. long. Capsules membranous, $\frac{1}{4}$ in. long, broadly oblong, turgid.

Northern Island: Mount Egmont, Ruahine range, etc., *Dieffenbach, Colenso,* etc. **Middle** Island: common in damp mountainous localities, Upper Wairau, alt. 2500–5000 ft., *Sinclair, Travers.*

2. **O. macrocarpa,** *Hook. f. Fl. N. Z.* i. 198. Erect, nearly glabrous. Rhizome prostrate, as thick as the little finger. Leaves all radical, with long stout petioles, 4–6 in. long; blade 1–2 in. long, orbicular broadly oblong or ovate-oblong, round at the tip, crenate, coriaceous, glabrous, except along the edges of the petiole, veins strongly reticulate. Scape very stout, 8–18 in. high; lower bracts in pairs, connate, oblong, crenate, upper whorled, all coriaceous. Flowers umbelled in several whorls. Pedicels stout, 1–3 in. long. Calyx 5-partite; lobes glabrous, $\frac{1}{2}$ in. long, linear-oblong, obtuse, very coriaceous. Corolla not seen. Capsule narrow-oblong, $\frac{1}{2}$ in. long.

Middle Island: Chalky Bay, *Lyall.* Well distinguished from *O. macrophylla* by its glabrous condition, stout habit, very coriaceous leaves, very large sepals and elongate capsule.

3. **O. Colensoi,** *Hook. f., n. sp.*—A small, erect, glandular-pubescent species, 2–6 in. high. Rhizome slender, creeping. Leaves $\frac{1}{2}$ in. long or less, petiole as long, oblong, obtuse, crenate, pubescent on both surfaces or glabrate. Scape 1–4-flowered, slender; bracts all in pairs, oblong, crenate. Flowers solitary or in pairs; pedicels slender, $\frac{1}{2}$–$\frac{2}{3}$ in. long. Calyx 5-partite; lobes $\frac{1}{6}$ in. long, narrow-linear, glandular-pubescent. Corolla white, nearly $\frac{2}{3}$ in. diam. Capsule shorter than the sepals, nearly orbicular.

Northern Island: top of the Ruahine range and Lake Taupo, *Colenso.* **Middle** Island: Wairau mountains, alt. 3–3500 ft., *Travers.*

4. **O. sessilifolia,** *Hook. f., n. sp.* Erect, hirsute, and glandular; rhizome creeping, as thick as a crow-quill. Leaves all radical, $\frac{2}{3}$–$1\frac{1}{3}$ in. long, broadly ovate, abruptly narrowed into broad short petioles, crenate-serrate, upper surface villous with glandular hairs, lower less so or glabrate. Scape stout, 3–4 in. high; bracts all opposite, oblong or obovate, crenate. Flowers in pairs; pedicels rather stout, $\frac{1}{2}$–1 in. long. Calyx 5-partite; lobes broad, linear, obtuse, glandular. Corolla large, nearly 1 in. diam.; lobes long, rounded, white. Capsule not seen.

Middle Island: Mount Brewster, alt. 5–6000 ft., *Haast.* Apparently a very distinct plant; I have only one good specimen.

5. **O. cæspitosa,** *Hook. f. Fl. N. Z.* i. 198. Stems creeping, stout, tufted, leafy, rather fleshy, with very short ascending branches, glabrous or loosely pilose or subtomentose. Leaves somewhat imbricate, patent, recurved, thickly coriaceous, $\frac{1}{4}$–$\frac{1}{3}$ in. long, obovate-spathulate, with recurved 2- or 3-lobed or crenate margins, sessile or narrowed into short, glabrous or ciliate, stout petioles. Scape 2–3 in. high, 1–6-flowered; bracts opposite, like the leaves. Flowers solitary or 2 or 3 together; pedicels stout. Calyx $\frac{1}{4}$ in. long, 5-lobed; lobes oblong, obtuse, glabrous. Corolla white, $\frac{1}{3}$–$\frac{1}{2}$ in. diam.; lobes short, broad. Capsule $\frac{1}{6}$ in. long, ovate-oblong.

Northern Island: tops of the Ruahine mountains, *Colenso.* **Middle** Island: Milford Sound, *Lyall;* Hurumui mountains, alt. 3–3500 ft., *Travers;* Southern Alps, common at 3500–6000 ft., *Haast;* Otago, lake district, alpine, alt. 6000 ft., *Hector and Buchanan.*

6. **O. glandulosa,** *Hook. f., n. sp.* Stems very stout, succulent, creeping, 3–6 in. long, glabrous. Leaves usually densely 2-fariously imbricate, $\frac{1}{3}$–$\frac{1}{2}$ in. long, obovate-spathulate, sessile by a broad base, spreading or recurved, thickly coriaceous, ciliated with stout closely articulate hairs, entire or obscurely crenate. Scapes stout, 1–2 in. long, with several pairs of opposite spathulate bracts, which as well as the scape and calyx are covered with glandular hairs. Flowers 1–3, pedicels slender. Sepals $\frac{1}{4}$ in. long, oblong, obtuse. Corolla white, $\frac{1}{3}$–$\frac{3}{4}$ in. diam.

Middle Island: Otago, lake district, alpine, alt., 5000 ft., forming large patches *Hector and Buchanan.*

10. EUPHRASIA, Linn.

Small or large, simple or branched, annual or perennial plants. Leaves small, opposite, toothed lobed or pinnatifid.—Calyx tubular or campanulate, 4-lobed, rarely 5-lobed. Corolla 2-lipped; upper lip concave, 2-fid, lobes broad spreading; lower spreading, 3-fid, lobes obtuse emarginate. Stamens 4, didynamous; anther-cells nearly parallel, one or both often spurred. Stigma dilated. Capsule oblong, compressed, 2-valved.

A genus of which one or two very small species are found throughout northern Europe and Asia, and as many in colder South America, but of which the majority are Australian and New Zealand. The species are difficult of discrimination and very variable.

Flowers $\frac{1}{3}$–$\frac{1}{2}$ in. long and broad.
Erect, 6–30 in. high. Leaves $\frac{1}{4}$–$\frac{2}{3}$ in., margins not revolute . . . 1. *E. cuneata.*
Erect, 3–6 in. high. Leaves $\frac{1}{6}$–$\frac{1}{3}$ in.; margins revolute . . . . . 2. *E. Munroi.*
Decumbent or tufted, much branched. Leaves $\frac{1}{12}$–$\frac{1}{6}$ in., margins cut and revolute . . . . 3. *E. revoluta.*
Flowers $\frac{1}{6}$–$\frac{1}{5}$ in. long.
Erect, 1–2 in. high, diffusely branched. Leaves cut; margins revolute 4. *E. antarctica.*
Creeping and rooting, 1–2 in. long. Leaves $\frac{1}{12}$ in., 3-fid; margins flat . . . . 5. *E. repens.*

1. **E. cuneata,** *Forst.;—Fl. N. Z.* i. 199. An erect, annual, branched, glabrous or puberulous, leafy, slender species, 6–30 in. high. Leaves in remote pairs or fascicled on short branchlets, $\frac{1}{4}$–$\frac{2}{3}$ in. long, obovate-oblong or spathulate or cuneate-oblong, rarely ovate, narrowed into stout or slender petioles, coarsely toothed; margins not recurved. Flowers numerous, shortly peduncled. Calyx campanulate, 4-lobed; lobes short, obtuse. Corolla $\frac{1}{3}$–$\frac{2}{3}$

in. long, pink purplish or yellowish; tube slender, funnel-shaped; lobes obovate, upper notched. Anthers pilose, mucronate, spurs of the posterior pair unequal. Capsule $\frac{1}{3}$ in. long, linear-clavate.

Northern Island: from the East Cape, southward, *Banks and Solander*, etc. **Middle** Island: common throughout. A very variable plant, allied to the Australian *E. collina*.

2. **E. Munroi,** *Hook. f., n. sp.* A short, erect, perennial (?), leafy species, 3–6 in. high, glabrous or minutely glandular and pubescent; branches ascending, leafy. Leaves rather crowded, spreading and recurved, $\frac{1}{5}$–$\frac{1}{3}$ in. long, sessile, broadly ovate-oblong or spathulate, obtuse, very coriaceous; margins recurved, thick, sparingly crenate. Flowers few, chiefly at the ends of the branches, very shortly peduncled. Calyx sometimes 2-lipped, lips erect, one 3-lobed the other 2-lobed or entire; lobes obtuse, short, with revolute edges. Corolla $\frac{1}{2}$ in. long; tube short, funnel-shaped; lobes short, retuse. Anthers hairy, anterior pair with 2 obtuse or shortly mucronate cells, posterior with 1 cell spurred, the other obtuse or acute. Capsule broadly oblong, retuse.

Middle Island: abundant on the alps, Dun mountain, *Munro;* Hurumui range and Discovery Peak, alt. 3500–5000 ft., *Travers;* Southern Alps, *Sinclair and Haast*. I fear that this may prove nothing but an alpine state of *E. cuneata*, but it looks very different. It is very near indeed to the Tasmanian *E. alpina*.

3. **E. revoluta,** *Hook. f. Fl. N. Z.* i. 199. A small, much branched, slender annual, 1–2 in. high, almost prostrate, sometimes tufted, glandular pubescent or glabrate. Leaves $\frac{1}{12}$–$\frac{1}{6}$ in. long, obovate-spathulate, sessile, obtuse, crenate or lobed; margins strongly recurved. Flowers solitary or very few at the ends of the branches; peduncles shorter or longer than the leaves. Calyx shortly 4-lobed; lobes obtuse; margins recurved. Corolla nearly $\frac{1}{2}$ in. long, and as much in diameter. Anthers nearly glabrous. Cells all shortly mucronate, those of the posterior pair unequally. Capsule small, broad, oblong, obtuse, retuse.

Northern Island: summit of the Ruahine range, *Colenso*. **Middle** Islands: Nelson mountains, alt. 5500 ft., *Travers;* Mount Brewster, *Haast;* Otago, lake district, alpine, forming patches, alt. 6300 ft., *Hector and Buchanan;* Dusky Bay, *Lyall* (leaves deeply cut). A very different-looking plant from *E. Munroi*, small, stout or slender, much branched, tufted or straggling, but some specimens are difficult to distinguish. The habit is that of *E. antarctica*. The flowers seem variable as to colour. Very near the Chilian *E. subexserta*.

4. **E. antarctica,** *Benth.;—Fl. N. Z.* i. 199. A small, slender, rarely robust, much branched, glabrous or glandular, puberulous annual, 1–2 in. high. Leaves sessile or very shortly petioled, $\frac{1}{10}$–$\frac{1}{6}$ in. long, deeply lobed or subpinnatifid, ovate or obovate; lobes obtuse; margins recurved. Flowers small, sessile or shortly peduncled, numerous towards the ends of the branches or solitary, often amongst crowded subterminal leaves. Calyx oblong; lobes very short, obtuse. Corolla $\frac{1}{6}$–$\frac{1}{4}$ in. long; upper lip short, arched; lobes short. Anther-cells all nearly glabrous and mucronate. Capsule broadly obovate-oblong, retuse.

Northern Island: summit of the Ruahine range, *Colenso*. **Middle** Island: abundant on all the alps, from 2–6000 ft. elevation, *Munro, Lyall, Travers, Sinclair, Haast, Hector and Buchanan*, etc. Hector traces three forms, corresponding to as many zones of elevation on Mount Alta; 1st, a slender, erect form, with peduncled flowers, alt. 1–2000 ft.; 2nd, a robust spreading form with the flowers amongst crowded, subterminal leaves, alt. 3–5000 feet; and 3rd, a most minute form, $\frac{1}{6}$ in. high, with a single flower, alt. 6000 ft. I have a very large robust form, from rivulets, Lake Tekapo, alt. 3500 ft. (*Haast*), with leaves and calyx $\frac{1}{3}$ in. long. This is a common Chili and Fuegian plant, and has a very nearly allied Australian alpine representative in *E. alsa*, F. Muell.

5. **E. repens,** *Hook. f. Fl. N. Z.* i. 200. A very small, slender, creeping, glabrous or puberulous plant, 1–2 in. long; branches prostrate, with fibrous rootlets. Leaves minute, sessile, $\frac{1}{12}$ in. long, in scattered pairs, cuneate, 3-lobed; lobes acute, erect; margins not recurved. Flowers axillary, shortly peduncled, large for the size of the plant, erect. Calyx oblong, campanulate, lobes short, acute. Corolla $\frac{1}{3}$ in. long; tube long; upper lip short; arched; lower with 3 short lobes. Anthers all nearly glabrous; cells all mucronate. Ovary pubescent; stigma elongate.

Middle Island: Bluff Island, *Lyall.* A curious little species, of which more specimens are much wanted.

ORDER LIV. GESNERIACEÆ.

(*Including* CYRTANDREÆ.)

Shrubs or herbs. Leaves opposite or alternate. Flowers hermaphrodite, usually irregular.—Calyx inferior, 5-toothed -lobed or -partite. Corolla usually 2-lipped, 5-lobed, imbricate. Stamens 2–4, inserted on the tube of the corolla, with sometimes the filament of a fifth; anthers 2-celled; cells sometimes confluent. Ovary superior, 2-celled; placentas 2, parietal, sometimes meeting in the axis; style simple, stigma 2-lobed; ovules numerous. Fruit capsular. Seeds small; albumen fleshy or 0.

A large Order, chiefly of tropical plants. The above character does not apply to many American and other genera, which have inferior ovaries and baccate fruit.

1. RHABDOTHAMNUS, A. Cunn.

A slender, hispid or pubescent shrub. Leaves opposite. Flowers solitary or in pairs.—Sepals 5, unequal, persistent. Corolla-tube campanulate; limb 2-lipped; upper lip 2-lobed, under 3-lobed. Stamens 4, the fifth rudimentary; filaments slender, arched; anthers cohering cruciately. Disk thin, annular, lobed. Ovary 1-celled, narrowed into a slender style; stigma small, obtuse, 2-lobed; ovules numerous, on 2-lobed prominent placentas. Capsule ovoid, beaked, 2-valved; valves 2-fid, separating from the placentas. Seeds very minute, albuminous.

1. **R. Solandri,** *A. Cunn.;—Fl. N. Z.* i. 186. Shrub 2–4 ft. high; branches opposite, hispid, terete. Leaves on slender petioles, broadly obovoid or orbicular, $\frac{1}{2}$–$\frac{2}{3}$ in. diam., coarsely toothed, harsh to the touch. Flowers axillary or terminal; peduncles slender, $\frac{1}{2}$–$\frac{3}{4}$ in. long. Sepals lanceolate, acuminate, $\frac{1}{3}$ in. long. Corolla $\frac{2}{3}$ in. long, orange with red stripes. Capsule shorter than the calyx.

Northern Island: from the Bay of Islands to Wellington, *Banks and Solander*, etc. The foliage somewhat resembles *Trophis aspera* and *Carpodetus serratus.*

Order LV. LENTIBULARIEÆ.

Marsh or aquatic herbs, often floating. Leaves various, sometimes 0, or reduced to minute bladders. Flowers on leafless scapes.—Calyx usually 2-lobed or of 2 sepals, rarely 5-lobed. Corolla irregular, 2-lipped, produced into a prominence or spur behind. Stamens 2, inserted at the base of the tube; filaments short incurved; anthers 1-celled. Ovary free, 1-celled; style short, stigma 2-lipped; ovules numerous, on a free central placenta. Capsule 2-valved, many-seeded. Seeds minute; albumen 0; embryo short.

An Order of two principal genera, one of which (*Utricularia*) is found in almost all parts of the world.

1. UTRICULARIA, Linn.

Herbs, often aquatic. Stems 0, or creeping or slender and floating; rhizome bearing minute air-bladders. Leaves 0 or cauline or radical, usually linear and quite entire or multifid. Scape slender, 1- or more-flowered.—Sepals 2, nearly equal, entire toothed or lobed. Corolla 2-lipped; lower lip gibbous or spurred at the base. Stamens 2; anthers adnate to the thick filament. Ovary globose; stigma 2-lipped.

A large genus, found in all parts of the temperate and tropical world. The New Zealand species want searching for and working up.

Stems floating. Leaves capillary, multifid. Flowers yellow . . 1. *U. protrusa.*
Stem 0 or creeping. Leaves linear.
Scape 3–4 in., 1–4-flowered. Upper lip of corolla white, cuneate 2. *U. novæ-Zelandiæ.*
Scape ½–1 in., 1-flowered. Upper lip of corolla purple, linear . 3. *U. monanthos.*
Scape 3–4 in., 1–4-flowered. Upper lip of corolla linear . . . 4. *U. Colensoi.*

1. **U. protrusa,** *Hook. f. Fl. N. Z.* i. 206. Stems floating, slender, a span long, covered with capillary multifid leaves and bearing minute bladders. Scape stout, erect, 2–4-flowered. Sepals oblong. Corolla yellow; upper lip 3-lobed, lower broader, subquadrate, its disk protruded; margins recurved. Spur short, obtuse.

Northern Island: bogs, Bay of Plenty, *Colenso.* The above description is taken from the 'Flora of New Zealand,' the specimens being lost.

2. **U. novæ-Zelandiæ,** *Hook. f. Fl. N. Z.* i. 206. A slender herb. Roots or rhizomes creeping, fibrous, covered with pedicelled bladders. Leaves all radical, few, linear-lanceolate, quite entire, 1-nerved, deciduous. Scape 3–5 in. high, simple, erect, 1–4-flowered at the very top. Flowers shortly pedicelled; bracts broadly ovate, obtuse. Upper sepal orbicular, obtuse, 2-lobed or retuse; lower concave, obscurely 3-toothed. Corolla white; upper lip wedge-shaped, retuse, lower broadly axe-shaped; margin entire. Spur prominent, obtuse.

Northern Island: wet rocks at Palliser Bay, *Colenso.* Allied to the Australian *U. dichotoma*, but much smaller.

3. **U. monanthos,** *Hook. f.*;—*Fl. Tasm.* i. 298. A small, very slender, erect plant, not 1 in. high. Rhizomes or roots creeping, bearing minute pedicelled bladders. Leaves narrow linear-lanceolate. Scape 1-flowered, ½–1

in. high. Sepals oblong, obtuse. Corolla purple; upper lobe linear, notched, lower axe-shaped. Spur short, broad, notched.

Middle Island: Rangitata range, *Sinclair and Haast.* Also found in Tasmania.

4. **U. Colensoi,** *Hook. f. Fl. N. Z.* i. 206. Altogether like *U. novæ-Zelandiæ*, but with upper lip of the corolla linear-oblong. 2-lobed, lower broadly cuneate, 3-lobed, middle lobe retuse; disk with 3 gibbous prominences.

Northern Island: east coast, *Colenso.* The only specimens were preserved in fluid with *U. protusa,* and lost with them.

ORDER LVI. VERBENACEÆ.

(*Including* MYOPORINEÆ.)

Shrubs or trees, more rarely herbs. Leaves opposite or alternate, exstipulate. Flowers hermaphrodite.—Calyx inferior, 4- or 5-toothed -parted or -lobed. Corolla regular, or irregular and 2-lipped, imbricate in bud. Stamens usually 4 or 5, inserted in pairs on the corolla, or equidistant and alternating with its lower lobes. Ovary entire or 4-lobed, rarely deeply, 2–4-celled; style slender, entire or 2-fid; ovules 1 or 2 in each cell, erect or pendulous. Fruit dry or fleshy, indehiscent, sometimes divided into cocci. Seed with little or no albumen; cotyledons thick, straight or conduplicate.

A large, chiefly tropical Order, found n all parts of the globe. The *Myoporineæ*, usually considered as a Natural Order, cannot be separated; they are chiefly confined to Australia and the Pacific islands.

Tree. Leaves opposite, 3–5-foliolate. Corolla 2-lipped. Drupe 4-celled . . . . 1. VITEX.
Shrub. Leaves opposite, simple. Corolla 2-lipped. Fruit dry, 4-partite 2. TEUCRIDIUM.
Tree, maritime. Leaves opposite, simple. Corolla regular. Seed germinating on the tree . . . . 3. AVICENNIA.
Shrub. Leaves alternate, with pellucid glands. Corolla regular . . 4. MYOPORUM.

1. VITEX, Linn.

Trees or shrubs. Leaves digitately 3–5-foliolate. Flowers in axillary or terminal cymes or panicles.—Calyx 5-toothed or -lobed. Corolla 5-lobed, 2-lipped. Stamens 4; filaments long, declinate. Ovary 4-celled; style 2-fid; ovules pedulous. Drupe with a 4-celled nut.

A large tropical and subtropical genus.

1. **V. littoralis,** *A. Cunn.;—Fl. N. Z.* i. 203. A large tree, 50–60 ft. high; trunk 20 in girth; wood hard. Leaves on petioles 2–4 in. long; Leaflets 3–5, petioled, oblong or obovate, acute, glabrous, 3–4 in. long. Panicles axillary, spreading, 4–8-flowered, dichotomously branched; peduncles and pedicels slender. Calyx cup-shaped, obscurely lobed. Corolla 1 in. long, pubescent, pink or dull red; upper lip arched, 2-fid; lower deflexed, 3-fid. Drupe obovoid, bright red.—Hook. Ic. Pl. t. 419, 420.

Northern Island: along the east coast, common. **Middle** Island: Canterbury, *Haast.* Wood much used, extremely hard, said to be indestructible under water.

2. TEUCRIDIUM, Hook. f.

Shrubs, with slender 4-angled branches. Leaves small, simple. Flowers axillary, solitary.—Calyx campanulate, 5-toothed, persistent. Corolla campanulate, 2-lipped; upper lip of 4 short lobes, lower longer. Stamens long, exserted, arching downwards; anthers 1-celled. Ovary 4-lobed, 2-celled; cells incompletely divided again; style 2-fid; ovules 4, pendulous. Fruit sunk in the calyx, small, 4-lobed, hispid, splitting into 4 nuts. Testa thin. Cotyledons large, ovate; inferior radicle short.

A genus of two species, one found in subtropical Australia (*T. sphærocarpum*, Muell.), the other the following. The lobed ovary is anomalous in the Order, and shows a tendency towards *Labiatæ*, but the reversed position of the flower at once distinguishes this.

1. **T. parvifolium,** *Hook. f. Fl. N. z.* i. 203. *t.* 49. A much-branched slender, twiggy shrub, 2–5 ft. high, forming thickets, dichotomously branched, more or less pubescent. Leaves orbicular, broadly ovate or spathulate, $\frac{1}{6}$ in. long, with petioles $\frac{1}{3}$–$\frac{1}{2}$ in. long. Peduncles short, 2-bracteolate. Calyx-teeth acute. Corolla hairy, blue, $\frac{1}{2}$–$\frac{1}{3}$ in. long.

Northern Island: Wairarapa valley, *Colenso.* **Middle** Island: Nelson, *Bidwill, Travers;* Akaroa, *Raoul;* Canterbury plains, *Travers.*

3. AVICENNIA, Linn.

Evergreen, littoral trees, hoary with down; roots branching over the mud. Leaves opposite, quite entire. Peduncles short, axillary, 3-chotomous. Flowers sessile, surrounded by bracts.—Calyx equal, 4- or 5-partite. Corolla small, coriaceous, campanulate; limb 4- or 5-fid; lobes equal or posterior larger, nearly valvate. Stamens 4, filaments short; anthers 2-celled. Ovary ovoid or conical, silky, 2-celled; style short or 0, stigmas 2 short erect finally diverging; ovules 2, collateral, pendulous in each cell. Fruit obliquely ovoid, compressed, coriaceous, 1-celled, 1-seeded. Seed consisting of an immense embryo, with scanty albumen and imperceptible testa; cotyledons cordate, very broad; radicle long, thick, woolly, descending from the fruit before it falls.

A small genus of littoral, tropical, and subtropical trees, abounding in brackish, muddy creeks and estuaries.

1. **A. officinalis,** *Linn.—A. tomentosa,* Jacq.;—Fl. N. Z. i. 214. A small tree. Branches spreading. Leaves petioled, 2–3 in. long, ovate or oblong, obtuse, coriaceous, turning black when dry. Flowers silky, $\frac{1}{4}$ in. long, in short, 3-chotomous, capitate panicles; bracts 3, ovate, silky. Style 0, or short.—*A. resinifera,* Forst. Prodr.

Northern Island: from the Thames river northward. **Chatham** Island, *Dieffenbach.* Forster erroneously supposed that this plant yielded an edible (?) gum, whence his name *A. resinifera:* it is abundant in Australia and throughout Asia, and very nearly related to the American species, if not a form of it.

4. MYOPORUM, Banks and Solander.

Shrubs, glabrous; branches sometimes viscid at the tips. Leaves alternate, quite entire or serrate, studded with pellucid glands. Flowers axillary, solitary or fascicled, white or purplish.—Calyx equal, 5-partite. Corolla campanulate; tube short; limb 5-lobed, lobes nearly equal 2 upper approximate. Stamens 4, nearly equal; anthers 2-celled, cells becoming confluent. Ovary ovoid, 2–5-celled; style slender, stigma obtuse; ovules usually solitary when the cells are more than 2, geminate in each cell when these are only 2. Drupe ovoid, 2–5-celled; cells 1- rarely 2-seeded. Seeds pendulous; albumen scanty, fleshy; embryo terete; radicle superior.

A large Australian and Pacific Island genus.

1. **M. lætum,** *Forst.;—Fl. N. Z.* i. 204. A shrub or small glabrous tree, 8–10 ft. high. Leaves 2–4 in. long, lanceolate or obovate-lanceolate, acute or acuminate, serrulate above the middle, narrowed into petioles, bright green and lucid. Flowers 2–6 in a tuft; peduncle $\frac{1}{3}$–$\frac{2}{3}$ in. long. Sepals subulate or narrow-lanceolate, variable in size. Corolla $\frac{1}{3}$–$\frac{2}{3}$ in. broad; lobes rounded, villous inside. Drupe $\frac{1}{4}$ in. long.—*Cytharexylon perforatum,* Forst.

Northern and **Middle** Islands: common on the shores, as far south as Otago. **Kermadec** Islands, *M'Gillivray.* Very closely allied to the Norfolk Island *M. obscurum* but the calyx is smaller. Kermadec Island specimens, however, are intermediate in this respect. It is also very near *M. serratum* of Tasmania, and some Pacific Island species.

Of Forster's *M. pubescens* nothing is known; it may be a *Scævola.*

Order LVII. LABIATÆ.

Herbs or shrubs, usually aromatic. Leaves opposite, rarely whorled. Flowers in small, sessile or stalked, opposite axillary cymes, or solitary or whorled, rarely panicled racemose or spiked, regular or irregular.—Corolla-tube usually long and cylindric or funnel-shaped; limb often 2-lipped. Stamens 2 or 4, inserted on the tube of the corolla, alternating with its lower lobes. Ovary 4-lobed, 2–4-celled; style filiform, stigma 2-fid; ovules 1 in each lobe, erect. Fruit of 4 small nuts enclosed in the calyx. Albumen little or 0; cotyledons flat.

A very large Order in most parts of the world, except the coldest, to which the Mint, Sage, Horehound, etc. belong. It is singularly rare in New Zealand.

Calyx nearly equal. Stamens equal . . . . 1. Mentha.
Calyx 2-lipped. Stamens didynamous . . . . 2. Scutellaria.

1. MENTHA, Linn.

Herbs, erect or procumbent, aromatic when bruised. Flowers solitary or in few- or many-flowered axillary whorled cymes. Bracts subulate.—Calyx tubular or campanulate, 5-toothed; throat naked or villous. Corolla-tube short, included in the calyx; limb equally 5-lobed or 2-lipped, the upper lip larger often 2-lobed. Stamens 4, equal, straight, erect; anther-cells 2, parallel. Style 2-fid at the tip. Nuts dry, smooth.

A very large European and Oriental genus, comparatively rare in other parts of the world. Several English species, as the Water-Mint, *M. aquatica,* L.; Peppermint, *M. piperita,* Sm.; Spearmint, *M. viridis,* L., etc., have been introduced into New Zealand.

1. **M. Cunninghami,** *Benth.;—Fl. N. Z.* i. 205. A fragrant, prostrate,

slender, diffusely branched, wiry herb; branches 2–10 in. long, often matted, pubescent. Leaves sessile or petioled, $\frac{1}{6}$–$\frac{1}{2}$ in. long, rounded or oblong-ovate, obtuse, quite entire, covered with pellucid dots. Flowers solitary, axillary, on short or long, slender pedicels. Calyx $\frac{1}{10}$–$\frac{1}{8}$ in. long, campanulate, villous externally, and on the teeth internally. Corolla lobes short. Stamens included in the corolla-tube of some flowers, exserted in others.—*Micromeria Cunninghamii*, Benth.

Northern and **Middle** Islands: abundant in rather dry places from the Bay of Islands to Otago, *Banks and Solander*, etc.

2. SCUTELLARIA, Linn.

Herbs or shrubs. Flowers racemed or whorled in the upper pairs of leaves. Bracts obscure or 0.—Calyx campanulate, 2-lipped; lips entire, upper with a flat shield or scale adnate to its upper surface, closed after flowering. Corolla-tube exserted, dilated above; limb 2-lipped, upper lip entire or notched, lower dilated 3-lobed. Stamens 4, exserted; anthers meeting in pairs, of longer stamens with unequal cells. Ovary on a long or short curved pedicel. Nuts dry, tubercled.

A very large genus, found in almost all quarters of the world.

1. **S. novæ-Zelandiæ,** *Hook. f. Fl. N. Z.* ii. 335. *S. humilis*, i. 205, not of Brown. A slender, sparingly branched, nearly glabrous herb, 6–12 in. high. Stem square, suberect or ascending. Leaves small, in distant pairs, $\frac{1}{8}$–$\frac{1}{2}$ in. long, petioled, ovate orbicular or subreniform, obscurely 3–5-lobed or crenate or quite entire, petioles $\frac{1}{4}$–1 in. long. Flowers in opposite pairs in the axils of the uppermost leaves; peduncles stout, $\frac{1}{12}$–$\frac{1}{6}$ in. long. Calyx short, puberulous; lips obtuse, rounded, scale shorter than the upper lip, but becoming much larger in fruit. Corolla $\frac{1}{4}$ in long; lobes of lips short and obtuse, villous. Anthers all glabrous.

Middle Island: Nelson, *Bidwill*; Fox Hill, *Munro*. Closely allied to the Australian *S. humilis*, Br.

Of *Plectranthus australis*, Br., an Australian and Pacific Island plant, introduced into Raoul's 'Choix de Plantes,' I know nothing.

Order LVIII. PLANTAGINEÆ.

Herbs, with radical tufted leaves, rarely having leafy stems. Flowers solitary or spiked on slender leafless scapes, inconspicuous, green or brown.—Sepals 4. Corolla scarious; tube cylindric; lobes 4, spreading, with incurved margins. Stamens 4, long, inserted on the corolla-tube, alternating with its lobes; anthers versatile. Ovary 1, 2- or 4-celled; style long, filiform, stigma downy; ovules 1–6 in each cell, inserted on the septum. Capsule usually bursting transversely. Seeds peltate; albumen dense; embryo generally cylindric.

A small Order, widely diffused.

1. PLANTAGO, Linn.

Herbs, with usually ribbed, rosulate, radical leaves. Scapes few- or many-flowered. Flowers usually densely spiked, green, hermaphrodite.—Sepals with broad, membranous margins. Capsule bursting across the middle. Seeds attached to either face of a free longitudinal septum, sessile, peltate.

A large genus, found in all temperate and many tropical countries, including the Plantains and other common weeds.

Scapes 1–3, *rarely* 4–5-*flowered.*

Small. Leaves villous at the base. Scape 1-flowered. Bracts and sepals acute	1. *P. uniflora.*
Small. Leaves nearly glabrous. Scape 3–5-flowered. Bracts and sepals obtuse	2. *P. Brownii.*
Small. Leaves very villous. Scape 2- or 3-flowered. Bracts and sepals subacute	3. *P. lanigera.*

Scapes many-flowered. Flowers spiked.

Spikes ½–1 in. long. Bracts and sepals pilose and ciliated	4. *P. spathulata.*
Spikes ¼–¾ in. Scapes slender. Bracts and sepals glabrous . . .	5. *P. Raoulii.*
Spikes 1–6 in. Scapes stout. Bracts and sepals glabrous	6. *P. Aucklandica.*

1. **P. uniflora,** *Hook. f. Fl. N. Z.* i. 207. Stems short, stout, ¼ in. high, tufted (?), villous at the crown. Leaves few, ½–1½ in. long, narrow-lanceolate, sinuate-toothed or quite entire, glabrous, villous at the base; nerve 1, obscure. Scape slender, as long as the leaves, 1-flowered, hairy. Sepals linear-oblong, acute, longer than the lower half of the capsule.

Northern Island: top of the Ruahine range, *Colenso.* Very near *P. Brownii*, of which it may be a variety, but the leaves are narrower, scapes more slender, flowers solitary, and sepals narrower and more acute. My specimens are indifferent and past flower.

2. **P. Brownii,** *Rapin.*—*P. carnosa*, Br.;—Fl. N. Z. i. 207. A small, tufted, rather fleshy species; root stout. Leaves very numerous, rosulate, ½–1½ in. long, lanceolate or oblong-lanceolate, acute, sinuate-toothed, glabrous or pilose, often villous at the base or in patches on the upper surface. Scapes stout, erect or prostrate, numerous, as long as or longer than the leaves, pilose, 3–5-flowered. Flowers in a small, dense head. Bracts and sepals broadly oblong, obtuse, nearly glabrous, with fleshy keel. Capsule short; cells 2–4-seeded.—Fl. Antarct. i. 65, t. 43.

Northern Island: summits of the Ruahine mountains, *Colenso.* **Middle** Island: hills and valleys, Hurumui, alt. 1200–1600 ft., *Travers.* **Lord Auckland's** group: on maritime rocks, common, *J. D. H.* Closely allied to the Antarctic American *P. pauciflora*, Lamb., but the habit is different. This is a common Tasmanian plant. The Auckland Island specimens have foliage nearly glabrous; the Ruahine mountain ones present patches of villous hairs on the upper surface of the leaves; the Tasmanian individuals are intermediate in this respect; Travers's specimens are broader-leaved, and resemble small states of *P. spathulata* in foliage. This, like the *P. maritima* and *P. Coronopus* of Europe, inhabits both the mountain-tops and sea-level; *P. paradoxa*, Hook. f., of Tasmania, is probably another state of it; *P. barbata*, Forst., of Fuegia, is its American representative.

3. **P. lanigera,** *Hook. f., n. sp.* Leaves densely rosulate, oblong-lanceolate, subacute, sessile, quite entire, ½–1 in. long, above densely woolly with matted, white, tortuous, jointed hairs, below glabrous or nearly so. Scapes short, inclined, tomentose, 1–3-flowered at the top. Bracts and sepals ovate-oblong, subacute.

Middle Island: Otago, lake district, in rocky alpine places, alt. 6000 ft., *Hector and Buchanan.* The flowers are long past in my specimens, in Buchanan's drawing they are large, the corolla much exceeding the calyx.

4. **P. spathulata,** *Hook. f. Fl. N. Z.* i. 208. Glabrous or villous with

weak, jointed, soft hairs. Leaves all radical, horizontally spreading, 1–4 in. long, spathulate or oblong- or lanceolate-spathulate, fleshy, quite entire or sinuate or toothed, petioles villous at the base. Scapes several or numerous, villous or pilose; spikes oblong, cylindric, obtuse, dense- and many-flowered, $\frac{1}{2}$–1 in. long. Bracts and sepals $\frac{1}{10}$–$\frac{1}{8}$ in. long, ovate, acute, with thick midrib, pilose and ciliate. Corolla-lobes ovate, acute. Stamens and style moderately long. Anthers large. Capsule short; cells 2-seeded.

Northern Island: east coast, near Pahawa, in gravel, rocks, and sand, *Colenso*. **Middle** Island: terraces and river beds in the Kowai valley, *Haast*. The spikes are sometimes compound.

5. **P. Raoulii,** *Decaisne;—Fl. N. Z.* i. 208. Pilose or almost hispid or villous. Leaves all radical, horizontally spreading, 2–10 in. long, rather flaccid, linear- or oblong-lanceolate, acute, coarsely irregularly sinuate-toothed. Petioles villous at the base Scapes slender, few or numerous, pilose. Spikes cylindric, $\frac{1}{4}$–$\frac{3}{4}$ in. long, obtuse, densely many-flowered. Bracts and sepals $\frac{1}{12}$ in. long, broadly ovate, obtuse, glabrous, with broad fleshy keel. Corolla-lobes very small. Capsule twice as long as the calyx, acute; cells 2-seeded.—*P. varia*, A. Cunn., not Brown.

Northern and **Middle** Islands: abundant in pastures and waste grounds. This is a representative of the common Australian *P. varia*.

6. **P. Aucklandica,** *Hook. f. Fl. Antarct.* i. 65, *t.* 42. Rhizome fleshy, stout, as thick as the thumb. Leaves densely crowded, 1–3 in. long, broadly ovate, obtuse, glabrous, fleshy, obscurely sinuate or quite entire, 7–10-nerved; petioles very short, broad, densely villous or woolly at the base. Scapes numerous, longer than the leaves, very stout, hairy. Spikes 1–6 in. long. Flowers small, crowded above the middle of the spike, in scattered tufts below it. Bracts and sepals $\frac{1}{16}$–$\frac{1}{12}$ in. long, broadly ovate, obtuse, glabrous, with broad fleshy centre. Corolla-lobes small, linear-oblong. Capsule twice as long as the sepals, acute; cells 1-seeded.

Lord Auckland's Island: common on the hilltops, in wet places, *J. D. H.*

Two European species of *Plantago* are now naturalized in New Zealand, viz. *P. major*, Linn., with large ovate or subcordate, long-petioled, 5–9-ribbed leaves, and a very long spike; and *P. lanceolata*, Linn., also a large species, with lanceolate, 5–7-ribbed leaves, and short, stout, dense-flowered spikes. Both are troublesome weeds in pastures.

ORDER LIX. NYCTAGINEÆ.

Trees shrubs or herbs. Leaves usually opposite, quite entire, exstipulate. Flowers usually hermaphrodite, often panicled.—Perianth tubular or funnel-shaped, 5-lobed, persistent, closing over the fruit. Stamens 1 or more, hypogynous, free or united at the base, equal or unequal. Ovary free, 1-celled; style filiform, stigma lobed or capitate; ovule 1, erect. Utricle enclosed in the hardened perianth-tube. Seed usually long; embryo with foliaceous cotyledons coiled round mealy albumen.

An Order of no great extent or importance, chiefly tropical, containing the common garden *Mirabilis*, or "Marvel of Peru."

1. PISONIA, Linn.

Trees or shrubs. Leaves opposite alternate or whorled, exstipulate. Flowers in terminal cymes or corymbs, green or reddish, usually small.—Flowers hermaphrodite or unisexual, with minute bracts at the base. Perianth cylindrical-campanulate; lobes 5, small, plaited. Stamens 6–10, included or exserted, unequal. Ovary elongate; style lateral or terminal, stigma entire lobed or plumose. Utricle enclosed within the thickened, smooth or costate, glabrous aculeate or glandular perianth.

A considerable genus, of chiefly littoral plants, found in tropical and subtropical parts of the world.

1. **P. Brunoniana.** *Endl. Prodr. Flor. Ins. Norf.* 43 —*P. Sinclairii.*

fascicles, often subspicate.—Flowers without bracts, hermaphrodite or unisexual. Perianth 3-5-parted, not enlarged in fruit. Stamens 1-5. Ovary globose or depressed; styles 2 or 3. Utricle depressed or erect. Seed horizontal or erect; embryo surrounding mealy albumen.

A very large, ubiquitous genus, comprising many weeds of waysides, dunghills, and cultivated grounds, several of which are no doubt introduced into New Zealand.

§ 1. *Seed horizontal, rarely vertical in* C. glaucum.

Leaves ¼–¾ in., quite entire, glaucous and pulverulent . . . . . . 1. *C. triandrum.*
Leaves 1–1½ in., triangular, toothed, not glaucous . . . . 2. *C. urbicum.*
Leaves ½–1 in., oblong or deltoid, mealy below . . . . 3. *C. glaucum.*
Leaves 1–2 in., glandular, aromatic, not glaucous . . . . 4. *C. ambrosioides.*

§ 2. *Seed vertical.* (*See* glaucum *in* § 1.)

Tall, erect. Perianth-segments much thickened . . . . 5. *C. carinatum.*
Small, much-branched, very slender. Perianth membranous . . . 6. *C. pusillum.*

1. **C. triandrum,** *Forst.;—Fl. N. Z.* i. 212. A small, much-branched, pulverulent herb, 6-12 in. high; branches very slender. Leaves opposite and alternate, ¼–⅔ in. long, with slender petioles, oblong-hastate, truncate cordate or cuneate at the base, rounded at the tip, quite entire. Flowers very minute, in axillary spikelets, and fascicled towards the ends of the branches. Fruiting perianth open. Stamens 2-4. Seed punctate, adhering to the utricle.

Northern and **Middle** Islands, from Auckland to Otago, frequent, *Banks and Solander*, etc. Apparently confined to New Zealand. (The plant bearing this name in A. Cunningham's Herbarium is *Euxolus viridis.*)

2. **C. urbicum,** *Linn.;—Fl. N. Z.* i. 213. A tall, erect or prostrate, coarse, branching, green herb, 2-3 ft. high, quite glabrous, not glaucous nor pulverulent; stems angled. Leaves 1–1½ in. long, with slender petioles, triangular or subcordate-hastate, margin irregularly notched, crumpled, lobulate. Flowers in dense axillary and terminal spikes. Seed punctate; margins obtuse.

Northern Island: Tanenuiarangi, *Colenso.* **Middle** Island: New River, *Hb. A. Richard;* Ashburton river and Rangitata range, *Haast.* A common weed and wayside plant in many parts of the world, perhaps not indigenous to New Zealand.

3. **C. glaucum,** *Linn.*, var. **ambiguum;**—*Fl. N. Z.* i. 213. Prostrate, much-branched, succulent; stems and branches 4-18 in. long, flaccid, glabrous. Leaves ½–1 in. long, lower petioled, upper sessile, ovate-oblong trapezoid or subhastate, quite entire lobulate or coarsely unequally toothed, obtuse, pulverulent below. Spikes short, glomerate, axillary. Perianth 3-5-parted. Seed erect or horizontal, punctate.—*C. ambiguum*, Br.

Northern and **Middle** Islands: not uncommon on muddy flats, shingle, etc., near the sea, *Banks and Solander*, etc. A common extratropical plant in the southern and also abundant in the northern hemisphere.

4. **C. ambrosioides,** *Linn.;—Fl. N. Z.* i. 213. Tall, erect, branched, herbaceous, 1-3 ft. high, glabrous or pubescent, everywhere glandular, very aromatic; stems and branches terete. Leaves 1-2 in. long, ovate- or oblong-lanceolate or lanceolate, acuminate, cuneate at the base, shortly petioled, coarsely obtusely or acutely toothed and cut. Flowers on short, axillary, spiciform, leafy branches. Stamens usually 5. Seed horizontal.

Northern Island: cultivated ground, *Colenso*, perhaps introduced. **Middle** Island:

Otago, Waitaki valley and lake district, *Hector and Buchanan.* An abundant tropical and subtropical weed; the seeds are described as sometimes vertical.

5. **C. carinatum,** *Br.;—Fl. N. Z.* i. 213. A branched, erect, glandular-pubescent, strong-scented herb; stem 1–2 ft. high. Leaves petioled, small, $\frac{1}{4}$–1 in. long, ovate-oblong, obtuse, sinuate-toothed. Flowers very minute, copious, in axillary glomerules, green. Perianth-segments 5, incompletely covering the fruit, oblong, with prominent, thick, but not succulent back. Stamen 1. Seed erect, compressed, brown, minutely punctate.—*C. Botrys*, A. Cunn. not Linnæus.

Northern Island: Bay of Islands, *Cunningham*, etc.; Auckland, *Sinclair.* I have seen no authentic Australian specimens of *C. carinatum*, Br., but suppose this to be it; it is the *C. glandulosum*, Moq. Tand., also a native of Australia.

6. **C. pusillum,** *Hook. f., n. sp.—C. Pumilio*, Br.; Fl. N. Z. i. 214. A small, much-branched, pubescent and glandular plant; stem very short; branches 3–6 in. long, very slender, leafy. Leaves petioled, $\frac{1}{12}$–$\frac{1}{6}$ in. long, oblong-ovate, obtuse, quite entire or obscurely sinuate, pubescent on both surfaces. Flowers in very minute axillary glomerules. Perianth 3–5-cleft, membranous, segments not thickened, not closing over the fruit. Stamen 1. Seed very minute, erect, compressed, brown, minutely punctate.

Northern Island: shores of the east coast and sandy shores of Lake Taupo, abundant in native cultivated ground, *Colenso.* This differs from Brown's *C. Pumilio* in the membranous perianth-segments.

2. SUÆDA, Forskal.

Erect or prostrate, succulent herbs, sometimes shrubby at the base. Leaves terete, fleshy. Flowers minute, clustered in the axils of the leaves, hermaphrodite; bracts 2, minute.—Perianth urceolate, 5-partite, fleshy, tumid in fruit. Stamens 5. Ovary truncate; styles 2–5. Utricle compressed, included in the tumid perianth. Seed free, vertical or horizontal; albumen little or 0; embryo spiral.

A small genus of chiefly maritime plants, scattered over the globe.

1. **S. maritima,** *Dumortier;—Fl. N. Z.* i. 316. Erect, branched, perfectly glabrous; stem shrubby at the base; branches slender. Leaves sessile, $\frac{1}{3}$–$\frac{1}{2}$ in. long, subcylindric, linear, glabrous or farinose. Utricle membranous. Seed punctate, horizontal or oblique; margin rather acute.—*Chenopodium australe*, Br.; *C. maritimum*, A. Cunn.; *Salsola fruticosa*, Forst.

Northern and **Middle** Islands: not uncommon in maritime swamps, muddy shores, etc., as far south as Otago, *Banks and Solander*, etc. A frequent plant on most temperate and many tropical coasts.

3. ATRIPLEX, Linn.

Herbs or shrubs, often fleshy, sometimes scaly or powdery.—Flowers small, green, crowded in clusters, which are axillary or sessile on the terminal naked tips of the branches, unisexual. *Male* ebracteate. Perianth 3–5-parted. Stamens 3–5. Ovary rudimentary. *Female* 2-bracteate. Perianth 0 or 5-partite. Ovary small; styles 2, united at the base. Bracts in the fruit much enlarged, erect, dilated, closely pressed together and enclosing the utricle Seed horizontal; albumen farinaceous; embryo annular.

A large genus, abounding in shores and waste places of temperate (more rarely in tropical) regions.

Plants covered with white scales . . . . 1. *A. cinerea*.
Glabrous or powdery, erect. Fruiting bracts rhomboid . . . . . 2. *A. patula*.
Glabrous, fleshy, and papillose. Fruiting bracts urceolate . . . . 3. *A. Billardieri*.

1. **A. cinerea,** *Poiret;—Fl. N. Z.* i. 214. A small shrub, 1–4 ft. high, everywhere covered with white glistening scales; branches angled, leafy. Leaves 1–2 in. long, narrow-oblong, obtuse, entire, narrowed into short petioles. *Male* fl.: clustered in dense globose or oblong spikes, which are panicled at the ends of the branches. *Female* fl.: axillary, clustered or solitary. Bracts of fruit $\frac{1}{6}$–$\frac{1}{4}$ in. long, broadly ovate, obtuse, corky, with thin margins.—*A. Halimus*, Br. Prodr. not Linn.

Northern Island: sandy shores of Palliser Bay, *Colenso*. **Middle** Island: Canterbury, *Haast* (nearly glabrous; leaves sinuate, perhaps different, but specimen imperfect). An abundant Australian and Tasmanian plant. Closely allied to the European *A. Halimus*, L.

2. **A. patula,** *Linn.;—Fl. N. Z.* i. 215. A glabrous or slightly powdery, usually tall and erect, branched, leafy herb, 2–4 ft. high; stems terete. Leaves 1–3 in. long, shortly petioled, narrow ovate oblong or hastate, quite entire or lobed, rarely laciniate, subacute or obtuse, uppermost often linear. Flowers in spikes, which are axillary, and terminate the slender branches. *Female* bracts $\frac{1}{10}$ in. diam., rhomboid in fruit, toothed, their back smooth or tubercled.

Northern Island: salt marshes on the east coast, plentiful, *Colenso*. Abundant in Australia, Tasmania, and many other parts of the old world.

3. **A. Billardieri,** *Hook. f. Fl. N. Z.* i. 215. A prostrate, branched, glabrous, succulent, papillose herb; stems angled, 6–12 in. long; branches ascending. Leaves small, $\frac{1}{4}$–$\frac{1}{3}$ in. long, oblong, obtuse, entire or sinuate, shortly petioled. *Male* fl.: fascicled, shortly pedicelled; *female*: solitary or 2 together, sessile; bracts combined into an urceolate, 2-lipped, fleshy cup, enclosing an erect utricle. Utricle nearly orbicular, with 2 rather slender styles. Seed erect, compressed, its edges opposite the 2 bracts (not parallel, as is usual in the genus).—*Theleophyton Billardieri*, Moq.-Tand. in A. DC. Prodr.

Northern Island, *A. Cunningham*; sandy places, Wangururu Bay, *Colenso*. A native of Tasmania.

4. SALSOLA, Linn.

Shrubs, or rigid, often spinescent herbs. Leaves small, subcylindric, fleshy, or rigid and pungent.—Flowers small, axillary, sessile, hermaphrodite, 2-bracteate. Perianth 4- or 5-partite, segments dilating greatly, closing over the fruit, and becoming transversely ridged, or membranous and broadly winged. Stamens 5; filaments often united at the base. Ovary depressed; styles 2. Seed horizontal; albumen 0; embryo spiral.

A large genus, especially in the saline districts, inland and maritime, of the temperate and subtropical regions of the globe.

1. **S. australis,** *Br.?;—Fl. N. Z.* i. 216. A rigid, woody, low, suberect or prostrate, much-branched plant, 1–2 ft. high; stems and branches

ribbed, rather rough. Leaves small, $\frac{1}{6}$–$\frac{1}{4}$ in. long, scattered, patent or recurved, rigid, ovate or broadly subulate, pungent. Flowers very inconspicuous, shorter than the bracts. Perianth 5-parted, segments after flowering expanded into a broad, horizontal, membranous, veined wing.

Northern Island: gravelly shores of Port Nicholson harbour, *Colenso*, who observes that it is perhaps introduced. A native of Australia, and very closely related to the common European *S. Kali*. My only specimen has shorter leaves than the Australian plant, which is sometimes pilose; it has not ripe fruit. I have from Canterbury what appear to me to be seedlings of this, sent by Mr. Travers.

5. SALICORNIA, Linn.

Herbs or shrubs; stems leafless, succulent, jointed.—Flowers minute, sometimes in approximate short joints, which form a sort of cone at the end of the branch, hermaphrodite, ebracteate, hidden in or between the tops of the joints of the stem. Perianth fleshy, turbinate. Stamens 1 or 2. Ovary ovoid; styles 2. Utricle compressed, included in the perianth. Seed vertical; albumen usually scanty; embryo annular or conduplicate.

Maritime or salt lake plants, found in all parts of the world.

1. **S. indica,** *Willd.?;—Fl. N. Z.* i. 216. Stems prostrate, 3–6 in. long, rather woody; branches ascending, 2–6 in., terminated by cylindrical cones. Joints very variable in length, $\frac{1}{3}$–$\frac{1}{2}$ in., rather compressed, dilated at the tip and obscurely 2-lobed, about as thick as a small quill. Cones $\frac{1}{3}$–2 in. long, rather thicker than the branches. Flowers numerous, sessile, whorled in the axils of the short joints of the cone. Perianth urceolate, fleshy, truncate, with a small central orifice. Stamen, 1 only seen. Fruiting perianth obpyramidal, with a flat top and closed orifice. Utricle membranous. Seed flattened, nearly orbicular; testa papillose, rather thick; albumen 0; embryo with very thick, plano-convex, pyriform cotyledons, and incumbent terete radicle.—*S. australis*, Forst.

Northern and **Middle** Islands: abundant on muddy shores and in sandy and rocky places, *Banks and Solander*, etc. This appears to be the same as the Indian plant figured in Wight's 'Icones,' t. 757, and quoted by Moquin-Tandon, and which is also found in Tasmania and Australia. The structure of the flower and fruit wants careful re-examination on living specimens. It does not agree with any description in Moquin's monograph of the genera in A. DC.'s 'Prodromus,' but is, I suppose, referable to his genus *Arthrocnemon.*

ORDER LXI. AMARANTHACEÆ.

Herbs, rarely shrubby. Leaves opposite or alternate; stipules 0.—Floral characters of *Chenopodiaceæ*, but the perianth is usually membranous or scarious, of 5 distinct leaflets; the stamens are most often united into a membranous cup with membranous expansions between the filaments, and the anthers are 1-celled in many. One tribe of the Order has several ovules in the ovary.

A large tropical and subtropical Order, including a few weeds of cultivation, as *Amaranthus (Euxolus) viridis*, which Cunningham has, under the name of *Chenopodium triandrum*, from the Bay of Islands; it resembles a *Chenopodium*, but has acute bracts and perianth-leaflets, and a minute hard nut-like fruit.

1. ALTERNANTHERA, Forskal.

Herbs, rarely shrubs; stems often branched, jointed, rooting. Leaves opposite. Flowers minute, white, in axillary or terminal clusters, 3-bracteate.—Perianth 5-parted, erect. Stamens 5; filaments united in a membranous cup; anthers 3–5, 1-celled; style very short, stigma 2-fid or capitate. Utricle erect, compressed, included in the perianth. Seed vertical, compressed.

A rather large tropical and subtropical genus.

1. **A. sessilis,** *Br.*;—*Fl. N. Z.* i. 212. Stems much branched from the root, herbaceous, prostrate, 2–6 in. long, with 2 lines of pubescence. Leaves ½–1 in. long, narrow obovate or oblong, obtuse, rather fleshy, pubescent in the axils. Flowers minute, white. Perianth-segments acuminate, glabrous. Anthers, 3 fertile, 2 imperfect.—*A. denticulata*, A. Cunn.

Northern Island: not unfrequent in boggy places. A very common tropical and subropical weed in the old world, and very variable.

ORDER LXII. PARONYCHIEÆ.

Herbs, often small. Leaves opposite or alternate, stipulate or exstipulate. Inflorescence various. Flowers small, regular, hermaphrodite.—Perianth 4- or 5-lobed or partite, persistent, and enclosing the fruit. Stamens 1–10, perigynous or hypogynous; filaments subulate; anthers small, 2-celled. Ovary ovoid, free, sessile, 1-celled; style 1, terminal, 2- or 3-fid, stigmas capitellate or subulate; ovule 1, erect. Utricle or nut included in the persistent calyx, 1-seeded. Seed erect or pendulous from a basilar funicle; testa coriaceous; embryo usually annular, surrounding farinaceous albumen.

A small Order, scattered all over the globe.

1. SCLERANTHUS, Linn.

Small, densely tufted, dichotomously branched, rigid herbs. Leaves connate by their base in opposite pairs, subulate, pungent, exstipulate. Flowers minute, solitary in pairs or cymose.—Perianth very coriaceous, 4- or 5-fid, very hard in fruit. Stamens 1–10, perigynous. Style 2-fid. Ovule pendulous from a basilar funicle. Embryo annular.

A small genus found in the temperate regions of both hemispheres.

1. **S. biflorus,** *Hook. f. Fl. N. Z.* i. 74. A very densely branched, mossy, rigid herb, 1–2 in. high, forming compact tufts. Leaves closely imbricated, $\frac{1}{10}$–½ in. long, subulate, serrulate. Flowers in pairs, with 4 bracts at their base placed crosswise, at length carried up on a short, rigid peduncle. Perianth 4-fid. Stamen 1.—*Mniarum biflorum*, Forst.; *M. pedunculatum*, Labill. Fl. Nov. Holl. t. 2; *M. fasciculatum*, Raoul, not Br.; *Ditoca muscosa*, Banks and Solander.

Dry, rocky, and sandy places throughout the islands, *Banks and Solander*, etc. Also found in Tasmania and Australia.

ORDER LXIII. POLYGONEÆ.

Herbs or shrubs. Leaves alternate, entire; stipules membranous, tubular, often lacerated. Flowers regular, small, solitary, spiked racemed or panicled. —Perianth 5- or 6-partite, imbricate, persistent, enclosing the fruit. Stamens usually 6–9, perigynous. Ovary free, 1-celled; styles 1–3, stigmas capitate; ovule 1, erect. Fruit a usually compressed or 3-gonous nut enclosed in the dry or fleshy, often enlarged perianth. Seed filling the nut; embryo cylindric, straight or curved; albumen mealy.

A very large tropical and temperate genus, containing various European weeds that may be expected to occur in New Zealand, besides the Buckwheat.

Flowers hermaphrodite. Perianth not succulent in fruit . . . . 1. POLYGONUM.
Flowers unisexual. Base of the perianth succulent in fruit . . . 2. MUHLENBECKIA.
Flowers hermaphrodite. Two or three inner lobes of perianth enlarged and closing over the fruit . . . . 3. RUMEX.

1. POLYGONUM, Linn.

Herbs, rarely shrubs; prostrate or erect, simple or branched. Flowers white or red, small, racemed spiked or axillary and solitary, spikes rarely panicled.—Flowers hermaphrodite. Perianth 5-partite, persistent. Stamens usually 6–8, rarely fewer. Ovary 3-gonous or compressed; styles 2 or 3, very short. Nut 3-gonous or compressed, included in or protruding from the withered, dry perianth.

A very large, tropical and temperate genus.

Flowers spiked. Nut flattened . . . . 1. *P. minus.*
Flowers axillary, 1–3 together. Nut 3-gonous . . . . 2. *P. aviculare.*

1. **P. minus,** *Huds.*, var. **decipiens;**—*P. prostratum*, A. Cunn.;—Fl. N. Z. i. 209, not Br. Stems herbaceous, prostrate at the base or suberect, glabrous or pilose, simple or sparingly branched. Leaves scattered, 2–8 in. long, linear-lanceolate, acuminate, eglandular, margins scabrous; stipules long, brown, with ciliated aperture. Spikes long, slender, terminal, simple or compound, 1–2 in. long. Flowers small, not crowded, reddish, $\frac{1}{12}$ in. long. Bracts truncate, glabrous or ciliate. Perianth lobes oblong, obtuse, eglandular, glabrous. Nut flattened, with obtuse edges.

Northern and **Middle** Islands: not uncommon, *Banks and Solander*, etc. This is a southern state of the common North-European *P. minus*; it entirely resembles *P. salicifolium*, Del., and *P. serrulatum*, Lag., except that the nuts are flattened and not 3-gonous; it abounds in Australia, tropical and subtropical India, Africa, and America, and is also found in Southern Europe; it differs from the typical *P. minus* in the more slender erect habit, and few, long, simple, long-peduncled spikes; it is the *P. decipiens*, Br., of Australia.

2. **P. aviculare,** *Linn.;—Fl. N. Z.* i. 210. Stems herbaceous, much branched, prostrate, woody at the base; branches spreading, 6–24 in. long, hard, wiry, grooved, rough or smooth. Leaves small, scattered, coriaceous, $\frac{1}{2}$–$1\frac{1}{2}$ in. long, linear-oblong or sublanceolate, obtuse or acute, margins recurved; stipules membranous, silvery, ragged, long or short. Flowers small, solitary or 2 or 3 together, axillary, pedicelled. Nut acute, 3-gonous, longer than the perianth.

Var. *β. Dryandri.—P. Dryandri*, Spr.; Fl. N. Z. i. 210. Smaller. Stipules shorter.

Northern Island: east coast, Ahuriri, *Colenso*. **Middle** Island: Akaroa, *Raoul;* var. *β*, east coast, Ruamahanga and Tuki-tukimiu, *Colenso;* Port Cooper, *Lyall;* Otago, covering acres of ground by roadsides, *Hector and Buchanan*. A very common plant in all temperate and some tropical parts of the world, perhaps introduced into New Zealand, where it is spreading with extraordinary rapidity.

2. MUHLENBECKIA, Meisn.

Shrubs and undershrubs, sometimes small, often rampant, and climbing with much-intertwined slender branches. Flowers small, whitish or greenish, spiked, or axillary and few or solitary; spikes sometimes panicled. Pedicels jointed below the perianth.—Flowers unisexual, monœcious or polygamous. Perianth 5-lobed, lower part becoming fleshy, often white in fruit. Stamens 8; filaments short; anthers oblong; very reduced thick and short in the female flowers with imperfect anthers. Ovary sessile, 3-gonous, with 3 very short, papillose or fimbriate stigmas. Nut ovoid, acuminate, 3-quetrous, black, enclosed in the perianth, whose basal part is succulent.

A small genus of Australian, New Zealand, and South American plants.

Leaves ½–2 in. long, broad, ovate or cordate. Spikes panicled . . . 1. *M. adpressa.*
Leaves ¼–½ in. long, broad, ovate or cordate. Spikes usually simple . 2. *M. complexa.*
Leaves $\frac{1}{10}$–¼ in. long, orbicular or oblong. Flowers subsolitary . . 3. *M. axillaris.*
Leaves 0, or minute and narrow. Male flowers spiked, female often solitary . . . . 4. *M. ephedroides.*

1. **M. adpressa,** *Lab.;—Polygonum australe,* A. Rich.;—Fl. N. Z. i. 210. A large, rambling and climbing, leafy bush, glabrous. Stem and branches often twining, flexuous, grooved; branchlets often minutely scaberulous. Leaves ½–2 in. long, petioled, cordate or broadly oblong and truncate at the base, obtuse acute or apiculate, glabrous, in young plants 3-lobed; stipules deciduous. Spikes panicled, many-flowered, glabrous; bracts obtuse, 1–3-flowered. Flowers small, unisexual. Stigmas plumose. Perianth fleshy in fruit. Nut 3-gonous, black.—*Coccoloba australis,* Forst.

Abundant throughout the islands, *Banks and Solander*, etc. Common in Norfolk Island, Australia, and Tasmania. *Meisner* (A. DC. Prodr. xiv. 146) distinguishes the New Zealand from the Australian plant by the stigmas plumose, not papillose, but I find the stigmas of the Australian plant to be the most plumose of the two. The Chilian *M. Chilensis*, Meisn., seems to be the same species.

2. **M. complexa,** *Meisn.—Polygonum complexum,* A. Cunn.;—Fl. N. Z. i. 210. Stems slender, prostrate or climbing over bushes, much interlaced, flexuous, scaberulous or glabrous, grooved, 1–5 ft. long. Leaves petiolate, quite glabrous or puberulous at the base and on the petiole, ¼–½ in. long, broadly obovate-cordate or orbicular, often contracted in the middle, sometimes dotted below, rarely acute; stipules deciduous; petiole long or short. Spikes simple or panicled, sometimes very short and reduced to 1 or 2 flowers, glabrous pubescent or tomentose. Bracts obtuse, 1–6-flowered. Perianth fleshy in fruit. Stigmas papillose. Nut 3-gonous, black.

Throughout the islands, abundant, *Banks and Solander*, etc. Some large states of this seem to pass into *M. adpressa*, Lab., and small ones into *M. axillaris*.

3. **M. axillaris,** *Hook. f.—Polygonum axillare,* Hook. f. Fl. N. Z. i. 211. A small species, 1–6 in. high; quite glabrous, except the branchlets and petioles,

which are sometimes puberulous. Branches slender, tufted, spreading from a woody stock. Leaves small, petioled, $\frac{1}{10}$–$\frac{1}{4}$ in. long, elliptic-oblong, obtuse, flat, glabrous; stipules short, truncate. Flowers solitary, axillary, pedicelled. Stigmas papillose. Perianth fleshy in fruit.

Northern Island: Wairarapa valley, *Colenso*. **Middle** Island: abundant on the mountains, ascending to 6500 ft., *Bidwill*, etc. Much smaller than either of the preceding, and of a very different habit, but possibly only a mountain variety; it is excessively variable, and found also in the Australian and Tasmanian alps, where it attains a larger size.

4. **M. ephedroides,** *Hook. f.—Polygonum ephedroides;*—Fl. N. Z. i. 211. A shrubby, prostrate, diffusely branched, nearly leafless species. Stems 6 in. to 2–3 ft. long, rigid, wiry, rush-like, deeply grooved, the twigs scaberulous. Leaves none or few and scattered, $\frac{1}{4}$–1 in. long, petioled or sessile, linear, subacute, base often dilated or subhastate; stipules short, truncate. *Male* fl. in simple, axillary, lax, quite glabrous spikes, with a few female flowers intermixed;—*fem*. fl. often axillary and solitary. Stigmas fimbriate. Perianth rather fleshy in fruit.

Northern Island: east coast, near the sea at Ahuriri, *Colenso*. **Middle** Island: Otago, Lower Waitaki, *Hector and Buchanan*. A remarkable species, resembling Rushes scattered on the ground. I suspect that the *Polygonum Cunninghamii*, Meisner, which forms impassable thickets in some parts of Australia, is the same plant as this.

2. RUMEX, Linn.

Herbs, rarely almost shrubby. Leaves usually long. Flowers hermaphrodite, usually pedicelled, pendulous, small, green, whorled on branched spreading or close panicles.—Perianth of 6 pieces; 3 inner enlarging, dry, veined, and closing over the fruit. Styles 3, short, stigmas fimbriate. Nut small, usually 3-gonous, enclosed in the much enlarged, dry, often toothed or ciliate perianth.

A large genus, found in all temperate and many tropical parts of the world.

1. **R. flexuosus,** *Forst.;—Fl. N. Z.* i. 211. Prostrate, glabrous, diffusely branched, 1–2 ft. long; branches flexuous, deeply grooved. Leaves 4–8 in. long, narrow linear-oblong, plane or waved at the margin, obtuse or acute, base truncate acute or cordate, upper nearly sessile. Flowers in distant 3–8-flowered whorls, lower whorls leafy. Peduncles curved, as long as the fruiting perianth. Inner lobes of fruiting perianth $\frac{1}{10}$ in. long, rhomboid, with long attenuated tips, veined, without thickened knobs, quite entire, or with 1–4 spines on each side, keeled in the middle, keel sometimes spinulose. —*R. Cunninghamii*, Meisn. in A. DC. Prodr. 14. 62; *R. Brownianus*, Raoul, Choix; *R. fimbriatus*, A. Cunn. not Br.; *R. cuneifolius*, *β*, Fl. Antarct. i. 67.

Northern and **Middle** Islands: common, *Banks and Solander*, etc. **Lord Auckland's** Island, *J. D. H.*; very nearly allied to the Australian *R Brownii*, but apparently differing in the prostrate habit. It is perhaps *R. dumosus*, A. Cunn., of New South Wales, which has the same habit.

Besides the above, the common English Docks *R. crispus*, L., and *R. obtusifolius*, L., together with the small *R. acetosa*, have been introduced into New Zealand, and are quasi-indigenous; some of them, indeed, are spreading at an enormous rate.

ORDER LXIV. LAURINEÆ.

Trees or shrubs, often aromatic; one genus is of climbing leafless herbs. Leaves alternate, rarely opposite, quite entire, exstipulate. Flowers usually hermaphrodite, small, green, fascicled panicled or umbelled.—Perianth of 4–8 (often 6), herbaceous, imbricate segments, rarely 0. Stamens usually 12–15, in 2 or 3 series, all fertile or the inner sterile, filaments naked or glandular at the base; anther-cells 2–4, opening in front or behind by upturned valves. Ovary free, 1-celled; style short, stigma simple or 3-lobed; ovule 1, pendulous, rarely 2. Drupe or berry free or enclosed in the perianth, 1-seeded. Seed with membranous testa, large plano-convex cotyledons, and no albumen.

A very large tropical Order, to which the Bay Laurel, Camphor, and Cinnamon belong.

Leafy shrub or tree. Flowers in a 4- or 5-leaved involucre . . . . . 1. TETRANTHERA.
Leafy trees. Flowers panicled . . . . 2. NESODAPHNE.
Leafless, twining, string-like herb . . . . 3. CASSYTHA.

1. TETRANTHERA, Jacquin.

Shrubs or trees. Leaves alternate, rarely opposite.—Flowers usually axillary, diœcious, umbelled within a 4- or 5-leaved, deciduous involucre. Perianth 0 or 4–8-parted. *Male* fl.: Stamens 6–15; all or the inner filaments glandular at the base; anthers 4-celled. Ovary rudimentary. *Female* fl.: Stamens rudimentary. Ovary oblong, with a dilated stigma. Berry ovoid.

A very large genus in the tropics of the Old World.

1. **T. calicaris,** *Hook. f. Fl. N. Z.* i. 216. A small, evergreen, glabrous, umbrageous tree. Leaves petioled, 3–4 in. long, ovate or oblong, obtuse, quite entire, sometimes glaucous below. Involucral leaves 4 or 5, concave, $\frac{1}{6}$–$\frac{1}{3}$ in. long, 4- or 5-flowered. Pedicels silky, as long as the involucral leaves. Perianth of 5–8 oblong segments. Stamens about 12; filaments slender, all 2-glandular; anthers dilated, 4-valved, bursting inwards. Berry ovoid, $\frac{3}{4}$ in. long; peduncle thickened at the top.—*Laurus calicaris*, A. Cunn.

Northern Island: from the Bay of Islands to the east coast, *Banks and Solander*, etc.

2. NESODAPHNE, Hook. f.

Forest trees, with evergreen, alternate leaves and axillary or terminal panicles.—Flowers hermaphrodite, panicled. Perianth 6-partite, deciduous. Stamens 12, in 2 series; 6 outer fertile opposite the perianth lobes, with eglandular filaments and introrse anthers; of the 6 inner, 3 are antheriferous with extrorse anthers, and glands opposite the bases of their filaments; the 3 others have imperfect anthers. Style short, stigma simple. Berry ovoid, top of the fruiting peduncle swollen.

A genus confined to New Zealand, as far as at present known.

Branches stout and petioles and panicles tomentose . . . . 1. *N. Tarairi.*
Branches slender, and petioles and panicles nearly glabrous . . . . . . 2. *N. Tawa.*

1. **N. Tarairi,** *Hook. f. Fl. N. Z.* i. 217. A lofty forest tree, 60–80 ft. high. Branches stout, petioles panicles and costa of leaf below densely

tomentose. Leaves 3–6 in. long, coriaceous, obovate-oblong, obtuse, glabrous above, puberulous and glaucous below. Panicles branched, 1–2 in. diam. Flowers shortly pedicelled, $\frac{1}{6}$ in. diam. Berry $1\frac{1}{2}$ in. long, ovoid, purple. —*Laurus Tarairi*, A. Cunn.

Northern parts of the **Northern** Island, *Banks and Solander*, etc. Wood white, splits freely, *A. Cunningham*. Berry much eaten by birds, and, when boiled, by man. Embryo said to be poisonous when raw.

2. **N. Tawa,** *Hook. f. Fl. N. Z.* i. 217. A lofty forest tree, 60–70 ft. high. Branches slender, youngest silky. Leaves variable, 3–4 in. long, lanceolate or narrow elliptic-oblong, acute, finely reticulated on both surfaces, glaucous below. Panicles slender, branches elongate. Flowers small, quite glabrous. Berries small.—*Laurus Tawa*, A. Cunn.; *L. Victoriana*, Colenso.

Northern parts of the **Northern** Island, *Banks and Solander*, etc. Wood poor, very destructible, used for spears. Berries eaten. Leaves very aromatic and pungent.

3. CASSYTHA, Linn.

Leafless, twining, herbaceous or half shrubby plants, attached by suckers to shrubs, etc., on which they are parasitical, as with *Cuscuta*. Stems and branches slender, terete, like whipcords. Flowers spiked capitate or panicled, hermaphrodite.—Perianth 6-partite; tube very short; segments erect, in 2 series. Stamens 12, in 2 series, the 3 interior opposite the inner segments of the perianth with imperfect anthers, the rest fertile, the filaments of some 2-glandular at the base. Anthers 2-celled; of outer row opening inwards, of inner row outwards. Fruit enclosed in the baccate perianth.

A large genus, abounding in Australia, with a few Indian, Pacific, and American species.

1. **C. paniculata,** *Br.;—Fl. N. Z.* i. 218. Branches glabrous, $\frac{1}{12}$–$\frac{1}{10}$ in. diam., smooth; tips silky; scales at the axils of the branches small, ovate, acuminate, membranous. Spikes 1–2 in. long, simple or branched. Flowers shortly pedicelled, distant, cylindrical, $\frac{1}{8}$–$\frac{1}{6}$ in. long, sometimes pubescent, with several orbicular, minute bracts at the base. Ovary glabrous.

Northern part of the **Northern** Island, *Dieffenbach, Colenso*. A native of New South Wales.

ORDER LXV. MONIMIACEÆ.

Trees or shrubs, often aromatic. Leaves opposite, rarely alternate, exstipulate. Flowers racemose or cymose, hermaphrodite or unisexual.—Perianth rotate or subcampanulate, 4–15-lobed. *Male* fl.: Stamens indefinite, all or most fertile; filaments short; anthers 2-celled, opening by slits or by ascending valves. *Female* fl.: Stamens 0 or reduced to scales. Ovaries numerous, 1-celled; style terminal or lateral, stigma simple; ovule 1, erect or pendulous. Fruit of numerous drupes, or of achenes with persistent plumose styles, often included in the urceolate tube of the perianth. Seed pendulous; embryo in the axis of oily and fleshy albumen; cotyledons very short, divaricating.

A small Order, native of the tropics of South America, South India, and its islands, and extratropical America, Tasmania, and Chili.

Anthers with valves. Ovule erect. Achenes with plumose styles . 1. ATHEROSPERMA.
Anthers with slits. Ovule pendulous. Drupes stipitate . . . . 2. HEDYCARYA.

1. ATHEROSPERMA, Lab.

Trees. Leaves opposite, aromatic. Flowers panicled, diœcious.—Perianth 5–8-fid; lobes in many series. *Male* fl.: Stamens 6–20; filaments 2-glandular at the base; anthers with 2 ascending valves. *Female* fl.: Stamens reduced to scales. Ovaries 5–20, villous. Achenes with long plumose styles, included in the urceolate perianth, which splits laterally.

A small genus, including the following, two Australian, and a Chilian species.

1. **A. novæ-Zelandiæ,** *Hook. f.—Laurelia novæ-Zelandiæ*, A. Cunn.;—Fl. N. Z. i. 218. t. 51. A tree, 150 ft. high, with buttressed trunk, 3–7 ft. diam., the buttresses 15 ft. thick at the base (*Bidwill*); bark white, wood soft, yellowish; branches whorled and petioles pubescent. Leaves petiolate, coriaceous, 1½–2½ in. long, ovate or oblong, obtuse, glabrous, obscurely serrate. Flowers in axillary racemes, silky, ¼ in. diam. Stamens 6–10. Fruiting perianth narrow-urceolate, elongate. Achenes 6–10, with their plumose styles 1 in. long, very narrow.

Northern and northern parts of the **Middle** Island, *Banks and Solander*. Wood used for boat-building.

2. HEDYCARYA, Forst.

Aromatic trees. Leaves evergreen, opposite, toothed or entire. Flowers panicled, diœcious.—Perianth rotate, 5–10-lobed. *Male* fl.: Anthers very numerous, sessile in the base of the perianth, opening by slits. *Female* fl.: Ovaries numerous; stigma sessile, obtuse; ovule pendulous. Drupes few, stipitate, seated on the perianth, which does not enlarge.

A small genus of Australian and New Zealand plants.

1. **H. dentata,** *Forst.;—Fl. N. Z.* i. 219. A small evergreen bush or tree, 20–30 ft. high; branches pubescent. Leaves 1–4 in. long, obovate or linear-oblong, obtuse or acute, coarsely serrate, rarely entire, glabrous or slightly pubescent. Panicles axillary, pubescent, shorter than the leaves. Perianth pubescent, ⅓ in. diam. Anthers hairy at the tip. Drupes oblong, obtuse, ½ in. long, red; endocarp coriaceous.—Raoul, Choix, t. 30; *H. dentata* and *H. scabra*, A. Cunn.; *Zanthoxylon novæ-Zelandiæ*, A. Rich.

Northern and **Middle** Islands: as far south as Akaroa, *Banks and Solander*, etc.

ORDER LXVI. PROTEACEÆ.

Shrubs or trees, usually rigid and dry. Leaves usually alternate, exstipulate. Inflorescence various. Flowers hermaphrodite.—Perianth of 4 narrow segments, usually connate below, their free portion revolute, valvate in bud. Stamens 4, inserted on the lobes of the perianth or below them; filaments short; anther adnate, linear, bursting by 2 slits. Hypogynous glands usually 4. Ovary free, 1-celled; style usually long, stigma simple; ovules solitary, geminate or numerous. Fruit a nut drupe samara or follicle, 1-celled or almost 2-celled by a false septum formed by the coats of the ovules, which when 2 are placed back to back. Seed various; albumen 0; embryo straight.

A large Australian and South African Order, rare elsewhere, but found in South India, Japan, the Malayan, some of the Pacific islands, and in South America.

Lofty, slender tree. Leaves serrate. Fruit a dry follicle . . . . . . 1. KNIGHTIA.
Small tree. Leaves narrow, entire. Fruit a drupe . . . . 2. PERSOONIA.

1. KNIGHTIA, Br.

Flowers densely racemed in subsessile cylindrical cones.—Perianth-segments cohering by their margins into a long club-shaped tube, ultimately separating. Stamens inserted towards the ends of the segments; anthers long, linear. Hypogynous glands 4. Ovary sessile, narrowed into a long, very stout style, stigma vertical; ovules 4. Follicle coriaceous, 1-celled. Seeds winged at the tip.

A small genus, containing a New Zealand and New Caledonian species.

1. **K. excelsa,** *Br.;—Fl. N. Z.* i. 219. A lofty, slender tree, 100 ft. high, with the habit of a Lombardy Poplar; branches very stout, woody, and as well as the inflorescence densely covered with rusty, velvety down. Leaves 4–8 in. long, hard, petioled, obovate- or linear-oblong, obtuse, coarsely toothed. Racemes sessile, 2–3 in. long, 2 in. diam. Flowers in densely crowded pairs, shortly pedicelled, 1–1½ in. long before expansion, $\frac{1}{12}$ in. diam. in the middle. Ovary tomentose; stigma clavate. Follicle woody, narrow linear-oblong, terete, downy, pedicelled, contracted into the stout, straight, persistent style. —Br. in Linn. Trans. x. 194. t. 2.

Northern Island: common in the forests, *Banks and Solander*, etc. Wood mottled red and brown, used for furniture and for shingles.

2. PERSOONIA, Smith.

Evergreen shrubs and trees. Leaves alternate, coriaceous, various in shape, etc. Flowers usually in short axillary spikes or racemes.—Perianth-segments cohering by their margins into a club-shaped tube, ultimately separating. Stamens on the middle of the segments. Hypogynous glands 4. Style slender, stigma obtuse; ovules 1 or 2. Drupe with a 1-or 2-celled nut.

A large Australian genus.

1. **P. Toro,** *A. Cunn.;—Fl. N. Z.* i. 219. Small evergreen tree; branches woody, glabrous. Leaves 3–8 in. long, coriaceous, very narrow, linear-lanceolate, gradually narrowed into the petiole, acuminate or obtuse and apiculate, quite entire, smooth and polished on both surfaces, lateral veins nearly parallel. Flowers ¼ in. long, very shortly pedicelled, in short, axillary, pubescent, 6–10-flowered racemes 1 in. long. Ovary glabrous, sessile.—Bot. Mag. t. 3513.

Northern Island: from Auckland northwards, *Banks and Solander*, etc.

ORDER LXVII. THYMELEÆ.

Shrubs with very tough bark, often acrid. Leaves opposite or alternate, simple, quite entire. Flowers usually in terminal corymbs, hermaphrodite.

—Perianth inferior, tubular campanulate or urceolate; limb 4-lobed, imbricate; throat naked or with glands. Stamens inserted on the tube of the perianth, usually 2 or 4, rarely more, then in 2 series, the upper opposite its lobes. Ovary 1-celled; style lateral or terminal, stigma capitate; ovule usually solitary, pendulous. Fruit a nut drupe or berry, 1-seeded. Seed pendulous; testa usually thin; albumen 0, scanty or abundant; cotyledons plano-convex.

A large European, Oriental, South African, Australian, and New Zealand Order (to which the Daphne and Mezereon belong), comparatively rare elsewhere. The bark yields excellent fibre; paper is made of it in India, China, and Japan. The above character does not include the tribe *Aquilarieæ;* Indian trees yielding Aloes- or Eagle-wood, in which the ovary is 2-celled and fruit capsular.

Stamens 2 . . . . 1. PIMELEA.
Stamens 4 . . . . 2. DRAPETES.

1. PIMELEA, Banks and Solander.

Shrubs, much branched, erect or prostrate. Leaves opposite, coriaceous, often imbricate; florals sometimes larger and whorled, forming an involucre. Flowers capitate, white rosy or yellow.—Perianth tubular, coloured; limb 4-lobed, throat naked. Stamens 2, opposite the outer perianth-lobes; filaments slender. Ovary ovoid; style lateral, slender. Fruit dry or baccate, naked or included within the perianth-tube. Albumen scanty or copious.—*Banksia*, Forst. Char. Gen.; *Passerina*, Forst. Prodr.; *Cookia*, Gmelin.

A very extensive Australian and New Zealand genus, not found elsewhere, extremely variable in foliage. In the New Zealand species the periauth is not jointed in the middle as in many Australian ones, and the nut is often enclosed in the persistent tube of the perianth, which is membranous or fleshy. The species are most variable, and difficult of discrimination. I have forms that appear intermediate between the best-marked species.

Erect shrubs. Branches and leaves perfectly glabrous.

Leaves 1–2 in., lanceolate. Flowers $\frac{1}{2}$ in. . . . . 1. *P. longifolia.*
Leaves $\frac{1}{3}$–$\frac{2}{3}$ in., oblong or lanceolate. Flowers $\frac{1}{4}$ in. : . . . . . 2. *P. Gnidia.*
Leaves $\frac{1}{6}$–$\frac{1}{4}$ in., oblong or obovate, obtuse. Florets large . . . . . 3. *P. Traversii.*

Erect, rarely prostrate shrubs. Branches silky villous or pubescent.

Leaves $\frac{1}{2}$–1 in., lanceolate, glabrate pilose or silky . . . . 4. *P. virgata.*
Leaves $\frac{1}{4}$ in., coriaceous, keeled, glabrous, floral large . . . . . . 5. *P. buxifolia.*
Leaves $\frac{1}{4}$–$\frac{1}{3}$ in., with appressed, shining, silky hairs below . . . . . 6. *P. arenaria.*

Procumbent or prostrate, rarely erect shrubs. Branches pubescent or villous.

Branches villous with white hairs. Leaves $\frac{1}{6}$ in., very thick, obtuse, glabrous . . . . 7. *P. Urvilleana.*
Branches grey. Leaves $\frac{1}{12}$–$\frac{1}{3}$ in., glabrous, oblong or lanceolate . 8. *P. prostrata.*
Branches silky. Leaves $\frac{1}{6}$–$\frac{1}{3}$ in., silky, oblong or lanceolate . . . . 9. *P. Lyallii.*
Branches villous. Leaves $\frac{1}{6}$–$\frac{1}{4}$ in., silky-villous above and below, linear-oblong, obtuse . . . . 10. *P. sericeo-villosa.*

1. **P. longifolia,** *Banks and Sol.;—Fl. N. Z.* i. 220. A small, erect shrub, 2–6 ft. high, perfectly glabrous, the inflorescence excepted. Leaves crowded, not imbricate, very shortly petioled, spreading, 1–2 in. long, $\frac{1}{4}$–$\frac{1}{2}$ broad, oblong- or linear-lanceolate, acuminate, flat, often glaucous below; veins distinct; floral similar or rather broader. Flowers numerous, silky, $\frac{1}{2}$

in. long, white, odorous. Stamens and style exserted. Nut crustaceous, enclosed in the base of the perianth.

Northern Island: common, *Banks and Solander*, etc.; and northern parts of the **Middle** Island: Nelson mountains, ascending to 2000 ft., *Travers*.

2. **P. Gnidia,** *Forst.*;—*Fl. N. Z.* i. 221. A short, erect, robust, glabrous (or nearly so) species, 1–5 ft. high; branches very stout; bark pale or dark, always glabrous. Leaves crowded, very coriaceous, shining above, petioled, $\frac{1}{3}$–$\frac{2}{3}$ in. long, oblong or oblong-lanceolate, acute or obtuse, almost or quite keeled, veinless; floral not very different. Flowers $\frac{1}{4}$ in. long, silky and villous.—*Passerina Gnidia*, Forst.; *Cookia Gnidia*, Gmel.

Northern Island: thickets near the top of the Ruahine range, *Colenso*. **Middle** Island: Dusky Bay, *Forster*, *Menzies*; South Island, *Lyall* (leaves longer, nearly flat). The var. *β. Menziesii*, of Fl. N. Z., is, I now believe, the true *P. Gnidia*, Forst., distinguished from *P. buxifolia*, which is confounded with it in that Flora, by the perfectly glabrous branches, and floral leaves not much differing from the cauline.

3. **P. Traversii,** *Hook. f.*, *n. sp.* A short, very robust, glabrous, alpine, usually erect species, 4–24 in. high, densely branched; branches very stout, often tortuous, always glabrous, tubercled with close-set scars at the insertion of fallen leaves. Leaves densely 4-fariously imbricate, very coriaceous, sessile, $\frac{1}{6}$–$\frac{1}{4}$ in. long, obovate oblong or almost orbicular, quite glabrous, obtuse, nerveless, midrib often obscure, yellowish when dry; floral large (sometimes 4 times as large), broader, often edged with purple, often verdigris-green when dry. Flowers very silky, $\frac{1}{4}$ in. long, white.

Middle Island: Macrae's Run, *Munro*; Hurumui and Wai-au-au mountains, *Travers*; Southern Alps, in various places, alt. 2–4000 ft., *Sinclair and Haast*; Otago, Waitaki valley, *Hector and Buchanan*. One of the most distinct species, though sometimes prostrate, and sometimes having a few hairs on the branches, when it approaches forms of *P. prostrata*.

4. **P. virgata,** *Vahl*;—*Fl. N. Z.* i. 220. A small, erect, dense shrub, 1–2 ft. high. Branches slender, strict, silky. Leaves spreading, not imbricate, $\frac{1}{2}$–1 in. long, linear- or oblong-lanceolate, obtuse acute or acuminate, glabrous or pilose or glabrous above and silky below; floral similar. Flowers small, about 8–10 in a head, $\frac{1}{4}$ in. long, silky; tube swelling below; lobes broad, obtuse. Nut obovate, enclosed in the dry or baccate base of the perianth, brittle.—*P. pilosa*, Vahl; *Passerina pilosa*, Forst. Prodr.; *P. axillaris*, Thunb.; *Banksia tomentosa*, Forst. Gen.

Abundant throughout the islands, *Banks and Solander*, etc. I think this passes into *P. prostrata*.

5. **P. buxifolia,** *Hook. f.*, *n. sp.* A small, stout, erect, rigid shrub, 1–5 ft. high, much branched; branches very stout, hirsute with short grey hairs; bark black, minutely verrucose. Leaves close-set, 4-fariously imbricate, very coriaceous, $\frac{1}{4}$ in. long, oblong-ovate, keeled, acute or obtuse, lateral nerves conspicuous when dry; floral half as large again and broader than the cauline, often green when dry. Flowers $\frac{1}{4}$–$\frac{1}{3}$ in. long, densely silky.

Northern Island, *Dieffenbach*; base of Tongariro and top of Ruahine range, *Colenso*. Very similar in habit to *Veronica buxifolia*. Closely allied to *P. Gnidia*, but the hairy branches and larger floral leaves at once distinguish it. The evident lateral nerves on the leaf are a good character. I have a small specimen from Colenso (stony banks, Ahuriri), which has the foliage of this, but habit of *P. prostrata*.

6. **P. arenaria,** *A. Cunn.;—Fl. N. Z.* i. 221. A very beautiful, erect, rarely prostrate, small shrub, 8–24 in. high, much branched, villous with white, shining, silky hairs; branches stout, much scarred, most densely villous. Leaves close-set, but not imbricating, spreading or reflexed, flat, $\frac{1}{4}$–$\frac{1}{2}$ in. long, broadly oblong or orbicular, obtuse, pilose above, very silky and shining below; floral rather larger and broader. Flowers $\frac{1}{4}$ in. long, villous or shaggy. Fruiting-perianth baccate.—Bot. Mag. t. 3270; *Passerina villosa*, Thunb.; *Gymnococca arenaria*, Fisch. and Mey.

Northern Island: especially on sand dunes, near the sea, *Banks and Solander*, etc. **Chatham** Island, *Dieffenbach*. Fruit eaten; bark used for cloth string, etc. Procumbent specimens are best distinguished from *P. Urvilleana* by the large flowers and silky leaves. Very closely allied to the Tasmanian *P. sericea*, and altogether similar to it, except that the perianth is not transversely articulate.

7. **P. Urvilleana,** *A. Rich.;—Fl. N. Z.* i. 221. A small, widely spreading, procumbent species; branches 2–12 in. long, horizontally extending, white with short silky wool. Leaves close-set, often imbricate, very thick and coriaceous, $\frac{1}{6}$–$\frac{1}{4}$ in. long, sessile, broadly oblong or obovate-oblong, obtuse, concave, nerveless, glabrous or sparingly hairy; floral rather larger and broader. Flowers very small, villous, $\frac{1}{6}$ in. long; lobes as long as the tube. —*P. prostrata, β*, Meisner in A. DC. Prodr. xiv. 517; *Gymnococca microcarpa*, Fisch. and Mey. (?), according to Meisner.

Northern Island: rocky places near the sea and inland, *Banks and Solander*, etc.; Bay of Islands, *A. Cunningham;* Mount Egmont, *Dieffenbach*. The bark of this was chewed and beaten out to make the cloth of which the top-knots of the chiefs were formed, etc. Meisner unites this with *P. prostrata*, not without some reason; but by the same rule most of the other New Zealand species must be united too, for there appear to be forms uniting many of them; this keeps its characters well, both on the coast and inland, and these are as marked as between any two consecutive species.

8. **P. prostrata,** *Vahl;—Fl. N. Z.* i. 220. A small, prostrate, spreading, extremely variable species; branches slender, 2–10 in. long, more or less hairy, but seldom white with villous hairs, as in *P. Urvilleana*. Leaves crowded and imbricated, erect patent or recurved, or sparse, sessile, not thickly coriaceous, $\frac{1}{12}$–$\frac{1}{3}$ in. long, rarely more, oblong-obovate or lanceolate, obtuse or acute, usually nerveless or with a stout midrib below, glabrous; margins often incurved when dry; floral usually a little larger and broader. Flowers $\frac{1}{8}$–$\frac{1}{6}$ in. long, villous silky or pilose; lobes of perianth shorter than the tube. Fruit often baccate.—*P. lævigata*, Gærtn.

Var. *α*. Stems stout, suberect, or branches ascending. Leaves oblong or lanceolate. Flowers very silky.

Var. *β*. Stems very diffusely branched, more slender; branches spreading. Leaves oblong or broadly obovate. Flowers very small, silky or nearly glabrous.

Var. *γ*. Stems suberect, black, tortuous, and scarred. Leaves erecto-patent, ovate or linear-lanceolate, acute, glabrous.

Throughout the **Northern** and **Middle** Islands: var. *α*, abundant, ascending to 4000 ft., *Banks and Solander*, etc.; var. *β* also abundant, especially in hilly districts; var. *γ*, alpine regions of the **Middle** Island, from Nelson (*Travers*) to Otago, *Hector*. A most abundant and most variable plant; alpine states, with slender, nearly glabrous branches, and short, obtuse, rounded leaves, look different, but pass insensibly into the ordinary states. The prostrate habit, glabrous, small foliage and small flowers, distinguish this from *P. virgata;* the grey (not snow-white) villous hairs of the branches, and less coriaceous leaves, from *P. Urvil-*

leana; the glabrous leaves and smaller flowers, from *P. Lyallii;* the var. γ looks different, but is certainly only an alpine form.

9. **P. Lyallii,** *Hook. f. Fl. N. Z.* i. 222. A small, prostrate or suberect, very pilose, rarely glabrescent species; branches short and suberect, or long and trailing, 2–20 in. long, covered with grey or silky pubescence, rarely glabrous, bark brown. Leaves usually close-set and imbricating, erect or patent, $\frac{1}{6}$–$\frac{1}{3}$ in. long, oblong or lanceolate, acute or acuminate, silky with long hairs chiefly on the lower surface, concave, nerveless; floral the same as the cauline. Flowers $\frac{1}{6}$–$\frac{1}{4}$ in. long, silky. Ovary pilose.

Northern Island: between the Ruahine range and Taupo, *Colenso.* **Middle** Island: Ruapuke Island and Port William, *Lyall;* Southern Alps, alt. 1500–2500 ft., *Sinclair and Haast;* Gordon's Nob, *Munro;* Wairau mountains, alt. 3–5500 ft., *Travers;* Otago, Waitaki valley, abundant, *Hector and Buchanan.* This resembles small specimens of *P. virgata,* but is of a totally different habit. From *P. prostrata* the silky leaves distinguish it. Travers and Buchanan send excellent series of forms, amongst which some have all but glabrous leaves, and hence run into *prostrata.* Haast sends from shingle flats on the Macaulay and Godley rivers, alt. 3000 ft., a stout, erect, small plant, with much the habit of states of *P. Lyallii,* but the leaves are more coriaceous and glabrous.

10. **P. sericeo-villosa,** *Hook. f., n. sp.* A small, prostrate, much branched, densely tufted species, densely villous with whitish shining silky hairs; branchlets very short, leafy. Leaves close-set, $\frac{1}{8}$–$\frac{1}{4}$ in. long, linear-oblong, obtuse, concave, equally villous above and below. Flowers few, densely silky, $\frac{1}{6}$ in. long. Ovary villous with long hairs.

Middle Island: Macrae's Run, *Munro;* Wairau mountains, *Travers;* Otago, Waitaki river, *Hector and Buchanan.* This, from all the above habitats, keeps its characters so perfectly, that it is difficult to suppose it to be a form of *P. Lyallii,* from which it differs in the more depressed habit, much shorter branchlets, and far more copious, silky clothing.

2. DRAPETES, Lamarck.

Small, tufted, moss-like herbs or suffruticose plants. Leaves small, linear, crowded. Flowers solitary or few together, inconspicuous, terminal.—Perianth tubular or funnel-shaped; limb 4-fid; throat sometimes furnished with 4 small glands or scales opposite the lobes. Stamens 4, alternate with the lobes; filaments subulate. Style terminal or lateral, filiform; stigma capitate. Nut small. Albumen copious.

A small southern alpine and Antarctic genus, containing a New Zealand, a Tasmanian, a Fuegian, and a Bornean species; it has been split into three genera by Endlicher and Meisner, according as the tube of the perianth is continuous or transversely articulate, and its throat naked or provided with glands.

Leaves linear or linear-oblong. Perianth funnel-shaped . . . . 1. *D. Dieffenbachii.*
Leaves ovate-oblong. Perianth campanulate . . . . 2. *D. Lyallii.*

1. **D. Dieffenbachii,** *Hook. Lond. Journ. Bot.* ii. 497. *t.* 17;—*Fl. N. Z.* i. 222. A small, densely tufted, moss-like plant; branches slender, 6–12 in. long. Leaves imbricated, $\frac{1}{6}$ in. long, linear, obtuse, bearded at the tip, keeled at the back. Flowers terminal, solitary or fascicled, as long as the leaves, very shortly pedicelled. Perianth not articulate; throat with 4 glands.

Northern Island: Mount Egmont, *Dieffenbach;* Tongariro, *Bidwill;* Ruahine range

Colenso. **Middle** Island: abundant on the alps, ascending to 5000 ft., from Nelson, *Bidwill*, etc., to Otago, *Hector and Buchanan.* This is, I think, the same with the Tasmanian *D. tasmanica* of my 'Flora Tasmaniæ,' which has also been found on the Australian alps by Mueller.

2. **D. Lyallii,** *Hook. f. Fl. N. Z.* ii. 336.—*D. muscosa*, Fl. N. Z. i. 223, not Lamarck. Stems short, densely tufted, rarely with long straggling branches. Leaves ovate-oblong, obtuse, ciliated at the tip. Perianth almost campanulate.

Middle Island: Nelson, Wai-au-ua valley, *Travers;* Southern Alps, *Sinclair and Haast;* Otago, lake district, alpine, *Hector and Buchanan;* Southern Island, *Lyall.* A very distinct plant from *D. Dieffenbachii*, to which it has been reduced by Meisner, of different habit, with shorter, broader, closer-set leaves, and shorter, broader perianth.

ORDER LXVIII. SANTALACEÆ.

Trees shrubs or herbs. Leaves usually alternate, quite entire, exstipulate. Flowers usually small, very inconspicuous, in heads cymes or spikes, hermaphrodite or unisexual.—Perianth wholly or partially superior; lobes 3–6, persistent or deciduous, valvate. Stamens as many as and inserted on the perianth-lobes; filaments usually very short; anthers 2-celled. Ovary inferior or superior, 1-celled; style usually short, simple or 3-fid; ovules 3–5, pendulous from an erect central placenta. Fruit a 1-seeded nut or berry, indehiscent. Albumen fleshy; embryo usually terete and small; radicle superior.

A widely-diffused Order in various parts of the globe, temperate and tropical.

Trees or shrubs, leafless. Flowers in minute spikes . . . . 1. EXOCARPUS.
Trees. Leaves long. Flowers in axillary cymes . . . . 2. SANTALUM.

1. EXOCARPUS, Labillardière.

Shrubs or trees, with naked, twiggy, leafless branches, bearing scales at the ramifications, rarely leafy. Flowers minute, green, in axillary spikelets or clusters, hermaphrodite or unisexual.—Perianth of 4–6 spreading, deciduous lobes. Stamens 4–6, short, inserted on the base of the lobes; filaments glabrous. Disk 5-lobed. Ovary superior, fleshy; style very short, conic; stigma subcapitate; ovules not ascertained. Nut oblong, seated on the swollen drupe-like tip of the peduncle or base of the perianth. Embryo minute, cylindric; cotyledons small.

An Australian, New Zealand, and Pacific Island genus.

1. **E. Bidwillii,** *Hook. f. Fl. N. Z.* i. 223. *t.* 52. A small, rigid, procumbent shrub, 8–16 in. high, much branched; branches stiff, short, grooved, terete, leafless, with minute triangular scales at the bases of the branches. Flowers 8–10, in short puberulous axillary spikes. Perianth 5- or 6-parted. Stamens 5 or 6. Nut black, drupaceous; peduncle red.

Middle Island: Warrau mountains, *Bidwill;* river beds of the Kowai and Waitaki rivers, alt. 2000–3500 ft., *Haast;* Hurumui mountains, *Travers.* Closely allied to the *E. humifusa* of Tasmania, and probably a variety of that plant, but the flowers are more spiked and 5- or 6-merous. I think two plants are confounded in the 'Tasmanian Flora' under *E. humifusus*, one the true plant of Brown, resembling this, the other much smaller, with compressed stems, which I originally called *E. nanus.*

2. SANTALUM, Linn.

Shrubs or trees. Leaves opposite or alternate. Flowers green, in axillary cymes, hermaphrodite.—Perianth-tube campanulate, with 4 or 5 spreading, ovate-triangular, deciduous leaflets, having a tuft of hairs at the inner base of each. Stamens 4 or 5; filaments short; anthers ovoid. Disk concave, 4- or 5-lobed. Ovary superior in the bud, afterwards inferior; style conic or cylindric, stigmas 2–4; ovules 2–4. Drupe inferior, globose or turbinate, crowned with the remains of the perianth, 1-seeded. Seed inverse; albumen fleshy; embryo cylindric; cotyledons very short.

A genus of few species, scattered through the tropics of Asia, Australia, and the Pacific islands. Several species produce Sandal-wood.

1. **S. Cunninghamii,** *Hook. f. Fl. N. Z.* i. 223. A small tree. Leaves excessively variable, alternate, opposite in young plants, shortly petioled, 2–4 in. long, from narrow linear-lanceolate to broad obovate, veined, minutely dotted. Flowers green, in short axillary cymes. Perianth $\frac{1}{8}$–$\frac{1}{4}$ in. long; tube hemispherical; lobes 4 or 5, deciduous. Drupe nearly $\frac{1}{2}$ in. long.—*S. Mida*, Hook. Ic. Pl. t. 563 and 575; *Mida salicifolia, eucalyptoides*, and *myrtifolia*, A. Cunn.

Northern Island: from the east coast, northwards, *A. Cunningham*, etc.

Order LXIX. EUPHORBIACEÆ.

Herbs trees or shrubs, usually abounding in milky juice. Leaves various, alternate or opposite, often stipulate. Inflorescence very various. Flowers always unisexual (incomplete males are crowded round a female in *Euphorbia*, within an involucre, which hence resembles a perianth).—Stamens few or many; anthers usually didymous. Perianth 0, or very various from a scale to an almost perfect perianth in 2 rows. Ovary usually 2- or 3-lobed, 2- or 3-celled; style 2- or 3-lobed; segments stigmatic along the inner face; ovules 1 or 2, pendulous in each cell. Fruit very various, a 2- or 3-valved capsule in the New Zealand genera, consisting of 3 carpels, finally separating and dehiscing dorsally. Seeds pendulous, albuminous; embryo with flat cotyledons and small radicle.

One of the largest and most important Orders of plants, containing a vast number of different forms, with flowers of very different structure, found in all parts of the world, but rare in cold countries. The Croton-oil, Castor-oil (almost naturalized in the Northern Island), and a vast number of other medicinal and economic plants belong to it.

Herbs. Flowers collected in a perianth-like involucre, ♂ of naked stamens, ♀ of a simple, naked pistil . . . . 1. EUPHORBIA.
Shrubs or trees. Flowers in long, slender spikes . . . . 2. CARUMBIUM.

1. EUPHORBIA, Linn.

Herbs, with abundant milky juice, rarely shrubs or subarboreous. Leaves opposite or alternate, quite entire or toothed. Flowers usually in terminal cymes.—Involucre (resembling a perianth) urceolate or cup-shaped, containing many stamens of unequal length, the filaments jointed in the middle, and 1 central stalked pistil. Each stamen is regarded as a pedicelled male flower, without perianth. The pistil represents a pedicelled female flower, also without a perianth. Ovary 3-celled; style 3-fid; ovules 1 in each cell. Capsule 3-lobed, separating from a central axis into 3 cocci. Testa crustaceous, polished, grey.

An enormous genus, found in all temperate and tropical parts of the world.

1. **E. glauca,** *Forst.;—Fl. N. Z.* i. 227. Perfectly glabrous, glaucous; stems herbaceous, 1–2 ft. high, rising from a woody rhizome, sometimes as thick as the finger, erect, umbellately branched at the top, very leafy. Leaves spreading, 1–4 in. long, from oblong-obovate to narrow- or lanceolate-obovate, obtuse or mucronate; floral leaves whorled below the divisions of the umbel, broadly oblong. Involucres hid amongst the foliage, shortly pedicelled, campanulate, $\frac{1}{4}$ in. diam., fleshy, with 4 or 5 flat purple lunate glands at the mouth. Capsule subglobose, quite smooth, about as large as a pea.

Throughout the islands, common on the beach, *Banks and Solander*, etc.

E. helioscopia, Linn., a common annual English weed, smaller than *E. glauca*, with shorter, toothed leaves, is introduced into some parts of New Zealand.

2. CARUMBIUM, Reinwardt.

Shrubs or small trees. Leaves stipulate, alternate, quite entire, glabrous; petioles long, 2-glandular at the tip. Flowers bracteate, in terminal spikes, monœcious.—*Male:* crowded. Perianth of 1 orbicular or 2 semi-orbicular leaflets, often connate, with a large gland on either side of the base of each. Stamens 6 or many; filaments short, more or less connate; anthers didymous, bursting laterally. *Female:* solitary. Perianth as in the male. Ovary sessile, 2- or 3-celled; style stout, 2- or 3-partite, stigmas 2 or 3 linear or oblong. Capsule 2- or 3-celled, 2- or 3-valved, 2- or 3-seeded.—*Omalanthus*, A. Juss.

A small genus, natives of tropical Australia and the Pacific islands.

1. **C. polyandrum,** *Hook. f., n. sp.—Omalanthus nutans*, Hook. f. in Journ. Linn. Soc. i. 127, not Guillemin. A small tree; branches terete, brittle, glaucous. Leaves 2–3 in. long, with petioles 2–4 in. long, triangular-ovate or nearly orbicular, acute, quite entire, membranous, somewhat undulate, glaucous below; stipules $\frac{1}{2}$ in. long, membranous. Raceme terminal, slender, nodding, 4–6 in. long. *Male* fl.: very numerous, rather distant, shortly pedicelled, $\frac{1}{10}$ in. long; bracts minute; perianth of 1 small orbicular scale; anthers very numerous, in a globose stipitate head;—*fem.* fl.: at the base of the raceme, shortly pedicelled; perianth of 2 leaflets; styles very short, 3-fid; stigmas linear, tortuous, minutely 2-fid at the tip and base.

Kermadec Islands, *MGillivray*. A very distinct species, not hitherto found elsewhere.

Order LXX. CUPULIFERÆ.

Shrubs or trees. Leaves alternate, with or without stipules. Flowers small, in catkins or fascicled or solitary, unisexual.—*Male:* Stamens 1 or more, surrounded by scales or in a 4–6-leaved or lobed perianth; filaments usually slender; anthers 2-celled. *Female:* solitary or 2–4 together, usually surrounded by scales, which often form a simple or lobed or parted perianth-like involucre. Perianth adnate with the ovary. Ovary 2–6-celled; styles 2–6, short, usually stigmatiferous on the inner surface; ovules usually 2, pendulous in each cell. Fruit usually of 1 or more 1-seeded nuts, seated in a cup or capsular involucre. Albumen 0; embryo various.

A very large Order, found in most countries, embracing the Oak, Beech, Chestnut, etc., unknown in tropical and South Africa, and in tropical America.

1. **FAGUS**, Linn.

Shrubs or trees; buds scaly. Leaves alternate, coriaceous, evergreen or deciduous, sometimes plaited, often unequal-sided; stipules deciduous, membranous. Flowers monœcious.—*Male:* Perianth campanulate, 5- or 6-fid. Stamens 8–12, inserted round a central disk; anthers 2-celled, apiculate. *Female:* 2 or 4, minute, sessile in a 4-lobed involucre, which is covered with adnate bracts. Perianth urceolate; tube adnate with the ovary; mouth laciniate. Ovary inferior, 3-celled; styles 3, filiform; ovules 1 in each cell, pendulous. Fruit 2–4 3-cornered nuts, enclosed in a 4-valved capsule-like woody lamellated involucre, each 1-seeded. Seed pendulous; testa thin; cotyledons thick, plaited, coherent.

A small genus, of 2 or 3 European and North American species, and several Antarctic-American, Chilian, New Zealand, and Tasmanian ones. Though true Beeches, these are usually called Birches in the southern hemisphere, because of their small foliage.

Leaves glabrous, doubly crenate, veinless . . . . 1. *F. Menziesii.*
Leaves glabrous or pilose, deeply serrate, veined . . . . 2. *F. fusca.*
Leaves white and downy below, oblong, entire . . . . 3. *F. Solandri.*
Leaves white and downy below, ovate rounded or cordate at the base, entire . . . . 4. *F. cliffortioides.*

1. **F. Menziesii,** *Hook. f. Fl. N. Z.* i. 229. A handsome tree, 80–100 ft. high; trunk 2–3 diam.; bark silvery, outer layers deciduous; branches tabular; twigs with fulvous pubescence. Leaves deep green, rigid, glabrous, $\frac{1}{3}$ in. long, rhomboid ovate or orbicular, obtuse, deeply doubly crenate; stipules linear oblong. Fruiting involucre puberulous, $\frac{1}{4}$–$\frac{1}{3}$ in. long; segments erect, with 5–7 tiers of soft spines having thickened glandular tips. Nuts downy, 2- or 3-winged; wings produced upwards into sharp points.—Hook. Ic. Pl. t. 652.

Northern Island: Ruahine range and Waikare Lake, *Bidwill, Colenso.* **Middle** Island: abundant from Nelson, alt. 3000 feet, *Bidwill,* to Dusky Bay, *Menzies.* This is the representative of the Tasmanian *F. Cunninghamii* and Fuegian *F. betuloides.* The "Birch" of Otago, and "Red Birch" of other colonies. The only species between Lake Wainaka and the west coast, *Haast.*

2. **F. fusca,** *Hook. f. Fl. N. Z.* i. 229. A handsome tree, 80–100 ft. high, sometimes 12 ft. diam.; branches pubescent. Leaves petioled, evergreen, young pilose above and glandular below, not very coriaceous, 1–1$\frac{1}{4}$ in. long, oblong-ovate, deeply serrate; stipules linear-oblong, very deciduous. *Male* fl.: 1–3 at the end of a viscid, short peduncle; perianth 5-toothed. Fruiting involucre broadly ovate; segments with entire or cut membranous scales at the back. Nuts winged; wings toothed at the tip.—Hook. Ic. Pl. t. 631.

Var. *β.* Leaves more coriaceous, teeth smaller and obtuser, *Hook. Ic. Pl. t.* 630.

Northern Island: common on the mountains, *Banks and Solander,* etc. **Middle**

Island: abundant, ascending to 3500 ft., *Bidwill*, etc. "Black Birch" of the colonists. I have a small specimen from Travers, without flower or fruit, in which the leaves are quite entire.

3. **F. Solandri,** *Hook. f. Fl. N. Z.* i. 230. A lofty, beautiful, evergreen tree, 100 ft. high; trunk 4 or 5 ft. diam.; bark when young white, smooth, old black, cracked; wood white, close, tough; twigs densely pubescent. Leaves shortly petioled, small, $\frac{1}{4}$–$\frac{3}{4}$ in. long, linear- or ovate-oblong, obtuse, quite entire, oblique and cuneate at the base, finely reticulated above, white and downy below; stipules very deciduous. *Male* fl. on short 1 flowered peduncles; perianth broad, shallow. Fruiting involucre glabrous or tomentose, $\frac{1}{4}$ in. long; segments with unequally toothed or entire scales.—Hook. Ic. Pl. t. 639.

Northern Island: abundant in mountain forests, *Banks and Solander.* **Middle** Island: ascending to alt. 3000–6000 ft., *Bidwill*, etc. "White Birch." I have what appears to be a form of this, with the leaves glabrous below, *Colenso* (Ruahine range) and *Sinclair.*

4. **F. cliffortioides,** *Hook. f. Fl. N. Z.* i. 230. Very similar indeed to *F. Solandri*, but a much smaller plant, with leaves ovate or oblong-ovate, rounded or cordate at the base.—Hook. Ic. Pl. t. 673.

Northern Island: top of the Raahine range, *Colenso.* **Middle** Island: abundant on the alps of Nelson and Canterbury, alt. 5–7000 ft., *Bidwill, Travers, Haast*, etc.; Otago, *Hector and Buchanan*, to Dusky Bay, *Menzies.* "White Birch." At first sight this looks like a variety of *F. Solandri*, but wherever found it retains its character, of the leaves round or cordate at the base, and I know of no intermediate forms.

ORDER LXXI. URTICEÆ.

Trees shrubs or herbs, juice often milky. Leaves and inflorescence various; stipules membranous. Flowers small, inconspicuous, usually cymose or fascicled, unisexual.—Perianth 1–5-lobed or -partite. Stamens usually as many as and opposite to the segments of the perianth; filaments often recurved, elastic; anthers 2-celled. Ovary free, 1-celled; style short or 0, stigma elongate or penicillate. Ovules 1 or 2. Fruit a small nut drupe or samara, 1-seeded. Seed with fleshy albumen or 0; radicle superior.

A very extensive and widely-diffused Order, including the Nettle, Fig, Hemp, Mulberry, and Breadfruit tree. Bark often very stringy, used as cordage, etc.

Tree. Juice milky. Male flowers spiked. Ovule pedulous . . . . 1. EPICARPURUS.
Shrubs or herbs. Juice watery. Flowers solitary or racemed or spiked. Ovule erect.
Perianth of male 4- or 5-parted. Stamens 4 or 5. Flowers spiked or racemed . . . . 2. URTICA.
Perianth cup-shaped. Stamen 1. Flowers solitary or few . . . 3. AUSTRALINA.
Perianth of male 4- or 5-parted, of female tubular, 4-fid . . . . 4. PARIETARIA.
Herbs. Flowers small, in a fleshy discoid receptacle. Ovule erect . 5. ELATOSTEMMA.

1. EPICARPURUS, Blume.

Trees or bushes. Leaves evergreen, harsh and rigid, alternate. Spikes axillary. Flowers diœcious.—*Male:* spiked or in catkins. Perianth of 4 spreading leaflets. Stamens 4, longer than the leaflets. *Female:* spiked.

Perianth of 4 leaflets. Ovary ovoid; styles 2, with subulate stigmas; ovule suspended. Fruit a drupe or nut, 1-seeded. Seed pendulous; albumen 0; cotyledons conduplicate; radicle curved upwards.

A small, tropical Asiatic genus.

1. **E. microphyllus,** *Raoul, Choix,* 14. *t.* 9;—*Trophis (?) opaca,* Banks and Sol.;—Fl. N. Z. i. 224. A large tree, 50–60 ft. high, variable in habit and foliage, abounding in milky sap; branches brittle, pubescent at the tips; bark brown. Leaves petioled, $\frac{1}{2}$–2 in. long, ovate-oblong, serrate, obtuse or acute, with reticulate veins on both surfaces. *Male* fl.: minute, in catkins $\frac{1}{2}$–$\frac{3}{4}$ in. long, which are solitary or panicled;—*fem.* fl. in much shorter-fewer-flowered spikes. Drupe small, red.

Northern and **Middle** Islands: as far south as Akaroa, *Banks and Solander.* "Milk-tree" of the colonists; the male spikes often become diseased, and present panicled branches covered with minute bracts.

2. URTICA, Linn.

Herbs or small shrubs, with stinging hairs, rarely glabrous. Leaves opposite. Flowers unisexual, glomerate, on simple or branched spikes.—*Male:* Perianth 4-partite. Stamens 4. Ovary imperfect. *Female:* Perianth unequally 4-partite. Ovary ovoid; stigma sessile; ovule erect. Fruit a small, dry nut. Seed compressed; albumen scanty; cotyledons plano-convex.

A considerable genus of tropical and temperate plants, including the common "Stinging Nettle," now probably naturalized in New Zealand.

Herbaceous. Stinging hairs copious. Leaves variable . . . . . 1. *U. incisa.*
Herbaceous, stout. Stinging hairs few. Leaves very broadly cordate 2. *U. australis.*
Shrubby. Stinging hairs most copious, $\frac{1}{4}$ in. long. Teeth of leaves bristle-pointed . . . . 3. *U. ferox.*
Herbaceous. Stinging hairs few. Leaves very pubescent . . . . 4. *U. aucklandica.*

1. **U. incisa,** *Poiret;—U. lucifuga,* Hook. f. Fl. N. Z. i. 225. A slender herb, sparingly covered with stinging hairs, 1–2 ft. high. Leaves with long petioles, extremely variable in form, length, and breadth, $\frac{1}{2}$–2 in. long, from narrow-lanceolate or linear to broadly ovate-cordate, acute, acutely deeply toothed; petiole very slender; stipules oblong or lanceolate, acute.

Northern and **Middle** Islands: common, *Banks and Solander,* etc. Also common in Southern Australia and Tasmania. The female perianth is sometimes tubular and 4-toothed.

2. **U. australis,** *Hook. f. Fl. N. Z.* i. 225. A stout, succulent herb, 1–2 ft. high, glabrous, except for a few scattered, weak, stinging hairs. Leaves large, upper sometimes ternate, 3–4 in. long, very broadly ovate- or almost orbicular-cordate, acute, deeply toothed or crenate; petioles 1–3 in. long; stipules large, 2-fid. Flowers racemose, monœcious?

Northern Island: southern extreme, *Bidwill.* **Lord Auckland's** Islands: in woods, *J. D. H.*

3. **U. ferox,** *Forst.;—Fl. N. Z.* i. 224. A tall, slender shrub, copiously covered with rigid stinging hairs $\frac{1}{8}$–$\frac{1}{4}$ in. long, stem woody; branchlets, petioles, and leaves below, puberulous. Leaves 2–5 in. long, narrow ovate-cordate or linear- or lanceolate-oblong, always broader at the cordate some-

times lobed auricled or hastate base, acuminate, very coarsely toothed, teeth ending in a rigid bristle; petioles ½–1½ in. long, bristling with rigid stinging hairs; stipules linear-oblong, entire, obtuse. Flowers racemed.

Northern and **Middle** Islands: not uncommon in woods from the east coast to Otago, *Banks and Solander*, etc. The pain of the sting sometimes lasts four days, *Colenso*.

4. **U. aucklandica,** *Hook. f. Fl. Antarct.* i. 68. Herbaceous, robust, softly downy all over, except the upper surface of the leaves, where there are a very few stinging hairs. Leaves broadly ovate-cordate, coarsely serrate or toothed, rather coriaceous or rigid; petioles stout, ½–1 in. long; stipules foliaceous, 2-fid or 2-partite. Flowers imperfect.

Lord Auckland's group: sandy shore and edges of the woods, *J. D. H.* My specimens are too young for a good description.

3. PARIETARIA, Gaudichaud.

Herbs, rarely shrubs, usually very flaccid and slender, often pubescent. Leaves quite entire, almost exstipulate.—Flowers polygamous, cymose, 1–3-bracteate.—*Male* and *hermaph.*: Perianth 4-partite, pubescent. Stamens 4. *Female:* Perianth tubular, 4-fid. Ovary ovoid or oblong; stigma capitate or spathulate, recurved. Ovule erect. Nut minute, included in the perianth. Seed exalbuminous; cotyledons plano-convex.

A small genus, abundant in waste places in almost all parts of the world.

1. **P. debilis,** *Forst.;—Fl. N. Z.* i. 226. A slender annual; stems 6–24 in. long, erect or decumbent, simple or branched. Leaves extremely variable in size, ¼–2 in. long, broadly ovate, acute, membranous, quite entire; petioles slender. Cymes dense or lax-flowered; bracts generally linear, not enlarging after flowering. Female flowers most numerous.

Common throughout the islands, *Banks and Solander*, etc. An abundant plant in Australia and in many tropical and temperate climates.

4. AUSTRALINA, Gaudichaud.

Small, tufted, slender, creeping herbs. Leaves opposite or alternate, crenate, 3-nerved, stipulate. Flowers minute, monœcious, in small, few-flowered, axillary clusters.—*Male:* pedicelled. Perianth boat- or funnel-shaped; limb unequally 2-lipped. Stamen 1. Ovary 0. *Female:* sessile. Perianth ventricose; limb somewhat 5-toothed. Stamen 0. Ovary ovate-lanceolate, narrowed into a straight style; stigma lateral, villous; ovule erect. Nut minute, ovate, included in the perianth. Seed exalbuminous; cotyledons plano-convex.

A small genus found in extratropical Australia, Abyssinia, and New Zealand.

1. **A. pusilla,** *Gaud.;—A. novæ-Zelandiæ*, Hook. f. Fl. N. Z. i. 226. Stems slender, intricate, rooting, rather pubescent, 4–8 in. long. Leaves alternate, ⅓ in. long, membranous, orbicular or broadly ovate, coarsely crenate; petiole as long as the blade; stipules subulate. *Male* fl.: in pairs from the upper axils, on a peduncle as long as the petioles;—*fem.* fl.: solitary or few together. Perianth compressed, flagon-shaped; style exserted.

Northern Island: damp woods or roots of trees, etc., Bay of Islands, Manawata river, etc., *Colenso*, etc. Also a native of Tasmania.

5. ELATOSTEMMA, Forst.

Herbs, rarely shrubby, usually succulent. Leaves distichous, opposite or alternate, unequal-sided; stipules axillary. Flowers minute, unisexual, densely crowded on axillary, fleshy, unisexual involucres.—*Male:* Perianth of 3 or 4 mucronate leaflets. Stamens 4 or 5. Ovary imperfect. *Female:* Perianth minute, imperfect. Stamens imperfect. Ovary ellipsoid; stigma sessile, penicellate; ovule erect. Nut minute, ovate or elliptic, compressed. Seed exalbuminous; cotyledons plano-convex.

A very large tropical genus in Asia, Australia, and the Pacific Islands.

1. **E. rugosum,** *A. Cunn.;—Fl. N. Z.* i. 304. Erect or decumbent, 1–2 ft. high, stout, succulent, glabrous or puberulous. Stem branching below. Leaves 4–10 in. long, alternate, sessile, narrow obovate-lanceolate or lanceolate, acuminate, curved, auricled at the base and half amplexicaul, deeply serrate or rugose; stipules membranous, lanceolate. Flowers diœcious. *Male:* receptacles discoid, fleshy, ½ in. diam., surrounded by adnate bracts, solitary or 2 together, sessile or shortly peduncled; perianth pedicelled, hidden amongst membranous bracteoles;—*fem.:* receptacle smaller, more pubescent; perianth nearly sessile; stigma capitate.

Northern Island: in dense moist woods, abundant, *Banks and Solander,* etc. Closely allied to an Australian and South Sea Island species.

ORDER LXXII. CHLORANTHACEÆ.

Herbs or small shrubs. Leaves evergreen, opposite, the petioles connate at the base, usually serrated, stipulate. Flowers minute, green, uni- or bisexual, in terminal and axillary spikes.—Bracts boat-shaped or 0. Perianth 0. Stamens 1–3, epigynous in hermaphrodite flowers; filament very short; anthers bursting laterally or inwards. Ovary sessile, 1-celled; stigma sessile, obtuse, deciduous; ovule 1, pendulous. Drupe small, fleshy, 1-seeded. Seed with a membranous testa, fleshy albumen, and minute embryo.

A very small, tropical and subtropical Order.

1. ASCARINA, Forst.

Flowers diœcious, loosely spiked; bract small.—Stamen 1. Anther linear-oblong, 2-celled, bursting laterally. Ovary ovoid.

1. **A. lucida,** *Hook. f. Fl. N. Z.* i. 228. A small tree or shrub, 12–14 ft. high, everywhere perfectly glabrous, evergreen and shining; branches terete. Leaves 1½–2½ in. long, oblong or oblong-lanceolate, acute or obtuse, coarsely obtusely serrate, rather glaucous below; stipules subulate. Spikes ¼–8 in. long, solitary or racemed.

Northern Island: Totara-nui, *Banks and Solander;* swamps in the Wairarapa valley, *Colenso.* **Kermadec** Islands, *M'Gillivray.* I find that the Kermadec Island plant with larger leaves, which I distinguished as *A. lanceolata,* (Journ. Linn. Soc. i. 129) is only a variety of the New Zealand one. Dr. Seemann refers a Fiji Island plant to the same.

ORDER LXXIII. PIPERACEÆ.

Herbs or shrubs, often aromatic. Leaves alternate opposite or whorled, with or without stipules. Flowers most minute, usually crowded on very slender catkins, amongst minute, angular, flat-topped scales.—Perianth 0. Stamens 2 or more; filaments very short. Ovary 1-celled; style 0 or very short, stigma entire or 2–6-lobed, capitate or plumose; ovule 1, erect. Berry 1-seeded. Seed with fleshy or horny albumen and minute embryo.

A considerable tropical and subtropical Order.

Small herb. Leaves succulent. Stigma deciduous . . . . 1. PEPEROMIA.
Shrub, aromatic. Leaves membranous. Stigmas 2–5 . . . . 2. PIPER.

1. PEPEROMIA, Ruiz and Pavon.

Succulent herbs. Leaves opposite alternate or whorled. Flowers hermaphrodite.—Bracts peltate. Stamens 2. Ovary sessile; stigma sessile, deciduous, pencilled. Berry sessile.

A large genus, found in all tropical parts of the world.

1. **P. Urvilleana**, *A. Rich.;—Fl. N. Z.* i. 228. A small, glabrous or puberulous, succulent herb, 4–10 in. high, creeping, branched. Leaves alternate, shortly petioled, $\frac{1}{2}$–1 in. long, broadly obovate or elliptic-oblong, obtuse, 3-nerved at the base. Catkin peduncled, axillary, solitary, erect, 1–1$\frac{1}{2}$ in. long.

Northern Island: common in damp woods, on mossy trees, rocks, etc., *Banks and Solander*, etc. **Kermadec** Islands, *M'Gillivray*. Also a Norfolk Island plant, and closely allied to some Pacific Island and Australian species.

2. PIPER, Linn.

Small trees, or climbing or erect shrubs. Leaves alternate, petioled; stipules adnate or deciduous. Flowers hermaphrodite or unisexual.—Bracts nearly sessile. Stamens 2. Ovary sessile, ovoid; stigmas 2–5.

A very large tropical genus, to which the Black and Betle Peppers belong, and the Kava of the South Sea Islands.

1. **P. excelsum**, *Forst.;—Fl. N. Z.* i. 228. A large bush or small tree, often 20 ft. high, quite glabrous, very aromatic. Stem flexuose, jointed. Leaves 3–5 in. long, broadly ovate-cordate, acuminate, 5–7-nerved at the base; petioles 1–2 in. long; winged by the adnate stipules at their bases. Catkins solitary or 2 together, axillary, erect, strict, slender, 1–4 in. long. Berries yellow.

Northern and **Middle** Islands: as far south as Canterbury, *Banks and Solander*, etc. **Kermadec** Islands, *M'Gillivray*. Also found in Norfolk Island and Lord Howe's Island. The plant which I referred to *P. macrophyllum*, in a short account of the vegetation of the Kermadec group (Linn. Journ. i. 127), is, I think, only a large state of this. The leaves are eaten, but seeds rejected. Leaves used as Tea, etc., to cure toothache.

ORDER LXXIV. BALANOPHOREÆ.

Stout, succulent, leafless root-parasites. Stem reduced to a tuberous, often lobed Potato-like rhizome, giving off simple, thick, erect scapes, often bearing concave scales, and at the top ovoid, spikes or spadixes of minute, unisexual, very imperfect flowers.—*Male* fl.: Perianth 0 or 3-cleft, lobes valvate. Stamens 1–3; filaments free or united, short or long; anthers 2–many-celled. Ovary 0. *Female* fl.: minute. Calyx adherent, with a lobed limb or 0. Ovary ovoid or globose, 1-celled, styles 1 or 2; stigmas simple; ovule 1, pendulous. Fruit a very minute nut or utricle enclosing an adherent seed, usually consisting of a homogeneous mass of granular albumen; embryo when found excessively minute, lodged in the albumen.

A small Order of chiefly tropical root-parasites; one is a native of the Mediterranean, *Cynomorium*, the "Fungus Melitensis" of the Crusaders. The only New Zealand genus is a most remarkable one, found nowhere else.

1. DACTYLANTHUS, Hook. f.

A fleshy root-parasite; rhizome subterranean, globular or misshapen, tubercled, giving off numerous club-shaped peduncles or stems, covered with concave, obtuse, imbricating scales, the upper larger and surrounding the spadixes.—Flowers minute, diœcious, crowded on erect, columnar spadixes. *Male:* a solitary, almost sessile, 2-celled anther. *Female:* Perianth adnate to the globose ovary, limb of 2 or 3 superior subulate lobes; style filiform, stigma simple.

1. **D. Taylori,** *Hook. f. in Trans. Linn. Soc.* xxii. 427. *t.* 75. Rhizome as big as the fist or smaller; stems 2–4 in. high; scales ovate, lower smaller, shorter, $\frac{1}{4}$ in. long; upper more oblong, larger, 1 in. long. Spadixes cylindric, numerous, crowded at the top of the stem, hidden by the uppermost scales, 1–1$\frac{1}{2}$ in. long; *male* covered by crowded anthers;—*fem.* by the erect, slender ovaries.

Northern Island: Wanga-nui, alt. 4000 ft., on roots of *Fagus* and *Pittosporum*, Rev. R. Taylor. Native name "*Pua reinga.*"

ORDER LXXV. CONIFERÆ.

Shrubs or trees, usually resinous. Leaves stiff, very various, often reduced to scales.—Flowers monœcious or diœcious, very minute and imperfect; males reduced to crowded naked stamens; anthers 2- or more celled. *Female* of one or more naked ovules, without ovary style or stigma, inserted on coriaceous scales, which are solitary or spiked, or collected into catkins or cones. Ripe seeds nut- or drupe-like; testa membranous, crustaceous or osseous; albumen copious; embryo usually terete; cotyledons 2 or more.

A very extensive Order, found in all parts of the world, to which the Yew, Juniper, Pine, Cypress, etc. etc., belong.

Leaves oblong, 1–3 in. Cones large, of many imbricating scales . . 1. DAMMARA.
Leaves small. Cone of few, erect, woody scales . . . . 2. LIBOCEDRUS.

Leaves linear or scale-like. Drupe inverted, adnate to a fleshy peduncle . . . . 3. PODOCARPUS.
Leaves linear or scale-like. Nuts erect, in fleshy cups . . . . . 4. DACRYDIUM.
Leaves fan-shaped. Nuts on the leaves, erect, in fleshy cups . . . 5. PHYLLOCLADUS.

1. DAMMARA, Linn.

Lofty trees. Leaves, when young distichous, flat, coriaceous, with numerous parallel veins. Cone large.—Inflorescence diœcious. *Male:* cylindric catkins of sessile imbricated stamens; anthers of 8–15 cells, pendulous from a peltate connective. *Female:* terminal, cones ovoid obovoid or globose, formed of many closely imbricating, woody or coriaceous, deciduous scales. Ovule solitary at the base of each scale, inverted. Seed unequally winged.—*Agathis*, Salisb.

A considerable genus, confined to Australia, New Zealand, the Malayan and Fiji islands, New Caledonia, and New Hebrides.

1. **D. australis,** *Lambert;—Fl. N. Z.* i. 231. A very large, lofty tree, 120 ft. high; trunk sometimes 10 ft. diam.; branches whorled in the young plant; bark thick, very resinous. Leaves coriaceous, sessile, lanceolate in young plants, 2–3 in. long, in old oblong or obovate, 1–1½ in. long, glaucous. *Male* cones 1 in. long;—*fem.* obovoid or club-shaped, 2–3 in. long.—*Podocarpus zamiæfolius*, A. Rich.

Northern Island: east coast, from Mercury Bay northwards, *Banks and Solander*, etc. The famous Kauri or Kaudi Pine. Though not now found south of Mercury Bay, the gum was stated (Mr. Haast thinks erroneously) to be dug up in the Middle Island. The great lumps now found in the Northern Island are said to be much larger than the existing trees produce.

2. LIBOCEDRUS, Endl.

Trees or shrubs. Branches compressed when young, 4-gonous when old, imbricate.—Inflorescence monœcious. *Male:* cylindric catkins of 6 or 7 stamens; anthers sessile, of 4 cells pendulous from a peltate connective. *Female:* terminal cones formed of 4 erect woody persistent scales, the alternate smaller. Ovules 2 at the base of each scale, erect. Seeds solitary at the base of each scale, compressed, unequally winged.

A small genus, consisting of two Chilian and two New Zealand species.

Branchlets all much compressed . . . . 1. *L. Doniana.*
Branches for the most part 4-gonous . . . . 2. *L. Bidwillii.*

1. **L. Doniana,** Endl.—*Thuja Doniana*, Hook.; Fl. N. Z. i. 231. A tree, 60–100 ft high and 8–10 ft. circumference; bark flaking off and stringy, wood fine-grained, heavy, dark-coloured; branches distichous, vertical and flat in young plants (like the Arbor-Vitæ), much compressed, $\frac{1}{10}$ in. broad, fastigiate, more 4-gonous in old plants. Leaves in 4 rows, of two sizes, the lateral larger, those in the upper and lower faces of the branches very small. Cones ovate, obtuse, ½ in. long; scales with a sharp curved spine at the back.—*Thuja Doniana*, Hook. Lond. Journ. Bot. i. 571. t. 18; *Dacrydium plumosum*, Don.

Northern Island: forests on the Bay of Islands, *Bennett, A. Cunningham*, etc. Hokianga, *Edgerley*. Wood said to be excellent both for planks and spars.

2. **L. Bidwillii,** *Hook. f.* Young leaves and branches similar to *L. Doniana*, but not so broad, old branches with fastigiate 4-gonous twigs, $\frac{1}{16}$ in. diam.

Northern Island: Ruahine mountain, *Colenso*. **Middle** Island: abundant from the Nelson mountains, where it ascends to 6000 feet, *Bidwill*, to Otago, where, at Dunedin, it descends to 2000 ft. Haast's Pass, alt. 1000 ft., *Haast*. I advance this species with much hesitation; it is difficult to suppose that a timber-tree described as having excellent wood, and growing at the Bay of Islands at the level of the sea (I gathered *L. Doniana* on the banks of the Kawa-kawa river), should be the same as one inhabiting the mountains of the Middle Island, and described by Buchanan as having soft, worthless wood; but I can find very little difference between the specimens. The fruiting and other branches of all my specimens of *L. Doniana* are flat; of some of those of *L. Bidwillii* also flat; but of most, including all the fruiting ones, tetragonous. This is a character however of little importance in these Conifers. My cones of *L. Bidwillii* are unripe, and therefore can only be compared with unripe ones of *Doniana*; which they entirely resemble. Buchanan says that the heartwood of *L. Bidwillii* is so soft that soap-bubbles may be blown through a foot length of it. Colenso says his Ruahine mountain plant is quite distinct from the Bay of Islands one, and is called "Pahautea," but gives no distinctive characters.

3. PODOCARPUS, L'Héritier.

Trees or shrubs. Leaves very various, scattered or imbricate or distichous, large or minute, often of 2 forms on each plant or branch of the plant.—Inflorescence diœcious or monœcious. *Male:* axillary or terminal, spiked racemed or solitary cylindric cones, formed of imbricating stamens; anthers sessile; cells 2, pendulous from a peltate connective. *Female:* a short rachis, with 1 or 2 scales, each bearing an inverted ovule adnate to its face. Drupe inverted, with a ridge on one side (the adnate scale), seated on a fleshy peduncle.

A considerable genus, found in various mountain-districts of the tropics, in Japan, and in the south temperate zone commonly, but not in Eastern Asia, Europe, or North America.

Leaves uniform, all linear. Male catkins solitary or 2 or 3.
- Leaves distichous, falcate, $\frac{1}{2}$–$\frac{3}{4}$ in., acute . . . . 1. *P. ferruginea.*
- Leaves imbricate, erect or recurved, $\frac{1}{4}$–$\frac{1}{2}$ in., obtuse . . . . . . 2. *P. nivalis.*
- Leaves distichous or imbricate, $\frac{1}{2}$–$1\frac{1}{2}$ in., rigid, pungent . . . . 3. *P. Totara.*

Leaves uniform, distichous, linear, obtuse. Male catkins spiked . . . 4. *P. spicata*

Leaves some $\frac{1}{6}$ in. distichous, falcate, others $\frac{1}{12}$ in., imbricate . . . 5. *P. dacrydioides.*

1. **P. ferruginea,** *Don;—Fl. N. Z.* i. 232. A lofty timber-tree, 50–80 ft. high; trunk 3 ft. diam.; bark rather scaling; wood brittle, close-grained, durable, reddish. Leaves distichous linear, acute, falcate, 1-nerved, $\frac{1}{2}$–$\frac{3}{4}$ in. long, red-brown when dry. *Male* catkins axillary, solitary, shorter than the leaves; connective obtuse. Drupe $\frac{3}{4}$ in. long, red-purple, glaucous.—Hook. Ic. Pl. t. 542.

Northern Island: common in woods, *Banks and Solander*, etc. **Middle** Island: Otago, ascending to 1000 ft. "Black Pine" of the colonists, *Hector and Buchanan.* Drupes taste of turpentine, greedily eaten by birds.

2. **P. nivalis,** *Hook. f. Fl. N. Z.* i. 232. A small, woody, densely branched shrub, 1–20 ft. high; trunk 3 ft. diam. Leaves not distichous, spreading or recurved, very thick and coriaceous, $\frac{1}{4}$–$\frac{1}{2}$ in. long, linear-oblong, obtuse, apiculate; midrib stout, green when dry. Male catkins short, often 3 together; connective obtuse. Drupe apparently dry.—Hook. Ic. Pl. t. 582.

Northern Island: on the mountains, Tongariro, *Bidwill;* Ruahine range, *Colenso.* **Middle** Island: Southern Alps, alt. 2500–5000 ft., *Sinclair and Haast;* forming much of the subalpine forest; Wairau mountains, alt. 4–5000 ft., *Travers.* Perhaps an alpine form of *P. Totara;* Otago, *Hector and Buchanan.* Very nearly allied to the Tasmanian *P. alpina.*

3. **P. Totara,** *A. Cunn.;—Fl. N. Z.* i. 233. A lofty and spreading tree, 60 ft. high; trunk 2–4 ft. diam.; bark rather flaking; wood red, close-grained, very durable. Leaves distichous or not so, very coriaceous, erect, spreading or recurved, straight or falcate $\frac{1}{2}$–$1\frac{1}{2}$ in. long, linear, acuminate,, pungent; midrib indistinct, pale-green when dry. *Male* catkins short, stout, obtuse, solitary or 2 or 3 together, bracteate at the base; connective toothed. Drupes solitary or 2, on a swollen peduncle as big as a cherry.—Hook. Lond. Journ. Bot. i. 572. t. 19.

Throughout the **Northern** and **Middle** Islands: *Menzies,* etc. Wood the most valuable in the islands. Drupe eaten. Bark used for roofing.

4. **P. spicata,** *Br.;—Fl. N. Z.* i. 232. A large tree, 80 ft. high; bark bluish-black, almost smooth; wood white, soft, close, and durable. Leaves more or less distichous, $\frac{1}{3}$–$\frac{1}{2}$ in. long, linear, straight or falcate, obtuse or apiculate, glaucous below. *Male* catkins numerous, spiked, horizontal. Drupes often spiked, very numerous, globular, $\frac{1}{3}$ in. diam.—Hook. Ic. Pl. t. 543; *Dacrydium taxifolium,* Banks and Sol.; *D. (?) Mai,* A. Cunn.

Northern and **Middle** Islands: abundant, *Banks and Solander,* etc. "Black Rue" of Otago. Drupe sweet, eatable.

5. **P. dacrydioides,** *A. Rich.;—Fl. N. Z.* i. 233. Trees gregarious, very lofty, 150 ft. high, 4 ft. diam.; wood white, soft, spongy. Leaves of 2 forms, of young trees and on twigs of old, distichous, $\frac{1}{6}$ in. long, linear, falcate, tip turned up and acuminate, nerveless; on old branches imbricated, small, subulate, $\frac{1}{12}$–$\frac{1}{6}$ in. long, keeled. *Male* catkins terminal, small, $\frac{1}{6}$ in., solitary, sessile; connective acute.—Drupes small, gibbous, on swollen peduncles.—*P. thuyoides,* Br.

Abundant throughout the **Northern** and **Middle** Islands: in swamps, *Banks and Solander,* etc. Twigs used for eel-baskets. Wood bad. Drupe eaten. The rootlets bear singular minute globular bodies, containing cells with spiral markings. The wood of Otago specimens is described as close-grained and heavy (*Buchanan*).

4. DACRYDIUM, Solander.

Trees or shrubs. Leaves linear and distichous or subulate and imbricating. —Inflorescence diœcious. *Male:* small, terminal, ovoid, solitary cones, formed of imbricating stamens; anthers sessile; cells 2, pendulous from a peltate connective. *Female:* a short rachis, with 1 or more scales, 1 or 2 of which bear an inverted ovule, which becomes erect as it ripens. Nut ovoid, erect, outer coat short, sheathing at its base, sometimes fleshy.

A small genus, confined to the Malayan and Pacific islands, Tasmania and New Zealand.

Lofty tree, branches weeping. Leaves subulate . . . . 1. *D. cupressinum.*
A tree, branches erect or spreading. Leaves obtuse . . . . 2. *D. Colensoi.*
A small, creeping bush. Leaves lax, obtuse . . . . 3. *D. laxifolium.*

1. **D. cupressinum,** *Soland.;—Fl. N. Z.* i. 233. Tree pyramidal, pale-green; branches weeping; trunk 80 ft. high, 4–5 ft. diam.; bark scaling; wood red, heavy, solid. Leaves closely imbricating all round; on the young

branches, rigid, 3-gonous, decurrent, subulate, curved, $\frac{1}{6}$ in. long, keeled; on old much smaller, more closely imbricated, $\frac{1}{12}$–$\frac{1}{10}$ in. long. Nuts ovoid, $\frac{1}{8}$ in. long, on the curved tips of the branchlets.—Richard, Conif. 827. t. 2.

Throughout the **Northern** and **Middle** Islands, abundant, *Banks and Solander*, etc. "Red Pine" of Otago colonists. Spruce beer was made of the young branches by Captain Cook; the wood is excellent, the fleshy cup of the nut eatable.

2. **D. Colensoi,** *Hook.;—Fl. N. Z.* i. 234. A tree, 12–40 ft. high, very variable in habit; bark whitish or pale-brown with white patches; wood light, yellowish; branches stout, woody. Leaves of various forms, some linear and spreading, $\frac{1}{3}$–$\frac{1}{2}$ in. long, obtuse, with stout costa, others small, densely 4-fariously imbricated, triangular, keeled, coriaceous, $\frac{1}{20}$–$\frac{1}{12}$ in. long. Male catkins terminal, solitary, sessile; anthers 4–6; connective obtuse. Nut small, on a horizontal resinous cup-shaped disk.—Hook. Ic. Pl. t. 548; *Podocarpus (?) biformis*, Hook. Ic. Pl. t. 544.

Northern Island: Tongariro and Ruahine range, *Colenso*. **Middle** Island: Nelson mountains, alt. 4–6000 ft., *Bidwill;* alps of Canterbury, alt. 2–4000 ft., *Sinclair and Haast;* Otago, alt. 3000 ft., *Hector and Buchanan.*

3. **D. laxifolium,** *Hook. f., Fl. N. Z.* i. 234. A small, weak, straggling, prostrate shrub; branches trailing, 6–12 in. long, flexible. Leaves as in *D. Colensoi*, but much smaller, linear ones spreading, $\frac{1}{12}$–$\frac{1}{10}$ in. long, imbricating ones very broadly ovate or trapeziform, oblong, keeled, $\frac{1}{24}$ in. long. Nuts small, erect, in red fleshy cups.—Hook. Ic. Pl. t. 825.

Northern Island: Tongariro, *Bidwill;* Ruahine range, *Colenso*. **Middle** Island: Nelson mountains, alt. 5000–6000 ft., *Bidwill;* Black Hills, alt. 4000 ft., *Haast;* Otago, *Hector and Buchanan.* Perhaps a small form of *D. Colensoi.*

5. PHYLLOCLADUS, Br.

Trees with whorled branches. Leaves of two forms, some minute and scale-like, others linear, seen only in young plants, but which in older are connate into flat fan-shaped or ovate coriaceous organs (phyllodes) resembling simple leaves, which bear the inflorescence on their edges.—Inflorescence monœcious; male and female close together. *Male* as in *Dacrydium*. *Female* a short rachis with a few scales, each bearing a solitary, sessile, erect ovule; girt at the base with a cup-shaped disk. Nut solitary, erect, girt by the fleshy disk and fleshy connate scales.

A small genus, natives of the mountains of Borneo, of Tasmania and New Zealand.

Flowers on the margins of the phyllodes . . . . 1. *P. trichomanoides.*
Flowers at the base of the phyllodes . . . . 2. *P. alpinus.*

1. **P. trichomanoides,** *Don;—Fl. N. Z.* i. 235. A slender tree, 60 ft. high; wood pale, close-grained. Phyllodes distichous, with scales (rudimentary leaves) at their base, very coriaceous, $\frac{1}{2}$–1 in. long, obliquely rhomboid cuneate or ovate, simple or pinnatifidly lobed; lobes truncate, erose; veins radiating from one central one outwards. Nuts compressed, solitary, on the margins of the phyllodes.—Hook. Ic. Pl. t. 549, 550, 551; *P. rhomboidalis*, A. Rich. Flora.

Northern Island: not rare in forests. *Banks and Solander*, etc. Bark used for dyeing red. Wood excellent, white, used for planks and spars.

2. **P. alpinus,** *Hook. f. Fl. N. Z.* i. 235. *t.* 53. A small, very rigid, densely-branched shrub or small tree, sometimes 2 feet diam. Leaves very much thicker, smaller and more coriaceous than in *P. trichomanoides*, sometimes linear-oblong and only ½ in. long, often glaucous below. Nuts at the base of the phyllodes, small.

Northern Island: Tongariro, *Bidwill;* Ruahine range, *Colenso.* **Middle** Island: Nelson, alt. 6000 ft., *Bidwill;* Wairau mountains, alt. 4–5500 ft., *Travers,* etc.; alps of Canterbury, *Sinclair and Haast;* alt. 2300–5000 ft.; Otago, *Hector and Buchanan.* Perhaps only a form of *P. trichomanoides*, but a very distinct one. Also very closely allied to the Tasmanian *P. aspleniifolia.*

Class II. MONOCOTYLEDONS.

Order I. ORCHIDEÆ.

Herbs, sometimes almost shrubby, either terrestrial with tubers or bulbs, or epiphytes with leafy branches, which are often thickened and called pseudobulbs. Leaves sheathing at the base. Flowers bracteate, extremely various, often beautiful, hermaphrodite.—Perianth superior, of 6 pieces in 2 series, 3 outer (*sepals*) usually nearly equal; of the 3 inner 2 are lateral (*petals*); and the innermost (*lip*) is either largest or differs in shape, direction, or surface; it is sometimes superior (or posticous), at others inferior (or anticous). Axis of the flower occupied by a *column* facing the lip, consisting of a stamen combined with the style and reduced apparently to 2–8 masses of *pollen* contained in a fixed or moveable, deciduous or persistent cap-like anther; pollen cohering in 2–8, often pyriform, waxy or granular masses, often attached in pairs by a caudicle to a gland, which is easily detached from the tip (*rostellum*) of the column. Ovary inferior, 1-celled; stigma a glandular depression or swelling on the front or base of the column, opposite the lip; ovules very numerous on 3 parietal placentas. Capsule 1-celled, 3-valved, many-seeded. Seeds very minute, light, with loose reticulated testa, and a solid embryo.

A very extensive Order, abounding in beautiful plants, the flowers always of curious structure. Theoretically the flower consists of 15 parts: viz. 3 sepals; 3 petals; 3 outer stamens, opposite the sepals, of which, that answering to the sepal opposite the lip is alone developed (the other 2 being suppressed and confluent with the lateral veins of the lip); 3 inner stamens all suppressed, of which 2 are theoretically confluent with the sides of the column, and the 3rd with the midrib of the lip); lastly, 3 stigmas, of which, that opposite the lip is alone developed. The correctness of this theory is supported by the presence of 2 lateral appendages which represent undeveloped stamens on the sides of the column of such genera as *Thelymitra*, *Prasophyllum*, etc., and by three crests or ridges on the lip, representing as many others (as *Chiloglottis*); also by the arrangement and direction of the vascular bundles in the ovary, column, and perianth. For the development of this view, and an account of the wonderful processes by which fertilization is effected in this Order by insects, see Darwin's work 'On the Fertilization of Orchids.'

1. ARTIFICIAL KEY TO THE GENERA.

A. *Perennial epiphytes, with evergreen leaves. Pollen waxy, except in* 1.

Stems slender, leafy. Sepals free. Disk of lip naked. Pollen granular . . . . 1. EARINA.
Stem slender, leafy. Lateral sepals adnate with column. Lip with crests on face . . . . 2. DENDROBIUM.
Leaves in pairs on pseudobulbs . . . . 3. BOLBOPHYLLUM.
Stems very short, leafy. Lip concave, middle lobe solid . . . . 4. SARCOCHILUS.

B. *Stems annual; roots terrestrial, bulbous, tuberous, or rarely creeping.*

α. *Leaf solitary, broad, membranous.*

Flower 1, sessile, purple, large. Sepals very long, filiform . . . 9. CORYSANTHES.
Flower 1, long-peduncled. Sepals broad, upper concave . . . . . 8. ADENOCHILUS.
Flowers few. Sepals and petals acuminate or awned . . . . . . 6. ACIANTHUS.
Flowers few. Sepals and petals linear, obtuse . . . . 7. CYRTOSTYLIS.

β. *Leaf solitary (rarely* 0 *in* Prasophyllum), *slender, flat or terete* (2 *broad in* Caladenia bifolia).

Leaf tubular. Flowers many, minute, with the lip below . . . 10. MICROTIS.
Leaf linear or terete. Flowers many, minute, with the lip uppermost 17. PRASOPHYLLUM.
Leaf flat, pubescent. Flowers 1–4, pink. Lip glandular . . . 11. CALADENIA.
Leaf terete, thick. Flowers 1 or more, yellow or blue. Lip sepals and petals all similar . . . . 15. THELYMITRA.

γ. *Leaves* 2 *or more, rarely* 0 *in* Pterostylis (*see* Caladenia bifolia, in β).

Flower solitary, large, green, galeate. Lip small, narrow . . . 12. PTEROSTYLIS.
Flower solitary. Upper sepal arched. Lip with large, purple glands 13. CHILOGLOTTIS.
Flowers several. Upper sepal galeate. Lip with 5 ridges . . . 14. LYPERANTHUS.
Flowers several. Upper sepal galeate. Lip 3-lobed . . . . . . 18. ORTHOCERAS.
Flowers numerous. Upper sepal oblong. Lip with crumpled edges 16. SPIRANTHES.

δ. *Leaves* 0. *Stem stout, with brown sheathing scales.*

Flowers numerous, brown. Sepals united in an urceolate tube . . 5. GASTRODIA.

2. NATURAL CLASSIFICATION OF THE GENERA.

1. Anther terminal, deciduous. Pollen masses granular, attached to a short caudicle :—1, EARINA.

2. Anther terminal, deciduous. Pollen masses waxy, attached to a broad caudicle :—2, DENDROBIUM; 3, BOLBOPHYLLUM: 4, SARCOCHILUS.

3. Anther terminal, deciduous. Pollen masses of large loose grains, without caudicle :—5, GASTRODIA.

4. Anther terminal, persistent. Pollen masses attached by their bases to the stigmatic gland: — 6, ACIANTHUS; 7, CYRTOSTYLIS; 8, ADENOCHILUS; 9, CORYSANTHES; 10, MICROTIS; 11, CALADENIA; 12, PTEROSTYLIS; 13, CHILOGLOTTIS; 14, LYPERANTHUS.

5. Anther parallel to the stigma, at the back of the column, persistent. Pollen masses attached by threads to the stigmatic gland :—15, THELYMITRA; 16, SPIRANTHES, 17, PRASOPHYLLUM; 18, ORTHOCERAS.

1. EARINA, Lindley.

Rigid, erect, tufted epiphytes. Stems leafy, simple, compressed. Leaves distichous, narrow linear, numerous. Flowers small, whitish, in terminal simple or branched bracteate spikes or panicles.—Sepals and petals nearly equal, spreading, ovate or oblong. Lip superior, sessile or shortly stalked, 3-lobed, disk eglandular, lateral lobes incurved. Column short. Pollen-masses 4, granular, united in pyriform pairs to a small short caudicle, which comes away with them.

A small genus, confined, as at present known, to New Zealand. The two species were confounded by Cunningham.

Leaves 4–6 in. long, acuminate. Panicles slender . . . . 1. *E. mucronata.*
Leaves 2–3 in. long, obtuse or acute. Panicle stiff . . . . 2. *E. autumnalis.*

1. **E. mucronata,** *Lindl.;—Fl. N. Z.* i. 239. Stems slender, 1–3 ft. long, two-edged. Leaves narrow linear or strap-shaped, 4–6 in. long, $\frac{1}{8}$ in. broad, acuminate. Panicle slender, sparingly branched. Flowers remote, $\frac{1}{4}$ in. diam.; bracts obtuse. Sepals and petals linear-oblong. Lip deeply 3-lobed, spotted.—Hook. Ic. Pl. t. 431.

Common throughout the **Northern** and **Middle** Islands, *Banks and Solander*, etc.

2. **E. autumnalis,** *Hook. f. Fl. N. Z.* i. 239. Stems stout, 1–1½ ft. high, slightly compressed. Leaves narrow linear, 2–3½ in. long, $\frac{1}{3}$ in. broad, rigid, nerved and striated. Panicle short, stiff, usually distichously branched. Flowers as in *E. mucronata*, but crowded, white, speckled; bracts short, obtuse, imbricated. Sepals broadly oblong. Petals ovate. Lip broad, obscurely 3-lobed, retuse.

Throughout the **Northern** and **Middle** Islands, not uncommon, *Banks and Solander*, etc.

2. DENDROBIUM, Linn.

Rigid, tufted, epiphytes. Stem stout or slender, leafy. Leaves distichous, numerous, usually narrow. Flowers panicled racemed or axillary and solitary, usually large and handsome.—Sepals spreading, the lateral adnate to the base of the column. Petals usually smaller. Lip inferior or superior, sessile and jointed on to the base of the column or adnate to it, usually large and 3-lobed, disk often with ridges or plates. Column semiterete, usually produced at the base. Pollen-masses 4, adhering in pairs to a strap-shaped caudicle.

A very large Asiatic, Australian, and Pacific Island tropical and subtropical genus, containing many most beautiful species, cultivated extensively in England; these differ greatly in habit, foliage, and inflorescence.

1. **D. Cunninghamii,** *Lindl.;—Fl. N. Z.* i. 240. Stems tufted, pendulous, slender, rigid, polished, branched, 1–2 ft. long. Leaves numerous, 1–1½ in. long, $\frac{1}{8}$ in. broad, linear, acuminate, pale green, striated, 3-nerved. Flowers in axillary slender 2- or many-flowered racemes, which are shorter than the leaves, $\frac{3}{4}$ in. diam., pale rose-coloured; pedicels slender; bracts short. Sepals acute, upper narrow, lower broad produced into a short spur behind. Petals as long, oblong, obtuse. Lip with a short claw, adnate to the lateral sepals and column, 3-lobed, lateral lobes small, middle broader than long, wavy, retuse or truncate; disk with 5 plates.—*D. biflorum*, A. Rich. not Swartz.

Throughout the **Northern** and **Middle** Islands. abundant, *Banks and Solander*, etc. Very nearly related to the South Sea Island *D. biflorum*.

3. BOLBOPHYLLUM, Thouars.

Very coriaceous, tufted epiphytes, with running matted stems bearing fleshy green tubers (pseudobulbs), each crowned with 1 or 2 leaves.—Flowers solitary or spiked on a scape rising from the base of the pseudobulb; their structure very similar to *Dendrobium*, but the column is very short, produced at the tip into 2 short horns.

A large genus, with the same distribution as *Dendrobium*.

1. **D. pygmæum,** *Lindl.;—Fl. N. Z.* i. 240. Minute, forming patches on mossy trunks of trees. Pseudobulbs as large as a pea. Leaf solitary, sessile, coriaceous, linear-oblong, $\frac{1}{4}$ in. long, grooved down the middle. Flowers minute, solitary, on a short bracteate scape. Upper sepal convex, subacute; lateral, broadly ovate, acute. Petals shorter, linear-oblong, obtuse. Lip ovate, obtuse, disk with thickened lines down the centre. Ovary gibbous, hairy.—*Dendrobium pygmæum*, Smith.

Throughout the **Northern** and **Middle** Islands: common, *Banks and Solander*, etc.

4. SARCOCHILUS, Br.

Small, coriaceous epiphytes. Stems short. Leaves distichous, linear-oblong. Flowers large or small, in bracteate spikes or racemes.—Perianth fleshy, open. Sepals nearly equal, obtuse, lateral adnate to the base of the lip. Petals smaller, obtuse. Lip continuous with the column, concave, fleshy; spur 0. Column short, erect. Anther terminal. Pollen-masses 4, waxy, cohering in globose pairs, attached to a broad strap-shaped caudicle, which is fixed to the gland of the stigma.

A small genus, native of subtropical Australia, the Fiji and Malay Islands.

1. **S. adversus,** *Hook. f. Fl. N. Z.* i. 241. Roots long, wiry, straggling. Stems short, 1–2 in. long. Leaves few, linear-oblong, obtuse or subacute, 1–2 in. long, jointed near the base when dry. Scape 1–2 in. long, subterminal, slender. Spike 1 in. long, 10–20-flowered, rachis thickened. Flowers minute, yellow-green, $\frac{1}{10}$ in. diam. Sepals oblong-ovate, obtuse. Petals linear-oblong, obtuse. Lip subquadrate, obscurely lobed.

Northern Island: Opuragi, *Banks and Solander;* Bay of Islands and Wairanaka Valley, *Colenso, Edgerley*. A very small-flowered species, compared with the Australian.

5. GASTRODIA, Br.

Tall, slender, leafless, whitish or brown terrestrial herbs. Root long, thick, fleshy, twisted, or tuberous. Stem with sheathing brown scales. Flowers racemed, pendulous, dirty-white or brownish.—Perianth gibbous at the base, subcampanulate or urceolate. Sepals united into a tube to near their tips, which are shortly reflexed; petals smaller, adnate to the tube, their tips reflexed exserted. Lip included, superior, clawed, linear-oblong; margins crenulate. Column short or long, scarcely winged, without lateral appendages. Anther terminal, horizontal, 2-celled, deciduous. Pollen-masses 4, united in pairs, curved, free, composed of very large grains. Stigma at the base of the column, with a free opening to the ovary. Ovary small.

A curious genus, found in Australia and Tasmania, New Zealand, and the Indian Islands.

1. **G. Cunninghamii,** *Hook. f., Fl. N. Z.* i. 251. Root sometimes 18 in. long, very stout. Stem 1–2 ft. high; scales scarious, short, distant, al-

ternate or opposite and connate. Racemes 6–10 in. long. Flowers 10–20, dirty-green, spotted with white; bracts short, scarious. Perianth fleshy, $\frac{1}{3}$–$\frac{1}{2}$ in. long. Claw of lip winged, blade linear-oblong, membranous, waved, with 2 thick ridges down the middle. Column very short.—*G. sesamoides*, Br.?, A. Cunn.

Northern and **Middle** Islands: in damp shaded woods, not uncommon, but easily overlooked; Bay of Islands, etc., *A. Cunningham;* Port Preservation, *Lyall;* Haast's Pass, Lake Wanaka, etc., *Haast*. The root is full of starch and mucilage, and was eaten by the natives. Odour of plant aromatic but disagreeable (*Haast*).

6. ACIANTHUS, Br.

Slender, small, flaccid herbs. Root of tubers at the end of long fibres. Leaf 1, sessile, cordate. Racemes few-flowered. Flowers rather large, green or brown, spreading.—Sepals and petals slender, acuminate or curved, upper sepal sometimes broad. Lip pendulous, or pointing forwards, entire, with 2 glands at the base, and sometimes a glandular disk or tip. Columns slender, terete, arched. Anther small, terminal. Pollen-masses 8, or 4 each 2-partite, powdery.

A small Australian, Tasmanian, and New Zealand genus.

1. **A. Sinclairii,** *Hook. f. Fl. N. Z.* i. 246. Stem 1–3 in. high, very delicate. Leaf broadly cordate, deeply 2-lobed at the base, acuminate, $\frac{1}{2}$–1 in. long. Flowers 2–6, green, $\frac{1}{6}$–$\frac{1}{4}$ in. diam. Bracts short, ovate, acute. Sepals linear-subulate, aristate, upper broad, 3-nerved; lateral narrower, 1-nerved, toothed towards the tip. Petals small, lanceolate. Lip ovate-lanceolate, with 2 glands at the base, and a thickened tip.

Common throughout the **Northern** Island. **Kermadec** Island: *M'Gillivray.*

7. CYRTOSTYLIS, Br.

Habit of *Acianthus*, but sepals and petals not acuminate or awned, and column winged above.

A small genus of Australian, Tasmanian, and New Zealand plants.

Leaf oblong . . . . 1. *C. oblonga.*
Leaf orbicular, deeply 2-lobed at the base . . . . 2. *C. rotundifolia.*

1. **C. oblonga,** *Hook. f. Fl. N. Z.* i. 246. Very slender, 1–3 in. high. Leaf oblong or oblong-cordate, obtuse or acute, cordate or rounded at the base. Flowers 1–3, nearly $\frac{1}{2}$ in. diam. Sepals and petals nearly equal, narrow linear, acute. Lip as long as the sepals, linear-oblong, obtuse or rather truncate at the tip, with 2 small glands at the base.

Northern Island and northern parts of the **Middle** Island: frequent in shaded woods, *Sinclair, Colenso, Travers,* etc.

2. **C. rotundifolia,** *Hook. f. Fl. N. Z.* i. 246. Quite similar to *C. oblonga*, but the leaf is orbicular, and often 1–1$\frac{1}{2}$ in. diam.

Northern Island: east coast, Raukawa ridge, Cape Kidnapper, etc., *Colenso*. My *C. macrophylla* is, I think, nothing but a large state of this, which may itself prove to be a variety of *C. oblonga*.

8. ADENOCHILUS, Hook. f.

A very slender glabrous herb. Stem simple, with one sessile, ovate, acute leaf in the middle, 1-flowered.—Perianth downy. Upper sepal concave, acuminate, adnate to the back of the column; lateral oblique, 3-nerved, lanceolate, placed under the lip. Petals suberect, linear-lanceolate. Lip shortly clawed, 3-lobed, middle lobe caudate, disk and middle lobe with about 4 rows of stipitate glands. Column slender, curved, winged; wings produced upwards into 2 toothed lobes. Anther hidden behind the stigma and between the lobes of the column, persistent. Ovary very long, erect.

A curious genus, of but one species, allied to *Caladenia*, *Chiloglottis*, and the Australian genus *Eriochilus*.

1. **A. gracilis,** *Hook. f. Fl. N. Z.* i. 246. *t.* 56 *A.* Stem 6–10 in. high. Leaves membranous, acute, ¾ in. long; sheath short. Stem above the leaf with one sheathing bract in the middle, and another at the base of the peduncle. Ovary ½–¾ in. long. Perianth ½ in. broad; segments all very acuminated, veined. Lip small, concealed within the perianth. Anther orbicular, mucronate.

Northern Island: Bay of Plenty, *Colenso.*

9. CORYSANTHES, Brown.

Very small, delicate, succulent, terrestrial herbs. Root of small tubers on slender caulicles. Leaf 1, very broad. Flower 1, large, purple, almost sessile on the leaf; peduncle elongating after flowering.—Upper sepal very large, long, curved forward from the base, narrow boat- or helmet-shaped; lateral free or connate at the base, very small, or long and filiform. Petals usually very small. Lip very concave or involute, usually fimbriate at the edges. Column short or slender. Anther terminal, 1-celled, persistent. Pollen-masses 4, powdery.

A remarkable genus of beautiful little plants, natives of the Malayan islands, temperate Australia, and New Zealand. The genus *Nematoceras*, established in the New Zealand Flora, I think merges into *Corysanthes* through the Malayan species. The species are difficult of discrimination, and perhaps are not all permanently distinct.

Leaf reniform, 3-lobed at the tip. Lip 2-partite, recurved . . . . 1. *C. triloba.*
Leaf oblong-ovate; petiole short. Lip convolute, truncate, toothed . 2. *C. oblonga.*
Leaf orbicular-cordate . . . . 3. *C. rotundifolia.*
Leaf ovate-oblong orbicular or cordate; petiole short. Lip tubular, recurved . . . . 4. *C. rivularis.*
Robust. Leaf large, cordate; petiole and scape very thick. Lip broad, recurved, undulate . . . . 5. *C. macrantha.*

1. **C. triloba,** *Hook. f.—Nematoceras*, Fl. N. Z. i. 250. Leaf ¾–1¼ in. diam., membranous, orbicular-reniform, 3-lobed at the very tip; middle lobe acute; petiole ½–3 in. long. Peduncle sometimes 8 in. long after flowering. Flower ⅓ in. long. Lateral sepals and petals filiform, sometimes 2 in. long, 5 times longer than the lip; upper dilated and obtuse at the tip. Lip very large, of 2 large recurved lobes, margins nearly entire. Column very small.

Northern Island: damp woods, east coast, Cape Palliser, Cape Titiokura, etc., *Colenso;* Auckland, *Sinclair;* Otago, *Hector and Buchanan.* The Middle Island specimens alluded to in Fl. N. Z., I think are rather referable to *C. macrantha*, which differs in the much larger size, very coriaceous, rarely 3-lobed leaf.

2. **C. oblonga,** *Hook. f.—Nematoceras*, Fl. N. Z. i. 249. t. 57 B. Leaf $\frac{3}{4}$–1$\frac{1}{2}$ in. long, sessile, membranous, ovate-oblong, apiculate, rarely subcordate at the base. Flower small, $\frac{1}{5}$–$\frac{1}{4}$ in. long, on a short scape. Upper sepal rather narrow; lateral and petals filiform, $\frac{3}{4}$–1 in. long. Lip involute and truncate, when spread out broadly cordate, deep blood-red; margin pale, toothed. Column short, curved.

Northern Island, *Edgerley*; interior, *Colenso*; Auckland, *Captain Haultain*. **Middle** Island: Nelson, *Travers*; Otago, *Hector and Buchanan*.

3. **C. rotundifolia,** *Hook. f.—Nematoceras*, Fl. N. Z. i. 251. Leaf sessile or petioled, $\frac{1}{3}$–1 in. diam., membranous, orbicular, acute or apiculate, cordate or 2-lobed at the base. Flower subsessile, very small, $\frac{1}{4}$–$\frac{1}{3}$ in. diam. Upper sepal rather narrow, lateral and petals filiform. Lip involute, truncate as in *C. oblonga*.

Northern Island: Manawata harbour, etc., *Colenso*. **Middle** Island: Nelson, *Travers*. **Lord Auckland's** group and **Campbell's** Islands, *J. D. H.* Perhaps a variety of *C. oblonga*.

4. **C. rivularis,** *Hook. f.—Nematoceras*, Fl. N. Z. i. 251. Leaf nearly sessile, orbicular ovate ovate-cordate or oblong-cordate, obtuse acute or acuminate, membranous, often 2-lobed at the base. Flower $\frac{1}{3}$–$\frac{1}{2}$ in. long. Upper sepal narrow, acuminate; lateral and petals filiform, 1–1$\frac{1}{2}$ in. long. Lip involute, when spread open trowel-shaped, recurved; tip retuse or apiculate; margins undulate.—*Acianthus rivularis*, A. Cunn.

Northern and **Middle** Islands: common on mossy trees, etc., in shady ravines. **Lord Auckland's** group, *Bolton*.

5. **C. macrantha,** *Hook. f.—Nematoceras*, Fl. N. Z. i. 249. Much the largest and stoutest species, often 6–10 in. high. Leaf usually with a long very stout petiole, 1–1$\frac{1}{2}$ in. broad, coriaceous or fleshy, oblong-orbicular, cordate or 2-lobed at the base, obtuse or apiculate, rarely 3-lobed, with a broad thickened margin, much reticulate. Flowers on short scapes at the base of the petiole, $\frac{1}{2}$–1 in. long, lurid purple. Upper sepal narrow, acuminate; lateral and petals very narrow linear, almost filiform, 1–2 in. long. Lip large, recurved, very broad, almost 2-lobed, strongly undulate.

Throughout the central parts of the **Northern** and all the **Middle** Island: not uncommon in damp shady woods, *Colenso*, etc. **Lord Auckland's** group, *Bolton*.

10. MICROTIS, Banks and Solander.

Erect, slender, green herbs. Root of oblong tubers. Leaves 1–2 in., very slender, sheathing at the base. Flowers minute, green, spiked, fleshy, spreading.—Sepals: upper connivent with the petals, and together with them forming a hood; lateral placed under the lip. Petals small, ascending. Lip pendulous, small, entire, sessile, with 2 thickened glands at the base. Column short, with 2 lateral lobes; anther terminal. Pollen-masses 4, powdery.

A genus of several Australian and a few New Caledonian, Java, and New Zealand species.

1. **M. porrifolia,** *Sprengel;—Fl. N. Z.* i. 245. Very variable in

stature, robustness, and number of flowers, 6–24 in. high. Leaf solitary, terete, tubular. Spike 1–6 in. long, 20–80-flowered. Flowers $\frac{1}{12}$ in. long; bracts very short. Upper sepal broadly ovate. Lip oblong, obtuse, crenate or crisped, rounded or obscurely 2-lobed at the tip, with 2 glandular lumps at the base and 1 towards the tip.—*M. Banksii*, A. Cunn.; *Epipactis porrifolia*, Swartz; *Ophrys unifolia*, Forst.

Throughout the **Northern** and **Middle** Islands: abundant, *Banks and Solander*, etc.

11. CALADENIA, Br.

Slender, small, pubescent pilose or villous herbs. Roots of small tubers terminating underground caulicles. Leaf solitary (rarely 2), radical or cauline, sheathed at its base. Scape 1–4-flowered; bract 1 or 0.—Perianth open, suberect, 2-lipped, glandular. Upper sepal erect or arched, lateral placed under the lip. Petals erect or spreading. Lip clawed, concave, 3-lobed, obovate or trowel-shaped; disk with rows of stipitate glands. Column slender, curved, winged above. Anther terminal, exposed, mucronate. Pollen-masses 4.

A large extratropical Australian, Tasmanian, and New Zealand genus.

Leaf solitary, linear. Scape very slender. Flower $\frac{1}{4}$–$\frac{1}{2}$ in.: . . . . . 1. *C. minor.*
Leaf solitary, linear. Scape stout. Flower $\frac{1}{2}$–1 in. . . . . 2. *C. Lyallii.*
Leaves 2, oblong. Scape stout. Lip undivided . . . . 3. *C.? bifolia.*

1. **C. minor,** *Hook. f. Fl. N. Z.* i. 247. *t.* 56 *B.* A small, very slender herb, 2–8 in. high, pilose with spreading hairs. Leaf very narrow, linear, at the very base of the stem. Flower 1, rarely 2, pink, $\frac{1}{4}$–$\frac{1}{2}$ in. diam. Sepals linear, obtuse, narrower than the petals. Lip broad, 3-lobed; lateral lobes broad, banded with purple; middle subulate, glandular at the margin; disk with 2 series of stipitate glands.

Northern Island: dry clay hills, abundant. **Middle** Island: Otago, *Lyall.*

2. **C. Lyallii,** *Hook. f. Fl. N. Z.* i. 247. Much stouter than *C. minor*, with longer spreading hairs, 4–8 in. high. Leaf narrow linear, sometimes $\frac{1}{4}$ in. broad. Scape stout, curved, 1–2-flowered. Flowers $\frac{1}{2}$–1 in. diam. Sepals obovate-oblong, acute or obtuse and apiculate; upper shorter, arched. Petals narrower. Middle lobe of lip subulate, recurved. Other characters as in *C. minor.*

Middle Island: Gordon's Nob, *Munro;* swampy places, Hurumui mountains, Lake Tennyson, and Mount Brewster, alt. 4–5000 ft., *Haast, Travers;* Otago, grassy places, *Lyall.* **Lord Auckland's** group, immature, *J. D. H.* Very closely allied indeed to the Tasmanian *C. carnea.*

3. **C. (?) bifolia,** *Hook. f. Fl. N. Z.* i. 247. Glandular-pubescent, 3–5 in. high, rather stout. Leaves 2, radical, oblong, spreading, $\frac{3}{4}$–1$\frac{1}{4}$ in. long, pubescent or ciliated. Scape stout, 1-flowered, ebracteate. Flower erect, $\frac{3}{4}$–1 in. broad. Sepals obtuse; upper linear-oblong, obtuse, suberect; lateral and petals linear, nearly equal. Lip broad, almost sessile, orbicular-obovate, membranous, quite entire, with 2 short narrow lines of glands near the base.

Northern Island, *Colenso.* **Middle** Island: Lake Tennyson, alt. 4400 ft., *Travers;* Mount Brewster, alt. 4–5000 ft, *Haast;* grassy hills, Otago, *Lyall.* **Lord Auckland's**

group, *J. D. H.* A singular plant, differing from *Caladenia* in several respects, but hardly generically.

12. PTEROSTYLIS, Br.

Usually slender, leafy, erect, simple, rather succulent, glabrous herbs. Roots of small tubers terminating underground caulicles. Leaves radical or cauline, sheathing at the base. Flowers 1 in the N. Zealand species, membranous, green.—Perianth closed or open. Upper sepal and petals combined or conniving and forming a very concave boat-shaped hood; lateral sepals connate, ascending or deflexed. Lip small, clawed, its tip exserted between the free tips of the lateral sepals, or wholly exposed, often irritable, furnished with an erect or curved appendage at the base. Column slender, curved, broadly winged above. Anther terminal, persistent. Pollen-masses 4.

A large temperate Australian, Tasmanian, and New Zealand genus, not found elsewhere.

A. *Lateral sepals erect. Lip glabrous, included, or the tip only exserted.*
Leaves all cauline, or cauline and radical, all similar or nearly so.
Flower 2–3 in. long. Leaves all linear, grass-like . . . . . 1. *P. Banksii.*
Like *P. Banksii*, but smaller. Flower $\frac{3}{4}$–1 in. long . . . . . 2. *P. graminea.*
Lower leaves oblong. Flower 1 in. long . . . . 3. *P. micromega.*
Radical leaves broad, cauline reduced to bracts.
Glabrous. Leaves oblong. Bracts large, sheathing . . . . 4. *P foliata.*
Glabrous. Leaves long-petioled, ovate-cordate . . . . . . 5. *P. trullifolia.*
Puberulous. Leaves short-petioled, ovate-cordate . . . . . 6. *P. puberula.*
B. *Lateral sepals deflexed. Lip filiform, plumose, exserted* . . . 7. *P. squamata.*

1. **P. Banksii,** *Brown; — Fl. N. Z.* i. 248. Tall, leafy, 6–18 in. high. Leaves numerous, alternate, sheathing the whole stem, rising above the flower, narrow linear-lanceolate, acuminate. Flower solitary, 2–3 in. long. Upper sepal arched forwards, and lateral produced into long slender tails. Lip linear, glabrous, its tip exserted; appendage linear, curved, villous at the tip.—Bot. Mag. t. 3172.

Var. β. Leaves broader, $\frac{1}{2}$–$\frac{2}{3}$ in. Sepals less produced into long tails. *P. australis*, Hook. f. Fl. N. Z. i. 248.

Abundant throughout the **Northern** and **Middle** Islands to Otago, *Banks and Solander*, etc. Var. β, **Middle** and **Southern** Islands: Port William and Thomson's Sound, *Lyall.*

2. **P. graminea,** *Hook. f. Fl. N. Z.* i. 248. Slender, leafy, 6–10 in. high. Leaves sheathing, narrow linear-lanceolate, 1–3 in. long. Flowers solitary, $\frac{3}{4}$–1 in. long, of the same form as *P. Banksii*, but much smaller, and the sepals less produced into slender tails.

Northern and **Middle** Islands, not uncommon: east coast, *Colenso*; Auckland, *Sinclair*; Otago, *Lyall.* Probably only a small state of *P. Banksii*, but approaching the Australian *P. præcox.*

3. **P. micromega,** *Hook. f. Fl. N. Z.* i. 248. Slender, 3–8 in. high, leafy. Lower leaves large, ovate-oblong or lanceolate, obtuse or acute, sessile or petioled, $\frac{1}{2}$–$1\frac{1}{2}$ in. long; cauline smaller, more acuminate, $\frac{1}{2}$–$\frac{3}{4}$ in. Flower solitary, suberect, 1 in. long. Upper sepal narrow, slightly arched, caudate-acuminate; lateral erect, with narrow slender points. Petals narrow, acuminate. Lip glabrous, its tip exserted; appendage villous.

Northern Island, *Edgerley*; east coast, and bogs near Wairarapa, *Colenso*; Coromandel gold-fields, *Jolliffe.* Approaching the Australian *P. cucullata*, but smaller.

4. **P. foliata,** *Hook. f. Fl. N Z.* i. 249. Rather stout. scapigerous,

2–10 in. high, quite glabrous. Leaves: radical petioled, oblong or elliptic-oblong, 1–2½ in. long, obtuse; cauline reduced to 1 or 2 large, erect, sheathing bracts, ¾–1¼ in. long. Flower solitary, erect, ½–¾ in. long. Upper sepal much curved towards the tip, acuminate; lateral erect, their points filiform, exceeding the upper. Petals as long as the upper sepal, obtuse or subacute. Lip glabrous, its tip a little exserted; tip of appendage villous.

Northern Island: east coast, bogs near Oroi; hillsides, Cape Palliser; Wairarapa valleys; and tops of the Ruahine mountains, *Colenso.*

5. **P. trullifolia,** *Hook. f. Fl. N. Z.* i. 249. Small, slender, glabrous, 2–5 in. high. Leaves radical, or towards the base of the scape, petioled, broadly ovate or orbicular-cordate or trowel-shaped, obtuse or acute, ¼–⅓ in. long; petiole slender. Bracts 3–5, spreading, lanceolate, acuminate, the lower sometimes petioled. Flower solitary, erect, ⅓ in. long. Upper sepal abruptly arched forward beyond the middle, acuminate; lateral erect, their points filiform, longer than the upper. Petals narrow, acuminate. Lip glabrous, its tip shortly exserted; tip of appendage villous.

Northern Island: Bay of Islands, *Edgerley, Colenso,* etc.

6. **P. puberula,** *Hook. f. Fl. N. Z.* i. 249. Small, slender, puberulous, especially below, 3–5 in. high. Leaves radical, crowded, small, short, shortly petioled, ¼–½ in. long, ovate-cordate, acute. Bracts numerous, erect, sheathing, lanceolate, acuminate. Flower erect, solitary, ½–¾ in. long. Upper sepal slightly arched, acute; lateral erect, thin; points filiform, as long as or exceeding the upper. Petals broad, as long as the upper sepal, tips broad, obtuse or truncate. Lip with the tip exserted; tip of appendage divided.

Northern Island, *Colenso;* Auckland, *Sinclair.* Very closely allied to the Tasmanian *P. nana.*

7. **P. squamata,** *Brown;—Fl. N. Z.* i. 249. Stout, glabrous, erect, 3–8 in. high. Leaves radical, crowded, sessile, erect, ⅕ in. long, ovate-lanceolate, acuminate; bracts 3–4, sheathing, acuminate. Flower solitary, ¾–1 in. long. Upper sepal slightly arched, acuminate; lateral deflexed, their tips linear, straight. Petals linear-subulate. Lip filiform, exserted, pendulous, plumose with long golden hairs, terminated by a large purple gland; appendage curved, villous at the tip. Wings of column each with a long erect tooth.

Northern Island: Auckland, *Sinclair.* Also a native of Tasmania and east and west temperate Australia.

13. CHILOGLOTTIS, Br.

Erect, rather stout, scapigerous, small herbs. Roots of small tubers at the end of underground caulicles. Leaves 2, at the base of the stem. Scape with 1 bract and 1 flower. Flowers suberect, lurid purple; peduncle lengthening after flowering.—Perianth 2-lipped. Upper sepal arched; lateral placed under the lip. Petals ascending or reflexed. Lip with a claw, with large glands or protuberances at the base and on the disk. Column long, 2-fid at the tip. Anther terminal, persistent. Pollen-masses 4.

A small south-east Australian and Tasmanian genus, with one Auckland Island species.

1. **C. cornuta,** *Hook. f. Fl. Antarct.* i. 69. A small herb, 2–4 in. high. Leaves 1–1½ in. long, linear-oblong, acute; veins reticulate. Scape

stout, lengthening much after flowering. Flower ½ in. diam. Upper sepal ovate-lanceolate, acuminate; lateral linear. Petals erect, short, ovate, acuminate. Lip trowel-shaped, shortly clawed, acuminate, with a horn-like projection at the base of the disk, one broader protuberance on each side of it, and 3 broad, flat tumid purple glands on the surface.

Lord Auckland's group, *Bolton*. **Campbell's** Island: mossy shady places, *Lyall*.

14. LYPERANTHUS, Br.

Erect, rather stout, leafy, glabrous, terrestrial herbs. Root tuberous. Leaves sheathing at the base; bracts large. Flowers spiked or racemed, lurid, rather fleshy.—Upper sepal large, arched, concave, boat-shaped; lateral small, deflexed. Petals similar to the lateral sepals, spreading or reflexed. Lip small, sessile, ascending, entire, oblong; disk with small ridges or glands. Column short; anther erect, persistent. Pollen-masses 4, granular.

A small Australian, Tasmanian, New Caledonian, and New Zealand genus.

1. **L. antarcticus,** *Hook. f. Fl. Antarct.* ii. 544. A span high, stout. Leaves 1–3, linear-oblong or oblong-lanceolate, 1–2 in. long, upper smaller. Flowers 2 or 3; bracts large, cucullate, ½–¾ in. long. Perianth horizontal, ⅓ in. long. Upper sepal very deeply arched, helmet-like, acute; lateral and petals linear-subulate, falcate, acute. Lip broad, ovate-oblong, obtuse, with 5 slender ridges on the disk. Column rather slender, arched.

Middle Island: open land, Waipori Creek, alt. 2500 ft., *Hector and Buchanan*. **Lord Auckland's** Islands, *Le Guillon, Bolton*.

15. THELYMITRA, Forst.

Erect, stout or slender, scapigerous, glabrous herbs. Roots of ovoid tubers. Stem with a membranous sheath below the leaf. Leaves 1 or rarely 2, narrow, much elongated, thick and coriaceous. Flowers few, spiked or racemed. —Perianth spreading. Sepals and petals all oblong ovate or obovate, equal and nearly similar. Lip like the petals, but rather smaller. Column hooded, 3-fid; lateral lobes (staminodia) erect or prominent, crenate or produced into feathery appendages; middle small, entire, notched or 3-fid Anther posticous, almost hidden between the lateral lobes of the column, attached to the middle one, persistent. Pollen-masses 4, fixed to the gland of the stigma, powdery.

A large Australian and Tasmanian genus, also found in New Zealand and the lofty mountains of Java. The species are most difficult of discrimination in a fresh state and impossible in a dried one; and the following all require revision. The only good characters I have been able to find are in the length and direction of the lateral lobes of the column.

Flowers blue or purple.

Column much longer than its short plumose appendages . . . . . 1. *T. longifolia.*
Column longer than its erect toothed appendages . . . . 2. *T. pulchella.*
Column as long as its erect 2-toothed appendages . . . . 3. *T. uniflora.*

Flowers yellowish, few, 1–3. Stems slender.

Column much shorter than its erect plumose appendages . . . . . 4. *T. Colensoi.*
Column as long as its crenate or fimbriate appendages . . . . . . 5. *T. imberbis.*

1. **T. longifolia,** *Forst. Char. Gen.*—*T. Forsteri*, Swartz;—*Fl. N. Z.* i. 243. Stout or slender, 8–16 in. high. Leaf variable, ⅙–1 in. broad, linear, very narrow and channelled or long linear-lanceolate, nerved and flat, coria-

ceous. Spike 2–10-flowered. Flowers excessively variable in size; sepals blue or purple, $\frac{1}{4}$–$\frac{3}{4}$ in. diam.; petals paler; both ovate-lanceolate, acute. Lip obovate. Column with rounded tip, the appendages excessively short, villous anteriorly.—*T. stenopetala*, Hook. f. Fl. Antarct. i. 69; *Serapias regularis*, Forst. Prodr.

Abundant throughout the **Northern** and **Middle** islands, *Banks and Solander*, etc. **Lord Auckland's** Island, *J. D. H.*, *Bolton*. A most variable plant in stature and robustness, the length and breadth of the bracts and leaves, and the number and size of the flowers. Apparently identical with the Tasmanian and Australian *T. nuda*, Br.

2. **T. pulchella,** *Hook. f. Fl. N. Z.* i. 244. A very handsome species, differing from *T. longifolia* in the usually broader sepals and petals, and the longer, erect, toothed appendages of the column. Flowers $\frac{3}{4}$–1 in. diam., fine blue-purple.

Northern Island, *Colenso*. **Middle** Island: Moutere hills, *Munro*; Otago, *Lyall*. Probably a state of *T. Forsteri*.

3. **T. uniflora,** *Hook. f. Fl. N. Z.* i. 70. Short, stout, 6–8 in. high. Leaf narrow-linear, fleshy, obtuse, channelled, curved, shorter than the scape. Flowers 1–3, pale blue, $\frac{1}{3}$–$\frac{1}{2}$ in. diam. Sepals and petals linear-oblong, acute or acuminate. Lip obovate-cuneate. Appendages of column erect, linear, 2-fid at the tip.

Middle Island: Milford Sound, *Lyall*. **Lord Auckland's** group, *Le Guillon*, etc.

4. **T. Colensoi,** *Hook. f.*—*T. pauciflora*, Fl. N. Z. i. 244, not Br. Very slender, 8–12 in. high. Leaf very narrow linear, flexuous. Flowers 1–3, yellowish, on slender pedicels, $\frac{1}{3}$ in. broad. Sepals and petals very narrow, linear-oblong, acute. Column very short; appendages very long, subulate, erect, plumose at the tip. Anther with a long point.

Northern Island, *Colenso*. This differs from all the other species, and from the Australian *T. pauciflora*, to which I had united it, in the very narrow sepals and petals, very short column, and very long erect appendages.

5. **T. imberbis,** *Hook. f. Fl. N. Z.* i. 244. Small, slender, 4–10 in. high. Leaf very narrow linear. Flowers 1 or 2, yellowish, about $\frac{1}{2}$ in. diam. Sepals and petals very broad, orbicular-ovate, obtuse. Column rather short; appendages curving upwards and forwards, with crenate fimbriate or slightly villous tips. Anther with a long broad point.

Northern Island, *Colenso*. Much better specimens of this and the preceding are wanted to establish their distinctness; this is very like the Tasmanian *T. carnea*, but the flowers are said to be yellow.

16. SPIRANTHES, Richard.

Terrestrial, erect, glabrous or puberulous herbs. Root of unbranched fibres or much divided tubers. Stem simple, leafy. Flowers spiked or racemed.—Perianth nearly closed. Lateral sepals usually saccate at the base; upper ascending. Petals similar but narrower. Lip inferior, shortly clawed, quite entire, concave, embracing the column at its base. Column short, without lateral appendages. Anther at the back of the column, persistent, stipitate. Pollen-masses 2, narrow, pyriform, powdery, fixed to the 2-fid rostellum above the stigma.

A considerable genus, found in various parts of the world.

1. **S. australis,** *Lindl.*;—*S. novæ-Zelandiæ*, Hook. Fl. N. Z. i. 242. A span high and upwards. Leaves 2 in. long, linear-lanceolate, acuminate narrowed into long petioles. Scape covered with loosely sheathing bracts. Spike spiral, 1–2 in. long. Flowers $\frac{1}{12}$–$\frac{1}{5}$ in. long, white; bracts ovate, acuminate, as long as the flowers. Sepals and petals narrow, ovate-oblong, obtuse. Lip oblong, disk thickened; margins crumpled and waved. Ovary glandular-pubescent.

Northern Island, *Colenso*. The lip is very variable in form, narrower in the New Zealand than Australian specimens. A very common Australian plant, also found in China, India, and Siberia, and perhaps not different from a European species. The lip of the New Zealand plant indeed quite agrees with that of the European *S. æstivalis*.

17. PRASOPHYLLUM, Br.

Terrestrial, glabrous herbs; roots of round tubers often coated with matted fibres. Stem with a membranous sheath below the leaf. Leaf solitary, radical, linear, fistular, rarely short or 0. Flowers small, spiked, with the lip above.—Sepals oblong or ovate, lateral free or combined, recurved. Petals nearly as large as sepals, unequal-sided. Lip superior, clawed, ascending, undivided, often with an adnate plate on the disk. Column erect, small, with lateral appendages (staminodia). Anther at the back of the column, persistent, obtuse or with a long mucro. Pollen-masses 2, powdery, fixed to the rostellum above the stigma.

A large Australian and Tasmanian genus, more rare in New Zealand. I have very insufficient specimens of all the species except *P. Colensoi.*

Leaf sheathing the scape halfway up. Perianth $\frac{1}{4}$ in. . . . . 1. *P. Colensoi.*
Leaf sheathing the scape to the top; blade very short or 0.
Perianth pointing forwards . . . . 2. *P. nudum.*
Perianth pointing downwards . . . . 3. *P. pumilum.*

1. **P. Colensoi,** *Hook. f. Fl. N. Z.* i. 241. Stout, erect, 4–10 in. high. Leaf exceeding the spike, its sheath seldom rising more than halfway up the scape. Spike 1–3 in. long, many-flowered; bract small, obtuse. Perianth horizontal, greenish yellow, $\frac{1}{4}$ in. diam., sweet-scented. Sepals ovate-oblong, acuminate, lateral free. Petals linear-oblong. Lip trowel-shaped, claw short, blade fleshy, subacute, thickened towards the tip. Column extremely short; staminodia broadly notched. Ovary obovoid, tumid.—Fl. Tasman. ii. 12. t. 112 A.

Northern and **Middle** Islands: abundant from Auckland southwards to Otago. I suspect that this is a variety of the sweet-scented Tasmanian *P. alpinum*, but the lip is shorter, more coriaceous, and thicker towards the tip.

2. **P. nudum,** *Hook. f.*;—*P. tunicatum*, Fl. N. Z. i. 242. Very slender, 8 in. high. Stem and leaf coated below with a fibrous lacerated sheath. Scape included throughout its length in the slender leaf-sheath. Blade of leaf very short, $\frac{1}{3}$ in. long. Spikes $\frac{3}{4}$–1 in. long; bracts minute, obtuse. Perianth horizontal, yellowish, $\frac{1}{12}$ in. long. Lateral sepals free, ovate-lanceolate, acuminate. Petals much smaller, same shape. Lip ovate-

lanceolate, with a flat, glandular, adnate plate; claw long. Column short; lateral lobes broad, acuminate, deeply 2- or 3-toothed. Anther apiculate. Ovary linear.

Northern Island: east coast, clay hills, Te Hawara, Port Nicholson, and Lake Taupo, *Colenso.* I have only three specimens. This entirely agrees with my Hobarton (Tasmanian) specimens described in 'Flora Tasmaniæ,' ii. 13, as *P. nudiscapum*, to which the name *nudum* should be transferred. The *P. nudum* of the same work (p. 14), differs according to Archer's specimens only in the rather longer teeth of the appendages of the column; but according to his figure (t. 113 C) in the more oblong, shorter, obtuse lip, with ciliate edges.

3. **P. pumilum,** *Hook. f. Fl. N. Z.* i. 242. Habit and leaf of *P. nudum*, but perianth bent downwards. Lateral sepals ovate, acuminate. Petals similar but smaller. Lip trowel-shaped, subacute, with a broad, glandular, adnate plate; claw rather long. Column short; its lateral lobes large, broad, obliquely truncate and toothed. Anther apiculate.

Northern Island, *Edgerley;* Auckland, *Sinclair.* Very similar to the Tasmanian *P. despectans*, H. f., if not a variety of it; but the perianth is shorter, its segments broader in proportion, and the lateral lobes of the column shorter.

18. ORTHOCERAS, Br.

Erect, glabrous, leafy, terrestrial herbs. Root of oblong tubers. Leaves narrow, filiform. Flowers rather large, racemed. Perianth closed.—Upper sepal suberect, hooded, fleshy, obtuse; lateral much longer, very narrow, almost filiform, quite erect. Petals minute, linear, 2-toothed at the tip. Lip inferior, small, 3-lobed, with small glands at its base, very shortly clawed. Column very short; lateral lobes subulate. Anther large, erect, persistent. Pollen-masses 2, powdery, attached to the rostellum above the stigma.

A genus of an Australian and a New Zealand species.

1. **O. Solandri,** *Lindl.;—Fl. N. Z.* i. 243. Stout, erect, 1–2 ft. high. Leaves filiform, with long sheathing bases, rarely linear. Raceme 1–6 in. long, many-flowered; bracts large, spathaceous, exceeding the ovary. Perianth $\frac{1}{3}$ in. long, greenish-yellow. Lateral sepals $\frac{1}{2}$ in. long.—*O. strictum*, A. Cunn. not Br.; *Diuris novæ-Zelandiæ*, A. Rich. Flor. 163. t. 25. f. 1.

Northern Island: abundant on clay-hills, etc., *Banks and Solander*, etc. **Middle** Island: Nelson, ascending to 4000 ft., *Bidwill.* Very near indeed to the Australian *O. strictum.*

ORDER II. IRIDEÆ.

Herbs with tuberous bulbous or creeping roots or rhizomes. Leaves mostly radical, alternately sheathing on opposite sides of the stem, linear, laterally flattened. Flowers usually large, racemed spiked or panicled, hermaphrodite.—Perianth superior, of 6 petaloid leaflets. Stamens 3, inserted at the base of the segments; anthers opening outwards. Ovary inferior, 3-celled; style simple, stigmas 3, various; ovules numerous. Capsule loculicidal, 3-valved. Seeds albuminous; embryo small.

A very large Order, especially in Europe and South Africa, less frequent elsewhere.

1. LIBERTIA, Sprengel.

Herbs. Flowers in subumbellate panicles.—Perianth spreading. Sepals linear-oblong, white or greenish. Petals larger, obovate, white. Stamens with filaments connate, or closely applied to the base of the style; anthers ovate, versatile. Stigmas filiform. Capsule obovoid or globose, coriaceous or membranous. Seeds angled and deeply pitted in the New Zealand species.

A small genus, also found in Australia, Tasmania, and Chili. The New Zealand species are very variable; perhaps two are confounded under *L. ixioides*, or perhaps it and *grandiflora* are but forms of one.

Leaves $\frac{1}{6}$–$\frac{1}{4}$ in. broad. Umbels panicled. Capsule $\frac{1}{4}$–$\frac{1}{2}$ in. . . . . . 1. *L. ixioides.*
Leaves $\frac{1}{3}$ in. broad. Umbels panicled. Capsule $\frac{1}{5}$–$\frac{2}{3}$ in. . . . . . 2. *L. grandiflora.*
Leaves $\frac{1}{16}$–$\frac{1}{10}$ in. broad. Umbels solitary. Capsule globose . . . 3. *L. micrantha.*

1. **L. ixioides,** *Sprengel;—Fl. N. Z.* i. 252. Variable in size, from 6 in. to 2 ft. Leaves rigid, narrow linear, acuminate, $\frac{1}{6}$–$\frac{1}{4}$ in. broad. Scape panicled above; branches alternate, arising from membranous spathes, bearing 2–10-flowered umbels; pedicels 1–2 in. long, with membranous bracts. Perianth white, 1 in. diam. or less; sepals oblong; petals much larger. Capsule $\frac{1}{4}$–$\frac{1}{2}$ in. long, yellow or brownish, oblong or narrow pyriform.—*Reichenbach*, Ic. Exot. t. 157.

Var *α*. Bracts all lanceolate. Capsule narrow, pyriform.
Var. *β*. Upper bracts ovate, acute. Capsule oblong.

Common throughout the **Northern** and **Middle** Islands, *Banks and Solander*, etc. Var. *β*. Canterbury, *Lyall*, *Travers*. Klatt ('Linnæa,' 31, 383) distinguishes this (erroneously called *vestioides*) by the margins of the leaves being scabrid, but in all my specimens of both varieties they are quite smooth, or scarcely perceptibly rough.

2. **L. grandiflora,** *Sweet*. Stout, 2–3 ft. high. Leaves $\frac{1}{3}$ in. broad. Scapes panicled above, branches alternate, sheaths, etc. as in *L. ixioides*, but petals usually larger in proportion to the sepals, and capsule very much larger, $\frac{1}{5}$–$\frac{2}{3}$ in. long, very turgid, broadly obovoid.—*L. macrocarpa*, Klatt, Linnæa, 31. 385; *Renealmia grandiflora*, Br.

Northern Island, *Banks and Solander;* Auckland, *Herb. Sinclair*. **Middle** Island: Port Cooper, *Lyall*. In the Fl. N. Z. I erroneously regarded this as a variety of *L. ixioides*.

3. **L. micrantha,** *A. Cunn.;—Fl. N. Z.* i. 252. Much smaller than the above, 4–6 in. high. Leaves rather membranous, $\frac{1}{16}$–$\frac{1}{10}$ in. wide. Scape as long as the leaves, pubescent above, bearing a solitary 3–8-flowered umbel. Perianth $\frac{1}{3}$–$\frac{1}{2}$ in. diam., its leaflets nearly equal; pedicels pubescent, surrounded by a many-leaved involucre. Capsule globose, membranous.

Northern and **Middle** Islands: common in damp, especially mountain woods, *Banks and Solander*, etc., ascending to 4000 ft. in the Nelson mountains.

ORDER III. HYPOXIDEÆ.

Herbs, usually perennial, with tuberous, bulbous, or tufted roots. Leaves parallel-veined, mostly all radical. Inflorescence various. Flowers hermaphrodite.—Perianth superior, of 6 equal leaflets. Stamens 6; anthers opening inwards. Ovary inferior, 3-celled; style simple, stigmas usually 3;

ovules numerous. Fruit a capsule or berry. Seeds with copious albumen and nearly straight embryo.

A large, chiefly subtropical family, most abundant in South Africa.

1. HYPOXIS, Linn.

Small herbs. Root bulbous or tuberous, covered with matted fibres. Leaves all radical, enclosed in a membranous sheath at the base. Scapes 1- or many-flowered. Flowers yellow.—Perianth leaflets nearly equal, yellow. Stamens 3; anthers erect. Stigmas 3. Capsule 3-celled. Seeds with a short beak at the hilum.

A large South African genus, also found in India, Australia, and America.

1. **H. pusilla,** *Hook. f. Fl. Tasman.* ii. 36. *t.* 130 B;—*H. hygrometrica*, Br.? Fl. N. Z. i. 253. Very small, 1–2 in. high. Leaves narrow linear, nearly glabrous. Scape shorter than the leaves, 1-flowered. Perianth glabrous, $\frac{1}{6}$ in. diam.; leaflets ovate-lanceolate. Ovary obovate, narrowed below.

Northern Island: east coast, Ahuriri, Te Hautotara, etc., *Colenso*. **Middle** Island, Banks's Peninsula, *Travers*. Also found in Tasmania.

Order IV PANDANEÆ.

Erect or climbing shrubs or trees, rarely stemless, often branched. Leaves usually long, narrow, and spinous-serrate. Spikes or heads often clustered, peduncled.—Flowers unisexual. Perianth 0, or rarely of 3 or 4 valvate leaflets. *Male* fl.: stamens numerous; filaments filiform; anthers inserted by their base, linear, 2–4-celled. *Female* fl.: ovaries 1-celled, usually numerous and variously connate; stigma sessile; ovules solitary or numerous, on parietal placentas. Fruit a drupe or berry. Seeds oblong; albumen fleshy or horny; embryo minute.

A rather large, tropical Order, containing many arborescent species.

1. FREYCINETIA, Gaudichaud.

Climbing shrubs, branched, leafy at the ends of the branches. Leaves sheathing at the base, narrow linear-subulate, margins prickly. Spikes fascicled, terminal, surrounded by bracts at the base of the peduncles. *Male* fl.: bundles of stamens surrounding a rudimentary ovary. *Female* fl.: several ovaries, variously combined, surrounded by imperfect stamens. Seeds numerous.

An Asiatic tropical genus of no great extent.

1. **F. Banksii,** *A. Cunn.;—Fl. N. Z.* i. 237. *t.* 54 *and* 55. A lofty climber. Leaves 2 ft. long, serrulate, concave, tip 3-gonous. Spikes cylindrical, 3–4 in. long, surrounded by white, fleshy bracts. Anthers 2-celled. Fruit an oblong, green spadix, consisting of a multitude of laterally compressed carpels, about $\frac{1}{3}$ in. long, clavate, truncate; the lower part soft, hollow, the walls of its cavity densely covered with pendulous seeds; upper part thickened, very hard, solid; the truncate tip crenate, the crenatures stigmatiferous. Seeds small, linear-oblong, with a cellular testa.

Northern Island: as far south as the east coast, *Banks and Solander*. The bracts and young spikes make a very sweet preserve; the leaves are used for basket-making.

(AROIDEÆ.)

CALADIUM *esculentum*, Vent. (*Colocasia esculenta*, Schott; *Arum esculentum*, Linn.), is the Tarro or Tullo of the natives, introduced by them into the islands, and still abundantly cultivated (Forst. Pl. Esc. p. 58); it is a staple article of food in many parts of the Old World.

ORDER V. TYPHACEÆ.

Erect, leafy, marsh or water-plants. Leaves linear, sessile, sheathing at the base. Flowers unisexual, rarely hermaphrodite, in dense heads or catkins. Perianth 0, or of irregular scales or hairs. Stamens densely crowded; filaments slender; anthers linear or ovoid, inserted by their bases. Carpels densely crowded, narrow, 1-celled, tapering into a slender style, with a narrow unilateral stigma; ovule solitary, pendulous. Fruit, small nuts or utricles. Seed pendulous; embryo straight, in copious albumen.

The following genera are the only ones known.

Flowers in long, dense, cottony, brown cylindric spikes . . . . . . 1. TYPHA.
Flowers in globular heads . . . . 2. SPARGANIUM.

1. TYPHA, Linn.

Tall, erect, marsh or water herbs; rootstock stout, creeping. Leaves all radical, long, linear, thick, flat, sheathing at the base.—Flowers in 1 or 2 long, dense, cylindric, terminal superimposed catkins; upper, or upper part if one only, male, lower female. Stamens mixed with hairs. Ovaries very minute, narrow, surrounded by hairs, which are thickened upwards, and form a copious brown, cottony mass. Nuts minute, slender, on slender stalks.

A very common tropical and temperate genus, of very few species, called Reed-maces and often Bulrushes.

1. **T. angustifolia,** *Linn.;—Fl. N. Z.* i. 238. Leaves 2–3 ft. long, $\frac{1}{3}$–$\frac{1}{2}$ broad. Scapes 4–8 ft. high, terete, solid, bearing 2 cylindric, brown catkins at the top, of which the upper or terminal is male, the lower female.—*T. latifolia*, Forster, not Linn.

Northern Island: in marshes and river banks, *Forster*, etc. Common in all parts of the world, including the Pacific Islands, Norfolk Island, and Australia. I refer this to *T. angustifolia* because the catkins, male and female, are separated by an interval, but the stature is that of *T. latifolia*, and I find the position of the catkins to vary extremely in tropical specimens of both. They are doubtless varieties of one plant. The present is extensively used for making walls and roofs of houses. The pollen is made into loaves of bread by the natives, as in Scinde.

2. SPARGANIUM, Linn.

Roots fibrous. Stems erect, simple or branched. Leaves alternate, sheathing at the base, linear, grass-like, erect or floating. Flowering-branches terminal, simple or branched.—Flowers in sessile, globose, spiked or panicled heads, the lower with leafy bracts at their base. Upper heads male. Stamens mixed with minute scales. Lower heads larger, female. Ovaries sessile, each surrounded by a perianth of 3–6, small linear scales. Nuts ovoid or obovoid, thick.

A small aquatic genus, found in all temperate climates.

1. **S. simplex,** *Huds.;—Fl. N. Z.* i. 238. An erect, rather flaccid herb, 1–2 ft. high. Leaves a foot long, $\frac{1}{8}$ in. broad, acute, channelled. Scape slender, erect, not branched. Nuts broadly obovoid, $\frac{1}{8}$ in. long, suddenly contracted into a slender style.

Northern Island: in watery places frequent, *Bidwill*, etc. "Simple Bur-reed" of England, a common plant in various parts of the New and Old World, including Australia, but not found in Tasmania.

ORDER VI. NAIADEÆ.

Floating, submerged, or erect marsh plants (in *Lemna* reduced to floating green scale-like fronds). Leaves alternate, linear or oblong, sheathing at the base. Stipules usually interpetiolar, membranous, sheathing. Inflorescence various, usually spiked.—Flowers minute, green, inconspicuous, unisexual or hermaphrodite. Perianth inferior, regular, of 4–6 green pieces in 2 series, or 0, or reduced to scales, or to a tubular sheath. Stamens 1, 4 or 6; filaments usually short; anthers 2- or 4-celled, inserted by their base, dehiscing longitudinally; pollen often of confervoid threads. Ovaries free, 1–6; styles short or long, stigmatiferous on the inner face; ovule 1, usually pendulous in each cell. Fruit of 1–6 indehiscent nuts, rarely drupes or capsules. Seed erect or pendulous, albumen 0; testa membranous; embryo with a very large radicle, usually curved or inflexed; plumule in a slit on one side of it.

An extensive family of water plants, scattered over the whole globe.

Small scale-like floating fronds . . . . 1. LEMNA.
Simple or branched marsh or water plants. Leaves linear or oblong. Flowers spiked.
Perianth 6-leaved. Stamens 6; anthers subsessile . . . . . 2. TRIGLOCHIN.
Perianth 4-leaved. Stamens 4; anthers subsessile . . . . . 3. POTAMOGETON.
Perianth 0. Stamens 4; anthers sessile. Ovaries 4. Nuts on long stalks . . . . 4. RUPPIA.
Flowers axillary. *Male* fl.: a long stamen. *Female:* 2–5 ovaries in a sheathing perianth . . . . 5. ZANNICHELLIA.

1. LEMNA, Linn.

Minute green scale-like fronds, covering the surface of ponds and ditches. Roots thread-like, suspended from the frond. Flowers most minute, appearing by threes in a cleft of the frond, surrounded by a thin membranous spathe.—Two of the flowers are males and consist of a single stamen each, with slender filament and 2-celled anther; the other is a female, and is a 1-celled 1- or 2-ovuled ovary, with a funnel-shaped stigma. Fruit a most minute 1–4-seeded utricle. Seed erect or horizontal; testa ribbed, membranous; albumen 0; cotyledon thick, farinaceous, including the radicle and plumule.

A genus of few species, scattered over the globe, rarely found in flower. "Duckweed."

Frond oblong or oval, flat below . . . . 1. *L. minor.*
Frond oblong or oval, convex below . . . . 2. *L. gibba.*

1. **L. minor,** *Linn.*;—*Fl. N. Z.* i. 239. Fronds ovate or oblong, green, about $\frac{1}{12}$–$\frac{1}{10}$ in. long, with a single capillary root. Flower not seen.

Middle Island: Port Cooper, *Lyall.* Probably abundant elsewhere (as throughout the world), but overlooked.

2. **L. gibba,** *Linn.*;—*Fl. N. Z.* i. 239. Like *L. minor*, but frond convex below. Flower not seen.

Northern Island: east coast, *Colenso.* Also a common plant in various parts of the world, but not so frequent as *L. minor.* Some of the Tasmanian specimens referred to *L. minor* in Fl. Tasm. are referable to this. Fronds of New Zealand and Tasmanian specimens are of the same size as those of *L. minor*, and much smaller than of English specimens.

2. TRIGLOCHIN, Linn.

Marsh herbs. Leaves all radical, filiform or rush-like, sheathing at the base. Scape naked, slender, bearing a terminal spike or raceme of small green flowers.—Flowers hermaphrodite, green, small, ebracteate. Perianth of 6, green, concave pieces in 2 series. Stamens 6; filaments very short; anthers bursting outwards. Ovaries 3 or 6, partially cohering; stigma sessile; ovule solitary, erect. Fruit of 3 carpels (or more, the alternate not ripening), separating from a central axis. Embryo straight, radicle inferior.

A widely diffused, but not large genus, in the temperate zones of both hemispheres.

1. **T. triandrum,** *Michaux*;—*Fl. N. Z.* i. 236. A very slender, glabrous herb, 3–10 in. high. Leaves filiform, grass-like, semiterete, as long as or longer than the scape. Flowers minute, green, pedicelled. Fruit globose, of 3 keeled carpels, alternating with as many or fewer imperfect ones. Stigmas recurved.—*T. flaccidum*, A. Cunn.; *T. filifolium*, Hook. Ic. Pl. t. 579.

Northern and **Middle** Islands: common in marshes, especially near the sea, *Banks and Solander*, etc. Also found in Australia, South Africa, and temperate North and South America.

3. POTAMOGETON, Linn.

Aquatic plants. Stems creeping, jointed and rooting. Leaves 2-ranked, usually alternate, all similar, or the lower narrow submerged, the upper broader floating; stipules membranous, free or united and sheathing. Spikes on axillary peduncles.—Flowers hermaphrodite, green, small, ebracteate. Perianth of 4 concave pieces, valvate in bud. Stamens 4; filaments very short. Ovaries 4; stigmas sessile or subsessile; ovule solitary, ascending. Fruit of 4 small drupe-like nuts, often compressed. Embryo curved.

A large and very abundant genus in all parts of the world, the species of which are very variable. Much better specimens of the New Zealand species are wanted.

Floating leaves oblong or lanceolate-oblong . . . . 1. *P. natans.*
Leaves, lower linear, grass-like, upper floating, oblong, lanceolate, acute 2. *P. heterophyllus.*
Leaves all narrow linear and grass-like. Stipules free, lacerate . . 3. *P. gramineus.*
Leaves all very narrow linear. Stipules united with leaf-margins . 4. *P. pectinatus.*

1. **P. natans,** *Linn.*;—*Fl. N. Z.* i. 236. Stems and branches long or short according to the depth of the water. Leaves on long petioles, floating, 1–3 in. long, oblong, obtuse or acute, often subcordate at the base, reddish-brown; submerged (if present) linear; stipules not winged, without ribs. Scape

not thickened upwards. Nut when fresh obtuse at the back, or obscurely 3-ribbed.

Northern and **Middle** Islands: abundant in ponds, rivers, etc. A most common aquatic herb in Europe and elsewhere throughout the globe.

2. **P. heterophyllus,** *Schreber?* Lower submerged leaves narrow-linear, membranous, 3–4 in. long, $\frac{1}{6}$–$\frac{1}{4}$ in. broad, margins waved or rather crisped, upper submerged leaves broader, passing into the floating, which are petioled, 1–1$\frac{1}{2}$ in. long, oblong, obtuse, coriaceous; stipules free.

Middle Island: Clethra river, Otago, *Lindsay*. Without flower or fruit; apparently the same as the Tasmanian plant, which is common in various temperate parts of the globe.

3. **P. gramineus,** *Linn.?—P. ochreatus,* Raoul, Choix, t. 7;—Fl. N. Z. i. 236, and ii. 336. Leaves all submerged, membranous, grassy, 1–3 in. long, $\frac{1}{6}$–$\frac{1}{4}$ in. broad, narrow-linear, 3–5-nerved; stipules free, lacerate. Peduncle very stout, 1 in. long. Spike oblong, continuous. Nuts compressed, obliquely broadly ovoid, with a mucronate style, $\frac{1}{10}$ in. diam.

Northern and **Middle** Islands: Akaroa and Bay of Islands, *Raoul, Colenso;* Auckland, *Sinclair*. A very common European, Australian, and Tasmanian plant; also found in many other temperate countries. In the N. Z. Flora I referred this to *P. compressus*, Linn., according to Fries' specimens: but Fries' plant is apparently not the true *compressus*, and considered by Prof. Oliver to be more probably a form of *pusillus*. My specimens from M. Raoul are ticketed as from the Bay of Islands, but in his 'Choix' he states it to be a native of Banks's Peninsula.

4. **P. pectinatus,** *Linn.* Leaves all submerged, membranous, grassy, very narrow-linear, 2–3 in. long, $\frac{1}{20}$ in. broad; stipules united with the base of the leaf into an entire sheath. Peduncle very long, slender. Flowers in distant fascicles. Nuts gibbous, compressed, obliquely very broadly ovoid, with a very short style.

Northern Island: lagoon at Tangloio and Hawke's Bay, *Colenso*.

4. RUPPIA, Linn.

A slender brackish-water plant. Stems extremely slender. Leaves filiform with sheathing bases. Flower-spikes first included in the leaf-sheaths, then lengthening, often twisted spirally, and ascending to the surface.—Flowers hermaphrodite, small, green, ebracteate, remote on a slender spadix, the stamens and pistils sometimes so wide apart that the flowers appear to be monœcious. Perianth 0. Stamens 4; anthers sessile, 1-celled, or 2 with separate anther-cells. Pollen a long curved cell. Carpels 4, sessile, with sessile stigmas and solitary suspended ovules. Fruit of 1–4 minute, long stipitate, obliquely ovoid, obtuse or pointed nuts. Embryo hooked.

An abundant plant in Europe and many parts of the world.

1. **R. maritima,** *Linn.;—Fl. N. Z.* i. 236.

Northern and **Middle** Islands: abundant in salt-water ditches, etc., as far south as Otago. A very frequent plant throughout the northern and southern hemispheres; abundant in Australia and Tasmania.

5. ZANNICHELLIA, Linn.

Slender, fresh- or brackish-water herbs. Stems forked, capillary. Leaves opposite or alternate, very slender, with sheathing membranous stipules.—Flowers unisexual, very minute, axillary, sessile, naked, usually in pairs (male and female). *Male*, a solitary stamen; filament very slender; anther 2–4-celled. *Female*, 2–5 carpels in a cup-shaped involucre; stigma peltate; ovule solitary, suspended. Fruit of usually 4 curved, stalked, narrow, beaked nuts. Embryo bent upon itself; cotyledon slender.

A genus of few species, scattered over the globe.

1. **Z. palustris,** *Linn.*;—*Fl. N. Z.* i. 237. A very slender aquatic, forming tangled masses in fresh or brackish water. Stems capillary, 3–12 in. long, branched. Leaves almost filiform, 2–3 in. long. Nuts shortly stalked, $\frac{1}{10}$ in. long, with filiform styles as long as themselves.

Northern Island: east coast, *Colenso*, in fresh water. Abundant in Europe and Australia, and generally in temperate as well as tropical countries.

Some other both fresh and salt-water plants of this Order may be found in New Zealand, as *Cymodocea* and *Posidonia*, both marine, and common in Australia.

ORDER VII. LILIACEÆ.

(*Including* SMILACEÆ, ASPHODELEÆ, *and* ASTELIEÆ.)

Herbs, rarely shrubs or trees, of very various habit, sometimes (*Rhipogonum*) climbing.—Flowers usually hermaphrodite and regular. Perianth inferior, of 6 coloured leaflets in 2 series, free or combined into a tube. Stamens 6, attached to the perianth or free from it; anthers bursting inwards, attached by a point. Ovary superior, 3-celled (1-celled in some *Asteliæ*); style simple, stigma 3-lobed; ovules 2 or more (rarely 1) in each cell. Fruit a membranous or coriaceous, 3-celled, loculicidally 3-valved capsule, or an indehiscent 1–3-celled berry. Seeds few or numerous, globose or flattened; testa crustaceous or membranous; albumen horny; embryo terete, straight or curved.

A very large Order, found in all parts of the world but the coldest.

A. *Fruit a berry. Seeds turgid; testa membranous. Leaf-veins netted* (Smilaceæ).

Climbing shrub. Ovary-cells 1-ovuled . . . . 1. RHIPOGONUM.
Small herb. Ovary-cells few-ovuled . . . . 2. CALLIXENE.

B. *Fruit a berry. Seeds globose or oblong; testa crustaceous.*

Glabrous. Stem often arboreous. Filaments subulate . . . . 3. CORDYLINE.
Glabrous herbs. Filaments curved, thickened upwards . . . 4. DIANELLA.
Herbs with villous silky or chaffy leaves. Flowers diœcious . . . 5. ASTELIA.

C. *Fruit a 3-celled, many-seeded capsule. Seeds flattened or angular; testa usually black.*

Flowers white, panicled, spreading; pedicels jointed in middle . . 6. ARTHROPODIUM.
Flowers racemed, yellow, spreading; pedicels not jointed . . . . 7. ANTHERICUM.
Flowers panicled, green and orange, tubular, irregular . . . . . 8. PHORMIUM.
Flowers solitary in spathes. Small herb . . . . 9. HERPOLIRION.

1. RHIPOGONUM, Forst.

Climbing wiry shrubs. Leaves opposite and alternate, petioled, 3-nerved, with netted venation. Flowers in spreading axillary and terminal racemes, hermaphrodite.—Perianth of 6 minute spreading leaflets. Stamens larger than the petals; filaments very short. Ovary 3-celled; style short; ovules 1 in each cell. Berry 1- or 2-seeded. Seeds hemispherical; testa membranous.

Only three species are known, the following and two Australian.

1. **R. scandens,** *Forst.*;—*Fl. N. Z.* i. 253. Stems very slender, knotted, forming interwoven wiry masses in the forest. Leaves 3–5 in. long, linear ovate or oblong, subacute, coriaceous. Racemes simple or branched, 1–2 in. long. Flowers ¼ in. diam. Berry scarlet.—*R. parviflorum*, Br.; *Smilax Rhipogonum*, Forst. Prodr.

Northern and **Middle** Islands: abundant as far south as Otago, *Banks and Solander*, etc. The long underground rootstocks have been used as Sarsaparilla by settlers, and the stems as cord and for basket-work by the natives.

2. CALLIXENE, Commerson.

Matted glabrous herbs, with wiry branching creeping stems, knotted at the joints, with membranous scales at the joints. Leaves alternate, distichous, with netted veins.—Flowers solitary, nodding, hermaphrodite. Perianth of 6 white spreading leaflets, with 2 glands at the base of 3 or all. Ovary 3-celled; style stout, 3-grooved; ovules few in each cell. Berry few-seeded; seeds globose, with a membranous testa.

A genus of 2 species, one a native of Fuegia and S. Chili, the other of New Zealand.

1. **C. parviflora,** *Hook. f. Fl. N. Z.* i. 254. Stems wiry, 8–10 in. long. Leaves ¾ in. long, with short twisted petioles, linear or oblong, acute or obtuse, 3–5-nerved. Flower ¾ in. broad, peduncle usually terminal. Perianth segments ovate-lanceolate, acute. Filaments glabrous. Berry globose.—Hook. Ic. Pl. t. 632.

Northern Island: mountainous districts, *Colenso*. **Middle** Island: in damp forests, ascending to 4000 ft., common.

3. CORDYLINE, Commerson.

Trees or rarely stemless herbs. Leaves alternate, very long, with broad bases, distichous, or crowded at the ends of the trunk or branches, nerves obliquely leaving the midrib, parallel. Flowers white, on much branched panicles, hermaphrodite.—Perianth campanulate, of 6 nearly equal persistent leaflets, more or less combined at the base. Stamens inserted on the petals, filaments subulate, glabrous. Ovary 3-celled; style simple, stigma 3-lobed; ovules numerous in each cell. Berry globose, 3-celled, few- or many-seeded. Seeds angular; testa black, crustaceous. Pedicels 3-bracteate; first and third bracts 1-nerved, intermediate one with 2 nerves.

A large genus in the southern hemisphere, most frequent in New Zealand, Australia, and the Pacific Islands.

Trunk arboreous.

Leaves 2 ft. × 1¼–1⅓ in., not petioled, striate. Bracts ½ as long as flower 1. *C. australis.*

Leaves 5–6 ft. × 1½–2 in., long petioled, nerved. Bracts extremely short 2. *C. Banksii.*

Leaves 2–3 ft. × 4–5 in., very thick. Bracts almost as long as flower 3. *C. indivisa.*

Stem 0, or slender and very short. Leaves 1–2 ft × ¼–⅓ in. . . . . 4. *C. Pumilio.*

1. **C. australis,** *Hook. f. Fl. N. Z.* i. 257; *Gard. Chron.* 1860, 792;

not Endlicher. Trunk arboreous, 10–40 ft. high, branched. Leaves ensiform, 2 ft. long, $1\frac{1}{4}$–$1\frac{1}{2}$ in. broad, slightly contracted above the broad base, striated with numerous parallel veins, all equally fine, midrib obscure. Flowers densely crowded, white, $\frac{1}{3}$ in. diam., sweet-scented; bracts from half as long to as long as the bud before expansion.—*Dracæna australis*, Forst. Prod. 151.

Northern and **Middle** Islands, *Banks and Solander*, etc, apparently common throughout. There are no specimens of this in the Banksian or Paris Herbariums from Forster himself, but his drawing identifies the plant; it is distinguished from *C. Banksii* by the shorter leaves not contracted at the base, without strong nerves, and by the large bracts. The *Dracæna australis* of Bot. Mag. has the short flowers of this species, but the minute bracts of the following, but differs from both in the much broader leaves. I find the same plant from Norfolk Island, whence I suspect Frazer procured it, and not from New Zealand.

2. **C. Banksii,** *Hook. f. in Gard. Chron.* 1860, 792. Trunk arboreous, 5–10 ft. high, simple or sparingly branched. Leaves very long, 5–6 ft. long by $1\frac{1}{2}$–2 in. broad, linear-lanceolate, gradually contracted into a petiole 1–2 ft. long, striated, and having 6 or 8 evident nerves on either side of the prominent midrib. Panicle lateral, drooping, 2–5 ft. long. Flowers white, longer than in *C. australis*, $\frac{1}{2}$ in. long, nearly sessile. Bracts small, not $\frac{1}{8}$ part as long as the flowers previous to expansion. Berry white.

Northern and **Middle** Islands: Port Nicholson and Ruahine mountains, *Colenso;* Canterbury, *Travers*, etc. There is a small specimen of this in the British Museum, collected by Banks and Solander, but not alluded to in their MSS. nor drawings; it differs from *C. australis* in the shorter trunk, much longer, narrower, petioled leaves, with evident strong veins, the minute bracts, and longer in the flowers, and (according to Colenso) in the drooping lateral panicle.

3. **C. indivisa,** *Kunth;—Fl. N. Z.* i. 258. Trunk simple, arboreous, 2–5 ft. high. Leaves excessively thick and coriaceous, 2–3 ft. long, 4–5 in. broad, scarcely contracted at the base, glaucous below; nerves numerous, yellowish, strong, midrib stout. Panicle very large, drooping. Flowers pedicelled, $\frac{3}{4}$–1 in. diam., most densely crowded, imbricate. Bracts nearly as long as the flower and pedicel.—*Dracæna indivisa*, Forst. Prod. n. 150; Plant. Esc. n. 33.

Northern Island: Ruahine range, *Colenso*. **Middle** Island: Thompson's Sound, *Lyall;* Dusky Bay, *Forster*. A magnificent plant, well distinguished by the broad, excessively thick leaves, glaucous below, and huge drooping panicle, covered with the large flowers. Flax from its leaves is used by the natives of the Northern Island to make a garment called Toii (*Colenso*).

4. **C. Pumilio,** *Hook. f. in Gard. Chron.* 1860, 792.—*C. stricta*, Fl. N. Z. i. 257. t. 58, not Endlicher. A small plant. Stem 0 or short, slender, not thicker than the thumb. Leaves 1–2 ft. long, $\frac{1}{4}$–$\frac{1}{3}$ in. broad, linear, grass-like, nerves few, midrib stout, prominent. Panicle erect, very lax, 2 ft. long, branches spreading, slender. Flowers few, scattered, pedicelled, white, $\frac{1}{6}$ in. diam.; bracts subulate, half as long as the flower and twice as long as the bracts.

Northern parts of the **Northern** Island: Bay of Islands, *A. Cunningham*. A very different-looking species from all the preceding. The *C. stricta* of Endlicher (*Dracæna stricta*, Bot. Mag. t. 2575; Bot. Reg. t. 965), is an arboreous species, with bluish flowers. The present was regarded by A. Cunningham as a young plant of *C. australis*.

4. DIANELLA, Lamarck.

Herbs with short tufted rhizomes. Leaves very long, linear, rigid, equitant or sheathing. Scapes bearing compound panicles. Flowers jointed on the rigid, curved pedicels, white or blue, drooping, hermaphrodite.—Perianth of 6 nearly equal, spreading, deciduous leaflets. Stamens 6; filaments incurved, thickened upward. Ovary 3-celled; style filiform, stigma simple; ovules numerous in each cell. Berry globose or oblong, many-seeded. Seeds globose, with metallic crustaceous testa.

A genus of no great number of species, natives of Madagascar, the South Sea Islands, India, Australia, and New Zealand.

1. **D. intermedia,** *Endl.;—Fl. N. Z.* i. 255. Rootstock woody, with underground runners. Leaves 1–5 ft. long, narrow linear-ensiform, rigid, rough on the edges. Panicle 10–18 in. long, much branched; peduncles and pedicels very slender, the latter curved. Flowers small, greenish-white, nodding, $\frac{1}{3}$ in. diam. Anthers linear. Berry $\frac{1}{3}$–$\frac{1}{2}$ in. long, dark blue.

Northern and **Middle** Islands: frequent in fern lands and in woods, *Banks and Solander*, etc. Also a native of Norfolk Island.

5. ASTELIA, Banks and Solander.

Large or small tufted herbs, usually more or less clothed with silky, shaggy, or chaffy hairs. Leaves all radical, long, narrowed from base to tip, sheathing the scape. Flowers numerous (rarely few) in branched racemes or panicles, white greenish or purplish, diœcious.—Perianth coloured, petaloid or rather dry in texture, campanulate or rotate, silky, 6-partite. Stamens 6. Ovary 3-gonous, 1 or 3-celled; style short or 0, stigma 3-lobed; ovules few or numerous, pendulous from the top of the cells, or attached to central or parietal placentas. Berry 1–3-celled, few- or many-seeded. Seeds oblong, angled or terete; testa black, brittle; albumen fleshy; embryo small.

A small genus, confined to the alps of Australia and Tasmania, New Zealand, the Pacific Islands, and Antarctic America. The New Zealand species are all diœcious, and, from growing usually on lofty forest trees, it is most difficult to match the sexes. I am not satisfied with the following determination of the species, which is that proposed in Fl. N. Z., since the publication of which no one has attempted to clear up their limits, as there urgently desired.

Perianth not enclosing the fruit. Berry 1-celled. Seeds terete.

Leaves 2–5 ft., silky and shaggy. Berry globose . . . . . . 1. *A. Cunninghamii.*
Leaves 1–6 in., glabrous or scaly. Berry ovoid . . . . 2. *A. linearis.*

Perianth usually enclosing the ripe fruit. Berry 3-celled. Seeds angled.

Leaves 1–3 ft. × $\frac{1}{2}$–$\frac{3}{4}$ in., strong, glabrous or silky; nerves 3 . . 3. *A. nervosa.*
Leaves 2–4 ft. × 3 in.; nerves 3. Filaments $\frac{1}{2}$ in. long . . . . 4. *A. Solandri.*
Leaves 2–5 ft. × $\frac{1}{3}$–$\frac{2}{3}$ in.; nerves obscure. Perianth rotate. Filaments subulate . . . . 5. *A. Banksii.*

1. **A. Cunninghamii,** *Hook. f. Fl. N. Z.* i. 259. A large, tufted, very silky herb. Leaves 2–5 ft. long, spreading and recurved, $\frac{1}{2}$–1 in. broad, plaited, silky and villous. *Male:* Scape 1–1$\frac{1}{2}$ ft., angled, flexuose, shaggy, much and widely branched. Flowers numerous. Perianth rotate, $\frac{1}{4}$ in. diam.; segments acuminate, red-brown when dry. *Female:* Scape 1 ft.; branches crowded, erect, 3–8 in. long. Perianth as in male. Ovary globose; style

short; ovules from the upper part of 3 parietal placentas, imbedded in mucilage. Berry globose. Seeds 6–8, linear, terete, curved; testa thick; in germination the embryo pushes before it a small black operculum.

Northern Island: common on limbs of forest-trees. **Middle** Island: Aorere valley, *Travers*. Very like *A. Banksii*, but differs in the larger flowers, ovary, fruit, and seeds.

2. **A. linearis,** *Hook. f. Fl. Antarct.* i. 76;—*Fl. N. Z.* i. 260. A small tufted herb, glabrous or more or less covered with appressed chaffy hairs. Leaves spreading, 1–6 in. long, narrow linear, acuminate, keeled, margins recurved, silky and villous at the base. Scape few-flowered, shorter than the leaves. Berry elongate, obtusely 3-gonous, red. Seeds obovoid, shining, not angled.

Northern Island: swamps on the summit of the Ruahine mountains, *Colenso*. **Lord Auckland's** group and **Campbell's** Island: boggy ground on the bare hillsides, *J. D. H.*

3. **A. nervosa,** *Banks and Solander*;—*Fl. N. Z.* i. 260. Densely tufted. Leaves 1–3 ft. long, narrow linear-subulate, $\frac{1}{2}$–$\frac{3}{4}$ in. broad, silky or glabrous, rigid, with 3 very strong, often red nerves. *Male:* Scape 1 ft. long, glabrate or very silky. Flowers scattered on long branches, pedicelled. Perianth $\frac{1}{3}$ in. diam., rotate with a hemispherical tube, cut to below the middle; segments broadly linear-oblong, acute. Stamens on the middle of the segments; filaments slender; anthers broadly oblong. *Female:* Scape shorter, stout, much branched. Flowers stoutly pedicelled. Perianth of the male. Ovary conical, 3-gonous. Berry 3-celled, with a stout style, sunk in the baccate tube of the perianth, yellow. Seeds black, shining, angled.

Northern Island: forming dense masses in alpine bogs, *Banks and Solander*; Taupo and Ruahine range, *Colenso*. **Middle** Island: Nelson mountains, alt. 5000 ft., *Bidwill*; Akaroa, *Raoul*; Canterbury, alpine meadows, amongst grass, alt. 2500–5000 ft., *Haast*; Otago, *Lindsay*.

4. **A. Solandri,** *A. Cunn.*;—*Fl. N. Z.* i. 260. A very large species. Leaves 2–4 ft. long, spreading and recurved, 3 in. broad at the base, and there clothed with dense, snow-white silky, villous wool, glabrous above, silky below, with 3 very strong nerves. *Male:* Scape stout. Panicle 6–18 in. long; branches, with the flowers on, 1 in. across. Flowers very crowded, shortly pedicelled. Perianth very large, $\frac{1}{2}$ in. long, membranous; segments linear, obtuse, silky externally. Filaments $\frac{1}{2}$ in. long; anthers linear-oblong. *Female:* Scape stout, curved. Panicle with very long slender (rarely short, stout) branches, 8–12 in. long. Flowers close-set, fascicled or whorled; pedicels slender, $\frac{1}{4}$ in. long. Perianth much smaller and more scarious than in the male; tube hemispheric; segments recurved. Ovary globose, 3-celled; style straight.

Northern and **Middle** Islands: common on trunks of trees, *Banks and Solander*, etc. I have no positive assurance of the plants here described as male and female belonging to one species, but have no reason to doubt it.

5. **A. Banksii,** *A. Cunn.*;—*Fl. N. Z.* i. 260. Habit and foliage as in *A. Cunninghamii*. Leaves 2–5 ft. long, $\frac{1}{3}$–$\frac{2}{3}$ in. broad, glabrous or silky. *Male:* Panicle densely silky; branches long, slender. Perianth glabrous, $\frac{1}{4}$ in. diam.; segments narrow, acuminate. Filaments subulate; anthers broadly

oblong. *Female:* Panicle shorter; branches stout, crowded, 3–4 in. long. Perianth broad, silky, not enclosing the berry. Ovary conical, 3-celled. Berry ovoid, yellowish, $\frac{1}{3}$ in. long; stigma sessile. Seeds 3–6, pendulous by slender funicles from the top of each cell, sharply angled; testa thick, hard.—*A. Richardi,* Kunth; ? *Hamelinia veratroides,* A. Rich. Flor. 158. t. 24.

Northern Island: on limbs of forest trees, probably common; Mercury Bay, *Jolliffe.* It is difficult to distinguish the male of this from that of *A. Cunninghamii.*

6. ARTHROPODIUM, Brown.

Herbs, with fleshy, fibrous roots. Leaves narrow, radical, flaccid. Flowers panicled or racemed on a long scape, white; pedicels jointed in the middle, hermaphrodite. Perianth of 6 spreading, persistent leaflets. Stamens 6; filaments bearded. Ovary 3-celled; style slender, stigma hispid; ovules numerous in each cell. Capsule subglobose, 3-valved, few-seeded. Seeds angular; testa black, membranous; embryo curved.

A small Australian and Tasmanian genus.

Scape 1–2 ft. Flowers $\frac{3}{4}$–1 in. diam. . . . . 1. *A. cirrhatum.*
Scape 4–8 in. Flowers $\frac{1}{4}$ in. diam. . . . . 2. *A. candidum.*

1. **A. cirrhatum,** *Br.;—Fl. N. Z.* i. 254. A tall herb, 1–2 ft. high, quite glabrous. Leaves ensiform-lanceolate, acuminate with attenuated points, 1 ft. long, $1\frac{1}{2}$ in. broad. Panicle much branched; bracts foliaceous. Flowers white, $\frac{3}{4}$–1 in. diam.; leaflets lanceolate, acuminate. Filaments bearded above the middle, 2-glandular at the base. Capsule $\frac{1}{3}$ in. diam. Seeds black, opaque, angular.—Bot. Mag. t. 2350; *Anthericum cirrhatum,* Forst.

Northern Island: abundant, *Banks and Solander,* etc.

2. **A. candidum,** *Raoul, Choix,* 14. *t.* 6;—*Fl. N. Z.* i. 254. A small, slender species, 4–8 in. high. Leaves very narrow linear, $\frac{1}{10}$–$\frac{1}{12}$ in. broad, flaccid and grass-like. Scape rarely branched, very slender, usually simple. Flowers solitary or in pairs, small, white, $\frac{1}{4}$ in. diam., on long secund pedicels; bracts long, leafy, lanceolate. Capsule small, globose, $\frac{1}{6}$ in. diam.

Northern and **Middle** Islands: mountain woods, frequent.

7. ANTHERICUM, Linn.

Herbs, with fleshy fibrous roots. Leaves all radical, linear, sheathing at the base. Scape radical. Flowers racemed, often yellow, hermaphrodite, rarely diœcious.—Perianth of 6 equal, spreading leaflets. Stamens 6, inserted at the base of the leaflets; filaments glabrous or bearded. Ovary 3-celled; style simple, stigma 3-lobed; ovules 2 or more in each cell. Capsule 3-celled, 3-valved, few- or many-seeded. Seeds usually compressed and 3-quetrous; testa black.

A large genus, found in various parts of both temperate zones.

Scape stout, 2–3 ft. high. Flowers diœcious . . . . 1. *A. Rossii.*
Scape 1 ft. high. Flowers hermaphrodite . . . . 2. *A. Hookeri.*

1. **A. Rossii,** *Hook. f.—Chrysobactron Rossii,* Hook. f. Fl. Antarct. i. 72. t. 44, 45. Very robust, 6 in.–3 ft. high; stem at the base often $1\frac{1}{2}$

in. diam. Leaves very numerous, spreading, 8–16 in. long, 1–1½ broad, very thick, concave, recurved, obtuse, nerveless. Scape stout, ¼–⅛ in. diam. Raceme very dense, 4–6 in. long, 1–2 diam. Flowers diœcious, bright yellow, very densely crowded; pedicels slender, erect, ½–¾ in. long, lower bracts nearly as long. Perianth ⅓ in. diam., expanded in the male plant, erect in the female; segments ovate-oblong, obtuse. Stamens glabrous in the male plant, with slender subulate filaments in the female. Ovary broadly ovoid in the female; style straight, small, stigma small lobed; ovules 1 or 2 in each cell; ovary in the male smaller, conical, with 3 subulate styles. Capsule ¼–⅓ in. long, ovate-oblong, 3-gonous. Seeds linear-oblong, 3-gonous, testa black.

Lord Auckland's group and **Campbell's** Island: abundant, *J. D. H.* A magnificent plant. I have seen a specimen 3–4 ft. high, with 3 crowns of leaves from one root, and 7 racemes of flowers, some of them 2-fid.

2. **A. Hookeri,** *Colenso.*—*Chrysobactron Hookeri*, Col.; Fl. N. Z. i. 255. Much smaller than *C. Rossii*, but most variable in stature. Leaves 8–10 in. long, ⅙–½ broad. Bracts very variable, sometimes as long as the pedicels. Racemes 3–5 in. long. Flowers ⅓ in. diam., always hermaphrodite.—Hook. Ic. Pl. t. 817; Bot. Mag. t. 4607.

Northern Island: between the Ruahine range and Lake Taupo, *Colenso.* **Middle** Island: apparently abundant throughout, in subalpine pastures.

8. PHORMIUM, Forster.

Tall, rigid herbs; stem swollen at the base, with stout, fibrous roots. Leaves all radical, linear-ensiform, distichous, coriaceous, excessively tough. Scape very tall, branched at the top, bracteate. Flowers large, panicled, erect, jointed on the pedicel, hermaphrodite.—Perianth tubular, curved, of 6 leaflets, outer erect, inner with spreading tips. Stamens 6, exserted; filaments unequal, 3 smaller. Ovary 3-celled; style stout, 3-gonous, stigma capitate; ovules numerous, 2-seriate in each cell. Capsule oblong or linear-oblong, obtusely 3-angled, coriaceous or somewhat membranous, loculicidally 3-valved, many-seeded. Seeds compressed; testa lax, black.

A remarkable genus, "New Zealand Flax," confined to New Zealand and Norfolk Island.

Leaves 3–6 ft., obtuse or subacute . . . . 1. *P. tenax.*
Leaves 2–3 ft., more acuminate . . . . 2. *P. Colensoi.*

1. **P. tenax,** *Forst.;—Fl. N. Z.* i. 256. A tall, coarse, dark-green, stout, densely-tufted plant. Leaves very thick and coriaceous, 3–6 ft. long, narrow-linear, rigid, erect, dark green above, paler and glaucous below; tip acute or apiculate (not acuminate), keeled, always slit when old. Scapes 5–15 ft. high, dark red, stout. Flowers usually lurid-red or yellow, but very variable in colour, 2 in. long. Perianth-segments nearly straight, obtuse. Pods very variable in length, 4–8 in., black, 3-gonous, coriaceous, straight or twisted.—Bot. Mag. t. 3199.

Throughout the **Northern** Island, abundant, *Banks and Solander*, etc. Also found in Norfolk Island. In the 'New Zealand Flora' I have gone at length into the question of the specific identity of this and the following plant. A re-examination of all the specimens does not alter my opinion of this and the following being races of one plant; they seem, however, to be permanently distinct, and as they differ in distribution somewhat, and so much in appearance as to be universally distinguished, I have thought it better to retain them as distinct. From Colenso's notes both appear very variable as to colour of flower, from dark-red to yellow, and size, form, and twisting of the capsule; but this is the coarser and less useful plant of the two. The synonymy of the two species, as given in the 'Flora of New Zealand,' has been altered to suit Mr. Colenso's notes on Le Jolis' *P. Cookianum*, which he asserts to be his *P. Forsterianum*, and consequently the following species, to which, however, the name *Forsterianum* cannot be applied, as Forster's plant is undoubtedly the *P. tenax*.

2. **P. Colensoi,** *Hook. f., in Raoul, Choix.—P. tenax*, *β*, Fl. N. Z. i. 256. Smaller in all its parts than *P. tenax*, and usually of a paler green. Leaves 2–3 ft. long, more acuminate, rarely split at the tip. Scapes 3–6 ft. high, usually green. Flowers 1–1½ in. long; inner segments of perianth acuminate, reflexed. Pod as in *P. tenax*, but smaller.—*P. Forsterianum*, Colenso; *P. Cookianum*, Le Jolis.

Northern and **Middle** Island, from the East Cape southwards, abundant, and often growing with *P. tenax*.

Dr. Hector informs me of a third species or variety of *Phormium*, from the south-west coast, with short, almost globular capsules.

9. HERPOLIRION, Hook. f.

Small alpine herbs, with wiry creeping stems, sending up short scapes clothed with linear leaves, sheathing at the base. Flower hermaphrodite, when in bud enclosed within 1–3 spathes or bracts.—Perianth tubular, campanulate, 6-partite; leaflets linear, nearly equal. Stamens 6; filaments filiform, glabrous or downy; anthers twisted a little. Ovary 3-celled, oblong; style filiform, stigma simple; ovules numerous in each cell. Fruit?

A small genus of one Australian and Tasmanian alpine species, and the following.

1. **H. novæ-Zelandiæ,** *Hook. f. Fl. N. Z.* i. 258. Leaves 1–2 in. long, spreading or recurved, narrow linear, acuminate, flat or longitudinally folded, striate, glaucous. Flowers waxy-white, tinged with blue in drying, almost sessile amongst the leaves, ¼–½ in. long; bract or spathe solitary. Filaments pubescent.

Northern Island: plains near Taupo, *Colenso*. **Middle** Island: swamps in the alps of Canterbury, of Nelson, *Travers, Sinclair, Haast*. This may prove the same with the Tasmanian plant.

ORDER VIII. PALMEÆ.

Trees or shrubs, with simple, erect or climbing stems, or stemless. Leaves simple pinnate or palmately divided, with sheathing bases. Flowers 3-bracteate, sessile in panicled spikes; panicles enclosed when young in one or more sheathing spathes, unisexual.—Perianth inferior, of 6 thick, coriaceous or fleshy leaflets, in 2 series, valvate or contorted in bud. Stamens usually 6; anthers 2-celled, versatile. Ovary 3-celled, or ovaries 3; stigmas 3, sessile; ovule 1, rarely 2, erect in each cell. Fruit or berry 1–3-celled, or 1–3 drupes.

Seed erect or laterally attached, embryo small, in a cavity of the horny albumen.

A very large tropical Order, including the Date, Cocoa-nut, etc., which reaches its highest south latitude in New Zealand.

1. ARECA, Linn.

Trees, with simple ringed trunks. Leaves pinnate, rarely entire. Inflorescence axillary.—Flowers monœcious, panicled; males numerous, females few; at the bases of the branches, or flowers ternate, and each female having a male on either side of it. Spathes 2- or 3-leaved. Stamens 6. Ovary 1–3-celled. Drupe with a fibrous inner coat. Embryo at the base of the albumen.

A large tropical, Indian, Australian, and Pacific Island genus, to which the Betel-nut Palm belongs.

1. **A. sapida,** *Soland.;—Fl. N. Z.* i. 263. *t.* 59 *and* 60. Trunk 6–10 ft. high; 6–8 in. diam.; cylindric, green. Leaves 4–6 ft. long, pinnate; leaflets very narrow-linear, margin replicate, nerves midrib and petiole covered with minute scales. Spathes 2 or 3, white, a foot long. Panicles densely branched, 18–24 in. long, white. Flowers very numerous, crowded, males and females mixed. Perianth-lobes acuminate, outer smaller. Ovary 3-celled; ovules lateral, pendulous. Drupe oblong, ½ in. long. Albumen surface smooth (not ruminated).—*A. Banksii*, Martius.

Northern and **Middle** Islands: as far south as Queen Charlotte's Sound, *Banks and Solander*, etc. Very closely allied to the Norfolk Island *A. Baueri*, which is a larger plant; young inflorescence eaten.

ORDER IX. JUNCEÆ.

Herbs, leafless or with stiff, rush-like or flat grassy leaves. Stems, or culms cylindric or compressed, bearing cymes heads or panicles of small, bracteate, hermaphrodite flowers.—Perianth of 6 dry, usually brown, lanceolate acuminate leaflets in 2 rows. Stamens 6, rarely 3, inserted at the bases of the leaflets. Ovary 3-angled, 1- or 3-celled; style short or long, stigmas 3, linear; ovules solitary or many in each cell. Capsule included in the persistent perianth, 1–3-celled, loculicidally 3-valved, 1–many-seeded. Seeds minute, albuminous; embryo minute.

A large Order, found in every part of the world, most frequent in the temperate zones.

Glabrous. Style short or 0. Ovary 3-celled, many-ovuled . . . . . 1. JUNCUS.
Glabrous. Style long. Ovary 1-celled, many-ovuled . . . . 2. ROSTKOVIA.
More or less hairy. Style short. Ovary 1-celled, 3-ovuled . . . . . . 3. LUZULA.

1. JUNCUS, Linn.

Herbs, glabrous, with annual, fibrous roots, on creeping, jointed rootstocks. Leaves slender or stout, sheathing at the base, terete or flat, jointed internally or not so, sometimes 0, or reduced to sheaths.—Flowers in terminal or lateral cymes heads or panicles, rarely solitary. Ovary 3-celled; style short, rarely distinct; ovules numerous.

A large genus, found all the world over. The species are variable and often difficult of discrimination.

1. *Rhizome stout, creeping. Culms terete.*

Leaves 0; *sheaths at the base of the culms, broad, open, obtuse.*

Culms 2–3 ft., very stout. Panicle pale. Stamens 6 . . . . 1. *J. vaginatus.*
Culms 1–3 ft., slender. Panicle pale. Stamens 3. . . . . . 2. *J. australis.*

Leaves few, terete, very long, pungent; sheaths below them appressed to the culm.

Culms 1–2 ft., rather stout. Panicle brown. Stamens 6 . . . . 3. *J. maritimus.*
Culms 1–3 ft., slender. Panicle pale. Stamens 3 . . . . . 4. *J. communis.*

2. *Roots fibrous* (*rhizome creeping in* 8). *Culms tufted, usually flattened, leafy at the base.*

Leaves not jointed internally (*see also* 10. J. novæ-Zelandiæ).

Leaves flat or concave, grass-like. Flowers $\frac{1}{12}$ in. long . . . . 5. *J. planifolius.*
Densely tufted. Leaves slender. Flowers $\frac{1}{4}$ in. long . . . . . 6. *J. bufonius.*
Minute. Leaves terete. Flowers 1–3, terminal . . . . 7. *J. antarcticus.*

Leaves terete or compressed, the pith jointed internally, often very inconspicuously in J. novæ-Zelandiæ.

Stout. Cymes divaricating. Flowers capitate, $\frac{1}{8}$ in. long . . . 8. *J. Holoschœnus.*
Slender. Leaves compressed. Flowers 2–8, terminal. Perianth as long as the acute capsule . . . . 9. *J. scheuzerioides.*
Slender. Leaves filiform. Flowers 3–5, terminal, brown. Perianth shorter than the black, shining, turgid capsule . . . . . 10. *J. novæ-Zelandiæ.*
Slender. Leaves capillary. Flowers 1–3, lateral, pale. Perianth as long as the acute, pale capsule . . . . 11. *J. capillaceus.*

1. **J. vaginatus,** *Br.;—Fl. N. Z.* i. 263. Very stout, tall, pale-coloured. Culms 2–3 ft. high, minutely striate, nearly $\frac{1}{4}$ in. diam. at the base, leafless, covered with large sheaths at the base. Panicle lateral, dense or loose, very variable in size, branches strict, compressed. Flowers $\frac{1}{10}$ in. long, fascicled; bracts short, ovate, acute; perianth segments lanceolate, acuminate. Capsules pale, obovoid, obtuse, rather longer than the perianth. Testa produced at each end.

Northern Island: wet clay soil near Wellington, *Colenso, Stephenson;* Waitaki, *Sinclair.* An abundant South Australian and Tasmanian plant, and apparently the same as the Chilian *J. procerus*, E. Meyer (*J. Valdiviæ*, Steud.).

2. **J. australis,** *Hook. f. Fl. Tasman.* ii. 67. *t.* 134 A. Quite similar to *J. vaginatus*, but much more slender, panicle very much smaller, and stamens only 3. Culms slender, 1–3 ft. high, leafless. Panicle pale, open or contracted and almost capitate.

Northern and **Middle** Islands: marshy places, probably common, *Colenso, Sinclair, Munro;* as far south as Otago, *Lindsay.* Also abundant in South Australia and Tasmania.

3. **J. maritimus,** *Lamarck;—Fl. N. Z.* i. 263. Stout, tall, rather dark-coloured. Culms 1–2 ft. high, leafless, or with 1 or 2 stout, long, terete, pungent leaves, surrounded with large sheaths at the base, finely striate. Panicle large, lateral, effuse, erect, much branched, often a span long, branches strict, compressed. Flowers fascicled, $\frac{1}{10}$ in. long; bracts as long as the flowers. Perianth segments lanceolate, acuminate. Stamens 6. Capsule ovoid, as long as the perianth. Testa produced at each end.

Northern and **Middle** Islands: abundant on the coasts. Equally abundant in Aus tralia, Tasmania, all temperate and many tropical countries. European and other specimens are as pale as *J. vaginatus* and *australis*, approaching them very closely.

4. **J. communis,** *E. Meyer*;—*J. effusus*, Linn.;—Fl. N. Z. i. 263. Culms slender, finely striate, 1–3 ft. high, clothed at the base with cylindrical, rather appressed sheaths. Leaves few, terete, pungent, finely striate. Panicle lax effuse and spreading or contracted, sometimes capitate. Flowers not fascicled, but often crowded, $\frac{1}{12}$ in. long. Perianth-segments lanceolate-subulate, acuminate, as long as the obovoid capsule. Stamens 3. Testa loose at each end of the seed.

Northern and **Middle** Islands: abundant in stiff, wet soil, etc. The slender habit close sheaths, and presence of leaves, distinguish this from *J. australis*. One of the commonest plants in all temperate countries.

5. **J. planifolius,** *Br*;—*Fl. N. Z.* i. 263. Culms 6–18 in. high, leafy at the base, stout or slender, compressed, striate. Leaves numerous, grass-like, flat or concave, membranous, $\frac{1}{20}$–$\frac{1}{4}$ in. diam.; sheaths open. Panicles terminal, capitate and dense, or open, in branched 3-chotomous cymes; involucral leaves longer or shorter than the panicles or 0. Flowers brown, fascicled, $\frac{1}{12}$ in. long. Perianth segments acuminate. Stamens 3–6. Capsule prismatic, mucronate, longer or shorter than the perianth. Seeds striate.

Abundant throughout the islands, *Banks and Solander*, etc. **Lord Auckland's** group, *Le Guillon*. A common Australian, Tasmanian, and Chili plant. Habit of a *Luzula*, but perfectly glabrous.

6. **J. bufonius,** *Linn.*;—*Fl. N. Z.* i. 264. A small, pale, annual species, 1–6 in. high, excessively branched from the base, roots fibrous. Leaves numerous, slender, compressed, linear-subulate, with long large sheaths, pith continuous inside. Flowers fascicled, usually in threes; fascicles lateral, sessile, solitary, or in loose long pedicelled sparingly-branched cymes, $\frac{1}{4}$ in. long. Perianth-segments narrow, subulate-lanceolate, long-acuminate, with broad membranous margins. Stamens 6. Capsule elongate, prismatic, longer than the perianth. Seeds ovate-globose, pale, shining.—*J. plebejus*, Br.

Damp clay soil, etc., throughout the **Northern** and **Middle** Islands. One of the commonest plants in the islands, and in all temperate countries.

7. **J. antarcticus,** *Hook. f. Fl. Antarct.* i. 79. *t.* 46. A minute, tufted species. Culms 1–2 in. high, branched at the base. Leaves $\frac{1}{2}$–1 in. long, subulate, nearly terete, obtuse; pith continuous inside; sheath short, rather broad. Flowers 1–4, fascicled at the top of the culm, $\frac{1}{10}$ in. long, sessile. Perianth-segments lanceolate, acuminate. Stamens 6. Capsule as long as the perianth, ovoid. Seeds ovoid or oblong, pale, smooth.

Campbell's Island: on the exposed mountain-tops, in wet places, *J. D. H.* This appears to be the same with a small Chilian species, (unnamed) collected by Gillies at San Pedro Nolasco, but the specimens are insufficient for an accurate comparison.

8. **J. Holoschœnus,** *Br.*;—*J. cephalotes*, Thunb.;—Fl. N. Z. i. 263. Culms erect, leafy, from a creeping perennial rhizome. Leaves and involucres slender, compressed, acuminate, pith jointed within. Flowers fascicled, $\frac{1}{6}$ in. long; fascicles collected into lateral cymes, with spreading, often divaricating

branches; involucral leaf solitary. Perianth-segments lanceolate, acuminate. Stamens 6. Capsule prismatic, as long as the perianth. Seeds linear-oblong; testa striate and rugose, produced at each end.

Northern Island: east coast, *Colenso*. I have only tops of culms, and so cannot certainly identify this plant, which is common in Australia and Tasmania. It is probably the same with Thunberg's *J. cephalotes*, which extends over Asia, Africa, Southern Europe, and parts of America.

9. **J. scheuzerioides,** *Gaud.;—Fl. Antarct.* i. 79 Culms short, tufted, leafy, fastigiately branched at the base, 1–8 in. high. Leaves compressed, linear-subulate, much longer than the culms; pith jointed inside. Flowers 2–8, in a small, terminal fascicle, $\frac{1}{10}$–$\frac{1}{6}$ in. long; bract setaceous or 0. Perianth-segments ovate-lanceolate, acuminate. Stamens 6. Style evident. Capsule as long as the perianth, ovate-oblong. Seeds ovoid, smooth; testa with a delicate outer membrane.

Middle Island: Otago, lake district, alpine, *Hector and Buchanan*. **Lord Auckland's** group and **Campbell's** Islands: in boggy places, *J. D. H.* Probably common in alpine bogs throughout the islands. Also found in the Falkland Islands, Fuegia, Chili, and the Andes of Ecuador.

10. **J. novæ-Zelandiæ,** *Hook. f. Fl. N. Z.* i. 264. Culms very slender, tufted, 2–6 in. high, leafy. Leaves longer or shorter than the culms, very slender, filiform, terete, striated; pith evidently jointed inside where the leaf is broad enough to observe it. Flowers 3–5, in a small terminal fascicle, $\frac{1}{12}$ in. long, rarely fascicles 2, superimposed; bracts setaceous or 0. Perianth-segments obtuse, red-brown, with white membranous margins. Stamens 6. Capsule broadly obovoid, very turgid, longer than the perianth, nearly black, shining. Seeds small, pale; testa scarcely produced at the ends.

Northern Island: bogs, east coast, Manawata and Waipona rivers, etc., *Colenso*. **Middle** Island: Otago, lake district, *Hector and Buchanan*. Probably common in subalpine bogs throughout the islands.

11. **J. capillaceus,** *Hook. f. Fl. N. Z.* i. 264. Culms very slender, tufted, 2–6 in. high, leafy. Leaves longer or shorter than the culms, setaceous or filiform, terete, striated; pith jointed inside. Flowers 1–3 together, pale, lateral, $\frac{1}{12}$ in. long. Perianth-segments acute or acuminate. Stamens 6. Capsule prismatic, narrow ovoid, pale, longer than the perianth. Seeds broadly oblong, smooth, minutely striate and reticulate.—Fl. Tasman. ii. 65 t. 134 B.

Northern Island: east coast, skirts of woods near Eparairua, *Colenso*. Easily distinguished from *J. novæ-Zelandiæ* by the pale flowers and capsule.

2. ROSTKOVIA, Desvaux.

Characters of *Juncus*, except that the flowers are large, solitary, terminal; the ovary 1-celled, and the style long; all have creeping rhizomes, and erect, rigid, terete culms and leaves. Testa very thick.

A small genus, native of the islands south of New Zealand and of Fuegia.

Flowers $\frac{1}{2}$ in. long . . . . 1. *R. magellanica.*
Flowers $\frac{3}{4}$–1 in. long . . . . 2. *R. gracilis.*

1. **R. magellanica,** *Hook. f. Fl. Antarct.* i. 81. Culms simple, tufted,

compressed below, 6–8 in. long. Leaves longer than the culms, numerous, strict, erect, rigid, concave in front, pungent, polished. Flowers $\frac{1}{4}$ in. long, 2-bracteate; lower bract subulate, scarcely equalling the perianth; upper 2–4 times longer, ovate-lanceolate below, then subulate. Perianth-leaflets chestnut-brown, linear-oblong, acute or acuminate, outer larger. Stamens included; anthers linear; connective unguiculate. Ovary oblong, 3-gonous; style stout; stigmas exserted. Capsule prismatic, acute, longer than the perianth, lenticular, obtuse, even.—*Rostkovia sphærocarpa*, Desvaux; *Juncus Magellanicus*, Lamarck.

Campbell's Island: bogs on the hills, *J. D. H.* Also found in the Falkland Islands, and Fuegia, and on the Andes of Quito, at 13,000 ft. elevation.

2. **R. gracilis,** *Hook. f. Fl. Antarct.* i. 83. *t.* 47. Much larger than *R. Magellanica.* Leaves 9–12 in. long, three times longer than the culms, very slender, terete. Flowers $\frac{2}{3}$–1 in. long.

Lord Auckland's group: rocky places and bare ground on the hills, *J. D. H.* Allied to the Fuegian *R. grandiflora.*

3. LUZULA, De Candolle.

More or less hairy, tufted, grass-like herbs. Leaves flat or involute, sheathing at the base. Flowers in terminal branched panicles cymes or heads, 2-bracteate at the base.—Stamens 6. Ovary 1-celled; style 1, stigmas 3, filiform; ovules 3, basal. Capsule 1-celled, 3-seeded.

A genus of temperate and frigid regions, not found in tropical climates. The species are excessively variable; the following are very possibly all referable to one.

Leaves usually flat, 1 or more in. long, ciliate with long hairs.

Culms slender. Bracts nearly entire or toothed, slightly ciliate . . . . 1. *L. campestris.*
Culms stout. Leaves broad, with cartilaginous margins. Bracts subciliate and toothed . . . . 2. *L. Oldfieldii.*
Culms stout. Leaves broad, with cartilaginous margins. Bracts lacerate and ciliate . . . . 3. *L. crinita.*

Leaves less than 1 in. long, subulate, very concave, not ciliate.

Perianth-leaflets acuminate . . . . 4. *L. pumila.*
Perianth-leaflets obtuse or acute . . . . 5. *L. Colensoi.*

1. **L. campestris,** *De Cand.;—Fl. N. Z.* i. 264. Culms slender, leafy, 2–16 in. high. Leaves $\frac{1}{2}$–12 in. long, flat and grass-like, with long white hairs on the margins and sheaths towards the base. Flowers fascicled, arranged in small heads on the long or short branches of irregular cymes, or collected into dense ovoid heads. Bracts membranous; margins slightly ciliated, entire or nearly so. Perianth $\frac{1}{16}$–$\frac{1}{6}$ in. long; leaflets varying from oblong to lanceolate, acuminate or almost aristate, white chestnut-brown or transparent with brown centre. Anthers longer than the filaments. Capsule broadly obovoid or nearly globose, 3-gonous or 3-lobed.

Var. *α.* Perianth-segments lanceolate-acuminate, chestnut-brown, with narrow white margins.

Var. *β. picta.* Perianth-leaflets lanceolate, long-acuminate, almost awned, with very broad white margins and narrow brown midrib.—*L. picta*, A. Rich.;—Fl. N. Z. i. 265.

Var. *γ. pallida.* Perianth-leaflets often white, shorter, broader, acute or acuminate. Inflorescence usually more capitate.

Northern and **Middle** Islands: *Banks and Solander*, etc. Var. β usually in woods. A most variable plant, found in all temperate parts of the globe; extremely common in temperate Australia and Tasmania. Var. β is the common Tasmanian form.

2. **L. Oldfieldii,** *Hook. f. Fl. Tasm.* ii. 68. More robust than *L. campestris*. Culm stout, 3–8 in. high, leafy. Leaves coriaceous, often $\frac{1}{3}$ in. broad, flat; margins cartilaginous, ciliate with very long hairs. Flowers in dense ovoid or subglobose, simple or compound, sessile or pedicelled heads, $\frac{1}{3}$–1 in. diam. Bracts ciliate, often lacerate. Flowers as in *L. campestris*, but rather larger.

Middle Island: Southern Alps, *Haast*; Otago, sand dunes, mouth of the Kaikorai, *Lindsay*; Waitaki valley, in swampy ground, *Hector and Buchanan*. Also a native of the alps of Tasmania. A very distinct form, but no doubt passing into *L. campestris* on one hand, and into *L. crinita* on the other.

3. **L. crinita,** *Hook. f.;—Fl. Antarct.* 85. *t.* 48. Much larger and stouter than *L. campestris*. Leaves much ciliated or glabrate, $\frac{1}{12}$–$\frac{1}{6}$ in. broad. Inflorescence a dense, chestnut-brown, broadly ovate, involucrate head, $\frac{1}{4}$–$\frac{3}{4}$ in. long, very woolly at the base. Bracts with fimbriate, ciliate edges. Perianth-leaflets subulate or lanceolate, much acuminate, with paler but not white borders.

Lord Auckland's group and **Campbell's** Island, *J. D. H.* **Macquarrie's** Island, *Frazer*. Closely allied to the Fuegian *L. Alopecurus*, Desv., and probably only a gigantic form of *L. campestris*, α, with broader leaves and more ciliated bracts.

4. **L. pumila,** *Hook. f., n. sp.* Small, densely tufted, nearly glabrous, 1–2 in. high. Leaves shorter than the culm, $\frac{1}{2}$–1 in. long, linear-subulate, obtuse, coriaceous, convex at the back, concave in front, Culm naked, 1–2 in. high. Flowers in a small, terminal, 4–10-flowered head, $\frac{1}{12}$ in. long. Bracts ovate, edges ragged. Perianth-leaflets subulate or lanceolate, long-acuminate, chestnut-brown, twice as long as the black capsule.

Middle Island: summit of Mount Torlesse and Mount Darwin, alt. 6–7000 ft., *Haast*; Otago, lake district, alpine, *Hector and Buchanan*.

5. **L. Colensoi,** *Hook. f., n. sp.* Very small, densely tufted, nearly glabrous, 1 in. high. Leaves longer than the culm, $\frac{1}{3}$–$\frac{2}{3}$ in. long, subulate, obtuse, coriaceous, convex at the back, very concave in front. Scape stout, sunk amongst the foliage. Flowers in small, terminal, 6–8-flowered heads, $\frac{1}{16}$ in. long. Bracts ovate, edges ragged. Perianth-leaflets ovate-oblong, obtuse or acute, white with brown centre, not much longer than the pale capsule.

Northern Island: summit of the Ruahine mountains, *Colenso*. Probably only an alpine state of *L. campestris*, γ, but totally different in habit and appearance.

ORDER X. RESTIACEÆ.

Grass-like or rush-like herbs. Leaves always narrow, sheathing below, the sheath usually split to the base. Flowers in terminal heads, or solitary in the sheaths of the leaves, or in spikelets terminating scapes, small, uni- or bisexual.—Perianth dry, often thin and membranous, of 2–6 leaflets in 2 series, or reduced to imbricating, flower-bearing scales (glumes). Stamens 1–6 (never 5); anthers 1-celled. Ovary 1–3-celled, or of 2 or more free or connate 1-celled carpels; style simple, entire, terminating in 1 or more filiform stigmas; ovules solitary and pendulous in each cell.—Fruit a small nut or utricle or 1–many-celled capsule, 1- rarely 3-seeded. Seeds usually oblong; testa membranous; albumen copious; embryo minute, lenticular.

A large Order of obscure plants, most abundant in Australia and South Africa, very rare in Europe, the rush-like ones differing from *Junceæ* in the 1-celled anthers and unisexual flowers; the grass-like differing from the Grasses in the leaf-sheaths usually split to the base, and in the anthers, habit, and fruit.

Flowers diœcious, in panicled spikelets. Perianth of 6 leaflets . . . 1. LEPTOCARPUS.
Flowers unisexual, few, hidden in the leaf-sheaths. Perianth of 6 leaflets 2. CALOROPHUS.
Flowers unisexual, within 2 glumes. Stamens 2. Stigmas 2 . . . 3. GAIMARDIA.
Flowers unisexual, within 2 glumes. Stamen 1 or 0. Stigmas or ovaries 2–18 . . . . 4. ALEPYRUM.

1. LEPTOCARPUS, Br.

Rush-like plants. Rhizome stout, creeping, scaly. Culms numerous, erect, terete, harsh, jointed, with sheaths at the joints. Male and female inflorescence often dissimilar.—Flowers diœcious, in cylindrical spikelets, which are panicled or fascicled. *Male:* Perianth of 6 dry leaflets in 2 series. Stamens 3, seated round a rudimentary ovary. *Female:* Perianth as in the male. Ovary 1-celled, 3-gonous; stigmas 3, deciduous. Nut enclosed in the perianth, 1-celled, 1-seeded.

A considerable genus, of which the following species is the only extra-Australian member.

1. **L. simplex,** *A. Rich.;—Fl. N. Z.* i. 265, *not Brown.* Variable in size, 1–3 ft. high. Rhizomes stout, creeping, scaly; scales chestnut-brown. Culms numerous, simple, slender or stout, terete, smooth. Sheaths distant, ¼ in. long, 1–3 in. apart. *Male:* Spikelets cylindrical, red-brown, peduncled or sessile, ¼–½ in. long, often in lateral panicles; peduncles glabrous or downy. Glumes ovate-acuminate or lanceolate, longer than the sessile flowers. Outer perianth-segments lanceolate, acuminate; inner half the size, oblong-lanceolate. *Female:* Spikes clustered in ovoid heads, sometimes ½ in. long, and as broad. Glumes ovate-acuminate. Outer perianth-segments lanceolate; inner smaller, oblong-lanceolate, retuse, apiculate. Stigmas 3, slender.

Northern and **Middle** Islands: abundant in marshy places, *Banks and Solander*, etc. Extensively used for thatching, etc. Very near the Tasmanian *L. Brownii*, Hook. f. (*L. simplex*, Br.), but the stigmas are much longer, and the inner perianth-leaflets of the female different.

2. CALOROPHUS, Labill.

Culms slender, long, flexuose, jointed, simple or excessively branched, with appressed distant sheaths, which sometimes terminate in small leaves at the joints.—Flowers monœcious or diœcious, in short, minute, few-flowered spikelets, that are hidden in the sheaths of the leaves, and surrounded by glumes. *Male:* Perianth of 6 long narrow leaflets. Stamens 3; anthers linear. *Female:* Perianth of 2–6 very short truncate leaflets. Ovary 1-celled;

stigmas 2 or 3, deciduous. Nut hard, surrounded at the base with the persistent perianth.

A small Australian, Tasmanian, and New Zealand genus.

C. elongata, *Lab.;—Fl. N. Z.* i. 267. Culms prostrate, 1–3 ft. long, excessively branched from the base; branches flexuose, slender, semiterete, not striated, $\frac{1}{24}$ in. diam.; sheaths $\frac{1}{4}$ in. long, with spreading, subulate points. *Male:* Spikelets 4–6-flowered; bract obtuse, bearded. Perianth of 6 linear leaflets, enclosed in pungent glumes. Anthers linear. *Female:* Spikelets 3-flowered. Flowers remote, lower with 2, upper with 6 leaflets. Styles long, tortuous.—Lab. Fl. Nov. Holl. ii. 7. t. 228.

Var. *β. minor.* Shorter, more slender. Male spikelet 2-flowered. Female usually 1-flowered.—*C. minor*, Fl. N. Z. i. 267.

Northern Island: Great Barrier Island, *Sinclair;* swamps at Wangerei, *Colenso.* **Middle** Island: Port Preservation, *Lyall.* Var. *β*, base of Tongariro, *Colenso;* Lake Tennyson, alt. 4400 ft., *Travers;* top of Morse mountain, alt. 6500 ft., *Bidwill.* Both varieties are Australian and Tasmanian plants.

In the 'New Zealand Flora' I have referred a third New Zealand plant from Chatham Island, collected by Dieffenbach, which is out of flower, provisionally to this genus; it has erect, much-branched stems, 2 ft. high, smooth and polished, neither striate nor flexuose; sheaths with acuminate points.

3. GAIMARDIA, Gaudichaud.

Very small, moss-like, densely-tufted plants, glabrous, bright green. Leaves imbricate, setaceous, with broad membranous sheaths.—Spikelets minute, on a short terminal scape, of 2 glumes enclosing 1 or 2 unisexual flowers. Perianth 0. Stamens 2. Ovaries 2, connate; stigmas 2, filiform. Capsule membranous, 2-celled, 2-valved, 2-seeded.

A small genus found in the New Zealand islands and Antarctic America.

Leaves setaceous, with acicular points; sheaths glabrous . . . . . . 1. *G. setacea.*
Leaves setaceous, obtuse; sheaths ciliated . . . . 2. *G. ciliata.*

1. **G. setacea,** *Hook. f. Fl. N. Z.* i. 267. Tufted, forming large moss-like patches; stems 1–2 in. high. Leaves setaceous, imbricating, $\frac{1}{4}$–$\frac{1}{2}$ in. long, with long acicular points, and broad, membranous, often lacerate sheaths produced upwards. Scape erect, $\frac{1}{2}$ in. long. Outer scale larger, $\frac{1}{12}$ in. long, convolute, linear-oblong, obtuse; inner narrow, with a terete tip.

Middle Island: Port Preservation, *Lyall.*

2. **G. ciliata,** *Hook. f.;—Fl. Antarct.* i. 86. Very similar to *G. setacea*, but the leaves are obtuse, and the membranous sheaths excessively ciliated.

Lord Auckland's group, forming dense moss-like tufts on the hills, *J. D. H.*

4. ALEPYRUM, Br.

Very small, tufted, moss-like herbs. Leaves subulate, soft, with broad, membranous white sheaths.—Spikelets minute, on a short terminal scape, of 2 glumes enclosing 1 or 2 flowers. Stamens 1 or 2; filament very long and slender. Ovaries 2–18, connate or free; stigmas as many, free or connate, filiform. Capsules 2–18, membranous, 1-seeded.

A small Australian, Tasmanian, and New Zealand genus, which should probably be united with *Gaimardia.*

1. **A. pallidum,** *Hook. f. Fl. N. Z.* i. 268. *t.* 62 C. A minute, tufted, moss-like, soft, pale green plant; stems ½–1 in. high. Leaves ¼ in. long, subulate; sheaths broad, membranous, transparent. Scape shorter than the leaves. Stamen 1. Ovary solitary or 2–4, united in pairs or threes (or 2 to each glume); styles very long.—*Gaimardia (?) pallida*, Fl. Antarct. 86.

Northern Island: tops of the Ruahine mountains, near the snow, *Colenso*. **Campbell's** Island, forming tufts on the hills, *J. D. H.*

ORDER XI. CYPERACEÆ.

Grass-like or rush-like herbs, with fibrous roots. Culms solid, usually 3-gonous, sometimes compressed flat or terete; sheaths not split to the base. —Flowers unisexual or hermaphrodite, in the axils of small scales (glumes), which form spikelets, being either imbricated all round a rachis or distichous. Glumes usually rigid, compressed, concave or convolute, obtuse cuspidate or awned, usually persistent, the lower empty or all floriferous. Perianth generally 0, sometimes of 3–6 or more hypogynous bristles or scales (the ovary of *Carex* and *Uncinia* is contained in a utricle or false perianth). Stamens 1–6; filaments flat, slender, sometimes elongating after flowering; anthers linear, inserted by the base, 2-celled, often with a claw at the tip. Ovary 1-celled; style 1, stigmas 1–3, filiform; ovule 1, erect, anatropous. Nut minute, compressed or 3-gonous. Seed filling the nut; embryo at the base of copious floury albumen.

An immense Natural Order, found in all parts of the world, distinguished from Grasses by the rarely round, solid culms; leaf-sheaths not split to the base; absence of evident ligule on the leaf; often 3-ranked leaves; anthers not versatile; separable, thick pericarp; stigmas not plumose; embryo not on one side of the base of the albumen but at the very base; and structure of the spikes and spikelets.

A. *Spikelets* 1- *or more flowered. Flowers hermaphrodite.*

Spikelets compressed. Glumes few or many, persistent, more or less distichous (rarely imbricated all round in Schœnus).

Glumes many, keeled, all floriferous. Bristles 0 . . . . 1. CYPERUS.
Glumes few, upper 1–3 floriferous. Bristles 0 or 3–6 . . . . 2. SCHŒNUS.
Glumes free, upper 1 floriferous. Bristles large, flat, plumose . . 3. CARPHA.

Spikelets terete. Glumes few or many, imbricate all round. (See Schœnus *in previous section.)*

Spikelets usually lateral, numerous. Glumes many, most floriferous. Stamens 1–3. Bristles 2–8 . . . . 4. SCIRPUS.
Spikelets solitary, terminal. Glumes as in *Scirpus*. Bristles 3–6 . 5. ELEOCHARIS.
Spikelets small, usually lateral. Glumes as in *Scirpus*. Bristles 0 6. ISOLEPIS.
Spikelets spirally arranged on the leafy culm. Glumes and flowers as in *Scirpus* . . . . 7. DESMOSCHŒNUS.
Spikelets umbelled. Glumes many, most floriferous. Stamen 1. Base of style tumid, hairy . . . . 8. FIMBRISTYLIS.
Spikelets fascicled. Glumes many or few, 1–3-floriferous. Bristles 0. Stamens 3, rarely elongating . . . . 9. CLADIUM.
Spikelets panicled. Glumes few, 1 floriferous. Stamens 3–6, much elongating. Bristles or scales 0 . . . . 10. GAHNIA.
Spikelets panicled. Glumes few, 1–2 floriferous. Stamens 3. Scales 6, minute . . . . 11. LEPIDOSPERMA.

Spikelets of 2 1-flowered deciduous glumes. Nut on a naked scape.

Tufted, rigid, low, dense herb, with subulate leaves . . . . . . 12. OREOBOLUS.

B. *Spikelets solitary or numerous, many-flowered. Flowers unisexual.*

Female utricle with an ovary and long hooked bristle . . . . . . 13. UNCINIA.
Female utricle with an ovary only, or rarely with one straight bristle 14. CAREX.

1. CYPERUS, Linn.

Rhizomes creeping or tufted. Culms erect, leafy or leafless. Leaves various.—Spikelets compressed, many-flowered, very variously disposed, solitary or numerous, often umbellate; umbels involucrate. Glumes numerous, distichous, much compressed, all or most floriferous. Stamens 3. Bristles 0. Nut 3-quetrous; style deciduous, not jointed nor tumid at the base. Stigmas 3.

A most extensive, chiefly tropical genus, extending sparingly into the temperate zone.

1. **C. ustulatus,** *A. Rich.;—Fl. N. Z.* i. 268. Tall, coarse, leafy, 2–4 ft. high. Culms usually 3-gonous, stout, smooth, $\frac{1}{4}$–$\frac{1}{6}$ in. diam. Leaves long, keeled, coriaceous, margins serrulate or rough. Involucre of many grassy leaves, 6–12 in. long. Umbels of 6–10 long or short rays. Spikelets $\frac{1}{4}$–1 in. long, densely arranged in oblong spikes, $\frac{1}{2}$–1 in. long, dark red-brown, sessile, suberect. Glumes 6–20, broadly oblong, obtuse or mucronate, sulcate, convex at the base, shining. Nut oblong, narrowed at both ends, dark-brown.—A. Rich. Flor. 101. t. 17.

Abundant in marshes throughout the **Northern** and **Middle** Islands, *Banks and Solander*, etc.

2. SCHŒNUS, Linn.

(*Including* CHÆTOSPORA, *Br.*)

Tufted or creeping plants, often rigid, with long or short culms and usually creeping rhizomes. Leaves narrow or 0.—Spikelets usually few, panicled fascicled or crowded, compressed, rarely nearly terete. Glumes few, distichous, rarely imbricate all round the rachis, usually convex at the back, often hard and brown, lower empty, upper 2 or 3 floriferous. Stamens 3. Bristles 0 or 3–6, usually scabrid, rarely plumose. Nut compressed or 3-gonous; style continuous with its top, not jointed nor swollen. Stigmas 2 or 3.

A very widely-diffused genus in the temperate zones of both hemispheres. The species with terete spikelets are with difficulty distinguished from *Scirpus*, except by habit and the fewer flowers.

Spikelets solitary, compressed. Glumes obviously distichous.

Small, usually creeping, leafy. Spikelets axillary, subsolitary . . . . . 1. *S. axillaris.*

Spikelets distinctly compressed, panicled fascicled or crowded.

Culms 1–2 ft. Spikelets panicled, 1 in. long. Bristles 0 . . . . . . 2. *S. tenax.*
Culms 1–2 ft. Spikelets panicled, $\frac{1}{4}$ in. long. Bristles very short . . 3. *S. Tendo.*
Culms 1–3 ft. Spikelets panicled, $\frac{1}{4}$ in. long. Bristles 6, long . . . 4. *S. pauciflorus.*
Culms 4–12 in. Spikelets crowded, $\frac{1}{8}$–$\frac{1}{10}$ in. Bristles 6, long . . . 5. *S. Brownii.*

Spikelets terete or obscurely compressed. Glumes obscurely distichous.

Culms 1–2 in. Spikelet solitary. Bristles 6, simple . . . . 6. *S. concinnus.*
Culms 2–10 in. Spikelets crowded. Bristles 6, plumose . . . . . 7. *S. nitens.*

1. **S. axillaris,** *Hook. f.—Chætospora axillaris,* Br. ;—Fl. N. Z. i. 274. *t.* 62 A. Tufted, small, flaccid, pale green, leafy. Culms branched, often creeping, 1–4 in. long, very slender, compressed. Leaves spreading, alternate, $\frac{1}{3}$–$\frac{2}{3}$ in. long, $\frac{1}{30}$ broad, very narrow linear, obtuse. Spikelets 1–3 together, axillary, shortly peduncled, $\frac{1}{12}$–$\frac{1}{8}$ in. long, pale brown. Glumes 4–6, lanceolate, acute, pale brown, 1 or 2 upper floriferous ; keel scabrid, green. Bristles 3–6, longer than the ovary. Stamens 3. Nut small, white, polished, broadly ovate, 3-gonous, tapering into the slender style. Stigmas 3.—*Scirpus foliatus,* Hook. f. in Journ. Bot. iii. 614.

Northern Island : abundant in wet places. Probably overlooked in the Middle Island. Also found in Australia and Tasmania.

2. **S. tenax,** *Hook. f.—Chætospora tenax,* Fl. N. Z. i. 273. Leafless. Culms rigid, tufted, erect, 1–2 ft. high, terete or compressed, polished ; sheaths at the base red-brown, with subulate blades 1 in. long. Spikelets compressed, $\frac{1}{3}$ in. long, lanceolate, pale brown, 2- or 3-flowered, in a contracted panicle 2–5 in. long ; branches slender, longer than the sheaths. Glumes 6–8, distichous, ovate-lanceolate, acute, pale-brown, not shining, concave, nerveless, subciliate. Bristles 0. Nut turgid, 3-gonous ; angles thick, transversely waved. Stigmas 2 or 3.

Northern and **Middle** Islands : dry hills, abundant, *Banks and Solander,* etc.

3. **S. Tendo,** *Banks and Solander.—Chætospora Tendo,* Fl. N. Z. i. 273. Very similar to *S. tenax,* but much more slender. Culms with 1 deep groove throughout their length. Spikelets only $\frac{1}{4}$ in. long, much darker. Bristles present, but few and short. Nut quite smooth.

Northern Island : Opuragi, *Banks and Solander ;* Clay hills, Bay of Islands, Auckland, etc., *Sinclair,* etc., abundant.

4. **S. pauciflorus,** *Hook. f. Fl. N. Z.* i. 273. Culms tufted, very slender, 1–3 ft. high, pale, leafy at the base, angled and grooved. Leaves subulate, grooved, rigid, erect ; sheaths long, grooved, dark chestnut-brown, shining. Panicle lateral, solitary, short. Spikelets few, slender, pedicelled, $\frac{1}{4}$ in. long, lanceolate, dark-brown, polished, 3- or 4-flowered. Glumes distichous, acuminate ; keel green. Bristles 6, capillary, as long as the very long style. Nut narrow-oblong, pale-brown, shining, striated with impressed dots. Stigmas 3.

Northern Island : marshes near Patea and base of Tongariro, *Colenso.* **Middle** Island : swamps, base of Mount Sinclair, alt. 2200 ft., *Sinclair and Haast ;* Otago, Lindis Pass, subalpine, alt. 2–4000 ft., *Hector and Buchanan.* The true " Snow Grass," according to Buchanan. Very near the South Chilian *Chætospora antarctica,* Hook. f., but the leaves are much shorter.

5. **S. Brownii,** *Hook. f.—Chætospora imberbis,* Br. ;—Fl. N. Z. i. 274. Culms densely tufted, very leafy, 4–10 in. high, grooved, slender, rather flaccid. Leaves alternate, very narrow linear, acuminate, channelled above, striate below longer than the culms ; sheaths red-brown, deeply grooved. Spikelets lateral, fascicled, 3–6 together, shortly pedicelled, $\frac{1}{10}$–$\frac{1}{8}$ in. long, 1- or 2-flowered, lanceolate. Glumes few, lanceolate-oblong ; upper obtuse, almost black, margins pale ; lower with hispid points. Bristles 6, slender.

Nut (in Tasmanian specimens) white, broadly obovate, 3-gonous, grooved, with impressed dots in the grooves.

Northern Island: east coast, *Colenso.* A very common Australian and Tasmanian plant, allied to *S. axillaris*, but very much larger with long leaves. The *Schœnus imberbis*, A. Cunn., is not the plant of Brown, which has no bristles.

6. **S. concinnus,** *Hook. f.—Chætospora concinna*, Fl. N. Z. i. 274. t. 62 B. Rhizome slender, creeping. Culms tufted, leafy, erect, very slender, 1–2 in. high, rigid, curved. Leaves shorter than the culms, setaceous, grooved; sheaths almost black. Spikelet nearly terminal, solitary, suberect, slightly compressed, $\frac{1}{6}$–$\frac{1}{5}$ in. long, 2- or 3-flowered, dark-brown. Glumes 5 or 6, scarcely distichous, oblong-lanceolate, obtuse, quite glabrous. Bristles 6, very slender. Nut pale-brown, smooth, with a dense brush of hairs at the base, broadly ovate, 3-gonous; tip and style scabrid. Stigmas 3.

Northern Island: foot of Tongariro, and moist bases of cliffs on the east coast, *Colenso.*

7. **S. nitens,** *Hook. f.—Chætospora nitens*, Br.; Fl. N. Z. i. 82. Rhizome slender, creeping. Culms densely tufted, 2–12 in. high, wiry, rigid, slender, leafy at the base. Leaves slender, erect, longer than the culms, semiterete, deeply grooved in front; sheaths black-brown, shining. Spikelets sessile, fascicled, 2–10 together, terminal or nearly so, $\frac{1}{6}$–$\frac{1}{4}$ in. long, ovoid, turgid, 2- or 3-flowered, dark-brown, shining. Glumes obscurely distichous, 4–6, broadly ovate, obtuse, nerveless, grooved. Bristles numerous, or 6 plumose at the base. Nut pale-brown, smooth, 3-gonous. Stigmas 3.

Northern Island: sandy flats, Porangahau, and Cape Palliser, *Colenso.* A very common Tasmanian and Australian plant, most closely allied to a South Chilian one. It is doubtful whether the bristles are 6 and divided to the base, or very numerous.

3. CARPHA, Banks and Solander.

Tufted, grass-like herbs. Leaves narrow linear, rigid, obtuse. Culm cylindric, obtusely 3-angled or compressed.—Spikelets in fascicles or corymbs, subtended by sheathing bracts, pale-yellow, compressed. Glumes distichous, few; lower small, empty; upper large, opposite, 1 floriferous; uppermost small. Bristles 3, or 6 in 2 series, as long as the glumes, plumose. Stamens 3. Nut 3-gonous, 3-ribbed; base of style long, conical, persistent.

A small genus, native of Australia, Tasmania, New Zealand, and Fuegia.

1. **C. alpina,** *Br.;—Fl. N. Z.* i. 273. Culms 1–12 in. high, longer or shorter than the leaves. Leaves grooved above, with broad, shining, smooth or slightly scabrid sheaths. Spikelets fascicled, $\frac{1}{3}$–$\frac{1}{2}$ in. long. Glumes 4–6, linear-oblong, glabrous, shining, concave. Bristles large, flat, feathery. Stigmas 2.

Northern Island: Taupo plains and top of the Ruahine range, *Colenso.* **Middle** Island: Nelson mountains, alt. 4–5000 ft., *Bidwill;* Chalky Bay, *Lyall.* Also a native of the alps of South-east Australia and Tasmania.

4. SCIRPUS, Linn.

Culms erect from creeping rhizomes, 3-gonous or 3-quetrous, stout or slender, leafy or leafless, sheathed at the base.—Spikelets fascicled umbelled panicled or corymbose, terete or angled, rarely compressed. Glumes numerous, imbricated all round the rachis, all or most floriferous. Bristles 2–8, usually scabrid. Stamens 1–3. Nut 3-gonous or compressed, tipped with the persistent, often swollen base of the style. Stigmas 2 or 3.

A very extensive genus, found in all parts of the globe, chiefly in watery places.

Culms 3-gonous, leafy. Leaves flat. Spikelets terminal. Involucre long, leafy . . . . 1. *S. maritimus.*
Culms terete, spongy, leafless. Spikelets lateral . . . . 2. *S. lacustris.*
Culms 3-quetrous, leafy at base. Leaves 3-quetrous. Spikelets lateral . 3. *S. triqueter.*

1. **S. maritimus,** *Linn.*;—*Fl. N. Z.* i. 269. Roots tuberous. Culms leafy, 2–6 ft. high, acutely 3-gonous. Leaves long, flat, keeled, edges scabrid. Umbels terminal, involucrate, irregular; rays 6–10. Spikelets 1–3, sessile, ovoid, pale brown, ½ in. long; involucral leaves long, grassy. Glumes numerous, glabrous or scaberulous, membranous, ovate, obtuse, entire or 2-fid, awned or mucronate. Nut 3-gonous or ovate, compressed. Bristles 2 or 3, retrorsely scabrid. Anthers twisted after flowering, apiculate. Stigmas 2 or 3.

Northern and **Middle** Islands: near the sea, *Banks and Solander*, etc., abundant. Root formerly eaten by the natives. Abundant in Australia, Tasmania, and most other temperate parts of the world.

2. **S. lacustris,** *Linn.*;—*Fl. N. Z.* i. 269. Rhizome stout, creeping. Culms terete, leafless, spongy, 2–8 ft. high, sheathed at the base. Spikelets in lateral irregular panicles, ovoid, ¼–⅓ in. long. Glumes broadly ovate, membranous, obtusely 3-fid, mucronate, ciliate. Bristles 6, retrorsely scabrid. Anthers apiculate. Nut 3-gonous, shortly obovoid.

Abundant in lakes, pools, etc., throughout the **Northern** and **Middle** Islands, *Banks and Solander*, etc. Also found in all temperate and many tropical countries. Used in Europe for coopering casks, making mats, packing bottles, etc.

3. **S. triqueter,** *Linn.*;—*Fl. N. Z.* i. 269. Rhizomes stout, creeping. Culms rather slender, 1–2 ft. high, 3-quetrous. Leaves few, like the culms. Spikelets 1 or more, fascicled, sessile, lateral, ¼–½ in. long, broadly ovoid, dark-brown. Glumes membranous, broadly obovate-oblong, often ciliate, bracteate, 2-fid, with a short stiff awn. Anthers with a rather long point, twisted after flowering. Nut, bristles, etc., as in *S. maritimus.*

Common throughout the **Northern** and **Middle** Islands, in wet places, *Banks and Solander*, etc. Also found in Australia, Tasmania, and many other parts of the world.

5. ELEOCHARIS, Br.

Culms tufted, erect, terete, usually simple and leafless, sheathed at the base.—Spikelet solitary, terminal, erect, terete or angled. Glumes numerous, imbricated all round the rachis, most floriferous. Bristles 3–6, scabrid. Stamens 3. Nut 3-gonous or compressed, swollen at the top. Style jointed at the base, deciduous.

A large genus in the northern hemisphere, temperate and subtropical, rarer in the southern.

Culms stout, hollow, septate . . . . 1. *E. sphacelatus.*
Culms slender, pith continuous . . . . 2. *E. gracilis.*

1. **E. sphacelata,** *Br.*;—*Fl. N. Z.* i. 269. Culms as thick as the little

finger, 1–2 ft. high, cylindrical, hollow, with distant joints, contracted towards the top, stoloniferous at the base. Spikelet 1–2 in. long, terminal, solitary, erect, cylindric. Glumes numerous, linear-oblong, obtuse, membranous, flat, $\frac{1}{4}$–$\frac{1}{3}$ in. long. Bristles 6–8, stout, retrorsely scabrid. Nut obovoid, compressed.

Northern Island: in marshy places, *Cunningham*, *Colenso*; Bluff Island, *Lyall*; Auckland, *Sinclair*. A common Indian, Pacific Island, and Australian plant, usually tropical, but extending into Tasmania.

2. **E. gracilis,** *Br.*;—*Fl. N. Z.* i. 270. Culms creeping, deeply grooved, slender, striate, cellular, 6–24 in. high, sheathed at the base, sheaths chestnut-brown, mucronate. Spikelet very pale, white or reddish, terete, $\frac{1}{4}$–$\frac{3}{4}$ in. long, narrow, ovoid, subacute. Glumes few, linear-oblong or obovate, obtuse, membranous, transparent, flat. Bristles 4–6, retrorsely scabrid. Nut broadly obovate, compressed, 3-gonous, shining. Stigmas 3.

Var. β. *gracillima.* Culms very slender. Spikelet short. Glumes ovate-oblong, lower longer. Bristles 5 or 6, longer than the nut.—*E. acicularis*, A. Cunn., not Linn.

Var. γ. *radicans.* Small, rhizome stout, creeping, almost woody. Culms filiform, 2–3 in. high. Spikelet $\frac{1}{10}$–$\frac{1}{8}$ in. long. Glumes few. Bristles 6.

Common in boggy places throughout the **Northern** and **Middle** Islands, *Banks and Solander*, etc. Var. β. Bay of Islands. Var. γ. Bay of Islands, in sandy places. Also abundant in Australia and Tasmania. Too nearly allied to the European and almost ubiquitous *E. palustris*, which has only 2 stigmas. The var. γ is a very peculiar-looking plant, and may be a different species.

6. ISOLEPIS, Br.

Usually very small, slender, tufted green herbs. Leaves slender or 0.—Spikelets small, terete, solitary or fascicled, lateral, rarely panicled or solitary and terminal, basal in *I. basilaris*. Glumes few or numerous, imbricated all round the rachis, most floriferous. Bristles 0. Stamens 1–3. Nut 3-gonous or compressed, usually tumid at the top; style jointed on to its top. Stigmas 2 or 3.

A very large genus, found in all parts of the globe.

Culms stout, leafless, pungent. Spikelets capitate . . . . . . 1. *I. nodosus.*
Culms slender, leafless, usually flaccid and proliferous . . . . . 2. *I. prolifer.*
Culms very slender, 1-leaved. Spikelets 1 or 2. Nut smooth . . 3. *I. riparia.*
Culms short, rigid, 1- or 2-leaved. Spikelets 1–3. Nut pitted . . 4. *I. cartilaginea.*
Culms slender, leafy. Spikelets 1 or 2. Nut white, smooth . . 5. *I. aucklandica.*
Culms short, leafy. Spikelets 1 or 2, basilar. Nut obovoid . . . 6. *I. basilaris.*

1. **I. nodosa,** *Br.*;—*Fl. N. Z.* i. 270. Culms stout, tall, leafless, 1–3 ft. high, compressed below, cylindric above, sheathed at the base, tips acute. Spikelets aggregated into dense globose heads, below the erect pungent top of the culm, ovoid, $\frac{1}{8}$ in. long. Glumes broadly ovate, obtuse, coriaceous, concave. Nut compressed, scarcely 3-gonous, smooth, polished, pale-brown, apiculate; stigmas 3.

Throughout the **Northern** and **Middle** Islands: abundant in marshy places, *Banks and Solander*, etc. **Kermadec** Islands, *M'Gillivray*. Also common in Australia and Tasmania, and very closely allied to the widely distributed *I. Holoschœnus*.

2. **I. prolifer,** *Br.*;—*Fl. N. Z.* i. 271. Very variable in size and habit. Culms stout or slender, weak, often filiform, striate, stoloniferous, usually leafless, terete or compressed, 2–12 in. long, sheathed at the base, usually obtuse at the tip. Spikelets small, $\frac{1}{12}$–$\frac{1}{6}$ in. long, ovoid, aggregated on

the side of the culm, few or many, rooting and throwing out culms from their base in wet places. Glumes small, coriaceous, red-brown, obtuse, concave in flower, keeled in fruit; striate, red-brown, with green nerves and margin. Stamen 1. Nut broadly obovate, striate, pale, dotted, compressed, 3-gonous. Stigmas 3.—*I. setacea*, A. Cunn. in part, not Br.

Throughout the **Northern** and **Middle** Islands, abundant, *Banks and Solander*, etc. Equally abundant and variable in Australia and Tasmania; also found in S. Africa, St. Helena, and I think the same plant occurs in Chili. More than one species (and perhaps states of some of the following) may be confounded under forms of this.

3. **I. riparia,** *Br.*—*I. setacea*, Fl. N. Z. i. 271 in part, not Br. Small, 1–4 in. high, densely tufted. Culms almost filiform, with 1 leaf at the base, tips acute. Leaf setaceous. Spikelet 1, rarely 2, lateral, ovoid, $\frac{1}{8}$–$\frac{1}{10}$ in. long. Glumes 5–8, broadly ovate, obtuse, keeled, green or brown. Stamens 3. Nut compressed, 3-gonous, smooth or minutely dotted. Stigmas 3.—Fl. Tasm. ii. 89. t. 145 C.; *I. setosa* (misprint for *setacea*), Raoul.

Common in marshy places throughout the **Northern** and **Middle** Islands. Also found in South Africa and Chili.

4. **I. cartilaginea,** *Br.;—Fl. N. Z.* i. 271. Culms short, densely tufted, rigid, leafy at the base, $\frac{3}{4}$–2 in. high, tips acute. Leaf subulate, channelled, keeled. Spikelets 1 or more, lateral, small, $\frac{1}{8}$ in. long. Glumes concave, grooved, thickly keeled, very coriaceous, with a thick obtuse mucro and chestnut-brown lines on each side. Stamens 2 or 3. Nut elliptic-oblong, obtusely 3-gonous, white, punctulate. Stigmas 3.—Fl. Tasm. ii. 89. t. 145 A, B.

Northern Island: wet places, probably common, but overlooked, *Colenso, Sinclair.* Abundant in Australia, Tasmania, and S. Africa; and I think also in Chili.

5. **I. aucklandica,** *Hook. f. Fl. Antarct.* i. 88. *t.* 50. A bright-green, flaccid, very leafy, densely tufted species. Culms 3–5 in. high, rather stout, chestnut-brown, much branched at the base, terete, striate, tips obtuse. Leaves similar to the culms and as long, plano-convex, obtuse. Spikelet 1, rarely 2, $\frac{1}{12}$–$\frac{1}{10}$ in. long, broadly ovoid, pale or brown. Glumes 6–8, ovate, concave, coriaceous, obtuse, with an obtuse excurrent thick keel. Stamens 3. Nut 3-gonous, white, quite smooth, not polished. Stigmas 3.

Northern Island: marshy places in the interior, *Colenso.* **Lord Auckland's** group and **Campbell's** Island: common in marshes. In many respects this so closely resembles some states of *I. prolifer*, that I should not be surprised if they proved identical, meanwhile the differences in habit are too great to admit of their union; the nut is usually larger and stamens 3. Some of the Chilian specimens named *I. pygmæa* in the 'Flora Antarctica' are certainly identical with this, the nut being rather smaller than usual. It is also very nearly allied to, if not the same as, a S. African plant.

6. **I. basilaris,** *Hook. f., n. sp.* Culms small, densely tufted, green, leafy, 1–2 in. high, much branched at the base. Leaves setaceous, plano-convex, channelled above, obtuse at the tip. Spikelets 1 or 2, almost radical, $\frac{1}{8}$–$\frac{1}{6}$ in. long, narrow oblong, pale green. Glumes rather numerous, oblong, obtuse, membranous, with an indistinct, thick, excurrent midrib. Stamens 2. Nut much smaller than the glume, broadly obovoid, compressed, not angled nor 3-gonous, pale-brown, smooth, not polished. Stigmas 2.

Northern Island: mud banks of Ngaruroro river, *Colenso.* A most distinct little species, with the almost radical spikelets of the Australian *I. acaulis*, and the Cape *I. humilis*, but differing in character from both.

7. DESMOSCHŒNUS, Hook. f.

Culms tall, rigid, obtusely 3-gonous, tufted, erect from a stout, woody, scaly rhizome. Leaves squarrose, very harsh, margin scaberulous.—Inflorescence a span long, of clusters of sessile, red-brown, globose spikelets, spirally arranged round the culm, subtended by rigid squarrose subulate leafy bracts. Glumes numerous, imbricate all round the rachis, all or most floriferous. Bristles 0. Stamens 3; anthers with a terminal awn. Nut compressed; style deciduous, not jointed at the base. Stigmas 3.

A very remarkable plant, agreeing with *Isolepis* in the floral characters, but of most peculiar habit.

1. **D. spiralis,** *Hook. f. Fl. N. Z.* i. 272. Culms 1–3 ft. high, leafy at the base, glabrous, surmounted by a sort of malformed interrupted catkin, bearing long squarrose leafy wiry bracts, with decurrent bases. Leaves very numerous, concave, keeled, 3-gonous towards the long harsh points; margins cutting with minute teeth. Inflorescence a span long. Spikelets ⅛ in. long. Glumes concave, striated, obovate, obtuse, shining. Nut quite smooth, broadly obovate, obtuse.—*Anthophyllum Urvillei*, Steud. Synops. Pl. Cyp. 160; *Isolepis spiralis*, A. Rich. Fl. 105. t. 19.

Northern and **Middle** Islands, *Banks and Solander*, etc., not uncommon in sandy seashores, as far south as Canterbury.

8. FIMBRISTYLIS, Vahl.

Culms tufted, usually rather soft, nearly terete or 3-gonous, generally leafy below. Leaves grasslike.—Spikelets usually in terminal involucrate corymbose or umbellate panicles, terete or angled. Glumes numerous, mostly floriferous, imbricated all round the rachis. Bristles 3. Stamen 1. Nut surmounted with the tuberous pilose or villous persistent base of the deciduous style. Stigmas 2 or 3.

A large tropical genus, rare in temperate climates.

1. **F. dichotoma,** *Vahl.*—*F. velata*, Fl. N. Z. i. 272, not Br. Culms densely tufted, annual, slender, leafy at the base, 4–8 in. high. Leaves very slender, shorter than the culms, flaccid, terete, grooved. Umbels with very many slender rays 1–2 in. long, surrounded with long involucral leaves. Spikelets ⅛–⅕ in. long, 3–5 on each ray, pedicelled or sessile, ovoid, pale-brown. Glumes numerous, spreading, narrow-lanceolate, acuminate; keel scabrid. Stamen 1. Nut nearly orbicular, compressed, margins thickened, crowned with the bulbous base of the style, from which long hairs descend over its face. Stigmas 2.

Northern Island: Bay of Islands, Auckland, etc., *Colenso, Sinclair*. A common Australian tropical and subtropical weed, perhaps introduced into New Zealand.

9. CLADIUM, Linn.

Tufted, rigid sedges, of various habit. Culms leafy or leafless, erect, slender or stout, flat terete angled or compressed.—Spikelets generally panicled, often with leafy or spathaceous bracts, small, terete, 1–3-flowered. Glumes few, imbricated all round the rachis, mostly empty. Bristles or hypogynous scales

0. Stamens 3; filaments rarely lengthening after flowering. Nut 3-gonous, usually terminated by the persistent, much enlarged base of the style, but triquetrous and cuspidate in *C. Sinclairii*. Stigmas 2 or 3.

A large and widely diffused genus, especially in tropical countries and temperate Australia; rarer in the north temperate hemisphere. The species differ from *Gahnia* more in habit than in any floral character.

Culms and leaves terete. Filaments not elongating. Nut with a tumid top.

Culms leafy. Bracts spathaceous. Glumes acuminate . . . . 1. *C. glomeratum.*
Culms leafy. Bracts small. Glumes awned . . . . 2. *C. teretifolium.*
Culms leafy. Pith jointed. Panicle very large . . . . . . 3. *C. articulatum.*
Culms leafy. Panicle very long, slender. Spikelets 1-flowered . 4. *C. Gunnii.*
Culms leafless. Sheaths very long, with small flattened tips . . 5. *C. junceum.*

Culms and leaves flat. Filaments elongating. Nut with a 3-quetrous cuspidate top . . . . 6. *C. Sinclairii.*

1. **C. glomeratum,** *Br.;—Fl. N. Z.* i. 275. Culms stout, tufted, 1–2 ft. high, glabrous, terete, soft, leafy; pith not jointed internally. Leaves terete, subulate. Panicle contracted, distantly much or sparingly branched, 2–6 in. long; bracts large, membranous, spathaceous, lower 1–2 in. long. Spikelets very numerous, fascicled, $\frac{1}{6}$ in. long. Glumes red-brown, ovate, acuminate, ciliate, glabrous at the back. Nut 3-gonous, polished; top very large, tumid, acute, puberulous.—*Fuirena rubiginosa*, Sprengel.

Northern and **Middle** Islands: clayey and marshy places, frequent, *Banks and Solander*, etc. A common Australian and Tasmanian plant.

2. **C. teretifolium,** *Br.;—Fl. N. Z.* i. 276. Habit, size, and foliage of *C. glomeratum*, but panicle shorter, 2–4 in. long, much more dense; primary branches shorter and closer together; bracts short, not spathaceous; glumes awned; nut corky, longitudinally grooved and wrinkled.

Northern Island: common in marshy places, *Banks and Solander*, etc. **Middle** Island: Canterbury, *Sinclair and Haast*. This is identical with *C. teretifolium* of Sieber's 'Agrostotheca,' No. 6 (*C. rigidifolium*, Presl, according to Steudel, Synops. Cyp. 152); it is a native of New South Wales.

3. **C. articulatum,** *Br.;—Fl. N. Z.* i. 276. Culms tufted, erect, 3–5 ft. high, very stout, terete; pith jointed internally. Leaves shorter than the culms, terete, pungent; pith jointed like the culms. Panicle very large, lax, effuse, 6–10 in. long, drooping; branches long. Bracts short compared with the panicles. Spikelets very numerous, pale-brown, $\frac{1}{8}$ in. long, 3- or 4-flowered. Glumes ovate-lanceolate, acute or very shortly awned, puberulous. Nut unripe, smooth with a large tumid top. Stigmas 2 or 3.

Northern Island: Lake Rotoetara, *Colenso*; Auckland, *Sinclair*. Also found in Australia and the Pacific Islands.

4. **C. Gunnii,** *Hook. f.—Lampocarya tenax*, Fl. N. Z. i. 277. Culms tufted, very slender, 2–4 ft. high, terete, leafy at the base, compressed or grooved towards the top. Leaves like the culms, with pungent, subulate points; sheaths long, narrow, red. Panicle very long, lax, slender, 6–24 in. long; branches few, distant, very slender, few-flowered. Spikelets few, alternate, not crowded or fascicled, $\frac{1}{6}$ in. long, 1-flowered. Glumes few, all long, lanceolate-subulate, very acuminate; upper awned, much larger, divaricating in fruit. Nut pedicelled, turgid, 3-ribbed, glabrous; top large, tumid. (Described chiefly from Australian specimens.)—Fl. Tasm. ii. 95, 148 B.

Northern Island: Opurago and Otago, *Banks and Solander;* Bay of Islands, *Cunningham;* Auckland, *Sinclair.* This is clearly the Australian plant, though the panicle is very much longer than in my Tasmanian specimens, being quite like Mueller's Victorian.

5. **C. junceum,** *Br.—Lepidosperma striata,* Fl. N. Z. i. 279, not Br. Culms ½–2 ft. high, densely tufted, rigid, rising from a woody scaly rhizome, terete, leafless, covered with a long appressed sheath; tip of the latter vertically flattened, sickle-shaped. Spikelets in a very short, simple or sparingly-branched, few-flowered spike, ½–1 in. long, 1-flowered; bracts small, obtuse, appressed. Glumes oblong-lanceolate, acuminate or awned, scaberulous at the back. Nut 3-gonous, obovoid; top tumid, hemispherical, puberulous.

Northern Island: abundant in fresh and brackish water marshes, etc., *Banks and Solander,* etc. Also a native of Australia and Tasmania.

6. **C. Sinclairii,** *Hook. f.—Vincentia anceps,* Hook. f. Fl. N. Z. i. 276. Culms tall, coarse, several ft. high, quite flat, ¼ in. broad, smooth. Leaves quite flat, striate, 1 in. broad; edges sharp but not cutting. Panicle 1 ft. long, soft, excessively branched, nodding; bracts sheathing; branches drooping, arising from smaller bracts. Spikelets small, fascicled, pale-brown, 2- or 3-flowered. Lower glumes and bracts awned, striated, scaberulous. Flowers about 3, of which 1 only seems to ripen fruit. Stamens elongating greatly after flowering. Nut attenuated at both ends; sharply 3-angled at the cuspidate top, crustaceous, mottled with red, pedicel with a tuberous base.

Northern Island: in swamps, *Banks and Solander;* east coast and interior, *Colenso;* Auckland, *Sinclair;* Mercury Bay, *Jolliffe.* I have referred this to *Cladium,* into which the genera *Baumea, Chapelliera,* and *Vincentia* must fall. There being a *Cladium anceps* renders it necessary to change the specific name of this.

10. GAHNIA, Forst.

(*Including* LAMPOCARYA, *Br.*)

Tall, coarse, leafy, rigid, scabrid sedges. Culm erect, terete or 3-gonous, stout. Leaves very long, involute, harsh and cutting, with scabrid edges and points.—Spikelets excessively numerous, small, terete, 1- or 2-flowered, disposed in a terminal branched, often effuse panicle. Glumes few, imbricated all round the rachis, the terminal (or last 2) minute, floriferous. Bristles and hypogynous scales 0. Stamens 3, 4, or 6; filaments usually lengthening much after flowering, and often twisting together and holding the nut suspended. Nut very hard, obscurely 3-gonous, sessile, obscurely thickened at the top. Stigmas 3 or 4, sometimes 2-fid.

A small genus, native of Australia, Tasmania, and the Malayan and Pacific Islands. Closely allied to *Cladium,* but very different in their more harsh habit and foliage.

Nut of stony hardness, transversely grooved inside. Stamens 4.

Spikelets black, ¼ in. long. Nut pale red-brown . . . . 1. *G. setifolia.*
Spikelets red-brown, ⅙–⅕ in. long. Nut pale-brown . . . . 2. *G. procera.*

Nut not transversely grooved internally. Stamens 3 or 4. (Lampocarya, *Br.*)

Leaves with very slender points. Stamens 4. Nut black . . . . . 3. *G. lacera.*
Panicle 2–3 ft. long. Branches very long, pendulous. Stamens 4.
Nut yellow . . . . 4. *G. xanthocarpa.*
Culms robust, angled, scabrous. Stamens 3. Nut black . . . . . 5. *G. arenaria.*

1. **G. setifolia,** *Hook. f. Fl. N. Z.* i. 279. A very large, tall, coarse, cutting sedge, 2–6 ft. high, forming huge tussocks in woods. Culms erect, terete, as thick as a goose-quill. Leaves very long, convolute, with cutting edges and long scabrid points. Panicle 1–2 ft. long, much laxly branched, very slender; bracts leafy, with long black sheaths. Spikelets pedicelled, ovoid, nearly $\frac{1}{4}$ in. long, 2-flowered. Glumes coriaceous, short, black-brown; outer downy, awned; inner acuminate. Upper flower perfect, lower male. Stamens 4; filaments lengthening much after flowering. Nut $\frac{1}{4}$ in. long, narrow elliptic-oblong, acute at both ends, when ripe red-brown, polished, grooved on one side, transversely furrowed within.—*Lampocarya setifolia*, A. Rich.

Northern Island: abundant in woods, *Banks and Solander*, etc. **Middle** Island: Nelson, *Travers* (panicle broader, excessively branched; nut smaller). Very near the Tasmanian *G. psittacorum.*

2. **G. procera,** *Forst.*;—*Fl. N. Z.* i. 278. Habit and foliage of *G. setifolia*, but smaller. Culms only 2–3 ft. high; panicle more slender, with fewer, shorter branches; bracts more spathaceous; spikelets much larger, on longer pedicels. Spikelets $\frac{1}{3}$–$\frac{1}{2}$ in. long, of a dark purple-brown colour. Glumes coriaceous, few; outer longer than the spikelet, acuminate and awned. Stamens 4; filaments very long. Nut pale-brown, scarcely shining (perhaps not ripe), transversely grooved within.

Middle Island: Dusky Bay, *Forster*; Port Preservation, *Lyall.*

3. **G. lacera,** *Steudel.*—*Lampocarya lacera*, A. Rich.;—Fl. N. Z. i. 277. A tall, coarse, harsh, densely-tufted sedge, 3–5 ft. high. Culms stout, leafy, as thick as a goose-quill. Leaves very long, involute, with cutting edges and long scabrid points. Panicle erect or inclined, 1–1$\frac{1}{2}$ ft. long, much branched, leafy. Spikelets alternate, pedicelled, pale-brown. Glumes membranous, downy, ovate, acuminate, awned, $\frac{1}{6}$–$\frac{1}{5}$ in. long. Stamens 4; filaments much elongating. Nut $\frac{1}{8}$ in. long, elliptic-oblong, black, shining, obscurely 3-gonous, not transversely furrowed within.

Northern and **Middle** Islands: abundant in woods, *Banks and Solander*, etc.

4. **G. xanthocarpa,** *Hook. f.*—*Lampocarya xanthocarpa*, Fl. N. Z. i. 278. A very stout, tall, densely-tufted, harsh, leafy sedge. Culms robust, 4–6 ft. high, as thick as the little finger. Leaves very long, involute, with scabrid, cutting edges. Panicle leafy, 2–3 ft. long, with numerous very long, pendulous and flexuous branches, a span long. Spikelets alternate, pedicelled, $\frac{1}{8}$–$\frac{1}{6}$ in. long, dark-brown. Glumes rather coriaceous; outer ovate, awned; inner acute. Stamens 4; filaments much elongating. Nut $\frac{1}{8}$ in. long, unripe pale-yellow, not transversely furrowed within.

Northern Island: east coast, *Banks and Solander*; Auckland, *Sinclair*; interior, *Colenso.* A noble plant.

5. **G. arenaria,** *Hook. f.*—*Lampocarya affinis*, Brong.;—Fl. N. Z. i. 277. A densely-tufted, harsh, very leafy, subsquarrose sedge. Culm leafy, 1–1$\frac{1}{2}$ ft. high, obtusely 3-gonous, scabrous. Leaves excessively numerous, much longer than the culms, narrow, involute, scabrid; sheaths very short. Panicle erect, leafy, rigid; branches distant, very short; bracts ex-

cessively long and foliaceous. Spikelets in dense ovoid fascicles, ½–1 in. long, narrow-lanceolate, very shortly pedicelled, turgid. Glumes rather coriaceous, obscurely distichous, very concave, ovate-lanceolate, acuminate, all awned, lower shorter. Stamens 3, filaments scarcely lengthening after flowering. Nut elliptic-oblong, subacute, black, polished, obscurely 3-gonous, not transversely grooved within.—*Morelotia Gahniæformis*, Gaudichaud, in Freyc. Voy. 416. f. 28.

Northern and **Middle** Island: common on hills, sandy shores, etc., *Banks and Solander*, etc. Very near a Sandwich Island species.

11. LEPIDOSPERMA, Labill.

Culms tufted, erect, from a stout creeping rhizome, very rigid, compressed terete angled or flat and leaf-like, with cutting edges, leafy at the base only. Leaves usually equitant, rigid, cutting.—Spikelets small, in rigid branched compressed spikes or panicles, 1- or 2-flowered, with an awned bract at the base. Glumes few, imbricating all round the rachis, brown, rigidly coriaceous, the upper 1 or 2 floriferous, lower flower alone fertile. Stamens 3. Hypogynous scales 6, thick, very small, adnate to the base of the nut. Nut coriaceous or osseous, terminated by the base of the style.

A very large Australian and Tasmanian genus, of which a few species are found in New Zealand and the Malay Islands, China, and Ceylon.

Culms and leaves 3- or 4-angled . . . . 1. *L. tetragona.*
Culms and leaves flat . . . . 2. *L. concava.*

1. **L. tetragona,** *Labill. Fl. Nov. Holl. t.* 17.—*L. australis*, Hook. f. Fl. N. Z. i. 279. Culms tufted, 1–1½ ft. high, irregularly 3- or 4-angled, very rigid, smooth. Leaves rigid, 2–8 in. long, 3- or 4-angled. Spikelets crowded, forming a short, oblong, terminal head ½ in. long; bract short, sheathing, with a subulate erect point. Spikelets nearly ¼ in. long. Glumes 6–8, very coriaceous, acuminate, lower awned, terminal floriferous. Stamens and stigmas 3. Scales connate into a small 6-lobed cup. Nut with a tumid top.—*Vauthiera australis*, A. Rich. Fl. 107. t. 20.

Northern Island: banks of lakes, etc, common, *Banks and Solander*, etc. **Middle** Island: Nelson, *Sinclair*; Akaroa, *Raoul.* A native of Australia and Tasmania.

2. **L. concava,** *Br*—*L. longitudinalis*, Hook. f. Fl. N. Z. i. 279, not Labill. Culms flat, 3–4 ft. high, ¼–⅕ in. broad, with scabrid cutting edges, thin, concave on one side. Leaves flat like the culms, acuminate. Panicle erect, rigid, sparingly branched, 2–3 in. long; branches distant; bract acute, spathaceous. Spikelets few, short, in appressed fascicles of 2 or 3 within an awned bract, 1-flowered. Glumes ovate, puberulous, rough on the back, awned. Stamens 3. Scales minute, connate. Nut obovate, terminated by the tumid base of the style.—Fl. Tasm. i. 91. t. 146 B; *L. elatior*, A. Cunn. not Br.

Northern Island: not uncommon in clay hills and Bay of Islands, *Cunningham*; Auckland, *Sinclair*. A native of Australia and Tasmania.

12. OREOBOLUS, Br.

A low, rigid, densely tufted plant, forming large cushion-like patches on the mountains. Culms short, densely compacted, leafy throughout, divided. Leaves subulate, distichous or imbricated all round.—Scape axillary, short, compressed, 1-flowered, strict, rigid, lengthening after flowering. Glumes 2, deciduous. Hypogynous scales 6, in 2 series, perianth-like, persistent. Stamens 3. Style deciduous, bulbous at the base; stigmas 3. Nut enveloped above the middle in the appressed scales, crustaceous, obovoid, 3-gonous, with a broad, terminal, depressed area.

A remarkable genus, confined to the alps of Tasmania, Australia, New Zealand, and Fuegia.

1. **O. Pumilio,** *Br.—O. pectinatus,* Hook. f. Fl. N. Z. i. 275. Culms 1–3 in. long, usually flattened. Leaves closely densely imbricated, distichous and spreading, or suberect and imbricating all round, curved, linear subulate, $\frac{1}{2}$–1 in. long, pungent; sheaths hard, striated. Spikelets minute. Scape an inch long after flowering. Glumes lanceolate, $\frac{1}{4}$ in. long. Scales lanceolate subulate.—*O. pectinatus,* Fl. Antarct. i. 87. t. 49.

Northern Island: mountainous regions, Taupo plains, and tops of the Ruahine range, *Colenso.* **Middle** Island: Morses mountain, alt. 6000 ft., *Bidwill;* Southern Alps, *Haast.* **Lord Auckland's** group and **Campbell's** Island, abundant, *J. D. H.* Found also in the Tasmanian and Victorian Alps.

13. UNCINIA, Persoon.

Tufted, usually grass-like sedges, with fibrous roots. Culms stout or slender, more or less 3-gonous. Leaves long, narrow, flat, involute or keeled.—Spikelet solitary at the end of the culm, linear-elongate, terete, many-flowered, upper part contracted, male. Glumes imbricated all round the rachis, all floriferous; upper *male* with 3 stamens, and no rudiment of scales or ovary; lower *female;* ovary contained in a compressed urceolate utricle (*perigynium* of authors); its style and 3 stigmas exserted; a long rigid exserted bristle, hooked at the tip, arises from the base of the ovary, also within the utricle. Nut 3-gonous, flattened, included in the utricle.

A considerable genus, native of Tasmania, Australia, New Zealand, Tristan d'Acunha, Abyssinia, Fuegia, and the Andes of S. America. The species have been elaborated by the late Dr. Boott for the 'New Zealand Flora,' which descriptions have in great part been embodied in the following pages. The New Zealand forms are most puzzling, *U. leptostachya, Banksii, Sinclairii, rubra,* and *ferruginea,* seem all very distinct, the rest are very difficult of discrimination.

Utricle scabrous towards the top.

Culms slender. Bract subulate or 0. Spikelet very slender . . . 1. *U. leptostachya.*
Culms short, stout. Bract subulate. Spikelet short, stout . . . 2. *U. Sinclairii.*

Utricle perfectly smooth. Glumes closely imbricate.

Culms short. Spikelet $\frac{1}{2}$–$\frac{3}{4}$ in. Bract 0 . . . . 3. *U. compacta.*
Culms long. Spikelet 3–6 in. Bract leafy. Glumes $\frac{1}{10}$ in. . . . 4. *U. australis.*
Culms long. Spikelet 3–6 in. Bract leafy. Glumes $\frac{1}{6}$–$\frac{1}{5}$ in. . . . 5. *U. ferruginea.*

Utricle perfectly smooth. Glumes lax, all or the lower distant.

Leaves longer than culm, green when dry.
Leaves $\frac{1}{14}$–$\frac{1}{10}$ in. broad. Spikelets 2–4 in. Bract leafy . . . . 6. *U. cæspitosa.*

Leaves $\frac{1}{10}$–$\frac{1}{16}$ in. broad. Spikelets 1–2 in. Bract 0. Glumes equalling utricle . . . . 7. *U. rupestris.*
Leaves $\frac{1}{40}$–$\frac{1}{30}$ in. broad. Spikelets $\frac{3}{4}$–$1\frac{1}{2}$ in. Bract 0. Glumes equalling utricle . . . . 8. *U. filiformis.*
Leaves $\frac{1}{40}$–$\frac{1}{30}$ in. broad. Spikelets 2–3 in. Bract 0. Glumes $\frac{1}{2}$ as long as utricle . . . . 9. *U. Banksii.*
Leaves shorter than culm, red when dry . . . . 10. *U. rubra.*

1. **U. leptostachya,** *Raoul, Choix,* 12. *t.* 5;—*Fl. N. Z.* i. 286. Culms filiform, scaberulous upwards, 1–2 ft. long. Leaves longer than the culms, narrow, $\frac{1}{12}$ in. broad. Spikelet $1\frac{1}{2}$–4 in. long, very slender, lax-flowered, upper male part filiform; bract setaceous. Glumes small, distant, ovate- or subulate-lanceolate, acute or obtuse, membranous. Utricle about as long as the glume, fusiform, nearly $\frac{1}{4}$ in. long, scabrid above, margins ciliate, faintly nerved.

Var. *β. distans.* Spikelet 1–2 in. long. Utricles as long as the acuminate glumes.—*U. distans,* Boott, Fl. N. Z. i. 285.

Var. *γ. scabra.* Spikelets 1–2 in. long. Utricle longer than the acute glumes.—*U. scabra,* Boott, Fl. N. Z. i. 286.

Northern and **Middle** Islands: probably common; top of the Ruahine range and Titiokura, *Colenso;* Akaroa, *Raoul.* I can hardly distinguish the above varieties as such.

2. **U. Sinclairii,** *Boott, mss.* Culms 4–8 in. high, stout, smooth. Leaves flat, grassy, $\frac{1}{10}$ in. broad, shorter than the culm. Spikelet $\frac{3}{4}$–1 in.; long, robust; male portion very short, narrow, cylindric; bract 0. Glumes imbricate, broadly ovate-oblong, obtuse, green with white membranous margins. Utricle longer than the glumes, $\frac{1}{4}$ in. long, narrow obovate-lanceolate, 3-gonous, scabrous above, edges ciliate; style tumid at the base.

Middle Island: Tarndale?, *Sinclair;* Lake Tennyson, alt. 4400 ft., *Travers.*

3. **U. compacta,** *Br.;* var. **divaricata.**—*U. divaricata,* Boott;—Fl. N. Z. i. 286. Culms short, stout, 4–6 in. high, smooth. Leaves much longer than the culms, flat, $\frac{1}{12}$–$\frac{1}{6}$ in. broad, rigid. Spikelet short, $\frac{1}{2}$–$\frac{3}{4}$ in., rather stout, $\frac{1}{4}$ in. diam., upper $\frac{1}{4}$ male; bract 0. Glumes closely imbricate, lanceolate, acute, deciduous, pale, the lower cuspidate. Utricle $\frac{1}{6}$ in. long, rather turgid, lanceolate, quite glabrous, spreading when old; bristle incurved, as long, stout.

Northern Island: top of the Ruahine range, *Colenso.* **Middle** Island: Milford Sound, *Lyall.* This differs from the Tasmanian plant only in the pale-green, not brown, glumes; it is also near the *U. Kingii* of Fuegia.

4. **U. australis,** *Persoon;*—*Fl. N. Z.* i. 287. Culms 6–12 in. high, rigid, smooth. Leaves very numerous, twice as long as the culm, flat, $\frac{1}{6}$–$\frac{1}{4}$ in. broad, smooth or scabrid. Spikelet 3–6 in. long, cylindrical, $\frac{1}{6}$ in. broad, upper $\frac{1}{3}$ or $\frac{1}{4}$ slender male; bract long, foliaceous. Glumes closely imbricating, ovate, subacute, $\frac{1}{10}$ in. long, pale-green, faintly browned. Utricle as long as or shorter than the glume, elliptical-oblong, or fusiform and turgid in the middle, equally narrowed at both ends.—*U. compacta,* A. Rich., not Br.; *Carex uncinata,* Forst.

Northern and **Middle** Islands: common in woods.

5. **U. ferruginea,** *Boott. in Fl. N. Z.* i. 388. *t.* 64 B. Very similar in habit, stature, and dimensions, to *U. australis,* but the glumes are much

longer, $\frac{1}{6}$–$\frac{1}{3}$ in. long, subulate-lanceolate, long acuminate, dark-brown with green keel, and the utricle is more attenuate upwards, and less at the base.

Northern Island: sides of streams, base of Ruahine range, *Colenso;* Wellington, *Ralphs.*

6. **U. cæspitosa,** *Boott, in Fl. N. Z.* i. 287. Culms 6–12 in. high, rather stout. Leaves much longer than the culms, slender, $\frac{1}{14}$–$\frac{1}{10}$ in. broad, flat or 3-gonous, rather scabrid. Spikelet 1–4 in. long, rather stout; bract leafy. Glumes rather close, lower remote, lanceolate, obtuse acute or acuminate, pale-green. Utricle as long as the glume, narrow ovate or fusiform, narrowed at both ends, 3-nerved.

Northern Island: frequent in mountainous districts, *Colenso,* etc. A very variable plant, small forms, with narrow leaves and small spikelet, are seen to pass into *U. filiformis,* others with broader leaves run into *U. rupestris,* and the large form with foliaceous bracts into *U. australis.*

7. **U. rupestris,** *Raoul, Choix,* 13. *t.* 5;—*Fl. N. Z.* i. 286. Culms slender, 3–6 in. high. Leaves very much longer than the culm, flat, $\frac{1}{16}$–$\frac{1}{10}$ in. broad, nearly smooth. Spikelet small, slender, 1–2 in. long; bract setaceous or 0. Glumes very laxly imbricate, $\frac{1}{6}$ in. long, narrow oblong-lanceolate, obtuse or acuminate, pale-brown. Utricle about as long as the glume, very narrow-ovate or fusiform, attenuate upwards, 3-nerved.—*U. Hookeri,* Boott, in Fl. Antarct. i. 91. t. 51; *U. riparia,* Hook. f. in Hook. Journ. Bot. iii. 417, not Br.

Northern Island: Wellington, *Stephenson.* **Middle** Island: Akaroa, *Raoul;* Canterbury, *Travers.* **Lord Auckland's** group and **Campbell's** Island, *J. D. H.* Extremely near the Tasmanian *U. riparia,* Br., and perhaps only a form of that plant.

8. **U. filiformis,** *Boott, in Fl. N. Z.* i. 286. Culm extremely slender, 4–10 in. high. Leaves filiform, much longer than the culm, $\frac{1}{40}$–$\frac{1}{30}$ in. broad, 3-gonous, grooved, scabrid. Spikelets filiform, $\frac{3}{4}$–$1\frac{1}{2}$ in. long; bract 0. Glumes approximate, $\frac{1}{8}$ in. long, lanceolate, acuminate, pale-green. Utricle as long as the glumes, lanceolate, 3-nerved.

Northern Island: top of the Ruahine mountains, *Colenso.* **Middle** Island, *Lyall.*

9. **U. Banksii,** *Boott, in Fl. N. Z.* i. 287. Habit and foliage of *U. filiformis,* but the spikelets are 2–3 in. long, male portion sometimes female at the top, the glumes distant, short, not half the length of the slender, almost terete, 3-nerved utricle.

Northern and **Middle** Islands: not uncommon in damp woods, *Banks and Solander,* etc.

10. **U. rubra,** *Boott, in Fl. N. Z.* i. 287. *t.* 64 A. Whole plant red-brown when dry. Culms rigid, 6–12 in. high, scabrid above. Leaves shorter than the culm, rigid, long, striate, concave, $\frac{1}{16}$–$\frac{1}{12}$ in. broad. Spikelet rigid, 1–2 in. long, male portion short, of few glumes; bract 0. Glumes few, distant, ovate-lanceolate, obtuse or acute, concave, rigid, persistent. Utricle as long as or shorter than the glume, oblong-lanceolate, terete, faintly nerved.

Northern Island: east coast, Tehawera, and summit of Titiokura, *Colenso.*

14. CAREX, Linn.

Tufted, usually grass-like sedges, with creeping rhizomes and fibrous roots.

Culms stout or slender, obtusely or acutely 3-gonous. Leaves long, narrow, flat involute or keeled.—Spikelets solitary or numerous, simple or compound, usually arising from the sheaths of long, leafy bracts, unisexual or bisexual, oblong or linear, terete, many-flowered. Glumes imbricated all round the rachis, entire or 2-fid, obtuse cuspidate or awned, all floriferous. *Male* fl: stamens 3; ovary 0. *Female:* an ovary contained in a sessile or stalked, urceolate, compressed utricle (*perigynium* of authors); style exserted; stigmas 2 or 3. Nut 3-gonous or flattened, included in the utricle, rarely accompanied with a slender bristle which is not hooked at the tip.

An immense genus, found in all parts of the world, but especially in mountainous regions; it has been a subject of special study by my friend the late Dr. Boott, F.L.S., who prepared the descriptions of the New Zealand species (and of *Uncinia*) for the 'New Zealand Flora.' On re-examining the specimens with additional ones received since, for the present work, I have found it necessary to unite *C. fascicularis*, Soland., with *C. Forsteri*, Wahl., and *C. secta*, Boott, with *C. virgata*, Soland.; and *C. Lambertiana*, and perhaps *vacillans* should also be united with *C. dissita*, Soland. There are, no doubt, many other species to be found in New Zealand; in the Middle Island they have been scarcely at all collected.

A. *Spikelet solitary, simple, small; upper part male. Stigmas usually* 3.

Culms 3–6 in., shorter than the flat leaves . . . . 1. *C. pyrenaica.*
Culms 2–6 in., and leaves wiry, terete, grooved . . . . . . 2. *C. acicularis.*

B. *Spikelets collected into a solitary compound spike, small, androgynous. Stigmas* 2.

Spikelets with male flowers below.

Leaves flattish. Spikelets 1–3, pale. Utricle ovate, beaked, serrate above . . . . 3. *C. inversa.*
Leaves involute, wiry. Spikelets 1–4, brown. Utricle broad, not beaked, serrate above . . . . 4. *C. Colensoi.*
Leaves keeled. Spikelets many, squarrose. Utricle spreading, longer than the glume . . . . 5. *C. stellulata.*

Spikelets with male flowers above. (*Utricle with serrate wings above.*)

Culm 1–2 ft., slender. Spike short. Utricle as long as glume . 6. *C. teretiuscula.*
Leaves 1–3 ft., $\frac{1}{8}$ in. diam., harsh. Spike very long and slender . 7. *C. virgata.*
Leaves 1–3 ft., $\frac{1}{4}$–$\frac{1}{2}$ in. diam., harsh. Spike very long and stout 8. *C. appressa.*

C. *Spikelets several, distinct, axillary, in long leafy bracts, usually peduncled.*

α. *Spikelets all unisexual, or the female with a few male flowers at top. Stigmas* 2. (*Utricle not serrate above in any.*)

Spikelets 3–6, peduncled, erect. Glumes obtuse, smaller than the flat utricle . . . . 9. *C. Gaudichaudiana.*
Spikelets 5–7, peduncled, erect. Glumes mucronate or cuspidate, smaller than the flat utricle . . . . 10. *C. subdola.*
Spikelets 15–24, peduncled, long, pendulous, brown. Glume awned . . . . 11. *C. ternaria.*
Spikelets 4, sessile, short, pale. Glumes cuspidate. Utricle shining . . . . 12. *C. testacea.*

β. *Spikelets all unisexual, or the females with male flowers below. Glumes cuspidate. Stigmas* 2, *rarely* 3 *in* C. lucida.

Spikelets 4–6, approximate, sessile, short . . . . 13. *C. Raoulii.*
Spikelets 4–8, distant, peduncled, cylindric . . . . 14. *C. lucida.*

γ. *Spikelets unisexual, or the females with a few male flowers above or below. Stigmas* 3.

† *Utricles spreading after flowering, glabrous.*

Culm shorter than leaves. Spikelets short, crowded, brown . . 15. *C. pumila.*
Culm tall, leafy. Spikelets cylindric, large, green . . . . . 16. *C. Forsteri.*

Culms short, leafy. Spikelets short, obtuse, green . . . . . . 17. *C. cataractæ.*
†† *Utricles erect, pubescent. Culms short* . . . . 18. *C. breviculmis.*

††† *Utricles erect, glabrous. Culms slender (except in* C. trifida), *long.*

Spikelets 6–10, stout, very large, brown, 2–4 in. Glumes 2-fid. 19. *C. trifida.*
Spikelets 6–10, slender, pedicelled, $\frac{3}{4}$–$\frac{1}{2}$ in. Glumes entire . . 20. *C. Neesiana.*
Spikelets 5–8, stout, suberect, $\frac{1}{2}$–1 in. Glume 2-fid . . . . . 21. *C. dissita.*
Spikelets 5–8, stout, suberect, 1–1$\frac{1}{2}$ in. Glumes 2-fid . . . 22. *C. Lambertiana.*
Spikelets 5–9, drooping, 1–2$\frac{1}{2}$ in. Glumes entire . . . . . . 23. *C. vacillans.*

1. **C. pyrenaica,** *Wahlenberg;—Fl. N. Z.* i. 280. Culms small, 3–6 in. high, 3-gonous, quite glabrous. Leaves flexuous, longer than the culm, coriaceous, flat, grooved below, edges scabrid. Spikelet terminal, solitary, $\frac{1}{3}$–$\frac{3}{4}$ in. long, linear-ovoid, pale red-brown; male flowers at the top. Glumes ovate, acuminate or obtuse. Utricle longer and narrower than the glume, stipitate, lanceolate, compressed, nerveless; beak pale, membranous, emarginate. Stigmas 3, rarely 2.

Northern Island: tops of the Ruahine mountains, forming clumps in snow runs, *Colenso.* A native of Europe and North America.

2. **C. acicularis,** *Boott;—Fl. N. Z.* i. 280. *t.* 63 C. Culms small, 2–6 in. high, rigid, wiry, very slender, nearly terete. Leaves as narrow as the culm, nearly terete, grooved in front, quite smooth. Spikelet solitary, ovoid, $\frac{1}{4}$–$\frac{1}{3}$ in. long, pale red-brown; male flowers at the top; bract long or short. Glumes few, lanceolate, acuminate, lower foliaceous; utricle stipitate, lanceolate, 3-quetrous, acuminate; edges serrate above; beak 2-fid. Nut 3-gonous, with a linear bristle at its side. Stigmas 3.

Northern Island: tops of the Ruahine range, *Colenso.*

3. **C. inversa,** *Br.;—Fl. N. Z.* i. 281. Small, rather slender, 1–10 in. high, quite smooth. Rhizome stout, creeping. Culms leafy below only, slender, 3-gonous, grooved. Leaves shorter, flat, keeled, very narrow, $\frac{1}{20}$ in. broad. Spikelets solitary, or broken up into 2 or 3, $\frac{1}{12}$–$\frac{1}{4}$ in. long, ovoid, pale; male flowers below; bracts very long, unequal. Glumes ovate, acuminate, cuspidate or awned. Utricle as long as the glume, oval, beaked, nerved, margins serrate above. Stigmas 2.

Northern Island: Hawke's Bay and Ruahine range, *Colenso.* Also a native of Tasmania and Southern Australia.

4. **C. Colensoi,** *Boott;—Fl. N. Z.* i. 281. *t.* 63 B. Small, slender, wiry, flexuous, 6–12 in. high; rhizome stout, creeping. Culms leafy at the base only, very slender, 3-gonous, grooved. Leaves longer or shorter, as narrow as the culms, involute, deeply grooved. Spikelets solitary or broken up into 2–5, ovoid, turgid, $\frac{1}{4}$ in. long; male flowers below; bracteate or not. Glumes ovate, acuminate, pale-brown; margins broad, white. Utricle shorter and narrower than the glumes, broadly ovate or oblong, not beaked, obscurely nerved; margins serrulate above. Stigmas 2.

Northern Island: dry plains, road from Patangata to Manawarakau, *Colenso.* **Middle** Island: Southern Alps, 3000–3500 ft., *Sinclair and Haast;* Acheron valley, 4000 ft., *Travers.*

5. **C. stellulata,** *Goodenough;—Fl. N. Z.* i. 281. Culms very slender, tufted, glabrous, leafy at the base only, 3-gonous, grooved. Leaves as long and narrow as the culms, or broader, flattish or involute, grooved. Spikelets

3 or 4, very small, $\frac{1}{12}$–$\frac{1}{10}$ in. long, very few-flowered, divaricating in fruit; male flowers below, uppermost male only; bracts 0. Glumes very few, ovate, acute, keeled, pale; margins broad, membranous. Utricle longer than the glume, ovate, acuminate, 3-gonous, nerved, serrate at the margin, beaked, 2-dentate. Stigmas 2.

Northern Island: bogs near Lake Taupo, *Colenso*. **Middle** Island: Acheron valley, 4000 ft., *Travers*. Also found throughout the temperate northern hemisphere, but only in New Zealand in the southern.

6. **C. teretiuscula,** *Goodenough;—Fl. N. Z.* i. 281. Culms tall, slender, erect, 1–2 ft. high, 3-quetrous, smooth or scabrid, grooved, leafy at the base only. Leaves shorter and broader than the culms, flat, deeply grooved; margins scabrid. Spikelets very small, $\frac{1}{6}$–$\frac{1}{4}$ in. long, collected into a linear head $\frac{1}{2}$–$\frac{3}{4}$ in. long, few-flowered; male flowers at the top; bracteate or not. Glumes pale-brown, ovate, acute; margins broad, membranous. Utricle about as long as the glume, ovate, plano-convex, nerved at the back, winged and serrate above; beak 2-dentate. Stigmas 2.

Northern Island: bogs, Tangoio village, Hawke's Bay, *Colenso*. **Middle** Island: watercourses near Lake Okau, *Haast*; Otago, lake district, *Hector and Buchanan*. Common throughout the north temperate zone, but not found elsewhere in the southern.

7. **C. virgata,** *Solander;—Fl. N. Z.* i. 283. Culms densely tufted, tall, harsh, leafy, 3-gonous, edges scabrid, 1–2 ft. high. Leaves rigid, flat, keeled, narrow, $\frac{1}{6}$ in. broad, much longer than the culms; margins scabrid. Spikelets very numerous, distantly spiked or panicled along the slender end of the culm, small, sessile, $\frac{1}{12}$ in. long, pale brown; male flowers at the top. Glumes ovate, acute, cuspidate or awned; margins broadly membranous. Utricle as long as the glume, 2-convex, oblong or broadly ovate; beak short, with serrate wings, obliquely 2-dentate; nerves strong. Stigmas 2.—Boott, Ill. Carex, t. 121, 122; *C. collata*, Boott, in Lond. Journ. Bot. iii. 447.

Var. *β. secta.—C. secta*, Boott, l. c. 283. Spike more lax slender and drooping. Utricle faintly nerved.—Boott, Ill. Carex, t. 123, 124.

Northern and **Middle** Islands: Var. *α* and *β*, common in marshes and bogs, abundant, as far south as Otago, *Colenso, Sinclair*, etc. Var. *β* forms tufts of roots sometimes 1–6 ft. high and 6–18 in. diam., like the stem of a Tree-Fern, *Buchanan*.

8. **C. appressa,** *Br.;—Fl. Antarct.* i. 90. A large harsh species, of the same habit and with nearly the same characters as *C. virgata*, but the leaves are upwards of $\frac{1}{4}$–$\frac{1}{2}$ in. broad, the culms acutely 3-quetrous, very stout and scabrid, the spike erect, more rigid, with shorter, stouter branches, and the utricles are plano-convex, with incurved margins and nerved faces.—Boott, Ill. Carex, t. 119, 120; Fl. Tasm. ii. 99.

Lord Auckland's group and **Campbell's** Island, abundant in woods near the sea, *J. D. H.* Also a most common Tasmanian and temperate Australian plant. This again I am inclined to regard as a form of *C. virgata*.

9. **C. Gaudichaudiana,** *Kunth*. Culms leafy, tufted, 2 in.–2 ft. high, nearly smooth, rather slender. Leaves shorter than the culms, flat, rather soft. Spikelets 3–6, erect, $\frac{1}{4}$–$\frac{3}{4}$ in. long, ovoid or cylindric, purple, shortly peduncled or the lower long-peduncled, the lower female or with very few male flowers at the top; bracts long, leafy. Glumes oblong or lanceolate-oblong, obtuse or rounded at the tip, dark-purple with green margin and

keel. Utricle broader and longer than the glume, ovate- or elliptic-lanceolate, compressed, strongly nerved, green, when ripe dotted with brown; beak short, entire or 2-dentate. Stigmas 2.—Fl. Tasm. ii. 99. t. 151 A.

Middle Island: Southern Alps, *Sinclair and Haast, Travers;* Otago, banks of the Clutha, *Lindsay;* lake district, *Hector and Buchanan.* All Sinclair's, Haast's, and Travers's specimens have an almost wholly unisexual spikelet. Lindsay's have male flowers at top. Buchanan remarks that it is a most important plant, filling up bogs, etc., and preparing the way for other vegetation. Also a native of South-East Australia and Tasmania.

10. **C. subdola,** *Boott;—Fl. N. Z.* i. 282. Very similar in habit and characters to *C. Gaudichaudiana,* but usually taller, 2–3 ft. high, and with the glumes mucronate or shortly aristate.

Northern Island: Bay of Islands, *Colenso,* etc. **Middle** Island: Acheron valley, alt. 4000 ft., *Travers.*

11. **C. ternaria,** *Forst.;—Fl. N. Z.* i. 282. A large, tall, very leafy species. Culms 2–4 ft. high, stout, leafy throughout; angles very scabrid. Leaves very long, flat, $\frac{1}{6}$–$\frac{1}{2}$ in. broad, smooth or scabrid. Spikelets very numerous, cylindric, 1–4 in. long, long-peduncled, nodding or pendulous, solitary or 2- or 3-nate, dark-brown, unisexual or the female with male flowers at top; bracts very long and leafy. Glumes ovate or lanceolate, truncate emarginate or acute, with hispid awns. Utricle shorter than the glume, flattened, broadly ovate- or elliptic-obovate, entire or emarginate, nerved, shortly beaked. Stigmas 2.—*C. polystachya,* A. Rich.; *C. geminata,* Schkuhr.

Northern Island, *Banks and Solander;* Great Barrier Island, *Sinclair;* Wellington, *Stephenson;* Hawke's Bay, *Colenso.* **Middle** Island, abundant. **Lord Auckland's** group, *J. D. H.*

12. **C. testacea,** *Solander;—Fl. N. Z.* i. 282. A rather small, perfectly smooth species. Culms 4–10 in. high, very slender, leafy. Leaves as long as or longer than the culms, very narrow, flat, $\frac{1}{12}$–$\frac{1}{10}$ in. broad. Spikelets 4, very close together, sessile, shortly cylindric, $\frac{1}{2}$–1 in. long, stout, uppermost very slender, male; bracts very long. Glumes broadly ovate, shortly awned, membranous, pale-brown. Utricle as large as the glumes, very broadly ovate, turgid, plano-convex, nerved, shining; beak short, usually broadly 2-cuspidate. Stigmas 2.

Northern Island: Tigada, *Banks and Solander;* Auckland, *Sinclair.* **Middle** Island: banks of the Clutha, *Lindsay;* Canterbury, *Sinclair and Haast.* Boott, who describes the utricles as serrated above,—but I do not find them so,—informs me that it is possibly a variety of *C. Raoulii;* but the 2-fid glumes of the latter readily distinguish it.

13. **C. Raoulii,** *Boott;—Fl. N. Z.* i. 283. Culms 2–3 ft. high, slender, leafy, smooth. Leaves long, narrow, $\frac{1}{6}$ in. broad, nearly smooth, flat. Spikelets 4–6, pale, close together or the lower distant, sessile, shortly cylindric, $\frac{1}{2}$–1 in. long, male flowers below, upper longer with more male flowers below; bracts very long and leafy. Glumes broadly ovate, 2-lobed, with a short hispid awn; margins broadly scarious and lacerate. Utricle as large as the glume, obovate, turgid, plano-convex, purplish, strongly nerved, toothed or entire on the edges above; beak emarginate. Stigmas 2.

Middle Island: Akaroa, *Raoul;* Southern Alps, *Sinclair and Haast.*

14. **C. lucida,** *Boott;—Fl. N. Z.* i. 283. A rigid, rather harsh, leafy species. Culms slender, leafy all the way up, 1–2 ft. high, slightly scabrid.

Leaves much longer than the culms, very narrow, $\frac{1}{14}$–$\frac{1}{10}$ in. broad, rigid, flat or keeled. Spikelets 4–8, distant, erect, cylindric, $\frac{1}{2}$–$1\frac{1}{2}$ in. long; upper male, more sessile, slender; lower male at the base only. Glumes ovate, acute obtuse or emarginate, cuspidate or shortly awned, pale-purple, shining; keel and margins pale. Utricle as long as the glume, ovate or elliptic, 2-convex, faintly nerved; margin scabrid or smooth, purple; beak short, minutely 2-dentate. Stigmas 2, rarely 3.—Boott, Ill. Carex, t. 173.

Northern Island: frequent in grassy places, *Banks and Solander*, etc. **Middle** Island: Upper Wairau, *Sinclair, Travers;* Dunedin, *Lindsay.*

15. **C. pumila,** *Thunberg;—C. littorea,* Labill. Fl. Nov. Holl. ii. t. 219; Fl. N. Z. i. 284. A harsh, short, rigid, very leafy species, 8–16 in. high; rhizomes stout, running in sand. Culms very short, stout. Leaves much longer than the culms, flexuous, rigid, $\frac{1}{12}$ in. diam., keeled, grooved, scabrid. Spikelets 4 or 5, hidden amongst the leaves, close together, short, $\frac{1}{2}$–$1\frac{1}{2}$ in. long, stout, erect, sessile or peduncled, upper male, slender; bracts very long, leafy. Glumes small, ovate, obtuse, acute or cuspidate, pale-brown; margins white, membranous. Utricle much larger than the glume, large, $\frac{1}{6}$ in. long, turgid, ovate, not nerved, smooth, pale-brown, opaque; beak short, 2-cuspidate. Stigmas 3.

Northern and **Middle** Islands: on sandy shores abundant, *Banks and Solander*, etc.; as far south as Port Preservation, *Lyall.* A native of Australia, Tasmania, and Japan.

16. **C. Forsteri,** *Wahlenberg;—Fl. N. Z.* i. 285. A rather flaccid, grassy, pale-green species. Culms 6–24 in. high, very leafy, slender or stout, slightly scabrid. Leaves flaccid and grassy, longer than the culms, $\frac{1}{6}$–$\frac{1}{2}$ in. diam., flat, not keeled, rather scabrid. Spikelets 3–8, very variable in size, $\frac{1}{2}$–$4\frac{1}{2}$ in. long, distant or close together, cylindric, pale, 1–3 upper male, the rest all female or with male flowers above or below, sessile or shortly peduncled, erect or nodding; bracts very long and leafy. Glumes very numerous, linear or lanceolate, cuspidate or shortly awned, membranous, pale red-brown, margins and nerve white. Utricle much larger than the glume, spreading, stipitate, elliptic-lanceolate, compressed, strongly nerved; beak long, 2-furcate. Stigmas 3.—Boott, Ill. Carex, t. 137; *C. debilis,* Forst.; *C. recurva,* Schkuhr; *C. punctulata,* A. Rich. Flora, t. 22.

Var. *β. minor.* Spikelets 3, $\frac{1}{2}$–1 in. long, sessile, crowded. Glumes often minute, subulate, and long-awned.

Var. *γ. fascicularis.—C. fascicularis,* Solander; Fl. N. Z. i. 283. Leaves $\frac{1}{4}$–$\frac{1}{2}$ in. broad. Spikelets with male flowers below.—Boott, Ill. Carex, i. t. 139, 140; *C. pseudocyperus,* Br.; Forst. not Linn.

Northern and **Middle** Islands: all varieties abundant in moist woods, etc., *Banks and Solander,* etc. Also common in Southern Australia and Tasmania. Very closely allied to indeed to the northern *C. pseudocyperus.*

17. **C. cataractæ,** *Br.* A rather small, flaccid, leafy, pale green species. Culms 6–10 in. high, stout, 3-gonous, grooved, smooth. Leaves much longer than the culms, flat, not keeled, striate, $\frac{1}{12}$–$\frac{1}{8}$ in. broad. Spikelets 6–8, very short, sessile or peduncled, erect, crowded, very pale green, $\frac{1}{4}$–$\frac{1}{2}$ in. long, stout, 3 upper male, lower female, or male at the top only. Glumes few, small, ovate, acute or obtuse, white, membranous. Utricle much larger than the glumes, spreading or deflexed, oval, turgid, green, strongly nerved; beak 2-cuspidate. Stigmas 3.—Fl. Tasm. t. 151 B.

Middle Island: Southern Alps, *Sinclair and Haast;* swampy gullies, Waitaki river, etc., *Haast;* Otago, lake district, *Sinclair and Haast*. Also a native of Tasmania.

18. **C. breviculmis,** *Br.;—Fl. N. Z.* i. 283. *t.* 63 A. A small, tufted, leafy species, 1–10 in. high, nearly smooth. Culms very short indeed. Leaves many times longer than the culms, very narrow, $\frac{1}{20}$–$\frac{1}{16}$ in., flat, keeled, grooved. Spikelets hidden amongst the leaves, few, crowded together, sessile, $\frac{1}{6}$–$\frac{1}{4}$ in. long, pale-green, male terminal, females with sometimes a few male flowers at the tip; bracts long, leafy. Glumes 6–8, nearly white, ovate, cuspidate or shortly awned. Utricle shorter than the glume, narrow elliptic-oblong, narrowed at both ends, scarcely beaked, nerved, green, pubescent. Stigmas 3.

Northern Island: in grassy places, Bay of Islands, Tarawara, Ahuriri, etc., *Colenso*, etc. Also a native of South-East Australia and Tasmania, Japan, China, and the Himalaya mountains.

19. **C. trifida,** *Cavanilles;—Fl. N. Z.* i. 284. A very large, tall, robust, leafy sedge, 3–6 ft. high. Culms stout, obtusely 3-gonous, leafy, smooth. Leaves very large and long, $\frac{1}{2}$ in. broad and more, flat, keeled, striate; margins scabrid. Spikelets 6–10 or more, approximate, dark-brown, shortly peduncled, erect or nodding, cylindric, very large, 2–4 in. long, $\frac{1}{2}$ in. broad, lower sometimes compound; bracts very long and leafy. Glumes linear-oblong or lanceolate, $\frac{1}{3}$ in. long, membranous, red-brown, 2-fid with a hispid awn at the tip. Utricle shorter than the glumes, oblong-obovate, strongly nerved; beak 2-fid. Stigmas 3.

Northern Island: Totara-nui, *Banks and Solander* **Middle** Island: Akaroa, *Raoul;* Dusky Bay, *Lyall;* Otago, *Lindsay*. **Lord Auckland's** group and **Campbell's** Island, *J. D. H.* Much the largest New Zealand species. Also a native of Fuegia, South Chili, and the Falkland Islands.

20. **C. Neesiana,** *Endl.—C. Solandri*, Boott; Fl. N. Z. i. 284. Culms tall, very slender, leafy, 1–2 ft. high. Leaves long, very narrow, $\frac{1}{10}$ in. diam., flat, grassy, keeled. Spikelets 6–10, distant, on long slender peduncles, nodding, $\frac{3}{4}$–1$\frac{1}{2}$ in. long, dark-brown, slender, cylindric, upper 1–4 male, lower sometimes compound; bracts long, leafy. Glumes ovate (rarely emarginate), cuspidate or shortly awned, very pale-brown. Utricle as long as the glumes, ovate- or elliptic-oblong, smooth, obscurely nerved, dark-brown, shining; beak short, sometimes scabrid or toothed at the edges, 2-cuspidate. Stigmas 3.—*C. Solandri*, Boott, Ill. Carex, i. t. 175.

Northern Island: Totara-nui, Opuragi, etc., *Banks and Solander, Colenso;* Wellington, *Ralphs, Stephenson*. A native of Norfolk Island.

21. **C. dissita,** *Solander;—Fl. N. Z.* i. 284. Habit of *C. Neesiana*, but more robust. Leaves broader, $\frac{1}{6}$ in. diam., flat, grassy, membranous, not keeled. Spikelets 5–8, distant, shortly peduncled, erect or nodding, short, $\frac{1}{2}$–1 in. long, stout, cylindric, pale-brown, uppermost one male, the rest female or male at the base only. Glumes broadly ovate, 2-fid, shortly awned, dark-brown; keel green Utricle as large as the glumes, like *C. Neesiana*, but shorter. Stigmas 3.—Boott, Ill. Carex, i. t. 176.

Northern Island: not uncommon in grassy places, Opuragi and Tigada, *Banks and Solander*. **Auckland** and **Great Barrier** Islands, *Sinclair;* east coast, Patea, *Colenso*.

Very near *C. Neesiana*, but spikelets much shorter and stouter, leaves broader, and glumes 2-fid.

22. **C. Lambertiana,** *Boott;—Fl. N. Z.* i. 284. Very similar in habit and characters to *C. dissita*, but larger, stouter, 2 ft. high, with coarser leaves ¼ in. diam., longer spikelets 1–1½ in. long, and always solitary male spikelets.—Boott, Ill. Carex, t. 177.

Northern Island: probably common, Totara-nui, *Banks and Solander*; Waitaki and Great Barrier Island, and Auckland, *Sinclair*; Bay of Islands, *J. D. H.* This appears to me a large form of *C. dissita.* I find no difference in the 2-cuspidate tip of the utricle of the two.

23. **C. vacillans,** *Solander;—Fl. N. Z.* i. 285. Culms very slender, 1 ft. high, scabrid, leafy. Leaves longer or shorter than the culms, rather rigid, ⅕–¼ in. broad, flat, striate, nearly smooth. Spikelets 5–9, approximate or distant, long peduncled, drooping, 1–3 in. long, cylindric, bright yellow-brown, uppermost 1–3 male, the rest male at the base only, lower lax-flowered at the base; bracts long and leafy. Glume ovate-subulate or lanceolate or almost linear-oblong, entire, shortly awned, pale red-brown. Utricle longer than the glume, fusiform, strongly nerved; beak long, 2-cuspidate. Stigmas 3.

Northern Island: common in moist woods, *Banks and Solander*, etc., *Sinclair*, *Colenso.* This again is very closely allied to the preceding two, but has the entire glumes of *C. Neesiana*, longer spikelets and peduncles, and much longer, narrower, strongly nerved utricles.

I have another, apparently very distinct species of *Carex*, collected near Lake Okau by Haast along watercourses, but in too immature a state for description. Leaves very coriaceous, flat, 10–12 in. long, ¼ in. broad. Male spikelets 5 or 6, shortly peduncled; female 6–8, very long-pedicelled, slender, cylindrical, with male flower at top. Glumes entire, cuspidate or short-awned. Stigmas 2.

Order XII. GRAMINEÆ.

Grasses, usually tufted plants, either perennial with often creeping rhizomes, or annual with fibrous roots. Culms hollow, closed at the joints, terete Leaves alternate, usually distichous, very long, flat convolute or keeled; sheaths split to the base, often having a membranous appendage (ligule), where the sheath joins the blade.—Flowers minute, hermaphrodite or unisexual, in the axils of imbricating 2-ranked scales (glumes), which are arranged in spikelets. Spikelets green, spiked panicled or fascicled. Glumes usually coriaceous, lanceolate, concave, keeled and ribbed, rarely flat, lower 2–4 empty, the rest often bearing a 2-nerved scale (pale) in their axis, between which and the glume is a hermaphrodite or unisexual flower; in one section (*Poaceæ*) there are only 2 empty (flowerless) glumes which are nearly opposite, and the rest are alternate on a slender rachis, the terminal being often imperfect; in the other section the uppermost glume is always hermaphrodite, and those below it have either male flowers or are flowerless. Perianth of 2 very minute, often oblique scales. Stamens 3 (rarely 1, 2, or 6, or more); filaments capillary; anthers versatile. Ovary with 2 feathery stigmas; ovule 1, erect. Fruit a grain, free or adhering to the pale. Seed firmly adhering to the membranous pericarp; albumen hard, floury; embryo small, nearly orbicular, on one side of the base of the albumen.

One of the largest and most important Orders of flowering plants, found in every region of the globe except the very coldest. The New Zealand species are in general very closely allied to Australian; like the *Cyperaceæ*, they have not been well collected either in the Northern or Middle Islands; Mr. Colenso alone having carefully sought for them. The determination of the genera is very difficult, and the species of some genera even more so.

A. *Spikelets on the spines of globose, polygamous, involucrate heads* 5. SPINIFEX.

B. *Spikelets sessile, in* 1 *or* 2 *series, on one or both sides of a flattened rachis. Empty glumes* 0 *or* 1–3 (*see* Festuca bromoides, *in* C. III. *β*).

Empty glumes 2 or 3, short; flowering solitary hard . . . . . . 6. PASPALUM.
Empty glume 1, margins connate; flowering solitary. Pale 0 . . . 9. ZOYSIA.
Empty glumes 2, lanceolate; flowering 3–16-awned . . . . . . 25. TRITICUM.
Empty glume 0 or 2 bristles; flowering 1–3-awned . . . . . . 26. GYMNOSTICHUM.

C. *Spikelets never sessile and distichous, pedicelled panicled or racemed.*

I. *Empty glumes* 3 *or more, below the solitary hermaphrodite flowering one.*

Glumes 5; 4 empty, acuminate, 1-flowering, obtuse . . . . . . 1. EHRHARTA.
Glumes 5; 2 empty minute, 2 empty awned, 1-flowering, acuminate . 2. MICROLÆNA.
Glumes 4, short, obtuse; 2 empty, 1 male, upper hermaphrodite . 8. ISACHNE.
Glumes 4, mucronate or awned; 2 empty, 1 male, upper hermaphrodite . . . . 7. PANICUM.

II. *Empty glumes* 2, *below the solitary flowering one.*

a. Panicle dense, cylindric, spike-like (*see* Poa anceps *and* Danthonia *in* III.).

Empty glumes equal, flattened. Pale 0. Panicle soft, spiciform . 3. ALOPECURUS.
Empty glumes rigid, acuminate. Panicle reduced to an ovoid, spinulose head . . . . 10. ECHINOPOGON.

β. Panicle effuse or contracted.

Fl. glume on a bearded pedicel, tip 2-fid, awned . . . . 11. DICHELACHNE.
Fl. glume pedicelled, ending in a long rigid awn . . . . 12. APERA.
Fl. glume sessile, short, acute. Seed loose in pericarp . . . . . 13. SPOROBOLUS.
Fl. glume sessile, truncate, awned at back or awnless . . . . . . 14. AGROSTIS.

III. *Empty glumes* 2, *below the* 2 *or more flowering ones. Glume* 1 *in* Festuca bromoides. *Flowering glumes rarely* 1 *in* Poa anceps, *in* Danthonia, *and* Deschampsia.

a. Flowering glumes awned, sometimes awnless in Hierochloe (*see* Festuca *in β*).

Fl. glumes 2–5, silky; awn at the 2-fid tip, slender . . . . . . 15. ARUNDO.
Fl. glumes 2–8, 2-cuspidate, with stout dorsal awn, and long hairs on sides and at base . . . . 16. DANTHONIA.
Fl. glumes 2–4, silky at base, 3-awned, middle awn longest . . . 19. TRISETUM.
Fl. glumes 2 or 3, shining, truncate, or 4-toothed . . . . 17. DESCHAMPSIA.
Fl. glumes 3, shining, obtuse, 2 lateral ♂ 3-androus, central, ⚥ 2-androus . . . . 4. HIEROCHLOE.
Fl. glumes 3–7, shining, 2-fid, with a short obtuse awn or 0. Panicle spiciform . . . . 18. KŒLERIA.
Fl. glumes 4–10, 2-fid, with an intermediate awn. Ovary villous at top . . . . 24. BROMUS.

β. Flowering glumes not awned except in some Festucas (*see* Hierochloe *and* Kœleria *in a*).

Fl. glumes 6–14, short, obtuse, green. Scales connate . . . . . 20. GLYCERIA.
Fl. glumes 2, short, truncate, erose, membranous . . . . 21. CATABROSA.
Fl. glumes 2–10, compressed, keeled, obtuse or acute . . . . . . 22. POA.
Fl. glumes 2–10, convex or keeled at back, often awned at the entire tip . . . . 23. FESTUCA.

ARRANGEMENT OF THE GENERA ACCORDING TO THE NATURAL SYSTEM.

(THE NON-INDIGENOUS GENERA ARE IN ITALICS.)

* *Spikelets with* 1 *fertile terminal flower, with or without a male or imperfect flower below it* (Paniceæ).

1. **Oryzeæ.** Flowering glumes hardening, and enclosing the grain. Empty glumes 4 or 5, unequal, laterally compressed, lower smaller.—1, EHRHARTA; 2, MICROLÆNA.

2. **Phalarideæ.** Flowering glume and pale hardening, and enclosing the grain. Empty glumes 2, equal, laterally compressed, keeled, longer than the flowering.—3, ALOPECURUS, *Phalaris;* 4, HIEROCHLOE, *Phleum, Anthoxanthum, Phalaris.*

3. **Paniceæ.** Flowering glume and pale hardening, and enclosing the grain. Empty glumes 2–4, outer smaller, often dorsally compressed.—5, SPINIFEX; 6, PASPALUM; 7, PANICUM; 8, ISACHNE.

4. **Andropogoneæ.** Flowering glume small, thin, transparent, or 0.—9, ZOYSIA, *Anthistiria, Andropogon, Apluda.*

** *Spikelets with* 1 *or more perfect flowers; the male or imperfect flowers, if present, above the perfect ones, the axis or rachis often ending in a point or bristle.*

5. **Agrostideæ.** Spikelets 1-flowered. Flowering glume awnless or with a simple awn. Grain free.—10, ECHINOPOGON; 11, DICHELACHNE; 12, APERA; 13, SPOROBOLUS; 14, AGROSTIS.

6. **Stipaceæ.** Spikelets 1-flowered. Flowering glume firm, with a simple or 3-cleft awn jointed on to its tip, closely enveloping the grain.—*Aristida.*

7. **Arundineæ.** Spikelets usually 2- or more-flowered, rachis with long silky hairs. Glumes all membranous, free.—15, ARUNDO.

8. **Chlorideæ.** Spikelets 1- or several flowered, sessile on the linear branches of a panicle or on a simple spike.—*Cynodon, Eleusine.*

9. **Avenaceæ.** Spikelets 2- or more flowered. Flowering glumes on a slender rachis, usually shorter than the empty ones, membranous, shining, split at the top with an intermediate awn that is oftent wisted at the base (rarely awnless).—16, DANTHONIA, *Holcus;* 17, DESCHAMPSIA; 18, KŒLERIA; 19, TRISETUM, *Avena.*

10. **Festucaceæ.** Spikelets usually 4- or more flowered. Flowering glumes usually longer than the empty ones, on a flexuous rachis.—20, GLYCERIA; 21, CATABROSA; 22, POA; 23, FESTUCA; 24, BROMUS.

11. **Hordeaceæ.** Spikelets 1- or more flowered (spiked), sessile on opposite sides of a simple rachis, solitary or 2 or 3 together, the glumes standing right and left to the axis of the spike.—25, TRITICUM; 26, GYMNOSTICHUM, *Hordeum, Lolium.*

1. EHRHARTA, Thunberg.

Culms branching. Leaves flat or concave, not involute. Spikelets panicled, 1-flowered.—Glumes 5, keeled, compressed, 4 lower empty, acuminate, flowering one terminal, obtuse. Pale linear. Scales 2-lobed. Stamens 2–6. Ovary sessile. Grain free within the hardened glumes.

An Australian, South African, and New Zealand genus.

1. **E. Colensoi,** *Hook. f. Fl. N. Z.* i. 288. *t.* 65 A. Tufted, glabrous, 4–8 in. high. Leaves suberect, distichous, 2–4 in. long, contracted at the sheath, ¼ in. broad, linear-subulate, scaberulous above, smooth below; nerves faint; ligule short, ragged. Panicle 1½–2 in. long, inclined. Spikelets on slender pedicels, compressed, linear-oblong, ¼ in. long. Glumes deeply striate or nerved; lower pair empty, acuminate; 2 following almost awned with silky hairs at the base; flowering shorter. Pale small, narrow, with 2 nerves, and a small bristle at its outer base. Stamens 2; anthers short.

Northern Island: forming large tufts on the tops of the Ruahine mountains, *Colenso.*

2. MICROLÆNA, Br.

Very slender, perennial grasses, simple or branched. Leaves small or long, flat or concave, not involute. Spikelets racemed or panicled.—Glumes 5 ; 4 lower flowerless, 2 lowermost opposite, minute; 2 following awned, much larger; terminal or flowering acuminate or awned. Pale short, linear, hyaline. Scales 2, glabrous. Stamens 2–4. Ovary sessile. Grain free within the hardened glumes.

A small genus (including *Diplax*, Br.) of Australian, Tasmanian, and New Zealand Grasses, natives generally of woods.

Two lowest glumes distant from the following . . . . 1. *M. stipoides.*
Two lowest glumes close to the following (*Diplax*).
Spikelets panicled. Leaves scaberulous, $\frac{1}{4}$–$\frac{1}{2}$ in. broad. Stamens 2 . 2. *M. avenacea.*
Spikelets racemed. Leaves smooth, $\frac{1}{12}$–$\frac{1}{6}$ in. broad. Stamens 4 . . 3. *M. polynoda.*

1. **M. stipoides,** *Br.;—Fl. N. Z.* i. 289. A slender grass, 18–24 in. high. Leaves glabrous or sparingly hairy, 2–6 in. long. Panicle slender, nodding, branched below. Lower spikelets on long pedicels, upwards of 1 in. long, awns included. Glumes: lowest pair very minute, acute, deciduous; 2 following seated at distant intervals on the bearded rachis, scabrid, awns as long as themselves; uppermost or flowering acuminate. Scales large. Stamens 4.—*Ehrharta stipoides*, Labill. Fl. Nov. Holl. i. 16. t. 118.

Northern Island: East Cape, Hawke's Bay, and Cape Palliser, *Colenso; Auckland*, *Sinclair*. A common Australian and Tasmanian plant.

2. **M. avenacea,** *Hook. f.—Diplax avenacea*, Raoul, Choix, ii. t. 3; Fl. N. Z. i. 289. A tall, handsome grass, 2–4 ft. high. Culms densely tufted, compressed and leafy at the base, simple and quite smooth above. Leaves 1$\frac{1}{2}$–2 ft. long, $\frac{1}{4}$–$\frac{1}{2}$ in. broad, margins scabrid. Panicle 10–15 in. long, with many long, capillary branches. Spikelets on capillary pedicels, $\frac{1}{2}$ in. long (awns included). Glumes: lowest pair very minute, unequal, persistent; 2 following close to the preceding, with long awns; upper or flowering acuminate. Stamens 2.

Northern and **Middle** Islands: abundant in woods, *Banks and Solander*, etc.

3. **M. polynoda,** *Hook. f.—Diplax polynoda*, Fl. N. Z. i. 290. Culms tufted, very long, 3–6 ft., slender, or stout and rigid, terete, branched, with knots at the joints, quite smooth. Leaves not scabrid, very narrow, 4–8 in. long, $\frac{1}{12}$–$\frac{1}{6}$ broad. Racemes simple, few-flowered. Spikelets on rather short pedicels, with the awns $\frac{1}{2}$ in. long. Glumes as in *M. avenacea*, but awns of the 2 upper flowerless ones, shorter. Stamens 4.

Northern Island: base of the Ruahine range and east coast, *Colenso*.

3. ALOPECURUS, Linn.

Creeping or erect, glabrous or downy grasses. Culms generally simple, leafy. Leaves flat. Panicles contracted into dense, cylindrical, obtuse spikes. Spikelets very shortly pedicelled, imbricated, 2-flowered.—Empty glumes equal, much compressed, keeled, usually connate at the base, erect, acute; flowering glume shorter, becoming hard, keeled; awn if present short, straight, dorsal. Pale 1-nerved or 0. Scales 0. Stamens 3. Grain free within the hardened glumes.

A considerable genus, confined to temperate or subtropical countries.

1. **A. geniculatus,** *Linn.;—Fl. N. Z.* i. 290. Culms prostrate and creeping at the base, then erect, 1–2 ft. high. Leaves glabrous, soft, flat; sheaths large, grooved; ligule long, membranous. Spike green, downy, 1½–2½ in. long, cylindric; rachis woolly. Empty glumes connate at the very base only, membranous, very flat, keel with long bristles; awn variable in length and position, recurved. Pale 0.

Northern and **Middle** Islands: marshy places, not unfrequent; east coast, *Colenso;* Canterbury, *Lyall;* Lake Okau, *Haast;* Otago, lake district, alpine, *Hector and Buchanan.* A very common European, North American, and North Asiatic Grass, also found in Victoria and Tasmania; the common "Foxtail Grass" of England.

A. agrestis, Linn., introduced into New Zealand, may be known by the outer glumes connate to the middle (*Wellington,* Stephenson, No. 111).

Phleum pratense, Linn., is another English naturalized Grass, distinguished from *Alopecurus,* which it closely resembles in characters, by the awned outer glumes.

Anthoxanthum odoratum, Linn., a very common and widely distributed European Grass, which gives the sweet smell to new-made hay, is widely dispersed in New Zealand, and has been gathered in a viviparous state at 3–4000 ft. on Mount Cook by Haast. It may be recognized by its scent (of *Hierochloe*) when drying, and the oblong spiked panicle, 2 outer empty glumes, 2 next also empty, both awned, and the fourth, or flowering, very short, awnless, with 2 stamens.

Phalaris Canariensis, Linn., another allied introduced Grass, has a large ovoid spiked panicle, and the glumes are large and winged at the keel, with green lines on each side; it is common in many places, and was gathered by Forster.

4. HIEROCHLOE, Gmelin.

Erect, tufted, glabrous, sweet-scented, leafy grasses. Leaves flat or involute, rather coriaceous. Panicles usually effuse.—Spikelets rather large, broad, pedicelled, flattened, shining, 3-flowered; 2 lower flowers male, terminal, hermaphrodite. Empty glumes nearly equal, oblong, keeled, obtuse or acute; flowering similar, awnless, or with a terminal or dorsal, short, not twisted awn. Pale 2-nerved. Scales 2, 2-lobed. Stamens 2 in the male fl., 3 in the hermaphrodite. Grain free, terete.

A beautiful genus of Grasses, natives of mountainous arctic and antarctic regions, and of South Africa.

Outer glumes as long as the flowering.
  Culms 2–3 ft. Leaves long . . . . 1. *H. redolens.*
  Culms 1–2 ft. Leaves short . . . . 2. *H. alpina.*
Outer glumes much longer than the flowering ones . . . . 3. *H. Brunonis.*

1. **H. redolens,** *Br.;—Fl. N. Z.* i. 300. Culms densely tufted, 2–3 ft. high, rather stout, very soft and smooth. Leaves flat, quite smooth or slightly scabrid; ligule broad, membranous. Panicle 6–10 in. long, nodding, branches capillary, slightly hairy, lower 2–3 in. long. Spikelets numerous, ¼ in. long and broad, pale, very shining. Empty glumes 1-nerved, or 3-nerved at the very base; inner 3-nerved to the middle; 2 lower flowering-glumes obtuse, bearded below, downy above, margins and keel ciliated, 5-nerved, shortly awned below the top.—*H. antarctica,* Br.; *Disarrhena Antarcticum,* Labill. Fl. Nov. Holl. ii. 83. t. 232.; *Torresia redolens,* A. Cunn.

Abundant in wet places throughout the islands, *Banks and Solander,* etc. **Campbell's** Island, *J. D. H.* Also common in Fuegia, Tasmania, and the Alps of South-West Australia.

2. **H. alpina,** *Rœm. and Schultes.—H. borealis,* Fl. N. Z. i. 300; Fl. Tasm. ii. 108, not R. and S. Culms tufted, 1–2 ft. high, slender, soft and smooth. Leaves short, 4–8 in. long, strict, quite smooth, flat. Panicle ovate, 2–3 in. long, branches few, spreading, capillary. Spikelets $\frac{1}{5}$ in. long and broad, shining. Empty glumes short, acute, 3-nerved; flowering pubescent; margins silky with long hairs; awn variable in length and position, usually inserted above the middle and exserted.

Northern Island: tops of the Ruahine mountains, *Colenso.* **Middle** Island: Nelson mountains, *Munro, Travers;* Hopkins' river, alt. 2–3500 ft., *Haast;* Otago, lake district, *Hector and Buchanan.* Very tall and stout; also a native of the Tasmanian mountains, and of northern and alpine Europe, Asia, and North America.

3. **H. Brunonis,** *Hook. f.;—Fl. Antarct.* i. 93. *t.* 52. Very similar to *H. redolens,* but the empty glumes are much larger, twice as long as the flowering, $\frac{1}{3}$ in. long, and the flowering have 2-fid acute tips, with awns as long as the glume between the lobes.

Lord Auckland's group: tufts on the hills, *J. D. H.*

5. SPINIFEX, Linn.

Very coarse, rambling, much branched, rigid, spinous, bushy, littoral grasses, glabrous or woolly. Leaves long, rigid, wiry, involute. Inflorescence diœcious.—Male spikelets spiked on rigid peduncles which are collected into umbels, with sheathing spathaceous leaves at their bases, 1- or 2-flowered; empty glumes 3, large; flowering glume membranous. Stamens 3; anthers very long. Female spikelets solitary or few in the sheathing bases of very long pungent leaves, which are extremely numerous and collected into very large globose masses, 1- or 2-flowered, 3 empty glumes as in the male, but larger, flowering glume coriaceous. Scales 2, fleshy. Grain free within the glume and pale.

A genus of two or three littoral plants, natives of the coasts of Tasmania and Australia, (where it is said to impede travelling on the western side in some places), and in India and China.

1. **S. hirsutus,** *Labill. Fl. Nov. Holl.* ii. 230, 231;—*Fl. N. Z.* i. 292. A strong-growing, silky or woolly grass; culms stout, knotted, creeping. Leaves 1–1½ ft. long, very coriaceous; lower sheaths shining; upper and back of leaf silky or villous. Male spikes with the rachis 1–3 in. long, numerous, peduncled, silky. Spikelet very shortly pedicelled, $\frac{1}{3}$ in. long. Glumes acute. Female spikelet at the membranous bases of leaves which terminate in rigid, slender spines 3–5 in. long. Glumes larger than the male; outer awned.—*Ixalum inerme,* Forst.

Northern Island: common on the coasts, *Banks and Solander.* **Middle** Island: Canterbury, *Travers.* A common Australian and Tasmanian, Indian and Pacific Island Grass.

6. PASPALUM, Linn.

Tufted or creeping grasses, of various habit. Leaves flat or involute.—Spikelets in the New Zealand species ovoid, much compressed, arranged in two rows on one side of a flat rachis, 1-flowered, short, acute or obtuse, without a callus at the base. Empty glumes 2 or 3, unequal, lower usually very small; flowering concave, hardening and enclosing the pale and grain;

all obtuse or acute, awnless. Pale like the flowering glume, but smaller and 2-nerved. Scales 2, short, fleshy. Stamens 3. Grain free within the hardened glume and pale.

A most extensive tropical and subtropical genus of grasses, often weeds of cultivation, of various forms, etc., not extending into Tasmania. The above character does not apply to many of the non-New-Zealand species.

Erect. Leaves flat. Spikelets obtuse . . . . 1. *P. scrobiculatum.*
Creeping. Leaves involute. Spikelets acute . . . . 2. *P. distichum.*

1. **P. scrobiculatum,** *Linn.;—Fl. N. Z.* i. 291. Glabrous, erect, 1–3 ft. high, stout, leafy. Leaves rather broad, flat or wrinkled, rough at the margin, sometimes hairy at the base. Spikes 2–6, alternate, 1–2 in. long, rachis flat, bristly at the base. Spikelets imbricate in two series, sessile, orbicular, $\frac{1}{12}$–$\frac{1}{10}$ in. long. Empty glumes thin, membranous, 1-nerved.—*P. orbiculare*, Forst.

Northern and **Middle** Islands: common, *Banks and Solander*, etc., forming pasture at the Bay of Islands. An abundant tropical and subtropical weed.

2. **P. distichum,** *Burmann;—Fl. N. Z.* i. 291. Creeping, perfectly glabrous. Culms branched, compressed, ascending, 4–10 in. high, covered with leaf-sheaths to the top. Leaves distichous, strict, involute. Spikes in pairs, 1 in. long; rachis narrow. Spikelets pale, loosely imbricate, glabrous, pedicelled, ovate, acute, $\frac{1}{8}$ in. long.—*P. littorale*, Br.

Northern Island: sandy, etc. places, generally near the sea; Bay of Islands, *Cunningham;* Auckland, *Sinclair*. A common tropical and subtropical grass.

7. PANICUM, Linn.

Erect decumbent or creeping grasses of very various habit. Leaves flat or involute.—Spikelets variously arranged, naked or with bristles at their base, spiked racemed or panicled, 1-flowered, or if 2-flowered the lower flower male. Glumes 4, awned or awnless, lowest small or minute, empty; 2nd, larger; 3rd, empty or male-flowered; uppermost, with a hermaphrodite flower, fainter-nerved, smooth, hardening, and enclosing the pale and grain. Pale like the glume but smaller, 2-nerved. Scales 2, truncate. Stamens 3. Grain free within the hardened glume and pale.

A most extensive tropical and subtropical genus, not extending into Tasmania.

1. **P. imbecille,** *Trinius.—Oplismenus æmulus*, Kunth;—Fl. N. Z. i. 292. Culms slender, prostrate, rooting, ascending, weak, sparingly branched, 6–10 in. long. Leaves 1–6 in. long, $\frac{1}{4}$–1 in. broad, lanceolate; sheaths and knots more or less pilose. Spikelets spiked in distant clusters of 2–6, nearly sessile, $\frac{1}{12}$ in. long, glabrous or nearly so, naked or with a brush of long hairs at their bases. Glumes 4, 3 empty concave, membranous, 3-nerved, green, pilose or glabrous; lower shorter, with a long, flexuous, stout, obtuse awn; 2nd, rather larger, acute or cuspidate, rarely awned; 3rd, acute; flowering one terete, nerveless, coriaceous, white, shining.—*Orthopogon æmulus*, Br.; *Hekaterosachne elatior*, Steudel.

Northern Island: frequent in woods and shady places, *Banks and Solander*, etc.

Middle Island: Canterbury, *Lyall*. Probably a form of some common Indian Grass; it is also found in Australia and the South Sea Islands.

The following grasses, allied to *Panicum*, are introduced as doubtful denizens.

Panicum (Digitaria) sanguinale, Linn., an annual, prostrate or erect grass, with more or less hairy leaves; panicle of 6–8, very slender, almost digitate, strict branches, 2–4 in. long; spikelets in pairs on one side of the branches, minute, 1 sessile, the other pedicelled; lowest glume very minute, 2nd concave, shorter than the 3rd, which is flat and 5-nerved.

P. colonum, Linn., one of the commonest of Australian and tropical grasses, has erect culms, 1–2 ft. high; leaves flat, glabrous; panicle of 6–10 short, curved, ascending, alternate, flat branches; spikelets green, crowded in 2–4 rows on one side of the branchlets; glumes hispid; outer, broad, short; 2 following, flat, acute or cuspidate.

P. gibbosum, Br., is indicated by Raoul as a New Zealand plant, but I suspect some mistake; it is a slender tropical Australian grass, with a solitary filiform spike, and bearded glumes, the flowering one gibbous.

P. glaucum, Linn., an excessively common, tropical, and temperate annual weed, has an ovoid panicle like *Echinopogon*, with many rough bristles on the pedicels of the spikelets, and the flowering glume wrinkled.

8. ISACHNE, Br.

Tufted grasses. Leaves generally flat; mouth of the sheath bearded.—Spikelets panicled, short, ovoid, 2-flowered, lower flower usually male, upper hermaphrodite. Empty glumes 2, nearly equal, often deciduous; flowering 2, nearly equal, hardening and surrounding the pale and grain. Pale nearly as large as the glumes, 2-nerved, also hardening. Scales 2, truncate. Stamens 3. Grain free within the hardened glumes and pale.

A small tropical and subtropical genus, common in Australia, but not extending into Tasmania.

1. **I. australis**, *Br.;—Fl. N. Z.* i. 291. Culms, 6–18 in. high, rather slender, decumbent and creeping at the base. Leaves scaberulous, 3–5 in. long, $\frac{1}{6}$–$\frac{1}{4}$ broad, flat. Panicle erect, ovoid, 1–2 in. long, lax; branches long, flexuous, sparingly divided. Spikelets few, pedicelled, obtuse, $\frac{1}{12}$ in. long; pedicels with a pellucid gland. Empty glumes glabrous; lower flowering sessile, glabrous; upper stipitate, pubescent.—*? Panicum gonatodes*, Steudel.

Northern Island: Bay of Islands, *A. Cunningham;* Auckland, *Sinclair;* Lake Taupo, *Colenso*. Common in Australia, India, and China.

9. ZOYSIA, Willdenow.

Small, creeping, usually littoral grasses, with rigid, running rhizomes. Leaves distichous, grooved or involute, filiform.—Spikelets few (1–10), sessile or shortly pedicelled, alternate and imbricating, on a stiff, erect, flattened flexuous rhachis. Empty glume 1, ovoid, terete, convolute, rigid, very coriaceous, glabrous, even, obtuse mucronate or awned; flowering glume solitary, sessile, included, membranous, convolute. Pale membranous or 0. Stamens 3. Grain free.

The genus consists of one variable species, common on the shores of the Indian, Australian, and Chinese seas, sometimes attaining 1 ft. high, with a compound spike.

1. **Z. pungens**, *Willd.;—Fl. N. Z.* i. 312. Culms 1–2 in. high, tufted, quite glabrous. Leaves spreading, filiform or subulate, involute, 1–3 in long; sheaths tumid, grooved; ligule 0. Spike $\frac{1}{8}$–$\frac{1}{2}$ in. long, often reduced to a solitary erect spikelet. Spikelets $\frac{1}{10}$–$\frac{1}{5}$ in. long. Glume ovoid or more elongated,

yellow, green towards the keeled tip, which is sometimes produced into an awn more than half its own length.—*Rottbœllia uniflora*, A. Cunn.

Northern Island: sandy and muddy places, generally near the sea, from the Bay of Islands southwards. **Middle** Island: Motucka valley, *Munro*.

Anthistiria australis, Br., the Kangaroo-grass of Australia, has been gathered by Sinclair near Auckland, but is not indigenous.

Andropogon refractus, Br., is an Australian Grass, of which I find a solitary specimen, without locality or ticket, in A. Cunningham's Herbarium, probably accidentally introduced there.

Apluda mutica, Linn. Of this, a common tropical Indian Grass, I find a single specimen, without ticket or locality, amongst Sinclair's Auckland plants.

10. ECHINOPOGON, Palisot.

Culms tufted, erect, simple, scabrid. Leaves flat. Panicle contracted to an oblong or ovoid, dense spike, bristling with rigid awns.—Spikelets subsessile, horizontal, 1-flowered, green. Empty glumes equal, as long as or longer than the flowering, rigid, acuminate; flowering one with a silky pencil of hairs at the base, 2-fid at the tip; awn terminal, long, rigid, not twisted. Pale as long as the glume, with a short stiff pedicel at its base. Stamens 3. Ovary bearded at the top. Grain free.

A genus of one species, differing from *Agrostis* in habit, inflorescence, the rigid glumes, terminal awn, and ovary bearded at top.

1. **E. ovatus,** *Palisot;—Fl. N. Z.* i. 298. A harsh, very scabrid Grass, 6–24 in. high; sheath of upper leaf long. Spike $\frac{1}{2}$–1$\frac{1}{2}$ in. long.—*Agrostis ovata*, Forst.;—Labill. Fl. Nov. Holl. i. 19. t. 21; *Hystericina alopecuroides*, Steudel.

Abundant throughout the **Northern** and **Middle** Islands, *Banks and Solander*, etc. Also abundant in Southern Australia, Tasmania, and Norfolk Island.

11. DICHELACHNE, Endlicher.

Erect, rigid, tufted, annual or perennial grasses. Leaves flat subulate or convolute. Inflorescence a contracted panicle.—Spikelets long and narrow, 1-flowered, shining. Empty glumes 2, membranous, acuminate; flowering one as long, on a bearded pedicel, scabrid or silky, 2-fid or entire at the tip, with a straight twisted or flexuous awn from the back or from between the lobes, which is not jointed nor thickened at the base. Pale shorter, linear, 2-fid. Scales 2. Stamens 3. Grain long, terete, free.

A small genus, natives of Australia, Tasmania, and New Zealand.

Perennial. Culms stout, 1–3 ft. high. Spikelets $\frac{1}{2}$–$\frac{3}{4}$ in. long . . . . 1. *D. stipoides.*
Annual. Culms slender. Panicle dense. Spikelets $\frac{1}{3}$ in. long . . . 2. *D. crinita.*
Annual. Culms slender. Panicle lax. Spikelets $\frac{1}{4}$ in. long . . . . 3. *D. sciurea.*

1. **D. stipoides,** *Hook. f. Fl. N. Z.* i. 294. *t.* 66. Densely tufted, rigid, erect, smooth, polished, handsome, 1–3 ft. high. Leaves longer than the culms, very slender, erect, involute. Panicle strict, erect, 4–6 in. long; branches few, short, capillary, erect. Glumes membranous, $\frac{1}{2}$–$\frac{3}{4}$ in. long, lanceolate, acuminate; flowering one shorter, covered with silky, spreading hairs. Awn curved, 1 in. long, glabrous.—*Agrostis rigida*, A. Richard, *Dichelachne rigida*, Steud.

Northern Island: on maritime banks, rocks, etc., east coast, *Banks and Solander;* Bay of Islands and Auckland, *Sinclair*, etc. Also found in Tasmania and Australia.

2. **D. crinita,** *Hook. f. Fl. N. Z.* i. 293. Culms annual, tufted, 1–3 ft. high, slender or stout, leafy, glabrous downy or rather scabrid. Leaves flat or involute. Panicle elongate, contracted, dense, spike-like, shining; branches almost hidden by the flexuous awns. Spikelet $\frac{1}{3}$ in. long. Empty glumes narrow, long-acuminate, exceeding the flowering, which is scabrid, acuminate, entire or 2-fid at the tip; awn capillary, inserted at the back above the middle, flexuous, not twisted, $\frac{3}{4}$–1 in. long.—*D. vulgaris* and *D. Forsteriana*, Trinius; *Muhlenbergia mollicoma*, Nees; *Agrostis crinita*, Br.; *Apera crinita*, Palisot; *Anthoxanthum crinitum*, Labill. Fl. Nov. Holl. ii. 115. t. 263.

Northern and **Middle** Islands: abundant in dry soil, *Banks and Solander*, etc. Very common in Australia and Tasmania.

3. **D. sciurea,** *Hook. f. Fl. N. Z.* i. 294. Very similar to *D. crinita*, but smaller, more slender; panicle much thinner, with fewer spikelets; empty glumes shorter, as long as the flowering; and awn twisted, not four times as long as the glume.—Fl. Tasm. t. 158 A; *D. Sieberiana*, Trin. and Rup.; *Agrostis sciurea*, Br.; *Stipa micrantha*, Nees; *Muhlenbergia Sieberiana*, Trin.

Northern Island: Bay of Islands and Auckland, *Cunningham*, etc. I suspect this is only a variety of the former; it is also found in Australia and Tasmania.

12. APERA, Adanson.

Slender, erect, annual or perennial grasses. Leaves involute. Panicles with very long, capillary, whorled branches.—Spikelets minute, 1-flowered. Empty glumes 2, nearly equal, membranous, longer than the flowering, which is terete, coriaceous, acuminate, quite entire at the tip, ending in a very slender, straight, not twisted awn. Pale membranous. Scales 2, membranous. Stamens 1–3. Grain terete, enclosed in the coriaceous glume, free.

A small European and North American genus.

1. **A. arundinacea,** *Hook. f. Fl. N. Z.* i. 295. *t.* 67. A tall, erect, densely tufted, glabrous, most graceful grass. Culms reed-like, with creeping, scaly rhizomes, erect, branched, rigid, strict, 2–5 ft. high. Leaves coriaceous, narrow; margins involute, slightly scabrid; sheaths long; ligule short, truncate. Panicle erect, 8–16 in. long; pedicels alternate on the long whorled branches. Spikelets minute, $\frac{1}{12}$–$\frac{1}{10}$ in. long, pale, shining. Empty glumes with a scabrid keel; flowering one sessile on a small glabrous callus, thickened and rough at the top; awn scabrid, deciduous, $\frac{1}{3}$ in. long. Scales linear. Stamen 1. Ovary pedicelled. Grain truncate.

Northern Island: Cape Turnagain, *Colenso;* Akaroa, *Raoul.* Also found in subtropical East Australia.

13. SPOROBOLUS, Br.

Culms stout or slender, erect, tufted, simple or branched. Leaves flat or involute. Panicles usually contracted, often spike-like, with erect, appressed, capillary branches. Spikelets minute, 1-flowered.—Empty glumes 2, unequal, awnless; flowering glume sessile, short, awnless. Pale small. Scales 2. Stamens 1–3. Grain free, terete, with a lax pericarp.

A large tropical and subtropical genus of grasses, which does not extend into Tasmania.

1. **S. elongatus,** *Br.;—Fl. N. Z.* i. 295. A stout, rigid, perfectly smooth, glabrous, perennial grass, 1–2 ft. high. Leaves spreading, narrow, involute; ligule short; sheath deeply furrowed. Panicle 6–12 in. long, much contracted, slender. Spikelets pedicelled, minute, $\frac{1}{20}$ in. long, pale-green.

Northern Island: Auckland, *Sinclair*. A common tropical Grass, possibly introduced into New Zealand.

14. AGROSTIS, Linn.

Slender, erect, tufted, usually annual, glabrous or scaberulous grasses. Leaves flat or involute. Panicles generally open, often effuse, with whorled capillary branches, sometimes contracted and spike-like.—Spikelets small, pedicelled, 1-flowered. Empty glumes 2, nearly equal, acuminate or acute, usually longer than the flowering; flowering glume sessile or shortly pedicelled, obtuse acute or truncate, awnless or with a shortly-twisted or straight, dorsal or basal awn. Pale membranous, often minute or 0, sometimes furnished at the base with the pedicel of a third glume. Scales 2. Stamens 2 or 3. Grain oblong, terete, free.

A very large genus, abounding in temperate and cold climates, absent in hot, forming a considerable proportion of pasture in Europe. One-flowered forms of several species of *Poa, Danthonia* and *Deschampsia* may be confounded with *Agrostis.*

I. *Glumes membranous, flowering one much shorter than the empty, truncate and jagged at the tip. Pale very membranous, much shorter than the flowering glume, or* 0, *without the pedicel of a second glume at its base.*

Flowering glume wholly glabrous, truncate. Panicle usually contracted.

Spikelets $\frac{1}{8}$ in., on hispid pedicels . . . . 1. *A. antarctica.*
Spikelets $\frac{1}{12}$–$\frac{1}{10}$ in.; pedicels scarcely scabrid. Branches of panicle whorled . . . . 2. *A. canina.*
Spikelets $\frac{1}{16}$–$\frac{1}{12}$ in.; pedicels scarcely scabrid. Branches of panicle few, opposite or 3-nate . . . . 3. *A. parviflora.*

Flowering glume silky. Panicle very broad.

Leaves narrow, usually involute and filiform . . . . 4. *A. æmula.*
Leaves broad, flat . . . . 5. *A. pilosa.*

II. *Glumes membranous; flowering much shorter than the empty, usually bearded at the base and jagged at the tip. Pale very membranous, shorter than the flowering glume, with the silky pedicel of a second glume at its base.*

Branches of panicle whorled, capillary. Spikelets $\frac{1}{8}$–$\frac{1}{4}$ in. long . . 6. *A. Billardieri.*

III. *Glumes hard, coriaceous; flowering nearly as long as the empty, often pedicelled, silky at the base. Pale hard, as long as the flowering glume, with a rigid, bearded pedicel of a second glume at its base.* (*Panicle contracted, branches very short.*)

Leaves filiform. Spikelets $\frac{1}{10}$–$\frac{1}{8}$ in. Awn exserted . . . . . 7. *A. setifolia.*
Leaves filiform. Spikelets $\frac{1}{8}$–$\frac{1}{4}$ in. Awn exserted . . . . . . 8. *A. avenoides.*
Leaves concave. Spikelets $\frac{1}{6}$–$\frac{1}{4}$ in. Awn very short . . . . . 9. *A. Youngii.*
Leaves concave. Spikelets $\frac{1}{6}$–$\frac{1}{4}$ in. Awn exserted . . . . . . 10. *A. quadriseta.*

1. **A. antarctica,** *Hook. f. Fl. Antarct.* ii. 374. Culms erect, tufted, 6–24 in. high, glabrous and smooth. Leaves involute, shorter than the culms; sheaths glabrous, smooth; ligule oblong, truncate. Panicle 1–4 in.

long, contracted, nodding, pale-green. Spikelets crowded on short, erect, hispid pedicels, $\frac{1}{6}$ in. long and upwards. Empty glumes nearly equal, lanceolate, long-acuminate; keel ciliate, sides scaberulous; flowering one sessile, half the length of the empty or shorter, glabrous, membranous, truncate; awn from the middle of the back, rather recurved, longer than the spikelet. Pale 0 or small.—*A. multicaulis*, Hook. f. Fl. Antarct. i. 95; perhaps *A. magellanica*, Lamarck.

Campbell's Island, on moist banks. The *A. multicaulis* of Campbell's Island is certainly only a small state of the common Chilian and Fuegian *A. antarctica*. Whether this be the same as Lamarck's *A. magellanica* is still doubtful; the Magellanic plant which I have referred to the latter in the 'Flora Antarctica' has a more open panicle, and a small beard at the base of the flowering glume.

2. **A. canina,** *Linn.;—Fl. N. Z.* i. 296. Culms tufted, leafy, 1–24 in. high, slender, smooth, and glabrous. Leaves flat or involute, glabrous; sheaths quite smooth; ligule oblong, obtuse, usually lacerated. Panicle open or contracted, usually ovate or lanceolate, 2–4 in. long; branches slender, lower whorled, and slender pedicels slightly scabrid. Spikelets $\frac{1}{10}$–$\frac{1}{12}$ in. long. Empty glumes nearly equal, lanceolate, acuminate, glabrous; keel slightly ciliate; flowering glume half the length of the outer, oblong, truncate; awn dorsal or 0. Pale usually 0.

Var. β? Culms shorter, 1–5 in. high. Panicle much contracted, 1–1½ in. long; lower branches very short.—*A. gelida*, F. Muell. mss.

Var. γ? Culms very short, most densely tufted and moss-like, ½–2 in. high. Leaves subulate, often flaccid. Panicle minute, hidden amongst the leaves. Awn 0. Pale 0.—*A. subulata*, Fl. Antarct. i. 95. t. 53; *A. parviflora*, *β. perpusilla*, Fl. N. Z. i. 296.

Mountainous parts of the **Northern** and **Middle** Islands: var. α, Milford Sound, *Lyall;* alps of Canterbury, common, alt. 2–4000 ft., *Sinclair and Haast*. Var. β, Ruahine range, *Colenso;* ascending higher than any other Grass on Mount Darwin, *Haast;* Otago, lake district, alpine, *Hector and Buchanan*. Var. γ. **Northern** Island: snow rills, top of the Ruahine mountains, *Colenso*. **Middle** Island: Lake Tennyson, alt. 4400 ft., *Travers*. **Campbell's** Island, *J. D. H.*

There can be no question, I think, of the identity of this plant with the common European *A. canina*, which also occurs commonly in Fuegia, and of which I have a specimen from Victoria, F. Mueller, labelled *A. parviflora*. I agree with Colonel Munro in thinking that *A. falklandica* (Fl. Antarct.) is a variety, and *A. tenuifolia* a third form. Of var. β I am not so sure. I have seen no European state of *canina* with so contracted a panicle, nor having such short lower branches. The foliage and flowers are, however, identical; it has been found on the Australian Alps. Var. γ is a very singular little plant, in a depauperated state. Colenso's and Travers's specimens entirely agree in size and habit. The Campbell Island ones are larger, and may be different; but much more numerous specimens are wanted of all.

3. **A. parviflora,** *Br.;—Fl. N. Z.* i. 296. Culms slender, erect, or prostrate at the base, tufted, 6–8 in. high, glabrous. Leaves very narrow, flat involute or setaceous; ligule oblong, truncate, lacerate. Panicle of few lax, long, capillary, 3-chotomous, scabrid branches, opposite or 2- or 3-nate. Spikelets minute, $\frac{1}{16}$–$\frac{1}{12}$ in. long, pale-green. Empty glumes spreading, nearly equal, glabrous, scabrid on the keel; flowering glume one-third shorter, quite glabrous, truncate; nerves faint; awn, if present, dorsal, included. Pale minute or 0.—Fl. Tasm. ii. 113. t. 158 B.

Northern Island: probably common, but overlooked; east coast, Patea village and Cook's Straits, *Colenso*. **Middle** Island: Otago, lake district, *Hector and*

Buchanan. Common in Tasmania and Australia, and very near *A. canina*, from which it is best distinguished by the very lax panicle and smaller spikelets. Nearly allied to the European *A. alpina*, but with a dorsal, never basal awn.

4. **A. æmula,** *Br.*—*Deyeuxia Forsteri*, Kunth;—Fl. N. Z. i. 298. A very elegant, glabrous, slender grass. Culms tufted, very slender, 6–24 in. high. Leaves very narrow, involute, often setaceous, rarely flat; sheaths glabrous; ligule narrow, oblong. Panicle large, open, spreading; branches whorled, capillary, 3–6 in. long, usually 3-chotomously branched; rachis often flexuous. Spikelets $\frac{1}{10}$–$\frac{1}{6}$ in. long, on very slender pedicels. Empty glumes equal, lanceolate, long-acuminate, smooth, keel scabrid; flowering one membranous, shorter by about one-third, sessile, truncate, silky with scattered long hairs, and bearded at the base; awn from the back about the middle, very slender, exserted. Pale 0 or small, without a pedicel at its base or with a very minute one.—*A. Forsteri*, Rœm. and Schultes; *A. Lyallii*, Fl. N. Z. i. 297; *A. leptostachys*, Fl. Antarct. i. 94; *Lachnagrostis Forsteri*, Trinius; *L. æmula*, Nees; *Deyeuxia æmula*, Kunth; *Avena filiformis*, Forst.

Throughout the **Northern** and **Middle** Islands, abundant, *Banks and Solander*. **Campbell's** Island, *J. D. H.* Equally common in temperate Australia and Tasmania. I find the pedicel at the base of the pale to be so often absent, and when present so minute, that I am obliged to remove this plant from the section or genus of *Deyeuxia*, and to unite with it my *A. Lyallii* and *A. leptostachya;* the latter (from Campbell's Island), however, has unusually large spikelets.

5. **A. pilosa,** *A. Rich. Flor.* i. 134. *t.* 23;—*Fl. N. Z.* i. 297. Quite similar to *A. æmula*, except in being more coarse and robust, with large flat leaves, sometimes $\frac{1}{2}$ in. broad, which are described by Richard as pilose and scabrid, but are only scaberulous in my specimens.

Northern Island: mountainous districts of the interior, *Colenso.* **Middle** Island: Astrolabe Harbour, *D'Urville;* Southern Alps, *Sinclair, Haast*, etc.; Otago, lake district, *Hector and Buchanan.* Perhaps only a large form of *A. æmula*, with which it entirely agrees in inflorescence and all floral characters; but is a different-looking and far more robust plant.

6. **A. Billardieri,** *Br.*—*Deyeuxia Billardieri*, Kunth; Fl. N. Z. ii. 298. Culms tufted, erect, leafy, 1–2 ft. high. Leaves 6 in. long, $\frac{1}{3}$–$\frac{1}{2}$ in. broad, flat, glabrous or scaberulous; ligule short, oblong. Panicle very lax, 4–10 in. long, effuse, scabrid; branches very long, whorled, capillary, 3-chotomous. Spikelets $\frac{1}{6}$–$\frac{1}{4}$ in. long, on very slender pedicels, green or purplish. Empty glumes lanceolate, acuminate, keel scabrid, sides glabrous or scabrid; flowering glume $\frac{1}{4}$ shorter, truncate, silky at the base, 4-nerved, the lateral nerves produced into short awns; awn on the middle of the back, bent, half to twice as long as the glume. Pale as long as the glume, longer than the silky pedicel at its back.—*A. vaginata*, Steudel; *Lachnagrostis Billardieri*, Trinius; *Avena filiformis*, Labill. Fl. Nov. Holl. i. 24. t. 31, not Forster.

Northern Island, *Banks and Solander*, abundant in some places, but apt to be confounded with large states of *A. Forsteri.* Also an abundant Southern Australian and Tasmanian grass.

7. **A. setifolia,** *Hook. f.*—*Deyeuxia setifolia*, Fl. N. Z. i. 299. t. 65 B. Culms erect, tufted, very slender, wiry, smooth, 6–8 in. high. Leaves very narrow, setaceous or filiform, involute, shorter than the culms, quite smooth;

ligule oblong; sheaths slender, short. Panicle 1–2 in. long, erect, contracted, very slender, branches few, suberect, short. Spikelets few, shortly pedicelled, erect, $\frac{1}{10}$–$\frac{1}{8}$ in. long, pale, shining. Empty glumes rigid, oblong-lanceolate, acuminate; flowering one hard, rather shorter, silky at the base, truncate, 4-toothed; awn on the middle of the back, exserted, rigid, recurved. Pale longer than the very silky pedicel, rigid.

Northern Island: Titiokura, top of the Ruahine range and Lake Waikare, *Colenso.*

8. **A. avenoides,** *Hook. f., n. sp.* Culms rigid, wiry, 8–12 in. high, quite smooth, slender. Leaves much shorter than the culms, involute, very slender, smooth; ligule short; sheath short, smooth. Panicle very narrow, erect, 2–4 in. long, much contracted, branches very short, erect, few-flowered. Spikelets $\frac{1}{6}$–$\frac{1}{4}$ in. long. Empty glumes oblong-lanceolate, acuminate, rigid, smooth and glabrous; flowering glume sessile, as long, hard, scabrid, 2-fid, silky at the base; awn dorsal, twisted, recurved. Pale as long as the glume; pedicel long, pilose.

Middle Island: grassy downs, Rangitata river, *Sinclair and Haast;* Otago, lake district, *Hector and Buchanan.* Closely allied to the Tasmanian *A. montana.*

9. **A. Youngii,** *Hook. f., n. sp.* Similar in habit to *A. avenoides,* but larger, 2–4 ft. high, more robust. Leaves flat, $\frac{1}{6}$–$\frac{1}{4}$ in. diam., panicle 4–6 in. long, very slender, flexuous; branches very short. Spikelets $\frac{1}{6}$–$\frac{1}{4}$ in. long. Empty glumes oblong-lanceolate, acuminate, rigid, smooth, glabrous, nerveless; flowering glume as long, pedicelled, hard, scabrid, 2–4-cuspidate; awn very short, almost terminal. Pale as long as the glume; pedicel stout, with long silky hairs.

Middle Island: dry hillsides, sources of the Waitaki, *Haast.* Allied to the Tasmanian *A. scabra.*

10. **A. quadriseta,** *Br;—Fl. N. Z.* i. 296. Culm 6 in. to 4 ft. high, smooth or rough, erect, stout or slender. Leaves shorter than the culm, flat or involute, often setaceous, glabrous or scabrid. Panicle 2–6 in. long, slender or stout, usually extremely dense cylindric and spike-like, rarely open below and pyramidal, or interrupted and as it were lobed, pale-green, shining, lower branches sometimes spreading, whorled. Spikelets pedicelled, variable, $\frac{1}{10}$–$\frac{1}{6}$ in. long, shining. Empty glumes lanceolate, keels scabrid; flowering glume shorter, on a short bearded pedicel, coriaceous, concave, scaberulous, obscurely 5-nerved; nerves usually ending in minute exserted points or awns, 2-fid with a stout dorsal awn. Pale as long as the glume.—*A. elatior,* Steud.; *Avena quadriseta,* Labill. Fl. Nov. Holl. i. t. 32; *Bromidium,* Nees; perhaps the *Agrostis montana,* Br., of Raoul, Choix, 39.

Abundant throughout the **Northern** and **Middle** Islands, *Banks and Solander,* etc. Also very common in Australia and Tasmania.

Aristida calycina, *Br.;—Fl. N. Z.* i. 293. Culms glabrous, wiry, tufted, annual?, 1 ft. high. Leaves subulate. Panicle 3 in. long, very slender. Spikelets few, $\frac{1}{3}$ in. long, shortly pedicelled, 1-flowered. Empty glumes of equal length, shorter than the flowering, which has a 3-cleft awn at its tip.

Northern Island: Bay of Islands?, *A. Cunningham.* I have seen no reputed New Zealand specimen of this, but its finder's; it is a native of tropical Australia, and not being found by any other collector in New Zealand, I suspect it may have got into A. Cunningham's herbarium by accident.

15. ARUNDO, Linn.

Tall, stout, leafy grasses. Leaves flat or involute. Panicles very large, effuse, with excessively numerous, much divided branches, usually nodding to one side.—Spikelets excessively numerous, white, very thin and membranous, silky or shining, 1–5-flowered. Empty glumes nearly equal, very long, lanceolate, acuminate; flowering glumes pedicelled, long, lanceolate, very silky, entire or 2-fid at the tip; the awn straight or twisted. Pale short. Scales 2, fleshy. Stamens 3. Grain free, long, terete.

Noble grasses, found in most tropical and temperate countries, especially in or near water.

1. **A. conspicua,** *Forst.*;—*Fl. N. Z.* i. 299. Culms 3–8 ft. high, growing in dense tussocks, from which rise a profusion of long curving leaves, and erect slender culms, with large white panicles. Leaves coriaceous, narrow, smooth or scabrid. Panicle 1–2 ft. long, branches drooping. Spikelets 1–3-flowered, white, pedicels capillary. Empty glumes ½ in. long; flowering half as long, surrounded with long silky hairs, acuminate, terminated by the slightly twisted, slender included awn.—*A. australis*, A. Rich.; *A. australis* and *A. conspicua*, A. Cunn.; *? A. Kakao*, Steud.; *Achnatherum conspicuum*, Palisot; *Gynerium (?) Zelandicum*, Steud.; *Calamagrostis conspicua*, Gmel.; *? Agrostis procera*, A. Rich.

Northern and **Middle** Islands: abundant in moist places, *Banks and Solander*, etc.; as far south as Otago, *Lindsay*. The largest New Zealand Grass, and confined to these islands. Culms used for thatching and lining houses with reedwork.

Cynodon Dactylon, Linn. An excessively common, small, tropical and temperate grass, much used for fodder in India (Doab Grass), is introduced into New Zealand; it creeps extensively, has involute leaves, digitate strict branches of the panicle; spikelets small, sessile, in 2 rows on one side of the branches, 1-flowered, empty glumes 2 keeled, flowering one pilose, thinner and broader; pale with a bristle at its base.

Eleusine indica, Gærtner, a very common tropical Grass, has been gathered near Auckland; it has the habit and inflorescence of *Cynodon*, but is larger and coarser, the glumes obtuse, and grain wrinkled.

16. DANTHONIA, Decandolle.

Tufted, usually harsh, rigid grasses. Leaves flat or involute. Panicles effuse or contracted.—Spikelets pedicelled, rather large, coriaceous or scarious, 2- or many-flowered (rarely 1-flowered). Empty glumes unequal or nearly equal, membranous, keeled, awnless; flowering 2 or more, pedicelled, silky or furnished with pencils of hairs on the sides, convex, broadly 2-fid, the divisions subulate cuspidate or awned; dorsal awn from between the divisions, long, slender or stout, filiform or flat and twisted at the base. Pale 2-fid. Scales glabrous or pilose. Stamens 3. Ovary stipitate, glabrous. Grain free.

A large South African, Australian, Tasmanian, and New Zealand genus, also found in the south of Europe. The species are very difficult of discrimination: when the flowering-glumes are reduced to two, the species may be confounded with *Trisetum;* and if to one, with *Agrostis*, but for the more deeply 2-cuspidate flowering-glume; from *Arundo* it is distinguished chiefly by habit, and the more deeply 2-fid flowering-glume; the long silky hairs on the sides of the flowering-glume generally distinguish it from all its allies.

I. *Empty glumes shorter than the spikelet. Flowering glumes with scattered long silky hairs at the base and sides.*

Awn subulate, not flattened nor twisted.
  Panicle open, large, effuse, branches 6–8 in. long . . . . . . 1. *D. Cunninghamii.*
  Panicle short, close, ovoid, branches ½–1 in. long . . . . . . 2. *D. bromoides.*
Awn flattened and twisted at the base.
  Panicle very lax, open. Leaves setaceous . . . . 3. *D. Raoulii.*
  Panicle very lax, open. Leaves flat, very coriaceous . . . . . 4. *D. flavescens.*

II. *Empty glumes longer than the flowering, and including them. Flowering glume with pencils of silky hairs on their sides and base.*

Spikelet ⅜ in. Awn longer than glume . . . . 5. *D. semi-annulari.*
Spikelet ⅕–¼ in. Awns included . . . . 6. *D. Buchanani.*
Spikelet ⅙ in. Awn very short indeed . . . . 7. *D. nuda.*

1. **D. Cunninghamii,** *Hook. f.*—*D. antarctica*, var. *β. laxifolia*, Fl. N. Z. i. 303. Culms 3–5 ft. high, stout, ¼ in. diam., quite glabrous or pilose. Leaves very coriaceous, concave, rigid, 1–2 ft. long, ¼ in. broad, glabrous or hairy on the upper surface; sheaths broad; ligule 0, or a line of long hairs. Panicle very open, lax, 8–12 in. long; branches many or few, in distant pairs, very slender, 4–8 in. long, pubescent. Spikelets alternate on the branches, shortly pedicelled, ¼–½ in. long, 2–8-flowered. Empty glumes unequal, lanceolate or linear-oblong, lower acute or obtuse; flowering ⅙–¼ in. long, narrow, glabrous, except at the base and sides where covered with long silky hairs, deeply 2-fid; awn recurved, not flattened nor twisted at the base, as long as the glume.—*Agrostis pilosa*, A. Cunn. not A. Rich.; *D. rigida*, Fl. N. Z. i. t. 69 A, not Raoul.

Northern Island: Bay of Islands, shady woods, Keri-Keri river, *A. Cunningham;* woods near the tops of the Ruahine mountains and Hawke's Bay, *Colenso.* **Middle** Island: uplands near Otago, *Lindsay;* lake district, *Hector and Buchanan.*

2. **D. bromoides,** *Hook. f. Fl. N. Z.* i. 303. *t.* 68 A. Culms and leaves as in *D. Cunninghamii*, but panicle much smaller, oblong or lanceolate, 4–6 in. long, contracted; branches 1–2 in. long, erect. Spikelets broader, glumes closer, more like those of *D. Raoulii.* Awn not flattened nor twisted, ¾ in. long.—*Bromus antarcticus*, Fl. Antarct. i. 97. t. 54, not *Danthonia antarctica*, Fl. N. Z.

Northern Island: hills near Wellington, *Stephenson.* **Lord Auckland's** group and **Campbell's** Island: abundant on the hills, *J. D. H.*

3. **D. Raoulii,** *Steud.*—*D. rigida*, Raoul;—Hook. f. Fl. N. Z. i. 303, not Steud. Habit and character of *D. Cunninghamii*, but leaves involute and filiform; spikelets longer, pedicelled, larger, ½–¾ in. long, broader; glumes larger, both empty ones long acuminate, the flowering closely imbricate, broader, and the awn is flattened and twisted in a corkscrew manner at the base.

Northern Island: from halfway up to the tops of the Ruahine range, and hills on the east coast, *Colenso.* **Middle** Island: Akaroa, *Raoul;* common in the Alps, ascending to 3000 ft., *Sinclair and Haast;* Milford Sound, *Lyall;* Tarndale, *Sinclair;* Otago, lake district, alt. 2000 ft., *Hector and Buchanan.* Very like indeed the *D. Cunninghamii*, in some respects, but I think quite distinct; it is more near *D. robusta* of the Victorian alps. The leaves in all specimens from all elevations are almost filiform.

4. **D. flavescens,** *Hook. f. n. sp.* A large coarse grass, of the size and

with the habit of *D. Cunninghamii*, from which it differs in the awn being broad, flattened, and twisted like a corkscrew at the base.

Middle Island: Southern Alps, *Sinclair and Haast;* Otago, lake district, alt. 2000 ft., *Hector and Buchanan.* One of the "snow grasses." This so much resembles *D. Cunninghamii*, that I advance it as a different species with great hesitation; the widely different awn and locality, however, seem very marked characters.

5. **D. semi-annularis,** *Br.;—Fl. N. Z.* i. 304. Culms tufted, slender, rigid, glabrous or pilose, 6–20 in. high. Leaves filiform or setaceous, involute, rarely flat; ligule 0; mouth of sheath with spreading hairs. Panicle 2–4 in. long, effuse or contracted, of few, erect, short branches. Spikelets few, $\frac{1}{3}$ in. long, 4–6-flowered. Empty glumes white or purplish, $\frac{1}{3}$ in. long, nearly equal, much longer than the spikelet; flowering glumes villous at the base, with several transverse series of tufts of hairs above the middle on their sides, deeply 2-fid at the top; lateral awns as long as the glumes; central twisted, black or pale, often twice as long as the glume and recurved.

Var. *α.* Leaves glabrous.—*D. semi-annularis*, Labill. Fl. Nov. Holl. i. 26. t. 33; *D. Unarede*, Raoul, Choix, ii. t. 4; perhaps *D. cingula*, Steudel.

Var. *β. pilosa.* More or less covered with long spreading hairs.—*D. pilosa*, Br.;—Fl. N. Z. i. 303; *D. Gunniana*, Nees.

Var. *γ. gracilis.* Pilose as var. *β.* Culms very slender. Leaves filiform. Fowering glumes fewer, awns smaller, included, or nearly so.—*D. gracilis*, Hook. f. Fl. N. Z. i. 304. t. 69 B.

Abundant throughout the **Northern** and **Middle** Islands: ascending to 4000 ft., *Banks and Solander*, etc. A most abundant grass in Australia and Tasmania also, very variable, especially in the size and number of the spikelets, and the relative and absolute length of the awns.

6. **D. Buchanani,** *Hook. f. n. sp.* Culms very slender, perfectly glabrous, 5–10 in. high. Leaves erect, involute, filiform. Panicle small, contracted, $\frac{1}{2}$–1 in. long; spikelets 6–8, erect, very pale green, $\frac{1}{5}$–$\frac{1}{4}$ in. long, on short pedicels. Empty glumes broader than in *D. semi-annularis*, longer than the spikelets; flowering 2-fid at the top, with a slender, scarcely twisted middle awn, which is barely longer than the glumes, with silky hairs at the base and sides.

Middle Island: Otago, lake district, *Hector and Buchanan.* This appears a distinct form, though closely allied to *D. semi-annularis*, from which its shorter, broader glumes, short awns, and less silky flowering-glume distinguish it; it is also very near the *D. pauciflora* of Tasmania.

7. **D. nuda,** *Hook. f. Fl. N. Z.* ii. 337. Culms prostrate at the base, 5–8 in. high, slender, leafy, glabrous, much branched below Leaves involute, filiform, glabrous and smooth; ligule ciliated; sheath with long hairs at the mouth. Panicle small, of 6–10 erect spikelets on short pedicels. Spikelets greenish-white, $\frac{1}{6}$ in. long, 3-flowered. Empty glumes longer than the spikelet; flowering shortly 2-fid at the top, with a very short intermediate awn, ciliated at the base, and with one tuft of hairs on each side near the margin.

Northern Island: mountains near the east coast, *Colenso.* A very distinct species: the glumes are very shortly 2-fid, and the awn not twisted, so that it is not a very characteristic species of the genus, except in its pencil of silky hairs on the glume.

Holcus mollis, Linn., one of the most abundant European grasses, occurs as an escape;

it is annual, usually softly pubescent, especially at the knots. Panicle erect, open. Spikelets 2-flowered; empty glumes concave, equal; flowering glumes very deciduous, upper male, awned, lower perfect, awnless.

17. DESCHAMPSIA, Palisot.

Tall, erect, tufted, usually perfectly smooth, shining grasses. Leaves flat or involute. Panicles large, branched.—Spikelets pedicelled, 2- or 3- rarely 1-flowered. Empty glumes 2, nearly equal, much compressed, keeled, awnless; flowering ones 2 or 3, and a terminal imperfect one, 4-toothed, or jagged at the truncate tip; awn dorsal, short, straight, obtuse. Pale 2-fid. Scales 2, entire. Stamens 3. Grain free.

A small genus of temperate Grasses. The obtuse awn distinguishes 1-flowered specimens from *Agrostis*.

1. **D. cæspitosa,** *Palisot;—Fl. N. Z.* i. 301. An elegant, tall, perfectly smooth, shining grass, rarely scaberulous. Culms 6–36 in. high, rather slender, leafy. Leaves involute, long or short; ligule long, acute; sheaths smooth and shining, glabrous or subpilose. Panicle 3–12 in. long, effuse, one-sided, branches whorled or fascicled, slender, rough, erect or spreading. Spikelets $\frac{1}{6}$–$\frac{1}{5}$ in. long, pale, shining, yellow-green or purplish. Empty glumes acute; flowering silky at the base, uppermost often reduced to a villous pedicel; awn basal, as long as the glume.—*Aira Kingii,* Fl. Antarct. 376. t. 135; *Triodia splendida,* Steudel; *? Aira australis,* Raoul, Choix, 12 (awn omitted); *Agrostis Aucklandica,* Fl. Antarct. i. 96 (form with 1-flowered spikelet).

Northern Island: low grounds on the east coast, Te Hawera and Hawke's Bay, common, *Colenso.* **Middle** Island: abundant, ascending to 3500 ft. in the alps. **Lord Auckland's** group, *J. D. H.* A very common and widely diffused Grass in the north and south temperate zones and mountains of Central Africa.

18. KŒLERIA, Persoon.

Erect, rather stout, annual or perennial grasses. Leaves flat or involute. Panicle spike-like, dense.—Spikelets shortly pedicelled, pale, shining, much compressed, 2–4-flowered. Empty glumes lanceolate, acute, unequal or nearly equal, as long as the spikelets or nearly so; flowering ones on a short rachis, like the empty, acute, entire at the tip or 3-toothed, with a short, straight, intermediate awn or 0. Pale as long as the glume. Scales 2. Stamens 3. Grain free.

A small European genus, equally allied to *Deschampsia* and *Poa,* and very difficult to distinguish generically from either.

3. **K. cristata,** *Persoon;—Fl. N. Z.* i. 305. Culms tufted, tall or short, rather stout, stiff, glabrous or downy, 1–3 ft. high. Leaves 6–8 in. long, narrow, flat or involute; ligule very short. Panicle strict, erect, 3–5 in. long, narrow, often interrupted or lobed below, branches very short. Spikelets crowded, erect, imbricated, white or purplish, shining, 2- or 3-flowered, $\frac{1}{6}$–$\frac{1}{5}$ in. long. Empty glumes unequal, acute; flowering ones, glabrous, rachis slender, pilose, nerveless, keeled, acuminate, sometimes minutely toothed and shortly awned at the tip. Pale as long as the glume.

Middle Island: Aglionby plains, *Munro;* Canterbury plains and Acheron valley, alt.

4000 ft., *Travers;* Otago, *Lindsay;* lake district, *Hector and Buchanan*, terraces on the Southern Alps, *Sinclair and Haast.* A very common European and northern temperate grass, widely diffused in Australia and the south temperate hemisphere generally, but possibly introduced only; it exists in none of the earlier New Zealand collections.

19. TRISETUM, Kunth.

Tufted grasses, usually slender, often downy. Leaves involute. Panicles open, or contracted and spiciform.—Spikelets 2–4-flowered. Empty glumes 2, keeled, awnless, shorter than the spikelet; flowering glumes 2–4, with a terminal imperfect one, 2-fid at the tip, divisions subulate; awn from between the divisions, twisted and recurved. Pale 2-nerved. Scales 2. Stamens 3. Grain free, glabrous.

Temperate, alpine, arctic and antarctic grasses, distinguishable from *Danthonia* by habit chiefly.

Glabrous, 1–2 ft. high. Panicle lax, spreading . . . . 1. *T. antarcticum.*
Downy, 1–6 in. high. Panicle spiciform . . . . 2. *T. subspicatum.*
Pilose, 1–3 ft. high. Panicle very slender, contracted . . . . . . 3. *T. Youngii.*

1. **T. antarcticum,** *Trinius;—Fl. N. Z.* i. 302. *t.* 68 B. A tufted, erect, slender, smooth, rarely pubescent, shining grass, 1–2 ft. high. Leaves flat or involute, usually narrow, long or short, sometimes setaceous, quite smooth or scaberulous; ligule truncate, very short, often silky. Panicle erect, slender, contracted or effuse, 2–10 in. long; branches short, very slender, suberect. Spikelets $\frac{1}{4}$ in. long, contracted, white, very shining or pale green, 3- or 4-flowered. Empty glumes unequal, acuminate; margins white; keel scabrid; flowering ones pedicelled, deeply 2-fid, with a pencil of long white hairs at the base, scabrous; awn not twisted, recurved, twice as long as the glume.—*Aira antarctica*, Forst.; *Avena antarctica*, Rœm. and Sch.; *A. Forsteri*, Kunth; *Danthonia antarctica*, Sprengel; *D. pallida*, A. Cunn.

Northern and **Middle** Islands: abundant, *Banks and Solander*, etc.; not found elsewhere. Haast sends a large form from Lake Okau, with pubescent leaves $\frac{1}{3}$ in. broad.

2. **T. subspicatum,** *Palisot;—Fl. Antarct.* i. 97 *and* 377. Small, rather stout, densely tufted, usually downy, 4–18 in. high. Leaves shorter than the culms; ligule short, truncate, silky. Panicle dense, subcylindric, spiciform, 1–3 in. long. Spikelets shortly pedicelled, imbricate, $\frac{1}{6}$ in. long, 2- or 3-flowered, pale greenish-white, shining, pubescent. Empty glumes shorter than the spikelet, unequal, obtuse; flowering ones on hairy pedicels, downy, 2-cuspidate; awn recurved, dorsal, as long as or longer than the glume, inserted below the 2-cuspidate tip.

Var. *β.* More glabrous. Spikelets rather narrower. Glumes nearly glabrous.

Middle Island: Upper Awatere valley, *Sinclair;* Otago, lake district, *Hector and Buchanan;* rivulets of the Hopkins river, alt. 2500 ft., *Haast.* **Campbell's** Island, *J. D. H.* Var. *β*, Upper Awatere and Wairau valleys, *Sinclair.* A native of arctic Europe, Asia, and America, the alps of the same continents, of South America, Australia, and Tasmania, and of Fuegia.

3. **T. Youngii,** *Hook. f., n. sp.* Culms slender, erect, 2–3 ft. high, glabrous, shining. Leaves flat, $\frac{1}{6}$–$\frac{1}{4}$ in. broad, and sheaths pilose; ligule truncate. Panicle slender, 3–6 in. long; branches very short, with few spikelets. Spikelets pale, $\frac{1}{6}$ in. long, shining, 1- or 2-flowered. Empty glumes

broad, oblong, acute, membranous, as long as the spikelets; flowering glumes 2-cuspidate, nearly glabrous, with a few short hairs at the base; awn dorsal, stout, recurved, exserted, inserted below the 2-cuspidate tip.

Middle Island: Macaulay valley, alt. 3-4000 ft., *Haast and Young*. A very distinct species, closely allied to a Victorian alps one, distinguished from *T. subspicatum* by its slender tall culm, long very slender panicle, and broad outer glumes. It may, however, prove to be a form of that species. It is named in compliment to Mr. W. Young, Mr. Haast's able and indefatigable assistant-surveyor, who paid much attention to collecting the Grasses of the Southern Alps.

Avena sativa, Linn., the cultivated Oat, has been found as an escape from cultivation in various parts of the islands.

20. GLYCERIA, Br.

Erect, glabrous, often aquatic grasses. Culms tufted or creeping below. Leaves flat or involute. Panicle long and contracted.—Spikelets pedicelled, linear, glabrous, herbaceous, many-flowered. Empty glumes unequal, concave, obliquely truncate, much shorter than the spikelet; flowering ones numerous, imbricated, oblong, obtuse, 7-nerved, awnless. Pale as long as the glume. Scales 1, or 2 and connate. Stamens 2 or 3. Grain free, glabrous.

A small genus, found in temperate regions of both the north and south hemispheres, the grain of one species of which has been used as food during famines.

1. **G. stricta,** *Hook. f. Fl. N. Z.* i. 304. Culms slender or stout, glabrous, 1–2 ft. high. Leaves glabrous, short, 2–4 in. long, involute, strict, almost filiform; sheath long, tumid, striate; ligule short, broad. Panicle very slender, 4–6 in. long; branches unequal, strict, whorled, erect, spreading after flowering. Spikelets terete, green, $\frac{1}{4}$–$\frac{1}{3}$ in. long, 6–14-flowered. Flowering glumes with obscure nerves. Scales connate, lobed at the broad tip.—Fl. Tasman. ii. 123. t. 162 B.

Northern Island: east coast, *Colenso*. **Middle** Island: Akaroa, *Raoul*. Also a native of Tasmania and South-East Australia.

21. CATABROSA, Palisot.

Tufted, glabrous, erect grasses.—Spikelets panicled, 2-flowered, scarious, shining. Empty glumes 2, short, unequal, awnless, convex at the back; flowering ones concave, truncate, erose, herbaceous, awnless, lower sessile, upper pedicelled. Pale 2-nerved. Scales 3. Stamens 3. Grain glabrous, free.

A small genus of temperate climates, not hitherto found in Australia or Tasmania.

1. **C. antarctica,** *Hook. f. Fl. N. Z.* i. 308. *t.* 56. Culms decumbent, very slender, 6–12 in. high, branched below, leafy. Leaves involute, longer than the culms, slender, almost filiform; ligule long, narrow, membranous; sheaths deeply furrowed. Panicle very slender, erect, contracted, 1–2 in. long; branches few, slender. Spikelets very few, small, flat, white, glistening, $\frac{1}{12}$ in. long. Empty glumes unequal, acute; flowering ones convex with an obtuse jagged tip, slightly webbed at the base; nerves obscure. Anthers broad, short.—Fl. Antarct. i. 102. t. 56.

Northern Island: summit of the Ruahine range, *Colenso*. **Campbell's** Island, on rocky ledges, *J. D. H.*

22. POA, Linn.

Creeping or tufted, soft (rarely harsh rigid or pubescent), often tall grasses. Panicle usually open, with whorled lower branches.—Spikelets green, herbaceous, pedicelled, 2–10-(rarely 1-)flowered. Empty glumes equal or unequal, compressed, keeled, acuminate, awnless, margin often membranous; flowering ones distant on a glabrous or villous rachis, upper often imperfect, compressed, keeled, obtuse acute or acuminate (never awned), glabrous or scabrid, naked or webbed at the base; edges often membranous. Pale 2-nerved, 2-fid. Scales 2. Stamens 2 or 3. Grain glabrous, free, rarely adherent to the pale.

A very large genus in all temperate and cold regions; the species are always most puzzling and difficult to discriminate from one another, and many may be put with *Festuca*, except for habit. *P foliosa* thus approaches forms of *F. littoralis* very closely, but is an evident congener of *P ramosissima* and others. *Kœleria* hardly differs when not awned, except in habit and the texture of the glumes.

Empty glumes not half the length of the flowering. (Leaves slender, flaccid. Spikelets $\frac{1}{12}$ in. long.)

Spikelets few. Flowering glumes oblong, obtuse; nerves faint . . 1. *P. imbecilla.*
Spikelets many. Flowering glumes, acuminate, nerved . . . . . 2. *P. breviglumis.*

Empty glumes more than half as long as the flowering.

Flowering glumes acuminate.

Panicle open. Culms naked below, rigid, branched. Leaves flaccid 3. *P. ramosissima.*
Panicle open. Culms leafy from the base. Leaves coriaceous . . 4. *P. foliosa.*

Flowering glumes obtuse.

Culms very minute, tufted. Spikelets few, nearly sessile . . . . 5. *P. exigua.*
Culms stout. Leaves coriaceous, flat or concave; ligule truncate . . 6. *P. anceps.*
Culms slender, polished. Leaves filiform; ligule 0 . . . . . . 7. *P. australis.*
Culms slender. Leaves filiform or subulate; ligule membranous . . 8. *P. Colensoi.*
Culms 2–4 in., slender. Leaves short, narrow, flat, green; ligule short . . . . 9. *P. Lindsayi.*

1. **P. imbecilla,** *Forst.*;—*Fl. N. Z.* i. 306. Culms weak, flaccid, decumbent, spreading, very slender, sometimes capillary, 6–12 in. long. Leaves very narrow, $\frac{1}{30}$–$\frac{1}{20}$ in. broad, flat, green; ligule membranous, short. Panicle very lax, open, elongate, with few alternate or 2-nate long capillary branches. Spikelets 1 or 2 on each branch, minute, $\frac{1}{12}$ in. long, green, 3–8-flowered. Empty glumes unequal, obtuse; flowering ones twice as long, remote, oblong, obtuse, glabrous, obscurely 3-veined, not webbed at the base. —*P. Sprengelii*, Kunth, according to Raoul, and probably *P. implexa*, Trinius, *P. australis*, Sieber, and *P. Sieberiana*, Sprengel.

Northern Island: abundant in woods, etc. **Middle** Island: probably also common, but overlooked; Milford Sound, *Lyall*. Also a native of Australia, but not of Tasmania. I have seen no specimens of Forster's plant, to which A. Cunningham first referred this. Banks and Solander referred the following to Forster's *P. imbecilla.*

2. **P. breviglumis,** *Hook. f. Fl. Antarct.* i. 101. Quite glabrous. Culms flaccid, slender, decumbent and much-branched below, 6–12 in. long, leafy. Leaves shorter than the culms, flat, very slender, flaccid, $\frac{1}{20}$–$\frac{1}{12}$ in.

broad; ligule oblong. Panicle lax, oblong, 3–5 in. long; branches in pairs or threes, capillary, suberect, with 2 or 3 pedicelled, green, 3- or 4-flowered spikelets $\frac{1}{12}$ in. long towards their tips. Empty glumes small, very unequal, oblong, obtuse, lower half the length of the flowering glume or shorter, upper larger; flowering glumes oblong or lanceolate, obtuse or acute, on nearly glabrous pedicels, green, strongly 3–5-nerved.

Northern Island, *Banks and Solander*. **Middle** Island: Akaroa, *Raoul;* Saddle Hill, Otago, *Lindsay*. **Campbell's** Island, grassy places near the sea, *J. D. H.* This differs from *P. imbecilla,* of which it may be a variety, in its stouter habit, the larger size of all its parts, and more acute flowering glumes, with much stronger nerves. The outer empty glume is shorter in Campbell's Island than in any New Zealand specimens, and sometimes toothed at the tip.

3. **P. ramosissima,** *Hook. f.;—Fl. Antarct.* i. 101. Culms densely tufted, forming naked, rigid, brown, branching, decumbent stems, 6–10 in. long, from which much-divided, flaccid, very leafy, slender branches, 2–6 in. long, ascend. Leaves most numerous, very narrow, flaccid, flat, $\frac{1}{6}$ in. broad, much longer than the culms; ligule oblong, truncate; sheaths slender. Panicle 1–2 in. long, narrow, green; branches quite glabrous, smooth, very short, $\frac{1}{4}$ in. long, interrupted. Spikelets $\frac{1}{6}$ in. long, very shortly pedicelled, glabrous, green, 3–5-flowered. Empty glumes lanceolate, acuminate, 3-nerved, nearly equal, as long as the flowering, which are narrower, glabrous, acuminate with incurved tips, obscurely 5-nerved, pedicel glabrous or a little webbed.

Lord Auckland's group and **Campbell's** Island, abundant everywhere, *J. D. H.* A grass of remarkable habit, from the long, naked, decumbent bases of the culms, which are excessively branched and leafy above.

4. **P. foliosa,** *Hook. f.—Festuca foliosa,* Fl. N. Z. i. 308. Culms stout, tufted, short or tall, 1–3 ft. high. Leaves flat, glabrous, coriaceous, shorter or longer than the culms, $\frac{1}{12}$–$\frac{1}{4}$ in. broad; ligule short, membranous; sheaths compressed. Panicle rather dense, oblong or ovate, inclined, 1–10 in. long; branches short, nearly glabrous, erect. Spikelets large, $\frac{1}{6}$–$\frac{1}{3}$ in. long, much compressed, green or purplish, 4–8-flowered, very shortly pedicelled. Empty glumes lanceolate, acuminate, tips often slightly incurved, as long as the flowering glumes, which are broader, acuminate, tips incurved, much compressed, slightly scabrid or glabrous, keel smooth or ciliate, pedicel webbed.

Var. *α.* Culms 2–3 ft. high, very leafy. Leaves longer than the culm. Panicle 6–10 in. long. Empty glumes longer in proportion to the spikelet.—*Festuca foliosa,* Fl. Antarct. i. 99. t. 55.

Var. *β.* Culms 6–12 in. high, leafy at the base only. Leaves much shorter than the culms. Panicle 1–6 in. long. Spikelets larger in proportion to the glumes.

Lord Auckland's group and **Campbell's** Island: var. *α*, common near the sea and on the hills. **Middle** Island: var. *β*, Milford Sound, *Lyall;* Southern Alps, *Sinclair, Haast, and Travers;* snow holes on Mount Darwin and Richardson glacier, alt. 4–6500 ft.; Otago, lake district, *Hector and Buchanan*. The very acute glumes with the tips usually more or less inclined to be incurved, well distinguish this from all but *P. ramosissima.* Some of Haast's and Hector's specimens have spikelets very large and broad.

5. **P. exigua,** *Hook. f. n. sp.* A very minute, tufted, glabrous, flaccid species, 1–1½ in. high. Leaves involute, erect, subulate, obtuse, $\frac{1}{2}$–$\frac{2}{3}$ in. long, sheaths membranous; ligule 0, or very short. Panicle almost reduced to an erect raceme, $\frac{1}{4}$–$\frac{1}{3}$ in. long, narrow, of few shortly-pedicelled, purplish, broadly

ovate spikelets $\frac{1}{12}$ in. long. Empty glumes obovate-oblong, very obtuse, purple, with broad membranous margins; upper nearly as long as the flowering, lower shorter; flowering glumes nearly orbicular when spread out, quite glabrous, purplish, with broad white margins, 5-nerved.

Middle Island: lake district, alpine, *Hector and Buchanan*. Apparently a very peculiar little plant, and quite unlike any *Poa* known to me, but considering that I have it but from one locality, and how variable its congeners are, I am by no means confident of its not proving a reduced form of perhaps *P. breviglumis*.

6. **P. anceps,** *Forst.*;—*Fl. N. Z.* i. 306. Culms rather stout, tufted, leafy, tall or short, smooth, $\frac{1}{2}$–3 ft. high, glabrous, often compressed at the base, simple or branched. Leaves distichous, coriaceous, strict or flexuous, flat or concave, longer or shorter than the culm; ligule coriaceous, very short. Panicle inclined, 1–12 in. long, ovate or lanceolate, effuse or contracted; branches whorled, long, slender, capillary (short or 0 in var. γ, δ, ϵ). Spikelets $\frac{1}{4}$–$\frac{3}{4}$ in. long, usually broad and flat, 1–6-flowered, green, rarely white. Empty glumes as long as the flowering, narrow, acute, glabrous or scabrid; flowering narrow, oblong, obtuse, scaberulous (smooth in var. ϵ), crowded or distant, webbed or naked at the base, 5-nerved.—*P. australis*, A. Cunn. not Br.

Var. *α. elata.* Culms 2–3 ft. Leaves $\frac{1}{4}$ in. broad. Panicle 4–8 in. long; branches long, slender. Spikelets $\frac{1}{2}$–$\frac{3}{4}$ in. Glumes spreading.

Var. *β. foliosa.* Culms much branched, 1–2 ft. Leaves numerous, strict. Panicle more contracted; branches long, slender. Spikelets smaller. Glumes broader, flowering 1–5, close together.—*P. affinis*, Br.;—Fl. N. Z. i. 307.

Var. *γ. breviculmis.* Culms 4–10 in. high, stout, short, compressed, leafy at the base only. Leaves 2–3 in. long, curved. Panicle ovate, contracted. Spikelets short, obtuse, 3- or 4-flowered. Glumes broad, rather acute, flowering close together.

Var. *δ. densiflora.* Culms 1–2 ft. Leaves 4–12 in., concave. Panicle 2–4 in., dense; branches very short, densely covered with 4–6-flowered spikelets.

Var. *ε. alpina.* Culms 6–8 in. Leaves 1–2 in., involute. Panicle much contracted, 1–2 in. long; branches very short. Spikelets short, glabrous, shining, white, 2–4-flowered (perhaps a different species).

Abundant in woods and open places; var. *α*, in the **Northern** Island chiefly; var. *β*, in both islands; var. *γ*, in dry pastures; var. *ε*, Southern Alps, snow holes on Mount Darwin, ascending to 6000 ft. on Mount Dobson, *Haast*. This is the common *Poa* of New Zealand, and is the same as a New South Wales plant, which I take to be Brown's *P. affinis*, and of which some states appear in that country to run into *P. australis*. Of Forster's *Poa cæspitosa* I find no specimens in the British Museum, but can scarcely doubt that it is a form of this or the following species.

7. **P. australis,** *Br.*, var. **lævis**;—Fl. N. Z. i. 307. Culms densely tufted, perfectly smooth, polished, slender, pale yellow when dry, 6–36 in. high. Leaves shorter than the culms, very slender, filiform, involute, erect, rigid; sheaths polished, smooth and shining; ligule 0. Panicle 1–4 in. long, ovate in outline, very lax; branches few, capillary, spreading, with few, pedicelled, pale, 4–6-flowered spikelets $\frac{1}{6}$–$\frac{1}{4}$ in. long, on each. Empty glumes scabrid, nearly equal, acute, shorter than the flowering, which are scaberulous, oblong-lanceolate, obtuse, webbed at the base.—*P. lævis*, Br.

Northern Island: dry plains, Taupo, Motukino, and woods at Tarawera, *Colenso*. **Middle** Island: abundant in upland and dry lowland situations. Intermediates between this and *P. anceps* may occur, but in their ordinary states the two plants are widely different, especially in habit. The absence of ligule at once distinguishes some forms of this from *P. Colensoi*.

8. **P. Colensoi,** *Hook. f.*, *n. sp.* Habit, colour, and foliage of *P. australis*, but usually not more than 1–6 in. high, and generally 6–8. Leaves the same, but the sheaths always very rigid, grooved, and coriaceous, terminating abruptly in a very large, white, membranous, sheathing ligule. Panicle 1–2 in. long, broadly ovate, of few, lax, capillary, spreading branches, each bearing 1 or 2 broad, flat spikelets, exactly as in *P. australis*. Flowering glumes webbed at the base, and sometimes villous at the back.

Northern Island: Taupo plains, and top of the Ruahine mountains, *Colenso*. **Middle** Island: Upper Awatere and Aglionby plains, *Sinclair, Munro, and Travers;* Hopkins river and Rangitata range, alt. 1500–2500 ft., *Sinclair and Haast;* Otago, lake district, alpine, *Hector and Buchanan*, a very small variety, 1–2 in. high, with much reduced panicle, and flowering glume villous at the back. So similar to *P. australis*, that I long confounded them, and am far from persuaded of their permanent differences; the large, white, sheathing ligules of this are, however, a most curious character.

9. **P. Lindsayi,** *Hook. f.*, *n. sp.* Culms numerous, densely tufted, erect, very slender, 3–5 in. high. Leaves $\frac{1}{2}$–2 in. long, flaccid, involute, subulate, green, $\frac{1}{20}$ in. diam.; sheaths very short; ligule oblong. Panicle lax, open, 1–2 in. long, ovate; branches 2- or 3-nate, horizontal, capillary, smooth, flexuous, lower $\frac{1}{2}$–$\frac{3}{4}$ in. long, bearing 1–3 subterminal, ovate, compressed, spreading, 4–6-flowered, brown-green spikelets $\frac{1}{12}$–$\frac{1}{8}$ in. long. Empty glumes nearly equal, oblong, subacute, as long as the flowering, which are similar, subacute, glabrous, nerveless; keel scabrid; rachis glabrous. Pale with scabrid nerves.

Middle Island: Otago, slopes of Saddle Hill, *Lindsay;* lake district, *Hector and Buchanan;* Canterbury, Acheron valley, alt. 4000 ft., *Travers;* Kowai valley, in crevices of rocks, alt. 2–3000 ft., *Haast*. A beautiful little species, of which all the specimens I have entirely agree in size, habit, and characters; it may be known by the densely-tufted habit, small, flat, narrow, green leaves, numerous capillary culms, broad, lax, open panicle, and spikelets like *P. imbecilla*, but the outer glumes very different.

P. annua, the commonest of English Grasses, is found by roadsides, etc.; it is a weak, flaccid grass, with flat, soft leaves, large membranous ligules, and green webbed glumes, like those of *P. anceps*, var. γ. *P. nemoralis*, Linn., *P. compressa* and *P. pratensis*, Linn., may also be expected to occur in New Zealand pastures, etc.

23. FESTUCA, Linn.

Tufted, often rigid or harsh, leafy grasses. Spikelets panicled racemed or spiked, pedicelled, 2- or more-flowered, green or pale, coriaceous, and usually rigid; rachis often jointed. Empty glumes 2 (rarely 1), unequal, acute, convex at the back; flowering ones convex at the back, acute, acuminate or awned at the tip, rarely minutely 2- or 3-toothed at the tip, naked or webbed at the base. Pale 2-nerved. Scales 2, acutely 2-fid. Stamens 1–3. Grain free or adherent to the pale, glabrous.

A very extensive genus in all temperate and mountainous countries, of which the species are as variable as those of *Poa*, and the two first are hardly generically distinguishable from that genus.

Glumes awnless.
Culms branched, 2–3 ft., rigid. Spikelets turgid . . . . 1. *F. littoralis.*
Culms simple, 6–24 in. Spikelets compressed . . . . 2. *F. scoparia.*
Glumes awned.
Culms very slender. Panicle effuse . . . . 3. *F. duriuscula.*

1. **F. littoralis,** *Br.*;—*Schenodorus littoralis*, Palisot;—Fl. N. Z. i. 310. A tall, rigid, perennial, densely tufted, perfectly smooth, much branched, polished grass. Culms 2–3 ft. high, leafy. Leaves erect, involute, terete, pungent, longer or shorter than the culm; ligule very short or 0. Panicle 3–10 in. long, slender, rarely pale; branches short, erect. Spikelets broad, turgid, 4–7-flowered, $\frac{1}{2}$–$\frac{3}{4}$ in. long, longer than the glumes. Empty glumes acuminate, glabrous; flowering ones longer, pubescent, 5-nerved, naked or webbed at the base, keeled, acute, obtuse or 3-toothed at the tip.—*Poa littoralis*, Labill. Fl. Nov. Holl. i. 22. t. 27; *Arundo triodioides*, Trinius.

Northern Island: abundant on rocks near the sea, *Banks and Solander*, etc.; East Cape, on sandhills, *Colenso*. Also a native of Tasmania and Australia.

2. **F. scoparia,** *Hook. f. Fl. Antarct.* i. 98; *Fl. N. Z.* i. 308. Densely tufted, 6–24 in. high, smaller than *F. littoralis* in all its parts, perfectly glabrous, polished, and shining. Leaves filiform, rigid, longer or shorter than the culm. Panicle $\frac{1}{2}$–2 in. long, ovoid or narrow-elongate and few-flowered; branches short, erect. Spikelets flattened, $\frac{1}{5}$–$\frac{1}{3}$ in. long. Empty glumes acuminate, shorter than the flowering, which are much webbed at the base, acuminate at the tip.

Northern Island: Auckland, *Sinclair*. **Middle** Island: Port William, *Lyall*. **Lord Auckland's** group and **Campbell's** Island: common in rocky places near the sea, *J. D. H.* I suspect that the Kerguelen's Land *Poa (Triodia) kerguelensis* is a starved form of this.

3. **F. duriuscula,** *Linn.*;—*Fl. N. Z.* i. 309. Very slender, densely tufted, glabrous. Culms 1–3 ft. high, leafy at the base chiefly. Leaves slender, involute, filiform or short and setaceous. Panicle 1–6 in. long, effuse or contracted; branches capillary, often flexuous, lower 2- or 3-nate. Spikelets few, 4–8-flowered, $\frac{1}{4}$–$\frac{1}{3}$ in. long. Empty glumes unequal, acute; flowering ones naked at the base, narrow-lanceolate, remote, scabrid, acuminate, terminating in a short, stiff awn.

Northern Island: in mountainous districts, Hawke's Bay, Wairarapa valley, Cape Turnagain, etc., *Colenso*. **Middle** Island abundant on the Alps, from Nelson to Otago, ascending to 4000 ft. A most common European grass, forming much of the mountain pasture; also found in Tasmania, Fuegia, and almost all temperate mountainous regions.

F. bromoides, *Linn*;—*Fl. N. Z.* i. 309, is certainly an introduced plant, and nowhere native; it may be recognized by its annual, extremely slender culms, 4–8 in. high, setaceous, involute, erect leaves, racemed or almost spiked spikelets, with 3–10 flowers; empty glumes very unequal, flowering ones ending in long slender awns. A most abundant European grass, introduced into all temperate parts of the southern hemisphere.

24. BROMUS, Linn.

Annual or perennial grasses. Leaves usually flat.—Spikelets panicled or racemed, many-flowered. Empty glumes 2, concave, convex at the back, not awned, rigid; flowering ones lanceolate, convex at the back, 2-fid, awned from between the lobes, not twisted, often recurved.—Pale 2-nerved; nerves ciliated. Scales 2, entire. Stamens 3. Ovary hairy at the top; styles generally distinct. Grain free, top hairy.

A very large genus found in all temperate countries.

1. **B. arenarius,** *Lab.*;—*Fl. N. Z.* i. 310. Annual, densely pubescent

or villous all over. Culms 3–24 in. high, leafy. Leaves flat, villous on both surfaces. Panicle 3–8 in. long, nodding, branches slender, 3–5-nate, spreading. Spikelets green, 1 in. long, on slender pedicels, villous, 5–7-flowered. Empty glumes lanceolate, long acuminate, much shorter than the flowering ones, tips and margins membranous; flowering glumes ciliate, narrow-lanceolate, strongly 7-nerved, 2-fid at the tip; awn as long as the glume. —Labill. Fl. Nov. Holl. i. 23. t. 28; *B. australis*, Br.; A. Cunn. Prodr.

Northern Island: rocky places near the sea, Bay of Islands, *Cunningham*; east coast, *Colenso*, etc. Also a native of Australia, but not of Tasmania; it is allied to the European *B. tectorum*, L.

Of introduced species, *Bromus mollis*, Linn., a small species with short ovoid panicles, and tumid villous spikelets, and broad glumes with short awns, has been gathered in the Acheron valley by Travers, alt. 4000 ft.

B. racemosus, Linn., a similar species, with glabrous spikelets, is introduced near Dunedin, Otago.

25. TRITICUM, Linn.

Annual tufted or perennial creeping grasses. Leaves involute or flat.—Spikelets spiked, usually jointed at the base, on alternate sides of an unjointed compressed rachis, sessile, parallel to the long axis of the rachis, 3–many-flowered. Empty glumes 2, often unequal, rigid, concave, keel often scabrid; flowering ones hard and coriaceous, obtuse acute or awned, concave, keel ciliate. Pale 2-nerved; nerves ciliate at the back. Scales 2, entire, often ciliated. Stamens 3. Ovary hairy at the top; styles distinct. Grain free, or adhering to the pale, hairy at the top.

A very large European and Oriental genus, also found in most temperate countries, to which the cultivated Wheat belongs.

Spikelets 8–16-flowered. Awns short . . . . 1 *T. multiflorum.*
Spikelets 6–10-flowered. Awns thrice as long as the glume, slender . 2. *T. scabrum.*
Spikelets 4–6-flowered. Awn 1½–2 in. long, five times as long as the glume, rigid, channelled, very stout . . . . 3. *T. Youngii.*

1. **T. multiflorum,** *Banks and Sol.;—Fl. N. Z.* i. 311. Culms annual, slender, tufted, erect or prostrate below, 1–4 ft. high, striate, quite glabrous. Leaves 4–8 in long, flat, rough on the upper surface. Spike 3–10 in long. Spikelets 6–12, 8–16-flowered, ¼–1½ in. long. Empty glumes linear-lanceolate, unequal, acuminate, nerved; flowering ones much longer, ¼–⅓ in. long, smooth, nerveless and pale below, green and nerved above; awn very short, rigid, scabrid.—*T. scabrum*, A. Cunn. Herb. not Br.; *T. repens*, A. Rich. Flora ?, not Linn.

Northern and **Middle** Islands: abundant, *Banks and Solander*, etc.; Cape Palliser, and elsewhere on the east coast and interior, *Colenso*; Auckland, *Sinclair*. Closely allied to the European *T. repens*, but annual.

2. **T. scabrum,** *Br.;—Fl. N. Z.* i. 311. Culms very variable, annual, tufted, slender, 3 in. to 3 ft. high, smooth. Leaves 1–4 in. long, flat or involute, usually scabrid on both surfaces. Spike 4–6 in. long. Spikelets 2–8 (rarely 1), with the awns 1½–2½ in. long, 6–10-flowered. Empty glumes lanceolate, often small, nerved, lower truncate; flowering as in *T. multiflorum*, but the awn is 3–5 times as long as the glumes, flexuous straight or recurved.

—*T. squarrosum*, Banks and Sol.; *Festuca scabra*, Labill. Fl. Nov. Holl. i. t. 26.

Northern and **Middle** Islands: common in dry ground, *Banks and Solander*, etc. ascending to 3000 ft. on the Hopkins, *Haast*. This may pass into the preceding, but the two plants are so different in their ordinary states that I hesitate to unite them. The present is abundant in Australia and Tasmania, and apparently the same is found in central Asia, Abyssinia, and Persia. Travers says that it is a most valuable fodder grass, growing freely up to 6000 ft., and ravenously eaten by all kinds of cattle.

3. **T. Youngii,** *Hook. f.* Habit of *T. scabrum*. Leaves quite glabrous below, slightly scabrid on the upper surface. Spike 2–3 in. long, of 3 or 4 very large spikelets 4 in. long, including the awns. Empty glumes $\frac{1}{3}$ in. long, acuminate, margins membranous; flowering ones nearly $\frac{3}{4}$ in. long without the awn, which is $1\frac{1}{2}$–2 in. long, very stout, rigid, scabrid, concave at the back, concave in front with scabrid edges, margin and sides of glumes scabrid and almost aculeate.

Middle Island: grassy flats, sources of the Waitaki, alt. 3000 ft., *Haast*. A remarkable plant, with few spikelets, almost twice as large as those of *T. scabrum*, and very long rigid awns. My specimens are imperfect, and some allowance must here be made for the description.

The Wheat, *Triticum vulgare*, Linn., is no doubt often found as an escape from cultivation, as is its ally the Barley, *Hordeum sativum*, L. The common *H. murinum*, Linn., of European roadsides, is also naturalized in Otago.

Lolium perenne, Linn., *temulentum*, Linn., and *arvense*, With., all common European "Rye-Grasses," are found occasionally near cultivation: the genus is known from *Triticum* by its solitary empty glume.

26. GYMNOSTICHUM, Schreber.

Characters of *Triticum*, but the outer empty glumes are absent or represented by a pair of rigid bristles, and 1–3-flowered; flowering glumes on a flattened rachis, seated on thickened calli.

The only other described species is a native of the United States.

1. **G. gracile,** *Hook. f., Fl. N. Z.* i. 312. *t.* 70. Perennial?, slender, erect, 3–4 ft. high, smooth. Leaves narrow, flat, upper surface rough, sheaths smooth. Spike 4–8 in. long, very slender, inclined; rachis flat, flexuous, edges ciliate. Spikelets 20–30, lax, sessile, solitary, $\frac{1}{4}$–$\frac{3}{4}$ in. long with the awns, 1–3-flowered. Empty glumes 0 or replaced by 2 persistent bristles, flowering 1–3, distant on a flattened rachis, each with a callus at its base, upper imperfect, lower shortly pedicelled, all lanceolate, 5-nerved, scabrid, awn straight, shorter than the glume. Scales 2-lobed, ciliate. Ovary villous, styles remote at the base.

Northern Island: woods at Patea and Tarawera, *Colenso*. **Middle** Island: Akaroa, *Raoul*; Otago, lake district, *Hector and Buchanan*. A very curious grass, closely allied to the North American *A. hystrix*, which has usually the spikelets in pairs, and the empty glumes deciduous; I have described the empty glumes as sometimes replaced by two rigid persistent bristles; these I take to be rudimentary spikelets.

The following grasses, which I find described as natives of New Zealand, are unknown to me, nor can I guess what they are.

Kampmannia Zeelandiæ, Steud. Synops. Gram. 35. Very imperfectly described, no habitat nor collector's name given. It is placed next to *Hystericina*, Steud., which I have referred to *Echinopogon* (p. 325).

Eragrostis eximia, Steud. l. c. 279, a New Holland tropical grass, allied to *Poa*, but with 6–12 flowering glumes; has never, that I am aware of, been found in New Zealand.

Stenostachys narduroides, Turcz. in Bull. Soc. Nat. Hist. Mosc. 35. t. ii. 331. As with *Staphylorhodos* (p. 57), so with this, all I can say is, that I know of no plant like it in New Zealand, nor does any such occur in Sir E. Home's original collections in the British Museum. Can it be *Triticum* badly described?

Class III CRYPTOGAMIA.

Order I. FILICES.

Herbs, rarely half-shrubby or arboreous plants, with fibrous roots, or most frequently with a perennial *rhizome*.—Rhizome short, stout, either forming an erect woody trunk, or prostrate, or slender and climbing, or creeping. Branches (*fronds*), tufted at the end of the rhizome or alternate upon it, continuous with it or jointed on to it, simple or more often pinnatifidly pinnately or 2- or 3-pinnately divided, the lower stalk-like portion (*stipes*) usually grooved on the upper side, as is its continuation (*rachis*); the fronds are sometimes of two kinds, barren and fertile (*Lomaria*, *Niphobolus*, etc.), at others the fertile portion of the frond is very distinct from the rest (*Ophioglossum*, *Botrychium*, *Osmunda*). Fructification consisting of microscopic *spores*, contained in minute *capsules* of various forms, usually placed on the under surface of the frond, but sometimes arranged in spikes or panicles. Capsules in most of the genera very minute, membranous, collected into brown masses (sori), often mixed with jointed hairs or imperfect club-shaped capsules, bursting by a transverse or longitudinal fissure; in a few genera at the end of the Order the capsules are much larger, coriaceous, either connate into 2-valved masses which open by pores (*Marattia*), or into a long spike (*Ophioglossum*), or are sessile and free on the branches of a panicle (*Botrychium*). Sori of various forms, globose oblong or linear; at the back or edge of the frond; on the tips or middle of the veins; naked or covered by the recurved edge of the frond, or by a special involucre (also called *indusium*); sessile or on a short or long, sometimes filiform (*Trichomanes*) *receptacle*. Involucre formed of the recurved edge of the frond, or of a scale attached by its centre base or sides to the side or centre of the sorus, membranous or coriaceous, simple or double, sometimes 2-lipped or 2-valved, one lip being the recurved edge of the frond. Spores usually obtusely 3-gonous, smooth or granular.

A very large, difficult, polymorphous and variable Order of plants, found in almost all quarters of the globe, but most abundant and beautiful in damp southern, tropical and temperate insular localities, where also the species attain their greatest size. To the above characters of the Order may be added, that the fronds of almost all are circinate or coiled inwards like the top of crozier when young; but this character does not hold good in the tribe *Ophioglosseæ*. Its mode of propagation is very curious; the microscopic spore (contained in the minute capsule), when it falls in a suitably damp place, bursts, and produces from its contents a minute, flat, green, membranous, cellular scale (*prothallium*) on the under-surface of which two kinds of organs appear, male and female. The male (*antheridia*) consist of cells (*sperm-cells*) containing spiral ciliated bodies endowed with active motion (*spermatozoids* or *antherozoids*); the female of cavities (*archegonia*), containing each a solitary free cell (*germ-cell*) at its base. Somehow, probably by the agency of water, one or more spermatozoids finds its way into the archegonium, and fertilizes the free cell at its base. The free cell forthwith begins to send rootlets downwards and a stem upwards, and becomes a Fern, the prothallium thereon withering away. To observe these parts and processes requires a good microscope; but the production of the prothallium may be easily observed by causing spores to germinate on damp earth under a tumbler.

Of the 120 Ferns described here, 104 were in the 'New Zealand Flora,' of the rest 3 were there considered as varieties or not properly discriminated, and the rest are new discoveries or additions from Lord Auckland's and Kermadec Islands. Of the whole number, 45 species and 1 genus are peculiar to the islands, 60 are common to Australia and Tasmania, and 9 to Britain. I have followed the 'Species Filicum' throughout in the sequence of the genera and species, as well as in the difficult matter of their limitation, feeling satisfied that it is, on the whole, far the best hitherto proposed.

I have given two Keys to guide the student to the determination of the genera of this very difficult Order. The most important characters of the natural arrangement are derived from the form and markings of the minute capsules, which cannot be detected without a strong lens and some practice.

1. ARTIFICIAL KEY TO THE GENERA.

1. Fructification on a long stipes, apart from leaf-like portion of the frond:—Tribe VIII. OPHIOGLOSSEÆ.

2. Fructification forming distichous spikelets which terminate the fronds or project from their surface:—Tribe VI. SCHIZÆACEÆ.

3. Fructification on the back or edges of the frond.

a. Fructification of hard 2-valved bodies which open by parallel slits on the inner faces of the valves:—29, MARATTIA.

β. Fructification of minute, reticulated, sessile or stalked capsules, collected into clusters (*sori*) of various shapes, bursting vertically or transversely.

§ Capsule-bearing fronds differing from the barren:—16, LOMARIA; *Polypodium rupestre, Doodia caudata.*

§§ Fronds all similar.

† Sori dorsal. Receptacle elevated. Usually tree-ferns:—Tribe II. CYATHEACEÆ.

†† Stem not arboreous. Frond with whorled or dichotomous branches, with buds in the axils:—1. GLEICHENIA.

††† Stem not arboreous. Receptacle not elevated. Frond simple or pinnately or pinnatifidly branched.

‡ Sori marginal, covered with an involucre. Capsules sessile on a long receptacle, with an oblique or horizontal ring:—Tribe III. HYMENOPHYLLEÆ.

‡‡ Sori marginal, covered with an involucre. Capsule stalked, with a vertical dorsal ring:—10, LINDSÆA; 11, ADIANTUM; 12, HYPOLEPIS; 13, CHEILANTHES; 14, PELLÆA; 15, PTERIS; various species of 19, ASPLENIUM; 24, NOTHOCHLÆNA.

‡‡‡ Sori distant from the margin of the frond, globose or punctiform:—8, CYSTOPTERIS; 9, DAVALLIA; 19, ASPIDIUM; 20, NEPHRODIUM; 21, NEPHROLEPIS; 22, POLYPODIUM.

‡‡‡‡ Sori distant from the margin of the frond, linear or oblong:—(1) Involucrate: 17, DOODIA; 18, ASPLENIUM. (2) Involucres 0: 22, POLYPODIUM *Grammitides*, var. *australis*; 23, GYMNOGRAMME; 24, NOTHOCHLÆNA.

KEY TO THE NATURAL ARRANGEMENT OF THE GENERA.

I. Capsules very minute, membranous, reticulated, bursting by an irregular transverse or longitudinal fissure, collected in brown masses or sori on the edge or back of the frond.

TRIBE I. **Gleicheniaceæ.** *Capsules* 1–6, *sessile, bursting longitudinally, completely girt by a transverse or oblique striated ring.*

Rhizome creeping Fronds rigid, coriaceous . . . . 1. GLEICHENIA.

TRIBE II. **Cyatheaceæ.** *Capsules numerous, sessile or stalked, forming a globose sorus, placed on an elevated receptacie, often mixed with jointed hairs, bursting transversely, half girt by a vertical striate ring.—Coriaceous Ferns; trunks arborescent in the New Zealand species.*

Involucre globose, first enclosing the capsules, then bursting irregularly and leaving a cup with torn edges . . . . 2. CYATHEA.
Involucre 0 . . . . 3. ALSOPHILA.
Involucre 2-valved, outer of the concave recurved tip of the frond, the inner concave, placed on the tip of a vein . . . . . . 4. DICKSONIA.

TRIBE III. **Hymenophyllaceæ.** *Sori on the edges of the frond. Capsules sessile, on a clavate or filiform, often very long receptacle, girt by a horizontal or oblique complete striate ring.—Fronds usually very delicate, membranous, transparent (coriaceous and rigid in* Loxsoma *and* Trichomanes Malingii). *Involucre 2-valved, or urceolate or campanulate; veins very rarely branched in the segments.*

Frond membranous. Involucres 2-valved . . . . 5. HYMENOPHYLLUM.
Frond usually membranous. Involucres tubular, urceolate or campanulate, 2-lipped . . . . 6. TRICHOMANES.
Frond coriaceous, opaque. Involucres urceolate, mouth truncate . 7. LOXSOMA.

TRIBE IV. **Polypodiaceæ.** *Sori on the edges or back of the frond. Capsules not raised on an elevated receptacle, stalked, furnished with an incomplete, vertical, dorsal, striate ring, bursting transversely where the ring is absent.*

A. *Sori covered with a more or less evident involucre.*

§ 1. *Sori globose, not on the edge of the frond. Involucre ovate or saccate, opening outwards.* (§ 5, ASPIDIEÆ, *differ in the involucre inserted by a point only.*)

Involucre attached by its base to the middle of a vein, membranous 8. CYSTOPTERIS.
Involucre attached by its base and sides to the tip of a vein . . 9. DAVALLIA.

§ 2. *Sori linear or oblong, close to or upon the edge of the frond. Involucre opening outwards* . . . . 10. LINDSÆA.

§ 3. *Sori linear or globose, on the edge of the frond. Involucre continuous with the edge of the frond, opening inwards.* (*See some* Asplenia, *in* § 4.)

Sori globose or oblong. Involucre reniform, of different texture from the frond . . . . 11. ADIANTUM.
Sori globose. Involucre formed of the green incurved tip of the segment of the frond . . . . 12. HYPOLEPIS.
Sori globose, confluent or linear. Involucre formed of the continuously inflexed margin of the frond (see NOTHOCHLÆNA, in B) 13. CHEILANTHES.
Sori continuous. Involucre continuous round the pinnæ, finally recurved. Frond simply pinnate . . . . 14. PELLÆA.
Sori and involucre as in PELLÆA, but the latter recurved. Frond 2- or 3-pinnate . . . . 15. PTERIS.
Sori linear. Involucre as in PTERIS, but sometimes intramarginal, and fertile fronds different from the barren . . . . . . 16. LOMARIA.

§ 4. *Sori linear or oblong, distant from the margin, except in some* Asplenia. *Involucre membranous, linear or oblong, opening towards the costa.*

Sori parallel to the costa. Involucre attached to confluent veins . 17. DOODIA.
Sori linear or oblong, usually oblique to the costa. Involucre membranous, attached laterally to a free vein. (Sori marginal in much-divided species.) . . . . 18. ASPLENIUM.

§ 5. *Sori globose, distant from the margin. Involucre orbicular or reniform, attached by a point.*

Involucre orbicular, peltate. Frond compound . . . . . . 19. ASPIDIUM.
Involucre reniform. Frond very compound . . . . 20. NEPHRODIUM.
Involucre reniform. Frond simply pinnate . . . . 21. NEPHROLEPIS.

B. *Sori naked, without any involucre.*

Sori separate, globose, rarely oblong . . . . 22. POLYPODIUM.
Sori linear or oblong, usually confluent, often in branching lines . 23. GYMNOGRAMME.
Sori continuous near the margin, sometimes partially covered by the recurved edge of the frond . . . . 24. NOTHOCHLÆNA.

TRIBE V. **Osmundeæ.** *Capsules much larger than in the preceding sections, distinct, clustered on the back of the frond, forming irregular sori, without involucre, sessile or shortly stalked, 2-valved vertically, with a short, transversely striate, lateral or subterminal areola.*

Frond opaque, coriaceous . . . . 25. TODEA.
Frond translucent, membranous . . . . 26. LEPTOPTERIS.

II. Capsules in spikelets or panicles, or confluent into lobed or many-celled masses.

TRIBE VI. **Schizæaceæ.** *Capsules distichously arranged on spikelets, ovoid, sessile, attached by their sides, with radiating striæ from their crown.*

Spikelets projecting from the margins of fertile flabellate pinnæ. Stipes scandent . . . . 27. LYGODIUM.
Spikelets distichous, pinnately arranged at the top of a simple stipes or of the flattened branches of a flabellate frond . . . . . . 28. SCHIZÆA.

TRIBE VII. *Capsules on the back of the pinnæ, very coriaceous, connate into a 2-valved, oblong or linear lobed sorus, with no ring or reticulations.*

Fronds very large, 2-pinnate . . . . 29. MARATTIA.

TRIBE VIII. **Ophioglosseæ.** *Capsules globose, coriaceous, 2-valved, separate or confluent in a spike, or on the branches of a compound panicle, which is borne on a stipes and is apart from the frond.—Vernation not circinate.*

Frond ovate, simple. Capsules on a simple terminal spike . . . 30. OPHIOGLOSSUM.
Frond pinnate or 2- or 3-pinnate. Capsules on a pinnate, or 2- or 3-pinnately-branched panicle . . . . 31. BOTRYCHIUM.

1. GLEICHENIA, Smith.

Rhizome wiry, rigid, creeping.—Stipes erect, often tall and slender. Frond dichotomously branched, with a terminal bud at each fork; branches sometimes whorled, simple forked or pinnate. Pinnæ pinnatifid; segments flat or concave, sometimes so much so as to form an involucre. Sori of 1–6 sessile capsules, which burst longitudinally, and are girt by a complete oblique ring.

A considerable tropical and southern genus of Ferns, not found in Europe or temperate North America, and only in subtropical and tropical Asia.

1. *Sori on the tips of the veinlets. Ultimate lobules of pinnæ small, short, broad.*

Lobules of frond flat or incurved. Sori exposed . . . . 1. *G. circinata.*
Lobules of frond pouch-like. Sori concealed in the pouches . . . 2. *G. dicarpa.*

2. *Sori on the middle or forks of the veinlets. Ultimate segments of pinnæ linear.*

Frond rigid, coriaceous; segments entire, glaucous below . . . 3. *G. Cunninghamii.*
Frond membranous; segments serrulate, not glaucous below . . . 4. *G. flabellata.*

1. **G. circinata,** *Swartz.*—*G. semivestita,* Labill.;—Fl. N. Z. ii. 5. Rhizome slender, chaffy. Fronds erect, very slender, 1–3 ft. high, dichotomously branched; stipes very slender, brown, more or less hairy scaly and chaffy or glabrous; branches very narrow, 6–12 in. long, forked and pinnate. Pinnæ $1\frac{1}{2}$–2 in. long, $\frac{1}{16}$–$\frac{1}{12}$ in. broad, uniformly pinnatifid to the base, the lobules semicircular flat or incurved, shining above, not pouch-shaped as in

the following, green or glaucous below, glabrous or covered below with long weak hairs; rachis and costa often chaffy when young, glabrate when old. Capsules 1–4.—Hook. Sp. Fil. i. 3, t. 2 A; *G. microphylla*, Br.

Var. *β. hecistophylla.* More chaffy; rachis often woolly; lobules of frond much incurved.—*G. hecistophylla*, A. Cunn.; Hook. Sp. Fil. i. 4. t. 2 B.

Abundant throughout the **Northern** and **Middle** Islands, *Banks and Solander*, etc. **Chatham** Island, *Dieffenbach*. Equally abundant in Australia, Tasmania, New Caledonia, and some of the Malayan and Pacific Islands. Young specimens of this have been sent from New Zealand as *G. speluncæ*, Br. Mrs. Jones observes that the natives affirm that this Fern is the first to appear after clearing the forest, and hence call it Matua Rarauhe (father of Ferns).

2. **G. dicarpa,** *Br.;—Fl. N. Z.* ii. 5. Rhizome slender, chaffy, and woolly. Fronds erect, of the same habit and character as *G. circinata*, but usually woolly, especially at the nodes, and the ultimate segments of the frond are more coriaceous and pouch-shaped, being very convex above, and presenting a small opening below, within which the capsules are seen.—Hook. Sp. Fil. i. 3. t. 1 C; Fil. Exot. t. 40.

Var. *β. alpina.* Smaller, 3–10 in. high, stouter in proportion, more densely woolly and chaffy.—*G. alpina*, Br.; Hook. Sp. Fil. i. 2.

Northern Island: var. *β*, Lake Taupo and mountainous districts of the interior, *Colenso*. **Middle** Island: var. *α* and *β*, probably common throughout to Foveaux Straits, *Lyall*.

3. **G. Cunninghamii,** *Heward;—Fl. N. Z.* ii. 6. *t.* 71. Rhizome stout, creeping. Fronds stout, erect, 1–2 ft. high, stiff, dichotomously and often proliferously branched. Stipes stout, pale, smooth, naked or covered with membranous bullate scales that extend to the rachis and costa; branches stout, curving, 6–18 in. long, flabellate, forked, usually woolly at the base, pinnate below, pinnatifid above, tips often caudate; pinnules $\frac{1}{2}$–1 in. broad, decurrent, linear, falcate, very coriaceous, $\frac{1}{3}$–$\frac{2}{3}$ in. long, $\frac{1}{6}$–$\frac{1}{4}$ broad, flat or with the margins recurved, quite entire, often glaucous and pilose below. Capsules 2–6, usually exposed.—Hook. Sp. Fil. i. 6. t. 6 B.

Common throughout the **Northern** Island, *Cunningham*, etc. **Middle** Island: Canterbury, *Travers*. **Stewart's** Island, *Lyall*.

4. **G. flabellata,** *Br.;—Fl. N. Z.* ii. 6. Habit of *G. Cunninghamii*, but taller, much more membranous, with often numerous tiers or whorls of erect or ascending branches; stipes, rachis, and branches without scales; pinnules glabrous or rarely pubescent below, serrulate towards the tip, green on both surfaces.—Hook. Sp. Fil. i. 6; Fil. Exot. t. 71.

Northern Island: in shady woods, Bay of Islands, *A. Cunningham*, etc. Also a native of New Caledonia, Australia, and Tasmania.

G. Hermanni, Br. (*Polypodium dichotomum*, Forst.), a common tropical plant, is erroneously introduced by Forster amongst his New Zealand plants, as a native of arid mountains. In his 'Esculent Plants' he states that the roots are roasted, pounded, and eaten by the natives; the plant does not exist in his or in any other New Zealand Herbarium.

2. CYATHEA, Smith.

Tree-ferns in New Zealand. Fronds very large, 2- or 3-pinnate. Sori distant from the margin, globose, enclosed in a spherical involucre, which bursts irregularly, leaving a cup all round, or shallow cup on one side. Capsules crowded on a short club-shaped receptacle, half girt with an incomplete vertical striate ring.

A large temperate and tropical genus, unknown in the north temperate zone. I have availed myself largely of Mr. Ralph's notes (Journ. Linn. Soc. Bot. iii. 163), in the descriptions of the habit, etc., of the species.

Fronds white and glaucous below . . . . 1. *C. dealbata.*
Fronds not white nor glaucous below.
Frond 10–20 ft., coriaceous, costa glabrous above, tubercled below; fertile pinnæ subpinnatifid . . . . 2. *C. medullaris.*
Frond coriaceous; rachis glabrous above, woolly below; fertile pinnæ entire, serrulate at the tip . . . . 3. *C. Milnei.*
Frond 6–9 ft., flaccid, membranous; costa strigose above; fertile pinnæ subpinnatifid . . . . 4. *C. Cunninghamii.*
Frond 8–9 ft., coriaceous; costa strigose above at the tips; pinnæ falcate, coarsely toothed . . . . 5. *C. Smithii.*

1. **C. dealbata,** *Swartz;—Fl. N. Z.* ii. 7. Trunk slender, branched, almost black, sometimes 40 ft. high. Fronds 8–10 or more, broadly oblong-lanceolate, 8–12 ft. long, dark-green above, milk-white below, 2-pinnate; stipes and rachis slender, pale, smooth, when young clothed with brown subulate scales; primary divisions 1–1½ ft. long, rachis and midrib covered with pale deciduous down; secondary linear-lanceolate, 2–4 in. long, acuminate or caudate, pinnate below, pinnatifid above; pinnules close-set, linear-oblong, obtuse, serrate. Sori numerous, small, pale; involucre membranous; receptacle pubescent.—Hook. Sp. Fil. i. 27; A. Rich. Flor. t. 10.

Abundant throughout the **Northern** and **Middle** Islands, *Banks and Solander*, etc. "Silver Tree-fern" of the settlers. Also found in Lord Howe's Island.

2. **C. medullaris,** *Swartz;—Fl. N. Z.* ii. 7. Trunk very stout, 12–40 ft. high, 4–8 in circumference, conical and densely covered with matted fibres below; above marked with hexagonal scars of fallen fronds; at the top rough with the projecting bases of the old fronds. Fronds very numerous, 30–40, 10–20 ft. long, erecto-patent, coriaceous, oblong-lanceolate, 3-pinnate, lanceolate-oblong, deep green above, pale below; stipes and rachis stout, covered with scattered tubercles; secondary divisions 4–6 in. long, sessile, linear, ¾–1¼ in. broad, acuminate, pinnate below, pinnatifid above, glabrous or covered with small ciliate or jagged scales; pinnules linear, falcate, acute, ½ in. long, $\frac{1}{10}$–⅛ broad, lobulate or subpinnatifid (barren and young broader and quite entire), lowest pinnatifid. Sori numerous, one on each lobe of the pinnule, which is often recurved over it; involucre cup-shaped, split from the base.—Hook. Sp. Fil. i. 27; *Polypodium medullare,* Forst.

Abundant throughout the **Northern** and **Middle** Islands, *Banks and Solander,* etc. "Black Fern" of the settlers. This differs from the Norfolk Island and Pacific island allied species in the fertile pinnæ being always lobulate, or almost pinnatifid. The thick mucilaginous pith was once an article of food with the natives.

3. **C. Milnei,** *Hook. mss.* Very similar to *C. medullaris,* but the rachis is not tubercled, is woolly below, and the fertile pinnæ are entire, ⅓ in. long, ⅛ broad, serrulate at the tip only.

Kermadec Islands, *Macgillivray*. This I took for *C. medullaris* in the account of Kermadec Island plants, published in the Linnæan Journal, Bot. i. 128. Trunk 20–30 ft. high, 9 in. diam.

4. **C. Cunninghamii,** *Hook. f. Fl. N. Z.* ii. 7. Trunk 12–20 ft. high, fibrous at the base and for 5 ft. up, covered with the persistent bases of the fronds. Fronds 20–30 in a crown, 6–9 ft. long, erecto-patent, flaccid, 3-pinnate, acuminate; stipes and rachis slightly warted, pale-coloured, pubescent and scaly as in *C. medullaris;* costa strigose above; rachis with linear, warty scars on each side; pinnules sessile, linear, pinnatifid, $\frac{1}{2}$ in. long, $\frac{1}{10}$–$\frac{1}{12}$ broad; segments rounded, quite glabrons. Sori numerous; involucres variously torn, sometimes irregularly from the top, at others from the base on one side, turning over and forming a shallow cup, as in *C. Smithii.*—Hook. Ic. Pl. t. 985.

Northern Island: Bay of Islands, *Cunningham*, etc.; east coast and interior, *Colenso;* Auckland, *Sinclair;* Port Nicholson, in dense forests, *Ralphs*. Very similar to *C. medullaris,* and perhaps only a variety of it, but a much more delicate and flaccid plant, with smaller pinnules and sori, and the rachis above usually covered with long brown hairs.

5. **C. Smithii,** *Hook. f. Fl. N. Z.* i. 8. t. 72. Trunk 20 ft. high, covered with the ragged naked stipites of the old fronds, densely fibrous at the base. Fronds 8-9 ft. long, lanceolate, not acuminate, 2-pinnate, bright pale green; stipes stout, dark-coloured, covered at the base with stiff, subulate, dark-brown scales 1–1$\frac{1}{2}$ in. long; rachis pale-coloured, quite glabrous and smooth, except toward the ends of the segments, where these and the costa are strigose above; primary divisions 12–15 in. long, 4–5 broad, glabrous above except the rachis and costa, pale beneath, secondary 2–2$\frac{1}{2}$ in. long, pinnate below, pinnatifid above; pinnules linear-oblong, acute, falcate, coarsely toothed. Sori on the forks of the veins; involucre bursting from the base on one side, turning over and forming a shallow cup.

Northern Island: mountainous districts in the east coast and interior, *Colenso;* Wellington, *Sinclair, Ralphs,* usually near streams. **Middle** Island: apparently common throughout. The most common species at Otago; trunk hard, close-grained, heavy, *Buchanan*. Mr. Ralphs observes that the young involucre never covers the sorus, and that this is hence a true *Hemitelia,* from which, however, it differs in habit and the narrow pinnules.

3. ALSOPHILA, Br.

Generally Tree-ferns.—Fronds very large, 2- or 3-pinnate. Sori distant from the margin, differing from those of *Cyathea* only in the absence of an involucre.

A large tropical genus, extending into Australia, Tasmania, and New Zealand; distinguished from *Polypodium* by its usually arboreous habit and tumid receptacle.

1. **A. Colensoi,** *Hook. f. Fl. N. Z.* ii. 8, *t.* 73. Trunk 4–5 ft. high (according to Colenso sometimes absent); young parts covered with lax fulvous or red-brown hairs, and tumid, fimbriate, membranous scales, covering a minute stellate pubescence. Fronds 2–4 ft. long, 2-pinnate; stipes clothed at the base with long subulate white scales $\frac{1}{2}$–$\frac{3}{4}$ in. long; rachis weak; primary divisions 1 ft. long, 4 in. broad, lanceolate, acuminate; pinnules 1$\frac{1}{2}$–2 in. long, acuminate, deeply cut into oblong, obtuse, and obtusely-toothed lobules. Sori numerous, prominent, on the middle of the veins.

Northern Island: Ruahine range, *Colenso*. **Middle** Island: probably common Southern Alps, *Haast;* Otago, alt. 2000 ft., *Sinclair, Hector,* and *Buchanan.*

4. DICKSONIA, L'Heritier.

Trunk often arboreous.—Fronds large, 2- or 3-pinnate. Sori on the margin, globose, enclosed in a 2-valved involucre; inner valve of involucre coriaceous, placed on the end of a vein; outer formed of the recurved concave tooth of the pinnule. Capsules on an elevated receptacle, half girt with an incomplete vertical striate ring.

A large genus of Ferns, most abundant in humid south temperate climates.

Trunk black. Stipes tubercled, black. Sori numerous on each segment . . . 1. *D. squarrosa.*
Trunk brown. Stipes smooth, pale, woolly at the base. Sori 4 or 5 on each segment . . . . 2. *D. antarctica.*
Trunk short or 0. Stipes smooth, woolly at the base. Sori 6–10 on each segment . . . . 3. *D. lanata.*

1. **D. squarrosa,** *Swartz;—Fl. N. Z.* ii. 9. Trunk 10–20 ft. high, slender, quite black, covered with the persistent bases of the fronds. Fronds few, 6–10 ft. long, young clothed with soft brown wool, rigidly coriaceous, 2- or 3-pinnate; stipes stout, black, tubercled, covered with deciduous hairs and chaff; primary divisions 10–30 in. long, 4–6 broad, with long points secondary often stipitate, 3–4 in. long, $\frac{1}{2}$–$\frac{3}{4}$ broad, linear, acuminate, fertile narrower; pinnules oblong, rigid, $\frac{1}{4}$–$\frac{1}{3}$ in. long, pungent, deeply toothed. Sori numerous, large.—Hook. Sp. Fil. i. 68; *Trichomanes squarrosum,* Forst.

Abundant throughout the **Northern** and **Middle** Islands, *Banks and Solander,* etc.

2. **D. antarctica,** *Br.;—Fl. N. Z.* ii. 10. Trunk 10–20 ft. high, stout, 1–2 ft. diam., covered with matted yellow-brown rootlets. Fronds few or numerous, 6–10 ft. long, lanceolate, 2-pinnate, coriaceous: stipes smooth, glabrous, pale-brown; rachis and midribs clothed with spreading deciduous hairs; primary divisions 10 in. long, $1\frac{1}{2}$ broad, narrowed into long tips; pinnules sessile, $\frac{3}{4}$–1 in. long, linear or narrow-oblong, acute, fertile pinnatifid, barren lobulate; segments oblong, acutely toothed. Sori about 4 on each segment.—Labill. Fl. Nov. Holl. t. 249; Hook. Sp. Fil. i. 66; *D. fibrosa,* Col. in Tasm. Journ.; Hook. Sp. Fil. i. 68. t. 23 B.

Northern Island: Wairarapa valley, Te Waiite and Mohaka, *Colenso.* **Middle** Island: Canterbury and Nelson, *Travers;* Otago, *Hector and Buchanan.* A most abundant Fern in Tasmania and South Australia, attaining there 40 ft. in height, and 4 in diam. Trunk used in building houses, *Colenso, Buchanan.*

3. **D. lanata,** *Col. in Tasm. Journ.;—Fl. N. Z.* ii. 10. Trunk usually stout or absent, rarely 4 ft. high. Fronds 1–7 ft. long, very coriaceous, broad lanceolate or ovate, 2- or 3- pinnate; stipes smooth, towards the base clothed with long soft yellowish or purplish silky hairs $\frac{3}{4}$ in. long, upper part and rachis glabrous; primary divisions 5–14 in. long, oblong-lanceolate, acuminate; pinnules sessile, $1\frac{1}{2}$–$3\frac{1}{2}$ in. long, linear, acuminate, fertile contracted again pinnate, barren lobulate or pinnatifid; segments short, oblong, obtuse or pungent. Sori about 4 on each segment, covering the whole under surface of the frond.—Hook. Sp. Fil. 69. t. 23 C.

Northern Island: Bay of Islands, east coast, and interior, abundant in many places, *Cunningham,* etc. **Middle** Island: Massacre Bay, *Travers.*

5. HYMENOPHYLLUM, Smith.

Rhizome very slender, wiry, creeping.—Fronds usually matted, small (½–10 in. long), pinnate or 2- or 3-pinnatifid, pale-green, pellucid, beautifully reticulated when magnified; segments with 1 stout, never branched midrib. Sori axillary or terminal, free or sunk in the segments, globose or oblong, enclosed in a 2-valved membranous involucre, whose lips open outwards. Capsules sessile on a stalked or filiform receptacle, depressed, girt by a complete horizontal striate ring.

A very large tropical and temperate genus, of most beautiful small Ferns. A few South American species have simple fronds. The species inhabit forests, often clothing stumps and trunks of trees.

I. Fronds glabrous; margins toothed or serrate.

Receptacle not exserted.

Frond pinnate, pinnæ pinnatifid. Invol. smooth, supra-axillary, valves jagged or toothed . . . . 1. *H. tunbridgense.*

Frond pinnate, pinnæ pinnatifid. Invol. free, smooth, supra-axillary, valves quite entire . . . . 2. *H. unilaterale.*

Frond minute, pinnatifid. Invol. terminal, free, with toothed backs and lips . . . . 3. *H. minimum.*

Frond 2- or 3-pinnatifid. Invol. terminal, base sunk in the frond, smooth; lips quite entire . . . . 4. *H. bivalve.*

Receptacle exserted. Involucre free . . . . 5. *H. multifidum.*

II. Fronds glabrous, or setose only on the stipes, rachis and costa; margins neither toothed nor ciliate.

Fronds pinnatifid, rarely pinnate below; rachis and generally the stipes also winged.

Rhizome and stipes capillary. Frond pendulous, flat, membranous. Invol. broad, flat, terminal, sunk in the frond . . . . . . 6. *H. rarum.*

Rhizome stout, bristly. Frond tall, broad. Invol. orbicular, free, terminal . . . . 7. *H. pulcherrimum.*

Rhizome stout, glabrous. Frond tall, 3-pinnatifid. Invol. orbicular, terminal, half sunk . . . . 8. *H. dilatatum.*

Frond 3-pinnatifid; segments and wing of rachis crisped or waved. Invol. ovate, free, terminal . . . . 9. *H. crispatum.*

Frond 3-pinnatifid; segments narrow, concave. Invol. ovate, free, axillary . . . . 10. *H. polyanthos.*

Frond pinnate below. Stipes not winged. Rachis winged above only.

Stipes and rachis glabrous. Frond bright-green. Invol. small, ovate 11. *H. demissum.*

Stipes and rachis bristly. Frond dark-green. Invol. orbicular . 12. *H. scabrum.*

Stipes woolly at the base only and stipes glabrous. Frond pale, glistening. Invol. small . . . . 13. *H. flabellatum.*

III. Costa, margins and sometimes both surfaces of the fronds covered with stellate hairs.

Frond pendulous, opaque, clothed with red-brown stellate hairs . . 14. *H. æruginosum.*

Frond pendulous, membranous, transparent; margins and costa with stellate hairs . . . . 15. *H. Lyallii.*

Frond rigid, erect, clothed with stellate hairs. See *Trichomanes Malingii*, p. 356.

1. **H. tunbridgense,** *Smith;—Fl. N. Z.* ii. 11. Small tufted or matted, quite glabrous. Rhizome slender, wiry. Frond 1–3 in. high, ovate or linear, pinnate below, pinnatifid above; segments close or distant, long or short, spreading or decurved, toothed. Involucres supra-axillary, orbicular, compressed, their base sunk in the segments (free in var. β), lips spinulose or toothed.—Hook. Sp. Fil. i. 95.

Var. β. *cupressiforme*. Frond longer, narrower, more rigid, erect; segments remote, decurved, very narrow. Involucres free.—*H. cupressiforme*, Labill. Fl. Nov. Holl. t. 250. f. 2; *H. revolutum*, Col.

Throughout the **Northern** and **Middle** Islands, abundant, on rocks and trees. **Lord Auckland's** group, *J. D. H.* A very common Fern in most humid parts of the globe, and found in England.

2. **H. unilaterale,** *Willdenow;—Fl. N. Z.* ii. 11. So closely allied to *H. tunbridgense*, that the entire lips of the involucre often alone distinguish it; but usually a narrow plant with denser cellular tissue, more like var. β. *cupressiforme* in habit and colour.—*H. Wilsoni*, Hook. Sp. Fil. i. 95; *H. novo-Zelandicum*, Van den Bosch (a small form).

Northern Island: Ruahine range, etc., *Colenso*. **Middle** Island: common, Southern Alps, *Haast; Otago, Hector and Buchanan*. Also found in England and various other parts of the world, often with *H. tunbridgense*, its real or supposed differences from which have given rise to endless discussions.

3. **H. minimum,** *A. Rich., Fl. t.* 14;—*Fl. N. Z.* ii. 12. Small, matted, quite glabrous. Frond 1–2 in. high, broadly ovate, falcate or recurved, pinnatifid, pinnate below; segments entire or cut into 2 or 3 linear, obtuse, ciliate-toothed, concave, rigid lobes. Involucres terminal, free, stipitate, obovate-cuneate; valves spinulose at the back; lips short, toothed.

Northern Island: on roots and stumps of trees, *D'Urville*, etc. **Middle** Island. Otago, *Hector and Buchanan*. **Lord Auckland's** group, *J. D. H., Bolton*.

4. **H. bivalve,** *Swartz;—Fl. N. Z.* ii. 12. Matted, quite glabrous. Rhizomes stout, wiry, creeping. Fronds 2–8 in. high, rather rigid, broadly ovate, 2- or 3-pinnatifid, often decurved, dark-green; segments narrow, often decurved, deeply toothed. Involucres terminal, broadly ovate, turgid, base sunk in the frond, 2-fid to the middle, smooth; lips quite entire; receptacle included.—Hook. Sp. Fil. i. 98. t. 35; *H. spathulatum*, Col. in Tasm. Journ., *Trichomanes bivalve*, Forst.; *T. pacificum*, Hedw.

Northern Island: east coast and hilly regions of the interior, *D'Urville, Colenso*, etc. **Middle** Island: abundant, *Forster*, etc.

5. **H. multifidum,** *Swartz;—Fl. N. Z.* ii. 12. Very similar in size, habit, and colour to *H. bivalve*, but usually more sharply toothed, rachis hardly winged above. Involucres axillary, often decurved or pendulous, quite free, large, urceolate or obovate, with the receptacle exserted.—Hook. Sp. Fil. i. 98; *Trichomanes multifidum*, Forst.

Throughout the islands, abundant, *Banks and Solander*, etc. **Lord Auckland's** group and **Campbell's** Island, *J. D. H., Bolton*.

6. **H. rarum,** *Br.;—Fl. N. Z.* i. 12. A very delicate, membranous, matted, pendulous fern, glistening and pale-green, quite glabrous, margins entire. Frond oblong or linear-oblong, 1–8 in. long, flat, flaccid, pinnatifid, pinnate at the base; stipes and rachis capillary; segments short, broad, flat. Involucres large, broad, compressed, ovate or rhomboid, sunk in the ends of the segments; lips very short, broad, quite entire.—Hook. Sp. Fil. i. 101; *H. semi-bivalve*, Hook. and Grev.; *H. fumarioides*, Bory; *H. imbricatum*, Col.

Abundant throughout the **Northern** and **Middle** Islands, clothing the trunks of tree-ferns. **Lord Auckland's** group, *J. D. H., Bolton*. Also common in Tasmania, Chili, South Africa, and some parts of India.

7. **H. pulcherrimum,** *Col.;—Fl. N. Z.* ii. 13. *t.* 74. Tall, stout, bright green, very handsome. Rhizome short, stout, clothed with stiff bristles. Frond glabrous, 12–18 in. high, ovate or linear-oblong, 2- or 3-pinnatifid; stipes winged to the base; segments linear, obtuse, membranous; ultimate alternate, margins entire. Involucres small, axillary or terminal, free, 2-lobed to the base; lips entire.—Hook. Sp. Fil. i. 103. t. 37.

Northern Island: mountains of the east coast and interior, *Colenso.* **Middle** Island: abundant, often pendulous from trees, *Lyall*, etc.

8. **H. dilatatum,** *Swartz;—Fl. N. Z.* ii. 13. Large, tall, handsome, bright-green. Rhizome long, stout, wiry, glabrous. Fronds 6–18 in. high, erect or decurved, broadly ovate or oblong-ovate, quite glabrous, 3-pinnatifid, stipes winged to the base; segments ovate or lanceolate, often cuneate at the base; ultimate linear, often elongate, obtuse, margins quite entire. Involucres orbicular, terminal, sunk in the tips of the segments, deeply 2-fid; lips entire.—Hook. Sp. Fil. i. 104; *Trichomanes dilatatum*, Forst.; *Leptocionium sororium*, Presl.

Abundant throughout the **Northern** and **Middle** Islands, *Banks and Solander*, etc. **Lord Auckland's** group, *J. D. H.* Also found in Java and the Fiji Islands.

9. **H. crispatum,** *Wallich;—Fl. N. Z.* ii. 13. Tufted or matted, dull green, crisped. Rhizome glabrous, wiry, creeping. Fronds erect or decurved, glabrous, rather rigid, 2–8 in. high, 2- or 3-pinnatifid; rachis with a broad crisped wing; pinnæ cut into narrow linear, quite entire, crisped or waved, rarely flat lobes $\frac{1}{10}$ in. broad. Involucres usually numerous, terminal, free, ovate, turgid, broader than the segments, 2-valved to the base; lips entire or jagged.—Hook. Sp. Fil. i. 105; *H. flabellatum*, Br. not Labill.; *H. flexuosum*, A. Cunn.; *H. atro-virens*, Col.

Northern Island: in woods, Bay of Islands, etc., but not common, *Logan, A. Cunningham*, etc. **Middle** Island: Canterbury, *Travers;* Otago, *Hector and Buchanan.* An abundant Tasmanian and Indian fern.

10. **H. polyanthos,** *Swartz.*—Var. *β. sanguinolentum*, Hook. Sp. Fil i. 106; Fl. N. Z. ii. 14. Matted, 2–6 in. high. Rhizome rather stout. Frond rather rigid, opaque, reddish, erect or decurved, broad ovate or oblong, 2- or 3-pinnatifid; stipes narrowly winged, glabrous or bristly; rachis broadly winged; segments spreading or decurved, narrow linear, flat or waved; ultimate short, quite entire; midribs flexuous, stout, dark. Involucres numerous, axillary and terminal, free, orbicular or ovate, broader than the segments, 2-valved to the base; lips obtuse, entire or jagged.—*H. sanguinolentum*, Swartz; *H. villosum*, Col.

Abundant throughout the **Northern** and **Middle** islands, *Banks and Solander*, etc. This fern has often strong peculiar odour when dry.

11 **H. demissum,** *Swartz;—Fl. N. Z.* ii. 14. Matted, forming large bright-green tufts. Rhizome glabrous, wiry. Frond membranous, 4–10 in. high, 3–4 broad, decurved deltoid or ovate-lanceolate, pinnate below, pinnatifid above; stipes rigid, glabrous, not winged; rachis winged above only; segments ascending, 2- or 3-pinnatifid; lobes narrow, $\frac{1}{20}$ in. broad, often elongate, quite entire. Involucres on the lateral segments, terminal, small, ovate, convex, 2-valved to the base; lips quite entire.—*Trichomanes*, Forst.

Abundant throughout the **Northern** and **Middle** Islands, *Banks and Solander*, etc. **Kermadec** Islands, *Macgillivray*. **Lord Auckland's** group, *J. D. H.*

12. **H. scabrum,** *A. Rich. Fl. t.* 14. *f.* 1;—*Fl. N. Z.* ii. 15. Rhizome stout. Frond rigid, 6–24 in. high, bright but dark-green, stout, erect or curved, elongate ovate or deltoid; stipes stout, not winged, covered with stout bristles, as are the rachis and often midribs; rachis winged above only; segments and involucres as in *H. dilatatum*, but the involucres are broader, with generally toothed lips.—Hook. Sp. Fil. i. 110; *Sphærocionium glanduliferum*, Presl.

Northern Island, *D'Urville;* Bay of Islands, *A. Cunningham;* east coast, *Colenso*. **Middle** Island: Nelson and Canterbury, *Travers, Haast;* Otago, *Hector and Buchanan.*

13. **H. flabellatum,** *Labill. Fl. Nov. Holl. t.* 250. *f.* 1, *not Brown;—Fl. N. Z.* ii. 15. Matted, densely clothing trunks of trees, with pale-green glistening fronds. Rhizome rigid. Fronds imbricate, erect or decurved, broadly ovate, or linear elongate, 2–6 in. long, pinnate below, pinnatifid above; stipes rigid, not winged, woolly at the base, glabrous above; segments 2-pinnatifid, flabellate, broadly cuneate at the base, glabrous; lobes linear, quite entire. Involucres small, terminal on lateral segments, orbicular or oblong; lips entire or toothed.—Hook. Sp. Fil. i. 111; *H. nitens*, Br.; *H. Hookeri*, Van den Bosch.

Abundant throughout the **Northern** and **Middle** Islands, *Banks and Solander*, etc. **Lord Auckland's** group, *Bolton*. Also found in the Philippine Islands.

14. **H. æruginosum,** *Carmichael;—Fl. N. Z.* ii. 15. Rhizome filiform. Fronds pendulous, ovate or long and linear, 3–10 in. long; pinnate below, pinnatifid above, red-brown when dry from the copious stellate hairs; stipes filiform, not winged; segments short, linear or ovate, 2-pinnatifid, quite entire; lobes linear, approximate. Involucres terminal, orbicular or broader than long, sunk in the frond; lips short, pilose.—Hook. Sp. Fil. i. 94. t. 34; *H. Franklinianum*, Col.

Northern Island: Waikare lake, *Colenso;* Wellington (*Mrs. Jones*). **Middle** Island: abundant in subalpine forests, as far south as Dusky Bay, *Menzies*, etc Also a native of Juan Fernandez, Chiloe, and Tristan d'Acunha.

15. **H. Lyallii,** *Hook. f., Fl. N. Z.* ii. 16. A small, pendulous species, like *H. rarum*. Rhizome capillary. Frond ½–2 in. long, orbicular or oblong, excessively membranous, pinnatifidly or digitately divided into linear segments, which are quite entire, obtuse, with stellate hairs on the margins and midrib. Involucres cuneate triangular or obcordate, sunk in the frond; lips concave, quite entire, retuse.

Middle Island: Thomson's Sound, on trees, *Lyall;* Otago, common in the sounds of the west coast, on overhanging rocks, etc., *Hector and Buchanan.*

6. TRICHOMANES, Smith.

Rhizome tufted or creeping, stout or slender.—Fronds erect or pendulous, usually small simple pinnate or 2- or 3-pinnatifid, pale or dark green, usually pellucid, beautifully reticulated when magnified; segments with one stout simple or branched midrib. Sori axillary or terminal, enclosed in a campanulate or tubular, elongate, 2-lipped, free or sunk involucre, opening outwards.

Capsules sessile on a filiform, elongate, often exserted receptacle, depressed, girt by a complete horizontal striate ring.

A large tropical genus of ferns, of which one species is found in various damp insular and mountain regions of the north temperate zone, and several in the south temperate.

Frond glabrous, erect, simple, reniform . . . . 1. *T. reniforme.*
Frond glabrous, erect, lurid-green, much divided.
Frond lanceolate, pale-green, membranous . . . . 2. *T. strictum.*
Frond ovate or deltoid, dark green, rigid . . . . 3. *T. elongatum.*
Frond glabrous, pendulous, very delicate, membranous, divided.
Margin of frond thickened. Midrib simple . . . . 4. *T. humile.*
Margin of frond not thickened. Midrib simple . . . . 5. *T. Colensoi.*
Margin of frond not thickened. Midrib branching . . . . 6. *T. venosum.*
Frond densely clothed with stellate down . . . . 7. *T. Malingii.*

1. **T. reniforme,** *Forst.;—Fl. N. Z.* ii. 16. Rhizome rigid, stout, creeping. Fronds remote, rigid, erect, 4–8 in. high, dark green, transparent when fresh, brown and horny when dry, reniform with a deep sinus, 2–4 in. broad, slightly decurrent on the stout glabrous stipes; veins repeatedly forked, rarely reticulated. Involucres numerous, often crowded along the edge of the frond, tubular or urceolate; receptacles club-shaped, exserted.—Hook. Sp. Fil i. 115; Fil. Exot. t. 2.

Abundant throughout the **Northern** and **Middle** Islands, *Banks and Solander*, etc. One of the most beautiful and singular ferns, confined to these islands.

2. **T. strictum,** *Menzies;—Fl. N. Z.* i. 17. Rhizome very short, stout, sending out many rigid roots, with many fronds tufted at its top. Fronds crowded, rigid, erect, dark green, 4–10 in. high, linear-oblong or lanceolate, pinnate; stipes stout, black, terete, slightly winged, with red-brown shining bristles at the base; pinnæ lanceolate, ascending or recurved, 2- or 3-pinnatifid or irregularly cut; segments very narrow, obtuse or retuse; midrib stout, unbranched. Involucres erect, free, pedicelled; receptacle included or exserted.—Hook. Sp. Fil. i. 136; *T. leptophyllum*, A. Cunn.; *T. Cunninghamii*, Van den Bosch.

Northern Island: Hokianga, *A. Cunningham.* **Middle** Island: Dusky Bay, *Menzies, Hector and Buchanan;* Massacre Bay, *Lyall, Travers.* Probably the same as a Fiji Island species.

3. **T. elongatum,** *A. Cunn.;—Fl. N. Z.* ii. 17. Rhizome short, horizontal, woody, with strong root-fibres and crowded ascending fronds. Fronds 4–10 in. high, rigid, erect, ovate or deltoid, 2-pinnate, lurid green, often covered with mosses, etc.; stipes rigid, stout, terete, not winged, slightly hairy at the base; pinnæ imbricate, crowded, oblong-cuneate, inciso-pinnatifid; segments broad, acute or notched at the tip; midrib stout, not branched. Involucres very numerous, supra-axillary, crowded, cylindric; lips short; receptacles long, rigid, exserted.—Hook. Ic. Pl. t. 701; Sp. Fil. i. 134.

Abundant in deep shaded woods throughout the **Northern** and **Middle** Islands, *Banks and Solander*, etc.

4. **T. humile,** *Forst.;—Fl. N. Z.* ii. 16. Rhizome capillary, creeping. Fronds pendulous, very pale, membranous, transparent, 1–4 in. long, linear-oblong, 1- or 2-pinnatifid; stipes slender, winged above; pinnæ ascending, pinnatifid or forked, quite entire, obtuse, glabrous, margin thickened; midrib

stout, unbranched. Involucres sunk in short lateral segments, urceolate, shortly 2-lipped; receptacle included or very long exserted, capillary.—Hook. Sp. Fil. i. 123; *T. Endlicherianum*, Presl; *T. aureum*, Van den Bosch.

Abundant throughout the **Northern** and **Middle** Islands, on trunks of trees, etc., *Banks and Solander*, etc. Also found in the Pacific and West Indian and Philippine Islands.

5. **T. Colensoi,** *Hook. f. in Ic. Plant. t.* 979;—*Fl. N. Z.* ii. 17. Rhizome capillary, creeping, hairy. Fronds pendulous, quite glabrous, linear-oblong, 1–3 in. long, 1- or 2-pinnate, dark green, very membranous; stipes and rachis filiform, glabrous; pinnæ 5–10 pairs, distant, shortly stipitate, pinnate below, pinnatifid above; segments linear, quite entire, $\frac{1}{16}$ in. broad, acute; margin not thickened; midrib stout, not branched. Involucres solitary at the base of the segments, free, erect, pedicelled, cylindric; mouth scarcely dilated; receptacle generally capillary, exserted.

Northern Island: dense forests, Waikare lake, *Colenso.* **Middle** Island: Nelson, *Travers;* Lake Wanaka, *Haast.*

6. **T. venosum,** *Br.*;—*Fl. N. Z.* ii. 17. Rhizome capillary, creeping, glabrous. Fronds pendulous, quite glabrous, very membranous, shining, transparent, linear, 2–5 in. long, pinnate. Stipes capillary, winged above; pinnæ remote, broadly linear-oblong, cuneate at the base; segments $\frac{1}{8}$ in. broad, obtuse or notched; margin waved, not thickened; midrib flexuose, giving off veins alternately. Involucres at the upper edge of the base of the pinnules, sunk in the frond or free, tubular or urceolate; mouth dilated, shortly 2-lipped; receptacle often capillary and exserted.

Abundant throughout the **Northern** and **Middle** Islands: clothing trunks of tree-ferns, etc., *Banks and Solander*, etc. Also common in Tasmania and South-east Australia.

7. **T. Malingii,** *Hook. Garden Ferns, t.* 64. Rhizome slender, filiform Frond 4–8 in. high, erect, rigid, narrow linear-oblong, 2–4-pinnate, red-brown beneath, everywhere covered with pale-brown stellate pubescence; stipes not winged; divisions all very narrow-linear, coriaceous, almost terete, obtuse, $\frac{1}{12}$ in. diam., quite entire. Involucres subglobose, terminating the segments, than which they are a little broader, rather turgid, concealed by the stellate pubescence; lips irregularly waved or crenate; receptacle included.

Northern Island: Mount Egmont (*Mrs. Jones*), **Middle** Island: mountains between Blind Bay and Massacre or Golden Bay, *Maling, Brunner;* Otago, Mount Cargill, near Dunedin, alt. 2000 ft., *Hector and Buchanan.* A very singular fern; I follow Sir W. Hooker in placing it in *Trichomanes*, though to me it appears most nearly related to *Hymenophyllum æruginosum.* For the fact of the Mount Egmont habitat I am indebted to a letter from Mrs. Jones, who however does not state the finder's name.

The supposed *Trichomanes*, from Manakau Bay, alluded to at vol. ii. p. 18 of the 'New Zealand Flora,' proves to be seedling fronds of *Polypodium tenellum*.

7. LOXSOMA, Br.

Rhizome stout, woody.—Fronds erect, coriaceous, pinnate, opaque. Sori marginal, enclosed in an urceolate coriaceous involucre, with a truncate mouth. Capsules shortly pedicelled, club-shaped, crowded on a long columnar exserted receptacle, mixed with jointed hairs, obliquely girt by an incomplete striate ring, bursting longitudinally.

1. **L. Cunninghamii,** *Br.;—Fl. N. Z.* ii. 18. Rhizome covered with long, curved, matted, red-brown hairs. Frond broadly triangular, 1–2 ft. high, bright green above, usually glaucous below, 3-pinnate; stipes and rachis glabrous, polished, channelled, pale brown; pinnæ ascending, lower opposite; secondary lanceolate, again pinnate or pinnatifid; segments linear-oblong, subacute, notched. Involucres in the notches of the segments, pointing backwards from the frond.—Hook. Sp. Fil. i. 86; Gen. Fil. t. 15; Garden Ferns, t. 51.

Northern Island: in woods, rare; Bay of Islands, at the falls of the Keri-Keri river, *Cunningham;* Wangarei river, *Sinclair;* Waitemata, *Mrs. Jones* (fronds never glaucous below); Coromandel, *Mrs. Jones* (fronds always white below). The only species of the genus, not found elsewhere.

8. CYSTOPTERIS, Bernhardi.

Delicate, flaccid ferns. Rhizome short, creeping.—Fronds tufted, pinnate or 2-pinnate. Veins pinnate and forked; veinlets free. Sori small, globose, on the back of the pinnules, inserted on the middle of a venule, distant from the margin. Involucre very membranous, attached by a broad base to the veinlet below the sorus, ovate or oblong, very convex, acute; margin jagged, at length reflexed. Capsules pedicelled, with a dorsal striate ring.

A small genus of ferns, natives of mountainous, cool, damp regions in the northern and southern hemispheres.

1. **C. fragilis,** *Bernhardi.* Fronds 3–6 in. high, lanceolate or oblong-lanceolate, pale green, pinnate or 2-pinnate; stipes slender, brittle, glabrous; rachis winged above, primary pinnæ rather remote, oblong or lanceolate, simple or again pinnate; pinnules incised and lobed, upper decurrent.—Hook. Sp. Fil. i. 198; Fl. Tasman. ii. 136. t. 166.

Northern Island: Mount Egmont, ranges (*Mrs. Jones*). **Middle** Island: Wairau and Kaikora mountains, alt., 3600 ft., *Travers, Sinclair;* Southern Alps, alt. 3000 ft., *Haast;* Otago, lake districts, *Hector and Buchanan.*

9. DAVALLIA, Smith.

Rhizome creeping in the New Zealand species, tufted in many others. —Frond compound, rarely simple. Veins pinnate, simple or dichotomous. Sori oblong or rounded, on or near the margin. Involucre coriaceous, superficial, arising from the tip of a vein, its base and usually its sides too, adnate to the surface of the frond, opening outwards. Capsules pedicelled, with a dorsal striate ring.

A large tropical and subtropical genus, of often very handsome ferns.

1. **D. novæ-Zelandiæ,** *Col.;—Fl. N. Z.* ii. 19. Rhizome creeping, woody, hispid. Frond 3–24 in. high, 3-pinnate, ovate or oblong, acuminate; stipes red-brown, polished, glabrous above, hispid below; rachis glabrous, polished, sometimes hairy in the axils; pinnæ stipitate, linear-oblong, acuminate; secondary ones oblong-lanceolate, with pinnatifid incised tips; ultimate stipitate, $\frac{1}{2}$ in. long, sharply cut. Involucre orbicular or broadly ovate, jagged, attached by a broad base.—Hook. Sp. Fil. i. 158. t. 152; Garden Ferns, t. 51; *Acrophorus hispidus*, Moore?

Northern Island: Bay of Islands, *Cunningham;* east coast and interior, *Colenso;* Wellington, *Stephenson, Jolliffe;* Port Nicholson, *Lyall.* **Middle** Island: Otago, common, *Hector and Buchanan.*

D. Lindleyi, Hook., alluded to at vol. ii. p. 18, of the 'New Zealand Flora,' proves to be a Fiji Island plant, erroneously supposed to have been sent from New Zealand.

10. LINDSÆA, Dryander.

Rhizome tufted or creeping.—Frond usually compound, coriaceous. Veins free or anastomosing. Sori linear, parallel with and close to the margin, continuous or interrupted. Involucre of two valves, opening outwards, the upper being the margin of the frond, the lower membranous and rising from the tips of the veins.

A large tropical and subtropical genus.

Frond linear, pinnate . . . . 1. *L. linearis.*
Frond lanceolate, 1-3-pinnate . . . . 2. *L. trichomanoides.*

1. **L. linearis,** *Swartz;—Fl. N. Z.* ii. 19. Rhizome stout, creeping, scaly. Fronds distant, erect, 2–18 in. high, linear, pinnate, coriaceous, bright green; stipes and rachis stout, glabrous, purplish; pinnæ distant, $\frac{1}{4}$ in. broad, cuneate or fan-shaped; margins entire or crenate, revolute when dry. Sori nearly continuous.—Hook. Sp. Fil. i. 206.

Common in rocky and stony places throughout the **Northern** and **Middle** Islands. Also abundant in temperate Australia and Tasmania.

2. **L. trichomanoides,** *Dryander;—Fl. N. Z.* ii. 19. Rhizome creeping, chaffy. Fronds tufted, erect, 2–4 in. high, ovate or linear-oblong, pinnate or 2- or 3-pinnate; stipes and rachis stiff, 3-gonous, polished, the former scaly at the base; pinnæ distant, rarely simple, oblong-lanceolate, usually pinnatifid or again pinnate; segments or ultimate pinnæ $\frac{1}{4}$–$\frac{3}{4}$ in. long, cuneate, rounded in fruit, with a deep continuous intramarginal sorus.

Var. *α.* Frond 2- or 3-pinnate.—*L. trichomanoides*, Dryander,—Hook. Sp. Fil. i. 218; *L. viridis*, Col.; *Adiantum cuneatum*, Forst.

Var. *β. Lessoni.* Frond pinnate or 2-pinnate below; pinnæ oblong-lanceolate, obtuse, entire lobed or pinnatifid.—*L. Lessoni*, Bory, in Duperrey, Voy. 287. t. 37. f. 2.

Abundant throughout the **Northern** and **Middle** Islands: in woods. Also found in Tasmania.

11. ADIANTUM, Linn.

Rhizome creeping.—Fronds usually tufted and very compound, with the pinnæ on slender stalks, membranous or coriaceous. Veins simple forked or netted. Sori marginal, rounded or oblong and parallel to the margin. Involucre of the reflexed, often kidney-shaped, white or dark margins of the frond, opening inwards, its surface veined, the veins continuous with those of the pinnæ.

A very large tropical and tempêrate genus of ferns, known as "Maidenhair."

Sori situated in the deep notches or crenatures of the pinnules.

Frond pedate. Rachis hispid. Pinnules coriaceous, striate, hispid . 1. *A. hispidulum.*
Rachis glabrous, polished. Pinnules membranous, sparingly setulose above . . . . 2. *A. affine.*
Rachis slender, glabrous, polished. Pinnules membranous, glabrous, orbicular . . . . 3. *A. æthiopicum.*

Sori situated on the lobules or teeth of the pinnules.

Rachis pubescent above. Pinnules rigid, glabrous . . . . . . 4. *A. formosum.*
Rachis glabrous. Pinnules rigid, glabrous, minutely rough below . 5. *A. Cunninghamii.*
Rachis pubescent above. Pinnules rigid, hirsute or setulose below . 6. *A. fulvum.*

1. **A. hispidulum,** *Swartz;—Fl. N. Z.* ii. 20. Frond broad, often 2-partite, 6–12 in. long, broadly fan-shaped or pedate, 2- or 3-pinnate; stipes dark-brown, shining, scabrid, 3–10 in. high; rachis hispid; branches spreading and rather recurved; pinnules stipitate, coriaceous, olive-green, striate, hispid, obliquely oblong, obtuse, often serrulate or acutely toothed, cuneate at the base, $\frac{1}{3}$–$\frac{1}{2}$ in. long. Sori numerous, on the upper margins of the pinnules, situated in the notches, red-brown, orbicular.—Hook. Sp. Fil. ii. 31; *A. pubescens,* Schkuhr; *A. pedatum,* Forst., not Linn.

Northern Island: common from the Bay of Islands to Cook's Straits. **Kermadec** Islands, *Macgillivray.* Also found in India, Africa, Australia, and the Pacific islands.

2. **A. affine,** *Willdenow;—Fl. N. Z.* ii. 20. Frond oblong or ovate in outline, pinnate or 2- or 3-pinnate, membranous, flaccid, sparingly branched; stipes 6–8 in. high and rachis slender, glabrous, polished, black; pinnules stipitate, $\frac{1}{2}$–$\frac{3}{4}$ in. long, dimidiate-oblong, falcate, truncate at the tip, crenate on the upper margin, dark green, with a few scattered hairs on the upper surface. Sori few, scattered, in the notches of the fronds; involucre reniform or truncate, pale.—Hook. Sp. Fil. ii. 32; *A. trapeziforme,* Forst.; *A. setulosum,* J. Smith.

Northern and **Middle** Islands: common, *Forster,* etc. Also found in Norfolk Island.

3. **A. æthiopicum,** *Linn.;—Fl. N. Z.* ii. 21. Rhizome tufted. Frond oblong in outline, 1–3 in. broad, pale green or yellowish, flaccid, membranous, 3- or 4-pinnate; stipes 4–10 in. high, slender and with the almost capillary flexuous rachis black, glabrous, polished; pinnules stipitate, orbicular, perfectly glabrous, with cuneate bases, $\frac{1}{4}$–$\frac{1}{3}$ in. broad, upper margin lobed. Sori 2–6, in the notches; involucre rather large, reniform or transversely oblong, pale.—Hook. Sp. Fil. ii. 37. t. 77 A; *A. assimile,* Swartz; *A. trigonum,* Labill. Fl. Nov. Holl. t. 248, f. 2.

Abundant throughout the **Northern** and **Middle** Islands, *Banks and Solander,* etc. Easily distinguished by its tufted rhizome, very slender glabrous habit, and rounded membranous pinnæ. A common fern in Australia, Tasmania, and many tropical countries.

4. **A. formosum,** *Br.;—Fl. N. Z.* ii. 21. Frond broad, deltoid, 4-pinnate or decompound, primary branches remote; stipes 1–3 ft. high, stout, scabrid, shining, black; rachis flexuose, pubescent on the upper surface, shining and glabrous below; pinnules small, stipitate, $\frac{1}{4}$–$\frac{1}{3}$ in. long, rigid, glabrous, oblong or obliquely rhomboid, obtuse, cuneate at the base, upper and outer margin crenate and toothed with retuse lobules, lower margin straight or arched. Sori upon the lobules or teeth of the pinnules; involucres transversely elongate, narrow.—Hook. Sp. Fil. ii. 51, t. 86 B.

Northern Island; banks of the Manganaitaka river, *Colenso;* Kaipara and Wangarei (*Mrs. Jones*). A native also of New South Wales and the Pacific islands.

5. **A. Cunninghamii,** *Hook. Sp. Fil.* ii. 52. *t.* 86 A;—*Fl. N. Z.* ii. 21. Rhizome creeping, scaly. Frond sparingly irregularly branched, deltoid

in outline, 2- or 3-pinnate, rarely pinnate; stipes 6–10 in. high, minutely scabrid below, smooth above, polished, black, as is the glabrous rachis; pinnules few, stipitate, ½ in. long, rigid, glabrous, often glaucous below or minutely roughened, obliquely oblong, obtuse, upper and outer margin crenate or lobed, lower margin entire. Sori numerous, small; involucre upon the lobes, reniform or suborbicular, with a narrow sinus.—*A. formosum*, A. Rich.; A. Cunn., not of Brown.

Northern and **Middle** Islands: common, *Banks and Solander*, etc. Specimens growing in dense woods have sometimes variously lobed pinnules.

6. **A. fulvum,** *Raoul;—Fl. N. Z.* i. 22. Rhizome stout, creeping, scaly. Frond 1–2½ ft. high, olive green, ovate-deltoid, sub-pedately 3- or 4-pinnate; stipes scabrous, black; rachis scabrous, strigose above; pinnules hard, coriaceous, glabrous above, with strigose stalks, hirsute or setulose below, obliquely dimidiate-oblong or subfalcate, obtuse, lobulate. Sori on the lobules, numerous, rather large; involucres orbicular-cordate.—Hook. Sp. Fil. ii. 53. t. 85 A.

Northern and **Middle** Islands: rather common in dry woods, *Raoul*, etc. The coriaceous hispid pinnules and larger sori distinguish this at once from *A. formosum;* and the ramification and sori on the lobules from *A. hispidum*, which it approaches very closely.

12. HYPOLEPIS, Bernhardi.

Rhizome stout, creeping.—Fronds usually large, 2–4-pinnate, often glandular-pubescent; veins forked, free, never netted. Sori globose, near the margin, on the tips of the veins, not confluent, covered more or less completely by an incurved tooth of the frond, which forms a spurious involucre.

This genus is, I think, only a section of *Polypodium*, the so-called involucre being merely the incurved tip of the segment on which the sorus is placed. In proof of this, the *H. tenuifolia* may be seen to pass directly into *Polypodium rugulosum*. As, however, these genera are kept distinct by all authors, I have retained *Hypolepis*. The following species may prove forms of one or two at the most.

Frond 2–5 ft., deltoid, 3- or 4-pinnate, glandular. Rhizome scaly . 1. *H. tenuifolia.*
Frond 6–12 in., deltoid, 3-pinnate, glabrate. Rhizome naked . . . 2. *H. Millefolium.*
Frond 6–18 in., rigid, ovate-elongate, glabrous. Rhizome scaly and woolly . . . . 3. *H. distans.*

1. **H. tenuifolia,** *Bernhardi;—Fl. N. Z.* ii. 22. Rhizome long, stout, creeping, scaly. Frond large, sometimes 2 ft. broad, deltoid, glandular-pubescent or glabrate, rather membranous or coriaceous, pale or dark green, 3- or 4-pinnate; stipes 1–2 ft. high, stout, erect, brown, scabrid or pubescent; rachis glandular-pubescent, rarely glabrous; primary branches spreading, ovate or oblong, acuminate, secondary and tertiary more lanceolate; pinnules sessile, linear-oblong, obtuse, pinnatifid, segments oblong, obtuse, crenate-toothed, the teeth forming reniform involucres.—Hook. Sp. Fil. ii. 60. t. 19; *Cheilanthes arborescens*, Swartz; *C. pellucida*, Colenso; *C. ambigua*, A. Rich.; *C. dicksonioides*, Endlicher; *Lonchitis tenuifolia*, Forst.

Northern and **Middle** Islands, abundant: *Banks and Solander*, etc. **Kermadec** Islands, *Macgillivray*. A common Australian and Tasmanian fern, found also in many other parts of the world. Forster erroneously described it as arborescent.

2. **H. Millefolium,** *Hook. Sp. Fil.* ii. 68. t. 95;—*Fl. N. Z.* ii. 23.

Rhizome slender, creeping, not scaly. Frond 8–16 in. long, glabrate, broadly ovate or deltoid, 3-pinnate; stipes and rachis sparingly glandular and pilose, pale; primary branches few, ovate or lanceolate, secondary $\frac{3}{4}$ in. long, shortly stipitate, linear-oblong, obtuse, tertiary also stipitate and similar, pinnatifid or lobulate, the segments crenate. Sori solitary on the lobes or crenatures of the pinnules.

Northern Island: shady places, top of the Ruahine mountains, *Colenso*. **Middle** Island: Nelson, Lake Rotniti, *Munro;* Southern Alps, *Sinclair and Haast;* Otago, *Lindsay*. This approaches very closely some forms of *Polypodium rugulosum.*

3. **H. distans,** *Hook. Sp. Fil.* ii. 70. *t.* 95 C;—*Fl. N. Z.* ii. 23. Rhizome rigid, dark-brown, scabrid and aculeate, woolly and covered with scaly brown hairs. Frond 6–12 in. long, rigid, sparingly branched, 2-pinnate, linear-ovate, acuminate; stipes 8–12 in. high, slender, and rachis red-brown and rough with scattered small prickles; primary branches distant, opposite, narrow linear-lanceolate; pinnules numerous, shortly stipitate, rather distant, linear-oblong, obtuse, $\frac{1}{4}$ in. long, glabrous, crenate lobed or pinnatifid. Sori minute.

Northern Island: near Cape Maria Van Diemen, *Edgerley;* Hokianga, Hutt Valley, and Tararua, *Colenso*; Manakau Heads (*Mrs. Jones*). **Middle** Island: Canterbury, *Travers;* Otago, *Hector and Buchanan.*

13. CHEILANTHES, Swartz.

Rhizome in the New Zealand species stout, short, scaly.—Fronds tufted, rigid, erect, 2- or 3-pinnate. Veins forked free. Sori small, rounded, numerous and close together, near the margins of the frond, terminating veins. Involucres formed of the continuously inflexed margin of the frond.

A considerable tropical and southern genus, chiefly distinguished from *Hypolepis* by the margin of the frond being continuously inflexed, and forming one long involucre over many sori. The only New Zealand species most closely resembles *Nothochlæna distans,* but is readily distinguished by its being glabrous.

1. **C. tenuifolia,** *Swartz;—Fl. N. Z.* ii. 23.—Var. **Sieberi.**—Rhizome 1–2 in. long, very stout, covered with long silky scales and old stipites of fronds. Fronds tufted, erect, linear-ovate or -oblong or deltoid, quite glabrous, coriaceous, contracted; stipes erect, 2–8 in. high, and rachis stout, brown, smooth, and shining or slightly hairy below; primary branches distant, erect; pinnules small, scattered, ovate- or linear-oblong, lobed or pinnatifid, lobes obtuse or obtusely crenate, all soriferous. Sori sometimes so close as to be almost continuous and cover the back of the pinnules.—Hook. Sp. Fil. ii. 82; *C. Sieberi,* Kunze; *Pteris humilis,* Forst.

Abundant throughout the **Northern** and **Middle** Islands, in stony and rocky places, *Banks and Solander,* etc. A large state of this is a common and widely-dispersed fern in the south hemisphere. The var. *Sieberi* is common in Australia.

14. PELLÆA, Link.

Rhizome usually creeping.—Fronds rather tufted, pinnate in the New Zealand species, pedate and 2- or 3-pinnate in others. Veins pinnate, forked, free, very obscure. Sori continuous round the edges of the pinnules. Involucre more or less membranous, formed of or continuous with the edges of the frond, opening inwards, continuous along the margin, at length recurved exposing the sorus.

A considerable tropical and temperate genus.

Erect. Pinnules lanceolate, falcate, acute . . . . 1. *P. falcata.*
Decumbent or prostrate. Pinnules orbicular . . . . 2. *P. rotundifolia.*

1. **P. falcata,** *Br.—Pteris,* Fl. N. Z. ii. 24. Rhizome short, wiry, rigid. Fronds tufted, 1–3 ft. high, rigid, erect, narrow-linear, pinnate; stipes black, hispid; rachis scaly and villous; pinnules quite glabrous, shortly stipitate, linear-lanceolate or oblong, $\frac{3}{4}$–$1\frac{1}{3}$ in. long, acute or mucronate, broadly obliquely cuneate at the base, which is sometimes gibbous above; veins very obscure. Sori broad, continuous round the pinnule; involucre very narrow. —Hook. Sp. Fil. ii. 135; *P. seticaulis,* Hook. Ic. Pl. t. 207.

Northern Island: Auckland, *Sinclair.* **Kermadec** Islands, *Macgillivray.* Also found in India, Tasmania, and Australia.

2. **P. rotundifolia,** *Forst.;—Pteris,* Fl. N. Z. ii. 24. Rhizome short, wiry, rigid. Fronds tufted, rigid, generally decumbent or prostrate, very narrow linear, 6–24 in. long, pinnate; stipes and rachis hispid and scaly; pinnules broadly oblong or rounded, obtuse or mucronate, obliquely truncate at the base, glabrous, dark green; veins very obscure. Sori often in interrupted lines.—Hook. Sp. Fil. ii. 136; Fil. Exot. t. 48.

Abundant throughout the **Northern** and **Middle** Islands, *Banks and Solander,* etc. Though, in its ordinary states, very unlike the preceding, intermediates occur.

15. PTERIS, Linn.

Rhizome usually creeping.—Fronds of various habit. Veins free forked or netted. Sori continuous along the edge of the frond. Involucre scarious or membranous, linear, continuous along the edge of the frond and confluent with its incurved margin, not recurved with age.

One of the largest genera of ferns, found in all quarters of the globe.

§ 1. EUPTERIS.—*Fronds 2- or 3-pinnate. Veins forked, free.*

Frond tall, rigid, glabrous or pubescent below. Pinnules decurrent. 1. *P. aquilina.*
Frond tall, membranous, quite glabrous. Pinnules oblong, large . 2. *P. tremula.*
Frond rigid, glandular-pubescent. Pinnules minute . . . . . 3. *P. scaberula.*

§ 2. LITOBROCHIA.—*Fronds 2- or 3-pinnate. Veins netted.*

Frond coriaceous, glaucous below . . . . 4. *P incisa.*
Frond membranous. Pinnules stipitate, coarsely serrate . . . . 5. *P. macilenta.*
Frond membranous. Pinnules serrate at the tips only . . . . . 6. *P. Endlicheriana.*

1. **P. aquilina,** *Linn.,* var. **esculenta,** Fl. N. Z. ii. 25. Rhizome subterranean, very stout, as thick as the thumb or more. Fronds solitary, sometimes 10 ft. high, 2–4 ft. broad, broad deltoid, 3- or 4-pinnate, rigid, coriaceous; stipes rigid, erect, grooved on one side, pale, glabrous, shining; branches spreading; pinnules linear, decurrent on the rachis, often hairy below; costa stout, often hairy; veins forked, free. Sori continuous, often all round the pinnule, and along their decurrent bases; involucre coriaceous.—Hook. Sp. Fil. ii. 196; *P. esculenta,* Forst.; Labill. Fl. Nov. Holl. t. 244.

Abundant throughout the islands, from the **Kermadec** to **Campbell's** Island: covering the hillsides over extensive areas, *Banks and Solander*, etc. The var. *esculenta*, distinguished chiefly by the decurrent pinnules, is common in the south temperate zone; the ordinary state of the plant is found over all other parts of the world. Rhizomes formerly roasted and eaten by the natives.

2. **P. tremula,** *Br.;—Fl. N. Z.* ii. 25. Frond 1–5 ft. high, rather membranous, broadly deltoid, 2- or 3-pinnate or decompound, quite glabrous, bright or pale-green; stipes and rachis perfectly smooth and polished; primary branches ascending; pinnules 1–2 in. long, $\frac{1}{4}$ broad, linear, obtuse, sessile, decurrent, crenate, subacute; costa shining; veins forked, free; fertile pinnules entire, sometimes narrow-linear, with the involucres almost meeting at the costa.—Hook. Sp. Fil. ii. 174; *P. affinis*, A. Rich.; *P. tenuis*, A. Cunn. (form with narrow pinnules).

Abundant throughout the **Northern** and **Middle** Islands, *Banks and Solander*, etc. Also frequent in Tasmania, Australia, and Chili.

3. **P. scaberula,** *A. Rich. Fl.* 82. *t.* 11;—*Fl. N. Z.* ii. 25. Rhizome stout, woody, hairy, and often scaly. Fronds 6–18 in. high, rigid, erect, coriaceous, yellow-green, glandular-pubescent or glabrate, ovate or linear-oblong, 2- or 3-pinnate; stipes and rachis stout, yellow-red, glandular-pubescent and scabrid; primary branches linear-lanceolate, stipitate; pinnules very small, $\frac{1}{12}$–$\frac{1}{4}$ in. long, substipitate, elliptic-oblong or obovate-oblong, crenate-serrate or entire, acute or obtuse, sometimes auricled at the base. Sori generally surrounding the whole pinnule.—Hook. Sp. Fil. ii. 174. t. 93 A.

Abundant throughout the **Northern** and **Middle** Islands: generally in woods, *Banks and Solander*, etc.

4. **P. incisa,** *Thunberg.—P. vespertilionis*, Labill. Fl. Nov. Holl. t. 245; Fl. N. Z. i. 26. Rhizome stout, glabrous. Fronds large, perfectly glabrous, glaucous below, broadly deltoid, 2–4 ft. high, membranous or rather coriaceous, 2- or 3-pinnate (rarely pinnate); stipes and rachis stout, quite glabrous, pale, polished, often glaucous; primary branches ovate-lanceolate; secondary linear-lanceolate, acute, often adnate at the base; pinnules broad-oblong or oblong-lanceolate, adnate or decurrent on the rachis, $\frac{1}{3}$–$\frac{2}{3}$ in. long, entire or obtusely lobed crenulate or pinnatifid; costa stout, flexuose; veins netted towards their bases.—*P. Brunoniana*, Endl.; *P. montana*, Colenso.

Abundant on skirts of woods throughout the **Northern** and **Middle** Islands, *Banks and Solander*, etc. **Lord Auckland's** group, *J. D. H.* An Australian and Tasmanian fern; also found in the East and West Indies, South Africa, Chili, and Brazil.

5. **P. macilenta,** *A. Rich. Fl. t.* 12;—*Fl. N. Z.* ii. 26. Fronds large, tall, 1–5 ft. high, broadly deltoid, very membranous, flaccid, pale green, quite glabrous, rarely puberulous below, 2–4-pinnate; stipes and rachis quite smooth, shining; primary and secondary branches on long slender stalks; pinnules scattered; the uppermost adnate and decurrent on the rachis; lower on slender stalks, ovate-oblong, acute, deeply coarsely toothed serrate or lobed, especially towards the tip; costa flexuous; veins forked, netted towards their base only. Sori in the notches, broad.—Hook. Sp. Fil. ii. 219.

Northern Island: abundant in shady woods. **Middle** Island: Canterbury, *Haast.* One of the most beautiful of ferns, not found elsewhere.

6. **P. Endlicheriana,** *Agardh.—P. comans*, Fl. N. Z. ii. 26; ? of Forst.

Frond large, tall, 1–5 ft. high, broadly deltoid, membranous, flaccid, glabrous or rarely puberulous, bright green, 2-pinnate; stipes and rachis glabrous, polished; primary branches ovate-lanceolate, acuminate; secondary narrower, pinnatifid, sometimes caudate; pinnules sessile, rarely stipitate, usually adnate and decurrent on the stipes, 1–1½ in. long, linear-oblong or linear, lobed or pinnatifid, the lobes serrate; costa straight; veins forked, netted at the base. Sori continuous or interrupted.—Hook. Ic. Pl. t. 973.

Northern Island: abundant, *Banks and Solander*, etc.; Waikate and Great Barrier Island, *Sinclair*. **Kermadec** Islands, *Macgillivray*. Also found in Tasmania, Australia, and Juan Fernandez. Closely allied to *P. tremula*, and best distinguished by the netted veins. I suspect it to be only a variety of Forster's *P. comans*.

16. LOMARIA, Willdenow.

Fronds usually tufted at the extremities of a stout prostrate or creeping rhizome, erect, rarely alternate and pendulous, of two forms, the outer barren or fertile at the base only, broader, the inner fertile narrower; veins free, simple or forked. Sori linear, close to the margin, continuous round the pinnules or lobes of the frond, often covering their whole lower surface. Involucre sometimes very obscure, linear, scarious, opening inwards, close to the margin and parallel to it, or continuous with it as in *Pteris*, sometimes reaching the costa.

A large southern and tropical genus, chiefly distinguished from *Pteris* by the habit, and the fertile fronds distinct from the barren.

a. Barren fronds pinnate. Pinnæ much contracted at the base or stipitate.

Rhizome very long, climbing. Fronds alternate, pendulous . . . 1. *L. filiformis.*

Rhizome prostrate or erect, short, stout. Fronds tufted, erect.

Fronds 1–4 ft., broad. Pinnules linear, very coriaceous . . . . 2. *L. procera.*

Fronds 8–18 in., very narrow. Pinnules many, oblong or rounded, membranous. Stipes and rachis scaly . . . . 3. *L. fluviatilis.*

Frond 6–10 in., narrow. Pinnules oval-oblong, membranous. Stipes and rachis naked . . . . 4. *L. membranacea.*

β. Barren fronds pinnatifid, or pinnate at the very base only; fertile pinnate. Pinnæ attached by a very broad base.

Rhizome wiry. Fronds 3–5 in., lanceolate. Pinnules oblong, membranous, crenulate . . . . 5. *L. pumila.*

Rhizome stout. Frond deltoid. Pinnules falcate, coriaceous, lower deflexed. 6. *L. Vulcanica.*

Rhizome short, stout, creeping. Frond pendulous, large, coriaceous. Pinnules broad . . . . 7. *L. elongata.*

Rhizome short, stout, erect. Frond linear. Pinnules oblong, rounded, membranous, green below . . . . 8. *L. lanceolata.*

Rhizome short, stout, erect. Frond linear. Pinnules oblong, rounded, coriaceous, brown below . . . . 9. *L. discolor.*

Rhizome long, creeping, scaly. Fronds alternate, linear. Pinnules coriaceous . . . . 10. *L. attenuata.*

Rhizome slender, creeping. Frond 2–12 in., linear, fertile longest. Pinnules coriaceous, linear-oblong . . . . 11. *L. alpina.*

Rhizome very short. Frond 2-12 in., linear, fertile shortest. Pinnules coriaceous, rounded or oblong . . . . 12. *L. Banksii.*

Rhizome very short. Frond 8–12 in., lyrate-pinnatifid. Pinnules short, broad, membranous, black . . . . 13. *L. nigra.*

γ. Fronds 2-pinnate or 2-pinnatifid.

Rhizome erect, slender. Frond ovate-lanceolate . . . . 14. *L. Fraseri.*

1. **L. filiformis,** *A. Cunn.—Stenochlæna heteromorpha*, J. Smith;—Fl. N. Z. ii. 46. Rhizome climbing lofty trees, stout, woody, covered with chaffy scales. Fronds alternate, pendulous, 6–24 in. long, linear, pinnate; stipes rigid, often chaffy; pinnules numerous, stipitate, jointed on to the rachis; barren glabrous, coriaceous, 2–3 in. long, linear-lanceolate, falcate, finely crenulate or serrulate, rounded at the base, tapering to a long point, in young plants small, membranous, orbicular or oblong, crenate or lobed; costa stout; veins forked, free, parallel; fertile pinnules filiform; involucres very inconspicuous.—Hook. Sp. Fil. iii. 333. t. 149; *L. propinqua*, A. Cunn.; *L. pimpinellæfolia*, Hook. f. in Lond. Journ. Bot. iii. 412.

Abundant throughout the **Northern** and **Middle** Islands, *Banks and Solander*, etc. A similar and perhaps identical species is found in the Fiji Islands; a very remarkable plant, differing in several points from the other *Lomariæ*.

2. **L. procera,** *Sprengel;—Fl. N. Z.* ii. 27. Rhizome short, stout, often woody. Fronds tall, rigid, very coriaceous, pinnate, 1–4 ft. high, extremely variable in stature and form; stipes stout, short or long, chaffy and scaly at the base; rachis naked, or with a few scales; barren pinnules 3–20 pairs, linear-oblong, lanceolate or ensiform, stipitate, or the upper adnate to the rachis, acute acuminate or caudate, narrowed truncate or auricled at the base, 2–12 in. long, minutely toothed; costa naked or chaffy; veins numerous; lowest pinnule sometimes orbicular or cordate, uppermost sometimes elongate and erect; fertile pinnules narrower, on separate fronds, or on the bases of the barren.—Hook. Sp. Fil. ii. 22; Ic. Plant. t. 407, 408; *L. latifolia*, Colenso; *Stegania procera*, Br.; *Blechnum procerum*, Labill. Fl. Nov. Holl. t. 247; *Parablechnum procerum*, Presl; *Osmunda procera*, Forst.

Var. *α.* Tall, robust, coriaceous; barren pinnules obliquely truncate or cuneate at the base.

Var. *β.* Tall, robust, coriaceous; barren pinnules auricled at the base, often 1 ft. long.

Var. *γ.* Tall, robust; barren pinnules narrowed at the base.

Var. *δ.* Fl. N. Z. ii. t. 75. Smaller, less coriaceous; pinnules few, barren truncate or auricled at the base, upper adnate.—Hook. Garden Ferns, t. 53; *Stegania minor*, Br.

Abundant throughout the islands, from the **Kermadec** Islands to **Campbell's** Island: in all situations, generally in humid, often marshy. Found throughout the southern hemisphere, and as far north as Mexico. One of the most variable of ferns; the varieties enumerated keep their characters under cultivation.

3. **L. fluviatilis,** *Sprengel;—Fl. N. Z.* ii. 28. Rhizome stout, often woody, chaffy. Fronds tufted, slender, narrow-linear, 8–18 in. high, 1–2 broad, pinnate; stipes chaffy at the base and often the rachis with long scales; pinnules numerous, barren ones rounded or linear-oblong, obtuse, waved or crenate at the tip, $\frac{1}{2}$–$1\frac{1}{2}$ in. long, lower shortly stalked, upper adnate, uppermost confluent; fertile pinnules narrow-linear, erect, obtuse, $\frac{1}{3}$–$\frac{2}{3}$ in. long.—Hook. Sp. Fil. iii. 34; *L. rotundifolia*, Raoul, Choix, t. 2 B, and Colenso, in Tasm. Phil. Journ.; *Stegania fluviatilis*, Br.

Abundant in hilly parts of the **Northern,** and throughout the **Middle** Island. Also found in Tasmania and South-eastern Australia.

4. **L. membranacea,** *Colenso.* Rhizome very short, stout, with wiry roots, scaly at the top. Fronds tufted, erect, membranous, green, 6–10 in.

high, linear-lanceolate, pinnate, subflexuose; stipes short, and rachis without scales; barren pinnules oval-oblong, broadly adnate, not decurrent, obtuse, dentate-serrate, uppermost confluent, lowermost suborbicular; fertile fronds longer than the barren, with a longer stipes, and remote, linear, sessile apiculate pinnules, the lowest minute.—Hook. Sp. Fil. iii. 35. t. 145.

Northern Island: Bay of Islands, common, *Colenso*, etc.; Waiheki Island, Auckland, *Jolliffe*. **Middle** Island: Otago, *Hector and Buchanan*. Closely allied to *L. fluviatilis*.

5. **L. pumila,** *Raoul, Choix, t.* 2 A;—*Fl. N. Z.* ii. 28. Rhizome slender, creeping, with wiry roots, and chaffy fulvous scales at the tip. Fronds tufted, membranous, lanceolate, rather obtuse, tapering to the base, 3–5 in. long, barren pinnatifid; stipes slender, chaffy at the base; rachis slender, glabrous; pinnules ovate or ovate-oblong, $\frac{1}{4}$–$\frac{1}{3}$ in. long, obtuse, entire or subcrenate: fertile fronds with longer stipes, pinnate; pinnules oblong, obtuse, mucronate, more or less decurrent at the base.—Hook. Sp. Fil. iii. 19.

Middle Island: Akaroa, *Raoul.*

6. **L. vulcanica,** *Blume;—Fl. N. Z.* ii. 29. Rhizome stout, woody, often as thick as the wrist. Fronds tufted, coriaceous, 4–18 in. long, ovate-lanceolate or deltoid, pinnatifid above, almost pinnate below; stipes slender, pale, crinite with blackish chaffy bristles at the base; barren pinnules close-set, $1\frac{1}{2}$–3 in. long, adnate by a broad base, lower decurved, falcate, acuminate, rarely obtuse, margin thickened, entire or irregularly crenate, glabrous or pubescent on the rachis and costa; fertile pinnules very narrow; involucres lacerate.—Hook. Ic. Pl. t. 969; Sp. Fil. iii. 13; *L. deflexa* and *deltoidea*, Colenso.

Northern Island: Tarawera, east coast, etc., *Colenso;* Auckland, *Sinclair;* Taranaki, etc. (*Mrs. Jones*). **Middle** Island: Nelson (*Mrs. Jones*): Otago, subalpine, *Hector and Buchanan.* Also found in Tasmania, Java, and some of the Pacific Islands. Colenso sends a small state with an enormous rhizome and small fronds (var. *nana*, Hook. Sp. Fil. l. c.).

7. **L. elongata,** *Blume;—Fl. N. Z.* ii. 29. Rhizome short, creeping. Fronds alternate, very coriaceous, glabrous, pendulous, 1–3 ft. long, smooth, shining, dark-green, entire lobed or pinnatifid; stipes stout, with a few large scales, winged; rachis broadly winged; barren fronds pinnatifid; pinnules few, remote, 1–2 in. broad, 2–5 in. long, linear-lanceolate, falcate, acuminate, often serrate at the tip, each decurrent and connate with the next below it; veins terminating in a depression within the margin; fertile fronds pinnatifid; pinnules very slender, 3–10 in. long. Involucre with torn edges.—Hook. Sp. Fil. iii. 3. t. 143; *L. Colensoi*, Hook. f. in Ic. Pl. t. 627, 628; *L. heterophylla*, Colenso.

Mountainous parts of the **Northern** Island: Waikare lake, *Colenso;* Port Nicholson, *Sinclair;* Mount Egmont (*Mrs. Jones*). **Middle** Island: Ship Cove, *Lyall;* Otago, *Hector and Buchanan.* Also found in Java, India, and the Malay and Pacific Islands.

8. **L. lanceolata,** *Sprengel;—Fl. N. Z.* ii. 29. Rhizome stout, erect or ascending, sometimes 2 ft. high. Fronds tufted, erect, pale-green, 1–2 ft. high, rather membranous, quite glabrous, narrow-linear or elongate, lanceolate, acuminate, narrowed below; barren fronds pinnatifid or pinnate at the very base; stipes short, stout, with long black subulate scales at the base; rachis and costa quite glabrous; pinnules of barren fronds numerous, close set,

oblong-lanceolate, subfalcate, obtuse acute or acuminate, quite entire or waved or crenate, lower orbicular and sometimes distant; fertile fronds shorter, with distant linear, acute or acuminate pinnules.—Hook. Ic. Pl. t. 429; Sp. Fil. iii. 11.

Abundant throughout the **Northern** and **Middle** Islands, *Banks and Solander*, etc. **Lord Auckland's** group and **Campbell's** Island, *J. D. H.* Also very common in Tasmania, South-east Australia, some of the Pacific Islands; also in South America if, as I believe, *L. blechnoides*, Bory, is the same plant.

9. **L. discolor,** *Willdenow;—Fl. N. Z.* ii. 30. Very similar to *L. lanceolata*, but usually larger, often 3 ft. high; rhizome the same; fronds forming an elegant crown, red-brown below, coriaceous, the pinnules narrower; those of the fertile fronds are often leafy at the base.—Hook. Sp. Fil. iii. 5; *L. lanceolata*, Fl. Antarct. i. 110, not Sprengel; *Onoclea nuda*, Labill. Fl. Nov. Holl. t. 246; *Osmunda discolor*, Forst.

Abundant throughout the **Northern** and **Middle** Islands, **Lord Auckland's** group and **Campbell's** Island, *J. D. H.* Very common in South-east Australia and Tasmania.

10. **L. attenuata,** *Willdenow.* Rhizome long, creeping, stout, densely clothed with chaffy, fulvous, shining scales. Fronds alternate, 1–2 ft. long, coriaceous; barren pinnatifid, pinnate at the very base, broadly lanceolate, very attenuate below; stipes short, scaly; pinnules horizontal, close together, with a narrow acute sinus, becoming gradually smaller downwards, acuminate or truncate and emarginate, entire or subserrate; fertile fronds with more remote, linear, sessile, apiculate or acuminate pinnules.—Hook. Sp. Fil. iii. 6.

Kermadec Islands, *Macgillivray.* Also found in Norfolk Island and in tropical countries of the Old and New World.

10. **L. alpina,** *Sprengel;—Fl. N. Z.* ii. 30. Rhizome creeping, chaffy, slender. Fronds tufted, coriaceous, quite glabrous, dark green, 2 in.–2 ft. high, linear, $\frac{1}{2}$–$\frac{2}{3}$ in. broad, pinnatifid or pinnate at the very base only, narrowed above and below; stipes and rachis stout, smooth, sometimes scaly here and there; pinnules of barren fronds very numerous, close, linear-oblong, obtuse, sessile by a broad base; fertile fronds longer than the barren, pinnate; pinnules remote, spreading or deflexed and curving upwards, linear, obtuse, lower remote, rounded, often without sori. Involucre scarious.—Hook. Sp. Fil. iii. 16; Fl. Antarct. t. 150; *L. linearis*, Colenso.

Common in subalpine districts throughout the **Northern** and **Middle** Islands. Not found hitherto in Lord Auckland's group and Campbell's Island, though common in the mountains of Tasmania, Chili, Fuegia, and the other antarctic islands.

12. **L. Banksii,** *Hook. f. Fl. N. Z.* ii. 31. *t.* 76. *f.* 1. Rhizome prostrate, stout, woody, covered with matted fibres, chaffy at the ascending tip. Fronds tufted, numerous, quite glabrous, coriaceous, 6–18 in. high, very narrow-lanceolate, pinnatifid, the fertile shorter, pinnate; stipes winged, scaly at the base and rachis, very stout, naked; pinnules numerous, close set, half orbicular or very broadly oblong, rounded at the tip, quite entire, sessile by a very broad base; the lower decurrent and often forming a wing to the stipes; fertile pinnules shorter and narrower, spreading, curved or straight. —Hook. Sp. Fil. iii. 17.

Common in dark woods, throughout the **Northern** and **Middle** Islands, *Banks and*

Solander, etc. The very broad coriaceous pinnules, altogether adnate, best distinguish this. A very similar and, I think, identical species, occurs in Chili.

13. **L. nigra,** *Colenso ;—Fl. N. Z.* ii. 31. Rhizome short, indistinct. Fronds tufted, small, 8–12 in. high, dark green, blackish and brittle when dry, lanceolate, obtuse, quite glabrous, or pubescent below and on the margin, rather membranous; stipes and rachis scaly; barren fronds often lyrate-pinnatifid or -pinnate; pinnules few, often interrupted, oblong, obtuse, sinuate-crenate, upper usually largest, lobed, lower sometimes also large stalked and deflexed; fertile fronds pinnate; pinnules few, distant, suberect, narrow-linear, tip subulate or apiculate, terminal often elongate.—Hook. Ic. Pl. t. 960; Sp. Fil. iii. 35; *Polybotrya nana*, Fée, Acrost. t. 38. f. 1.

Northern Island: in dark moist woods, by watercourses, etc.; east coast and interior, *Colenso, Sinclair*, etc. **Middle** Island: Milford Haven and Bligh's Sound, *Lyall, Hector and Buchanan.*

14. **L. Fraseri,** *A. Cunn. ;—Fl. N. Z.* ii. 31. Rhizome slender, erect, like that of a small tree-fern, sometimes 2–3 ft. high, as thick as the thumb, black, covered with the bases of old stipes. Fronds tufted, very numerous at the top of the rhizome, 8–18 in. high, ovate or ovate-lanceolate, 2-pinnatifid, quite glabrous, rather membranous; stipes scaly at the base, quite glabrous, as is the rachis, both are interruptedly winged, the wing sharply angled and lobed; primary branches 2–4 in. long, $\frac{2}{3}$ broad, tapering to a long point; pinnules crowded, linear-oblong, acute, quite entire or serrate; fertile frond narrower than the barren, sometimes 2-pinnate, caudate; segments sometimes stipitate. Involucres marginal, recurved.

Northern Island and northern parts of the **Middle** Island: Bay of Islands, *Fraser*, etc.; Auckland, *Sinclair;* Massacre Bay, *Lyall.* Not found except in New Zealand.

17. DOODIA, Br.

Fronds tufted on a very short rhizome, simply pinnate, erect or prostrate, the inner often the most fertile. Veins free, simple or forked, uniting again only where the sorus is placed. Sori oblong or reniform, distant from the margin, parallel to the costa. Involucres of the same shape, membranous or scarious, attached by one side to a transverse vein, opening towards the costa.

A small genus, confined to the southern Pacific Islands, Australia, Java, and Ceylon. The following species, I think, all pass into one another.

Fronds erect, pinnatifid in upper $\frac{1}{2}$–$\frac{2}{3}$. Pinnules 3–6 in. . . . . . 1. *D. connexa.*
Fronds erect, pinnatifid in upper $\frac{1}{2}$ or $\frac{1}{3}$. Pinnules 1–2 in. . . . . . 2. *D. media.*
Frond usually decurved or prostrate, pinnate nearly to the top. Pinnules $\frac{1}{2}$–1 in. . . . . 3. *D. caudata.*

1. **D. connexa,** *Kunze.* Fronds tufted, 1–2 ft. high, erect, lanceolate, rather membranous, pinnate below, pinnatifid above the middle or lower one-third; stipes stout, black; rachis smooth, naked; pinnules close-set, horizontal, 3–6 in. long, very narrow linear-lanceolate, attenuate at the tip, sharply serrate, bright green, lowest truncate or auricled at the base. Sori in 1–3 series on each side the costa.—Hook. Sp. Fil. iii. 75.

Kermadec Islands, *Macgillivray.* Also found in the Society and other Pacific groups.

2. **D. media,** *Br.* Rhizome short, stout, ascending. Fronds tufted, 6–12 in. high, suberect, lanceolate, rather rigid, pinnatifid above, pinnate below; stipes black, 3–5 in. long, with subulate scales at its base, sometimes scabrid; rachis smooth, naked; pinnules close together, spreading, linear-oblong, obtuse, upper decurrent at the base, lower auricled, uppermost generally elongate and subcordate, all rigid, spinulose or toothed. Sori in one series.—Hook. Sp. Fil. iii. 74; *D. Kunthiana*, Gaudichaud in Freycinet's Voy. t. 14.

Northern Island: on clay hills, etc., common. Very common in Australia, Norfolk Island, and some of the Pacific Islands.

3. **D. caudata,** *Br.;—Fl. N. Z.* ii. 37, *excl. synonyms.* Very closely allied to *D. media*, and probably a variety of it, but more flaccid, the fronds often prostrate, pinnate nearly to the top, fertile more distinct from the barren and longer.—Hook. Sp. Fil. iii. 75; Exot. Flora, t. 25.

Northern Island: common, *A. Cunningham*, etc. The fragrant fern which occurs in the Wellington Valley appears to be a variety of this. A common Australian and Tasmanian fern.

18. ASPLENIUM, Linn.

Fronds tufted on a usually very short rhizome, or with no rhizome, usually green, coriaceous, flaccid, pinnate 2- or 3-pinnate or decompound (never entire in New Zealand); venation various. Sori linear or oblong, parallel to and upon the veins, oblique with regard to the costa, distant from the margin, except when the frond is decompound. Involucre membranous, of the same form as the sorus, attached laterally to a vein, opening towards the costa.

One of the largest and the most difficult genera of ferns, because the fronds of many of the species vary extremely from simple to decompound, and from almost membranous to coriaceous, etc. None of the New Zealand species have quite simple fronds, nor have any the double involucre of the *Diplazium* section. Of the species here described, numbers 4, 5, and 12 are quite distinct from one another and from all the others; 6 and 7 pass into one another, but not into others; 1, 2, and 3 pass into one another and into 8, 9, 10, and 11, which together form a network of varieties, of which the characters given are those of prevalent forms only; thus there are only 5 universally separable forms in the Island.

A. Euasplenium. Veins free. Involucre flat.

a. Fronds pinnate, sometimes 2-pinnate below in 3. Sori several on each pinnule, not close to the margin (see *A. Hookerianum*, in *β*).

1. *Rhizome stout or* 0.

Frond 3 in.–3 ft., stout. Pinnules oblong, crenate or serrate. Sori oblong or linear . . . . 1. *A. obtusatum.*

Frond 1–2 ft., very stout. Pinnules narrow-lanceolate, very coriaceous, serrate or pinnatifid. Sori short . . . . 2. *A. scleroprium.*

Frond 1–2 ft., stout. Pinnules oblong-lanceolate, serrate. Sori linear, very long . . . . 3. *A. lucidum.*

Frond 2–12 in., slender, erect. Pinnules ¼ in., shortly oblong or obovate. Veins distant. Sori in 2 series . . . . 4. *A. Trichomanes.*

Frond 2–18 in., slender, prostrate, rooting at the naked tips. Pinnules rounded. Veins flabellate. . . . . 5. *A. flabellifolium.*

2. *Rhizome creeping, scaly, slender.*

Frond 1–2 feet. Pinnules 2–4 in., lanceolate, caudate. Veins very close-set . . . . 6. *A. caudatum.*

Frond 1–2 ft. Pinnules 4–6 in., lanceolate, acuminate. Veins very close-set . . . . 7. *A. falcatum.*

β. Fronds 2- or 3-pinnate, segments usually narrow. Sori generally 1 on each segment and close to its margin.

Frond 4–12 in., erect, membranous, oblong, pinnate or 2-pinnate. Pinnules stalked . . . . 8. *A. Hookerianum.*
Frond flaccid, erect or pendulous, 1–3 ft., 2- or 3-pinnate, ovate-lanceolate, often proliferous . . . . 9. *A. bulbiferum.*
Frond 3–5 in., stout, erect, ovate, acuminate. Pinnules sessile . . 10. *A. Richardi.*
Frond 1–2 ft., very coriaceous, ovate linear or lanceolate. Pinnules very narrow, linear or lanceolate . . . . 11. *A. flaccidum.*

B. Athyrium. Fronds 2- or 3-pinnate. Veins forked. Involucre cylindric, often irregularly torn, several on a segment.

Frond tall, very broad, deltoid, membranous . . . . 12. *A. australe.*

1. **A. obtusatum,** *Forst.;—Fl. N. Z.* ii. 33. Rhizome short, thick, woody. Fronds very coriaceous, 3 in.–3 ft. high, erect or pendulous, linear or oblong-linear, pinnate; stipes very stout, with shining lanceolate scarious scales at its base; rachis stout, compressed, margined, glabrous or with a few weak hairs; pinnules 1–4 in. long, stipitate, oblong or linear- or oblong-lanceolate, obtuse or acute, truncate cuneate or rounded at the base, coarsely crenate or serrate, uppermost confluent. Veins usually simple. Sori linear or oblong, oblique.—Hook. Sp. Fil. iii. 96; Fil. Exot. t. 46; Labill. Fl. Nov. Holl. t. 242.

Var. *β. obliquum.* Pinnæ generally longer, more acute. Sori linear, more numerous. *A. obliquum,* Forst.; Labill. l. c. t. 242. f. 1. *A. apicidentatum,* Homb. and Jacq. Voy. au Pôle Sud, Bot. t. 1 A.

Abundant, both varieties, throughout the islands, especially on maritime cliffs, from **Kermadec** Islands, *Macgillivray,* to **Campbell's** Island, *J. D. H.* Common throughout the south temperate zone.

2. **A. scleroprium,** *Homb. and Jacq. Voy. Pôle Sud, Bot. t.* 1 D. Very similar to *A. obtusatum,* but more thick and coriaceous, with narrow-lanceolate, caudate-acuminate, entire serrate or almost pinnatifid pinnules. Sori extending to the margins of the segments.—Hook. Sp. Fil. iii. 98; *A. flaccidum, β. aucklandicum,* Fl. Antarct. 109.

Lord Auckland's group: common in woods near the sea, *Hombron and Jacquinot, J. D. H., Bolton.* This is a transition form between *A. obtusatum* or *lucidum* and *A. flaccidum.*

3. **A. lucidum,** *Forst.;—Fl. N. Z.* ii. 33. Habit and characters of *A. obtusatum,* but larger, more flaccid, bright green, shining; fronds sometimes 2 ft. long and broad; pinnules usually oblong-lanceolate, serrate, often 6 in. long, upper base rounded incised or lobed. Sori numerous, linear, very long.—Hook. Sp. Fil. iii. 98.

Var. *β. paucifolium.* Dwarf; pinnules 3–7, terminal much elongate. Hook. l. c. 99.

Var. *γ. Lyallii,* Fl. N. Z. ii. 33. t. 77. Lower pinnules again pinnate, the next deeply lobed.

Abundant from **Kermadec** Islands, *Macgillivray,* to Otago, *Banks and Solander,* etc. Var. *γ.* Otago, *Lyall.* This passes on one hand into *A. obtusatum,* and on the other into *bulbiferum,* through *γ. Lyallii,* and into *flaccidum* through *scleroprium,* and others.

4. **A. Trichomanes,** *Linn.* Rhizome short, thick, fibrous. Fronds

numerous, linear, erect, 3–12 in. long, pinnate; stipes smooth, polished, margined, black; pinnules numerous, rather coriaceous, small, horizontal, scarcely stalked, $\frac{1}{4}$ in. long, oblong or obovate, obliquely cuneate at the base, upper margin near the base rounded truncate or auricled, lower excised, entire or crenate; veins distant oblique, usually forked. Sori oblique, in 2 series; involucre pale brown, membranous, entire or jagged.—Hook. Sp. Fil. iii. 136.

Middle Island: Kaikoras Mountains, *Sinclair*; Acheron valley, *Travers*; Canterbury, Southern Alps, *Sinclair and Haast*; Otago, lake district, *Hector and Buchanan*. A very common fern throughout the northern temperate hemisphere; also found in South Africa, Australia, Tasmania, the Sandwich Islands, and the Andes of South America.

5. **A. flabellifolium,** *Cavanilles;—Fl. N. Z.* ii. 33. Rhizome 0. Fronds tufted, flaccid, spreading decumbent or pendulous, often rooting at the long tips, 4–18 in. long, linear, pinnate; stipes and rachis smooth, naked or with a few small black scales, slender, green; pinnules numerous, very variable, $\frac{1}{4}$–$\frac{2}{3}$ in. long, shortly stalked, bright green, orbicular or rhomboid, cuneate or reniform at the base; outer margin coarsely crenate or lobed; veins flabellate. Sori 3 or 4, radiating from the base of the pinnules; involucre pale brown, membranous.—Hook. Sp. Fil. iii. 146; *A. flabelliforme*, Hook. Exot. Flora, t. 208.

Abundant throughout the **Northern** and **Middle** Islands, *Banks and Solander*, etc., on rocks and in stony places. Also frequent in Australia and Tasmania.

6. **A. caudatum,** *Forst.* Rhizome stout, terete, creeping. Fronds broadly lanceolate, acuminate, 1–2 ft. long, very coriaceous, pinnate, pinnatifid towards the tip; stipes below, and young rachis brown and clothed with large, falcate, imbricate, subulate, shining scales; pinnules rather remote, 2–4 in. long, stalked, lanceolate from an obliquely cuneate subrhomboid base, long acuminate, upper base rounded or auricled, lower more or less excised, coarsely serrate or pinnatifid, lower serratures again cut; veins erecto-patent, generally forked. Sori linear, long, almost parallel to the costa, often confluent; involucre firm and membranous.—Hook. Sp. Fil. iii. 152; Schkuhr, Filices, t. 77.

Kermadec Island, *Macgillivray*. Also found in Australia, the Pacific and Malay Islands, and India. Apparently a form of *A. falcatum*.

7. **A. falcatum,** *Lamarck.—A. polyodon*, Forst.;—Fl. N. Z. ii. 34. Rhizome creeping, clothed with large brown scales. Fronds lanceolate or linear-lanceolate, coriaceous, 8 in.–2 ft. long, erect, pinnate; stipes long, brown, and rachis villous with slender scales; pinnules horizontal, long-stalked, 4–6 in. long, $\frac{1}{2}$–$\frac{3}{4}$ in. broad, lanceolate, cuneate at the base, gradually narrowed to an acuminate point, lobed or pinnatifid, upper base broader auricled, lower excised; veins close, erecto-patent with forked branches. Sori linear, numerous, giving a striate appearance to the pinnules; involucre narrow, firm, membranous.—Hook. Sp. Fil. iii. 160.

Common throughout the **Northern** Island, *Banks and Solander*, etc. A most abundant fern in all tropical and warm south temperate latitudes, but not extending into Tasmania. It passes into *A. caudatum*.

8. **A. Hookerianum,** *Colenso.—A. adiantoides*, var. *a*, Fl. N. Z. ii.

35. Rhizome short, fibrous. Frond 2–12 in. long, erect, oblong and pinnate or ovate and 2-pinnate, rather membranous; stipes slender, pale brown, with subulate scales at the base and narrower deciduous scales above; primary pinnæ 1–2 in. long, long stalked; pinnules rhombeo-subrotund, stalked, $\frac{1}{6}$–$\frac{1}{4}$ in. long, lobed toothed or pinnatifid; veins subflabellate, dichotomous. Sori 3–5, oblong, on the disk of the pinnule; involucre membranous.—Hook. Sp. Fil. iii. 194; *A. udiantoides*, Raoul, Choix, t. 1, not of *Raddi;* Hook. Ic. Pl. t. 983.

Var. *β. Colensoi*, Moore. Pinnules shortly stalked, deeply cut into narrow segments, with the sori on their margins.—*A. Adiantoides, β Colensoi*, Hook. f. in Ic. Pl. t. 984; *A. Colensoi*, in Lond. Journ. Bot. iii. 26.

Northern Island: mountains of the east coast and interior, *Colenso.* **Middle** Island: Nelson and Canterbury alps, *Travers;* Akaroa, *Raoul, Lyall;* Otago, *Hector and Buchanan.* Var. *β*, Wairarapa valley, etc., *Lyall, Colenso;* Nelson, Canterbury, and Otago, *Sinclair, Travers*, etc. This passes into *A. Richardi* and *flaccidum.*

9. **A. bulbiferum,** *Forst.;—Fl. N. Z.* ii. 197. Rhizome stout, horizontal or oblique, scaly at the tip. Fronds flaccid, erect pendulous or declinate, bright-green, thinly coriaceous, 1–3 ft. long, ovate-lanceolate, pinnato-pinnatifid or 2- or 3-pinnate; stipes green, semiterete or compressed, scaly at the base and deciduously so above; pinnules often proliferous, 6 in. long, long-stalked, lanceolate, acuminate; segments $\frac{1}{2}$–2 in. long, ovate-oblong, pinnatifid. Sori on the disk of the more entire pinnules, and sides of the lobes of the more cut ones, oblique, short, oblong; involucre firm, greenish.—Hook. Sp. Fil. ii. 196; Ic. Pl. t. 423; Homb. and Jacq. Voy. au Pôle Sud, Crypt. t. 3.

Var. *β. laxa.* Fronds pendulous, decompound; segments narrow.—*Cænopteris appendiculata*, Labill. Fl. Nov. Holl. t. 243; *A. triste ?*, Raoul, Choix.; *A. laxum*, Br.; Homb. and Jacq. Voy. au Pôle Sud, t. 3 *bis.*

Var. *γ. tripinnata.* Fronds tripinnate, as in *A. flaccidum*, but more membranous.—*A. tremulum*, Homb. and Jacq. l. c. t. 3 *bis.*

Abundant in damp woods throughout the **Northern** and **Middle** Islands, on the ground, and on rocks and trees, *Banks and Solander*, etc. Also frequent in South-east Australia and Tasmania, and in some parts of India; it is united by varieties with all the other species of this section.

10. **A. Richardi,** *Hook. f. Fl. N. Z.* ii. 197. Rhizome short, stout, tipped with subulate scales. Fronds erect, 3–5 in. long, rather flaccid, ovate, acuminate, 2-pinnate; stipes stout; pinnæ stalked, 1–1$\frac{1}{2}$ in. long, close set, lanceolate; pinnules crowded, sessile, ovate or obovate, obtuse, $\frac{1}{6}$–$\frac{1}{3}$ in. long, pinnatifid; segments obtuse, upper confluent; veins one to each segment, forked, thickened at the tip. Sori solitary, broad, oblong, close to the margin of each segment; involucre membranous.—Hook. Sp. Fil. iii. 197, excl. var. *β*; *A. adiantoides*, var. *Richardi*, Hook. f. in Ic. Pl. t. 977; *A. Raoulii*, var. *Richardi*, Mettenius.

Middle Island: New River, *Herb. A. Richard.* A more erect plant than *A. Hookerianum*, with an ovate frond, closer pinnæ, and crowded sessile pinnules; its habit is that of erect states of *A. flaccidum*, but it is very much more membranous. I suspect, however, that it is nothing but a form of one of these species. In the 'Species Filicum' Sir W. Hooker refers *β Colensoi* of *A. Hookerianum* to this, but is now disposed to keep it where I had placed it; some specimens of this var. *β* appear to me to be absolutely intermediate.

11. **A. flaccidum,** *Forst.;—Fl. N. Z.* ii. 35. Rhizome stout, erect, crowned with large, subulate-lanceolate, membranous scales. Fronds usually pendulous, polymorphous, coriaceous, 6 in.–3 ft. long, generally greenish-white when dry, lanceolate, acuminate, pinnate or 2-pinnate; stipes stout, rather short, compressed or 3-gonous, scaly at the base; rachis winged; lower primary branches of frond 4–10 in. long, lanceolate, often caudate, stalked, pinnatifid or pinnate, middle branches pinnatifid; upper forked, or all much divided into linear-oblong, slightly incurved obtuse pinnules $\frac{1}{3}$–$\frac{2}{3}$ in. long; veins one in each segment. Sori oblong, near or on the very margins of the segments; involucre firm.—Hook. Sp. Fil. iii. 205; *A. heterophyllum*, A. Rich.; *Cænopteris flaccida*, Thunberg; Schkuhr, Fil. t. 82; *C. novæ-Zelandiæ*, Sprengel; *Darea flaccida*, Smith.

Var. *α.* Fronds pendulous, 2-pinnatifid; pinnæ very narrow, distant, cut into deep narrow lobes.

Var. *β.* Fronds pendulous, pinnate; pinnæ incised or toothed; lobes linear, obtuse. Sori marginal.

Var. *γ.* Fronds erect or pendulous, pinnate; pinnæ falcate-lobed. Sori dorsal or half-dorsal. (Passing into *A. scleroprium.*)

Var. *δ.* Fronds erect, rigid, deltoid, 2-pinnate; pinnæ ovate or lanceolate; pinnules shortly stipitate.

Var. *ε.* Fronds 2–6 in., short, stout, ovate, 2- or 3-pinnate, like those of *A. Richardi*, but very thick and coriaceous. (An alpine form.)

Var. *ζ.* Fronds large, lax, 3-pinnate.—*A. Shuttleworthianum*, Kunze; Hook. Sp. Fil. iii. 210.

Abundant throughout the **Northern** and **Middle** Islands: *Banks and Solander*, etc. **Lord Auckland's** group, *J. D. H.* Var. *ε.* Alpine districts of both Northern and Middle Islands. Var. *ζ.* **Kermadec** Islands, *Macgillivray*. Also a native of the more humid parts of Australia and Tasmania. One of the most variable ferns known.

12. **A. australe,** *Brackenridge.—A. Brownii*, Smith, in Fl. N. Z. ii. 36. Rhizome very stout, creeping. Frond very large, broad, flaccid, membranous, pale-green, 2–3 ft. high, broadly deltoid, spreading, 2- or 3-pinnate; stipes long, slender, smooth, polished, scaly at the base; rachis slender, flexuous, glabrous; primary branches linear-oblong, acute or acuminate; secondary pinnules 1–2 in. long, oblong-lanceolate, shortly stalked, pinnatifid or pinnate; pinnules linear-oblong, obtuse, crenate or incised, rarely entire; veins simple or forked. Sori several on each segment, nearer the costa than the margin; involucre cylindric, attached by both margins and bursting with a torn edge down one side or the middle.—Hook. Sp. Fil. iii. 233; *A. Brownii*, Hook. Ic. Pl. t. 978; *Allantodia australis* and *A. tenera*, Br.

Northern Island: in damp woods frequent, *Banks and Solander*, etc. **Middle** Island: Nelson, *Travers*. A very beautful fern, also found in Norfolk Island, in South-eastern Australia and Tasmania, and under a slightly different form in India, the Malay Islands, Madeira, the Canaries, and various other parts of the world. In New Zealand this species becomes smaller and coriaceous in exposed localities.

19. ASPIDIUM, Swartz.

Rhizomes slender or stout and creeping, or forming a short, very stout, erect or inclined trunk.—Fronds in the New Zealand species erect, 2- or 3-pinnate, alternate on the slender rhizome or tufted on the end of the short trunk or stock; venation various, simple and free in the New Zealand species.

Sori on the back of the pinnæ, distant from the margin, on the middle of a vein, globose. Involucre orbicular, membranous, flat or convex, peltately attached to the centre of the sorus.

One of the largest genera of ferns, found in all parts of the globe. The *A. hispidum*, Swartz, is referred to the genus *Nephrodium*.

1. *Rhizome short, stout, woody, and scaly. Fronds tufted. Involucre flat or concave.*

Frond 2-pinnate, narrow, attenuate below. Scales on stipes large, broad, membranous, mixed with hairs . . . . 1. *A. aculeatum.*
Frond rigid, 1- or 2-pinnate, oblong or ovate, not attenuate below. Scales on stipes narrow, black, rigid. Involucre with dark disk . . 2. *A. Richardi.*
Frond rigid, 2-pinnate, not attenuate below. Scales on stipes rigid, black with white edges. Involucre with large black disk . . . . 3. *A. oculatum.*

Involucre very convex, bullate, subglobose.

Frond flaccid, narrow. Scales very large, pale, membranous . . . . 4. *A. cystostegia.*

2. *Rhizome creeping. Fronds rather membranous.*

Frond 2–4-pinnate, deltoid. Pinnules aristate . . . . 5. *A. aristatum.*
Frond 2–4-pinnate, deltoid. Pinnules obtuse . . . . 6. *A. coriaceum.*
Frond hispid with spreading bristles. (See *Nephrodium hispidum*.)

1. **A. aculeatum,** *Swartz;* var. **vestitum,** *Hook.—Polystichum vestitum,* Presl;—Fl. N. Z. ii. 38. Rhizome very stout, woody, erect or ascending, sometimes 2–4 ft. high. Fronds tufted, rather flaccid, forming a spreading crown, 1–3 ft. high, coriaceous, linear or ovate-oblong, 2-pinnate, gradually narrowed below, sometimes proliferous; stipes stout, covered with large, brown, ovate-lanceolate, acuminate, membranous, often lacerate, straight or curved scales, which are sometimes 1 in. long, and margined with white; rachis woolly with brown hairs, glabrate in age, and having a few scales also; primary divisions of frond very numerous, horizontal, linear-lanceolate, acuminate; pinnules numerous, ovate-oblong, pungent, sharply toothed or pinnatifid, segments shortly stipitate, pungent, lower outer margin auricled. Sori numerous; involucre orbicular, flat, stalked.—Hook. Sp. Fil. iv. 22; *A. pulcherrimum* and *A. waikarense*, Colenso; *A. venustum*, Hombr. and Jacq. Voy. au Pôle Sud, Bot. t. 4 and 5.

Abundant in the mountainous parts of the **Northern** Island, and throughout the **Middle** to **Campbell's** and **M'Quarrie's** Islands. Also very common in South-east Australia, Tasmania, and Fuegia; this differs a good deal from the prevalent European forms of *A. aculeatum*, but is connected by innumerable intermediates.

2. **A. Richardi,** *Hook.—Polystichum aristatum,* Fl. N. Z. t. 78. f. 5, not of Presl. Rhizome short, stout, scaly. Fronds 8 in.–2 ft. high, tufted, rigid, ovate-oblong or almost deltoid, not narrowed below, 1- or 2-pinnate, finely acuminate, glabrous or with white woolly hairs below; stipes and rachis slender, rigid, rough with scattered, narrow, subulate, black, rigid, deciduous scales, that have fimbriate bases; primary branches patent or ascending, lanceolate, acuminate, deeply pinnatifid or again pinnate; pinnules numerous, lanceolate or ovate-lanceolate, serrate, pungent, sessile or stalked, glabrous or woolly and scaly like the stipes below. Sori in 2 rows on each segment; involucre flat, with a small dark disk.

Var. *α.* Frond broader, 3-pinnate; pinnules spreading.—*A. coriaceum,* var. *acutidentatum,* A. Rich.

Var. β. Frond pinnate; pinnules ascending, pinnatifid at the base only.—*A. Richardi*, Hook. Sp. Fil. iv. 23, t. 122.

Abundant in woods, etc., throughout the **Northern** and **Middle** Islands. Var. β at the Bay of Islands only, *J. D. H.* This differs from *A. aristatum* in the rhizome not creeping, and much smaller, less pinnate fronds; from *A. oculatum* in the small scales, and wanting the large black disk to the involucre; and from *A. aculeatum* in the rigid habit, much smaller, broad, more coriaceous fronds, of an ovate form, not tapering to the base, and the free, less numerous primary pinnæ, and especially in wanting the large membranous scales of that plant. It seems to be confined to New Zealand and the Fiji Islands.

3. **A. oculatum,** *Hook. Sp. Fil.* iv. 24. *t.* 228 Rhizome absent. Fronds 10–20 in. long, coriaceous, ovate-oblong, acuminate, 3-pinnate, pale and clothed with woolly hairs below; stipes stout, straw-coloured, covered with rigid, large, subulate, brown scales margined with white; rachis with fewer, softer scales, and lax woolly hairs; primary divisions of the frond 2–4 in. long, narrow ovate-lanceolate, acuminate, stalked, not close together; secondary also lax, $\frac{2}{3}$–1 in. long, sessile or stalked; pinnules alternate, sessile, decurrent, $\frac{1}{4}$ in. long, obtuse or mucronate, obtusely toothed or subpinnatifid. Sori abundant over the whole under surface, 2–4 on each segment; involucre orbicular, shortly stalked, with a large black disk and narrow reddish margin.

Northern Island: Wairarapa valley, *Colenso.* **Middle** Island: Akaroa, *Raoul.*

4. **A. cystostegia,** *Hook. Sp. Fil.* ii. 30. *t.* 127. Rhizome short, small. Fronds tufted, soft, rather flaccid, 4–10 in. high, pale green, narrow oblong-lanceolate, pinnate or 2-pinnate; stipes and rachis stout but brittle, clothed with copious, very large, convex, membranous, ovate, pale scales, $\frac{1}{4}$–$\frac{2}{3}$ in. long, with subulate tips; rachis similarly clothed; primary divisions distant below, crowded above, $\frac{1}{2}$–$1\frac{1}{2}$ in. long, spreading, pinnatifid or pinnate with a flat or winged rachis, segments or pinnules $\frac{1}{4}$ in. long, ovate-lanceolate, obtuse or acute, entire or toothed, with a few long hairs below. Sori numerous and large; involucres bladdery, very membranous, with the edge turned down all round, forming a sort of globe.

Northern Island: Tongariro, *Dieffenbach;* Mount Egmont (*Mrs. Jones*). **Middle** Island: Discovery Peaks, alt. 5800 ft., *Travers;* Wairau gorge, alt. 4400 ft., *Sinclair;* between Lake Tennyson and the west coast, *Maling;* Canterbury on the Southern Alps, alt. 5–6000 ft., *Haast;* Otago, Mount Ida, *Dr. Buchanan*; lake district, alpine, *Hector*.

5. **A. aristatum,** *Swartz;* not *Polystichum aristatum*, Fl. N. Z. Rhizome long, stout, creeping, clothed with narrow, subulate, rusty scales. Fronds alternate, 1–$2\frac{1}{2}$ ft. high, deltoid-ovate, acuminate, 3- or 4-pinnate, coriaceous, glossy; stipes stout and rachis crinite with subulate scales; primary divisions stipitate, lanceolate, finely acuminate, lowest with the lower secondary pinnæ much elongated and 2- or 3-pinnate; pinnules obliquely ovate or rhomboid-lanceolate, subfalcate, mucronate, serrate. Sori generally small, and often in two rows on each pinnule; involucres small, flat, orbicular or slightly reniform.—Hook. Sp. Fil. iv. t. 27; Schkuhr, Fil. t. 42.

Kermadec Islands, *Macgillivray.* A very common Pacific Island and Indian plant, not found in Africa or America.

6. **A. coriaceum,** *Swartz.—Polystichum coriaceum*, Fl. N. Z. iv. 32. Rhizome stout, creeping, very long, clothed with tawny silky scales. Fronds

distant, 1–4 ft. high, rigidly coriaceous, ovate-deltoid, acuminate, 3-pinnate; stipes stout, and rachis covered with deciduous scales; primary branches erecto-patent, stipitate, lower obliquely deltoid, acuminate, with the lower secondary divisions much elongated and divided; pinnules 1–2 in. long, ovate or lanceolate, entire or obtusely serrate or pinnatifid; segments oblong, obtuse, rarely acute, entire or obtusely serrate; veins sunk, veinlets close. Sori in 2 rows, nearer the costa than margin; involucre large, orbicular, with a deep sinus.—Hook. Sp. Fil. iv. 82; Schkuhr, Fil. t. 50; *A. Cunninghamianum*, Colenso; *Polypodium adiantiforme*, Forst.

Northern and **Middle** Islands: common in woods, etc., *Banks and Solander*, etc. An abundant fern throughout the south temperate zone.

20. NEPHRODIUM, Br.

Rhizome stout, erect creeping or ascending.—Fronds tufted or distant, 1–3-pinnate in the New Zealand species. Veins forked and free, rarely united in the New Zealand species. Sori on the back of the frond, distant from the margin, globose. Involucre kidney-shaped, fixed by the sinus to the centre of the sorus, membranous, often deciduous.

A very large tropical and temperate genus. Involucre sometimes absent in *N. velutinum*, often orbicular and peltate in *N. hispidum*.

1. Eunephrodium.—*Primary veins pinnate; secondary angularly united.*

Frond lanceolate, pinnate, pubescent . . . . 1. *N. molle.*

2. Lastrea.—*Veins and veinlets all free.*

Frond lanceolate, pinnate, glabrous, with a few bullate scales . . 2. *N. thelypteris, β.*

Frond deltoid or 5-angular, 2–4-pinnate, softly tomentose, brown when dry . . . . 3. *N. velutinum.*

Frond deltoid or ovate, 2–4-pinnate, glabrous or pubescent, green when dry . . . . 4. *N. decompositum.*

Frond rigid, 3- or 4-pinnate, pubescent below. Stipes, rachis, and costa hispid with stiff spreading bristles . . . . 5. *N. hispidum.*

1. **N. molle,** *Desvaux.* Rhizome short, stout, horizontal, densely rooting. Fronds 1–3 ft. high, rather membranous and flaccid, oblong-lanceolate, abrupt or narrowed to the base, pinnate, pubescent on both surfaces; stipes and rachis green; pinnæ numerous, horizontal, sessile, 3–5 in. long, oblong or oblong-lanceolate, pinnatifid; pinnules variable in length, oblong, broad or narrow, obtuse, falcate; lowest veins uniting and sending forth a veinlet which is prolonged to the sinus, the rest free, simple or forked. Sori usually in 2 rows; involucres villous or pubescent.—Hook. Sp. Fil. iv. 67; *Polypodium molle*, Forst., and *Aspidium nymphale*, Forst.; Schkuhr, Fil. t. 34.

Northern Island?, *Forster*; Rotomahana (*Mrs. Jones*). **Kermadec** Islands, *Macgillivray*. One of the commonest of tropical and subtropical ferns, also found in New South Wales. The Rotomahana habitat I copy from Mrs. Jones's little work.

2. **N. thelypteris,** *Desvaux*; var. *β*, **squamulosum,** *Schlechtendal, Fil. Cap. t.* 11.—*Nephrodium squamulosum*, Fl. N. Z. ii. 39. Rhizome slender, creeping?, very long, black. Fronds scattered, 1–2 ft. high, membranous but firm, lanceolate, glabrous below, pinnate; stipes slender, pale, black at the base, naked or with the costa covered with scattered, pale, membranous, very convex scales; pinnæ subopposite, sessile, spreading, 1–1½ in.

long, linear-oblong, acuminate, deeply pinnatifid; pinnules ⅛ in. long, linear-oblong, obtuse or acute, quite entire; veinlets simple, the lower forked. Sori numerous, towards the margin; involucres cordate-reniform, glandular.—Hook. Sp. Fil. iv. 88; *Aspidium invisum*, Forst.?

Northern Island, *Forster;* Bay of Islands and east coast, *Colenso.* The var. *β* is a native also of South Africa; the typical plant, which wants the bullate scales, is common in the north temperate zone.

3. **N. velutinum,** *Hook. f.—Lastrea velutina*, Fl. N. Z. ii. 39. t. 80. Rhizome unknown. Frond rather membranous, deltoid or 5-angled, 1–2 ft. high and broad, acuminate, downy with silky hairs on both sides, often glandular beneath, 2- or 3-pinnate or 4-pinnate at the base, red-brown when dry; stipes long, slender, pubescent, clothed with subulate scales at the base; primary pinnæ, 4 in.–1 ft. long, broadly oblong, acuminate, long-stalked, lowest inferior secondary pinna deflexed, most compound; pinnules sessile, oblong, acute, pinnatifid, segments ovate, obtuse, serrate or crenate; veinlets simple or forked. Sori small, one on each segment; involucres hairy, red-brown, often fringed with glands, sometimes minute or absent.—*Aspidium velutinum*, A. Rich.

Abundant throughout the **Northern** and **Middle** Islands, *Banks and Solander*, etc.

4. **N. decompositum,** *Br.;—Fl. N. Z.* ii. 39. *t.* 79. Rhizome short, stout, creeping, black. Fronds as in *N. velutinum*, but very variable, often smaller, always green when dry, glabrous or pubescent, segments more acutely serrated, sometimes only pinnate, pinnules often elongate or caudate; stipes with fewer scales at the base.—Hook. Sp. Fil. iv. 146; *N. glabellum*, A. Cunn.

Abundant throughout the **Northern** and **Middle** Islands, *Banks and Solander*, etc. **Kermadec** Islands, *Macgillivray.* Also found in Tasmania, Australia, Norfolk Island, and the Pacific islands.

5. **N. hispidum,** *Hook. Sp. Fil.* iv. 150.—*Polystichum hispidum*, J. Sm.; —Fl. N. Z. ii. 38. Rhizome long, stout, creeping, covered with subulate brown scales. Fronds 1–2 ft. high, rigid, ovate-acuminate, 3- or 4-pinnate, pubescent beneath; stipes stout, and rachis and costa hispid with long rigid spreading or deflexed bristles with a swollen base; primary pinnæ 3–6 in. long, stipitate, ovate, acuminate, lowest pair deltoid; pinnules decurrent, narrow lanceolate, ¼–½ in. long, deeply serrate, segments pungent; veinlets solitary in each segment. Sori solitary on the segments; involucre flat, orbicular or reniform, dark brown.—*Aspidium hispidum*, Schkuhr, Fil. t. 49; *Polypodium setosum*, Forst.

Throughout the **Northern** and **Middle** Islands: abundant, *Banks and Solander*, etc. I have followed the 'Species Filicum' in referring this to *Nephrodium*, but the involucre is more that of *Polystichum.*

Aspidium Serra, Forst. Nothing is known of this plant, which is probably the common tropical *Nephrodium propinquum*, Br., or *N. unitum*, Sieb., which Forster gathered in the South Sea Islands. Raoul enters it into his catalogue as having been collected at Akaroa by himself, probably confounding it with something else.

Mrs. Jones (Handbook of N. Z. Ferns) mentions another species of *Nephrodium* as having been found near to hot springs, which differs from *N. molle* in the pinnæ tending upwards, and the position of the sori.

21. NEPHROLEPIS, Schott.

Rhizome creeping or short or 0.—Fronds coriaceous, pinnate. Pinnæ jointed on the rachis, deciduous, with minute white waxy dots on the upper surface. Sori dorsal, on the tip of the upper branch of a vein. Involucre flat, cordate reniform or lunate, attached by a broad base.

A small tropical or subtropical genus.

1. **N. tuberosa,** *Presl.* Rhizome scarcely any; roots of long wiry fibres, bearing ovoid or oblong scaly tubers. Fronds tufted, 1–3 ft. long, linear-lanceolate, acuminate, pinnate; stipes stout and rachis with deciduous scales; pinnules numerous, close set, horizontal, ½–1 in. long, oblong, obtuse, truncate cordate or obliquely auricled at the base, straight or falcate, lower shorter. Sori halfway between the costa and margin; involucre firm, reniform, brown, black towards the base.—Hook. Sp. Fil. iv. 151.

Northern Island: hot springs near Waikati, *Dr. Hochstetter and Sinclair.* A most abundant tropical fern, also found in the warmer parts of Australia. The tubers are eaten in India.

22. POLYPODIUM, Linn.

Rhizome erect or creeping or 0.—Fronds tufted or not, simple or compound. Venation various. Sori globose (sometimes oblong in *P. Grammitidis*). Involucre 0.

As above defined, this is the largest genus of ferns. It is broken up into many genera by most authors, but these are united by so many ambiguous species, that it is difficult to retain any of them. Many of the species are naturally referable to genera in which there is always an involucre; thus *P. rugulosum* passes into *Hypolepis tenuifolia; P. pennigerum* is a *Nephrodium,* without the involucre; and *P. sylvaticum* approaches *Aspidium aculeatum,* very closely, in habit and other respects.

I. Veins usually free. Sori rounded, linear or oblong, placed on the ends, rarely on the middle of the veinlets.

a. Frond simply pinnate or pinnatifidly divided.

Rhizome short, creeping, scaly. Frond quite entire. Sori oblong or linear . . . . . 1. *P. australe.*
Rhizome very short, scaly. Frond pinnatifid. Sori round or oblong 2. *P. Grammitidis.*
Rhizome long, slender, climbing. Fronds pendulous, pinnate, glabrous . . . . . 3. *P. tenellum.*

β. Frond 2–4-pinnate.

Rhizome short, stout, erect. Fronds tufted, erect, lanceolate; pinnules spinulose . . . . . 4. *P. sylvaticum.*
Rhizome long, stout, creeping. Frond deltoid, glandular; pinnules obtuse . . . . . 5. *P. rugulosum.*

II. Veins all free except those next the margin, which meet or almost meet (*Goniopteris*).

Rhizome stout, erect. Frond pinnate, membranous, glabrous . . 6. *P. pennigerum.*

III. Veins anastomosing, hidden in the frond which is densely covered with stellate hairs (*Niphobolus*).

Rhizome creeping, scaly. Frond entire . . . . . 7. *P. rupestre.*

IV. Veins anastomosing, with free veinlets in the areoles. Fronds glabrous (*Phymatodes*).

Rhizome very short. Fronds tufted, lanceolate, acuminate, quite entire . . . . . 8. *P. Cunninghami.*

Rhizome slender, scandent, squarrose with scales. Frond simple or pinnatifid . . . . 9. *P. pustulatum.*
Rhizome stout, glaucous, creeping or scandent, scales appressed. Frond simple or pinnatifid . . . . 10. *P. Billardieri.*

1. **P. australe,** *Mettenius.—Grammitis australis*, Fl. N. Z. ii. 44. Rhizome short, creeping, scaly. Fronds densely tufted, ¼ in.–1 ft. high, coriaceous, simple, linear- or obovate-lanceolate, acute or obtuse, glabrous pubescent pilose or ciliate, quite entire, narrowed into a winged stipes. Sori oblong or (being confluent) linear, in 1 series on each side of the obscure midrib.—Hook. Sp. Fil. iv. 167; *Grammitis Billardieri*, Willd.; *G. ciliata*, Colenso; *G. australis, rigida*, and *humilis*, Homb. and Jacq. Voy. au Pôle Sud, t. 2. f. F. G. H.

On trees, rocks, etc., throughout the **Northern** and **Middle** Islands: abundant, *Banks and Solander*, etc., ascending the mountains to 5000 ft. **Lord Auckland's** group and **Campbell's** Island, *J. D. H.* Equally abundant in Australia, Tasmania, Chili, Fuegia, and Tristan d'Acunha.

2. **P. Grammitidis,** *Br.;—Fl. N. Z.* ii. 41. Rhizome very short, scaly, roots fibrous. Fronds densely tufted, 1–10 (rarely 15) in. long, coriaceous, glabrous, linear-oblong ovate or lanceolate, deeply pinnatifid, or almost pinnate, narrowed below into a short winged stipes, which is scaly at the base; segments or pinnules distant, linear, obtuse, quite decurrent, entire lobed or pinnatifid; segments sometimes caudate. Sori round or oblong.—Hook. Sp. Fil. iv. 230; *Grammitis heterophylla*, Labill. Fl. Nov. Holl. t. 239.

On trunks of trees, etc., throughout the **Northern** and **Middle** Islands, *Banks and Solander*, etc. **Lord Auckland's** group and **Campbell's** Island, *J. D. H.* Also found in Australia.

3. **P. tenellum,** *Forst.;—Arthropteris tenella*, J. Sm. in Fl. N. Z. ii. 43. t. 82. Rhizome scandent, long, slender, rigid, wiry, scaly. Fronds scattered, pendulous, 1–2 ft. long, lanceolate, pinnate, rather membranous, bright-green; stipes slender, smooth and polished, jointed above the base; pinnules alternate, distant, 2–3 in. long, jointed on the rachis, linear-lanceolate, attenuated with an obtuse tip, sometimes forked, obliquely cuneate at the base; margin sinuate; costa slender; veins oblique, once or twice forked; veinlets with thickened tips. Sori globose, in 1 series on each side the costa, on the tip of a veinlet near the margin.—Hook. Sp. Fil. iv. 217; Schkuhr, Fil. t. 16; *A. (?) filipes*, Moore, of J. Sm. in Gard. Chron.

Northern Island: climbing lofty trees, abundant, *Banks and Solander*, etc. Also a native of New South Wales, Lord Howe's, Norfolk Island and other Pacific Islands.

4. **P. sylvaticum,** *Colenso;—Fl. N. Z.* ii. 41. *t.* 81. Rhizome stout, erect. Frond tufted, rather coriaceous, 1–2 ft. long, lanceolate, acuminate, 2-pinnate, glabrous, except on the costa; stipes stout, and rachis villous with deciduous subulate soft ferruginous scales mixed with rigid black curved ones; primary divisions distant, stalked, 2–4 in. long, oblong, acuminate; pinnules ½–¾ in. long, sessile, ovate-lanceolate, with an obliquely cuneate base, pinnatifid; lobes spinulose, subincurved; veins once or twice forked. Sori globose, equidistant between the costa and margin.—Hook. Sp. Fil. iv. 249.

Northern Island: Ruahine and other ranges, *Colenso*; Port Nicholson, *Lyall.* **Middle**

Island: not uncommon as far south as Otago. Very similar indeed to *Aspidium aculeatum*, and with similar scales, but the pinnules are more stipitate, longer, and narrower.

5. **P. rugulosum,** *Labill. Fl. Nov. Holl. t.* 241;—*Fl. N. Z.* ii. 41. Rhizome stout, creeping, rigid, villous or hispid and scaly. Fronds scattered, $\frac{1}{2}$–3 ft. high, rather membranous, covered with glandular viscid pubescence, often rufous when dry, ovate-lanceolate or deltoid, 2–4-pinnate; stipes and rachis bright red-brown, glandular pubescent and tubercled; primary divisions in distant stipitate pairs, oblong, acuminate; pinnules $\frac{1}{2}$–$\frac{3}{4}$ in. long, oblong or linear-oblong, obtuse, pinnatifid with rounded lobes or angled at the margin; veins once or twice forked. Sori on the veinlets, sometimes very abundant and covering the under surface of the pinnæ.—Hook. Sp. Fil. iv. 272; *P. viscidum*, Sprengel; *Cheilanthes ambigua*, A. Rich.

Abundant in woods throughout the **Northern** and **Middle** Islands, *Banks and Solander*, etc. **Lord Auckland's** group and **Campbell's** Island, *J. D. H.* A very common tropical and southern temperate fern, in all quarters of the globe, passing into *Hypolepis tenuifolia*.

6. **P. pennigerum,** *Forst.*;—*Goniopteris pennigera*, J. Sm.; Fl. N. Z. ii. 40. Rhizome erect, 6–12 in. high, stout, woody, covered with the bases of the old stipes. Fronds 2–3 ft. high, glabrous, oblong, lanceolate, membranous, pinnate, pinnatifid at the top; stipes and rachis stout, quite glabrous and smooth, scaly at the very base; pinnules often opposite, sessile, truncate or auricled at the base, linear-elongate, 4–8 in. long, $\frac{2}{3}$ in. broad, long acuminate or caudate, pinnatifid to the middle, lobes short, ovate, obtuse, quite entire; veins 6–8 pairs, free except the lowest branch, which meets the next above it at the sinus between the lobes. Sori globose, numerous, on the middle of the veins, nearer the costa than the margin.—Hook. Sp. Fil. v. 7; Schkuhr, Fil. t. 22; *Aspidium pennigerum*, Swartz.

Abundant throughout the **Northern** and **Middle** Islands to Akaroa (*Raoul*), *Banks and Solander*, etc. I have seen no Otago specimens; nor is the plant found in the Kermadec Islands or Norfolk Island.

7. **P. rupestre,** *Br.*;—*Niphobolus rupestris*, Hook. and Grev. Ic. Fil. t. 93; Fl. N. Z. ii. 44. Rhizome slender, creeping, tortuous, scaly, branched, scales narrow. Fronds distant, erect, of 2 forms, simple, tapering into a short stipes which is jointed on to the rhizome, very coriaceous, covered with rusty-brown or white stellate hairs and down; barren fronds 1–4 in. long, $\frac{1}{2}$–1 in. broad, obovate spathulate or orbicular, obtuse; fertile 3–6 in. long, linear or linear-oblong or lanceolate; veins irregularly anastomosing, sunk in the substance of the frond. Sori large, numerous, protruding through the tomentum, often confluent, irregularly placed, but chiefly on the upper half of the frond.—Hook. Sp. Fil. v. 46; *P. stellatum*, A. Rich.; *P. serpens*, Forst.; *P. elæagnifolium*, Bory.

Abundant throughout the **Northern** and **Middle** Islands, *Banks and Solander*, etc. **Kermadec** Islands, *Macgillivray*. Also a native of Australia and the Pacific Islands. The stellate hairs of this plant, horizontally placed on a jointed stalk, are beautiful microscopic objects.

8. **P. Cunninghami,** *Hook. Sp. Fil.* v. 58.—*Dictymia lanceolata*, J. Sm.; Fl. N. Z. ii. 43. Rhizome very short, covered with broad imbricating

scales, and woolly roots. Frond glabrous, tufted, 2–12 in. long, $\frac{1}{2}$–$\frac{3}{4}$ in. broad, lanceolate, acuminate, quite entire, tapering into a short margined stipes, which is not jointed on to the rhizome; costa distinct; veins sunk in the substance of the frond, anastomosing, free veinlets 0. Sori large, sunk in pits of the frond, globose or oval, in 1 series on each side of and near the costa. Capsules mixed with jointed hairs, on long pedicels.—Hook. Sp. Fil. v. 58; *P. attenuatum*, A. Rich., not Brown; Hook. Ic. Pl. t. 409.

Northern and **Middle** Islands: on trunks of trees and rocks, as far south as Akaroa, *Banks and Solander*, etc. Also found in the New Hebrides, but not in Australia, where the *P. attenuatum*, with which this was long confounded, takes its place.

9. **P. pustulatum,** *Forst.;—Phymatodes pustulata*, Presl; Fl. N. Z. ii. 42. Rhizome long, branched, slender, creeping, squarrose with subulate scales. Fronds distant, stipitate, numerous, pendulous, rather membranous, flaccid, 6–18 in. long, 1–6 in. wide, tapering into a slender glabrous stipes, lanceolate or oblong, acuminate, entire or pinnatifid to the rachis; segments distant, 2–3 in. long, $\frac{1}{2}$–1 wide, tapering from a broad base to an obtusely acuminate point; venation lax, anastomosing, with free included veinlets. Sori oval or globose, prominent, scarcely sunk in the frond, forming a series parallel with the margin and near it.—Hook. Sp. Fil. v. 80; Schkuhr, Fil. t. 10; *P. membranifolium*, Br.; *P. scandens*, Forst.

Abundant throughout the **Northern** and **Middle** Islands, *Banks and Solander*, etc. Climbing lofty trees. Plant very fragrant; also found in Norfolk Island and Australia.

10. **P. Billardieri,** *Br.;—Phymatodes Billardieri*, Presl;—Fl. N. Z. ii. 42. Rhizome stout, creeping, usually glaucous, more or less covered with appressed scales. Fronds numerous, polymorphous, erect or pendulous, distant, coriaceous, perfectly smooth and glabrous; narrowed into a stout smooth stipes, which is jointed on to the rhizome, 3–18 in. long, lanceolate or oblong, obtuse or acute, entire and acuminate or broader and pinnatifid; margin thickened; pinnules broad or narrow-oblong, remote or distant, obtuse or acute; costa stout, prominent; veins anastomosing, with a few free veinlets in the areoles. Sori partially sunk in the frond, numerous, large, globose or oblong, in one series on each side of the costa.—Hook. Sp. Fil. v. 82; *P. scandens*, Labill. Fl. Nov. Holl. t. 240, not Forst.; *Niphobolus glaber*, Kaulf. of A. Rich. Flora (?).

Abundant throughout the **Northern** and **Middle** Islands, *Banks and Solander*, etc. **Kermadec** Islands, *Macgillivray*. **Lord Auckland's** group and **Campbell's** Island, *J. D. H.* Also abundant in Australia, Tasmania, and the Pacific Islands.

23. GYMNOGRAMME, Desvaux.

Ferns of various habit; the New Zealand species are small and tufted.—Fronds pinnate or 2- or 3-pinnate. Veins free or anastomosing. Sori oblong or linear, usually spreading in branching lines over the under surface of the pinnæ, either on the veins or between them. Involucre 0.

A large genus of temperate and tropical ferns, chiefly distinguished from *Polypodium* by the more linear sori, which often branch.

Glabrous, annual, very membranous, 2- or 3-pinnatifid . . . . . . 1. *G. leptophylla.*
Villous and glandular, perennial, pinnate . . . . 2. *G. rutæfolia.*

1. **G. leptophylla,** *Desvaux;—Fl. N. Z.* ii. 45. Annual; roots fibrous; fronds tufted, 1–8 in. high, perfectly glabrous, flaccid, membranous, pale green, shining, oblong-lanceolate, 2- or 3-pinnatifid; outer barren, smaller than the inner which are more fertile; stipes slender, brittle; pinnules small, obovate-cuneate, 2- or 3-fid or lobed, lobes obtuse, decurrent; veins dichotomous. Sori simple or confluent, oblong.—Hook. Sp. Fil. v. 136; Hook. and Grev. Ic. Fil. t. 25; *Grammitis novæ-Zelandiæ*, Colenso.

Northern Island: hills on the east coast, *Colenso, Sinclair;* summit of Mount Wellington (*Mrs. Jones*). A very widely diffused fern, found in various parts of all the continents and in many of the oceanic islands.

2. **G. rutæfolia,** *Br.;—Fl. N. Z.* ii. 45. Rhizome very short, stout, ascending, perennial. Fronds tufted, 1–3 in. high, densely glandular and villous, rather membranous, linear-oblong, obtuse, pinnate; stipes brittle, rather slender; pinnules alternate, rather distant, $\frac{1}{2}$–$\frac{3}{4}$ in. long, obovate or obliquely rhomboid or flabellate, cuneate at the base, and tapering into a short petiole, variously lobed or pinnatifid, segments cuneate; veins flabellate, dichotomous. Sori oblong or linear, simple or forked.—Hook. Sp. Fil. v. 137; Fil. Exot. t. 5.

Northern Island: Cook's Straits, on cliffs, very rare, *Colenso.* A very common Australian and Tasmanian plant, also found in the Pyrenees and in Bourbon, but hitherto in no other countries.

G. involuta, Don (*Grammitis scolopendrina,* Bory), is stated by Bory to be a native of New Zealand, but erroneously; in the Flora of New Zealand I had supposed it to be the same with *Polypodium australe* (*G. australis*).

24. NOTHOCHLÆNA, Br.

Rhizome short, or long and creeping, scaly.—Fronds 2- or 3-pinnate, usually small, stiff, erect. Pinnæ small. Sori marginal, rounded, oblong or linear, confluent, often partially covered by the recurved margin of the frond, veins simple or forked. Involucre 0.

A considerable genus of tropical and temperate ferns, very closely resembling *Cheilanthes,* but with no true involucre.

1. **N. distans,** *Br.;—Fl. N. Z.* ii. 45. Rhizome very short, stout, suberect or prostrate, scaly; roots matted. Fronds tufted, erect, rigid, 4–10 in. high, coriaceous, covered with subulate, ferruginous, piliferous scales below, hirsute above, linear-oblong, obtuse, 2-pinnate; stipes stout, dark red-brown, shining; primary divisions stipitate, opposite, $\frac{1}{2}$–1 in. long, erecto-patent, deltoid, ovate, the lower distant; pinnules in few pairs, ovate-oblong, obtuse, lower pinnatifid at the base, margins recurved. Sori continuous.—Hook. Sp. Fil. vi. 114, Ic. Pl. t. 980; Labill. Sert. Nov. Caled. t. 8.

Northern Island: common on basaltic rocks, *Colenso,* etc. A native of Australia, Tasmania, and New Caledonia.

25. TODEA, Willdenow.

Rhizome very stout, erect.—Fronds tufted at the top of the rhizome, large, 2-pinnate, very coriaceous. Veins simple or forked. Sori on the under-surface of the lower pinnæ of each branch of the frond, inserted on the veins, large, subglobose. Capsules large, shortly stipitate, subglobose, vertically 2-valved, with a short transverse, lateral or subterminal, transversely striate areola. Spores oblong, with a dark nucleus. Involucre 0.

The following is the only species, but the genus should perhaps be united with *Osmunda;* it differs from *Leptopteris* in the coriaceous fronds.

1. **T. africana,** *Willdenow.* Fronds 4–8 ft. high, very coriaceous, 2-pinnate; quite glabrous, oblong-lanceolate; stipes and rachis very stout, pale brown, quite smooth and glabrous; primary branches linear, a span and more long; pinnules alternate on the branches, narrow, linear-oblong or lanceolate, acute or acuminate, crenate or serrate, $\frac{3}{4}$–$1\frac{1}{2}$ in. long, sessile by a broad, often decurrent base.—Fl. Tasman. t. 168.

Northern Island: Mount Carmel, *Jolliffe;* Hokianga (*Mrs. Jones*). Also a native of Australia, Tasmania, and South Africa.

26. LEPTOPTERIS, Presl.

Fronds erect, membranous, pellucid, deep green, 2- or 3-pinnate. Capsules scarcely collected into sori, scattered over the under-surface of the pinnæ, upon the veinlets, pedicelled, subglobose, gibbous, the gibbous part with a small transversely striate areola. Spores depressed, with a dark spot.

A small genus of Australian and New Zealand ferns, as transparent as *Hymenophyllum,* and differing in this respect only from *Todea.*

Frond truncate below, the lowest pinnæ long . . . . 1. *L. hymenophylloides.*
Frond narrowed below, the lower pinnæ becoming gradually smaller 2. *L. superba.*

1. **L. hymenophylloides,** *Presl;—Fl. N.Z.* ii. 48. Rhizome short, stout, creeping. Fronds 6 in.–3 ft. high, ovate-lanceolate or deltoid, 2-pinnate; stipes slender or stout, and rachis glabrate, or covered with rather woolly red-brown tomentum; primary divisions spreading, linear-lanceolate, acuminate, stalked, the lower not becoming gradually very small; pinnules numerous, crowded, shortly stipitate, oblong, obtuse, deeply pinnatifid, segments narrow, generally forked.—*Todea hymenophylloides,* Presl;—A. Rich. Flor. t. 16; Hook. Garden Ferns, t. 54; *T. pellucida,* Hook. Ic. Pl. t. 8.

Abundant throughout the **Northern** and **Middle** Islands, *Banks and Solander,* etc. Closely allied to the New South Wales *L. pellucida.*

2. **L. superba,** *Hook. Ic. Pl. t.* 910; *Fl. N. Z.* ii. 48. Fronds forming a crown on the top of the rhizome, lanceolate, narrowed at the base, by the pinnæ gradually becoming very small, and often produced to the very rhizome, stout, erect, 2–3 ft. high; stipes and rachis very stout, woolly; primary pinnæ more numerous and close together than in *L. hymenophylloides,* not stipitate, the basal pinnules often overlapping the primary rachis; pinnules excessively numerous, densely crowded, ovate, often crisped, and concealing the rachis.—*Todea superba,* Colenso.

Forests of the mountainous parts of the **Northern** Island, *Colenso;* more common in the **Middle** Island, *Banks and Solander, Forster,* etc. **Lord Auckland's** group, *Bolton.* A most splendid fern, but I suspect that it passes into the preceding.

27. LYGODIUM, Swartz.

Rhizome creeping.—Stipes climbing, very long, branched, slender, tough, like whipcord. Frond dichotomously branched; veins forked, free in the New Zealand species; barren pinnæ linear-oblong, fertile flabellately divided or similar to the barren. Sori distichous, forming short spikelets projecting from the frond. Involucres imbricating, distichous, each containing one ovoid capsule. Capsule attached by its side, pointing downwards and inwards, reticulated, the top with radiating striæ. Spores obtusely 3-angled.

A very common tropical genus, of which the New Zealand species is found in no other part of the world except Norfolk Island.

1. **L. articulatum,** *A. Rich. Flor. t.* 15;—*Fl. N. Z.* ii. 47. Perfectly glabrous. Rhizome slender, covered with brown, slender scales. Stipes very numerous, angular, 50–100 ft. long, forming pendulous matted screens in the forests, jointed here and there. Fronds dichotomously palmatipartite, very much branched; barren pinnules stalked, divaricating, jointed on the rachis, 1–4 in. long, linear or oblong, obtuse or acute, cuneate at the base, often glaucous below; costa slender; fertile pinnules small, lobed, cuneate or fan-shaped, lobed; lobes small, short, with adnate spikelets towards their tips.

Abundant throughout the **Northern** and **Middle** Islands, *Banks and Solander*, etc.

28. SCHIZÆA, Smith.

Rhizome short, creeping.—Stipes rigid, erect, flat or nearly terete. Frond simple forked or flabellate; segments terminated by a short pinnatifid fruiting comb-like limb, the incurved divisions of which are covered with imbricating capsules. Capsules in 2 parallel series, sessile, naked or partially covered by the incurved margins of the divisions, ovoid, reticulated, the top with radiating striæ, bursting laterally. Spores obtusely 3-angular.

A common tropical genus, rare in the northern temperate zone, more common in the southern.

Frond flattened, flabellately divided above . . . . 1. *S. dichotoma.*
Frond terete or semiterete, simple or forked . . . . 2. *S. bifida.*

1. **S. dichotoma,** *Swartz;—Fl. N. Z.* ii. 47. Frond 6–24 in. high; stipes flat or compressed below, above suddenly expanding into a fan-shaped limb, of numerous narrow flat segments, each about 2 in. long and $\frac{1}{12}$–$\frac{1}{10}$ in. broad. Fruiting limb broad, short, $\frac{1}{4}$ in. long, of 4–10 pairs of pinnules of equal length, with laciniate or fimbriate margins.—Hook. and Grev. Ic. Fil. t. 17.

Northern Island: in marshy places, Bay of Islands, *Cunningham;* Manakau Bay, *Colenso.* A common East Indian, Australian, Pacific Island, and African plant.

2. **S. bifida,** *Swartz;—Fl. N. Z.* ii. 47. Fronds crowded, erect, 6–18 in. high, rigid, wiry, semiterete, grooved down one side, simple or forked, smooth or slightly rough to the touch. Fruiting limb short, broad, $\frac{1}{2}$–$\frac{3}{4}$ in. long, of 8–10 pairs of pinnules, with crinite margins.—*S. propinqua,* A. Cunn.

Var. β. *australis.* Frond 1-3 in. high.—*S. australis,* Gaudichaud; *S. pectinata,* Homb. and Jacq., Voy. au Pôle Sud, Bot. t. 4 Z.

Northern Island: common in wet clayey places. **Middle** Island: probably common, but overlooked; Otago, *Hector and Buchanan.* **Lord Auckland's** group, *Hombron and Jacquemont, J. D. H., Bolton.*

29. MARATTIA, Smith.

Rhizome large, tuberous, very thick.—Fronds numerous, very large and long, 2- or 3-pinnate; stipes very stout, jointed on to the rhizome, with adnate stipules at its base. Sori on the under surface of the pinnules, near the margin, terminating the veins; each sorus consists of 2 parallel oblong bodies (formed of connate coriaceous capsules), with plane opposite faces and convex backs, the faces marked with transverse slits leading to cavities containing the spores. Spores elliptical, very minute. A narrow fimbriate involucre fringes the sorus.

A tropical genus of few, very variable species.

1. **M. salicina,** *Smith;—Fl. N. Z.* ii. 49. Fronds 6–10 ft. high, coriaceous, dark green, deltoid, 2- or 3-pinnate; stipes very stout, green; pinnules sessile or stalked, jointed on to the rachis, 3–7 in. long, lanceolate or linear-oblong, acuminate, rounded at the base, serrate; costa stout, glabrous or hairy; veins parallel, free, simple or forked. Sori brown, ⅛ in. long.

Forests in the northern and eastern parts of the **Northern** Island, *Cunningham,* etc.; Taranaki (*Mrs. Jones*). The same species is found in Australia, India, South Africa, South America, and the Pacific Islands.

Angiopteris evecta, Hoffm., a tropical plant allied to *Marattia,* has been stated, on insufficient authority, to be a native of New Zealand.

30. OPHIOGLOSSUM, Linn.

Root of fleshy fibres.—Stipes solitary, erect, bearing about the middle one oblong, erect, leaf-like frond, with reticulated venation, and terminated by a linear flattened spike of fructification. Spike consisting of 2 opposite rows of 6–30 globose, connate, coriaceous capsules, each bursting transversely. Spores very minute, rounded and 3-gonous.

A genus of but few species, found in all parts of the world, usually in grassy pastures.

1. **O. vulgatum,** *Linn.;—Fl. N. Z.* ii. 50. Quite glabrous, 1 in.–1 ft. high. Frond ¼–3 in. long, very variable in length, breadth, and texture.

Var. β. *costatum.* Frond ovate or lanceolate, with usually a distinct costa and evident veins.—*O. costatum,* Br.; *O. elongatum,* A. Cunn.

Var γ *gramineum.* Frond ovate or lanceolate, acute; costa 0; veins indistinct.—*O. gramineum,* Willdenow.

Var. δ. *lusitanicum.* Frond linear-lanceolate or linear-oblong; costa 0; veins indistinct.—*O. lusitanum,* Willdenow; *O. coriaceum,* A. Cunn.

Var. ε. *minimum.* Small, 1-2 in. high. Frond rhomboid- or oblong-ovate, acute.

Common in grassy places throughout the **Northern** and **Middle** Islands, *Banks and Solander,* etc. A native of England and most temperate parts of the world.

31. BOTRYCHIUM, Linn.

Root of thick fleshy fibres.—Stipes erect, bearing 1 pinnate or much-divided frond, and a branched spike or panicle of fructification. Capsules globose, distichously arranged on the branches, separate from one another, bursting transversely. Spores very minute, 3-lobed.

A small genus, found in most temperate and warm regions of the globe, but rare in very hot ones.

1. **B. cicutarium,** *Swartz.* Stout, rarely slender, 3–18 in. high, glabrous or slightly pilose. Frond broadly ovate or deltoid, 2- or 3-pinnatifid or 3-nately decompound; pinnæ variable in shape, oblong, obtuse, crenate, obscurely veined; peduncle of the fruiting panicle radical or nearly so, 2-pinnately branched above.

Var. *α.* Frond stout, fleshy.—*B. virginicum*, Fl. N. Z. ii. 50, not Linn.; Fl. Tasman. ii. t. 169 B; *B. australe*, Br.; *B. lunarioides*, Swartz.

Var. *β. dissectum.* Frond slender, much more finely divided.—*B. dissectum*, Muhlenberg; *B. lunarioides*, var. *dissectum*, A. Gray.

Northern and **Middle** Islands: var. *α*, abundant, *Banks and Solander*, etc.; var. *β*, Whangarei, near Auckland, *G. Burnett, Esq.* The var. *α* is a common plant in many parts of the globe, temperate and tropical, including Australia and Tasmania. Var. *β* is also frequent in North America and in some parts of Asia and Europe, but seems to be very scarce in New Zealand: it looks remarkably different, but is united by intermediate forms in the northern hemisphere. Its discoverer in New Zealand observes that var. *α* is abundant in the same neighbourhood where var. *β* itself is so scarce.

Order II. LYCOPODIACEÆ.

Erect or prostrate or creeping, rarely climbing, simple or branched plants, with usually rigid stems. Stems and ribs of the leaves with bundles of vascular tissue, consisting of wood-fibres and spirally-marked and barred vessels.—Leaves imbricated all round the stem or distichous or 4-fariously arranged, small, usually coriaceous, subulate, and nerveless, sometimes flattened and 1-nerved, sometimes of 2 forms, the larger distichous, smaller stipule-like. Capsules sessile in the axils of the leaves or of the scales of terminal sessile or peduncled, cylindric or 4-gonous cones, 2- or 3-valved, 1–3-celled, of two kinds: 1st, compressed, often reniform, coriaceous capsules, 2-valved, full of microscopic, obtusely 3-gonous spores; 2nd, larger capsules, containing 3 or 4 much larger spores, each marked with 3 radiating lines at the top.

Germination has been observed in the large spores of the genus *Selaginella* only, which does not occur in New Zealand. In this genus the contents of the large spore develope a small cellular expansion within its coat, under the position of the three radiating lines; upon this (the pro-embryo or prothallium) cellular papillæ (archegonia) appear, each containing an open cavity, at the base of which is a free cell, from which the future plant is afterwards developed. The cellular papillæ are produced in abundance on the prothallium along three radiating lines corresponding to those on the coat of the spore, but only one gives origin to a young plant. The small spores of *Selaginella* produce antheridia, which are cells containing a spiral thread endowed with motion (spermatozoa); these no doubt gain access to the papillæ on the prothallium, but no one has proved this to be so. Nothing is known of the process of fertilization or reproduction in any of the New Zealand genera, to which the attention of the student should be directed.

Lycopodiaceæ are found in all situations and all quarters of the globe except the driest; the New Zealand species are the largest of the Order, and present nearer affinities to the fossil *Lycopodiaceæ* of the Coal period than any other existing plants.

Leaves linear, radical. Cone borne on a solitary radical peduncle . 1. PHYLLOGLOSSUM.
Stems leafy, branched. Leaves small. Capsules 1-celled, in the upper leaf-axils or in cones . . . . 2. LYCOPODIUM.
Stems leafy, simple. Leaves large. Capsule on the leaf base, 2-celled . . . . 3. TMESIPTERIS.
Stem leafless, branched, 3-gonous. Capsules lateral on the branches, 3-celled. . . . . 4. PSILOTUM.

1. PHYLLOGLOSSUM, Kunze.

Root of 2 ovoid tubers, with long fleshy fibres from the crown.—Leaves all radical, 6–10, terete, subulate. Scape simple, erect, terete, terminated by a terete or ovoid spike of imbricating, 3-gonous, pedicelled scales. Capsules solitary in the axils of the scales, 1-celled, 2-valved. Spores most minute, obtusely tetrahedral, with 3 radiating lines at the top.

A remarkable little plant, with the habit of *Ophioglossum*. Also found in Australia and Tasmania.

1. **P. Drummondii,** *Kunze ;—Fl. N. Z.* ii. 51. Green, quite glabrous, 1–3 in. high. Leaves ½ in. long. Scales of spike sub-3-angular, suddenly contracted upwards into an obtuse beak, and downwards into a short rounded auricle.—Hook. Ic. Pl. t. 908.

Northern Island: in grassy places, clay banks, etc., Bay of Islands, Auckland, etc., *Sinclair, Colenso,* etc.

2. LYCOPODIUM, Linn.

Rhizome tufted or creeping.—Stems erect or pendulous, simple or branched, covered with distichous 4-ranked or imbricating, small leaves. Capsules solitary in the axils of the upper stem-leaves, or in the scales of slender terminal spikes, coriaceous, usually reniform, 1-celled, bursting transversely into 2 valves, full of excessively minute 3-gonous or tetrahedral spores.

An immense genus, found in all parts of the world, and in all climates but the very dry.

I. *Leaves imbricated all round the stem. Capsules axillary in the upper leaves, or in the scales of sessile, terminal, 4-gonous spikes.*

Stems 4–8 in., tufted, erect, stout. Leaves all similar, broad, subulate, the upper with capsules . . . . 1. *L. Selago.*
Stems 6–18 in., stout, curving, erect. Leaves linear, decurrent, obtuse. Capsules in terminal drooping spikes . . . . 2. *L. varium.*
Stems 2–4 ft., slender, pendulous. Leaves linear. Capsules in long, branched, pendulous spikes . . . . 3. *L. Billardieri.*

II. *Leaves imbricated all round the stem. Spikes sessile, cylindrical.*

Stems stout, erect, densely branched. Leaves very variable, subulate. Spikes terminal ; scales squarrose . . . . 4. *L. densum.*
Stems slender, erect, rarely branched. Spikes lateral ; scales erect . 5. *L. laterale.*
Stems stout, creeping ; branches erect, fastigiate ; tips incurved. Spikes short ; scales toothed . . . . 6. *L. cernuum.*

III. *Leaves imbricated all round the stem. Spikes terete, peduncled.*

Stems creeping. Peduncles lateral, long. Scales of spike with spreading points . . . . 7. *L. carolinianum.*
Stems creeping, with erect fastigiate branches. Peduncles terminal. Scales of spike recurved . . . . 8. *L. clavatum.*

IV. *Leaves distichous. Spikes terminal, terete.*

Stems creeping, 1–2 ft. Spikes solitary or 2-nate . . . . 9. *L. scariosum.*
Stems climbing, 2–10 ft. Spikes in branched panicles . . . . . . 10. *L. volubile.*

1. **L. Selago,** *Linn.;—Fl. N. Z.* ii. 52. Rhizome stout, creeping. Stems 4–8 in. high, stout, cylindric, obtuse, $\frac{1}{4}$–$\frac{1}{3}$ in. diam. Leaves densely imbricated, green or reddish, $\frac{1}{6}$–$\frac{1}{3}$ in. long, usually erect or incurved, rarely squarrose or spreading, subulate-lanceolate, acute or acuminate, quite entire or obscurely serrate. Capsules all axillary in the upper leaves.

Middle Island: Nelson mountains, *Bidwill;* Southern Alps, *Haast;* Otago, lake district, *Hector* A very common alpine and arctic plant, found in most cool mountainous damp regions of the globe.

2. **L. varium,** *Br.;—Fl. N. Z.* i. 52. Rhizome short or 0. Stems tufted, stout, erect or pendulous from trees, 6–18 in. high, simple or branched. Leaves closely imbricated all round the stem, $\frac{1}{4}$–$\frac{1}{2}$ in. long, erect or spreading, linear, obtuse, decurrent, quite entire, dark green; midrib obscure. Spikes terminal, sessile, 4-gonous, simple or branched, usually drooping; scales short, ovate, keeled, obtuse, sometimes foliaceous.—Hook. and Grev. Ic. Fil. t. 112.—*L. sulcinervium*, Spring.

Northern and **Middle** Island: mountainous districts, abundant, *Dieffenbach, Colenso,* etc. **Lord Auckland's** and **Campbell's** Islands, *J. D. H.* A common Tasmanian plant, which, I suspect, passes into *L. Selago* on one hand and *L. Billardieri* on the other; it is scarcely distinguishable from a South African species.

3 **L. Billardieri,** *Spring.;—Fl. N. Z.* ii. 53. Rhizome short. Stems tufted, pendulous from trees, 2–4 ft. long, slender, rigid, cylindric, grooved, flexuous, branched. Leaves $\frac{1}{4}$–$\frac{3}{4}$ in. long, scattered below, above close-set, imbricating all round, decurrent, linear-oblong or ligulate, obtuse or acute, the lower with sometimes an evident midrib. Spikes slender, flaccid, several inches long, dichotomously branched, 4-gonous; scales 4-fariously imbricate, broadly ovate, obtuse, keeled, often smaller than the capsules.—*L. flagellaria*, A. Rich., not of Bory; *L. Phlegmaria*, A. Cunn., not of Linn.

Abundant throughout the **Northern** and **Middle** Islands, *Banks and Solander,* etc. This almost passes into various tropical forms, but is not exactly similar to any.

4. **L. densum,** *Labill. Fl. Nov. Holl. t.* 251;—*Fl. N. Z.* ii. 53. Rhizome stout, creeping. Stems tall, erect, woody, rigid, 1–3 ft. high, densely fastigiately branched; branches erect or patent. Leaves imbricated all round, subulate, acuminate or piliferous, $\frac{1}{16}$–$\frac{1}{8}$ in. long, appressed spreading or squarrose, sometimes 6-fariously imbricated. Spikes terminal, solitary, sessile, $\frac{1}{2}$–$\frac{3}{4}$ in. long, cylindric, obtuse; scales scarious, spreading or squarrose, peltate, 3-angular, jagged. Spores hispid.

Northern Island: abundant, *Banks and Solander,* etc. **Chatham** Island, *Dieffenbach.* Also a native of Norfolk Island, Tasmania, Australia, and New Ireland.

5. **L. laterale,** *Br.;—Fl. N. Z.* ii. 53 Rhizome slender, creeping. Stems ascending or erect, 3–10 in. high, simple or divided, obtuse. Leaves imbricating all round, spreading and squarrose, $\frac{1}{3}$ in. long, narrow-subulate. Spikes solitary, axillary, sessile, erect, $\frac{1}{4}$–$\frac{2}{3}$ in. long, terete, obtuse; scales 4-farious, coriaceous, broadly ovate or rounded, suddenly contracted to a stiff erect point; margins scarious, jagged.—Labill. Sert. Nov. Caled. t. 15.

Northern and **Middle** Islands not rare in wet places, amongst grass, etc. A native of Australia, Tasmania, and New Caledonia.

6. **L. cernuum,** *Linn.;—Fl. N. Z.* ii. 54. Stem stout, creeping, 2–3 ft. long, leafy, with erect, stiff, much divided branches 6–12 in. high. Leaves numerous, spreading, squarrose or incurved, $\frac{1}{6}$ in. long, acerose or subulate, keeled, inserted all round the stem and branches, quite entire. Spikes short, terminal, sessile on short incurved branchlets, cylindric; scales imbricated all round, ovate or ovate-lanceolate, with long serrate acuminate points, margins scarious and ciliate.

Northern parts of the **Northern** Island: Bay of Islands, *Cunningham*, etc. A most abundant plant in all hot and subtropical climates.

7 **L. carolinianum,** *Linn.;—Fl. N. Z.* ii. 54. Stem creeping and rooting, 2–6 in. long; branches few, also creeping, never erect or ascending. Leaves imbricated all round, curved, ascending, subulate, $\frac{1}{5}$–$\frac{1}{4}$ in. long. Spike 1–2 in. long, cylindric, on an erect lateral stiff peduncle which is 1–4 in. high and covered with erect subulate leaves. Scales peltate, in about 6 rows, stiff, broadly ovate below, with rigid spreading points, and scarious toothed rarely entire margins.

Northern Island: Bay of Islands and east coast, *Colenso*, etc. A common plant in Australia, Tasmania, and in many tropical and subtropical parts of the world.

8. **L. clavatum,** *Linn.;*—var. *magellanicum*, Fl. N. Z. ii. 55. Stems stout, creeping below, rigid, 4–10 in. long; branches short, stout, erect, much fastigiately branched, 3–12 in. high. Leaves imbricating all round, spreading incurved or squarrose, sometimes subsecund, linear-subulate, quite entire, acuminate, not hair-pointed. Spikes 1–3 in long erect, cylindric, on solitary or twin, terminal, stout, leafy peduncles; leaves on the peduncles shorter, often whorled. Scales peltate, trapezoid, toothed at the base, with long recurved points. Spores granular.—*L. magellanicum*, Swartz; *L. fastigiatum*, Br.; *L. pichinchense*, Hook. Ic. Pl. t. 85.

Abundant in mountainous situations throughout the **Northern** and **Middle** Islands, *Banks and Solander*, etc. **Lord Auckland's** group and **Campbell's** Island, *J D. H.* Most abundant in subalpine South America, Tasmania, Australia, and the Antarctic islands. This is the southern variety of *L. clavatum*, a very common plant in the northern and some parts of the southern hemisphere, which has usually more or less serrulate and hair-pointed leaves, but which is connected with it by numerous intermediate forms.

9. **L. scariosum,** *Forst ;—Fl. N. Z.* ii. 55. Stems long, stout, creeping, often 2 ft. long, sparingly leafy, sending up prostrate or erect, flattened, flabellately divided branches 6–10 in. long and $\frac{1}{4}$ in. broad. Leaves of 2 forms, the larger 2-farious, coriaceous, falcate, ovate lanceolate, acute or acuminate, laterally flattened, sessile, decurrent, $\frac{1}{8}$ in. long, dark green above, pale or glaucous below; smaller leaves appressed to the under surface of the branches, more numerous, subulate. Spikes 1–3 in. long, solitary or geminate, cylindric, on long or short peduncles, which are covered with imbricate subulate leaves. Scales somewhat 6-farious, ovate; points broad, recurved, margins toothed. Spores areolate.—Hook. Ic. Pl. t. 966; *L. decurrens*, Br.; *L. Jussieui*, Desv.; Hook. Ic. Pl. t. 186; *L. Lessonianum*, A. Rich.?

Open mountainous regions throughout the **Northern** and **Middle** Islands; more rarely in woods, *Banks and Solander*, etc. **Lord Auckland's** group, *Lyall* Also found in Tasmania and throughout the Andes of South America to Fuegia, and on the mountains of Jamaica.

10. **L. volubile,** *Forst.*;—*Fl. N. Z.* ii. 55. A lofty climber, festooning trees, etc. Stems slender, wiry, many feet long, much branched, sparingly leafy; branches spreading, compressed, dichotomously and flabellately divided. Leaves of 2 kinds, exactly as in *L. scariosum*, but with more acuminate points. Spikes in branched dichotomous panicles, slender, ½–2½ in. long, cylindric, or obscurely 4-farious, pedicelled. Scales small, orbicular, suddenly contracted into a subulate point, margins toothed or entire. Spores broadly pyriform, broad end granular, the other transparent.—Hook. and Grev. Ic. Fil. t. 70; *L. D'Urvillei*, A. Rich.?

Abundant throughout the **Northern** and **Middle** Islands, *Banks and Solander*, etc. Also found in Java, but hitherto in no other part of the world. The most beautiful species of the genus.

A specimen of *L Phlegmaria*, Linn., a tropical species, having been mixed with Menzies New Zealand collections, has been erroneously described as a native of these islands. Cunningham's specimens so named are *L. Billardieri*.

3. TMESIPTERIS, Bernhardi.

Roots fibrous.—Fronds pendulous, flaccid, simple or rarely branched; stipes angled. Leaves vertical, sessile, falcate, decurrent, oblong or ensiform, obtuse truncate or acute, rarely acuminate, coriaceous; costa stout, often excurrent, oblique; veins 0, fertile ones stalked, 2-lobed or 2 together with a capsule seated in the fork between them. Capsule transversely oblong, coriaceous, 2-lobed, 2-valved, bursting by a vertical slit. Spores most minute, oblong, curved.

A curious but very variable plant, found in Australia, the Pacific Islands, and California.

1. **T. Forsteri,** *Endlicher*;—*Fl. N. Z.* ii. 51. Stems 6–24 in. long. Leaflets ½–1 in. long, pale green. Capsules ⅙ in long.—*T. Tannensis*, Labill. Fl. Nov. Holl. t. 252.

Abundant throughout the **Northern** and **Middle** Islands: often epiphytic on tree-ferns, *Banks and Solander*, etc. **Lord Auckland's** group, *Bolton*.

4. PSILOTUM, Swartz.

Rhizome short, stout.—Fronds tufted, rigid, 3-quetrous, dichotomously branched above, leafless or with minute scattered scale-like leaves. Capsules in the axils of the minute scales, large, coriaceous, 3-lobed, 3-celled. Spores very minute, bursting when placed in water, oblong, hyaline, with a central pale nucleus.

1. **P. triquetrum,** *Swartz.* Fronds erect or pendulous, 2–12 in. long, green, slender, simple below, dichotomously branched above. Branches 3-gonous. Scales or leaves scattered, ⅛ in. long. Capsules ⅙ in. diam.—Hook Fil. Exot. t. 63.

Northern Island: Motuhona Island, Bay of Plenty, *Jolliffe*. A common fern in very many parts, especially of the southern hemisphere, both temperate and tropical, also found in California.

Order III. MARSILEACEÆ.

Creeping or floating, marsh or water plants. Stems with very slender vascular bundles, like those of *Lycopodiaceæ*.—Vernation circinate. Capsules of two kinds, one containing a single spore, the other many minute spores, enclosed in a receptacle (or common capsule) with valvular dehiscence; in some both kinds of capsule are enclosed in a common involucre, in others each kind is in a separate involucre. Of these spores the larger, which are solitary, reproduce the plant; the smaller, which are very numerous, contain cells with ciliated filaments.

The reproductive process in *Marsileaceæ* is analogous to that of ferns in many respects. The receptacles burst, and both kinds of spores are emitted; the larger developes a prothallium at its top, which terminates in one perforated conical papilla (archegonium), the perforation leading down to a cell in the body of the prothallium. This cell gives origin to the new plant; the small capsules (antheridia) emit cells containing each a spiral thread endowed with motion (spermatozoa), of which one, no doubt, enters the perforation of the papilla and fertilizes the cell at its base. Only one genus, *Azolla*, has been hitherto found in New Zealand; but as 2 others are common in Australia, these may be mentioned here, viz. *Pilularia*, which has creeping stems, subulate leaves, and globose involucres at the base of the leaves; and *Marsilea*, with a creeping stem, 4-foliolate leaves, and leaflets like Clover. The latter is the Nardoo of Australia, upon the starch contained in the capsules of which, Burke and his companion-explorers subsisted in the desert for some time.

1. AZOLLA, Linn.

Plant floating, forming small red patches, consisting of pinnate fronds covered with minute imbricating leaves. Roots of solitary simple threads.—Stem consisting of a central cellular axis with a few spiral threads, surrounded by a circle of air-cells. Receptacles very minute, pendulous from the under surface of the branches; the larger sort female, ovoid, bursting irregularly, full of spherical stalked capsules, each containing a few globular spores; smaller sort bursting transversely, containing rounded antheridia? peltately borne on the sides of a central erect column.

A genus of few species found in the warmer temperate and tropical zones, as far north as the United States; the following is the only one that inhabits cold countries. I have never examined the fructification, the generic description of which is taken from A. Gray's Manual of the Botany of the North United States.'

1. **A. rubra,** *Br.*;—*Fl. N. Z.* ii. 56. Frond red, ovate orbicular or somewhat triangular, 1–2 in. diam., pinnate; branches close-set, almost imbricating, somewhat palmate. Leaves densely imbricate, ovate, entire, obtuse, $\frac{1}{20}$ in. long, smooth on the upper surface. Roots longitudinally plumose beyond the middle.

Northern Island: pools on the east coast and interior, *Colenso*. **Middle** Island: apparently common. Also found in Tasmania, Australia, and South America, and perhaps not specifically different from an African and North American plant.

Printed in the United States
By Bookmasters